AVIAN FORAGING: THEORY, METHODOLOGY, AND APPLICATIONS

Michael L. Morrison, C. John Ralph,
Jared Verner, and Joseph R. Jehl, Jr., editors.

Proceedings of an International Symposium of the
COOPER ORNITHOLOGICAL SOCIETY
held at Asilomar, California,
December 18–19, 1988

Sponsors:

Pacific Southwest Forest and Range Experiment Station, U.S.D.A. Forest Service
Cooper Ornithological Society
Department of Forestry and Resource Management, University of California, Berkeley
Western Foundation of Vertebrate Zoology
San Francisco Bay Area Chapter, The Wildlife Society

Studies in Avian Biology No. 13

A PUBLICATION OF THE COOPER ORNITHOLOGICAL SOCIETY

Cover drawing by John Schmitt

STUDIES IN AVIAN BIOLOGY

Edited by

Joseph R. Jehl, Jr.
Sea World Research Institute
1700 South Shores Road
San Diego, California, 92109

CONSULTING EDITORS FOR THIS ISSUE

William M. Block	Richard L. Hutto	Brian A. Maurer
Barry R. Noon	Martin G. Raphael	Harry F. Recher
John T. Rotenberry	Kimberly G. Smith	

EDITORIAL ASSISTANTS

| William M. Block | Lorraine M. Merkle |
| Linnea S. Hall | Suzanne I. Bond |

Studies in Avian Biology is a series of works too long for *The Condor,* published at irregular intervals by the Cooper Ornithological Society. Manuscripts for consideration should be submitted to the editor at the above address. Style and format should follow those of previous issues.

Price $29.50 including postage and handling. All orders cash in advance; make checks payable to Cooper Ornithological Society. Send orders to Jim Jennings, Assistant Treasurer, Cooper Ornithological Society, Suite 1400, 1100 Glendon Ave, Los Angeles, CA 90024.

ISBN: 0-935868-47-X

Library of Congress Catalog Card Number 90-081089
Printed at Allen Press, Inc., Lawrence, Kansas 66044
Issued 18 April 1990

Copyright © by the Cooper Ornithological Society 1990

CONTENTS

LIST OF AUTHORS vii

INTRODUCTION .... M. L. Morrison, C. J. Ralph, and J. Verner   1

SECTION I: *Role of Birds in Natural Ecosystems and the Quantification of Resources*

SECTION OVERVIEW:
Quantifying food resources in avian studies: present problems and future needs .... K. G. Smith and J. T. Rotenberry   3

Quantification of Resources

Ecological and evolutionary impact of bird predation on forest insects: an overview .... R. T. Holmes   6

Predation by birds and ants on two forest insect pests in the Pacific Northwest ... T. R. Torgersen, R. R. Mason, and R. W. Campbell 14

Measuring the availability of food resources .... R. L. Hutto 20

Arthropod sampling methods in ornithology .... R. J. Cooper and R. C. Whitmore 29

Food availability for an insectivore and how to measure it .... H. Wolda 38

Quantifying bird predation of arthropods in forests .... D. L. Dahlsten, W. A. Cooper, D. L. Rowney, and P. K. Kleintjes 44

Factors influencing Brown Creeper (*Certhia americana*) abundance patterns in the southern Washington Cascade Range .... J. M. Mariani and D. A. Manuwal 53

Food resources of understory birds in central Panama: quantification and effects on avian populations ...... J. R. Karr and J. D. Brawn 58

Spatial variation of invertebrate abundance within the canopies of two Australian eucalypt forests .... J. D. Majer, H. F. Recher, W. S. Perriman, and N. Achuthan 65

Quantifying abundance of fruits for birds in tropical habitats .... J. G. Blake, B. A. Loiselle, T. C. Moermond, D. J. Levey, and J. S. Denslow 73

Quantification of Diets

Approaches to avian diet analysis .... K. V. Rosenberg and R. J. Cooper 80

Diets of understory fruit-eating birds in Costa Rica: seasonality and resource abundance .... B. A. Loiselle and J. G. Blake 91

Dietary similarity among insectivorous birds: influence of taxonomic versus ecological categorization of prey .... R. J. Cooper, P. J. Martinat, and R. C. Whitmore 104

Foraging and nectar use in nectarivorous bird communities .... B. G. Collins, J. Grey, and S. McNee 110

SECTION II: *Foraging Behavior: Design and Analysis*

SECTION OVERVIEW:
Biological considerations for study design ....
.... M. G. Raphael and B. A. Maurer 123
Analytical considerations for study design ....
.... B. R. Noon and W. M. Block 126

Observations, Sample Sizes, and Biases
Food exploitation by birds: some current problems and future goals
.... D. H. Morse 134
A classification scheme for foraging behavior of birds in terrestrial habitats .... J. V. Remsen, Jr. and S. K. Robinson 144
Proportional use of substrates by foraging birds: model considerations on first sightings and subsequent observations ....
.... G. W. Bell, S. J. Hejl, and J. Verner 161
Sequential versus initial observations in studies of avian foraging ...
.... S. J. Hejl, J. Verner, and G. W. Bell 166
Analysis of the foraging ecology of eucalypt forest birds: sequential versus single-point observations ..... H. F. Recher and V. Gebski 174
Use of radiotracking to study foraging in small terrestrial birds ......
.... P. L. Williams 181
Influence of sample size on interpretations of foraging patterns by Chestnut-backed Chickadees .... L. A. Brennan and M. L. Morrison 187
Precision, confidence, and sample size in the quantification of avian foraging behavior .... L. J. Petit, D. R. Petit, and K. G. Smith 193
Interobserver differences in recording foraging behavior of Fuscous Honeyeaters .... H. A. Ford, L. Bridges, and S. Noske 199

Intraspecific, Spatial, and Temporal Variation
Within-season and yearly variations in avian foraging locations .....
.... S. J. Hejl and J. Verner 202
The importance and consequences of temporal variation in avian foraging behavior .... D. B. Miles 210
Seasonal differences in foraging habitat of cavity-nesting birds in the southern Washington Cascades ....
.... R. W. Lundquist and D. A. Manuwal 218
Yearly variation in resource-use behavior by ponderosa pine forest birds
.... R. C. Szaro, J. D. Brawn, and R. P. Balda 226
Variation in the foraging behaviors of two flycatchers: associations with stage of the breeding cycle .... H. F. Sakai and B. R. Noon 237
Seasonal changes in foraging behavior of three passerines in Australian eucalyptus woodland .... H. A. Ford, L. Huddy, and H. Bell 245
Geographic variation in foraging ecology of North American insectivorous birds .... D. R. Petit, K. E. Petit, and L. J. Petit 254
Geographic variation in foraging ecologies of breeding and nonbreeding birds in oak woodlands .... W. M. Block 264
Sex, age, intraspecific dominance status, and the use of food by birds wintering in temperate-deciduous and cold-coniferous woodlands: a review .... T. C. Grubb, Jr., and M. S. Woodrey 270

Effects of unknown sex in analyses of foraging behavior ....
.... J. M. Hanowski and G. J. Niemi 280
Differences in the foraging behavior of individual Gray-breasted Jay
flock members .... L. M. McKean 284

Analytical Methods

Use of Markov chains in analyses of foraging behavior ....
.... M. G. Raphael 288
A comparison of three multivariate statistical techniques for the analysis
of avian foraging data .... D. B. Miles 295
An exploratory use of correspondence analysis to study relationships
between avian foraging behavior and habitat ....
.... E. B. Moser, W. C. Barrow, Jr., and R. B. Hamilton 309
Analyzing foraging use versus availability using regression techniques
.... K. M. Dodge, R. C. Whitmore, and E. J. Harner 318
Analyzing foraging and habitat use through selection functions ......
.... L. L. McDonald, B. F. J. Manly, and C. M. Raley 325

SECTION III: *Specialists Versus Generalists*

SECTION OVERVIEW:
Specialist or generalist: avian response to spatial and temporal changes in
resources .... H. F. Recher 333
When are birds dietarily specialized? Distinguishing ecological from
evolutionary approaches .... T. W. Sherry 337
Behavioral plasticity of foraging maneuvers of migratory warblers: multiple selection periods for niches? .... T. E. Martin and J. R. Karr 353
Dead-leaf foraging specialization in tropical forest birds: measuring resource availability and use .... K. V. Rosenberg 360
Bird predation on periodical cicadas in Ozark forests: ecological release
for other canopy arthropods? ....
.... F. M. Stephen, G. W. Wallis, and K. G. Smith 369
Influence of periodical cicadas on foraging behavior of insectivorous
birds in an Ozark forest ....
.... C. J. Kellner, K. G. Smith, N. C. Wilkinson, and D. A. James 375
The influence of food shortage on interspecific niche overlap and foraging behavior of three species of Australian warblers (Acanthizidae)
.... H. L. Bell and H. A. Ford 381

SECTION IV: *Energetics and Foraging Theory*

SECTION OVERVIEW:
Studies of foraging behavior: central to understanding the ecological consequences of variation in food abundance .... R. L. Hutto 389

Energetics of Foraging

Digestion in birds: chemical and physiological determinants and ecological implications .... W. H. Karasov 391
Birds and mammals on an insect diet: A primer on diet composition
analysis in relation to ecological energetics .... G. P. Bell 416
Energetics of activity and free living in birds .... D. L. Goldstein 423

Behavioral and Theoretical Considerations

A functional approach to foraging: morphology, behavior, and the capacity to exploit .... T. C. Moermond 427

Ecological plasticity, neophobia, and resource use in birds .... R. Greenberg 431

Food availability, migratory behavior, and population dynamics of terrestrial birds during the nonreproductive season ...... S. B. Terrill 438

Foraging theory: up, down, and sideways .... D. W. Stephens 444

Extensions of optimal foraging theory for insectivorous birds: implications for community structure .... B. A. Maurer 455

Meeting the assumptions of foraging models: an example using tests of avian patch choice .... J. B. Dunning, Jr. 462

LITERATURE CITED 471

LIST OF AUTHORS

N. ACHUTHAN
School of Mathematics and Statistics
Curtin University of Technology
GPO Box U987
Perth, W.A. 6001, Australia

RUSSELL P. BALDA
Dept. of Biological Sciences, Box 5640
Northern Arizona University
Flagstaff, AZ 86011

WYLIE C. BARROW
School of Forestry, Wildlife, and Fisheries
Louisiana Agricultural Experiment Station
Louisiana State University Agricultural Center
Baton Rouge, LA 70803

GARY P. BELL
The Nature Conservancy
Santa Rosa Plateau Preserve
22115 Tenaja Rd.
Murrietta, CA 92362

GRAYDON W. BELL
Dept. of Mathematics
Box 5717
Northern Arizona University
Flagstaff, AZ 86011

HARRY L. BELL
Dept. of Zoology
Univ. of New England
Armidale, N.S.W. 2351
Australia

JOHN G. BLAKE
Dept. of Zoology
Univ. of Wisconsin
Madison, WI 53706
(Present address: Center for Water
 and the Environment
Natural Resources Research Institute
University of Minnesota
Duluth, MN 55811)

WILLIAM M. BLOCK
Dept. of Forestry and Resource Management
University of California
Berkeley, CA 94720
(Present address: Pacific Southwest Forest and Range
 Experiment Station
USDA Forest Service
2081 E. Sierra Ave.
Fresno, CA 93710)

JEFFREY D. BRAWN
Smithsonian Tropical Research Institute
P.O. Box 2072
Balboa, Republic of Panama, 34001-0011

LEONARD A. BRENNAN
Dept. of Forestry and Resource Management
University of California
Berkeley, CA 94720

LYNDA BRIDGES
Dept. of Zoology
Univ. of New England
Armidale, N.S.W. 2351
Australia

ROBERT W. CAMPBELL
Dept. of Environmental Forestry and Biology
New York State University
College of Environmental Sciences and Forestry
Syracuse, NY 13210

BRIAN G. COLLINS
School of Biology
Curtin University of Technology
Kent Street
Bentley, W.A. 6102, Australia

ROBERT COOPER
Forest Arthropod Research Team
Division of Forestry
West Virginia University
Morgantown, WV 26506
(Present address: Dept. of Wildlife
Humboldt State University
Arcata, CA 95521)

WILLIAM A. COPPER
Division of Biological Control
University of California
Berkeley, CA 94720

DONALD L. DAHLSTEN
Division of Biological Control
University of California
Berkeley, CA 94720

JULIE S. DENSLOW
Dept. of Biology
Tulane University
New Orleans, LA 70118

KEVIN M. DODGE
Forest Arthropod Research Team
Division of Forestry
West Virginia University
Morgantown, WV 26506
(Present address: McGarrett Community College
McHenry, MD 21541)

JOHN B. DUNNING, JR.
Dept. of Zoology
University of Georgia
Athens, GA 30602

HUGH A. FORD
Dept. of Zoology
University of New England
Armidale, N.S.W. 2351
Australia

VAL GEBSKI
Statistical and Computing Laboratory
Macquarie University
North Ryde, N.S.W.
Australia

DAVID L. GOLDSTEIN
Dept. of Biological Sciences

Wright State University
Dayton, OH 45435

RUSSELL GREENBERG
National Zoological Park
Smithsonian Institution
Washington, D.C. 20008

JAMES GREY
School of Biology
Curtin University of Technology
Kent Street
Bentley, 6102 W.A., Australia

THOMAS C. GRUBB, JR.
Dept. of Zoology
Ohio State University
Columbus, OH 43210

ROBERT B. HAMILTON
School of Forestry, Wildlife and Fisheries
Louisiana Agricultural Experiment Station
Louisiana State University Agricultural Center
Baton Rouge, LA 70803

JOANN M. HANOWSKI
Center for Water and the Environment
Natural Resources Research Institute
University of Minnesota
Duluth, MN 55811

E. JAMES HARNER
Dept. of Statistics and Computer Science
West Virginia University
Morgantown, WV 26506

SALLIE J. HEJL
Pacific Southwest Forest and Range Experiment Station
USDA Forest Service
2081 East Sierra Ave.
Fresno, CA 93710
(Present address: USDA Forest Service
Intermountain Research Station
P.O. Box 8089
Missoula, MT 59807)

RICHARD T. HOLMES
Dept. of Biological Sciences
Dartmouth College
Hanover, NH 03755

LEONIE HUDDY
Dept. of Zoology
University of New England
Armidale, N.S.W. 2351
Australia

RICHARD L. HUTTO
Biological Sciences
University of Montana
Missoula, MT 59812

DOUGLAS A. JAMES
Dept. of Zoology
University of Arkansas
Fayetteville, AR 72701

WILLIAM H. KARASOV
Dept. of Wildlife Ecology

University of Wisconsin
Madison, WI 53706

JAMES R. KARR
Smithsonian Tropical Research Institute
APO, Miami, FL 34002-0011
(Present address: Dept. of Biology
Virginia Polytechnic Institute and State University
Blacksburg, VA 24061-0406)

CHRISTOPHER J. KELLNER
Dept. of Zoology
University of Arkansas
Fayetteville, AR 72701
(Present address: Dept. of Biology
Virginia Polytechnic Institute and State University
Blacksburg, VA 24061-0406)

PAULA K. KLEINTJES
Division of Biological Control
University of California
Berkeley, CA 94720

DOUGLAS J. LEVEY
Dept. of Zoology
University of Florida
Gainesville, FL 32611

BETTE A. LOISELLE
Dept. of Zoology
University of Wisconsin
Madison, WI 53706
(Present address: Center for Water
and the Environment
Natural Resources Research Institute
University of Minnesota
Duluth, MN 55811)

RICHARD W. LUNDQUIST
Wildlife Science Group
College of Forest Resources AR-10
University of Washington
Seattle, WA 98195
(Present address: Raedeke Associates Scientific
Consulting
5711 Northeast 63rd
Seattle, WA 98115)

JOHNATHAN D. MAJER
School of Biology
Curtin University of Technology
GPO Box U1987
Perth, W.A. 6001, Australia

BRYAN F. J. MANLY
Dept. of Mathematical Statistics
University of Otago
Dunedin, New Zealand

DAVID A. MANUWAL
Wildlife Sciences Group
College of Forest Resources, AR-10
University of Washington
Seattle, WA 98195

JINA M. MARIANI
Wildlife Sciences Group
College of Forest Resources, AR-10
University of Washington

Seattle, WA 98195
(Present address: USDA Forest Service
White River Ranger District
857 Roosevelt Ave., E.
Enumclaw, WA 98022)

THOMAS E. MARTIN
Arkansas Cooperative Fish and Wildlife Research Unit
Dept. of Zoology
University of Arkansas
Fayetteville, AR 72701

PETER J. MARTINAT
Forest Arthropod Research Team
Division of Forestry
West Virginia University
Morgantown, WV 26506
(Present address: Chemical and Agricultural
  Products Division
Abbott Laboratories
14th and Sheridan
North Chicago, IL 60064)

RICHARD R. MASON
USDA Forest Service
Pacific Northwest Research Station
Forestry and Range Sciences Lab
LaGrande, OR 97850

BRIAN A. MAURER
Dept. of Zoology
Brigham Young University
Provo, UT 84602

LYMAN L. MCDONALD
Dept. of Statistics and Zoology
University of Wyoming
Laramie, WY 82071-3332

LAURIE MCKEAN
Dept. of Ecology and Evolutionary Biology
University of Arizona
Tucson, AZ 85721
(Present address: Dept. of Zoology
Box 7617
North Carolina State University
Raleigh, NC 27695-7617)

SHAPELLE MCNEE
School of Biology
Curtin University of Technology
Kent Street
Bentley, 6102 W.A., Australia

DONALD B. MILES
Dept. of Zoological and Biomedical Sciences
Ohio University
Athens, OH 45701

TIMOTHY C. MOERMOND
Dept. of Zoology
University of Wisconsin
Madison, WI 53706

MICHAEL L. MORRISON
Dept. of Forestry and Resource Management
University of California
Berkeley, CA 94720

DOUGLASS H. MORSE
Graduate Program in Ecology and Evolutionary Biology
Division of Biology and Medicine
Brown University
Providence, RI 02912

EDGAR BARRY MOSER
Dept. of Experimental Statistics
Louisiana Agricultural Experiment Station
Louisiana State University Agricultural Center
Baton Rouge, LA 70803

GERALD J. NIEMI
Center for Water and the Environment
Natural Resources Reserch Institute
University of Minnesota
Duluth, MN 55811

BARRY R. NOON
Redwood Sciences Laboratory
USDA Forest Service
1700 Bayview Drive
Arcata, CA 95521

SUSAN NOSKE
Dept. of Zoology
University of New England
Armidale, N.S.W. 2351
Australia

W. S. PERRIMAN
School of Mathematics and Statistics
Curtin University of Technology
GPO Box U987
Perth, W.A. 6001, Australia

DANIEL L. PETIT
Dept. of Zoology
University of Arkansas
Fayetteville, AR 72701

KENNETH E. PETIT
348 Church St.
Doylestown, OH 44230

LISA J. PETIT
Dept. of Zoology
University of Arkansas
Fayetteville, AR 72701

CATHERINE M. RALEY
Dept. of Zoology
University of Wyoming
Laramie, WY 82071

C. JOHN RALPH
Redwood Sciences Laboratory
USDA Forest Service
1700 Bayview Drive
Arcata, CA 95521

MARTIN G. RAPHAEL
Forestry Sciences Laboratory
USDA Forest Service
222 South 22nd Street
Laramie, WY 82070
(Present address: Forestry Sciences Laboratory
3625 93rd Ave. SW
Olympia, WA 98502)

HARRY F. RECHER
Dept. of Ecosystem Management
University of New England
Armidale, N.S.W. 2351
Australia

J. V. REMSEN, JR.
Museum of Natural Science
Foster Hall 119
Louisiana State University
Baton Rouge, LA 70803

SCOTT K. ROBINSON
Illinois Natural History Survey
607 East Peabody Drive
Champaign, IL 61820

KENNETH V. ROSENBERG
Museum of Natural Science and Dept. of
 Zoology and Physiology
Louisiana State University
Baton Rouge, LA 70803

JOHN T. ROTENBERRY
Dept. Biological Sciences
Bowling Green State University
Bowling Green, OH 43403

DAVID L. ROWNEY
Division of Biological Control
University of California
Berkeley, CA 94720

HOWARD F. SAKAI
Redwood Sciences Laboratory
USDA Forest Service
1700 Bayview Drive
Arcata, CA 95521

THOMAS W. SHERRY
Dept. of Biological Sciences
Dartmouth College
Hanover, NH 03755
(Present address: Dept. of Biology
2000 Percival Stern Hall
Tulane University
New Orleans, LA 70118)

KIMBERLY G. SMITH
Dept. of Zoology
University of Arkansas
Fayetteville, AR 72701

FREDERICK M. STEPHEN
Dept. of Entomology
University of Arkansas
Fayetteville, AR 72701

DAVID W. STEPHENS
Dept. of Zoology
University of Massachusetts
Amherst, MA 01003
(Present address: School of Biological Sciences
University of Nebraska
Lincoln, NE 68588)

ROBERT C. SZARO
Rocky Mountain Forest and Range Experiment Station
Forestry Sciences Laboratory
Arizona State University
Tempe, AZ 85287
(Present address: USDA Forest Service
P.O. Box 96090
Washington, D.C. 20090-6090
ATTN: Room 610 RP-E)

SCOTT B. TERRILL
Max-Planck-Institut fur Verhaltensphysiologie
Vogelwarte Radolfzell
D-7760 Schloss Moggingen
Federal Republic of Germany
(Present address: Dept. of Biological Sciences
Siena College
Loudonville, NY 12211)

TOROLF TORGERSEN
USDA Forest Service
Pacific Northwest Research Station
Forestry and Range Sciences Lab
LaGrande, OR 97850

JARED VERNER
Pacific Southwest Forest and Range Experiment Station
USDA Forest Service
2081 East Sierra Ave.
Fresno, CA 93710

GERALD W. WALLIS
Dept. of Entomology
University of Arkansas
Fayetteville, AR 72701

ROBERT C. WHITMORE
Forest Arthropod Research Team
Division of Forestry
West Virginia University
Morgantown, WV 26506

NOMA C. WILKINSON
Dept. of Zoology
University of Arkansas
Fayetteville, AR 72701

PAMELA L. WILLIAMS
Museum of Vertebrate Zoology
University of California
Berkeley, CA 94720

HENK WOLDA
Smithsonian Tropical Research Institute
APO, Miami, FL 34001-0011

MARK S. WOODREY
Dept. of Zoology
Ohio State University
Columbus, OH 43210
(Present address: Dept. of Biological Sciences
Univ. of Southern Mississippi
Hattiesburg, MS 39406)

INTRODUCTION

MICHAEL L. MORRISON, C. JOHN RALPH, AND JARED VERNER

Studies of foraging behavior and food resources comprise part of an overall attempt by biologists to associate behavior, distribution, and abundance of birds to their biotic and abiotic environments This is part of a natural progression. Inferences about bird-habitat relationships lead to questions involving environmental requirements, including those of food availability and the birds' use of that food. Studies of foraging in this century began with qualitative descriptions of habitat and foraging locations and advanced to more quantitative analyses of food habits and foraging behavior. Field work in the 1970s and 1980s emphasized quantification of rates of movement, intersexual and interseasonal changes in resource use, and even experiments designed to assess the impact of birds on their prey. In the early 1980s, an increasing number of studies related the "use" of prey or substrates to their "availability," because theoretical developments suggested the importance of these factors for assessing interactions within and among species.

Exploration of any biological process, including foraging, often proceeds logically from a theoretical framework to study design, data collection, data analysis and interpretation, and, finally, publication. It seems to us that contemporary biologists have given much attention to the theoretical framework (e.g., habitat selection, foraging theory, competitive interactions) for their studies, as theories have received extensive attention in the literature. Furthermore, biologists are gaining an appreciation for the value of proper statistical analyses of their data. Unfortunately, much less attention has been given to the intermediate step of study design: duration, temporal and spatial scale, number and training of observers, needed sample size, independence of observations, and the usually complex interactions among these and other factors. Conclusions based on poorly designed studies are suspect, and usually such studies cannot be rescued by statistical manipulations.

We believe that careful attention to study design is an essential precursor to every investigation, and a primary objective of this symposium is to focus attention on those aspects that pertain to foraging studies. As numerous papers in this symposium show, rigorous design features required for an adequate study have seldom been met in the past. While this is not a reason to discard all previous literature on avian foraging behavior, it does require researchers to decide critically which previous literature meets the standards that current research shows to be necessary.

This symposium emphasizes *what, when, where,* and *how* data on avian foraging behavior should be collected. It is not merely a compilation of natural history notes, although much good natural history will be found here. The various papers deal with aspects of sampling methods, foraging behavior, food resources, foraging theories, sources of bias, needed sample sizes, and so on. Specifically, these proceedings have been divided into six major subject areas:

Role of Birds in Natural Ecosystems and the Quantification of Resources
    Quantification of Resources
    Quantification of Diets
Foraging Behavior: Design and Analysis
    Observations, Sample Sizes, and Biases
    Intraspecific, Spatial, and Temporal Variables
    Analytical Methods
Specialists Versus Generalists
Energetics and Foraging Theory
    Energetics of Foraging
    Behavioral and Theoretical Considerations

Alert readers will soon realize that many problems bedevil studies of avian foraging behavior. As shown herein, the challenges of sampling variable food supplies; accounting for observer variability; phenological, seasonal, and annual variability; geographic variability; sex and age-class variability; and the extraordinary sample sizes often needed, all result in high costs in time and money, and will put extreme demands on our ability to design and execute future studies. These considerations must be recognized in advance of initiating any study.

It is probably wise for us all to admit that it may be impossible to conduct many of the studies we would like to, given the many factors—and their interactions—that influence bird foraging. Critical here is the clear statement of objectives, followed by careful evaluation of how each type of variability will be addressed and the number of samples necessary to attain those objectives. Attempting to address multifaceted objectives with inadequate sampling effort gives results with little or no predictive ability; and predictability is one of the goals of scientific research! The result is paper after paper presenting empirical results, but no concomitant refinement of theory. Without increased attention to, and discussion of, study design, progress in this and other aspects of avian ecology will be slow.

SECTION I

ROLE OF BIRDS IN NATURAL ECOSYSTEMS AND THE QUANTIFICATION OF RESOURCES

Overview

QUANTIFYING FOOD RESOURCES IN AVIAN STUDIES: PRESENT PROBLEMS AND FUTURE NEEDS

KIMBERLY G. SMITH AND JOHN T. ROTENBERRY

A major goal of avian ecological research is to determine both the role of birds in determining structure and functioning of ecological communities (*sensu* MacMahon et al. 1981), and how distribution and abundance of resources provided in those communities influence dynamics of populations and interactions among species (Wiens 1984b). Thus, with renewed interest in ways in which bird populations influence and react to changes in food availability, many avian ecologists are now attempting to quantify available food resources. Sampling food resources may seem like a simple problem involving only techniques borrowed from other disciplines. However, as papers in this section show, the problem is complex, and pitfalls associated with some sampling techniques make them of little use to ornithologists. Indeed, in some cases, avian ecologists now are asking questions for which standard sampling techniques do not exist.

PRESENT PROBLEMS

The basic problem associated with quantifying food resources in the context of their exploitation by birds is that two different distributions are being sampled simultaneously, each of which (Fig. 1) may be affected by independent processes. Thus, within a given habitat, one finds both a pattern of food availability that is likely controlled by a battery of environmental factors (e.g., Stephen et al., this volume) and a pattern of food exploitation that is likely a result of biological interactions (e.g., Torgersen et al., this volume). Investigators have often assumed that relatively simple processes link those two patterns, such that food exploitation is more or less directly related to food availability (and vice versa), and that this relationship directly reflects fitness of individual consumers. However, a variety of ecological and behavioral "filters" may be interposed between distributions of potential food resources in the environment and the ultimate fitness of birds, and the mapping between the two may often be complex and difficult to describe accurately (Wiens 1984b). Indeed, elucidation of that mapping is the goal of this symposium.

Even without the complication of considering dynamic feedbacks between foraging behaviors of birds and distribution of their prey, the papers in this section point out the variety of problems that confound accurate quantification of food resources. Although compendia of detailed arthropod sampling techniques exist (e.g., Southwood 1978), avian ecologists have difficulty applying those methods, because they often need to characterize entire arthropod communities, whereas most techniques efficiently sample only certain arthropod taxa (Cooper and Whitmore, this volume). Arthropod sampling is further complicated due to patchy distributions that vary substantially in time and space (Majer et al., this volume). Also, different conclusions may be reached concerning relative importance of taxa depending on level of taxonomic identification of arthropod prey items (Cooper et al., this volume), a problem that may be common to many studies where prey items are not identified to species (Green and Jaksić 1983).

Although much of the emphasis of the symposium is on arthropods, sampling plant resources also may present problems. For example, plant ecologists have been relatively uninterested in quantifying fruit abundance, leaving avian researchers to develop their own methods. Blake et al. (this volume) discussed sampling fruits in tropical communities where diversity of both fruits and fruiteaters is high, and where defining a fruit (or at least what part of a plant a particular bird consumes) can be a problem. Standard methods for sampling nectar resources have been established with the help of avian researchers interested in pollination ecology (e.g., Collins et al., this volume).

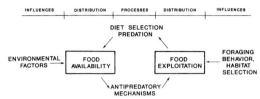

FIGURE 1. A diagrammatic view of the basic problem associated with quantifying food resources when two distributions are being sampled simultaneously. Researchers assume that those two distributions are linked such that food exploitation influences food availability through such processes as diet selection and predation, and that food availability influences food exploitation through antipredatory mechanisms such as crypsis and unpalatability. However, food availability also is influenced by environmental factors and food exploitation is influenced by biological interactions, affecting such things as foraging behavior and habitat selection.

FUTURE NEEDS

Papers in this section present many suggestions for future studies. Some offer general comments concerning ecological studies, while others are directed at specific problems associated with resource sampling. We suggest that the most profitable avenue is one that operates at what we perceive to be the level of the basic problem, that of the dynamic interface between distribution of arthropods and distribution of avian foraging behavior. We recognize, however, that most researchers, either by inclination or training, will tend to emphasize one distribution over the other. For avian ecologists, how exploitation of food resources ultimately affects fitness is a question that all researchers should be interested in, but one that rarely is addressed explicitly.

Several authors pointed out the need for detailed study of bird behavior in relation to specific arthropod prey. In particular, Holmes (this volume) proposed that the two "goals" of a caterpillar are to accumulate biomass and to avoid predation. It accomplishes the first by interacting with a plant and the second by not interacting with a predator. He suggested that predation by birds on canopy arthropods, by numerically reducing prey abundance, has acted as a strong evolutionary selective force, influencing caterpillar foraging behavior, crypsis, and life history patterns. Future studies considering bird-insect interactions also should consider ecological constraints and benefits (e.g., incorporation of secondary substances from plants as a defense mechanism) arthropod prey obtain from insect-plant interactions. Wolda (this volume) identified a need for avian researchers to consider more closely behavior and microhabitat selection of arthropod prey.

Hutto's (this volume) suggestion that changes in foraging behaviors of birds may indicate changes in arthropod abundance is refreshing in its originality, but remains to be confirmed. He also raised old questions that must still be considered: How does one know whether food availability has been adequately measured? How can existing techniques be verified when independent data sets do not exist? How does one know the proper scale of measurement to assess accurately a bird's perception of a food resource? Nonetheless, Hutto's approach explicitly incorporates an examination of the dynamic feedback between avian foraging behavior and distribution of arthropods.

Future studies need to focus on the relative importance of different predator guilds or functional groups (*sensu* MacMahon et al. 1981) on prey populations, and competitive effects of predators on each other. Changes in foraging behavior and habitat distribution of birds in the absence of an avian competitor have been reported (e.g., Sherry 1979, Williams and Batzli 1979b), suggesting that interactions between avian predators might alter patterns of prey exploitation. Researchers working with sessile organisms, such as plants and marine invertebrates, appear to be making progress in delineating fundamental (i.e., preinteractive) and realized (i.e., postinteractive) niches (e.g., Grace and Wetzel 1981). It now remains for clever ecologists to devise experimental methods for teasing apart fundamental and realized food niches of birds in terrestrial communities.

More emphasis must also be placed on experimental approaches. Recent studies that demonstrate the relative importance of different predator groups on an arthropod food resource (Torgersen et al., this volume; Pacala and Roughgarden 1984; Steward et al. 1988b) are especially persuasive because of the experimental designs that were used.

We strongly agree with Dahlsten et al. (this volume) that ornithologists should consult with entomologists about arthropod sampling, as new techniques are continually being developed. It seems as presumptuous for avian researchers to devise arthropod sampling techniques as for entomologists to invent techniques for censusing birds.

A problem common to many arthropod sampling techniques is that they only measure standing crop (Hutto, this volume; Cooper and Whitmore, this volume; Wolda, this volume), which may reveal very little about arthropods that are important to birds (see Martin 1986). Another problem seldom discussed is that researchers and arthropod predators are simultaneously sampling the same distribution, so that what is really sampled is the residue of predation. Both problems seem to lend themselves to experimental manipulation, as demonstrated in the exclosure study by Mariani and Manuwal (this volume).

Future studies must address components of variation found in food resource populations. As shown by Majer et al. (this volume), statistical analyses can be designed to handle variations within and between intraspecific and interspecific distributions. Geographic variation in arthropod communities or patterns of exploitation by bird communities is another topic that is rarely addressed (Wolda, this volume). The study of spatial and temporal variation in fruit abundance in relation to exploitation patterns of birds also has just begun to receive the attention that it deserves (Loiselle and Blake, this volume).

Deciding how to analyze arthropod samples can be a sticky problem (Cooper et al., this volume), particularly because most ornithologists cannot identify arthropods to species. Although one might like to have that level of precision, it is often only necessary to know how many different species are present (Wolda; Stephen et al.; Cooper et al.; this volume). In those cases, we suggest that researchers consider the use of operational taxonomic units (Vandermeer 1972), since arthropod species can just as easily be given numbers as names. We have found that seemingly difficult arthropod groups such as spiders can usually be identified on the spot (e.g., Smith et al. 1988). In cases where it is necessary to identify individual species, ornithologists must rely on their entomologist colleagues, with whom collaboration can be stimulating and productive (e.g., Stephen et al., this volume; Steward et al. 1988a, 1988b).

A general conclusion from this section is that sampling avian food resources in a meaningful manner is a difficult problem that, in some cases, seems nearly impossible and intractable, particularly in complex communities. However, there appear to be steps that researchers can take to alleviate some of those problems. In some cases, examining relatively simple communities may lead to greater insights concerning interactions between predators and their exploitation patterns of a food resource (e.g., Pacala and Roughgarden 1984). Studies can be designed that have a broad geographical scope, yet examine only a few species on a local basis (e.g., Wiens and Rotenberry 1979). Initially focussing on a single bird species (e.g., Mariani and Manuwal, this volume) or a few bird species may be another way to gain information concerning avian exploitation patterns in complex communities. Finally, situations where many species of birds are exploiting the same food resource may hold some promise for gaining insights into ways in which food availability can influence exploitation patterns (e.g., Collins et al., this volume; Hutto, this volume; Kellner et al., this volume; Loiselle and Blake, this volume).

ACKNOWLEDGMENTS

C. Kellner, T. Martin, D. Petit, and L. Petit offered suggestions on an early draft of this summary. Smith's participation in this symposium was made possible by an award from the Dean of the J. William Fulbright College of Arts and Sciences, University of Arkansas, Fayetteville, and National Science Foundation grant BSR 84-08090. Rotenberry was supported by the Graduate School, Bowling Green State University.

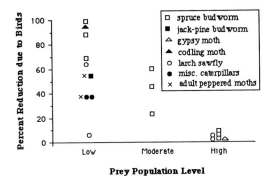

FIGURE 1. The impact of predation by forest birds on Lepidoptera as a function of prey density (see Table 1 for references and text for further explanation).

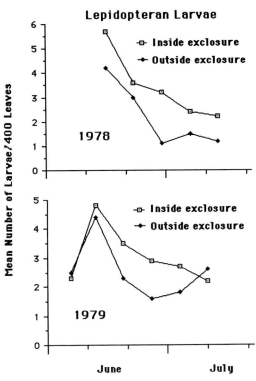

FIGURE 2. Densities of Lepidopteran larvae on foliage inside and outside of 10 exclosures in 1978 and 1979 in the Hubbard Brook Experimental Forest, N.H. Data from Holmes et al. (1979c) and Holmes and Schultz (unpublished).

tion technique in a detailed and rigorous way, reasonably well controlled exclusion experiments, or studies of predation rates on released adult moths (Table 1). Because most such studies were done on prey populations that regularly undergo periodic irruptions (and often cause economic damage), I have classified the data from each study as being obtained during periods of low, moderate, or high population levels of the prey, based largely on the authors' assessments. High levels generally represent periods of insect outbreaks in which defoliation is extensive, moderate levels are those in transition before or after peak irruptions, and low levels reflect "normal," nonoutbreak conditions.

Comparison of results of studies listed in Table 1 reveals (Fig. 1) two major points. First, it seems that birds take only a small percentage of the available insects when they are present in high densities. Although they exhibit both numerical and functional responses to increasing prey densities (Morris et al. 1958, Sloan and Coppel 1968, Mattson et al. 1968, Gage et al. 1970, Holmes and Sturges 1975, Crawford and Jennings 1989), birds seem unable to respond sufficiently to influence the continued rise in the abundance of these prey (McFarlane 1976, Otvos 1979). Although birds cannot keep up with a rapidly expanding defoliator population, their relatively strong impact at endemic levels (Fig. 1) suggests that such predation could delay the onset of an outbreak, as suggested previously (e.g., Morris et al. 1958, McFarlane 1976, Otvos 1979). Indeed, modeling of spruce budworm populations suggests that predation by birds may be a significant factor in maintaining endemic population levels of this species (Peterman et al. 1979, see also Crawford and Jennings 1989).

The second point from Figure 1 is that the impact of bird predation is proportionately much greater when insects are at low densities. This is further illustrated by experiments conducted by me and colleagues at the Hubbard Brook Experimental Forest in New Hampshire (e.g., Holmes et al. 1979c). In 1978 and 1979, we excluded birds from patches of understory vegetation and measured densities of all leaf-dwelling insects inside and outside of these exclosures. We moved exclosures to different patches of vegetation in 1979. In both years, the numbers of Coleoptera, Hemiptera, and spiders were not significantly different inside and outside of the exclosures, probably because these more mobile arthropods could readily move through the approximately 2-cm mesh netting. For Lepidopteran larvae, which are more sedentary, the numbers outside the exclosures were significantly reduced in several of the sampling periods (Fig. 2). Because other predators of these larvae, such as wasps or possibly ants, were not excluded by the netting, the reduction can be attributed almost entirely to birds. In the two years, birds reduced larval numbers by 20 to 63%, varying with the sampling period during the season; the average reduction in each season was 37%. The periods of greatest impact of bird predation were in late June and early July in both seasons (Fig. 2), which were times when birds were feeding nestlings and fledglings and thus when food demand was probably greatest.

These results, along with those in the literature (see Table 1), suggest that birds can have significant numerical effects on insect populations at endemic levels. This finding is particularly significant in view of the fact that most forest-dwelling Lepidoptera and similar species in temperate forests typically occur at low densities and rarely if ever exhibit population irruptions (Morris 1964, Mason 1987b). Even the few species that irrupt become abundant for only short periods and then decline to low population levels for several years (Berryman 1987, Wallner 1987). Moreover, when outbreaks occur, they are often geographically patchy (Campbell 1973, Martinat 1984). The result is that any one forest stand may only occasionally experience an outbreak. For northern hardwood forests, this may be once every 10–20 years (Holmes 1988), much longer than the lifetime of most individual birds. Consequently, birds probably lack highly evolved systems for detecting and responding to such temporal and geographic variability, although a few species may do so (e.g., MacArthur 1958, Morse 1978b). Hence, while outbreaks provide a locally abundant food in some years and places, the endemic population levels of most Lepidoptera and other arthropods provide the majority of the food source for birds most of the time.

Available data, such as those in Table 1 and Figure 1, suggest that the low abundances of insect species may be maintained at least in part by heavy predation pressure from birds, although wasps (Steward et al. 1988b), ants (Campbell et al. 1983), small mammals (Smith 1985), as well as viral and other disease organisms, are undoubtedly involved in various combinations. This general importance of natural enemies in the regulation of herbivorous insects, while controversial (Hassell 1978, Dempster 1983), is also supported by studies of the prey organisms employing key factor analysis and other demographic techniques (e.g., Varley et al. 1973, Pollard 1979, Mason and Torgersen 1987; also see Strong et al. 1984).

It is difficult to generalize about the numerical impact of birds on groups other than Lepidoptera, largely because of the lack of detailed or experimental studies. However, Gradwohl and Greenberg (1982b) showed through an exclusion experiment that tropical antwrens (*Mymotherula fulviventris*) reduced arthropods in dead leaf clusters by about 44%. Likewise, Askenmo et al. (1977) and Gunnarsson (1983) showed that birds removed 17–50% of spiders on spruce foliage over the course of the winter. Other examples of birds reducing local abundances of insects are given by Stewart (1975), Bendell et al. (1981), Loyn et al. (1983), and Takekawa and Garton (1984), and many anecdotal records are cited by Murton (1971), McFarlane (1976), and others. Finally, numerous studies, some manipulative, have found significant effects of bird predation on the abundances of bark beetles and other bark-burrowing insects (see review by Otvos 1979). Taken together, these findings suggest that birds probably have significant numerical effects in a wide variety of habitats and ecosystems.

Finally, contrary to generalizations by Fretwell (1972) and Wiens (1977) that limitation of many temperate bird populations may occur primarily in the winter, evidence is accumulating that food may often limit insectivorous bird populations in the temperate summer (Martin 1987). Recent studies at Hubbard Brook in New Hampshire, for example, indicate that food becomes abundant only during insect outbreaks, which occur sporadically and infrequently (Holmes et al. 1986, Holmes 1988). Birds in these deciduous forests depend heavily on non-irrupting prey, whose abundances they further depress during the breeding period (Holmes et al. 1979c; see above) at a time when the growth and survivorship of newly hatched young are greatly affected (Rodenhouse 1986). Birds in this temperate deciduous forest appear to experience prolonged periods of food limitation (Rodenhouse and Holmes, in prep.) partly because of the strong numerical effect exerted by the birds themselves.

I conclude that birds in temperate forests may exert a strong numerical impact on their arthropod prey, and that this may occur most often during the height of the breeding period. The effect may be to depress or maintain insect numbers at low levels and, in the case of prey species that exhibit population irruptions, to extend the periods between such events. This is consistent with the syntopic population model developed by Southwood and Comins (1976) in which an "endemic ridge" is separated from an "epidemic ridge" by a "natural enemy ravine." More large-scale experiments on the impact of birds and other enemies on endemic prey populations will clarify the extent and influence of such interactions. Extending such studies of the impact of bird predation on defoliators to tropical or other ecosystems, or to other kinds of arthropod prey, should be an important priority.

THE EVOLUTIONARY IMPACT OF BIRDS ON THEIR INSECT PREY

In the long term, the important effect on insect prey of intensive foraging by birds will be evolutionary. For example, the 37–57% predation rates recorded by Kettlewell (1955, 1956, 1973; see Table 1) on the peppered moth have been generally accepted as evidence of strong selection by birds for the evolution of morphological and behavioral traits in this insect (Cook et al. 1986, Endler 1986; but see Lees and Creed 1975). Since available evidence indicates that predation at this level by birds may be common (e.g., Table 1, Fig. 1), it seems likely that birds could have had, and continue to exert, a strong selective influence on their prey. The possibility that birds and other predators have an evolutionary impact on patterns of crypsis and other supposed predator-avoidance traits in insects has long been recognized (e.g., McAtee 1932, Cott 1940) and seems to be more or less taken for granted by many biologists (but see Endler 1986). However, ramifications of bird predation go beyond the evolution of crypsis or other antipredator traits that have not, in my opinion, been adequately considered. These include influences on the life-styles, feeding patterns, and other characteristics of these insects, which in turn affect their involvement and role in ecosystem processes, as I discuss below.

BIRDS AS SELECTIVE AGENTS ON INSECT MORPHOLOGY AND BEHAVIOR

Birds have long been implicated as a major agent of selection for aposematism (Harvey and Paxton 1981) and mimicry (Wickler 1968, Robinson 1969), as well as for nonmimetic polymorphisms in various prey populations (e.g., Cain and Shepherd 1954, Allen 1974, Wiklund 1975, Mariath 1982). Differential predation by birds affects the sex ratio of their prey (Bowers et al. 1985, Glen et al. 1981). Baker (1970) proposed a variety of ways in which predation by birds may have influenced evolution of the sizes, shapes, colors, and behavior of larval and pupal stages of *Pieris* butterflies, and Sherry and McDade (1982) inferred importance of bird predation on the shapes and sizes of tropical insects. Also, the evolution of spines, hairiness, and other similar features of insects and other prey are usually considered to be anti-predator adaptations (Root 1966, Edmunds 1974). Waldbauer and associates (Waldbauer and Sheldon 1971, Waldbauer and LaBerge 1985) proposed that the early-season occurrence of certain hymenopteran-mimicking Diptera was due primarily to strong selection pressures by inexperienced birds foraging in midsummer. Relevant to all of these examples, however, Robinson (1969) pointed out the paucity of experimental evidence concerning the adaptiveness and selective forces influencing such presumed anti-predator traits. Two decades later, this still appears to be the situation.

Nevertheless, passerine birds have been shown to be able to distinguish between shape (Brower 1963), color (Jones 1932, Schmidt 1960, Brower et al. 1964, Bowers et al. 1985), and pattern (Blest 1956, Sargent 1968), which gives them the potential for being discriminate foragers (Curio 1976a). In some early experiments, Ruiter (1952) showed that birds could distinguish geometrid caterpillars from similar inanimate objects (twigs), although movement of the prey was often required for this process to occur. Further, Pietrewicz and Kamil (1977) showed that Blue Jays (*Cyanocitta cristata*) could discriminate cryptic *Catacola* moths on bark, and Mariath (1982) demonstrated that predation rates by birds varied with the proportion and spatial distribution of two morphs of a geometrid caterpillar and with the color of the plant background. Jeffords et al. (1979) painted diurnally flying moths to look like swallowtail and monarch butterflies, and showed that predators, mostly birds, distinguished among the different colors and patterns. Moreover, Chai (1986) showed that jacamars (*Galbula ruficauda*) discriminated among tropical butterflies on the basis of color and of taste, supporting the hypothesis that birds exert strong selection pressures influencing the evolution of mimicry patterns in butterflies. Not all evidence is positive, however. Lawrence (1985), for example, found that European Robins (*Erithacus rubecula*) and Great Tits (*Parus major*) did not easily learn to detect cryptic prey.

The degree to which an insect or other prey item is detectable probably depends most strongly on its choice of substrate and on its movement patterns. Those that choose an inappropriate substrate or that move at the wrong time should be more subject to predation. Wourms and Wasserman (1985) showed experimentally that prey movement influences birds' feeding choices, and Sherry (1984) described how the behavior of certain insects, including their movement patterns, makes them differentially susceptible to bird predators. Since most birds in terrestrial habitats are diurnally active predators that hunt by visual means, they will be actively searching for and taking prey from a variety of substrates, and any prey organism on the wrong background, moving actively, or being otherwise conspicuous will be quickly removed. With many different bird species occupying a single habitat, each with different searching techniques and methods of prey capture (Smith 1974b, Robinson and Holmes 1982, Gendron and Staddon 1983, Lawrence 1985, Holmes and Recher 1986a) and each being fairly opportunistic and catholic in its prey preference (MacArthur 1958, Rotenberry 1980a, Robinson and Holmes 1982, Sherry 1984), the risk of predation is potentially high. Among the bird species in northern hardwoods forests, for example, some closely scrutinize nearby substrates as they move along branches and twigs, some examine undersurfaces of branches and leaves, while others move rapidly and flush prey from the foliage and twigs (Robinson and Holmes 1982). Furthermore, some forest birds differentially search and take prey from upper versus lower leaf surfaces (Greenberg and Gradwohl 1980, Holmes and Schultz 1988) and from particular plant species (e.g., Holmes and Robinson 1981, Holmes and Schultz 1988). They also may use leaf damage caused by chewing insects as prey-finding cues (Heinrich and Collins 1983) or develop search images (Tinbergen 1960) and other forms of learning (Orians 1981) to locate potential prey. All of these factors make it difficult for the prey to go undetected, and likely have led to the evolution of the observed antipredator traits.

The main points are that birds are discriminate foragers and that they use the appearance and behavior of their prey as major cues for locating those prey. These findings, coupled with the possibility that birds are often food-limited and that they can depress the numbers of their prey (except during insect outbreaks), implicates birds as important and significant selective forces that influence the evolution of many antipredator traits found among insects and other prey organisms.

ECOLOGICAL CONSEQUENCES OF THE EVOLUTION OF ANTIPREDATOR TRAITS BY INSECTS

As reviewed above, most considerations of the evolutionary effects of predators on prey have focussed on the morphological (e.g., size, shape, color, hairiness) and behavioral (e.g., background choice, startle responses) traits of the prey. However, other equally interesting and important consequences or ramifications of such traits affect the life-styles and ecology of these prey organisms. For instance, consider a caterpillar that mimics a twig. It must remain motionless on its correct substrate for its crypsis to be effective, and any movement or change in substrate, at least during the day, is likely to increase the probability of its being detected by a foraging bird. Its feeding may therefore be restricted to night hours when its risk of predation by birds is lowest. These constraints in turn affect the ways in which the caterpillar feeds, and hence its pattern of herbivory. Herbivorous insects in temperate forests typically consume <10% of annual leaf production per year (Mattson and Addy 1975, Schowalter et al. 1986); this low level may result in part from the constraints imposed on the major herbivores, namely caterpillars, by their antipredator adaptations (i.e., indirectly by bird predation) and partly by their interactions with the variable quality of the green leaves on which they feed (see below). The hypothesis that I want to develop here is that bird predation, acting in concert with the host plant and other factors, produces selective forces that act to organize and consequently influence the life history patterns—particularly feeding schedules—of leaf-chewing forest insects. The arguments are similar to those of Price et al. (1980), but focus specifically on bird-insect-plant interactions in forest habitats.

Because caterpillars do not mate, defend territories, or feed young (Schultz 1983a), their main "goals" are to accumulate biomass as rapidly as possible and to avoid being killed by natural enemies (i.e., parasites, disease, and invertebrate predators as well as foraging birds; Heinrich 1979c, Schultz 1983a). Means of achieving these goals may conflict. As argued by Schultz (1983a), maximizing feeding time and food quality should involve feeding throughout the day and night and because of variable food quality (see below), the larva may need to move frequently in search of new feeding places. At the same time, to avoid predation, the insect should minimize exposure during feeding, which, if diurnally hunting predators are important, might be done by feeding only at night or at least by restricting movement during daylight hours (Schultz 1983a).

The situation is complicated because the quality of leaves for herbivorous insects varies seasonally (Feeny 1970, Schultz et al. 1982), from tree to tree, from one leaf to another (Schultz 1983a, b), and even among different parts of a single leaf (Whitham and Slobodchikoff 1981). On sugar maple (*Acer saccharum*) and yellow birch (*Betula allegeniensis*) trees at Hubbard Brook, for instance, adjacent leaves on a single branch differ in chemical and physical properties important to herbivorous insects (Schultz 1983b). Since caterpillars are capable of discriminating among chemical cues (Dethier 1970) and of making behavioral "choices" of places to feed (Schultz 1983a), they should be able to respond to such local variation, although this has not been well documented (see below). Furthermore, short-term changes in phenolics and other defensive compounds can be induced by physical damage to the leaves, such as that caused by tearing or chewing (Haukioja and Niemala 1977, Schultz and Baldwin 1982, Baldwin and Schultz 1983, West 1985, Bergelson et al. 1986, Hunter 1987). Silkstone (1987) found that larvae fed less on damaged leaves, while Bergelson et al. (1986) showed that simulated damage to single leaves resulted in a significant increase in phenolic compounds within several days and that larvae moved away from these areas, grew more slowly, and took longer to reach the pupal stages.

Such short term induction of defensive chemicals, if widespread, implies that the longer a caterpillar stays on a leaf, the higher the probability that it will become less palatable. Thus, to optimize feeding and growth, caterpillars may need to move periodically to new leaves in search of higher-quality feeding sites. This results in a trade-off situation: if it feeds and moves extensively during the day, it would be subject to high predation; if it feeds only at night and remains motionless through the day, it would probably not only grow more slowly but also take longer to reach the pupal stage. The latter is important because longer development means the larvae will be exposed longer to natural enemies, including parasites and disease (Pollard 1979, Schultz 1983b, Dammon 1987). Also, in temperate zones, night temperatures in spring and early summer are often cool, which might increase the energetic costs of searching at night, as well as further slowing metabolic processes and therefore growth.

If this scenario is correct, one would expect some relationship between feeding behavior and the antipredator traits of the prey. Surprisingly, little quantitative or experimental data exist on the ecology and behavior of caterpillars with respect to food choice and predation risk, and most of what does exist is anecdotal. Heinrich (1979c) reported that the feeding strategies and time budgets of palatable caterpillars were consistent with their need to minimize predation. The species he observed either fed only at night or stayed on the underside of leaves, and often moved from feeding sites after eating only small amounts of leaf tissue. They also often clipped off partially eaten leaves after feeding on them, which he proposed was an antipredator trait reducing the chances that birds would find the larvae by using leaf-damage cues (Heinrich 1979c, Heinrich and Collins 1983). Unpalatable larvae did not cut off partially eaten leaves, and were often seen exposed while resting and feeding on leaf surfaces during daylight hours (Heinrich 1979c). Bergelson and Lawton (1988) found that larvae of two Lepidopteran species moved relatively little in response to foliage damage, but became more vulnerable to predation by ants, but not by birds, when experimentally forced to move.

Schultz (1983a) found that caterpillars are often specific in their choices. He also described observations of feeding caterpillars that appeared to taste (mandibulating leaf edges) and often reject feeding sites. Lance et al. (1987) report similar behavior by gypsy moth (*Lymantria dispar*) larvae. These observations suggest that some of the partial chewing of leaves reported by Heinrich (1979c) may in fact have represented food choice and later rejection by the caterpillar rather than a predator avoidance trait. On predation risk, Dammon (1987) showed that pyralid caterpillars survived better in leaf rolls than when exposed openly on leaf surfaces, and those on the undersides of leaves survived better than those on upper surfaces. In addition, the risk factor was apparently so important that the larvae chose leaves that were low in food quality. Hairy caterpillars, which are generally less preferred by avian predators (Root 1966, Whelan et al. 1989) might be expected to survive better or to have different feeding patterns from smooth-skinned, cryptic larvae. However, I am unaware of any study that has made such a comparison.

Caterpillars of some species in the forest at Hubbard Brook differ in feeding schedules and patterns of crypsis, which appear to reflect different evolutionary responses to predation risk (Schultz 1983a). For example, *Pero honestaria* (Geometridae) remains motionless all day on large twigs and branches far from feeding places, where it closely matches the background; at night it moves long distances from its resting sites to feeding areas and feeds during the dark hours. A closely related geometrid species, *Anagoga occiduaria*, feeds during both the day and night, but possesses a cryptic pattern that matches the small twigs and petioles near the leaves where it feeds; it is then able to "lean" over and take bites out of leaves during the day with only minimal body movement (Schultz 1983a). Another geometrid, *Cepphis armataria*, matches its own feeding damage on the leaves and thus remains in feeding position throughout the day and night; it feeds around the clock. Thus, different patterns of crypsis seem to allow insects to exploit their food in different ways. This comparison of closely related species is all the more interesting because they co-occur on the same host plant, striped maple (*Acer pensylvanicum*).

Although many of these ideas need experimental verification, and more information is needed on the interactions between bird foraging and prey defenses and feeding, the implications from the hypotheses developed here are that the evolutionary impact of bird predation, although indirect, has important ramifications on the life styles of the prey organisms and affects the structure and functioning of other ecosystem components. Birds are therefore not simply frills in ecological systems, as suggested by Wiens (1973), but exert through their foraging activities important influences in communities on both ecological and evolutionary time scales.

ACKNOWLEDGMENTS

My research at Hubbard Brook, which led to this review, has been supported by grants from the National Science Foundation, with the logistic support of the Northeast Forest Experiment Station, U.S. Forest Service, Broomall, PA. I thank Dr. Jack C. Schultz for his stimulating discussions and collaborations concerning the interactions between birds and insects, and the many colleagues and students who have contributed to the Hubbard Brook studies. J. T. Rotenberry, J. C. Schultz, T. W. Sherry, and B. B. Steele read and constructively criticized early versions of the manuscript. I greatly appreciate their efforts.

Quantification of Resources

PREDATION BY BIRDS AND ANTS ON TWO FOREST INSECT PESTS IN THE PACIFIC NORTHWEST

TOROLF R. TORGERSEN, RICHARD R. MASON, AND ROBERT W. CAMPBELL

Abstract. We used artificial stocking techniques, specialized prey-census methods, and selective exclosures and sticky barriers to identify and quantify bird and ant predation on Douglas-fir tussock moth (*Orgyia pseudotsugata*) and western spruce budworm (*Choristoneura occidentalis*). Fourteen species of birds preyed on tussock moth larvae. We observed losses of 0.08 larvae/m²/day. Six species of birds preyed on tussock moth pupae, among which we observed 6–47% losses from predation. Bird predation was implicated in reductions of 43–71% in egg survival.

Birds and foliage-foraging ants were the dominant predators of budworm larvae and pupae. Predation was studied using bird exclosures around tree branches 2–20 m above the ground, and around entire 9-m-tall trees. Sticky barriers kept ants off branches or trees. When exclosures or sticky barriers were used to protect larvae from predation, 2–15 times as many budworm survived to the pupal stage. At high larval densities survival of protected larvae was about double that of unprotected larvae. At low densities survival was 10–15 times higher among protected larvae. Predation was influenced by crown stratum; ants were most effective in lower strata, and birds excelled higher in the crown. Survival of pupae protected by branch-cages and sticky barriers was four times higher than unprotected pupae.

Predatory ants and many of the insectivorous birds identified in this study are influenced by the availability of standing or down dead wood, or stumps. Forest plans that provide for retention and recruitment of snags or logs can affect the ability of stands to support populations of these beneficial predaceous birds and ants.

Key Words: Predation; insectivorous birds; predaceous ants; exclosure techniques; Lymantriidae; Tortricidae.

The two most important forest-defoliating insects in the Pacific Northwest are the Douglas-fir tussock moth (*Orgyia pseudotsugata*) and the western spruce budworm (*Choristoneura occidentalis*). Their preferred host species are Douglas-fir (*Pseudotsuga menziesii* var. *glauca*) and grand fir (*Abies grandis*). Outbreaks of either species often extend over hundreds of thousands and even millions of hectares. In this paper we summarize studies that describe the population behavior of the tussock moth and budworm, and consider management strategies for preventing or minimizing damage; we also review studies of the possible role of predation in the dynamics of these two important pests. The methods used to identify and quantify predation included specialized prey-census methods, artificial stocking techniques, and selective exclosures and sticky barriers.

STUDIES ON DOUGLAS-FIR TUSSOCK MOTH

Population dynamics

Before starting the predation studies, we had monitored populations of the tussock moth near Crater Lake, Oregon, for several years (Mason and Torgersen 1987). For sampling, we used a pole-pruner and basket to collect tussock moth stages on 45-cm, mid-crown, branch tips (Paul 1979). Branch tips are roughly triangular, so area was calculated as the product of length and width divided by two. Tussock moth density was expressed as the number of larvae, pupae, or egg masses/m² of foliage (Mason 1979). The samples showed that average population density declined over 90% between the early larval stage and the pupal stage late in the season. We knew what proportion of these stages were parasitized, but we could not account for the disappearance of larvae and pupae.

Identifying predation

Larval stocking trials. To identify the causes of these losses, we stocked lower crown branches of host trees with known numbers of larvae. Under one set of branches were drop-trays to catch larvae falling from the foliage. A sticky, polybutene substance prevented escape. Larvae on another set of branches were protected by fine-mesh nylon bags to prevent predation or other losses. By the end of larval development, losses of larvae on the unprotected branches were eightfold higher than on branches protected by mesh bags. We had not actually observed predation or the source of these losses, which we attributed to "arthropod predation," based on the mangled appearance of the dead larvae, "dispersal" when larvae fell to the tray, and "disappearance." Disappearance of small, early larvae was attributed to spiders and predatory insects that left unidentifiable remains. Disappearance of large, late larvae was suspected to be caused by birds (Mason and Torgersen 1983).

To confirm our suspicions regarding bird predation on larvae and to quantify possible predation on pupae and egg masses, we continued artificial stocking trials using tussock moth larvae and, later, pupae and egg masses. The next set of larval stocking trials consisted of cohorts of five larvae each, placed on clusters of four branches with drop-trays below. Each of the three clusters of branches was observed for 4 or 5 hours every third day from a blind about 10 m away. Before each observation period branches were examined for missing larvae, which were replaced as necessary. Foraging visits and observations of apparent predation by birds were recorded. The observer counted the larvae on the branch and in the tray after each visit by a potential predator to confirm predation or dislodging of the prey. We directly observed nine species of birds eating tussock moth larvae, and recorded "suspected" predation by 14 others. In the latter cases birds visited trial branches and appeared to be foraging. Immediately after they departed, one or more larvae had disappeared. Late in the season, after some larvae had pupated, six bird species were also observed preying on pupae (Table 1).

Predation of stocked larvae was expressed as loss per exposure day. The daily loss rate was used to compare mortality between exposure periods of different lengths and examine the relation between predation rates and bird densities. For the 2840 exposure days when 228 stocked larvae disappeared, we calculated a mean, daily, larval loss rate of $0.08/m^2$. We tested differences in mean loss per exposure day among periods, sites, and sites by period. The analysis showed that peak losses occurred during 1–7 August, followed by a general decline in losses toward the end of the season (Fig. 1). Losses of larvae were closely correlated with the total number of about 30 species of birds classed as high-potential predators of tussock moth. Simple correlation analysis indicated that estimated bird density accounted for about 78% ($r = 0.885$, $P < 0.01$) of the variation in loss rate (Torgersen et al. 1984b).

Kendeigh (1970) suggested that birds generally seek prey of a size that produces a food value at least equal to the energy expended for locating and consuming it. Tussock moth larvae apparently do not reach this size—that is, about fourth instar—until late July or early August. The rate of larval loss from the stocked branches was probably influenced by the number of available large tussock moth larvae in the natural population. The observations of Curio (1976), who suggested that birds maintain search images of preferred prey during certain periods, could account for the onset of heavy predation losses. In

TABLE 1. AVIAN SPECIES OBSERVED OR SUSPECTED OF PREYING ON DOUGLAS-FIR TUSSOCK MOTH LARVAE AND PUPAE (TORGERSEN ET AL. 1984B)

Species	Number of prey	
	Larvae	Pupae
Observed predation		
Dark-eyed Junco (*Junco hyemalis*)	4	3
Red-breasted Nuthatch (*Sitta canadensis*)	3	1
Mountain Chickadee (*Parus gambeli*)	3	1
Golden-crowned Kinglet (*Regulus satrapa*)	2	1
Western Tanager (*Piranga ludoviciana*)	2	—
Nashville Warbler (*Vermivora ruficapilla*)	1	1
Black-headed Grosbeak (*Pheucticus melanocephalus*)	1	—
Black-capped Chickadee (*Parus atricapillus*)	1	—
Chipping Sparrow (*Spizella passerine*)	1	—
Subtotal	18	7
Suspected predation		
Red-breasted Nuthatch (*Sitta canadensis*)	10	—
Dark-eyed Junco (*Junco hyemalis*)	7	—
Ruby-crowned Kinglet (*Regulus calendula*)	6	1
Bushtit (*Psaltriparus minimus*)	5	—
MacGillivray's Warbler (*Oporornis tolmiei*)	5	—
Yellow-rumped Warbler (*Dendroica coronata*)	3	—
Chestnut-backed Chickadee (*Parus rufescens*)	2	—
Lincoln's Sparrow (*Melospiza lincolnii*)	2	—
Pine Siskin (*Carduelis pinus*)	2	—
Veery (*Catharus fuscescens*)	2	—
Wilson's Warbler (*Wilsonia pusilla*)	2	—
Cassin's Finch (*Carpodacus cassinii*)	1	—
Solitary Vireo (*Vireo solitarius*)	1	—
White-crowned Sparrow (*Zonotrichia leucophrys*)	1	—
Subtotal	49	1
Total	67	8

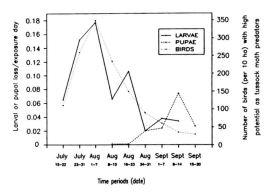

FIGURE 1. Tussock moth larvae and pupae lost per exposure day, and density (per 10 ha) of known or presumed avian predators of the tussock moth, by period, from 15 July to 14 September 1977, Fort Klamath, Oregon (from Torgersen et al. 1984b).

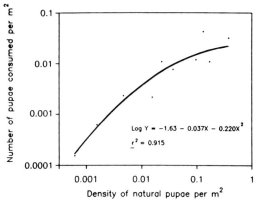

FIGURE 2. Relation of the number of pupae consumed to density of natural Douglas-fir tussock moth pupae. Data from sites near Placerville, California, and Fort Klamath and Malin, Oregon (from Torgersen et al. 1983).

this study, such losses appeared to coincide with the presence of large larvae and peak bird density when birds were foraging both for themselves and for their young.

Rates of predation on stocked branches were higher than those in the natural population. Losses were also higher than those estimated for stocked larvae in a previous study at the same location, where larvae that disappeared or dropped off the foliage were not replaced (Mason and Torgersen 1983). Natural tussock moth larval density was less than $0.05/m^2$ at the time of this study, so that density on stocked branches (about $20/m^2$) was considerably higher than natural densities. Even so, we saw no patterns of losses suggesting that birds or other predators were returning to the trial branches and systematically taking most or all the stocked larvae (Torgersen et al. 1984b).

Pupal stocking trials. The larval stocking study suggested that avian predation might also be a significant mortality factor among tussock moth pupae. To quantify predation, we stocked pupae in the same and one other Oregon site, and at two California sites. Cocooned pupae were produced in the laboratory (Thompson and Peterson 1978) and individually wired to the underside of foliated branches of white fir (*Abies concolor*) to simulate naturally occurring pupae. Pupae were stocked, one to a tree, on branches about 2 m above the ground. Trees were spaced at about 10-m intervals according to the method and plot design described by Torgersen and Mason (1979). We stocked 46–136 pupae at each site for 1–6 years, for a total of 11 place-years.

Two types of predation were observed: either the entire cocoon was missing, leaving only the attachment wire, or, more commonly, the cocoon was torn open and the pupa was missing or only fragments of it remained. When the entire cocoon was missing, or the cocoon was torn open and the pupa was missing, we presumed avian predation. Subsequent observations indicated that some of the predation in which only pupal fragments remained was caused by ants (*Camponotus* probably *modoc*). Pupal mortality of this kind was also observed by Dahlsten and Copper (1979).

Predation of stocked pupae varied from about 6–49% and was inversely correlated ($r = -0.725$; $P < 0.05$) with the estimated density of naturally occurring pupae. In terms of absolute numbers, the maximum number of pupae consumed by predators was less than $0.1/m^2$. Density of natural pupae at each site was estimated directly by sampling branches for cocoons (Mason 1977, 1987a). With increasing prey density, the absolute number of pupae preyed on (natural prey density times percent predation) increased, but at a decreasing rate to a maximum of about $0.04/m^2$. This occurred at a natural pupal density of $0.13/m^2$ (Fig. 2).

Egg-mass stocking trials. In sampling tussock moth egg masses, we noticed that some masses and associated cocoons had been disturbed. Some cocoons seemed to have only a partial complement of eggs, and the remaining portion of the egg mass and the cocoon were tattered. Dahlsten and Copper (1979) suggested that avian predators might account for such partial, tattered egg masses. They also reported predation on egg masses by Mountain Chickadees.

We undertook a stocking study to examine the incidence of both partial loss and complete removal of egg masses at nine sites in Oregon, Idaho, and California. From 1977 to 1981, we collected predation data on these sites for a total of 17 plot-years. Overwintering losses of entire egg masses, presumably from predation, were 5–33% and averaged about 14%. Among surviving egg masses, about two-thirds lost some eggs, and more than half lost about 50% of their eggs. Analyses showed that among the masses that remained in the spring, only about 60% of the original egg complement survived. Thus, in combination, partial predation and complete removal of egg masses resulted in reductions in egg survival of 43–71%, averaging about 52%.

We attributed major egg losses to predation by resident, foliage-gleaning birds. The capture of a Red-breasted Nuthatch in a snap-trap baited with an egg mass, and individual observations of a Dark-eyed Junco and a Nashville Warbler preying on egg masses partly verified our suspicions.

One observation was made of a foliage-foraging ant (*Camponotus* probably *modoc*) pulling apart an egg mass and carrying off an egg (Torgersen and Mason 1987). Dahlsten and Copper (1979) also suggested that ants might be preying on eggs.

Avifaunal censuses. The patterns of predation we observed in the artificial stocking trials may be correlated with avian density or species composition, or with other unknown factors influenced by habitat differences among the sites. Unfortunately, we do not have comparative avifaunal censuses for all sites, but censuses were done on the Oregon sites during the 1977 field season. Avian species composition and density/10 ha were determined from nine straight-line censuses (Emlen 1971) from mid-July to mid-September. These censuses indicated that six of the known avian predators of larvae, pupae, or eggs—Red-breasted Nuthatches, Mountain Chickadees, Dark-eyed Juncos, Golden-crowned Kinglets, Black-capped Chickadees, and Nashville Warblers—numerically dominated the area.

Because larvae, pupae, and egg masses were installed only on lower crown branches, the predation we recorded does not necessarily represent that occurring in other strata. However, we think our values provide a relative index of avian predation on the Douglas-fir tussock moth. They also suggest that such predation is an important component among the mortality factors that keep numbers of this pest low for long periods (Mason and Torgersen 1987).

STUDIES ON WESTERN SPRUCE BUDWORM

Population dynamics

Budworm sampling. In 1979 we began studies to examine budworm population behavior. Four study sites were established in the upper Methow River valley in northcentral Washington. In 1980 we added two study sites about 50 km away on the Okanogan Highlands. Sampling for larvae, pupae, and egg masses was done much as described for the tussock moth. Density of each stage was expressed as number/m² of foliage based on insect counts and measurements of foliage (Srivastava et al. 1984).

Because of our studies on the tussock moth, we were interested in the role birds might have in the population dynamics of the budworm. The literature also suggested that birds were a potentially important source of mortality.

Selective exclusion methods on branches

Identifying predation on larvae. The first set of exclusion trials was done on the three population sampling sites in northcentral Washington in 1979. Our first experimental design used 3/4-m³, single-branch exclosures with frames of 13-mm polyvinylchloride (PVC) pipe covered with 1-cm × 2-cm polypropylene garden mesh. The exclosures were placed on branches of Douglas-fir and grand fir at about 2 m and 5 m above the ground (Campbell et al. 1981). These branches were compared with unprotected control branches at the same heights, but accessible to all predators. The exclosures were installed when budworm larvae had completed spring dispersal and bud- and needle-feeding had begun. Protected and control branches were left undisturbed until all larvae had pupated, when surviving pupae were counted to compare predation among treatments.

Budworm survival was about twice as high on protected branches as on unprotected branches. Survival was significantly higher on protected branches at 2 m, but not at 5 m. Most of the differences, however, were accounted for by two of our three study sites. On these two sites, survival on protected branches was about triple that on unprotected ones. Differences in predation were possibly related to differences in natural budworm densities among the sites. The two sites where survival among treatments was pronounced had budworm densities of about 16/m²; the site where no significant difference was discernible had a density of about 32/m² (Torgersen and Campbell 1982).

Assessing ant predation on pupae. Because we were interested in processes that might maintain sparse budworm populations, we chose an additional Washington study site in 1979, where host trees showed little evidence of either prior defoliation or current budworm activity. We stocked branches with clipped twigs containing pupae inside their web shelters. Fine wire was used to attach the twigs to trial branches. Because numerous colonies of a potentially predaceous ant (*Formica haemorrhoidalis*) inhabited this site, half of the trial branches had a sticky barrier applied to the base of the branch to prevent access by ants. Equal numbers of treatment and control branches in the same whorl were stocked with 5, 10, or 20 pupae at 2 m and 5 m.

Where no barrier had been applied, 84% of the stocked pupae were missing or reduced to fragments after three days, whereas only 8% were missing or in fragments on branches protected by a barrier. Few direct observations of predation by ants were made in 1979, but in repeat experiments in 1980 we observed nearly 100 instances of ants investigating or eating stocked budworm pupae. Other work in our study sites in 1981 and 1982 identified nine species of ants that preyed on budworm pupae (Youngs and Campbell 1984).

Assessing bird predation on pupae. Five branches each were stocked with 5, 10, or 20 pupae at 2 m and 5 m; each branch was protected from predation by birds by a single-branch exclosure. Every branch with an exclosure was accompanied by three unprotected control branches in the same whorl and stocked with the same number of pupae.

About 98% of the pupae installed on the control branches disappeared or were reduced to fragments after 12 days, vs. 84% on branches protected from birds ($P < 0.001$). The status of pupae on protected branches at 2 m differed only slightly from those at 5 m ($P < 0.05$). The relatively small differences emphasized the possible importance of predaceous ants.

Selective exclosure trials. We conducted further experiments to clarify the roles of birds vs. ants as predators of pupae (Campbell and Torgersen 1982). Treatments and a control were randomly assigned to equal numbers of branches stocked with 5, 10, or 20 pupae. We used both sticky barriers and whole-branch exclosures, or sticky barriers alone to exclude both ants and birds, or only ants at 2 m and 5 m.

Survival of budworm pupae was nearly four times higher (49% vs. 13%) on branches with both birds and ants excluded than on unprotected branches. Survival on branches with ants excluded was about three times higher than on controls (36% vs. 13%). Analysis of survival among treatments between crown strata was more complicated. Apparently the sticky barriers offered the pupae little or no protection on branches at 2 m. Occasionally, we watched ants drop from one branch to another, and enough ants may have fallen from higher branches to those at 2 m to confound results on branches with sticky barriers, whether in exclosures or not.

Exclusion trials on whole trees

Selective exclosure methods for whole trees. Results from the single-branch exclosures prompted us to design exclosures for whole trees up to 9 m tall. In 1980, exclosure trials were done on two sites in northcentral Washington and four sites near McCall, Idaho. In 1981, we established four sites near Seely Lake and Potomac in northwestern Montana, where we conducted both population sampling and exclosure trials. In northeastern Oregon, we established five sites for population sampling, only two of which were used for exclosure trials.

Our experimental design was expanded to include four treatments: birds excluded, ants excluded, both excluded, and neither excluded. At each site on grand fir, Douglas-fir, or both, the four treatments were completely randomized, and each treatment was done twice. Birds were excluded by polypropylene garden net attached to a 9-m-tall hexagonal framework of 13-mm PVC pipe reinforced with wooden 2 × 4's (Campbell et al. 1981). Ants were excluded from treatment trees by applying a 50-cm-wide sticky barrier below the base of the live crown. These trials were installed after completion of spring budworm dispersal and before budworm emergence from the host shoots. The exclosures were removed after adult moth emergence. Hence, insects in exclosures were protected during the interval from instar IV to adults.

Beginning density in each trial site was determined from samples of 45-cm branch tips from the upper, middle, and lower crown thirds of the trial trees and 25 additional trees in the site. Plot density based on this sample was determined from equations developed by Srivastava et al. (1984). At the end of the developmental period — that is, when most budworm moths had emerged — the trial trees were dissected. Every branch of each treatment tree was removed. The foliated area of every third, fourth, or fifth branch (depending on the year of study) was calculated, and all pupal remains were counted. Posttreatment density based on dissection of trial trees was expressed as number of surviving budworm per square meter of foliage.

The results of the 1980 trials in Washington and Idaho indicated that at the lowest initial budworm density — about $1.7/m^2$ — 10 to 15 times as many budworms survived on trees protected from both birds and ants as on control trees. Even when density was high, about $25/m^2$, survival continued to be fully twice as high on the doubly protected trees as on the controls. In the 1981 trials in Montana and Oregon, a similar strong inverse relation was apparent between budworm density and the effects of birds and ants. This predation was consistently adequate to reduce survival to about 5% in populations with budworm densities near $1/m^2$. When birds and ants were excluded, survival increased to about 40%. Birds and ants displayed different patterns of predation among crown strata of the trial trees. Ants were most effective in the lower third of the crown; birds were most effective in the upper third (Campbell and Torgersen 1983b, Campbell 1987).

At the lower densities, in both years and all areas, birds alone or ants alone were usually sufficient to greatly dampen the high survival observed when both groups were excluded. In fact, the contribution of either birds or ants largely compensated for the absence of the other guild in the single exclusions. Little or no evidence of further mortality was found after birds and ants were excluded. Thus, during the period from early foliage-feeding larvae through the pupal stage at the densities where we worked, other mortality-causing factors played minor roles (Campbell and Torgersen 1983a).

Single-branch exclosures in tall trees. Based on the apparent differences in predation by birds or ants among crown strata, we hypothesized that birds would continue to be important budworm predators even in trees much taller than 9 m, and that ants would play a decreasing role as tree height increased. Accordingly, we attempted to test our hypothesis on higher branches in tall trees. Because whole-tree exclosures were out of the question, we used single-branch exclosures at two sites in Montana at about 2 m and 20–25 m above the ground in Douglas-fir and Engelmann spruce (*Picea engelmannii*). The sites had widely different budworm densities ($0.28/m^2$ and $23.1/m^2$). A truck-mounted, 27-m hydraulic lift was used to install and remove exclosures, apply sticky barriers, and stock branches with pupae. Pupae were individually wired to branch tips, five to a branch.

Results paralleled those on smaller trees. Across all treatments, predators had relatively minor effects on the high-density site, confirming that predation of both the budworm and the tussock moth by ants, birds, or both was inversely related to insect density (Campbell et al. 1983, Torgersen et al. 1983). On the site with low natural budworm densities, mortality among pupae on both high and low branches protected from birds and ants was about 40%, as compared with 72% on controls. Birds appeared to be more effective predators than ants high in the trees, but were about equally effective in low branches. These results left little doubt that birds and ants, separately or together, were at least as effective predators on high branches of old-growth trees as on branches or trees up to 9 m tall (Campbell and Torgersen 1983b).

Identification of avian predators. In concurrent studies designed to observe and identify avian predators on several of our study sites, Langelier and Garton (1986) and Garton (1987) identified several species of birds that were eating the budworm. Observations and stomach analyses confirmed that about two dozen species of birds were preying on the budworm in these sites. Half of these were also on our list of bird predators of the Douglas-fir tussock moth.

CONCLUSIONS

These studies showed that insectivorous birds and foliage-foraging ants are major predators of two of the most important forest insect pests in the Pacific Northwest. Management-induced habitat changes can influence the abundance and diversity of these predators and other natural enemies of these pests. For example, forest plans that provide for retention and recruitment of snags can affect the ability of stands to support populations of predaceous birds and ants. Almost all of the ants, and many of the birds that prey on the tussock moth and the budworm, are influenced by the availability of standing or downed dead wood. Even birds that are not cavity nesters will use snags for foraging, perching, roosting, or singing.

The need to reduce damage to forests from insect pests suggests that managers view these and other natural enemies as a resource to be conserved and enhanced. One of the great challenges for land-management professionals today is to use new knowledge to broaden their perspectives and expand their management alternatives to maintain and improve forest health. We hope the results reported here will focus more attention on the beneficial role of natural enemies of insect pests in forest ecosystems.

ACKNOWLEDGMENTS

The research reported here was funded in part by two USDA programs: the Douglas-fir Tussock Moth Research and Development Program, and the Canada-USA Spruce Budworms Program. The authors thank the following for reviewing earlier drafts of this paper: M. Brookes, E. Bull, M. Henjum, W. McComb, J. Rotenberry, P. Shea, A. Torgersen, and J. Verner. G. Paul prepared the graphics.

MEASURING THE AVAILABILITY OF FOOD RESOURCES

RICHARD L. HUTTO

Abstract. To assess the role of food supply in the biology of forest birds, available food density must be measured with precision. In reviewing 155 recent papers that deal with the role of food supply, I found that most authors justify use of a particular sampling method by intuitive arguments and numerous assumptions. An intuitive approach may be inadequate, however, because (1) we do not perceive food availability in the same manner that birds do, (2) we ignore scale-of-measurement problems, and (3) we measure only standing crop. To avoid those potential problems, I suggest using quantitative measures of behavioral acts that are necessarily correlated with variation in food abundance as a "check" on the reliability of measurements of food availability. These might include a bird's temporal and spatial attack rate, its mean stop-to-stop movement length, or the proportion of its daily time budget spent foraging. Future studies may be strengthened if such behaviors are used to confirm that a given measure of food availability is appropriate.

Key Words: Food availability; prey density; stomach contents; functional response; feeding rates; search tactics; time budgets.

INTRODUCTION

Of biological parameters that might influence the evolution of adaptations among species, the distribution and abundance of food, predators, and mates are especially important (Krebs and Davies 1987). Virtually every aspect of the life cycle of an individual has been molded to some degree by those variables, as Crook (1964) began to demonstrate in his classic studies of social organization of weaver finches.

Information on food availability alone has contributed to our understanding of numerous life history characteristics and their population- and community-level consequences. The importance of food availability as a hypothesis to explain various biological patterns is reflected in the large number of studies that deal with this issue. For example, in a perusal of a dozen ecological and ornithological journals published since 1978, I located 155 articles on landbirds that dealt specifically with the relationship between food supply and several ecological patterns, including timing of annual cycles, territoriality, habitat selection and territory placement, diet, mating system, clutch size, reproductive success, population size, geographic distribution, and community structure.

The role of food supply in a few of those cases has become clear, either because of an unusual ability to measure food availability precisely (e.g., territoriality in nectarivores, or use of space by ground-feeding shorebirds and insectivores), or because of the ability to manipulate food supply experimentally (e.g., optimal foraging, or clutch-size experiments). The role of food in other arenas of investigation (e.g., timing or occurrence of various annual cycles) has become dogmatically accepted, despite the lack of careful measurements of food resources. The role of food availability for still other (mostly population- and community-level) phenomena remains unresolved and controversial.

The inability to resolve whether food is important often results from difficulty knowing whether food availability has been measured adequately. Often these measures are of questionable relevance to the organisms involved. For example, several authors reported that food density and habitat use by raptors were not well correlated (Wakeley 1978, Baker and Brooks 1981, Bechard 1982), but vegetation structure was related to habitat use. Therefore, vegetation structure was deemed to be more important than food as a factor influencing habitat use, even though the importance of vegetation lay with its effect on food availability. In fact, after converting rodent density (as estimated from trap data) to rodent "availability" (as estimated by multiplying rodent density by the fraction of incident light at ground level), Bechard (1982) concluded that food availability *was* related to habitat use. If the researchers had measured prey availability as perceived by hawks at the outset, then the correspondence between food supply and habitat use would have been more readily apparent.

At the population level, Pulliam and Dunning (1987) argued that local population density of sparrows over a series of years was independent of food abundance, when abundance exceeded some threshold level. They based their conclusion on a lack of correlation between sparrow density and seed availability, as estimated by counting seeds that fell into small traps. However, seed traps may not accurately reflect food availabile to sparrows (especially in view of the unmeasured seed stores that must have been present in the soil). As these two examples suggest, measures of food availability undoubtedly have contributed to the conflicting results and disagreements that surround the more controversial arenas of investigation. Such conflicts have, consequently, led to pleas for greater care in the measurement of food availability (Wiens and Rotenberry 1979, Wiens 1983, Morrison et al. 1987b).

But how can we measure food availability in a biologically meaningful manner? Even if one samples selected prey types from a single microhabitat, the relative prey abundance between sites can differ significantly among sampling methods (Majer et al., this volume). To learn more about the factors that should be considered when measuring food availability, I searched through the current literature for patterns in the way biologists justify their sampling methods. In this paper, I synthesize results of this search, and suggest how we might begin to test whether our measures of food availability are appropriate.

METHODS

After cataloguing the ways by which biologists measure food availability in the field, I chose to concentrate on the arguments given to justify use of a given measure. In addition to including some references published prior to 1978, I searched through all issues of *American Naturalist, Animal Behaviour, Auk, Behavioral Ecology and Sociobiology, Condor, Ecology, Ecological Monographs, Ibis, Journal of Animal Ecology, Journal of Field Ornithology,* and *Wilson Bulletin* published after 1978 for articles involving the impact of food availability on biology of landbirds and shorebirds.

I conducted field studies on the relationship between food availability and bird behavior in western Montana Douglas-fir (*Pseudotsuga menziesii*) forests in 1985 and 1986. In most coniferous forests of the western United States, the western spruce budworm (*Choristoneura occidentalis*) is the most widely distributed and destructive defoliator (Carolin and Honing 1972, Fellin and Dewey 1982). It is also an important prey species for forest birds during the nesting season. The use of systemic insecticide implants (Reardon 1984) was tested in northwestern Montana in 1985 by USDA Forest Service personnel as a method to reduce foliage and cone crop loss. I watched groups of trees that contained both experimentally treated and adjacent untreated trees to discover whether artificially reduced budworm levels on treated trees would affect the probability of a bird visiting a tree, the length of a given visit, or a bird's feeding rate.

Thirty Douglas-fir trees were selected for experimentation by Forest Service personnel associated with the Northern Region Cooperative Forestry and Pest Management Division, and 15 of those trees were randomly chosen for treatment with insecticide implants (Reardon 1984). I used 14 of their treated trees, their 15 control trees, and an additional 20 trees as controls, so that I sampled nine groups of five to seven trees. Each group had at least one, but no more than two treated trees. Three groups of trees were in Lubrecht Experimental Forest of the University of Montana [46°52'N, 113°27'W] within a mixed conifer forest that was dominated by Douglas-fir, and six groups were on Champion International Paper Co. land [46°48'N, 113°33'W] on a pure Douglas-fir site that was commercially thinned in 1980.

Trees were treated with implants on 18 April 1985, and I sampled late-instar budworm larvae on 29 June 1985 by clipping two or three 45-cm terminal branch tips from the lower to middle crown of each tree using a 9-m pole pruner affixed with a collecting bag. Contents were emptied into plastic bags and transported to the laboratory where I sprayed them with a pyrethrin-based insecticide to reduce the activity of budworm larvae. Branch samples were then placed on white cardboard, and foliage surface area was estimated by compacting foliage into the smallest single-layered space possible and measuring length and width of the area to the nearest cm. Each branch sample was searched carefully for budworm larvae and other arthropods, which were then removed and "rinsed" of debris in a wash bowl containing 70% alcohol before being dried through contact with a paper towel and weighed on an electronic balance to the nearest 0.01 g.

From one observation point, each group of focal trees formed a slight semicircle (concave toward the observer) and fell within a 120° arc. Consequently, all trees could be watched simultaneously for bird activity. The observer (myself or an assistant) observed for 90 min before moving to another group of trees. From 18 June to 1 July 1985, we recorded bird activity between 07:30 and 11:00. Observation times were rotated so that each group of trees was watched for 180 min during each half of the morning.

When a bird landed in an experimental tree, we recorded the tree number, time of day, bird species, duration of its stay in the tree (in sec), its activity (feeding, singing, or perching), and when possible, its foraging attack rate (recorded as number of pecks/sec of observation). On rare occasions, when several birds were present at the same time in a group of trees, we noted the identity of each visitor, and estimated the duration of stay for each bird.

In 1986, we studied avian foraging behavior in a Douglas-fir stand 5 km southeast of Missoula, Montana [46°50'N, 113°56'W]. The 5-ha site was traversed in a systematic fashion on a daily basis from mid-June through mid-July. An observer recorded the identity and height of every bird encountered.

RESULTS AND DISCUSSION

Literature review

Food not measured directly. Twenty percent of the authors did not attempt to measure food because their comparison obviously involved relatively food-rich vs. relatively food-poor conditions. For example, Tryon and MacLean (1980) interpreted the use of space by Lapland Longspurs (*Calcarius lapponicus*) in terms of food availability, which was assumed to be greater at times of "cranefly pupation" and when "the tundra was aswarm with adult Diptera." Strehl and White (1986) studied reproduction of Red-winged Blackbirds (*Agelaius phoeniceus*) during years that had and years that did not have a periodical cicada outbreak.

Food measured directly but relevance not addressed. Forty-seven percent of the authors measured food density but made no explicit assumptions about relevance of their measures in terms of food availability. Their implicit assumptions were so reasonable that most of us would not think to question the measures. For example, Baird (1980) and McPherson (1987) measured fruit availability to frugivorous birds by counting fruits on trees in their study areas. Similarly, biologists who have studied nectar-feeding birds generally counted flowers but did not explicitly assume that such samples adequately reflected food available to birds (e.g., Carpenter and MacMillan 1976, Kodric-Brown and Brown 1978, Feinsinger and Swarm 1982).

Food measured directly and relevance addressed. Twenty-four percent of authors took a simple measure and explicitly assumed that it was correlated with food availability. For example, after describing a vacuum sampling technique, K. G. Smith (1982) stated that his "meadow samples reflect actual abundances available to birds." Or, Blancher and Robertson (1987) trapped "flying insects between ground level and 1 m" because that height range represented food availability for Western Kingbirds (*Tyrannus verticalis*). Conner et al. (1986) stated that their sweep samples were not a direct measure of food for Northern Cardinals (*Cardinalis cardinalis*) but that they would "give a relative index of overall food availability." Dunning and Brown (1982) assumed that food resources available to wintering sparrows were "closely and positively correlated" with what they chose to measure: the quantity of precipitation during the previous summer.

Food measured, then adjusted to be more relevant. In still other instances (9% of the studies reviewed), researchers "adjusted" their measures of food density on the basis of some intuitive argument before making the explicit assumption that their adjusted measure accurately reflected food availability. Hutto (1980, 1985a), for example, derived an "adjusted insect density" by multiplying number of insects trapped on sticky boards by a measure of vegetation density. Adjusted density was assumed to be better correlated with food availability to foliage-gleaning insectivorous birds than was either flying insect density or vegetation density alone. Greenlaw and Post (1985) determined the "food value" of Seaside Sparrow (*Ammodramus maritimus*) territories by multiplying volume of potential prey in each of several patch types by a factor that accounted for both relative use and relative abundance of that patch type within the territory. The most common adjustment, however, involved a refinement of food types considered on the basis of stomach contents of the bird species. For example, Bryant (1975a), Zach and Falls (1979), Smith and Anderson (1982), and Smith and Shugart (1987) eliminated prey types from the sample if they were not present in stomachs.

Problems with current methods

I cannot judge the accuracy of any of these methods but, clearly, no current method of measuring food *abundance* is immune to the criticism that it may be an unreliable measure of food *availability*. Baker and Baker (1973) warned that "the food density for shorebirds as revealed by ordinary sampling techniques is related to the food density experienced by the bird by some often complex functions or may be entirely unrelated." Their warning applies equally to forest birds (see Martin 1986, and Wolda, this volume). At least three categories of potential problems would apply to any of the sampling methods outlined above, as discussed next.

We lack the bird's perception and do not know its feeding constraints. Even for relatively simple fruit and nectar systems, all fruits or flowers may not be equally available (as assumed by simple counts). In general, sampling the "kind" of food a bird eats probably falls short of a meaningful measure because the animal's perception screens items in a manner that differs from that of a sticky board (Seastadt and MacLean 1979; Hutto 1980, 1985; Cody 1981), sweep net (Wilson 1978; Wittenberger 1980; Fischer 1981, 1983; Folse 1982; Laurenzi et al. 1982), vacuum cleaner (Craig 1978, K. G. Smith 1982, Smith and Anderson 1982, Ault and Stormer 1983), suction trap (Bryant 1975a, Holmes et al. 1978, Turner 1982), snap trap (Wakeley 1978, Baker and Brooks 1981, Bechard 1982), or visual count (Salomonson and Balda 1977; Holmes and Robinson 1981; Schluter 1982a, b; McFarland 1986a). Items will be sampled differentially because of mechanical and perceptual differences between a given sampling technique and a bird (Robinson and Holmes 1982, Heinrich and Collins 1983, Sherry 1984). Moreover, lacking a bird's perception, we do not know which prey items it would ignore because of the prey's crypticity (Janzen 1980a), inaccessibility (Kantak 1979, Moermond and Denslow 1983, Avery and Krebs 1984), difficulty of capture (Hespenheide 1973a), mechanical defenses (Davies 1977b, Sherry and McDade 1982, Heinrich and Collins 1983) or chemical defenses (Eisner 1970, Janzen 1980a).

Some believe that these perceptual problems can be solved by adjusting sampling methods to match stomach contents. They reason that if the sample has the same prey types as stomach contents, the sample will be relevant. Although stomach contents can help refine one's definition of available prey types, differential digestibilities among prey types (Custer and Pitelka 1975), variation in times of collection, and differences between diets of adults and the young they feed will cause biases in the estimate of what the bird actually takes from the field. It is not a simple matter to determine a bird's diet. That issue aside, a mere listing of the contents of both stomachs and field samples to show that they are "more or less the same" (Terrill and Ohmart 1984, Blancher and Robertson 1987), or adjusting the measure of food availability by eliminating what is not in stomachs (Feinsinger et al. 1985, Smith and Shugart 1987), does not necessarily solve the perception problem. Unless diet and field samples have the same proportions of item types, they are not likely to have been sampled with the same perceptual "filter." Even if fruit species A were the only prey type sampled from the environment and the only prey type found in stomachs, not all fruits are equally accessible; a mere tally of the appropriate food type may be an inadequate representation of food availability. In short, without accounting for a bird's perception, simple biomass measures (even "adjusted" ones) are probably poor reflections of actual food abundance available to birds (see also Moermond, this volume).

We ignore scale-of-measurement problems. Scale problems of tremendous magnitude occur when determining food availability, and these seem to be routinely ignored by researchers. Consider the following hypothetical problem. Suppose we want to test whether number of feeding trips/nestling/hour is related to food availability. Food availability would have to be measured and averaged over a unit at least as large as a territory—the unit searched by the bird for food. One could not use a single trap on each territory to represent conditions over the whole territory unless variation among traps within a territory was known to be less than variation between territories. Similarly, imagine a system where the ranking of areas by food density (measured as amount of food per branch) differs markedly from a ranking of those same areas when food density is measured as amount of food/leaf, or food/tree (Fig. 1). Holmes and Robinson (1981) measured food availability in terms of numbers of arthropods per cm^2 of leaf area after counting 400 leaves. Would number/leaf be a better indication of value of the tree to the bird, or perhaps number/tree? Such problems are not trivial because number/leaf cannot be extrapolated to number/study area (and vice versa) unless food is distributed uniformly throughout. Since food is not so distributed, one's density estimate will vary with the scale of measurement. So which scale of measurement is correct?

We measure standing crop only. Most of our measures of food availability are equated with standing crop volume, number, or biomass (Carpenter 1987), even though bird behavior can depend on whether a patch of food is depletable (Kamil and Yoerg 1985). With the exception of nectar resources (Gill and Wolf 1979, van Riper 1984, Feinsinger et al. 1985), attempts to measure (or even discuss the effects of) renewal rates for continuously renewing food resources are rare (notable exceptions include Zach and Falls 1976b, 1979; Davies 1977a; Davies and Houston 1981, 1983). Yet, an area with two food items/m^2 that is restored to the same density within a second after removal of an item has much greater food availability than another with 20 items/m^2 and no renewal. Furthermore, a place with greater food density at the time of sampling is assumed to have more later, even though some food resources (e.g., fruit and seeds) are not continuously renewing.

Toward the validation of food availability measures

Given the potential problems, do authors ever attempt to confirm the appropriateness of their chosen method, beyond the use of intuitively logical assumptions and adjustments? They generally do not, based on my literature search. Occasionally authors will compare two methods of sampling food and presume that agreement between the two means that either is valid. For example, Ault and Stormer (1983) vacuumed the soil and got the same seed types that scraped samples produced, so they concluded that any dietary deviation from the sample would reflect a food "preference" by birds. A correlation between the abundance measure of two samples does not, however, validate either as an adequate measure of food availability. Not only has the animals' perception been ignored but, also, identical sample contents from two methods do not guarantee the correctness of either.

Most of us would consider stomach contents to be one way to validate sampling methods, but stomach contents can only guide one's "adjustment" of a measure to be closer to what the bird actually experiences. Samples that match stom-

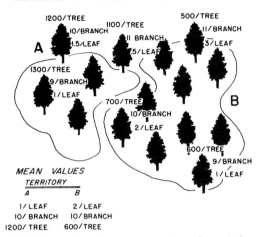

FIGURE 1. Hypothetical examination of the relative availability of food in two areas. Results depend on whether the unit area used to estimate food density is a single leaf, a branch, or a tree.

ach contents still do not address the perception problem or the other two classes of problems outlined earlier. Is it possible, then, to confirm that one's sampling method is meaningful? How do we know when we have measured actual food availability? At least one possibility deserves consideration.

Why not use patterns of bird behavior to confirm that our measures of food availability are appropriate? In fact, because conditions good for one individual may be poor for another, we cannot afford to measure food availability independently from bird behavior. Even the same food abundance can change in "availability," for example, as the thermal load of a bird changes (Clark 1987). If variation in some behavioral act were necessarily correlated with food availability, we might be able to use that behavior to "check" the validity of a food availability measurement made for some other purpose. Figure 2 depicts the essence of this argument. Normally, to understand whether food availability affects some biological phenomenon, we measure food by one of the four approaches categorized earlier, and then interpret results. I suggest that we simultaneously monitor a behavioral act, the rate of which is known to be influenced by food availability, and check the (partial) correlation between our food measure and the behavior. A significant correlation between our chosen measure and an act that is known to be related to food availability would strengthen the argument that we have measured food availability adequately.

Food availability surely affects some aspects of foraging behavior in predictable ways (Robinson and Holmes 1982, 1984). Indeed, birds can rapidly adjust their foraging behavior in response to prey availability (Paszkowski 1982, Pienkowski 1983). But which behaviors have been shown to be universally correlated with variation in food abundance under well-controlled experiments, such that we might use them to find a meaningful sampling method?

To find such a behavior, we must look at systems in which food availability can be undeniably ranked independently from bird behavior. Laboratory systems and field systems in which vegetation structure is relatively simple and available prey types are limited in number should allow one to measure food availability as accurately as possible. For example, in western Montana, Douglas-fir often occurs in homogeneous, nearly monospecific stands. Little other than western spruce budworm is available as a food source in early summer. On the basis of foliage samples taken from a series of 48 trees in June 1985, spruce budworm larvae comprised 72% of the 1035 arthropods that I collected. The predominance of spruce budworm larvae was most evident in the biomass measurements, however, where they comprised 96% of the total. Analyses of stomach contents from mixed-conifer forests in both Washington and Montana confirm that most forest passerines depend heavily, if not exclusively, upon budworm larvae for food from May through July (DeWeese et al. 1979). Remarkably, species that are known to feed on the ground during most other times of the year [American Robin (*Turdus migratorius*), Chipping Sparrow (*Spizella passerina*), and Dark-eyed Junco (*Junco hyemalis*)] fed extensively on larvae in trees from mid-June to mid-July; the entire insectivorous bird community appeared to rely on this single food source during the breeding season. Recognition that forest birds depend heavily upon lepidopteran larvae at this time is nothing new (MacArthur 1959, Robinson and Holmes 1982), but the preponderance of western spruce budworm larvae in both field samples and diets means that food availability should be exceptionally easy to estimate in that habitat type at that time of year.

The mean density of late-instar budworm larvae was significantly less on trees treated with systemic pesticide implants than on control trees during the year of treatment (Table 1). Twelve bird species visited the experimental control tree groups, and individuals of each species were observed eating or gathering budworm larvae. Limited sample sizes prohibited a meaningful species-by-species analysis, but results pooled across species showed that neither the probability of a

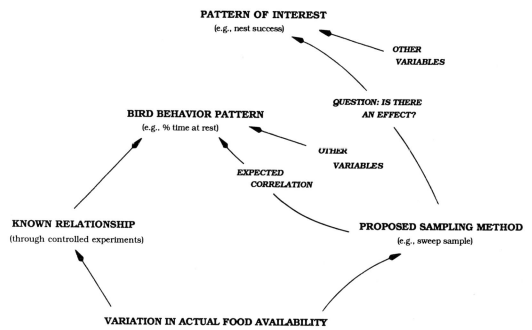

FIGURE 2. Diagrammatic illustration of the way bird behavior might be used to "screen" possible sampling methods, so that an appropriate measure of food availability is selected.

bird entering a tree (visits/hour) nor the mean duration of a bird's stay was significantly greater in trees that harbored more food. The same held true if I considered feeding observations only, although the trend in all cases was to spend more time in trees with higher food densities. In contrast, mean attack rate of birds that foraged in systemically treated trees was significantly less than mean attack rate in control trees.

In an effort to uncover a series of easily quantified behavioral variables (such as attack rate) that might be unquestionably related to food availability, I searched through the literature for additional laboratory or field studies that bore on the relationship between behavior and food availability. I found information on the following behavioral acts:

Temporal attack rate (number of attacks/unit time). Based on the well-studied functional responses of animals to prey density (Holling 1965, 1966), feeding rate of a predator should be proportional to food density until it can increase no further because of satiation or handling limitations. Linear (Type I) and exponential (Type II) responses have been shown to exist for birds that feed, respectively, on invertebrate or seed resources in the wild (Schluter 1984). Therefore, providing that we record foraging observations

TABLE 1. A Comparison of Mean Values of Bird Use between Systemically Treated and Control Trees

Measure	Untreated trees			Treated trees			U[a]	P
	N	$\bar{X}$	SE	N	$\bar{X}$	SE		
Budworm density								
No. budworms/m^2	34	135.50	14.3	14	36.70	6.9	58	0.000
No. budworms/tip ($\times 100$)	33	7.87	0.7	14	3.13	0.6	82	0.000
No. budworms/g ($\times 100$)	33	6.19	1.3	14	2.04	0.7	59	0.000
Bird use								
No. visits/hr	37	0.78	0.2	14	0.74	0.2	256	0.950
Duration of visit (sec)	160	58.24	6.4	58	48.10	7.2	4275	0.373
No. pecks/sec ($\times 100$)	30	4.80	0.8	13	1.90	0.6	106	0.016

[a] Mann-Whitney U-statistic.

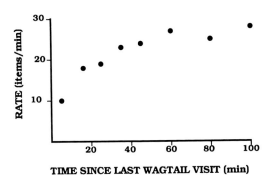

FIGURE 3. Pied Wagtails (*Motacilla alba*) exploit a renewing food supply. The feeding rate of an individual within a patch depends on the time since that patch was last depleted by the same or another wagtail. Foraging attack rate is clearly related to food availability to the birds (redrawn from Davies and Houston 1981).

during periods of active feeding prior to satiation, we should expect a more or less linear relationship between food density and feeding rate.

My experimental decreases in spruce budworm density on selected trees produced decreased attack rates by foliage-gleaning birds. Many other studies have produced similar results (Goss-Custard 1977a, Greenwood and Harvey 1978, Tinbergen 1981, Paszkowski 1982, Pienkowski 1983, Robinson and Holmes 1984, Schluter 1984, Maron and Myers 1985, Marcotullio and Gill 1985). Perhaps the strongest field study of this sort is that of Davies and Houston (1981), who worked with a relatively simple two-dimensional system. They found a Type II relationship between peck rate of ground-feeding wagtails on a patch and the time since last visit to the patch (Fig. 3). The relationship between prey availability and feeding rate seems irrefutable in this instance.

Neither Davies (1977a), Morse (1981), Møller (1983), nor Roland et al. (1986) found correlations between their measures of food availability and feeding rate. Careful examination of methods, however, revealed that food availability was not measured well or was not the only variable likely to have influenced feeding rate. Specifically, Davies measured prey availability by using a cumulative-total trap method, which may not have reflected food availability accurately over a smaller portion of the day. Morse did not measure food directly. Møller compared attack rates among seasons, over which time period the food types changed dramatically. And Roland et al. used number of larvae per cluster as a measure of food availability, which may not have been the best scale of measurement for determining food availability because a tree could have only a few leaf clusters despite a high density of larvae per cluster.

Spatial attack rate (number of pecks/unit distance). Intuitively, it seems that number of items taken per unit distance traveled should be greater in relatively food-rich areas. Goss-Custard (1977c) showed such a response for Redshanks (*Tringa totanus*) feeding on large worms in the mud, and Hendricks (1987) used this measure after assuming it to be well correlated with food availabile to Water Pipits (*Anthus spinoletta*). The relationship deserves further study.

Rate of progression (distance/unit time). The number of steps that a shorebird takes following a successful capture is generally less than the number following an unsuccessful capture (Baker 1974). Thus, movement rate might decrease when a bird is in a relatively food-rich area. Area-restricted searching would also predict a slower rate of beeline progression with an increase in prey availability. In apparent contrast with these expectations, Baker and Baker (1973), Baker (1974), and Zach and Falls (1976c) found that movement rate was positively correlated with temporal attack rate (= food availability?). Goss-Custard (1970) found no relationship between the number of paces/min and prey density, whereas Zach and Falls (1979) found rate of progression (based on beeline distances) to be positively related to food supply.

Search velocity (hops or perch changes/unit time). Search velocity has been shown to be well correlated with temporal attack rate (Robinson and Holmes 1982). Because it may be easier to measure than attack rate for birds that forage in dense vegetation, search velocity might be more useful.

Average stop-to-stop movement length (hops/unit distance). We might expect a greater number of hops per unit distance in relatively food-rich areas because of area-restricted searching, which has been shown to occur after a successful capture (Croze 1970; Krebs 1973; Smith 1974b; Zach and Falls 1976b, c). Smith (1974b), in fact, showed that average move length by a foraging thrush decreased after a prey capture.

Search tactics. Birds may change search tactics with variation in prey availability. For example, several species have been shown to perform proportionately more aerial flycatching maneuvers as flying insects become more abundant (Davies 1976, 1977b; Davies and Green 1976; Greenwood and Harvey 1978; Holmes et al. 1978; Robinson and Holmes 1984). These changes probably reflect shifts in relative availability of one prey type over another, however, and not necessarily a change in overall prey availability.

TABLE 2. Some Foraging Behaviors Likely to be Influenced by Food Availability Levels

Behavior	Expected relationship with food availability	No. studies consistent with trend[a]	No. studies inconsistent with trend[a]	Other variables to control[b]
Temporal attack rate (No. pecks/sec)	Positive	11	4	1–7
Spatial attack rate (No. pecks/m)	Positive	2	0	11
Rate of progression (cm/sec)	Negative	0	5	9
Search velocity (No. hops/sec)	Positive	1	0	7
Mean move length (No. hops/m)	Negative	6	0	11
Search tactic change (glean, sally . . .)	Change	0	6	1, 7, 8
Food delivery rate (No. trips/hr)	Positive	3	1	10
Rate of aggression (No. attacks/hr)	Positive	3	0	8
Percent time feeding/resting (daily time budget)	Negative/positive	6	0	6, 8, 9

[a] Specific references are cited in the text.
[b] The other variables are: (1) time of day or season; (2) quality or quantity of food per peck; (3) prey type consumed; (4) foraging tactic used; (5) success rate; (6) physiological condition of bird; (7) foraging microhabitat; (8) sex, individual identity; (9) weather; (10) clutch size; and (11) none reported yet.

Therefore, this is not likely to be a measure that accurately reflects changes in food availability.

Nestling food delivery rate (trips/unit time). In several instances, food delivery rates by aerial foragers have been positively correlated with food density (Zammuto et al. 1981, Turner 1982, Blancher and Robertson 1987). Strehl and White (1986), however, recorded fewer trips/hour by Red-winged Blackbirds during times of high food (periodical cicada) density. The latter result was a consequence not only of a change in prey types available but also of a change in foraging locations used by adults. Therefore, the positive relationship between food availability and delivery rate seems to be consistent among recent studies.

Rate of aggression (number of supplanting attacks/unit time). Rates of aggression may increase when food availability decreases, as Hinde (1952), Gibb (1954), and McFarland (1986a) have reported for forest tits and honeyeaters. Although results are consistent, the difficulty associated with obtaining large sample sizes in most instances will almost certainly render this measure useless as an index of food availability.

Percent time feeding (from time budget information). We might expect a bird in a food-poor area to spend more time feeding and less time resting, relative to a bird in an area of high food availability. Davies and Lundberg (1985) added food to Dunnock (*Prunella modularis*) territories three months prior to the breeding period. The birds not only bred earlier, but spent significantly more time perching (resting) (20% vs. 7%) and interacting (9% vs. 1%), and less time feeding (62% vs. 89%) than control females that were without supplemental food. Similar patterns have been shown nonexperimentally for tits (Gibb 1954), hummingbirds (Gill and Wolf 1979), ducks (Hill and Ellis 1984), shorebirds (Maron and Myers 1985), and honeyeaters (McFarland 1986a).

CONCLUSIONS

By using bird behavior to confirm that a measure of food availability is biologically meaningful, we can probably avoid the three major problems discussed earlier. The bird's perception of food availability is no longer ignored, scale-of-measurement questions are automatically resolved, and renewal rates are also automatically integrated. Nonetheless, potential problems remain. In particular, search tactics (behavioral acts), patterns of locomotion (rates), and time budgets may change independently of food availability because of changes in (micro)habitat (Robinson and Holmes 1984), time of day (Davies and Green 1976, Holmes et al. 1978), season (Root 1967), weather (Grubb 1978), and physiological condition of the bird (Moore and Simm 1985, Clark 1987). Many of these aspects of foraging behavior are also sex-specific (Holmes et al. 1978; Smith 1974a, b).

As an example, foraging attack rate should vary not only with food availability but also with quality and quantity of food/peck (McFarland 1986a), prey type or size consumed (Goss-Custard 1977a, b; Paszkowski and Moermond 1984; Robinson 1986), foraging tactic used to acquire food (Baker and Baker 1973), probability of success for a given attack (Goss-Custard 1970, Baker and Baker 1973), and physiological condition of the bird (Paszkowski and Moermond 1984, Moore and Simm 1985). Thus, one would need to control those additional variables before using attack rate as an index of food availability. That can be accomplished by restricting comparisons to time periods and locations in which such changes should be minimal, and by recording only number of successful captures/sec. Even in the absence of control for those variables, however, they will only add variance to the relationship between attack rate and food availability and decrease the chance of observing a significant correlation. Finding a significant correlation in the face of such scatter would only strengthen the argument that the measure is a reliable estimate of food availability.

For each of the foraging behaviors considered here, I have summarized (Table 2) whether the behavioral act is likely to be reliable as an indicator of food availability (based on the consistency of published results where both the behavior and prey density were measured). I have also included a list of nonfood-related variables found to affect a given behavior, so that they might be controlled as much as possible.

It is important to note that the behavioral acts outlined here are those for which published information exists. Undoubtedly, other behavioral measures (e.g., pecks/stop) might be sensitive to variations in food supply. Researchers working with systems that afford accurate measurement of food availability could record bird behaviors to test the usefulness of those measures. Meanwhile, temporal and spatial attack rates, mean stop-to-stop movement length, and percent time feeding are probably the most promising behaviors to record.

Finally, it may be practical to use foraging behavior to validate a measure of food availability when one's goal is to determine whether food availability is important in explaining observed biological differences among individuals. Indeed, behavior alone might be an adequate index of food availability in such instances. If, on the other hand, one wishes to determine whether food supply is important in explaining why some parcels of land are used and others are not used by individuals of a given species, the problem is more difficult. Even if sweep net samples provide a perfect measure of food availability (as evidenced by a perfect correlation with variation in some behavioral act), one cannot assume that sweep samples from occupied and unoccupied areas will be comparable because the correlation between bird behavior and food abundance will have been based entirely on data taken from occupied areas. Occupied and unoccupied areas may differ significantly in physical structure such that food might not be perceived the same way in those locations. Another possibility is that predators or competitors may occur in areas that are avoided by the subject species. Thus, measures of food abundance could be similar between occupied and unoccupied areas, but food could still be less available in the unoccupied areas.

To compare food availability between occupied and unoccupied areas, we must know the constraints on what is *possible* for the bird to use. Just as we must know about the subset of prey types and sizes that should be excluded from estimates of food availability, we need to know the subset of (micro)habitats that should be excluded for comparisons of occupied and unoccupied areas. This problem will stand as a fundamental obstacle to our eventual understanding of the relationship between habitat use and food availability.

ACKNOWLEDGMENTS

I thank Jed Dewey and Larry Stipe for letting me tie into their research on effects of systemic insecticide treatments, and for their loan of tree pruning equipment. Hank Goetz granted permission to work in Lubrecht Forest and John Mandzak granted permission to work on Champion International land. The field work was funded in part by an NSF-MONTS grant. I thank Sue Reel for help with the budworm sampling and counting, and Paul Hendricks, Don Jenni, C. J. Ralph, Andy Sheldon, Kim Smith, and Jerry Verner for suggestions that improved the manuscript.

… Studies in Avian Biology No. 13:29–37, 1990.

ARTHROPOD SAMPLING METHODS IN ORNITHOLOGY

ROBERT J. COOPER AND ROBERT C. WHITMORE

Abstract. We review common methods used by entomologists and ornithologists for sampling terrestrial arthropods. Entomologists are often interested in one species or family of insects and use a trapping method that efficiently samples the target organism(s). Ornithologists may use those methods to sample a single type of insect or to compare arthropod abundance between locations or over time, but they are often interested in comparing abundances of different types of arthropods available to birds as prey. Many studies also seek to examine use of prey through simultaneous analyses of diets or foraging behavior. This presents a sampling problem in that different types of prey (e.g., flying, foliage dwelling) must be sampled so that their abundances can be compared directly. Sampling methods involving direct observation and pesticide knockdown overcome at least some of these problems. Trapping methods that give biased estimates of arthropod abundance can sometimes be related to other methods that are less biased (but usually more expensive) by means of a ratio estimator or estimation of the biased selection function.

Key Words: Arthropods; insects; prey abundance; prey availability; sampling.

Numerous techniques exist for sampling insects (e.g., Southwood 1978); many have been used in ornithology. Most ornithological studies use insect sampling to determine types, numbers, and distribution of insects available to birds as prey; many also are designed to examine the use of those prey through simultaneous studies of diet and foraging behavior. Most techniques, however, effectively sample only a portion of the total insect fauna available to birds, and estimates of total arthropod abundance using these techniques will be biased accordingly.

Our objectives are (1) to review a portion of the literature on sampling techniques commonly used in entomological field studies, (2) to describe the advantages and disadvantages of those techniques, (3) to review their use and misuse in ornithological studies, and (4) to make recommendations concerning arthropod sampling methods in light of the objectives of field ornithologists. Our review of the entomological literature is largely limited to sources in which arthropod sampling is a focal point of the paper; it excludes techniques that sample arthropods normally unavailable to passerines, such as light trapping, and techniques that focus on a single species, such as capture-recapture sampling. An excellent review of these techniques is contained in Southwood (1978). Other reviews of interest include Morris (1960) and Strickland (1961).

DEFINING ARTHROPOD AVAILABILITY

The usual objective of most ornithological studies that sample arthropods is to relate some aspect of bird behavior or ecology (e.g., diet, foraging behavior, territory size, productivity) to arthropod abundance and distribution (availability). Simple arthropod abundance, however, may not reflect the prey actually available to birds, because not all arthropods in a bird's foraging area will be eaten by the bird. The size, life stage, palatability, coloration, activity patterns, and other characteristics of arthropods influence the degree to which they are located, captured, and eaten. These are the "translators" (Wiens 1984b) or factors that translate simple arthropod abundance into availability (e.g., see Hutto, this volume; Wolda, this volume). The problem is one of perception; the researcher must assess availability as the bird does. This, of course, is impossible. One approach is to use dietary data to determine availability (Sherry 1984, Wolda, this volume). Using this approach, the prey available to the bird are those it commonly eats. This approach is useful in some types of investigations (e.g., foraging behavior), but it is nonsensical in others (e.g., dietary preference).

Usually, as in this paper, when arthropods are sampled to determine availability, the sampling is designed to estimate the types, numbers, or distribution of some or all arthropods in the foraging area of one or several species of birds. Ornithologists are often interested in locations (e.g., tree species, heights) where birds forage in relation to arthropod availability. In this case, availability would be defined only in terms of the specific prey types in the location(s) of interest. Dietary data can be used to narrow the focus of the sampling effort. If a species eats large percentages of caterpillars a sampling technique specific to caterpillars should be used. Most methods described here are designed to sample one type of arthropod (e.g., flying, foliage dwelling). Such an objective is generally easier to achieve than sampling different types of arthropods for comparison (see Sampling Problem).

If a study seeks to assess "preference," one must estimate the numbers and types of all arthropod prey in the foraging area and compare prey eaten and not eaten by the birds. Reasons for their choices may be found in the ecologies of predator and prey. For example, Robinson and Holmes (1982) and Cooper (1988) found that prey types eaten by different species reflected differences in searching and attack strategies. Aviary studies involving feeding trials (Whelan et al., 1989) or simulated ecosystems (Heinrich and Collins 1983) can also provide information about why certain prey are eaten or avoided.

THE SAMPLING PROBLEM

Estimates of arthropod abundance are either relative or absolute. Relative measures provide only indices of abundance, such as numbers per surface area of sticky trap in a given time period. They have limited utility in studies of arthropod abundance and availability. Absolute measures, on the other hand, permit estimates of arthropod density that can be used for interspecific comparisons and comparisons among different habitats and seasons. Absolute measures usually require an intermediate sampling effort to quantify the density of the unit (plant, leaf, branch, and so on) used to assess arthropod numbers. Arthropod sampling, as used in this paper, covers both types of measures.

An obvious problem is comparing results from one or more methods involving different sampling units. This is especially the case if one wishes to compare arthropods eaten by birds with those available. Because various foraging sites and maneuvers differ among bird species, many sorts of arthropods may have to be sampled simultaneously in a given habitat, including flying insects, foliage-dwelling forms (e.g., spiders, Homoptera, some beetles), caterpillars, and litter-dwelling insects (e.g., ground beetles, ants). Because most of these groups require a unique sampling technique, one cannot easily relate results from one group to another. For example, how might we compare a frequency of two caterpillars/100 leaves with a density of 10 ichneumonid wasps/400 cm^2 of sticky-trap surface per week? The problem requires either (1) a technique that equivalently samples all types of arthropods in a given habitat (defined as unbiased measurement of the abundance of all arthropod taxa on experimental units that completely cover the habitat of those arthropods), or (2) a method that permits unbiased comparison of results from one technique with those of another. The first solution can sometimes be achieved in arthropod sampling by using methods that allow estimates of density, instead of indices of relative abundance. The second solution involves relating one estimator to another, as commonly done in entomology and other fields by using a ratio or regression estimator (Cochran 1977), which takes the form

$$\bar{y} = \left(\frac{\bar{y}_1}{\bar{x}_1}\right) \bar{x}_2$$

where $\bar{x}_1$ is the estimate of mean abundance from sampling technique 1, $\bar{y}_1$ is the estimate of mean abundance from sampling technique 2 when applied to the same study site, and $\bar{x}_2$ is the estimate of mean abundance from sampling technique 1 when applied by itself in a second study site.

Usually x is less costly and less accurate than y. Both x and y are measured on several sampling units and the relationship between them is expressed as a ratio ($\bar{y}_1/\bar{x}_1$). Then a larger number of samples is taken measuring only x. An estimate of y for the entire study area can then be obtained by using the general formula above. This method is mentioned frequently in the following sections. McDonald and Manly (1989) consider an alternative to ratio (regression) estimation in which an attempt is made to calibrate a biased sampling procedure by estimating a selection function.

ESTIMATING RELATIVE ABUNDANCE (INDICES)

The following techniques are designed to give indices of relative abundance, not to estimate absolute abundance or density of arthropods. Most of the methods use a trapping device, such that the sampling unit is insects per trap or time period.

Sticky traps

Sticky traps of various design (see Southwood 1978:250–252) are commonly used to sample flying insects inexpensively; an insect settles on or strikes an adhesive surface and is trapped. Trap size and shape (Heathcote 1957a, b; Younan and Hain 1982) and color of the trap surface (Purcell and Elkinton 1980; Weseloh 1972, 1981) are important. The traps are messy. Temperature can affect the consistency and effectiveness of the adhesive, and large insects tend to bounce off or escape.

Sticky traps have been compared with other flying insect traps or sampling techniques mostly with unfavorable results, in that certain insect taxa, or sizes, or both, were underrepresented (Trumble et al. 1982, Younan and Hain 1982). Because trap color alters the effectiveness of the traps for many insects, between-species comparisons of abundance may be biased. However, strong correlations were documented between sticky traps and absolute counts (Heathcote et al. 1969) and suction traps (Elliott and Kemp 1979), suggesting that sticky traps could be used in a ratio or regression estimation scheme, given a common sampling unit.

Despite the considerable shortcomings, sticky traps have been used widely in ornithological field studies, probably because of their simplicity and low cost. Cody (1981) used them to study the relationship between precipitation patterns and the insects available to birds. Given the extremely low sampling intensity (5 days/year), results should be viewed with caution. Blake and Hoppes (1986) used them to determine prey abundance in treefall gaps, and Hutto (1980, 1981a, 1985b) used them in several habitats. These authors recognized that their sampling schemes did not sample the same arthropods that foliage-gleaning insectivores capture, but assumed that the numbers of insects captured in the traps were correlated with actual prey availability. Given that sticky traps do not capture larval Lepidoptera, an important prey source for birds in many areas, that assumption is questionable.

Moreover, because sticky-trap catches cannot be meaningfully related to a sampled area, and because comparisons of catches between arthropod taxa are biased for various reasons, we doubt that sticky traps are useful in ornithological studies, especially because more reliable methods exist.

Malaise traps

The Malaise trap is an interceptive device made of fine-meshed netting that uses a series of baffles to herd insects into a closed chamber that may or may not contain a killing fluid (see Steyskal 1981 for an excellent bibliography). Malaise traps have been used effectively to sample a variety of flying insects in a variety of habitats (Evans and Murdoch 1968, Matthews and Matthews 1970, Walker 1978). Results indicate that these traps perform well for larger Hymenoptera, adult Lepidoptera, and some Diptera, but they are unsatisfactory for Coleoptera and Hemiptera, which tend to be less common in collections than expected, because they usually drop when encountering obstacles (Juillet 1963, Tallamy et al. 1976, Reardon et al. 1977). Although more comparative studies are needed, the trap's advantages are clear. It samples most flying insects except the Coleoptera and Hemiptera with roughly equal intensity. Collections are funneled into a jar that is easy to handle and process. The jars may be removed if only a portion of the day is of interest. However, they are expensive (approximately $300/trap), transportation is difficult, and they often must be operated for some time to obtain large numbers of insects.

Malaise traps have been used to sample flying insects assumed to be available to a variety of aerial-hawking birds. Often diets of such species have been compared with availability as determined solely from Malaise trap captures (e.g., Beaver and Baldwin 1975, Davies 1977b), or in concert with sticky traps and direct observation (Blancher and Robertson 1987), or together with sweep net samples (Blendon et al. 1986). At least Davies (1977b) recognized that Coleoptera and perhaps other taxa were underrepresented but assumed that trapping results were acceptable because the flycatchers he studied seldom eat beetles. Robust analysis of use versus availability (Johnson 1980) can be helpful when one is unsure of including questionable prey items.

A viable alternative to Malaise traps that operates on a similar principle (i.e., interception) is the stationary tow net, a large net that swivels around to face into the wind. Quinney and Ankney (1985) and Quinney et al. (1986) used them to assess use versus availability of flying insects by Tree Swallows (*Tachycineta bicolor*). These nets have an advantage over Malaise traps because they capture insects that fall when striking an object. Such insects may also be sampled with a window trap—basically a sheet of glass held vertically with a fluid-filled collecting trough below (Chapman and Kinghorn 1955).

Beating or shake-cloth methods

These methods have been in use for a long time in a variety of situations. Typically, a cloth supported by a frame is placed underneath a branch or plant. The vegetation is then shaken or beaten to dislodge insects, which collect in the cloth below. The technique is seldom considered to result in an accurate estimate of absolute density, although the number of leaves in a selected plant or branch can be counted to arrive at a density estimate. Boivan and Stewart (1983) found that while most individuals were dislodged from struck branches, many missed the cloth or moved off too quickly to be counted. Similarly, Rudd and Jenson (1977) found that the technique did not sample highly mobile species efficiently. Frequently, therefore, this method is used together with a more expensive but accurate technique in the form of a regression estimator (Bechinski and Pedigo 1982, Linker et al. 1984), although Marston et al. (1976) were not satisfied with the results of the ratio estimators they derived using shake-cloth sampling. Ornithologists that have used versions of shake-cloth sampling to obtain a measure of relative arthropod abundance include Boag and Grant (1984) and Brush and Stiles (1986).

Sweep-net sampling

The sweep net is probably the most widely used device for sampling arthropods from vegetation. Its advantages are simplicity and speed. Sweep netting has been used in numerous ecosystems where plants of interest are short. Strong positive correlations between sweep netting and more accurate but expensive procedures suggest that the technique may be useful in a regression estimation scheme (Bechinski and Pedigo 1982, Fleischer et al. 1982, Linker et al. 1984). However, others have found regression estimators employing sweep-net sampling to be generally unacceptable (Byerly et al. 1978, Purcell and Elkinton 1980, Ellington et al. 1984). Marston et al. (1982) found sweep netting to collect some groups of insects more efficiently than others, so resulting ratio estimators varied in precision. They also provide some sample size guidelines for sweep netting in ratio estimation schemes.

Sweep netting does not provide a measure of absolute density and it is biased in several ways. It collects only arthropods located in the upper portions of plants. The method is ineffective in tree foliage or extremely short vegetation. The taller a plant is, the smaller the proportion of the plant that is adequately sampled, so arthropods differing in their vertical distributions cannot be compared using sweep netting. Because of the effect of foliage height on the efficiency of sweep netting, it is not useful for comparing the abundance of arthropods between different habitats or between seral stages on the same site (Southwood 1978:240–242).

Sweep-net sampling is extremely popular in ornithology, undoubtedly because it is easy. Murphy (1986) used sweep netting to relate breeding biology of Eastern Kingbirds (*Tyrannus tyrannus*) to food availability, noting that arthropods commonly eaten by kingbirds were located in the upper portion of the field vegetation. Many ornithologists have also used sweep netting in woody vegetation, either to track abundance of arthropods in the same location over time (Sealy 1979, 1980; Biermann and Sealy 1982; Rosenberg et al. 1982; Boag and Grant 1984), to compare abundance of arthropod prey between different areas (Blenden et al. 1986), or to compare availability and use of arthropod prey (Root 1967, Beaver and Baldwin 1975, Busby and Sealy 1979). Because sweep netting of woody vegetation undoubtedly captures more active prey and relatively few caterpillars, which adhere to leaves and twigs more readily than active prey, use-versus-availability estimates using sweep-nets are probably biased, perhaps severely so.

Pitfall traps

Pitfall traps are designed to capture surface-dwelling arthropods, especially such active forms as spiders (Uetz and Unzicker 1976, Doane and Dondale 1979) and ground-dwelling beetles (Thomas and Sleeper 1977, Shelton et al. 1983). The pitfall trap is a receptacle (e.g., cup, jar, can), usually with killing or preserving fluid, sunk into the ground with its opening level with the ground surface. One improvement provides a cover to prevent rain from filling the receptacle (Shubeck 1976), and another uses plastic cups placed one inside the other to prevent escape (Morrill 1975). Barriers leading to the receptacles can increase captures significantly (Durkis and Reeves 1982). Like other trapping techniques discussed in this section, absolute population density cannot be estimated from pitfall traps alone. Frequently, if a single species is of interest, pitfall trapping is used as part of a capture-recapture study (Rickard and Haverfield 1965, Brown and Brown 1984).

The method has been seldom used in ornithological studies, probably because it effectively samples only actively crawling arthropods, and not larvae in the litter layer. Pitfall traps have been used to compare numbers of different arthropod taxa that were known prey of insectivorous birds in pesticide treated and untreated areas (Johnson et al. 1976, Sample 1987), and to compare abundance of surface-dwelling arthropods among Wheatear (*Oenanthe oenanthe*) territories (Brooke 1979), objectives for which the technique is appropriate.

ESTIMATING ABSOLUTE ABUNDANCE (DENSITY)

Methods that allow density estimates are usually labor intensive and expensive and differ in several ways from those previously discussed. First, they depend upon instantaneous measures, whereas most trapping methods measure relative abundance over a period of time. Second, results can be expressed in numbers per unit area, volume, or weight. An intermediate sampling step is usually needed to relate the sampling unit to area sampled, so arthropod counts can be converted to a density estimate. Unlike measures of relative abundance, of course, density estimates allow direct comparisons between different taxa in the same habitats, or between the same or different taxa in different habitats. Certain ways of sampling arthropods in vegetation and in the air also allow density estimates.

Sampling arthropods in vegetation often involves collecting all or part of a plant, with determination of the number of arthropods per leaf, leaf area, shoot, branch, or plant. Arthropods may also be collected from whole plants without collecting the vegetation as well (e.g., by fumigation or careful examination of plants and physical removal of organisms). If the collection technique is efficient, a reasonable estimate of numbers per plant or other unit of vegetation can be obtained. In some cases, arthropods can be counted directly on foliage without collecting vegetation or removing the organisms. In all of these instances, knowledge of the density of the collection unit then allows conversion of arthropod counts to density estimates. Another approach that allows density estimation uses suction traps to capture flying insects, with counts being expressed in terms of a given volume of air.

Collecting vegetation

Counting arthropods on collected whole plants is usually restricted to relatively small plants, frequently crops such as cotton and soybeans. Because it can be time consuming, whole plant assessment is often used as a basis of comparison for other sampling techniques, such as vacuum and sweep-net sampling (Smith et al. 1976, Byerly et al. 1978), shake-cloth and vacuum sampling (Fillman et al. 1983), and shake-cloth and sweep-net sampling (Linker et al. 1984).

Pole pruning or branch clipping is similar to whole plant sampling, but the vegetation of interest (usually trees) is too large to allow collecting the entire plant; branches are pruned and collected instead. This is often done with a pole pruner, featuring a cutting device at the end of one or several extendable poles that is operated from the ground. The cut branch is either collected in a basket suspended beneath the cutter or it crashes to the ground, usually onto a tarpaulin, where it and any expelled arthropods are collected.

Because more active arthropods often escape or are expelled when a branch is disturbed, pole pruning is largely restricted to use with caterpillars and other relatively sedentary arthropods. It has been used widely in ornithological research to study bird-insect relationships associated with caterpillar populations (e.g., Morris et al. 1958; Tinbergen 1960; Buckner and Turnock 1965; Royama 1970; Morse 1973, 1976a; and Emlen 1981). It seems to be the preferred technique for mid- to upper-canopy caterpillar sampling and has been used to sample larval stages of spruce budworm (*Choristoneura fumiferana*) (Carolin and Coulter 1971, Torgersen et al. 1984a), Douglas-fir tussock moth (*Orgyia pseudotsugata*) (Mason and Overton 1983), gypsy moth (*Lymantria dispar*) (Martinat et al. 1988), leaf miners (Pottinger and LeRoux 1971) and others (Markin 1982, Martinat et al. 1988).

A variation of this technique involves placing a plastic bag over a branch and clipping it with shears. The sample is then fumigated and the arthropods are collected. Majer et al. (this volume) found that few arthropods escaped using this method. Schowalter et al. (1981) used a long-handled insect net fitted with a closeable plastic bag and a long-handled pruning hook to cut the sample.

Few studies have compared the effectiveness of pole pruning with other sampling methods. Mason (1970, 1977), who developed sampling techniques for the Douglas-fir tussock moth, concluded that pole pruning at midcanopy was an ineffective technique when populations were low, because of the small sample sizes. His preferred method involved beating lower canopy branches over a shake cloth on which dislodged larvae could be counted. Majer et al. (this volume) compared branch clipping with pesticide knockdown for sampling canopy arthropods in eucalypt forests. Branch clipping gave a much better representation of sessile arthropods, such as psyllids, caterpillars, and web-spinning spiders, but was inadequate for sampling mobile arthropods.

The value of pole pruning depends on study objectives. The technique is appropriate for determining caterpillar abundance, but not for determining use versus availability of all prey by birds. Further, pole pruners are difficult to operate at heights >15 m, thus precluding sampling of taller forest canopies. Those problems can largely be overcome by bagging, clipping, and fumigating samples, but the investigator must gain access to canopy foliage. Schowalter et al. (1981) used platforms to reach canopy foliage, and Majer et al. (this volume) used a mobile cherry picker.

In addition to collecting live foliage by pole pruning, researchers have measured arthropod fauna available to birds by collecting dead foliage. Gradwohl and Greenberg (1982b) collected dead leaves inside and outside of exclosures to determine the effect of avian predation on dead leaf arthropods. Smith and Shugart (1987) related prey abundance and territory size of ovenbirds (*Seiurus aurocapillus*) by collecting litter samples within a circular hoop and sorting arthropods from the litter. Berlese funnels considerably facilitate this process (Southwood 1978: 184–186).

Stationary suction traps

First developed by Johnson (1950) and Taylor (1951), stationary suction traps vacuum flying insects into a collection device in a fixed spot. The trap usually features an electric fan that pulls or drives air through a fine gauze cone, which filters out insects. The trap may be fitted with a device that separates the catch by time intervals. Taylor (1955, 1962) standardized air flow and trapping results of numerous suction traps, and estimated their absolute efficiency. Based largely on those results, Southwood (1978) considered the suction trap to sample a fixed unit of habitat and thus provide an estimate of absolute abundance. Because they are believed to sample most flying insects in an unbiased fashion, suction traps are a substantial improvement over sticky and Malaise traps. The primary disadvantage is cost.

Suction traps have been used to sample aphids, lacewings, Coleoptera, Diptera, and Hymenoptera (Taylor 1951, 1962). Johnson (1950) compared his suction trap with sticky traps and tow-nets (Broadbent 1948) and found that it performed best for aphids and other small, airborne insects. Elliott and Kemp (1979) also used suction traps for aphids and developed regression estimators to compare them with sticky-trap results.

Suction traps have been used effectively to determine abundance of flying insects in several studies. Holmes et al. (1978) used them to determine diurnal change in flying insect abundance and response in foraging behavior by American Redstarts (*Setophaga ruticilla*). Catchability bias associated with insect size was calibrated using Taylor's (1962) correction factors. Bryant (1973, 1975b) used suction traps to assess use versus availability of flying insect prey of House Martins (*Delichon urbica*). Bryant (1975a) also used suction traps to relate breeding biology of House Martins to food supply. Because suction trapping is efficient and its bias can be calibrated, we believe the above procedures resulted in good estimates of the flying insects available to the birds of interest.

Portable vacuum sampling

Also called suction sampling, this procedure uses a portable vacuum. It was first applied by Johnson et al. (1957) and Dietrick et al. (1959). Dietrick's model was later improved (lightened to approximately 27 lbs.) and is now known as the d-Vac sampler (Dietrick 1961).

The d-Vac has been used widely in agricultural and other ecosystems, such as for Homoptera in flooded rice fields (Perfect et al. 1983) and cherry orchards (Purcell and Elkinton 1980), weevils in thistle plants (Trumble et al. 1981), mosquitos in salt marshes (Balling and Resh 1982), aphids on peaches (Elliott and Kemp 1979), and various arthropods in cotton (Leigh et al. 1970, Smith et al. 1976, Byerly et al. 1978) and soybeans (Bechinski and Pedigo 1982, Culin and Yeargan 1983).

Portable vacuum samples are closely correlated with direct counts (Ellington et al. 1984) and have even exceeded whole-plant visual sampling of thistles (Trumble et al. 1981). Although they have been used to estimate densities (Perfect et al. 1983), Wiens (1984b:404) found that d-Vac sampling of arthropods on sagebrush was only 55% efficient, and that different taxa were sampled with differing effectiveness. Leigh et al. (1970) also concluded that suction sampling alone cannot estimate density; they recommended using the d-Vac with a sampling cube for such an estimate.

Portable vacuum sampling has not been used extensively in ornithological research. The cost of suction samplers (about $1000; Dietrick, pers. comm.) precludes their use in many studies, and their bulk makes them unsuitable in certain situations, such as forest canopy sampling. Suction samplers are especially efficient in shrubby or field-like habitats. For example, K. G. Smith (1982) used a portable vacuum sampler with a sampling cube to collect herbaceous and understory arthropods in a standardized area and time period in a study of drought-induced changes in a bird community. Rotenberry (1980b) used a portable vacuum sampler and a quicktrap (Turnbull and Nicholls 1966) to sample shrubsteppe arthropods.

Direct observation

Occasionally it is feasible to count arthropods directly. Use of more than one observer introduces observer bias, and direct observation is especially time consuming and requires well-trained observers. Furthermore, not all types of arthropods are equally observable, due to activity or crypsis. However, the method has the major advantage that all observable arthropods are measured in the same units (e.g., insects per leaf, leaf area, or plant). Also, many ancillary data (location, substrate, plant species association, or escape behavior of arthropods) can be recorded, most of interest to ornithologists (e.g., Greenberg and Gradwohl 1980, Holmes and Robinson 1981, Cooper 1988, Holmes and Schultz 1988).

Direct observation is rarely used in entomological field studies to sample arthropods of forest canopies. The objective of much entomological research is to sample populations of one species or family of arthropod, which can generally be sampled more efficiently by using one or a combination of the previously mentioned methods. Ornithologists, however, are usually more interested in the entire arthropod community in terms of its availability to birds as prey. Often relative numbers of different prey taxa are compared with the frequency of those taxa in bird diets. Thus, a sampling method is required that targets all arthropod taxa. This is accomplished with direct observation methods, if performed carefully. Not surprisingly, then, the method has been used frequently in ornithological research to count arthropods on herbaceous vegetation (Schluter 1984, Blancher and Robertson 1987), tree trunks (Cooper 1988), understory tree foliage (Holmes et al. 1979c), dead leaves (Gradwohl and Greenberg 1982b), and mid- to upper-canopy foliage (Greenberg and Gradwohl 1980, Holmes and Schultz 1988). Access to canopy foliage has been done using towers (Greenberg and Gradwohl 1980) or tree-climbing gear (Cooper 1989, Holmes and Schultz 1988).

MISCELLANEOUS TECHNIQUES

Many other arthropod sampling methods do not fit under the above categories but have been used in ornithological research. Some are designed especially to sample a single species. For example, frass traps have commonly been used to sample larval Lepidoptera (e.g., Betts 1955). They are funnel-shaped structures placed on the forest floor to collect arthropod excrement, providing an index of abundance. If mean daily production of excrement can be calculated, absolute abundance can be estimated (Liebhold and Elkinton 1988). Use of frass traps requires prior study of frass from target species, so it can be distinguished in the field from that of other insects.

Pheromone traps are commonly used for adult Lepidoptera of pest species and were used by Crawford et al. (1983) to sample spruce budworms in a study of avian predation. Some species, such as the gypsy moth, lay conspicuous masses of eggs that overwinter and can be counted as an index of abundance (Smith 1985). Burlap bands wrapped around trees have also been used to count late instar gypsy moth larvae, which hide beneath the burlap during the day (Campbell and Sloan 1977).

Emergence traps, which are cone-shaped nets erected with the circular end flush to the ground, were used to estimate the emergence rate of periodical cicadas (Homoptera: Cicadidae) in a study of avian response to this superabundant prey source (K. Smith, pers. comm.). Buckner and Turnock (1965) trapped emerging larch sawfly (*Pristiphora erichsonii*) adults and Orians and Horn (1969) trapped emerging damselflies in similar studies using emergence traps.

A method that seems to be gaining popularity among ornithologists working in forest habitats is the pan or water trap (Southwood 1978:252–253). These are plastic containers filled part way with water and a preserving solution (e.g., salt or antifreeze) and placed on the ground or hung in the canopy. They effectively capture many arthropod taxa (Morrison et al. 1989). Although pan traps are undoubtedly biased against certain types of arthropods and are likely to be affected by trap color, they are an inexpensive way to assess canopy arthropod abundance over time or between locations.

Another method, pesticide knockdown, can be used to sample all types of arthropods in a less biased manner than many of the previously mentioned techniques. Using a fogging machine and a pyrethroid pesticide, which has strong knockdown ability but breaks down quickly and has low vertebrate toxicity, the forest canopy can be fogged in a systematic fashion. Pyrethrin killed virtually all arthropods in patches of foliage examined before and after fogging (Cooper, unpubl. data). Majer et al. (this volume) found that pesticide knockdown missed some types of sessile arthropods obtained in branch clipping samples. Some flying insects were also able to escape at the time of spraying. Dead insects fall to the ground and are collected in jars at the bottoms of funnels made of canvas or plastic (Wolda 1979; Majer et al., this volume) or on collecting cards (Raley 1986). The percent composition of each arthropod taxon can then be computed and compared with the percent of each taxon in bird diets. The drawbacks of this method are that arthropod densities are difficult to compute and that arthropods are not observed until they are collected, so an understanding of their location and behavior must be obtained in some other way. Foggers and pesticides are also expensive. A major advantage is that large numbers of arthropods are collected per sample in a short period of time. Also, in forests with extremely tall canopies, such as tropical rain forests, fogging may be the only way to sample arthropods from the upper layers (Wolda 1979).

DISCUSSION

If the objective is to measure the abundance of a particular arthropod taxon or overall abundance of all arthropods in the same location over time, or to compare the abundance of a particular taxon or all arthropods between different locations, then almost any of the techniques described above will suffice, because the inherent biases of a sampling method against certain prey taxa should be more or less constant. However, if the objective is to assess the relative abundance of different prey taxa available to birds, the method of choice must sample all relevant arthropods with equal intensity. This is relatively easy for some bird species, such as swallows and some flycatchers, which feed almost entirely on flying insects that can be sampled with a stationary suction sampler or stationary tow nets (Bryant 1973, 1975b; Quinney and Ankney 1985).

Because most bird species do not entirely feed on one type of arthropod, but use a variety of foraging behaviors and different substrates to capture several types of arthropod prey, this presents a formidable sampling problem; different types of arthropod prey (i.e., flying, foliage-dwelling, bark-dwelling) must be sampled in a consistent fashion that allows the researcher to compare the abundances of all types.

TABLE 1. SUMMARY OF SOME ARTHROPOD SAMPLING TECHNIQUES COMMONLY USED IN ORNITHOLOGY

Method	Arthropods sampled	Advantages	Disadvantages
Sticky trap	Flying or otherwise active	Inexpensive; able to cover large area	Messy; influenced by trap color, temperature; small interceptive surface
Malaise trap	Flying	Easy to maintain; large interceptive surface	Expensive, bulky; biased against Coleoptera; few catches per unit time
Stationary tow-net	Flying	Inexpensive; captures most flying insects with equal probability	Small interceptive surface
Shake-cloth	Foliage-dwelling	Inexpensive; good for sessile arthropods	Active arthropods can escape; hard to sample in canopy
Sweep-net	Foliage-dwelling	Simple, inexpensive; good for active arthropods	Biased by foliage height and against sessile arthropods
Pitfall trap	Ground-dwelling	Simple, inexpensive; can estimate density of single population using capture-recapture	Biased against inactive litter arthropods; captures affected by density and type of ground cover
Pesticide knockdown	Foliage-dwelling	Samples many types of arthropods with approximately equal probability	Foggers, pesticide expensive; affected by wind; can miss attached or extremely active arthropods
Frass traps	Caterpillars	Field methods simple, inexpensive; absolute density estimable	Requires arthropods be kept in captivity
Emergence traps	Arthropods emerging from soil or water	Inexpensive; can estimate density of emerging arthropods	Large number often required to adequately cover area of emergence
Pole pruning	Sessile, foliage-dwelling	Inexpensive method of reaching forest canopy	Biased against active arthropods; few arthropods per sample
Branch-clipping	Foliage-dwelling	Captures many arthropods missed by pole pruning; inexpensive but must gain access to forest canopy	Biased against active arthropods; few arthropods per sample
Suction	Foliage-dwelling	Gives good estimates of abundance when used with sampling cube or quick trap	Expensive; can miss some arthropods
Stationary suction	Flying	With correction factors gives good estimates of abundance; can sort samples by time	Expensive; difficult to sample large area
Direct observation	Foliage-dwelling	Can directly compare abundances of different arthropod taxa; many ancillary data on arthropod ecology collected; arthropods "collected" quickly on tape recorder	Observability bias likely for both arthropods and observers; must gain access to forest canopy, strenuous; identification to species level often difficult

In field-like ecosystems, for example, sweep netting is often used. It is fast, simple, and efficient, but it is biased against arthropods located near the ground. This bias can be corrected by using a more accurate method, such as portable vacuum sampling, on a subset of the units sampled by sweep netting and relating them by means of a ratio or regression estimator. Of course, if possible for all samples, portable suction samplers would be more desirable than sweep netting.

In forest ecosystems, canopy-dwelling, foliage-gleaning birds feed upon a variety of arthropods associated with bark, foliage, and air. Most of the aforementioned sampling methods work best on only one of these substrates and would be inappropriate for assessing use versus availability. Two methods, direct observation and pesticide knockdown, are effective for comparing relative frequencies of different arthropod taxa available to and used by foliage-gleaning birds. While those methods are not unbiased, they are preferable for sampling canopy arthropods.

Most ornithological studies have used arthropod sampling in an effort to relate some aspect of avian feeding ecology to arthropod availability. Yet few studies have done this adequately. Typical shortcomings include inadequate sample sizes and inappropriate extrapolation from a specialized technique to make inferences about total arthropod availability. Sample size is more likely to be a problem with methods like branch clipping or direct observation that obtain only a few arthropods per sample than with methods that obtain large numbers of arthropods per sample (Gibb and Betts 1963, Cooper 1989; Majer et al., this volume). Many types of arthropods have clumped distributions, which can greatly inflate variance estimates using methods that sample a small volume of foliage or airspace. Sample-size problems like these are offset by the larger number of samples usually obtainable per unit time and the shorter amount of sorting time required using branch clipping and direct observation. Sorting time can reach mountainous proportions in techniques that obtain large numbers of arthropods per sample (Table 1).

Not surprisingly, many studies that have meaningfully associated some aspect of avian ecology with arthropod availability have either been done in structurally simple habitats such as shrubsteppes or pine plantations or have involved birds known to feed almost exclusively on one type of insect. Other studies have concentrated on a single type of insect known to be especially important to the bird species of interest.

The few meaningful studies (see Root 1967, Holmes and Robinson 1981, Robinson and Holmes 1982, Rosenberg et al. 1982, Holmes et al. 1986, Holmes and Schultz 1988) have several things in common. First, most lasted three or more years. Second, sampling procedures were frequently directed towards only one type of prey such as caterpillars (Holmes and Schultz 1988) or cicadas (Rosenberg et al. 1982). Sampling methods used for assessing arthropod availability, such as sweep-net sampling of woody vegetation (Root 1967, Rosenberg et al. 1982) were often biased. However, the methods were sufficient to demonstrate seasonal changes in arthropod abundance, which is often all that is needed. Third, the authors all performed dietary analyses to convincingly establish which prey birds were selecting, and which strengthened the conclusions concerning arthropod availability and foraging or reproductive behavior.

Studies of bird-insect relationships have long been of interest to avian ecologists and seem to be gaining popularity. Because this area of ecology involves insects as much as birds, knowledge of insect ecology and behavior is important, as it clarifies how and why birds capture and eat certain types of prey. Sampling methods involving direct observations can be particularly insightful.

Virtually all techniques in this paper have been developed by entomologists. Comparison of methods, advantages and disadvantages, calibration of biases, and required sample sizes appear in the entomological literature. Other methods, such as direct observation and pesticide knockdown, which are likely to gain favor with ornithologists, should be similarly assessed (e.g., Majer et al., this volume). No single, magic sampling method exists. Each has strengths and weaknesses, and each is biased to some extent. Bias can be tolerated in certain situations and corrected in others. An appropriate sampling design depends upon study objectives and the scale of investigation. In general, more time and effort should be devoted to arthropod sampling in ornithological research than has been done in the past. We encourage ornithologists to investigate, compare, and report the efficacy of different methods and designs as they pertain to the objectives of ornithological research.

ACKNOWLEDGMENTS

This research was supported by funding through the McIntire-Stennis program and, in part, by USDA Forest Service Cooperative Agreements 23-972, 23-043, and 23-144, and a Sigma Xi grant-in-aid of research. We thank L. Butler, L. L. McDonald, M. L. Morrison, C. J. Ralph, J. T. Rotenberry, K. G. Smith, and J. Verner for reviewing the manuscript. P. W. Brown, K. M. Dodge, and P. J. Martinat provided many helpful comments. Published with the approval of the Director, West Virginia University Agriculture and Forestry Experiment Station as Scientific Article No. 2120. This is contribution No. 7 of the Forest Arthropod Research Team.

species in a pine wood with frass collectors. Frass pellets were identified to species and instar. These data could be used to estimate absolute abundances, because they were calibrated by simultaneously measuring the densities of caterpillars, by species and instar, on the trees.

GROUPING SPECIES INTO HIGHER TAXA

Counting or weighing insect samples as a whole, or after classification to higher taxa, is far easier and much less time consuming than classifying them at the species or morpho-species level, which makes the procedure very popular. This is understandable because of limited time and funds. But for reasons given below, I am convinced that in most cases studies of insect availability are irrelevant unless analyses are done at the (morpho-)species level.

SELECTIVITY OF SAMPLING TECHNIQUES

Insect collecting techniques differ in their efficiency in capturing a given species and this efficiency varies among species. As a consequence, the relative frequencies of insect species in a sample depends on the collecting techniques used (e.g., Fenton and Howell 1957, Mikkola 1972, Tallamy et al. 1976, Service 1977, Dowell and Cherry 1981, Zelazny and Alfiler 1987, D'Arcy-Burt and Blackshaw 1987, Mizell and Schiffhauer 1987, Cooper and Whitmore, this volume, Majer et al., this volume). It is doubtful that any collecting method can produce an unbiased picture of the faunal segment under study (Cooper and Whitmore, this volume; Hutto, this volume; Majer et al., this volume). If the nature of the bias is known, correction factors can be applied, as is done for suction-trap samples (Taylor 1962) and a few other cases (Weseloh 1987). Normally, however, both direction and magnitude of bias are unknown. Relative abundances of the species in a sample may have very little predictive value for those in the field.

ESTIMATES OF TOTAL ABUNDANCE FROM SAMPLES

Because relative abundances of species in a sample may have little relation to relative abundances in nature, they provide unreliable estimates of total abundance. For example, a hypothetical fauna of 10,000 individuals comprises five equally abundant species, A–E (Fig. 1). A given collecting effort obtains a sample of 100 individuals, among which the five species are unequally represented because of different capture probabilities. One species (A) was not captured at all, while another (E) made up 60% of the sample. (Such differences in capture probabilities are probably commonplace.) If any of these five species increases fivefold, from 2000 to 10,000 individuals, the total fauna increases to 18,000 individuals. How this would be perceived in the sample, however, depends on capture frequency (Fig. 2). Here increase in total number of individuals varies from 0 to 240%, depending on which of the five species increased. If individual species were not counted separately, the sample would present a very distorted picture of the natural situation. One can make the model more realistic by allowing some species to increase and others to decrease. A real increase in the total fauna might then very well translate to a decrease in the number of individuals in a sample and vice versa. Unless samples are analyzed at the species level, conclusions about abundance are likely to be erroneous.

THE INSECTIVORE'S VIEW

An insectivore is likely to be at least as selective as an entomologist (Hespenheide 1975, Belwood and Fullard 1984, Sörensen and Schmidt 1987). Probabilities of encounter, detectability, or acceptability are usually different for different insect species. In Figure 3 I show the same hypothetical fauna used in Figure 1. Species are arbitrarily assigned different distributions in the habitat such that different proportions of populations occur in the correct microhabitat and, accordingly, have different probabilities of being encountered by a bird. This results in an "available" fauna different from the total fauna. Similarly, among-species differences in probabilities of being detected and being accepted result in detected and accepted "faunas" with a species composition that is very different from the fauna as a whole (Fig. 3). Diet composition, affected by still more probabilities, may be different again.

A faunal increase of 80%, from 10,000 to 18,000 individuals, with only one of five equally abundant species increasing fivefold (as in Fig. 2) is perceived by a bird as an increase in the number of acceptable prey items, which varies from 0% to 250% depending on which species experienced the increase (Fig. 4). Again, if insects are analyzed at the species level, data provide an accurate picture of which species underwent an increase. If not, the apparent relationship between diet and fauna may be difficult to explain.

SELECTION OF COLLECTING TECHNIQUES

The central problem in selecting a collecting technique is to determine both distribution and abundance of potential prey items. With a collecting technique that monitors potential prey only in appropriate microhabitats, one could directly measure availability, or changes in availability. Some instances seem to approach this

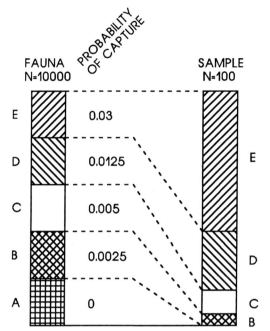

FIGURE 1. Effect of species-specific capture probabilities on the relative abundances of the species in an insect sample of 100 individuals taken from a fauna of 10,000 individuals consisting of five equally abundant species, A–E. Representation of a species in the sample depends on its capture probability.

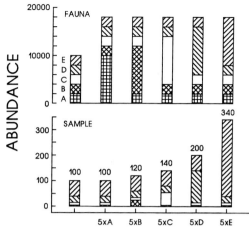

FIGURE 2. Effect of a fivefold increase in abundance of one species of the fauna of Figure 1 on total insect abundance in the fauna as well as that in the sample using the capture probabilities of Figure 1. How the real increase in abundance is perceived in the sample depends on the capture probability of the species that underwent the increase.

ideal. Birds specializing in insects that hide in aerial leaf litter are an excellent example (Rosenberg, this volume). Aerial nets placed at appropriate heights may directly measure availability of prey insects to predators of flying insects (Hespenheide 1975, Bryant and Westerterp 1981, Quinney and Ankney 1985, Hussell and Quinney 1987). Visual inspection of foliage in the understory of a forest may approximate availability of insects for an understory foliage gleaner (Graber and Graber 1983, Karr and Brawn 1988). Often, however, such direct measurements of insect abundance in the correct microhabitats are impossible. In such cases one should select the monitoring technique that comes closest to that ideal, and one that does not select against species important to an insectivore. A splendid example is given by Castillo and Eberhard (1983), who used artificial webs to measure insect prey available to a spider. Finally, one should attempt to calibrate abundances in samples against those in the field (Tinbergen 1960, Cooper and Whitmorte, this volume).

DETAIL OF TAXONOMIC ANALYSES

The level of taxonomic detail needed to analyze an insect fauna and the diet of a bird is determined by the ecology of both bird and insects. Insect taxa that are perceived identically by a bird can safely be pooled, if one can determine the bird's perception. Two species of flies may be alike in appearance, but if one concentrates in microhabitats used by the bird and the other does not, pooling their measures of abundance is unjustified. If they occur in the same microhabitats, but have a different probability of being captured, they are again not identical to a bird.

Whether "similar" prey from our standpoint are identical for a given bird can be determined only if their relative frequencies are determined both in the diet and in the suitable microhabitat. Initial classification into "potential prey" and "nonprey" items depends to a large extent on guesswork. If birds do not take ants, counting ants can be avoided. However, for many taxa the decision is not obvious. A bird may be known to take "small beetles," but it actually may take only a few kinds and avoid others.

When insect samples can be analyzed only at a coarse taxonomic level, extreme care should be taken to interpret the results. One should not expect changes in abundance in one set of species to be "representative" of those in another (cf. Hutto 1985b). In cases where previous studies have established that a bird feeds only in a well-defined microhabitat, takes only a certain category of prey, and does not discriminate among those items, an analysis at the species level is a waste of time. In general, however, one should

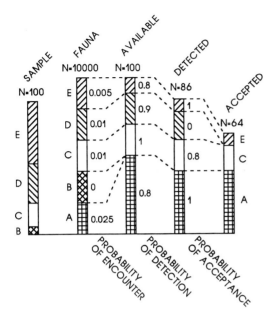

FIGURE 3. Perception of relative abundances of the five species in the fauna of Figure 1 by an insectivore, given species-specific probabilities of encounter, of detection, and of acceptance. Probabilities of pursuit, capture, and being eaten similarly affect the species composition in the diet, but are not included here for sake of simplicity. The sample "taken by the investigator" (Fig. 1) is given for comparison.

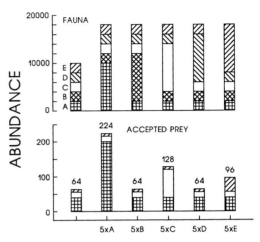

FIGURE 4. Effect of a fivefold increase in abundance of one of the species of the fauna of Figures 2 and 4 on total insect abundance as well as that in the category of prey items acceptable to the bird (using the probabilities given in Fig. 4). How the real increase in abundance in the fauna is perceived by the bird depends on the (encounter, detection, etc.) probabilities of the particular species which underwent the increase.

err on the safe side initially and perform the analysis at as detailed a level as possible, preferably at the species level. Pooling of taxa can be done later, but splitting taxa requires a reexamination of the same samples.

In most cases, it is unnecessary to get too involved in insect taxonomy. With the help of an insect reference collection, using code numbers instead of names if convenient, one can classify individuals at a "morphospecies" or "operational taxonomic unit" (OTU) level. The goal is to work at the level of real species, but one should avoid getting bogged down in problems of sibling species or the analysis of genera that can be analyzed only by a specialist. If two individuals cannot be separated on relatively simple external characters, they can be classified as belonging to the same morphospecies. This facilitates the task considerably, and it can be accomplished even in diverse habitats. My assistants and I have been doing it for years in tropical forests in Panama. This procedure has the potential for errors of the kind mentioned above, but I believe its advantages outweigh the disadvantages. The advice of competent insect taxonomists is invaluable when deciding which characters to use in the classification.

A common objection to the (morpho)species approach is economical. However, lack of funds is no excuse for an unsound study. With proper planning, acceptable procedures can be designed that can be executed at a reasonable cost, and defended in grant proposals.

CONCLUSIONS

To obtain reliable results from studies of insect availability for birds, one must be prepared to take time away from birds and spend a considerable proportion of available resources on the insects. Simply putting up some sticky traps or taking some sweepnet samples, and then scoring insects at best at the ordinal level, is insufficient. It is preferable to estimate absolute abundances of potential prey insects. However, measures of relative abundance, using some carefully selected standard monitoring technique, may often provide sufficient information, especially when results are calibrated against field abundance. The more detailed the analysis of the insect samples, the more reliable and the more informative the results. If insects are tallied only at a coarse taxonomic level, the best one can hope for is that nonprey species do not dominate the sample to the extent that existing correlations between insect availability and bird performance are obscured. Loiselle and Blake (this volume) and Blake et al. (this volume) clearly demonstrate the need to classify fruits in birds' diet to the species level.

Whenever possible, one should take the analysis to the morphospecies level and pool species only if it is known that this can be done with impunity.

If, instead of one species, an entire guild of insectivores is under investigation, the situation becomes more complex. Different insectivores are likely to be different in hunting characteristics, and so are likely to have different values of availability, detectability, and so on, for each prey taxon. The overlap in insect taxa constituting "potential prey" and in "microhabitats used" may be small. Composition of the diet of a guild of insectivores is a complex, composite picture of selections made by the component species. Under these conditions an approach that does not distinguish among potential prey species is unlikely to produce useful results.

Trying to avoid problems introduced by an inappropriate collecting technique or a coarse taxonomic analysis by using a variety of techniques simultaneously may be self deception. The assumption that errors will cancel out (e.g., Fenton and Howell 1957) is wishful thinking. Executing a proper analysis of the importance of food for an insectivorous bird is a formidable but rewarding challenge.

ACKNOWLEDGMENTS

I am grateful to Jim Karr, David Manuwal and two anonymous reviewers for critically reading the manuscript.

QUANTIFYING BIRD PREDATION OF ARTHROPODS IN FORESTS

Donald L. Dahlsten, William A. Copper, David L. Rowney, and Paula K. Kleintjes

Abstract. Sampling insects and other arthropods in forest environments is complicated because of the unique attributes of this ecosystem. Entomologists have used many techniques to quantify forest arthropods, some of which are applicable for quantifying the impact of bird predation, as we illustrate in studies of several defoliators and bark beetles. We describe sampling methods for a defoliator, Douglas-fir tussock moth (*Orgyia pseudotsugata*), and a bark beetle, western pine beetle (*Dendroctonus brevicomis*). We discuss the types of information that can be obtained for insect populations from these methods, the time or cost for different levels of sample error, and the application of these methods for evaluating bird predation on arthropods.

Key Words: Sampling; predation; defoliators; insectivorous birds; forests; conifers; western pine beetle; Scolytidae; Douglas-fir tussock moth; Lymantriidae.

Forest entomologists have struggled with the quantification of arthropod abundance for many years. Much work has been done by applied biologists interested in population dynamics of certain species, efficacy of treatments, or the impact of insects on resources. Quantitative studies are more complicated in forests than in other environments where insects are of economic importance (such as agriculture; Dahlsten 1976), because forests are vast, continuous regions composed of different tree species of different ages, and a mosaic of stocking (density) patterns (Pschorn-Walcher 1977). The advantage of forest ecosystems is that they generally encounter less perturbation than agricultural systems and probably have a more stable arthropod community. Outbreak species (those that reach very high densities periodically) are relatively rare in forests (Berryman 1986).

Most studies of forest insect populations deal with single species; associated insects such as natural enemies, inquilines, and organisms in the same feeding guild are often ignored. Regional or forest-type arthropod faunistic or community studies are rare and typically more qualitative than quantitative. Yet, population information gathered by entomologists may be useful in assessing the impact of birds on a single insect species.

Meanwhile, ornithologists desire quantitative population information about arthropod species eaten by birds. Birds typically feed on several different species and at different heights in the foliage. To quantify an adequate number of prey items on several substrates is costly and time-consuming, however, so compromises and stratifications are required.

Based on work by our laboratory, we believe that better quantification of arthropod prey for birds is possible. We have had a long-term interest in the impact of natural enemies on forest insects, particularly insectivorous cavity-nesting birds (Dahlsten and Copper 1979). In this paper we discuss the types of sampling we have used to assess avian impact on insects on two substrates in the forest, foliage and bark, and also what it costs to obtain useful information.

FOLIAGE SAMPLING

LODGEPOLE NEEDLE MINER

The lodgepole needle miner (*Coleotechnites milleri*; Lepidoptera: Gelechiidae), because of its cyclic availability, is a suitable species for studying the role of birds in its dynamics. The adult moths appear only in alternate years and have a short period of activity, whereas the larvae and pupae are available for a long period. The insect has a discrete 2-year life cycle, passing the first winter in an early larval instar and the second winter in the fifth instar. As the birds feed only on larger larvae, this food source is available only in alternate years. In addition, because the insect is a needle miner, the birds must open needles to obtain the larvae, leaving evidence of their feeding. Finally, the distribution of immature needle miners in the trees has been studied and a sampling method developed (Stark 1952, Stevens and Stark 1962). The method is similar to that for tussock moths, discussed below, and involves sampling the tips of lodgepole pine branches.

At a study site in the Inyo National Forest, Telford and Herman (1963) found that Mountain Chickadees (*Parus gambeli*) concentrated their feeding efforts in alternate years on the needle miner larvae and that the chickadees exhibited a functional response to prey density. The chickadees peeled needles in a characteristic way, leaving evidence of their feeding, and Cassin's Finches (*Carpodacus cassinii*) also fed on needle miners by clipping the ends of the needles

(Dahlsten and Herman 1965). Nest boxes were later placed in areas infested and not infested with needle miners. Mountain Chickadees increased in density in the infested areas, both during the breeding and postbreeding periods.

The needle miner-chickadee system has great potential for evaluating the impact of a bird on a single insect species. Because the insect is cryptic during the stage eaten, evidence of chickadee feeding can be easily detected. The system is ideal for studying the functional response of chickadees, because the prey is available only in alternate years, and nest boxes and avian census techniques permit study of the numerical response of the predator to its prey.

BUD-MINING SAWFLIES

Bud-mining sawflies (*Pleroneura* spp.; Hymenoptera: Xyelidae) are also well suited for evaluating avian predation. Four species mine new buds on expanding shoots of white fir (*Abies concolor*) in California; three have been studied in detail (Ohmart and Dahlsten 1977, 1978, 1979). The species of early instar larvae and adults can be distinguished, but the late larval instar (the stage most likely to be eaten by birds) cannot be separated to species.

The three species were treated as a single group in an analysis of within-crown distribution and the development of sampling methods at Blodgett Forest, El Dorado County, California (Ohmart and Dahlsten 1978). Over 94% of the infested buds occurred in the outer portion of the crown, coinciding with the foraging area of several birds at Blodgett, particularly the Mountain and Chestnut-backed (*P. rufescens*) Chickadees. Also, the chickadee nesting period coincided with the late larval instars of the sawflies, late May to early June (Ohmart and Dahlsten 1977).

We did not learn how birds open buds to remove larvae, or if they leave characteristic evidence. However, mortality of the *Pleroneura* fifth larval instar was substantial (Ohmart and Dahlsten 1977), seemingly because of avian predation, as chickadees were observed and photographed by nest box camera units bringing numerous *Pleroneura* larvae to their young.

PINE SAWFLIES

Larvae that feed in the open, like sawflies, are often fed upon by birds, but no evidence is left on the foliage when they are removed. However, birds often remove sclerotized portions of insects, such as the elytra of beetles and the head capsules of larvae, before eating them or feeding them to nestlings. Sawfly larvae, in particular, exude a brownish substance from their mouth when threatened by a parasitoid or predator. This substance is probably distasteful (Eisner et al. 1974).

In studying the population dynamics of a pine-feeding sawfly in the *Neodiprion fulviceps* complex at Mt. Shasta, California, Dahlsten (1967) watched Evening Grosbeaks (*Coccothraustes vespertinus*) feeding on their larvae. Ten trees, 2–4 m in height, were sampled in each of three study areas at different elevations in a plantation. All sawfly stages, starting with eggs, were counted. Drop cloths were placed beneath each sample tree. The cloths did not catch cocoons, but they did catch head capsules and thoraxes of larvae, which were discarded by grosbeaks. Some larval remains were also stuck to foliage; counts on and beneath the trees showed a total of 166 sawflies— 10% of all the larvae on the study trees in one area (Dahlsten 1967).

Because the birds were feeding on a known population, the portion taken was known, at least from the sample trees. Area-wide estimates can be made from such samples. This is a labor-intensive technique, limited to smaller trees where foliage-feeding insects could be counted and larval remains could be found on foliage or drop cloths.

DOUGLAS-FIR TUSSOCK MOTH (DFTM)

The tussock moth (*Orgyia pseudotsugata*; Lepidoptera: Lymantriidae), because of its economic importance in western North America, has been the focus of many studies, including the role of insectivorous birds in its population dynamics (Brooks et al. 1978; Torgersen et al., this volume). The tussock moth overwinters as eggs in masses on top of female cocoons. Both male and female cocoons are commonly spun on foliage, although cryptic sites such as cavities in trees are also used. The cocoons and egg masses, in particular, are suitable for stocking studies. Egg masses can be sampled and then examined for evidence of predation, or they can be stocked on branches or trunks of trees at different known densities and predation evaluated (Dahlsten and Copper 1979, Torgersen and Mason 1987). Pupal stocking showed that most predation was due to birds, although some was due to ants (Dahlsten and Copper 1979, Torgersen et al. 1983).

SAMPLING ARTHROPODS ON WHITE FIR

This study illustrates how the distribution of a community of organisms on a given tree species can be determined. Sampling programs can then be developed for any species known to be eaten by birds. Relationships among sampling error, time spent sampling, and cost are shown, so that the researcher can better manage available financial resources.

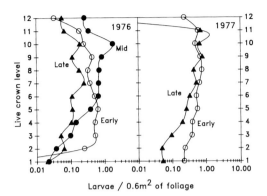

FIGURE 1. Distribution of Douglas-fir tussock moth (*Orgyia pseudotsugata*) on white fir, in 12 equal levels of the crown in different sample periods in 1976 and 1977, El Dorado and Modoc counties, California.

Methods

Two areas in California were selected for sampling, based on Douglas-fir tussock moth activity in previous years. Three plots were established in each area, at Yellowjacket Springs, Tom's Creek, and Roney Flat in Modoc County, and at Iron Mountain, Plummer Ridge, and Baltic Ridge in El Dorado County.

A road ran lengthwise through each plot, which was 2–5 km long. Each plot was divided into quarters; two spots were randomly selected in each quarter. At each spot, the nearest white fir between 9–12 m in height became the first sample tree. Sample spots were permanent and were revisited each subsequent sampling period; since the sampling was destructive, on each subsequent visit the 9–12 m white fir nearest to the originally selected sample tree was chosen.

Eight trees, one from each of the eight spots in a plot, were sampled in each of the six plots during each sample period, giving a sample size of 48 trees per period. Five periods during the DFTM generation were sampled in 1976: Period 1 = late spring–early summer for cocoons and egg masses laid by the previous generation; Period 2 = early larval stage; Period 3 = midlarval stage; Period 4 = late larval stage; and Period 5 = a final sample in early to late fall for the cocoon-egg mass stages. The five trees in each spot therefore spanned the development of the DFTM generation and gave phenological information for the DFTM defoliator guild, and for its predators and parasites.

For each sample tree, all live branches were numbered beginning from the lowest north-side branch. Computer-generated random number lists were used to select a sample of one-third of all branches on the tree. All branches were cut from the tree; branches selected for sampling were caught in large canvas bags and beaten over a large canvas on the ground. All insects and spiders were recorded, as was the branch number, dimensions (for foliage area), and other characteristics. Some insects were retained for rearing or identification. A crew of three or four, processing from two to four trees per day, was needed for the intensive sampling procedure.

During periods 1 and 5 (spring and fall) sampling was supplemented by a 100% search for DFTM cocoons and egg masses, as these occur in relatively low numbers. These data were kept apart from the regular sample.

In the second year of sampling (1977), some modifications were made. Because cocoons and egg masses were rare in 1976, only the two plots with the most cocoons and egg masses in 1976 were sampled during the first and fifth periods of 1977, and no sampling was done during period 3 (medium larval stage).

Field data sheets were designed for direct keyboard entry, and computer programs were written to produce summaries of each insect species' density by whole trees, plots, areas, and by each of 12 equal crown levels. Another program was written to simulate sampling in different ways, such as two midcrown branch samples, two branches at each of three crown levels, and so on. This program gave variance, bias, and cost figures necessary to sample a plot at any level of precision for each sampling method.

Results

Foliage distributions of DFTM, *Neodiprion*, *Melanolophia* sp., and associated insects were calculated by 12ths of the live crown from the whole-tree sample of 48 trees per period, with both areas combined. Numbers of egg masses and cocoons in periods 1 and 5 were too low to estimate meaningful distributions. Many empty cocoons were found, presumably a result of avian predation.

Distributions of small, medium, and large larval DFTM differed by crown level and by years (Fig. 1). Early summer (small larvae) distribution was relatively constant across levels in 1976 except for the lower and upper foliage, whereas in 1977 density increased steadily from the lower to the upper one-fourth of the foliage. Late summer (large larvae) distributions tended to increase by a factor of 10 or more from the lower one-third to upper one-third of the trees, with the 1977 trees showing considerably higher density in the upper crown. The unpredictable changes indicate the need for multilevel crown sampling to avoid biased estimates.

Live crown densities of *Neodiprion* larvae for late spring were very low ($<0.2/0.6$ m^2) in 1976 and almost zero in 1977 (Fig. 2). In early sum-

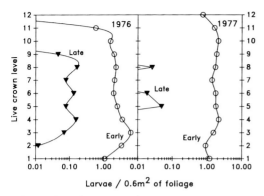

FIGURE 2. Distribution of sawfly larvae (*Neodiprion* species) on white fir in 12 equal levels of the crown in different sample periods in 1976 and 1977, El Dorado and Modoc counties, California.

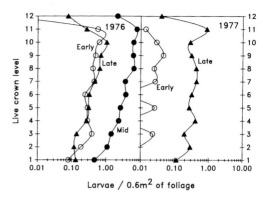

FIGURE 3. Distribution of greenstriped forest looper larvae (*Melanolophia imitata*) on white fir in 12 equal levels of the crown in different sample periods in 1976 and 1977, El Dorado and Modoc counties, California.

mer, the density jumped to high levels, especially in 1976 (peak of >6/0.6 m²). The distributions of early summer populations varied markedly between the two years, with higher densities in the lower one-third crown in 1976, but relatively even distributions across levels in 1977 (Fig. 2). Sampling for this insect would require a multi-level technique to reduce bias to an acceptable level. A sample of the lower one-third crown level would estimate that the 1976 density was 3–4 times higher than in 1977, whereas the whole-tree density of the intensively sampled trees indicated 1976 was only about 1.5 times higher. This insect also illustrates the timing problem in estimating prey density; its density increased about 20 times between late spring and early summer and then dropped to near zero by mid- to late-summer (not shown).

Another known chickadee prey, the greenstriped forest looper, *Melanolophia imitata,* a common geometrid larva on white fir, did not appear in significant numbers until early summer in 1976 and 1977. Densities rose from about 0.5/0.6 m² in early summer to about 5.0/0.6 m² in midsummer, and then dropped to about 0.5/0.6 m² by late summer of 1976 (Fig. 3). Distributions were biased toward the upper third of the live crown during all periods. In 1977, density in early summer was about ten times lower than in 1976, but in late summer was similar to 1976 (no midsummer sample was taken). Possibly a single level sample, probably at midcrown, could be used with minimal bias, if the low/middle/upper ratios seen in these two years were consistent over a number of years.

If the objective of sampling is to estimate total prey availability in foliage, a multilevel sample will be required for relatively precise, unbiased estimates. To illustrate this, we used computer-generated sub-sampling of the original data from all trees, under a variety of sampling rules, to compare their DFTM density estimates to those using the complete intensive sample. We then used estimated cost figures to determine the most efficient methods for given total error levels.

The computer sampling program simulated these sampling methods: two branches taken at random from the lowest two meters (lower two meter sampling method); two branches from the middle ⅓ of the crown (midcrown sampling method); two, three, or four branches from the whole crown at random (whole crown–two branch method, whole crown–three branch method, etc.); two, three, and four equal crown levels, with sets of two, three, or four branches from each level (giving nine methods, for example the two level–two branch per level method, three level–three branch per level method, and four level–four branch per level method). For each of these methods the program calculated tree mean densities using means per level weighted by the average proportion of foliage per level.

Within-tree sampling error (WSE) was the square root of the variance of the density estimates for all possible samples. Between-tree errors (BSE) were calculated from the mean squared differences between area means and individual tree means. Bias was found by subtracting the density mean (SM) of the samples chosen by the program from the "actual" (intensive sample) tree mean density (AM). Total standard error (TSE) for a sampling method with n sample trees was then calculated as:

$$\text{TSE} = \sqrt{((\text{BSE}^2 + \text{WSE}^2)/n + \text{BIAS}^2)}$$

where BIAS = AM − SM.

It is important to use a sampling method with low and stable bias, because bias cannot be re-

TABLE 1. DOUGLAS-FIR TUSSOCK MOTH SAMPLING SIMULATION: PERCENT MEAN BIAS[a] OF DIFFERENT SAMPLING METHODS FOR DOUGLAS-FIR TUSSOCK MOTH FOR PERIOD 2 (SMALL LARVAE), 1976 AND 1977

Number of branches per division	Year	Divisions					
		Lower 2 m only	Mid-crown only	Whole crown	2 Level	3 Level	4 Level
2	1976	−94.3	58.1	4.9	0.9	1.4	3.1
2	1977	−85.3	59.5	10.3	8.7	6.1	11.6
3	1976			3.0	2.3	2.3	4.0
3	1977			6.6	7.7	5.1	10.5
4	1976			2.0	2.9	2.7	4.4
4	1977			4.7	7.2	4.7	10.0

[a] Percentages of unbiased means of 0.333/0.6 m² (1976) and 0.434/0.6 m² (1977).

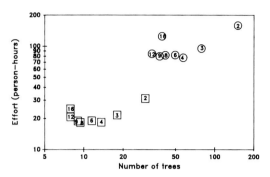

FIGURE 4. Effort to sample Douglas-fir tussock moth (*Orgyia pseudotsugata*) on different numbers of trees with varying numbers of branches per tree (numbers are the number of branches from 2–16) with 20% standard error (circles) and 40% standard error (squares). Based on 1977 period 2 small larvae sampling, El Dorado and Modoc counties, California.

duced by increasing sample size. The methods tried above using two, three, or four branches from two or three levels generally yielded the lowest percent bias figures (Table 1 shows DFTM small larvae for two years). The percent bias for both the lower two meter method and the mid-crown method was high and unstable.

Comparisons between methods may be made by selecting an acceptable level for TSE and calculating the number of trees and total branches required for a given mean density and its associated BSE, WSE, and BIAS. Labor costs may then be calculated from the estimated time to locate a tree and sample a branch. A conservative estimate is 15 min per tree, plus three min per branch for a crew of three people.

For example, in 1977 the mean density of small larvae (Period 2) was 0.434/0.6 m², the WSE varied from 0.181 to 0.804, and the BSE varied from 0.409 to 0.441, depending on the sampling method. Total trees and effort needed to determine the mean with a TSE of 20%, 40%, or 60% of the mean were calculated for each method, and trees were plotted vs. effort for different methods at two error levels (Fig. 4). Only the low bias methods and more efficient of any two methods that used the same number of branches per tree are shown.

For any error rate, the minimum point for curves in terms of effort indicates the most efficient sampling for the time assumptions used. The three-level, two-branch-per-level method is a good choice, as it is easy for field crews to divide a crown by eye into three levels, and it ensures a relatively representative sample, even if the branches chosen in each level are not random. Methods using greater numbers of branches are more likely to cause significant damage to the tree.

Tree and effort figures were calculated for all the sample periods in both years. Relationships between methods for other periods were similar to those for Period 2, 1977. However, the numbers of trees necessary for a given proportional sample error increased significantly for sample periods with lower mean densities. Using the three-level, two-branch-per-level method, the number of trees necessary for standard errors of 20%, 40%, and 60% of the mean was calculated and plotted vs. density, along with least squares regression lines for each error level (Fig. 5). This figure can be used to plan a low-level population sampling program, given the degree of precision required and an estimate of the populations in an area, perhaps from the previous year's population or a pilot study. Such methods are costly, but they can provide estimates of prey species abundance with reliable error rates and low bias.

BARK SAMPLING

Sampling of the bark substrate by our laboratory has mostly been below the surface of straight-boled conifers for species such as bark beetles (Scolytidae) and scales (Margarodidae). This group of cryptic, bark-inhabiting arthropods has special advantages for evaluating avian predation. One is that bark foragers and gleaners can be excluded by screening. Another is that birds usually leave evidence of feeding on insects in the phloem-cambial region, such as flaked or holed bark. However, sampling is often labor-intensive and costly. Below are examples of specific attempts to evaluate avian predation and of costs of sampling programs.

WESTERN PINE BEETLE (WPB)

The biology and control of the western pine beetle (*Dendroctonus brevicomis*; Coleoptera: Scolytidae) has been a problem for over 80 years

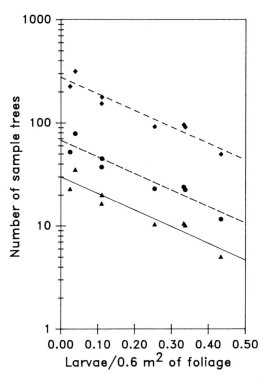

FIGURE 5. Number of sample trees needed for sampling Douglas-fir tussock moth (*Orgyia pseudotsugata*) at different densities for standard errors = 20% of mean (diamonds), 40% of mean (circles), and 60% of mean (squares), using the three-crown-level, two-branch-per-level method.

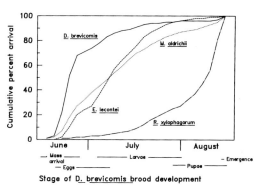

FIGURE 6. First generation arrival patterns of western pine beetle (*Dendroctonus brevicomis*) (8334 individuals) and three representative species of associates totalling: 3480 *E. lecontei*, 1684 *M aldrichii*, and 2728 *R. xylophagorum*. Data are mean cumulative frequencies from five trees at Blodgett Forest in 1970 and 1971. The mean collection interval was 2.5 (±0.06) days, and the mean trapping period was 62.8 (±1.6) days. The approximate stages of pine beetle within-tree brood development are shown (from Stephen and Dahlsten 1976b).

(Miller and Keen 1960, Stark and Dahlsten 1970). Their attack and colonization of ponderosa pine (*Pinus ponderosa*) has three phases (Wood 1972): (1) dispersal from the overwintering generation and selection of new susceptible trees in early spring (May and June); (2) concentration (mass attack) by feeding females; and (3) establishment that is associated with mating, excavation of egg galleries, and brood development. This same sequence occurs for a second generation that is usually prolonged, and which may overwinter as late larvae or pupae. However, in warm years a third generation may develop in October–November. In each generation, starting with the mass-attack phase and throughout the establishment phase, numerous other arthropods, parasites, and predators are attracted to the developing brood in a sequential pattern (Fig. 6).

In order to obtain information on the arrival pattern of pine beetles (Stephen and Dahlsten 1976a) and the subsequent arrival of associated arthropods, it is necessary to find trees just as they are under mass attack (Stephen and Dahlsten 1976b). Because locating sample trees was difficult, we induced mass attack by using female-infested bolts (logs) hung in trees (about 6 m from the ground), or by using synthetic attractants hung in trees.

We trapped insects continuously at the bark surface at three heights (1.5 m, 4.5 m, and 7.5 m) of the bole. A pulley system was installed so that a series of Stickem® coated traps could be removed and replaced easily. Traps were changed every other day during the concentration and establishment phases, and every fourth day during brood development. Traps were cleaned in warm kerosene to dissolve the Stickem®. Insects were separated from the solution by fine mesh screens and placed in alcohol.

Estimates of attack densities, gallery length, eggs laid, and brood development were recorded for correlation with arrival patterns of associated arthropods. Since the western pine beetle develops within the bark, an X-ray technique was used to count larvae, pupae and adults, along with some predators, parasitoids, and associates. Also, predation by woodpeckers was estimated visually; see Berryman and Stark (1962), Stark and Dahlsten (1970), and Dudley (1971) for details.

We found that initial beetle attack occurs at midbole, then spreads down and more slowly upward (cf. Miller and Keen 1960, Demars 1970). Height appears to influence brood distribution within trees more than aspect. Also, differences in trapping densities and generations (season) indicate a faster developmental time during the first generation and a higher concentration of

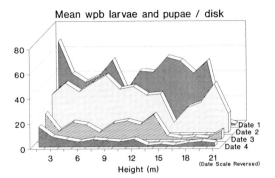

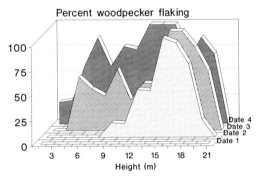

FIGURE 7. Changes in western pine beetle (wpb) (*Dendroctonus brevicomis*) larvae and pupae (UPPER) and percent bark flaking by woodpeckers (LOWER) by height and sample dates (1 = 16 Sept., 2 = 4 Oct., 3 = 10 Nov., 4 = 16 May, date scale reversed for visibility, UPPER only). Three trees combined, Blodgett Forest, California, 1967–1968.

broods in the lower portion of the bole in the second generation.

WESTERN PINE BEETLE AND WOODPECKERS

Because the bark beetle larvae develop within the bark during the later life stages, radiographs (X-rays) of bark samples made larvae easy to count; in many cases predators and parasitoids could also be counted (Berryman and Stark 1962, Berryman 1964). Otvos (1965, 1970) used this technique to determine the combined effect of the four main species of woodpeckers by comparing samples from caged and uncaged portions of trees and by examining bark samples (Stark and Dahlsten 1970). Otvos (1965) first examined all beetle-killed trees (438, from years 1961–1963) in the study area to determine generation and species of beetles killing the trees. He also determined that 53% of the trees had been drilled by woodpeckers, with the most activity occurring on the overwintering (second generation) broods.

Otvos' radiograph data (1962–1964 generations) showed 31.8% brood consumption by woodpeckers. A more significant benefit of woodpecker activity was increased parasitism. Otvos estimated that a 3–10 fold increase in parasitism may result from reduction in bark thickness by providing parasites with shorter ovipositors a larger area of oviposition.

Otvos (1970) also measured the western pine beetle broods removed by woodpeckers by X-raying bark strips and plotting positions occupied within the bark by larvae. Among 379 larvae, 220 (58%) were located within the woodpecker-flaked portion of the bark. Additional larval mortality in the thinner bark could also be caused by desiccation and by freezing during the winter months.

A similar study of an overwintering generation of bark beetles in 1967 (Dahlsten, unpubl. data), corroborates Otvos' (1965) results. In this case, bark thickness and percent of woodpecker activity were taken directly from bark samples of infested trees.

We removed pairs of 88 cm^2 bark disks on opposite sides of the bole at 1.5 m intervals from the base to the top of WPB infestation. The first sample date was shortly after the peak of attacks and adult gallery construction, and subsequent samples were spaced through larval stages to the emergence of brood adults. Each sample was X-rayed so that insects within could be identified and counted quickly without dissection of the sample, and the proportional area of bark surface flaked by woodpeckers was recorded.

We found the lowest density of WPBs per disk later in the sampling season when the percent of bark with woodpecker flaking was highest (Fig. 7). Data were from a single generation (1967 overwintering) and represented the mean at each height for three trees close together at Blodgett Forest, California. This pattern is common in the overwintering bark beetle populations. The initial bark beetle attacks probably took place between 7.5 and 12.0 m and fill-in attacks occurred between 1.5 and 18 m. By the second sampling date, woodpeckers had become active high in the tree and the beetle brood showed the reduction at that level. Woodpecker activity continued down the bole on the next two sampling dates, and the beetle brood declined further.

Decline of the beetle brood (in this case brood is the offspring of all females attacking the tree) was not entirely due to woodpeckers. Predatory insects were present prior to woodpecker activity and began to increase at heights below peak woodpecker activity (Fig. 8). (Woodpeckers no doubt feed on predaceous insects also.) WPB larvae infected with parasites attained their highest densities in the upper portion of the tree during

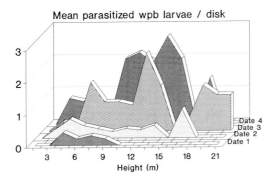

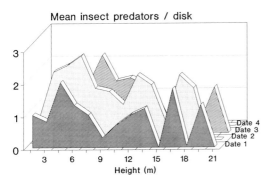

FIGURE 8. Changes in numbers of parasitized western pine beetle (wpb) (*Dendroctonus brevicomis*) larvae (UPPER) and insect predators (LOWER) by height and sample dates (1 = 16 Sept., 2 = 4 Oct., 3 = 10 Nov., 4 = 16 May). Three trees combined, Blodgett Forest, California, 1967–1968.

the last sample date. Parasitization was also shown to be enhanced by woodpecker activity in an earlier Blodgett study (Otvos 1965).

To evaluate woodpecker-prey relationships in this system, at least two sample dates are required per WPB generation—one shortly after the peak of the WPB egg stage and prior to woodpecker activity to measure initial larval and egg densities, and another near the emergence stage for brood adults (Table 2). Because woodpecker activity and larval density vary by location, sampling to be representative must include at least four heights along the infested bole. The X-ray technique is probably the fastest method to determine bark beetle numbers within the bark, but it requires some special equipment and training. The cost for this type of sampling is shown in Table 2 on a per tree basis.

MOUNTAIN PINE BEETLE

Larvae of mountain pine beetles (*Dendroctonus ponderosae*; Coleoptera, Scolytidae), unlike those of western pine beetles, develop at the bark–wood interface, not in the bark. The mountain pine beetle has been recorded from many host species (McCambridge and Trostle 1972), and the parasite-predator complex differs by host and location. Dahlsten and Stephen (1974) began to record the associated fauna of mountain pine beetles from sugar pine (*P. lambertiana*) in California. One tree had numerous woodpecker strikes that could be associated with a larval or pupal chamber when the bark was peeled back; 436 individual woodpecker strikes were recorded from the sample bolts, 70% in the upper half of the tree. Because the mountain pine beetle pupae and larvae are beneath the bark, woodpeckers make individual strikes. Cost estimates for different sample sizes were developed for sampling mountain pine beetles in another study (Table 3).

CONCLUSIONS

We have shown that sampling forest arthropod populations is difficult. It can be labor intensive, time consuming, and expensive; moreover, results may or may not help determine the impact

TABLE 2. ESTIMATED TIME AND COSTS FOR SAMPLING WESTERN PINE BEETLE WITHIN TREE DEVELOPMENT STAGES AND ASSOCIATED ARTHROPODS AND WOODPECKER ACTIVITY. ASSUMPTIONS ARE: TWO TRAINED PERSONNEL, FOUR SAMPLE HEIGHTS PER TREE, TWO SAMPLE DISKS CUT PER HEIGHT PER SAMPLE DATE, AND TWO SAMPLE DATES

	Person-hours
Field sampling	
Locate sample tree (highly variable)	
First sample date (includes setup, limbing, installing ladders, and so on)	2.0
	6.0
Second sample date (includes removal of emergence cartons, measurement for woodpecker bark flaking)	4.0
Field total:	12.0
Lab analysis	
First sample date (eight sample disks)	
Count attacks, eggs, gallery length	4.0
X-ray samples, measure bark thickness	0.5
Read X-rays twice	2.0
Place disks in rearing cartons, periodically check over 6-week period	3.0
Second date (eight sample disks, eight emergence cartons)	
X-ray samples, measure bark thickness	0.5
Read X-rays twice	2.0
Place disks in rearing cartons, periodically check over 6-week period	3.0
Emergence cartons: count known arthropods	2.5
Lab total	17.5
Grand total per tree	29.5

TABLE 3. ESTIMATED TIME NEEDED TO SAMPLE MOUNTAIN PINE BEETLE POPULATIONS[a]

Sample	Hours to remove bark samples	Hours to analyze samples
1000-cm² rectangle, six/tree	1.50	0.90
500-cm² rectangle, six/tree	1.00	0.60
250-cm² rectangle, six/tree	0.70	0.40
100-cm² circular disk, six/tree	0.25	0.25

[a] Cost of locating, felling, and measuring tree and infestation parameters about $80.00.

of avian predators upon their prey. Estimates of arthropod populations can be made, but a proportion of arthropod prey will not be found by any sampling technique.

A decision must be made whether to examine the impact of one bird species or the entire forest bird community upon one or several forest arthropods. It may be easier to obtain more accurate quantitative results when working with only one insect species; yet, all lifestages must be included. A continuous annual study should be attempted to produce good results from this type of investigation.

Another approach may be to intensively sample an entire forest arthropod community occupying a single species of tree. Arthropod samples could then be compared with arthropods found in a bird's diet, which can usually be determined from feces or stomach samples, by visual observation, and in photographs from cameras attached to nestboxes. Correlations could then be made between arthropods within a bird's diet, location of the same arthropod species on a sampled tree, and the locations where the bird spends most of its time foraging on the tree. Avian impacts on arthropod prey could then be assessed by plotting the percent of time birds forage vs. the abundance of specific arthropods at foraging locations.

In general, sampling a limited prey resource quantitatively is the most feasible method for measuring the impact of a predator upon its prey.

ACKNOWLEDGMENTS

We thank Dorothy DeMars and Nettie Mackey for typing and the editors of this volume for numerous constructive comments and suggestions.

FACTORS INFLUENCING BROWN CREEPER (*CERTHIA AMERICANA*) ABUNDANCE PATTERNS IN THE SOUTHERN WASHINGTON CASCADE RANGE

JINA M. MARIANI AND DAVID A. MANUWAL

Abstract. During the spring of 1984, we sampled arthropods in three young (65–80 years old), three mature (105–130 years old), and three old-growth (375 years old) forest stands in the western hemlock zone of the southern Washington Cascade Range. Crawl traps, designed to collect arthropods crawling upwards on the bark surface of tree boles, and flight traps, designed to catch arthropods alighting on tree boles, were installed on 45 live Douglas-fir trees. Brown Creeper abundance was correlated significantly and positively ($P < 0.01$) with the abundance of spiders (6–11 mm) estimated from the crawl traps. Spiders were found in all six creeper digestive tracts we examined. Spiders of all sizes and soft-bodied arthropods (≥ 12 mm) were the only arthropod variables that were significantly and positively associated with bark furrow depth, which is highly correlated with tree diameter. A quantitative method for estimating bark surface area as it changes with diameter, height, and bark furrow depth was designed to evaluate how arthropod abundances differed with changes in bark structure. We discuss the limitations and usefulness of these arthropod sampling methods.

Key Words: Tree trunk arthropods; Douglas-fir trees; Brown Creepers; bark surface features.

Several species of bark-foraging birds use some tree species and sizes disproportionately as foraging substrates (e.g., Jackson 1979; Morrison et al. 1985, 1987; Lundquist and Manuwal, this volume). Differential use of foraging substrates may partly be attributed to the composition and availability of arthropods (Jackson 1979), which vary in response to the suitability of microclimatic conditions created by bark structure (Jackson 1979, Nicolai 1986).

Characteristics of tree-trunk bark differ both interspecifically (Travis 1977) and intraspecifically with respect to tree size and age (Jackson 1979). In the western hemlock zone, southern Washington Cascades, Douglas-fir (*Pseudotsuga menziesii*) trees have the most rugose bark structure of any tree species and the furrow depths become substantially deeper as the trees increase in diameter. In both old-growth and second growth forest stands of the western hemlock zone, the trunks of large Douglas-fir trees (≥ 50 cm at diameter breast height) are the only substrates used disproportionately as foraging sites by Brown Creepers (*Certhia americana*) during spring and winter (Lundquist and Manuwal, this volume). Brown Creepers typically begin foraging at the base of a tree and proceed up the bole searching for prey.

Our study was designed primarily to determine the degree of association between Brown Creeper and arthropod abundance on Douglas-fir trunk surfaces in three forest age classes. We also evaluated the association between arthropod abundance and changes in bark structure. To achieve these objectives, we designed a method for calculating the bark surface area of tree boles by measuring bark furrow depth. In this paper we compare the arthropod survey techniques we employed.

STUDY AREA AND METHODS

We worked in the U.S. Forest Services' Wind River Ranger District, in coniferous forest stands of the southern Washington Cascade Range. Our study sites were in the low elevation western hemlock zone (Franklin and Dyrness 1973), where western hemlock (*Tsuga heterophylla*) is the primary regenerating tree species in old-growth forest stands. Stands selected for this study originated from natural disturbances and had no silvicultural treatments applied throughout their development. The nine study sites comprised three young (65–80 years old), three mature (105–130 years old), and three old-growth (all 375 years old) forest age classes. Elevations ranged from 420 to 710 m.

Douglas-fir and western hemlock were the most abundant tree species in all forest age classes. Western red cedar (*Thuja plicata*), Pacific yew (*Taxus brevifolia*), western white pine (*Pinus monticola*), and several true fir species (*Abies* spp.) were present in varying amounts in the old-growth stands. The common deciduous tree species included big-leaf maple (*Acer macrophyllum*), red alder (*Alnus rubra*), and black cottonwood (*Populus tricocarpa*). Specific details of the plant associations and stand structure of forests in the western hemlock zone are found in Topick et al. (1986).

The annual temperature regime is considered moderate, and most of the precipitation, averaging about 154 cm annually, occurs from October through May. Summers are typically dry and warm (Topick et al. 1986).

Brown Creeper abundance

We counted Brown Creepers by using the variable circular plot (VCP) method (Reynolds et al. 1980). Twelve permanent VCP stations were established at

FIGURE 1. Arthropod traps as they were installed on the trunks of 45 Douglas-fir trees in nine forest stands of the southern Washington Cascades. The crawl trap consisted of three basic parts: a removable collecting tray, a cover, and a netting girdle. The flight trap (to the upper right) consisted of a 30 × 30 cm² piece of plexiglass suspended by wire clips in a 36 × 7.5 × 5 cm plastic tray.

150-m intervals along a rectangular transect within each stand. Six censuses were conducted in each stand from 25 April to 30 June 1984. We avoided conducting surveys on days with precipitation or high winds. All visual and aural bird detections were recorded for a period of 8 min at each count station. A 1-min pause time followed our arrival at a count station to allow for resumption of normal bird activity. We recorded the estimated horizontal distance from the observer at the plot center to the birds detected. Abundance estimates of Brown Creepers were calculated with the program TRANSECT (Laake et al. 1979) as described by Burnham et al. (1980). Creeper abundances are expressed as birds/40 ha.

Tree abundance

All trees were counted in circular plots centered at each VCP count station. Each tree was identified to species and assigned to one of four size classes measured at diameter breast height. Trees 1–10, 11–50, and 51–99 cm were counted in 0.05 ha plots, and trees ≥100 cm were counted in 0.20 ha plots.

Arthropod sampling

We sampled arthropods from the bark surface of five Douglas-fir tree trunks in each of nine forest stands. All sample trees were within a size range (diameter measured at breast height) known to be average for forest stands of that age class (T. Spies, pers. comm.). We randomly selected five of the 12 VCP bird count stations that had been established in each stand, and within a radius of 25 m of the count station centers one tree was randomly selected on which to install the traps.

Two types of arthropod traps were attached to each tree bole at 1.5 m from the ground. One trap was designed to collect arthropods crawling upward on the bark surface. It consisted of a removable collecting trap, a cover, and a netting girdle (Fig. 1). The netting girdle was attached around the circumference of the tree and followed the contours produced by the bark furrows. The girdle acted as a funnel for arthropods climbing upward on tree trunks by guiding them into the collecting tray. For specific details of the materials, design, and installment, see Moeed and Meads (1983).

The other trap was designed to collect air-borne arthropods that alighted on the tree bole. This flight trap consisted of a 30 × 30 cm² piece of plexiglass suspended by wire clips in a 36 × 7.5 × 5 cm plastic tray (Fig. 1). These traps were attached to tree boles by two nails located in the back of the tray, and the tray had two small holes at each end (located 1 cm from the bottom and covered with mesh) to prevent overflow from rainfall.

We began collecting samples from the crawl traps on 9 May 1984 and from the flight traps on 16 May 1984, and collected samples from both traps weekly through 1 August 1984. We collected 165 flight trap samples and 195 crawl trap samples from each forest age class. A total of 495 samples was collected from the flight traps and 595 samples from the crawl traps. Antifreeze was used in the collecting trays of all traps to capture and preserve arthropods, which were removed from the antifreeze and stored in vials containing 70% alcohol.

Bark structure and area

We recorded several measurements on each of the trees sampled for arthropods and on 16 additional (randomly selected) trees in each of the nine forest stands to evaluate changes in bark structure in relation to tree diameter and bark furrow depth. The following measurements were made at diameter breast height on each tree bole: (1) four bark furrow depth measurements equally spaced around the tree bole, (2) tree bole circumference without accounting for furrow depth, and (3) bark circumference taking into account the larger area produced by the depth of bark furrows. We took the last measurement by molding electrical wire around the tree to conform to the contours produced by furrows. Measuring the length of the stretched wire then equaled the circumference of the tree at diameter breast height, accounting for bark furrow depth.

Prey composition

We collected two Brown Creepers from stands in each of the three forest age classes in June. All were shot from the trunks of live Douglas-fir trees after watching them feed. The entire digestive tract was extracted immediately and stored in 70% formaldehyde.

DATA ANALYSIS

Arthropod classification and abundance

We sorted, counted, and classified to Order and Family the arthropods from each sample. All insects were grouped into one of six categories defined by exoskeleton condition (hard or soft) and body length: small (1–5 mm), medium (6–11 mm), and large (≥ 12 mm). The longest insect measured was 27 mm. Spiders were grouped into the same size classes defined above but maintained as separate variables. Size classes were determined by examining the frequencies of individuals measured lengthwise from several randomly chosen samples. Our categorization was based on the assumption that there may be constraints imposed by the morphology of the Brown Creepers' bill for obtaining or ingesting very large arthropods or those with very hard exoskeletons.

To evaluate differences in the types of arthropods collected in each trap, we calculated dry weight biomass of arthropods by body condition (spiders were included in the soft-bodied estimates) and calculated Pearson correlations between weight and abundance for each arthropod category identified.

We calculated Spearman Coefficients of rank-order correlation to examine the various associations of abundance (e.g., creeper and arthropod abundances) or relationships (arthropod abundance and bark furrow depth) being investigated. In most analyses, correlation coefficients were derived using stand level abundance estimates, and the sample sizes equaled nine. We used nonparametric rank-order correlations because we have only indices of abundance, which represent ordinal scale data (Zar 1984:3). All data sets were analyzed using SPSS (Nie et al. 1975).

Estimates of arthropods calculated from crawl traps are expressed as numbers per m^2 of bark surface area, and those from flight traps as numbers per 30 cm^2 (the area encompassing the plexiglass plate).

Bark surface area

To estimate arthropod abundance from the crawl traps, we calculated the bark surface area, including furrow depth sampled under the traps, to express arthropod abundance per unit area.

We used tree circumference, without measures of bark furrow depth, as an independent variable (X), and bark circumference including bark furrow depth as the dependent variable (Y) in two least squares regression models to generate slope and intercept coefficients. One model used measurements taken on 120 trees in young and mature stands (referred to as second growth); the other used measurements taken on 60 trees in old-growth stands. A BASIC computer program was written to calculate the bark surface area of Douglas-fir trees at any given diameter and height. The program incorporated both the slope and intercept coefficients produced by our linear regression models, and taper curve coefficients derived for second and old-growth Douglas-fir trees in British Columbia (D. Briggs, pers. comm., Kozak et al. 1969). Area of bark surface was calculated at 0.5-m intervals to account for changes in diameter and furrow depth.

Spider abundance and bark surface area

Bark surface area encompassing the lower two-thirds of the tree bole was calculated for representative young, mature, and old-growth Douglas-firs. The upper one-third of the tree bole was not included in the analyses because pronounced taper and the presence of limbs introduces additional and less predictable error into bark area calculations (D. Briggs, pers. comm.). We calculated the number of medium (6–11 mm) spiders occurring on a bole (daily and weekly) based on their abundances in the crawl traps. We used spiders of this size because their abundance was correlated most positively and significantly with creeper abundance. We assumed that spider abundance did not vary with height on a bole. We have no quantitative estimate of spider distribution and abundance with tree height so the degree to which this assumption is violated is unknown. The abundance of spiders (6–11 mm) per tree size was used only for considering the potential energy to be derived by creepers from foraging on trees of various sizes.

RESULTS

The probability of incurring Type I errors increases when numerous simple correlations are computed. We attempted to lessen the chance of incurring those errors by focusing only on those correlations significant at the $P < 0.01$ level.

Weekly arthropod abundance and biomass (N = 13) were significantly correlated ($r = 0.84$, $P < 0.01$) from the crawl traps only. Of the correlations between bird and arthropod abundances (Table 1), creeper abundance was significantly and positively correlated with the abundance of medium (6–11 mm) spiders measured in the crawl traps only. Brown Creeper abundance was correlated positively with very large (≥ 100 cm dbh) Douglas-fir trees ($r_s = 0.73$). No significant correlations were found between the abundance of creepers and any other tree species.

The correlation between tree diameter and bark furrow depth was highly significant ($r = 0.92$, $P < 0.0001$). Bark furrow depth was correlated significantly with the abundances of small ($r_s = 0.35$, $P < 0.01$), medium ($r_s = 0.77$, $P < 0.001$), and large spiders ($r_s = 0.66$, $P < 0.001$), and soft-bodied large arthropods ($r_s = 0.49$, $P < 0.001$).

Because of the low sample size (N = 6), we have only a qualitative assessment of prey capture by Brown Creepers. Spiders were present in the digestive contents of all six creepers and one creeper also contained numerous spider eggs. Unidentified larvae and pupae of the order Lepidoptera were found in three creepers, and soft-bodied adult arthropods of the orders Diptera

TABLE 1. SPEARMAN COEFFICIENTS OF RANK–ORDER CORRELATION MEASURING THE DEGREE OF ASSOCIATION BETWEEN BROWN CREEPER (BIRDS/40 HA) AND ARTHROPOD ABUNDANCE. ARTHROPODS WERE SAMPLED FROM 9 MAY THROUGH 1 AUGUST 1984 IN NINE FOREST STANDS OF THE SOUTHERN WASHINGTON CASCADE RANGE

Arthropod variables	Trap type	
	Crawl	Flight
Spiders		
Small (1–5 mm)	−0.18	0.68**
Medium (6–11 mm)	0.82***	0.49
Large (≥12 mm)	0.14	−0.27
Soft-bodied types		
Small (1–5 mm)	−0.63**	0.07
Medium (6–11 mm)	0.28	0.20
Large (≥12 mm)	−0.14	−0.39
Hard-bodied types		
Small (1–5 mm)	−0.25	−0.10
Medium (6–11 mm)	−0.08	0.56
Large (≥12 mm)	−0.64**	0.48
Total arthropod abundance	−0.65**	0.08

*** Significant at $P < 0.01$; ** $P < 0.05$.

(1), Neuroptera (1), Tricoptera (1), Lepidoptera (3), Hemiptera (2), and Homoptera (1), were found in the digestive tracts of four creepers. Coleoptera were found in the digestive contents of two creepers.

One Douglas-fir tree (112 cm dbh and 53 m tall) had 125 m^2 of bark surface area encompassing two-thirds of the height, a mature tree (67 cm dbh and 44 m tall) had 61.4 m^2, and a young tree (29 cm dbh and 30 m tall) had 18 m^2. We multiplied these areas by the average number of spiders found daily on trees in young, mature, and old-growth forests. We found that a creeper would have to fly to 13 young trees (29 cm dbh) or 3.3 mature trees (67 cm dbh), to obtain the same number of spiders available on one old-growth tree 112 cm dbh. Average daily spider estimates were 0.26/m^2 in old-growth, 0.17/m^2 in mature, and 0.14/m^2 in young stands.

DISCUSSION

Surveying even one substrate may require using more than one trapping technique because of the high temporal and spatial variability associated with arthropod abundance. The two traps we used were designed primarily to capture arthropods that use different types of locomotion. Both sampled an unknown amount of air space; the crawl traps also sampled an unknown area of forest floor surrounding the tree. Biomass of arthropods captured in the flight traps was more variable than those captured in the crawl traps. The flight traps often captured swarming arthropods (e.g., Diptera: Chironomidae) whose weights were slight relative to numbers. Both traps captured spiders; some of the spiders in the flight traps may have been young that "balloon" to colonize new substrates (R. Gara, pers. comm.). In general, the flight traps were ineffective for establishing relationships between creeper and arthropod abundances.

We did no observations of capture efficiency (i.e., the proportion of arthropods encountering a trap and subsequently caught) for either trap. Moeed and Mcads (1983) found that for crawl traps only a few cockroaches (Blattodea) and ground beetles (Coleoptera) avoided capture by climbing over the netting girdle, and some Collembolla and mites (Acari) passed through the 1.5 mm mesh of the netting girdle. They observed spiders residing in down-traps (designed to capture arthropods crawling downward on tree trunks) on three occasions and on up-traps on one occasion but concluded that these were isolated instances and likely had no effect on capture rates for other insects.

We installed up-type (crawl) traps only and never observed spiders residing in them. Periodically checking and cleaning our traps between scheduled sampling periods was not feasible because our sites were not readily accessible.

Our study was an exploratory analysis of associations between Brown Creepers and certain habitat characteristics, including potential food resources. Whole prey items found in creeper digestive tracts were never larger than our medium-sized category, but arthropods with both hard and soft body conditions were present. Although not conclusive, bill morphology may not limit creepers' use of food items, as we had assumed.

We did not compare Brown Creeper use of prey items in comparison to the relative abundance of prey, but our results suggest that spiders may have been an important food item for creepers during the 1984 breeding season. The significant relationship between creeper abundance and very large trees may have been mediated to some extent by the deep bark furrows on such trees. Large trees or those with deeper furrows tend to have high densities of spiders (New Zealand—Moeed and Meads 1983; Europe—Nicolai 1986; USA—this study) and large, soft-bodied arthropods (this study). Spiders apparently comprise a major food source for creepers (e.g., Martin et al. 1951, this study), and Kuitunen and Tormala (1983) found that 90% of the food items (by number) brought to Treecreepers (*Cethia familiaris*) in Finland were spiders. Finally, spiders have a higher protein content than insects (Hurst and Poe 1985), perhaps making them a premium food item for small birds, and especially for creepers, which expend considerable energy climbing upward on tree boles (Norberg 1986).

Bark furrow depths, which are significantly correlated with tree size, increase available foraging substrate without substantially increasing the actual area over which the bird has to move to search for prey. Based on our calculations of bark surface area and the number of spiders (6–11 mm) potentially occurring on trees of various sizes, we hypothesize that creepers may be able to increase their energy intake by foraging on one large diameter Douglas-fir tree versus numerous small trees.

We conducted this study during a short time frame and our methods enabled us to conduct only descriptive types of analyses. Arthropod abundance and composition on tree trunks are affected by a combination of several factors including the microclimatic conditions produced by bark features (Nicolai 1986), the presence of fungi and epiphytes on bark, the proximity and composition of surrounding vegetation (Jackson 1979), and the tree species' relative abundance throughout recent geological history (Southwood 1961).

More comprehensive and intensive sampling efforts of arthropod populations are needed within and among seasons and on a long term basis. This information would be especially useful if collected in the context of examining the effects that habitat alterations have on food resource availability.

ACKNOWLEDGMENTS

We acknowledge the assistance of John M. Stone and Anne B. Humphrey with trap construction and field work. Laurie Goodpasture classified thousands of arthropods and David Briggs, of the College of Forest Resources, University of Washington, wrote the computer program for calculating bark surface area. Thomas Spies, Pacific Northwest Forest and Range Experiment Station, Forestry Sciences Lab., Corvallis, Oregon, suggested techniques for examining bark surface features and provided critical forest stand data. Stephen D. West provided valuable advice and support throughout the duration of the study, and we thank Michael L. Morrison, Stephen D. West, John F. Lehmkuhl and Jared Verner for reviewing this manuscript. Funding was provided by the USDA Forest Service, Pacific Northwest Forest and Range Experiment Station, Forestry Sciences Laboratory, Olympia, Washington under contract number PNW-83-219. This paper is contribution No. 35 of the Old-growth Wildlife Habitat Project.

FOOD RESOURCES OF UNDERSTORY BIRDS IN CENTRAL PANAMA: QUANTIFICATION AND EFFECTS ON AVIAN POPULATIONS

JAMES R. KARR AND JEFFREY D. BRAWN

Abstract. Habitat associations of birds inhabiting the understory of tropical moist forests vary in time and space. We investigated whether this variation was related to changes in resource abundance. Foliage and litter arthropod abundances were estimated at about 60 sampling sites in central Panama from 1983 through 1985. Bird activity was also determined with mist nets at these sites. Activity (i.e., capture rates) of about 20 species and five foraging guilds revealed widely varying consumer-resource associations. Certain species were positively correlated with variation in arthropod abundances, whereas others were less common when and where their presumed resources were comparatively abundant. Microclimate (i.e., humidity) influenced the nature of bird-food interactions; dry sites appeared to be unsuitable habitat for certain species despite sometimes abundant arthropods. We conclude that habitat associations of birds in central Panama are not solely food-resource mediated.

Key Words: Understory birds; food resources; Panama; arthropods; tropical forest.

Understanding the nature of consumer-food interactions is critical to the study of tropical avian ecology. For example, the notably high species richness in many neotropical habitats may stem from the variety of available food resources, associated feeding locations, and the tendency for many species to be omnivorous. At least 20–35% of tropical forest species consume some combination of fruits, insects, and nectar (Karr 1975, Karr et al. in press). Two factors make observation of foraging behavior especially difficult for a large proportion of Neotropical forest species. First, many species are rare, secretive, or both (Karr 1971, Terborgh 1985). Second, even if a species is common, its mobility can impede observations of behavior in tropical forest habitats (Remsen 1985). Large frugivores, such as parrots and toucans, and many insectivores travel over large areas, often in mixed-species flocks. Thus, complete description of "community foraging space" of birds is clearly difficult in tropical moist forests.

These logistical problems have, understandably, led to a research emphasis on long-term studies of selected species or guilds with relatively sessile and quantifiable resources or easily observed foraging behavior. Studies of nectarivores (Wolf et al. 1976, Feinsinger 1978, Stiles 1978), frugivores (Snow 1981, Moermond and Denslow 1983), army ant followers (Willis and Oniki 1978), and flycatchers (Fitzpatrick 1980, Sherry 1984) are examples.

Problems in estimating consumer and resource abundances in tropical moist forests exacerbate the difficulty of studying avian foraging ecology. Estimating avian abundances is laborious and time consuming (e.g., mist-netting), or demands bird identification skills that can take considerable time to develop in tropical forests (e.g., spot-mapping, Terborgh 1985). Standardized protocols for estimating variation in food resources of, say, insectivorous birds have not been established (Wolda, this volume). Estimating arthropod abundance is tractable for certain groups of consumers (e.g., arthropods in hanging litter; Gradwohl and Greenberg [1982b]), but not for others (e.g., canopy-dwelling birds). Arthropod abundances on understory foliage and in litter can be estimated, but the method used can strongly affect results (Wolda and Wong 1988). Further, the often high diversity of plant species with specialized leaf-eating arthropods present formidable sampling difficulties. Finally, the natural histories of many neotropical birds are so poorly documented that even the elementary step of choosing which resources to study may be problematic (Remsen 1985). For example, hummingbirds routinely feed on arthropods in many neotropical habitats (Remsen et al. 1986), but few efforts to apply foraging theory to nectarivores have considered the importance of arthropods in their diet (Karr 1989).

Much theory in ecology proceeds from assumptions about the relationships among organisms and their foods. We believe that the aforementioned uncertainties justify caution in acceptance of general, often paradigmatic, statements about the habitat, population, and community ecology of tropical birds (e.g., narrow niches, the stability-diversity dogma).

The diversity of tropical birds and their resources precludes detailed study of all groups, so we adopted a compromise between the number of species studied and the level of detail of the study. We sought to estimate variation in resource availability and its influence on habitat use by selected undergrowth birds in moist lowland forests of central Panama. We consider two topics that are integral to the understanding of relationships among birds and their food resources: (1) methods used to estimate variation in leaf litter and foliage arthropods; and (2) relationships between arthropod abundances and avian activity, including examination of the effects of variation in understory microclimate.

METHODS

SAMPLING

Birds

Bird activity was sampled with mist nests (Northeastern Bird Banding Association ATX, 12.0 × 2.6 m, 36 mm mesh) at over 60 net sites in the undergrowth of forest in Parque Nacional Soberania in central Panama (Karr and Freemark 1983). Nets were operated at each site for 3–6 days in March (dry season) and July (wet season) from 1979 through 1986. Additional samples were collected in January and May 1983. Sampling effort included 45,008 net hours and yielded 6896 captures. We analyzed activity of the 20 most commonly netted species and of five foraging guilds each composed of 3–4 species (Table 1). Several guilds represented by only one species were not included in the latter analysis. A total of 1754 captures ($\bar{X}$ = 88/species, range = 36–382) are used here from data collected during 1983–1985. Assignments of species to guilds was based on similarities in food type and foraging location (Karr 1971, pers. obs.; Stiles 1983a).

Activities of species and guilds were estimated as number of captures/100 net hours. Capture rates for each guild were based on total number of captures for all species in the guild, not the average of individual species capture rates. We assumed that capture rates estimated intensity of activity at our sample sites (Karr and Freemark 1983). Nets were open all day, except during heavy rains, so variation in activity as a function of time of day was not a concern. Nets used in this way assessed changes in avian activity in a mosaic of habitats and thus reflected a dynamic habitat selection process (see Karr and Freemark 1983:1489).

The objective of evaluating associations between measures of bird activity and resource availability presents a problem of selecting the appropriate sampling method to detect ecological relationships. General indexes of food availability can indicate broad connections between birds and resources (Martin and Karr 1986b), but stronger inferences are likely with direct measures of food resources (Blake and Hoppes 1986). Moreover, direct estimates of resource abundance are most useful when derived from samples that coincide in space and time with the collection of bird data.

Accordingly, we sampled bird activity and resources thought or known to be consumed by birds at the same sites and in the same months. All resource sampling was done within a 10- × 25-m quadrat adjacent to each net site. The choice of quadrat size was somewhat arbitrary but reflected our goal of sampling a reasonably sized area associated with each mist net. The center line of each quadrat's long axis ran perpendicularly from the center of each net. Quadrats were successively alternated from the right to the left sides of nets along a line of net sites. Three general categories of resources were sampled: leaf-litter arthropods, undergrowth-foliage arthropods, and undergrowth fruit. Other types of food resources were too ephemeral (e.g., fruit fallen to the ground) or difficult to sample efficiently (e.g., bark arthropods) to justify estimation of availability. Only the arthropod data will be presented in this paper. Foliage arthropods and birds were sampled four times in 1983 and twice in 1984; litter arthropods were sampled in these periods and during one additional period in 1985.

Foliage arthropods

Our goal in sampling foliage arthropods was to mimic the search methods of foliage-gleaning birds. Therefore, we used an approach mentioned briefly by Janzen

TABLE 1. WEIGHTS, FORAGING-GUILD ASSIGNMENTS, AND NUMBER OF CAPTURES FOR EACH OF THE 20 MOST FREQUENTLY CAPTURED SPECIES, PARQUE NACIONAL SOBERANIA, PANAMA, JANUARY 1983 TO MARCH 1985

Species	Number of captures	Weight (g)	Foraging guild[a]
Geotrygon montana (dove)	47	128	GRFR
Phaethornis longuemareus (hermit)	105	6	NI
Dendrocincla fuliginosa (woodcreeper)	70	41	ANTF
Automolous ochrolaemus (foliage-gleaner)	40	40	FGIN
Sclerurus quatemalensis (leaftosser)	86	34	GRIN
Thamnophilus punctatus (antshrike)	49	22	FGIN
Myrmotherula axillaris (antwren)	45	8	FGIN
M. fulviventris (antwren)	41	10	FGIN
Gymnopithys leucaspis (antbird)	151	30	ANTF
Hylophylax naevioides (antbird)	95	17	ANTF
Phaenostictus mcleannani (antbird)	58	51	ANTF
Formicarius analis (antthrush)	40	57	GRIN
Pipra coronata (manakin)	75	10	UNFR
P. mentalis (manakin)	382	15	UNFR
Terenotriccus erythrurus (flycatcher)	54	7	FLIN
Myiobius sulphureipygius (flycatcher)	44	12	FLIN
Platyrinchus coronatus (spadebill)	91	9	FLIN
Mionectes oleaginea (flycatcher)	159	10	UNFR
Cyphorhinus phaeocephalus (wren)	86	20	GRIN
Cyanocompsa cyanoides (grosbeak)	36	32	UNOM

[a] Foraging guilds: GRFR = ground frugivore (1 species); NI = nectarivore insectivore (1); FGIN = foliage-gleaning insectivore (4); GRIN = ground insectivore (3); ANTF = ant follower (4); UNFR = undergrowth frugivore (3); FLIN = flycatching insectivore (3); UNOM = undergrowth omnivore (1).

(1980a). We conducted timed visual surveys within each sampling quadrat by counting arthropods on leaves from 0.5 to 2.0 m high during unpatterned walks through each quadrat. All samples were done between 07:30 and 14:00 during periods with little or no cloud cover. A flashlight was occasionally used to aid in detection of arthropods on the undersides of leaves. Each arthropod observed was categorized according to taxonomic group (Order, sometimes Family), size (<5 mm, 5–15 mm, and >15 mm), and leaf surface (upper or lower). Taxonomic groups used for adults were: Arachnida, Coleoptera, Hemiptera, Homoptera, Lepidoptera, Diptera, Formicidae, and Orthoptera. Uncommon taxa were lumped as "other." Small numbers of cryptic insects may have been missed despite our best efforts. We do not include detailed analyses of arthropod taxa here. As more specific data on diets of tropical birds (e.g., Sherry 1984) become available, analysis of patterns between individual bird species and insect groups might be useful. We excluded ants from our analyses because adult ants are rarely consumed by the species commonly captured in our net samples (J. R. Karr, pers. obs.).

During January 1983, variation among four observers in simultaneous counts varied by less than 10% for abundances and taxonomic assignments. Consequently, we made no additional effort to evaluate variation among observers. All observers conducting these counts had training in insect identification.

Leaf-Litter arthropods

We followed procedures established by Willis (1976) for sampling litter arthropods. One sample was collected at each net site at randomly determined coordinates within each sampling quadrat. Samples were collected by placing a 29- × 34-cm (0.10 m²) plastic container on the forest floor and sliding a plexiglass sheet beneath to gather the litter. Samples were than placed in plastic bags, arthropods immobilized with Kahle's Solution (Borror et al. 1976), weighed, and sorted. Although we were unable to obtain dry weights of our samples, other studies (S. Levings, unpubl. data) provided estimates of moisture content of litter in wet and dry periods. Moisture content of leaf litter in central Panama averaged 25% in late March and 53% in July (S. Levings, unpubl. data). All weight-based analyses of litter arthropods were standardized to estimate dry weights of litter and were expressed as number of arthropods/100 gm of litter. Each litter arthropod was classified according to size (2–5 mm, 5–10 mm, 10–15 mm, >15 mm) and taxonomic group (Phalangidae, Acarina, Arachnida, Isopoda, Diplopoda, Blattaria, Orthoptera, Isoptera, Hemiptera, Homoptera, Neuroptera, Coleoptera, Lepidoptera, Diptera, Formicidae, and other Hymenoptera, "other," and unknown). Cast skins and arthropods <2 mm were not counted.

ASSOCIATIONS OF BIRDS WITH ARTHROPODS ALONG A MOISTURE GRADIENT

Microclimate in the undergrowth varied little among net sites during the wet season, but appreciable variation among sites existed in temperature and humidity during the dry season (Karr and Freemark 1983). Microclimate at our sample sites was influenced by local topography, presence of nearby streams, and vegetation structure (e.g., canopy openness). Each sample location was assigned to one of four moisture classes (1 = driest, 4 = wettest) based on these factors, and temperature and humidity data were collected with sling psychrometers during several dry seasons (Karr and Freemark 1983). All net sites were sampled hourly (07:00–17:00) on the same day during both wet and dry season sample periods.

STATISTICAL ANALYSES

Our approach in analyzing relationships between avian activity and variation in arthropod abundances was primarily correlative and exploratory. For temporal variation, we combined (i.e., averaged) data from all sampling sites (net locations) or those within a moisture class and calculated Spearman's rank-order correlations between mean arthropod abundance (foliage or litter) and capture rates of species or guilds. We combined sampling sites because of uncertainties in independence of observations among nets; thus, our results are conservative.

RESULTS

RELATIONSHIPS AMONG ABUNDANCES OF BIRDS AND ARTHROPODS

Variability in abundances of understory foliage and litter arthropods

Abundances of arthropods varied considerably among sampling periods (Table 2). Foliage arthropod abundances were relatively low from January thru July 1983, but increased sharply in 1984. Abundances of foliage arthropods in March 1984 were, on average, about 125% greater than those observed during the previous year's March sample. About 45% of the 1983 to 1984 increase was due to increased numbers of small adult Diptera.

Litter arthropods displayed the same general pattern of temporal variation in abundance as the foliage arthropods; numbers were higher in 1984, a result consistent with another study of litter arthropods in central Panama (Wheeler and Levings, in press). The patterns were not always concordant, however. For example, the peak period in abundances of litter arthropods (July 1984) lagged behind that of foliage arthropods (March 1984).

Arthropod abundances and capture rates of five foraging guilds

No correlations between capture rates by guild and either foliage or litter arthropod abundances were significant ($P > 0.05$; critical values = 0.83 for foliage and 0.75 for litter arthropods), but differences among the guilds were striking (Fig. 1). Spearman rank correlations ranged from 0.61 for the ground foraging insectivores with litter arthropods to -0.54 for flycatchers with foliage arthropods. Ground-foraging insectivores and flycatchers had the most positive and negative

TABLE 2. ABUNDANCES OF UNDERSTORY FOLIAGE AND LITTER ARTHROPODS IN CENTRAL PANAMA

Sampling period	Foliage (No. observed/hour) X̄ (SE)	Litter (No. individuals/ 100 g litter) X̄ (SE)
January 1983	37.8 (2.3)	5.4 (4.3)
March 1983	53.5 (4.8)	4.6 (1.0)
May 1983	54.4 (3.1)	5.4 (1.0)
July 1983	63.9 (2.8)	3.7 (0.3)
March 1984	120.4 (16.1)	11.2 (3.5)
July 1984	108.0 (4.9)	14.1 (2.6)
March 1985	No data	6.7 (1.2)

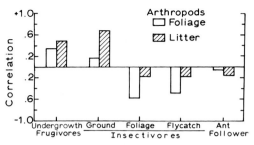

FIGURE 1. Correlations (Spearman's Rho) of capture rates for five avian foraging guilds with abundances of foliage and litter arthropods for 1983 to 1985 in central Panama.

associations with arthropod abundances, respectively.

Arthropod abundances and capture rates of the 20 most common species

The associations of individual species with arthropod abundances also varied (Fig. 2). Spearman correlations between capture rates and foliage arthropod abundances (Fig. 2A) averaged 0.02 and ranged from 0.61 for the Black-faced Antthrush (*Formicarius analis*) to −0.71 for the Sulphur-rumped Flycatcher (*Myiobius sulphureipygius*).

Correlations of individual species capture rates with abundances of litter arthropods (Fig. 2B) averaged 0.18 and ranged from −0.32 for Ochre-bellied Flycatcher (*Mionectes olivaceus*) to 0.77 for *Formicarius analis* ($P < 0.05$). Associations of bird activity with litter arthropods were therefore generally weak, but more positive than those with abundances of foliage arthropods. Moreover, species within guilds were more consistent regarding litter arthropods; all correlations for the ground insectivores were positive and all those of the flycatchers were negative.

Associations of species capture rates and foliage arthropod abundances were especially heterogeneous within certain foraging guilds (Fig. 3). For example, two antfollowers, Ocellated Antbirds (*Phaenostictus mcleannani*) and Spotted Antbirds (*Hylophylax naevioides*), had Spearman correlations of 0.60 and −0.54, respectively. The three-member ground-foraging guild was the most consistent. Wald-Wolfowitz runs test on the Spearman's rank correlations suggested a systematic difference between flycatchers and ground gleaners in their associations with litter and foliage arthropods ($P < 0.05$).

Arthropod abundances and capture rates of birds along a moisture gradient

In the wet season foliage arthropod abundances were similar at dry and wet sampling sites, but during dry periods moist sampling sites had lower abundances than more xeric sampling sites. Litter-arthropod abundances were less variable along the moisture gradient, but tended to be higher at comparatively wet and dry sites.

We found no consistent pattern of covariance (Table 3) between capture rates of undergrowth bird species and arthropod abundances along the moisture gradient. Capture rates did not increase or decrease systematically within any guild along the moisture gradient. Bird-arthropod associations were somewhat more positive at relatively mesic sampling sites (e.g., Moisture class 2 for litter arthropods). Moreover, at the most xeric sites, capture rates of all guilds were negatively associated with abundances of foliage arthropods (Table 3). Correlations of capture rates with abundances of litter arthropods were more positive, especially for ground insectivores. Activities of flycatchers and foliage gleaners were negatively associated with foliage-arthropod abundances at all moisture conditions. The association of foliage gleaners with foliage arthropods at the driest sites was distinctly negative.

DISCUSSION

The clear differences in arthropod abundances between 1983 and 1984 coincided with an extremely dry dry season, possibly caused by the severe El Niño in 1983 (Brawn and Karr, unpubl. data). The dry season in 1983 was the longest and driest recorded in central Panama since 1929. Only 26 mm of rain were recorded at nearby Barro Colorado Island from January to March (40-year mean ± 1 SD = 122 ± 96 mm [Karr and Freemark 1983]). During early 1983 many trees and shrubs exhibited signs of moisture stress (e.g., wilting, excessive leaf abscission [J. Karr, pers. obs.]). Moreover, rarity of intermittent dry-season rains in 1983 delayed development of new leaves, flowers, and fruits of many tree species in central Panama (D. Windsor, pers. commun.); thus, phenological differences in availability of

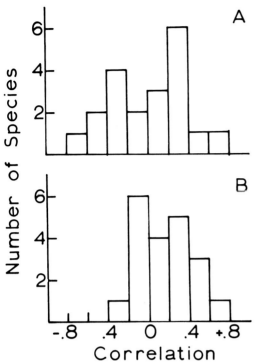

FIGURE 2. Frequency histograms of Spearman's Rho correlations between capture rates and abundances of (A) foliage arthropods and (B) litter arthropods based on the numbers of species that exhibited given levels of correlations for the 20 most common species for 1983 to 1985 in central Panama.

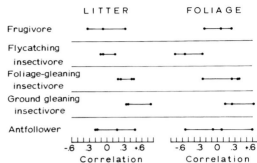

FIGURE 3. Distributions within five foraging guilds of Spearman's Rho correlations for capture rates and arthropod abundances for 1983 to 1985 in central Panama.

resources for insects or direct moisture effects on insects may have indirectly influenced resource availability for birds. The influence of the 1983 drought may have been more direct on litter arthropods; observational and experimental evidence indicates that numbers of litter arthropods in central Panama are enhanced by soil moisture (Levings and Windsor 1984).

Overall, the direction of associations between birds and arthropods was not consistent with the notion that birds were "tracking" food resources. For example, the strong positive correlation of ground-foraging birds with litter arthropods suggests a bird-food association, but contrasts with the negative associations between capture rates of foliage-gleaning birds and foliage arthropods. Positive correlations between undergrowth frugivores and both foliage and litter arthropods are puzzling.

Karr and Freemark (1983) observed that patterns of habitat selection by undergrowth birds in central Panama are partly explained by inter- and intraspecific variation in activities of birds along microclimatic gradients. Each species exhibited some preference among the range of moisture conditions and many species altered their primary habitat association over time as they appeared to track changing microclimate conditions. Karr and Freemark suggested two underlying mechanisms for nonrandom distribution of activity along a moisture gradient: birds seek micro-climatic optima for physiological reasons, or they track food resources whose abundances are directly related to moisture conditions, or both. Our results do not show a clear association between bird abundances and their food resources. Therefore, as hypothesized by Karr and Freemark (1983), physiological factors may impede these species from exploiting sometimes abundant food resources at dry sites.

Results of correlative analyses can be discussed for their biological significance or judged critically owing to perceived problems in analytical issues such as validity of sampling method. Certain biases are inherent in sampling with mist nests as with any survey or census method (Karr 1979, 1981). However, mist nets minimize problems associated with detecting species that are difficult to observe or that vocalize rarely, problems that introduce unknown biases into more conventional census procedures.

Variations in foraging activity and mobility among species and even among sex and age classes of the same species (Karr 1971, 1979, 1981) yield capture rate variation among species. High capture rates of very mobile species such as Red-capped Manakin (*Pipra mentalis*) do not necessarily reflect higher densities than those of seldom-captured species such as Spotted Antbirds. High recapture rates in our study (consistently 50–70% of captures), however, suggest that birds do not learn to avoid nets. In addition, we find no variation in recapture rates of sedentary

TABLE 3. CORRELATIONS OF FORAGING GUILD CAPTURE RATES WITH ARTHROPOD ABUNDANCES AT DIFFERENT MOISTURE CLASSES FROM DRY (1) TO WET (4) FOR 1983 TO 1985 IN CENTRAL PANAMA

Arthropod group	Moisture class	Undergrowth frugivores	Ground insectivores	Foliage-gleaning insectivores	Flycatching insectivores	Antfollowers
Litter	1	0.27	0.57	−0.57	−0.11	0.18
	2	0.27	0.57	0.33	0.32	0.79
	3	−0.46	0.36	−0.18	0.21	−0.14
	4	0.57	−0.17	−0.01	−0.39	−0.05
Foliage	1	−0.46	−0.14	−0.83	−0.29	−0.43
	2	0.64	0.17	−0.49	−0.37	−0.09
	3	−0.03	−0.77	−0.02	−0.43	0.09
	4	0.71	−0.49	−0.02	−0.31	0.02

species in several guilds, suggesting that different guilds do not vary systematically in their ability to detect and avoid nets.

We also note that imprecision should not be confused with systematic error, a distinction that is critical when evaluating results of field studies. Our estimates of bird activity and resource abundance were derived from sampling a full range of microclimates and vegetation structure within a 90 ha area. In addition, counts of birds and their presumed food were done in the same plots over short time periods, a goal that has rarely been attained over so many sample plots (about 60 net sites).

Our data on arthropod abundances and avian activity are, admittedly, "blunt instruments" for determining the effects of food availability on habitat selection by insectivorous birds. All the arthropods detected in our abundance estimates were not potential prey items for birds; some may be unpalatable or require excessive time or energy for capture (Martin 1986; Wolda, this volume). Sherry (1984) demonstrated that, for flycatchers in Costa Rica's moist lowland forests, what is or is not a food item for an insectivore is a function of a predator's foraging technique as well as prey distribution and body size.

Associations between capture rates of birds and arthropod abundances vary among species and guilds, suggesting that foraging mode may determine how "opportunistic" a given species can be. Ground insectivores (by species and as a guild) appear to track temporal variation in resource availability more precisely than species in other guilds regardless of environmental conditions. In contrast, flycatchers seem to be *less* common when and where arthropod abundances are high. Unfortunately, we have no information about the relationship between foliage and flying-insect abundances. Other studies of Neotropical flycatchers suggest that a species' diet can be influenced by time and energetic-physiological constraints (Fitzpatrick 1981, Sherry 1984). Our results support these suggestions and add that such constraints may be more rigid in dry areas. A study incorporating physiological and energetic measurements evaluated along environmental gradients would be useful in clarifying habitat use and resource ecology of birds of tropical forest undergrowth.

As empirical evidence accumulates, it seems that the factors responsible for ecological pattern vary among populations and communities. Accordingly, the value of pluralistic theory, though not a new idea, is gaining acceptance (Schoener 1986a). The expectation of finding valid univariate explanations is thus naive. Our results clearly indicate that species and guilds do not respond in concert to variation in environmental factors. Even within a species, the influence of food availability, physiological conditions, and predation may vary in importance with vegetation structure, macroclimate, and microclimate. The concept of a "normal" or typical bird species, population, or community is simply inappropriate (Wiens and Rotenberry 1987).

We believe that pluralism is also appropriate in the field. The complexity of tropical forest avifaunas and their food resources invites an especially high diversity of valid approaches to understanding ecological patterns. Further, the system being examined can influence the types of questions that can be effectively addressed. For example, experimental manipulations of avian abundances or supplemental feeding experiments, formidable in any habitat (Wiens et al. 1986a), would be difficult for most insectivores in tropical forests. Similarly, supplemental feeding might be possible in the case of frugivores, but the scale of habitat use by many tropical frugivores may make tractable manipulations inappropriate for evaluating bird-food associations in the real world.

Two final points are important—one specific to our study and one a general observation. First, the methods and results described here considered general patterns, but also served to identify specific aspects of the foraging ecologies of certain species and guilds that merit more detailed examination (e.g., more detailed analysis of the insect taxa consumed by specific bird species, the comparative stability of fruit vs. insects as food resources, or the mechanisms of omnivory that allow survival through crunch periods). Second, tropical species have long been considered to be ecological specialists (Klopfer and MacArthur 1961). Many species are habitat or food specialists, but many also repeatedly exhibit an ability to alter their behavior (foraging and habitat selection) in response to changing environmental conditions. An exploration of the temporal reliability of resources and the evolution of plasticity to exploit a broader range of resources, although a difficult task, could help clarify the role of food resource availability in governing the ecology of species and the development of assemblages of species.

ACKNOWLEDGMENTS

Financial support for this study came from Earthwatch, National Science Foundation (DEB #82-06672), and the Smithsonian Tropical Research Institute. Panama's Natural Resource Agency (IN.RE.NA.RE.) kindly provided permits to work in Soberania National Park. N. Smith, E. Leigh, S. Rand, T. Martin, H. Wolda, and R. Fraga provided comments on an early draft of this paper.

Studies in Avian Biology No. 13:65-72, 1990.

SPATIAL VARIATION OF INVERTEBRATE ABUNDANCE WITHIN THE CANOPIES OF TWO AUSTRALIAN EUCALYPT FORESTS

J. D. Majer, H. F. Recher, W. S. Perriman, and N. Achuthan

Abstract. We compared branch-clipping and pyrethrin-knockdown methods for estimating relative abundances of arboreal invertebrate taxa in two Australian eucalypt forests, one at Karragullen in Western Australia, the other at Scheyville in New South Wales. Branch clipping was designed to sample sessile invertebrates and galls. The pyrethrin method sampled mainly mobile invertebrates and those associated with bark and branches. Invertebrates were sampled in subcanopy (1–7 m) and the canopy (7.1–20 m) layers. Both methods indicated higher levels of invertebrates at Scheyville than at Karragullen. Adjusted for exceptionally abundant taxa, both techniques showed invertebrates to be more abundant in the upper canopy of all tree species and, within a particular forest, most abundant on marri (*Eucalyptus calophylla*) and narrow-leaved ironbark (*E. crebra*). Differences between tree species were more pronounced with branch-clipping than with pyrethrin data; however, branch-clipping data were insufficient for computing variances. Pyrethrin data showed that within-tree variation was generally greater than that between trees. Variances were generally greater for invertebrates from the subcanopy. In both forests, tree species with the highest invertebrate abundance had the highest variances in counts of invertebrates. Each method has its limitations. We recommend using both together to measure relative abundances of invertebrates in forest ecosystems.

Key Words: Invertebrates; insects; invertebrate sampling; relative abundance; arboreal.

INTRODUCTION

Understanding processes that govern the dynamics of bird communities requires information on the abundance and variability of food resources. For many reasons well known to avian ecologists, estimates of the absolute abundances of invertebrates available to birds are seldom available. However, indices of invertebrate abundance can be used to interpret seasonal patterns of avian abundance (e.g., Bell and Ford, this volume; Recher et al., 1983), year-to-year variation in numbers (e.g., Bell and Ford, this volume), and the timing of reproductive activities (e.g., Nix 1976, Recher et al. 1983). In these instances, avifauna respond to large changes in the abundance of invertebrate prey. Indices of invertebrate abundance are less useful for interpreting differences in population densities and community composition between habitats or differences between species in the use of particular substrates. Understanding the reasons why birds select particular substrates, and explaining small differences in species abundance or community composition, require precise measures of the abundances of individual prey items on specific substrates.

Before such measurements can be obtained, we need sampling procedures with predictable levels of variance. Two inherent sources of error confound all invertebrate sampling procedures: variation in the application of the procedure itself, and spatial and temporal variation in the distribution and abundance of the invertebrates being studied. Sources of error due to temporal variations in invertebrate numbers can be controlled by taking samples on consecutive days, sampling at the same time of day, and restricting samples to specific weather conditions (i.e., temperature, cloud cover, wind speed, incidence of rain). Assuming that sampling procedures can be standardized, the major sources of error in invertebrate samples result from variation in spatial patterns of distribution and abundance of invertebrates and also the substrates on which they occur.

Our objectives in this study were (1) to examine the patchiness of invertebrate abundance within and between tree species in forests at two localities, and (2) to compare two different methods for measuring the relative abundances of various invertebrate taxa in tree canopies.

METHODS

Study Sites

Sampling was done during February 1987 at Scheyville, New South Wales (56°05'S, 150°51'E), where we sampled invertebrates on narrow-leaved ironbark (*Eucalyptus crebra*) and grey box (*E. mollucana*); and during April 1987 at Karragullen, Western Australia (32°04'S, 116°07'E), where we sampled jarrah (*E. marginata*) and marri (*E. calophylla*). The forest at Scheyville was dominated by narrow-leaved ironbark (51% of trees, 42% of tree foliage) and grey box (40% of trees, 51% of tree foliage) with smaller numbers of forest red gum (*E. tereticornis*) (7% of trees) and thin-leaved stringybark (*E. eugenoides*) (<1% of trees). Canopy cover was 40-45%, with the canopy averaging 15-18 m in height. Individual trees emerged above the canopy to 25 m (Recher and Gebski, this volume). The understory consisted of eucalypt saplings; grasses and forbs comprised the ground cover. At Karragullen, jarrah (92% of trees and foliage) dominated the forest; marri comprised only 8% of all trees. Canopy cover was 60%, and mean canopy height was 15–18 m, with individual trees to 30 m. Karragullen had a more diverse understory than the forest at Scheyville, with a dense subcanopy of eucalypt saplings, sheoak (*Allocasuarina fraserana*), and banksia (*Banksia grandis*). The site had a rich ground vegetation.

Climate at the Karragullen site is Mediterranean—cool, wet winters and hot, dry summers. Mean annual rainfall is 1241 mm with most rain falling between May and October. At Scheyville, although spring (August–October) tends to be drier than other seasons, rain falls fairly evenly throughout the year. Mean annual rainfall is 874 mm, summers are warm and winters are cool with occasional frosts.

INVERTEBRATE SAMPLING PROCEDURES

Two methods were used to sample invertebrates: branch clipping and chemical (pyrethrin) knockdown. The efficiency of these methods for sampling canopy arthropods is reviewed by Cooper and Whitmore (this volume). Majer and Recher (1988) described both methods and compared their effectiveness and costs (in time) for sampling invertebrate communities in eucalypt forests. In brief, branch clipping sampled sessile foliage invertebrates, which could be expressed as sample weight or numbers/leaf area. Pyrethrin knockdowns sampled mobile invertebrates on leaves and bark, with abundances expressed per unit area of canopy.

For each procedure 10 trees of each species were sampled at each of two height ranges—the subcanopy layer (1–7 m) and the canopy (7.1–20 m). Flowering trees were avoided. Mature trees (>15 m in height) were selected for sampling canopy vegetation. These were reached with a trailer unit with an extendable arm and bucket capable of lifting two people to a height of 13 m. Ten samples were taken from each tree. We sometimes had difficulty taking 10 samples from saplings; in such cases we sampled from a monospecific cluster of trees. A ladder was used to place nets in the subcanopy.

Pyrethrin-knockdown samples

Cotton, funnel-shaped nets with a surface area of 0.5 m² were used to collect pyrethrin samples. Each net was fitted with a sleeve that held a 100-ml plastic tube. Nets were held about 60–70 cm below the vegetation. A swivel-and-line arrangement allowed movement in the wind but kept nets from slipping vertically. Within a given tree (or cluster of saplings), nets were suspended at different heights according to the distribution of suitable branches for attachment, so that no nets overlapped. As nearly as possible, net positions were selected to equalize the amount of foliage (determined by visual inspection) in the column directly above the nets. The height of each net was recorded. Nets were positioned the afternoon prior to spraying, to allow disturbed invertebrates to return, although we detected no case (e.g., insects flying away) of disturbance during this process.

The morning (07:00–10:00) of the following day, the canopy to a height of 7 m above each net was sprayed with 0.2% synthetic pyrethrin pesticide, synergized with piperonyl butoxide, using a motorized-knapsack mist-blower. Two liters of diluted (10:1) pesticide were used per tree, and trees were left for 30 min to allow silk-attached invertebrates to drop into nets. The canopy was then shaken to dislodge remaining invertebrates, and specimens were brushed into the collecting tubes and preserved with 70% ethanol.

We sampled five trees (50 nets) each morning for 10 days (10 high and 10 low trees of each of two species = 400 nets) to sample each height range at each site. The canopy was sampled first, then the subcanopy layer. Spraying was done only when it was dry and calm. In the event of poor weather, nets were left in place and we sprayed on the first suitable morning (usually the next day).

Branch clipping

At the same time that nets were hung, 10 small branches (<10 mm in diameter) were clipped from the outer foliage of each tree. Samples weighed from 25 to 125 g and contained at least 40 leaves. Branches with numerous seed capsules were avoided. Samples were inserted into a plastic bag prior to clipping; bags were sealed and frozen until processed. We never saw invertebrates leaving samples before bagging. In the laboratory, bags were weighed and samples vigorously shaken prior to removal. Invertebrates dislodged by shaking were identified and counted. Forty leaves were taken randomly from each sample and inspected on both surfaces for sessile invertebrates and those in webs or cocoons; these were identified and counted. Each 40-leaf sample was weighed to the nearest 0.1 g. Mean leaf area of each tree species at each height range at each site was estimated from the mean of a randomly selected subsample of 150–200 leaves, using a Li-Cor® portable area meter.

DATA SUMMARY AND ANALYSIS

Pyrethrin samples

The objective of the analysis was to compare mean levels of each taxon in each stratum and tree species, with an assessment of the relative variability of taxon counts attributable to "between-tree" and "within-tree" (between-net) variation. The experimental design involved selection of a random sample of 10 trees (experimental units) from each stratum and subsample of 10 nets/tree. Means and variances of the numbers of each invertebrate taxon in each stratum of each tree, and on all 10 trees of a given species and stratum, were computed using the SPSS computer package. Three independent comparisons were made for each individual taxon: (1) between strata for each tree species; (2) between jarrah and marri for each stratum; and (3) between grey box and narrow-leaved ironbark for each stratum. Analyses were restricted to common invertebrate taxa—those occurring in >80% of samples.

To compare between strata for each tree species, we denoted by Y_{ijk} the observed value of each taxon (invertebrate count per net) for the kth net from the jth tree in the ith stratum and assumed the following nested design model for Y_{ijk}:

$$Y_{ijk} = m + S_i + t_{ij} + n_{ijk}$$

where i = 1 (lower) or 2 (upper) for comparisons (1); j = 1, 2, ... 10 (trees); and k = 1, 2, ... 10 (nets).

TABLE 1. Means and Standard Deviations of the Numbers of Invertebrates Sampled per Tree by the Pyrethrin Method for Both Canopies and Subcanopies of Jarrah (*Eucalyptus marginata*) and Marri (*E. calophylla*) in Western Australia. The Number of Invertebrates per Tree Was Based on 10 0.5-m^2 Nets within Each Tree

	Taxon	Jarrah				Marri			
		Canopy		Subcanopy		Canopy		Subcanopy	
		$\bar{X}$	SD	$\bar{X}$	SD	$\bar{X}$	SD	$\bar{X}$	SD
Arachnida	Pseudoscorpionida	1.8	4.7	0.0	0.0	0.3	0.7	0.0	0.0
	Acarina	7.5	5.1	11.9	9.9	7.7	5.7	10.2	13.9
	Araneae	26.1	10.5	26.1	13.4	31.7	10.7	18.7	9.4
Crustacea	Isopoda	1.0	3.2	0.3	0.5	0.1	0.3	0.4	0.7
Collembola		3.0	3.3	32.4	37.1	1.2	1.0	109.6	94.6
Insecta	Thysanura	4.3	5.6	1.3	1.6	5.2	11.3	0.8	1.2
	Odonata	0.0	0.0	0.0	0.0	0.0	0.0	0.0	0.0
	Blattodea	4.9	5.0	0.8	0.6	7.6	9.4	1.2	1.0
	Mantodea	0.6	1.6	0.1	0.3	0.2	0.4	0.0	0.0
	Dermaptera	1.1	0.9	1.5	2.6	2.8	2.6	0.2	0.4
	Orthoptera	0.0	0.0	0.2	0.4	0.1	0.3	0.2	0.4
	Phasmatodea	0.0	0.0	0.1	0.3	0.1	0.3	0.1	0.3
	Embioptera	0.5	0.9	0.3	0.7	0.0	0.0	0.3	0.7
	Psocoptera	13.3	10.2	3.7	2.2	14.7	8.5	8.4	8.8
	Hemiptera (psyllids)	8.0	6.7	4.5	4.4	4.1	2.3	4.9	4.8
	(others)	16.1	8.3	5.5	6.0	15.5	4.8	10.2	11.5
	Thysanoptera	5.1	2.3	3.5	3.7	16.1	9.6	4.6	5.5
	Neuroptera (adults)	0.2	0.4	0.1	0.3	0.1	0.3	0.0	0.0
	(larvae)	0.0	0.0	0.1	0.3	0.0	0.0	0.0	0.0
	Coleoptera (adults)	40.4	10.3	18.2	10.8	48.0	28.5	20.5	15.5
	(larvae)	2.1	0.7	1.7	1.5	4.7	4.2	1.5	1.8
	Diptera (adults)	12.0	6.9	11.0	6.9	13.9	7.7	12.1	6.4
	(larvae)	3.2	2.0	6.1	3.7	0.4	0.8	1.3	1.5
	Lepidoptera (adults)	0.7	1.0	1.4	1.6	1.4	2.6	0.5	0.9
	(larvae)	1.4	1.2	1.7	3.7	3.2	4.8	1.3	1.2
	Hymenoptera (ants)	44.7	40.5	6.2	7.9	25.3	28.4	4.8	5.1
	(others)	44.3	16.7	20.4	11.3	48.6	21.8	20.4	10.9
Totals (excluding ants)		197.6	50.9	152.9	72.7	227.7	72.3	227.4	125.5

Algebraically, the model states that the observed invertebrate count Y_{ijk} was equal to the overall mean (m), plus the deviation (S_i) of the ith stratum mean (m + S_i) from the overall mean, plus a random deviation (t_{ij}) representing global variation between trees, plus an independent random deviation (n_{ijk}) representing local (within-tree) variation between nets. Additionally, t_{ij} and n_{ijk} were assumed to be independently distributed with zero means and variances σ^2_t and σ^2_n, respectively. Thus the model implies that the Y_{ijk} (invertebrate counts per net) were distributed about a mean of m + S_i with total variance $\sigma^2_{tot} = \sigma^2_t + \sigma^2_n$. The analysis of variance (ANOVA) for this model gave estimates of m + S_i as the stratum means, with assessment of the statistical significance of the difference between stratum means, along with estimates of the components of variance σ^2_t and σ^2_n that show relative variability "between-trees" and "within-trees," respectively.

A similar nested-design model was used for Y_{ijk} to compare between trees in each stratum:

$$Y_{ijk} = m + T_i + t_{ij} + n_{ijk}$$

where i = 1 (jarrah) or 2 (marri) for comparison (2), or i = 1 (grey box) or 2 (narrow-leaved ironbark) for comparison (3), and T_i = deviation of the ith species mean from the overall mean m. The t_{ij} and n_{ijk} have the same interpretation as in comparison (1).

Branch-clipping samples

Because counts of invertebrates were so low in the branch-clipping samples, we could not analyze the data statistically. Instead we computed the number of invertebrates/g of sample (bag contents) and the number/cm^2 of leaf area for each tree species and stratum. In the latter case, numbers were halved to allow for invertebrates on upper and lower leaf surfaces.

RESULTS

TOTAL INVERTEBRATES IN PYRETHRIN SAMPLES

Twenty-seven taxa were sampled at both sites, but counts of invertebrates were generally much higher at Scheyville than at Karragullen (Tables 1 and 2). The most abundant taxa on trees at Karragullen were Araneae (spiders), Psocoptera (booklice), Hemiptera (sucking bugs other than psyllids), Coleoptera (adult beetles), Diptera (adult flies), and Hymenoptera (ants). At Scheyville, the most abundant taxa were Acarina

TABLE 2. MEANS AND STANDARD DEVIATIONS OF THE NUMBERS OF INVERTEBRATES SAMPLED PER TREE BY THE PYRETHRIN METHOD FOR BOTH CANOPIES AND SUBCANOPIES OF GREY BOX (*Eucalyptus mollucana*) AND NARROW-LEAVED IRONBARK (*E. crebra*) IN NEW SOUTH WALES. THE NUMBER OF INVERTEBRATES PER TREE WAS BASED ON 10 0.5-M^2 NETS WITHIN EACH TREE

| | | Grey box | | | | Narrow-leaved ironbark | | | |
| | | Canopy | | Subcanopy | | Canopy | | Subcanopy | |
	Taxon	$\bar{X}$	SD	$\bar{X}$	SD	$\bar{X}$	SD	$\bar{X}$	SD
Arachnida	Pseudoscorpionida	0.0	0.0	0.0	0.0	0.0	0.0	0.0	0.0
	Acarina	24.6	26.1	23.0	16.1	18.0	14.9	31.0	16.9
	Araneae	65.8	29.6	68.8	30.0	70.7	39.4	119.7	36.8
Crustacea	Isopoda	0.0	0.0	0.0	0.0	0.1	0.3	0.0	0.0
Collembola		0.2	0.6	1.5	2.4	0.0	0.0	1.1	1.3
Insecta	Thysanura	0.2	0.6	0.0	0.0	0.2	0.4	0.1	0.3
	Odonata	0.0	0.0	0.3	0.5	0.1	0.3	0.2	0.4
	Blattodea	10.5	6.5	17.4	11.6	0.3	0.5	1.3	1.7
	Mantodea	0.0	0.0	0.0	0.0	0.2	0.4	0.3	1.0
	Dermaptera	0.1	0.3	0.0	0.0	0.0	0.0	0.0	0.0
	Orthoptera	0.4	0.5	1.6	1.6	0.4	0.7	1.5	1.7
	Phasmatodea	0.0	0.0	0.0	0.0	0.0	0.0	0.0	0.0
	Embioptera	0.3	0.7	0.0	0.0	0.1	0.3	0.1	0.3
	Psocoptera	0.8	1.9	3.4	2.3	1.3	1.6	4.3	6.9
	Hemiptera (psyllids)	164.6	111.4	67.8	37.0	637.0	587.7	306.0	295.2
	(others)	56.1	32.4	40.3	8.0	37.9	28.0	29.7	11.6
	Thysanoptera	13.7	11.8	16.1	11.2	25.1	22.0	26.5	31.4
	Neuroptera (adults)	1.6	1.3	1.6	2.6	3.3	2.9	3.4	4.3
	(larvae)	2.6	4.3	1.9	3.3	0.8	1.9	1.3	2.3
	Coleoptera (adults)	83.2	27.4	77.8	40.5	70.2	20.0	94.5	39.2
	(larvae)	13.7	11.9	14.3	10.8	8.9	6.2	17.7	13.8
	Diptera (adults)	23.3	11.5	15.8	6.8	25.2	22.9	26.4	15.7
	(larvae)	26.9	17.3	15.5	12.5	18.8	18.8	12.2	9.5
	Lepidoptera (adults)	7.3	4.9	5.7	4.1	2.1	1.5	5.1	5.5
	(larvae)	8.0	4.6	7.4	4.8	9.5	5.7	5.5	2.5
	Hymenoptera (ants)	168.1	176.8	105.8	98.7	218.6	253.8	206.8	136.3
	(others)	81.4	48.3	51.2	33.2	117.4	88.7	61.8	43.6
Totals (excluding ants)		585.3	173.5	431.4	151.7	1047.6	730.7	749.7	441.8

(mites), Araneae, Blattodea (cockroaches), Hemiptera (psyllids and other families), Coleoptera (adults and larvae), Diptera (adults and larvae), and Hymenoptera (ants and wasps).

The invertebrate count was higher on marri than on jarrah (excluding ants), regardless of stratum (Table 1) and higher on narrow-leaved ironbark than on grey box (Table 2). Counts were higher in the upper than the lower strata of jarrah, grey box, and narrow-leaved ironbark. This was also the case for marri, if the high count of Collembola (springtails) for the lower stratum was taken into account (Table 1).

STATISTICAL ANALYSIS OF PYRETHRIN DATA

Influence of tree species and stratum

Mean counts of invertebrates differed significantly between tree species and strata [P < 0.05, F statistics from ANOVA of comparisons (1), (2), and (3)]. More taxa were more abundant on either the upper or lower stratum of marri than on the corresponding stratum of jarrah. However, only counts of insect larvae were significantly different, with more larvae found on the lower stratum of the jarrah than of marri [P < 0.05, F statistics from ANOVA of comparison (2)].

At Scheyville, spiders were significantly more abundant on narrow-leaved ironbark (lower stratum), as were psyllids (both strata) and total invertebrates (lower stratum). Other sucking bugs were significantly more abundant on the lower stratum of grey box than of narrow-leaved ironbark [P < 0.05, F statistics from ANOVA of comparison (3)].

Many taxa were significantly more abundant in samples from upper than lower strata (e.g., spiders on marri; booklice on jarrah, other sucking bugs on jarrah; psyllids on grey box; beetles, ants, and other hymenopterans on jarrah and marri). Only two taxa had significantly higher counts on lower foliage—spiders on narrow-leaved ironbark and booklice on grey box [$P < 0.05$, F statistics from ANOVA of comparison (3)].

Variability within and between trees

Values pooled across tree species for a particular forest showed that variability within trees (σ^2_n) was generally greater than that between trees (σ^2_t). The average percentage contribution of within-tree variance (σ^2_n) to total variance (σ^2_n) was 80% (range 65–96%) at Karragullen and 66% (range 35–100%) at Scheyville. Little difference was evident in within-tree variability for the tree species in a particular forest (percentage contribution averaged 80% and 77% for jarrah and marri, respectively). Again the lower within-tree variability at Scheyville was evident (percentage contribution 66% and 67% for grey box and narrow-leaved ironbark, respectively).

The coefficient of variation (CV = $\sigma^2_{tot}/\bar{X}$) of each tree species and stratum revealed some interesting trends. At Karragullen, the CV was greatest for each invertebrate group on the lower stratum of the tree (mean = 1.09 and 1.65 for upper and lower strata, respectively). It was also greatest on the lower stratum for six of the nine groups tested at Scheyville (mean = 1.24 and 1.40 for upper and lower strata, respectively).

Mean CV on marri (1.36) was slightly higher than on jarrah (1.18), with six of the nine common invertebrate groups exhibiting the highest CV on marri. At Scheyville, invertebrates generally had a higher CV on narrow-leaved ironbark (1.59) than on grey box (1.05), with seven of the invertebrate groups having the highest CV on narrow-leaved ironbark. Thus, in both forests the trees with the highest invertebrate abundance also had the greatest variability in counts of invertebrates. Note that CVs in the two forests were similar. Thus, invertebrates in the two forests generally exhibited the same degree of patchiness, although individual taxa exhibited differences between forests.

TOTAL INVERTEBRATES IN BRANCH-CLIPPING SAMPLES

This method detected only 12 taxa at Karragullen and 17 at Scheyville, compared with the 27 taxa obtained in the pyrethrin samples at both locations. The most abundant invertebrates in branch-clipping samples at Karragullen were spiders, psyllids, other sucking bugs, adult beetles and ants. At Scheyville, the most abundant were spiders, psyllids, other sucking bugs, adult beetles, moth larvae, and ants (Tables 3 and 4).

As with the pyrethrin samples, many more invertebrates were detected per cm² of foliage at Scheyville than at Karragullen (Tables 3 and 4). This was also true on a per g basis, except for grey box. The trends between tree species and strata at Scheyville were similar for both the branch-clipping and pyrethrin samples. However, in this case, the numbers of invertebrates per g of foliage were between 5 and 10 times greater on narrow-leaved ironbark than on grey box, depending on which stratum was considered. This compares with a differential of only 1.7–1.8 times for the pyrethrin method (Table 2). Similarly, the differential in number of invertebrates per cm² of foliage was also more pronounced by the leaf-clipping method, with about five times the number being observed on narrow-leaved ironbark than on grey box foliage.

Invertebrates were more abundant per g and per cm² of leaf area in the upper than the lower stratum of grey box and per cm² in the upper than the lower stratum of ironbark (Tables 3 and 4).

The data from tree species at Karragullen did not give the same trends as observed by the pyrethrin procedure. More individuals from branch clipping were found on the upper than the lower stratum by a factor of 3.5–5.0, using the leaf-area measure. Similarly, 1.7 times more invertebrates on a per g basis were on the upper than the lower stratum of jarrah. Slightly fewer invertebrates were on the upper than the lower stratum of marri, and this seemed to be associated with the high count of spiders on the lower stratum.

The between-tree species trends for branch clips were most at variance with results obtained from pyrethrin samples. No differences were observed between the lower stratum of marri and jarrah on a per g basis. Twice as many invertebrates were observed on the upper stratum of jarrah than on marri. This differential was exaggerated with the leaf-area measure, with 7–10 times more invertebrates observed on jarrah than marri, depending on stratum. This compares with 1.1–1.4 times more invertebrates on marri than on jarrah by the pyrethrin method.

DISCUSSION

The two methods produced similar results: both yielded more invertebrates in Scheyville than in Karragullen. We do not know whether this was due to a real difference in invertebrate abundance or to different seasonal patterns of abundance between the forests, although preliminary analyses suggest that invertebrates were more abundant in all seasons at Scheyville. At Scheyville, both techniques indicated similar differences in invertebrate abundance between tree species. However, the excess of invertebrates on narrow-leaved ironbark vs. grey box was exaggerated by

TABLE 3. Means and Standard Deviations of Numbers of Invertebrates Sampled by Branch-clipping Method for Both Canopies and Subcanopies of Jarrah (*Eucalyptus marginata*) and Marri (*E. calophylla*) in Western Australia. The Number of Invertebrates per Bag Was Based on 10 Clips of 40 Leaves Each Taken from Each Tree

Taxon or measure			Jarrah						Marri						
			Subcanopy sample (N = 6)				Canopy sample (N = 6)				Subcanopy sample (N = 6)				Canopy sample (N = 6)	
			Bag		40-leaf		Bag		40-leaf		Bag		40-leaf		Bag		40-leaf
			$\bar{x}$	SD	$\bar{x}$	SD	$\bar{x}$	SD	$\bar{x}$	SD	$\bar{x}$	SD	$\bar{x}$	SD	$\bar{x}$	SD	$\bar{x}$	SD
Arachnida	Araneae		2.3	1.0	0.2	0.4	2.8	1.6			3.2	1.2			2.0	0.9		
Collembola			0.2	0.4											0.2	0.4		
Insecta	Blattodea														0.2	0.4		
	Dermaptera														0.5	0.8		
	Hemiptera	(psyllids)	1.0	1.3	1.5	1.8	1.2	0.8	2.8	3.1	0.7	0.8	0.7	1.6	0.5	0.8	5.3	10.2
		(others)	0.2	0.4	5.5	3.9	0.3	0.5	28.2	14.2			1.7	2.3			2.5	2.5
	Coleoptera	(adults)	0.2	0.4			3.2	3.4			0.8	1.2			1.0	0.9		
		(larvae)							0.5	0.6								
	Diptera	(adults)									0.2	0.4						
	Lepidoptera	(larvae)			0.2	0.4	0.2	0.4	0.2	0.4								
	Hymenoptera	(ants)	0.5	1.2			2.5	3.8			0.5	0.8			0.3	0.8		
		(others)	0.8	0.5			0.7	0.8			0.2	0.4			0.2	0.4		
	Unidentified				6.3	6.3			26.0	36.5							1.7	3.3
Mean biomass (g/40 leaves)			50.0	10.0			69.0	18.7			58.9	7.8			61.3	14.5		
Mean leaf surface (cm^2)					17.8				15.1				21.4				27.4	
Number of bags without invertebrates (40 leaves)			6.7	1.0			4.2	1.2			5.8	1.0			6.7	2.3		
Number of leaves without invertebrates					37.7	1.4			36.1	1.5			27.1	7.8			32.5	2.4
Number of invertebrates/1000 g of sample			9.1				15.7				9.3				7.0			
Number of invertebrates/100,000 cm^2 of leaf surface					96.2				469.4				13.4				47.0	

TABLE 4. MEANS AND STANDARD DEVIATIONS OF NUMBERS OF INVERTEBRATES SAMPLED BY BRANCH-CLIPPING METHOD FOR BOTH CANOPIES AND SUBCANOPIES OF GREY BOX (*Eucalyptus molluscana*) AND NARROW-LEAVED IRONBARK (*E. crebra*) IN NEW SOUTH WALES. THE NUMBER OF INVERTEBRATES PER BAG WAS BASED ON 10 CLIPS OF 40 LEAVES EACH TAKEN FROM EACH TREE.

		Grey box								Narrow-leaved ironbark							
		Subcanopy sample (N = 10)				Canopy sample (N = 10)				Subcanopy sample (N = 10)				Canopy sample (N = 10)			
		Bag		40-leaf		Bag		40-leaf		Bag		40-leaf		Bag		40-leaf	
Taxon or measure		x̄	SD	x̄	SD	x̄	SD	x̄	SD	x̄	SD	x̄	SD	x̄	SD	x̄	SD
Arachnida	Acarina					0.1	0.3					0.1	0.3			0.2	0.6
	Araneae	4.2	2.6	0.7	0.7	4.3	4.9	0.8	2.2	5.8	3.6	0.8	0.8	4.5	3.9	0.3	0.6
Insecta	Blattodea	0.2	0.4	0.1	0.3	0.1	0.3										
	Mantodea	0.2	0.4														
	Orthoptera																
	Hemiptera (psyllids)	1.2	2.0	10.3	11.0	2.7	2.7	17.9	23.7	2.2	3.0	16.8	16.7	4.2	3.1	25.0	16.3
	(others)	0.5	1.3	0.1	0.3	2.0	2.1	0.1	0.3	1.3	2.1	0.4	1.0	0.9	1.4	1.3	3.1
	Thysanoptera			0.2	0.6			0.6	0.7			0.1	0.3			0.2	0.4
	Neuroptera (larvae)					0.1	0.3										
	Coleoptera (adults)	0.5	1.1	0.1	0.4	1.4	1.4	0.4	0.9	0.7	0.3	0.1	0.3	1.7	0.9	0.3	0.6
	(larvae)			0.3	0.7	0.1	0.3	0.2	0.6	0.1	0.3	0.2	0.4	0.1	0.3		
	Diptera (adults)					0.1	0.3							0.4	0.9		
	(larvae)					0.1	0.3	0.5	1.2					0.1	0.3		
	Lepidoptera (adults)	0.1	0.3			0.2	0.4			0.4	0.5			0.3	0.4		
	(larvae)	0.6	0.7	2.4	1.1	0.6	1.0	2.4	2.5	3.3	3.1	1.9	2.6	3.1	3.7	1.4	1.5
	Hymenoptera (ants)	2.0	1.5	1.2	1.0	1.5	3.1	0.1	0.3	0.1	0.3			0.4	0.9		
	(others)					0.2	0.4	0.3	1.0			0.1	0.3	0.2	0.4	0.2	0.4
	Unidentified			0.1	0.3	0.6	1.8									0.1	0.3
Mean biomass (g/40 leaves)		111.2	29.3	—	—	126.3	3.5	17.0	7.7	26.7	9.7	—	2.3	50.9	22.7	—	2.0
Mean leaf surface (cm²)		—	—	17.2	10.0	—	—	—	—	—	—	5.0	—	—	—	3.5	—
Number of bags without invertebrates (40 leaves)		2.7	2.1	—	—	2.3	1.9	—	—	1.6	1.6	—	—	0.7	1.1	—	—
Number of leaves without invertebrates		—	—	37.6	2.4	—	—	36.6	3.5	—	—	37.8	2.0	—	—	37.3	2.4
Number of invertebrates/1000 g of sample		13.0	—	—	—	22.0	—	—	—	138.0	—	—	—	107.0	—	—	—
Number of invertebrates/100,000 cm² of leaf surface		—	—	194.0	—	—	—	369.9	—	—	—	1102.5	—	—	—	1850.0	—

branch clipping. The two techniques at first seemed to produce conflicting trends between tree species at Karragullen. Although more invertebrates were obtained from pyrethrin samples of marri than of jarrah, the reverse was true of branch clipping. This reversal was exaggerated when comparing the number of invertebrates per cm^2 of leaf area. However, Table 3 shows that the preponderance of invertebrates on jarrah was tied up with numbers of sucking bugs (excluding psyllids) and certain other small, unidentified invertebrates. The trend between tree species is reversed if these categories are deleted.

Branch clipping produced fewer invertebrates from a narrower range of taxa than the pyrethrin method (cf. Majer and Recher 1988). However, branch clipping gave a good representation of sessile invertebrates such as web-spinning and leaf-rolling spiders, moth larvae, psyllids, and certain other families of sucking bugs that remain attached to leaves. Suitability of this technique for sampling sessile invertebrates has also been pointed out by Cooper and Whitmore (this volume). Some of these invertebrates were obtained only by branch clipping. However, the clipping method was less efficient at obtaining rare or mobile invertebrates.

We conclude that branch clipping was more susceptible to "noise" caused by the abundance of one or a few types of invertebrates, perhaps because some of the most abundant sessile invertebrates are colonial (e.g., psyllids) and their distribution may be patchy. Because it samples sessile fauna less effectively, the pyrethrin method is less vulnerable to this problem. Pyrethrin sampling gives larger samples of a wider range of invertebrates. Variations in the distribution of individual taxa may therefore cancel out, producing a "more uniform" sample. However, the pyrethrin technique did have limitations, because some invertebrates flew away at the time of spraying and some remained attached to trees. In addition, wind may have blown dying organisms away from collection nets. This problem may be mitigated by keeping the drop distance from 0.5–1.0 m.

The greater variance in numbers of invertebrates within trees than between trees was due in part to differences in the volume of foliage above each net, despite our efforts to standardize canopy volumes. Differences in the amount of foliage sampled may be relatively high for nets hung within the same tree or cluster of saplings, but this effect tends to even out between trees. A way to compensate for this would be to use more, smaller nets but that would increase maintenance time.

Sampling of two tree species at a site required three persons for two weeks. The clipping and pyrethrin methods could be done concurrently. At Scheyville, the laboratory phase took one person two weeks to sort the clip samples and four weeks to sort the pyrethrin samples. Because of the smaller samples, lab work at Karragullen required one and three weeks, respectively. Field time did not change appreciably when branch clipping was omitted, although the time needed to process samples decreased markedly.

Because both techniques underestimate the true abundance of canopy invertebrates, we recommend using them together and interpreting results with an understanding of each method's limitations.

ACKNOWLEDGMENTS

We thank Greg Gowing, Stuart Little, and Nick Carlile for assistance with field work and sorting of specimens in N. S. W. and Sean Kelly, Andrew Steed, and John van Schagen for similar assistance in W. A. Ian Abbott and Hugh Ford provided helpful comments on an earlier draft of the manuscript.

QUANTIFYING ABUNDANCE OF FRUITS FOR BIRDS IN TROPICAL HABITATS

JOHN G. BLAKE, BETTE A. LOISELLE, TIMOTHY C. MOERMOND, DOUGLAS J. LEVEY, AND JULIE S. DENSLOW

Abstract. Inherent biases in different sampling techniques influence our interpretations of fruit-frugivore interactions. We review three general methods for sampling fruits: phenological studies based on repeated sampling of individual plants, fruit fall traps, and area-based sampling techniques. Phenological studies provide the least amount of quantitative information on fruit abundance. Fruit fall traps sample an unknown area, do not adequately sample all types of fruits dispersed by birds, and measure a residual quantity (that which is not eaten). Area-based samples frequently will be the best approach for many bird studies. Unripe fruits are used by birds under certain circumstances and provide information on future availability of ripe fruit. Therefore, both ripe and unripe fruits should be included in samples of fruit abundance, but as separate categories.

Key Words: Fruits; frugivores; tropics; sampling.

INTRODUCTION

Approximately one-third of the resident bird species in many neotropical forests are frugivorous (Terborgh 1980a, Stiles 1985b, Blake et al. in press, Karr et al. in press); the percentage of species that at least occasionally eat fruit is much larger. An estimated 50–90% of trees in neotropical forests and up to 98% of neotropical understory shrubs produce fruits whose seeds are dispersed by animals (Howe and Smallwood 1982), including birds (Gentry 1982, Stiles 1985b).

The method by which fruit abundance is estimated is critical to assessment of fruit as a resource for birds. In this paper we critique three commonly used techniques of quantifying fruit abundance, reviewing those as they might be or have been applied to bird studies. First, we consider studies that determined phenological patterns of plant species, which provide a general description of the seasonal availability of fruit. Second, we review use of traps to collect fallen fruits; such data have been used to estimate fruit abundance, seasonality, and diversity. Third, we discuss use of actual or estimated counts of fruits, fruiting plants, and species over a predetermined area. These three general methods are not necessarily mutually exclusive. For example, a phenological study can be area-based, and fruit abundance can be assessed on the basis of actual numbers, biomass, energy content, or some other factor (e.g., nutrient content).

METHODS USED TO COUNT FRUITS

PHENOLOGICAL STUDIES

The classic method of documenting phenological patterns is to record flowering and fruiting activity of plants over time (e.g., Frankie et al. 1974). Phenological patterns may be determined from collections made for taxonomic studies (e.g., Croat 1969, 1975), but more detailed information is obtained when reproductive activities of a series of individually marked plants are recorded at some repeated interval (Table 1). (Use of marked plants reduces such observer errors as overlooking unfamiliar or cryptic fruits.) Presence or absence of fruits (and flowers) or a simple index or estimate of abundance (e.g., "none, few, many"; Frankie et al. 1974) is noted. When conducted over a number of years, a general understanding of seasonal phenological patterns emerges. Those results, however, provide little quantitative, comparative data and are of limited value in studies on influences of fruit abundance on bird populations.

Phenological studies also may be designed to determine fruit production of a selected, small set of species (Table 1). For example, Howe and Vande Kerckhove (1979) analyzed fecundity and seed dispersal in 65 *Casearia corymbosa* (Flacourtiaceae). Total fruit counts were made over a 2-day span to determine crop sizes; fruits on 17 trees were counted daily to determine rates of fruit removal. Intermediate between community- and species-oriented studies are those that follow fruit production in a group of plant species that are important to a particular bird species (Worthington 1982, Wheelwright 1983) or to the frugivore community (Wheelwright 1986a). For example, Worthington (1982) sampled plant species that were known to produce fruit eaten by two species of manakins (Golden-collared, *Manacus vitellinus*; Red-capped, *Pipra mentalis*). Crop sizes were counted at biweekly intervals and were used, in combination with data on relative abundance of plant species, to provide an estimate of total fruit production. Because she worked on a small (18 ha) island, Wor-

TABLE 1. SELECTED STUDIES DESCRIBING PHENOLOGICAL PATTERNS OF TROPICAL PLANTS

Study length (months)	No. species	No. plants	Census interval	Count type[a]	Reference
			Community-oriented studies		
24	185	468	1 m	index	Frankie et al. 1974
14	113	1137	1 m	index	Frankie et al. 1974
36	154	?	6 wk	index	Opler et al. 1980
36	95	?	1 m	index	Opler et al. 1980
36	51±	145	1 m	index	Van Schaik 1986
108	44	61	2 wk	p/a	Medway 1972
13	?	?	?	p/a[b]	Charles-Dominique et al. 1981
21	?	?	2 wk	p/a[b]	Sabatier 1985
120	2	104	2 wk	p/a	Milton et al. 1982
24	13	109	?	p/a	Gautier-Hion et al. 1981
			Species-oriented studies		
12	21	210	2 wk	count	Worthington 1982
84	16	265	2 wk[c]	index	Wheelwright 1986
12	3	77	—[d]	count	Murray 1987
48	1	30–60	—[e]	count	Fleming 1981
4	1	5	—[e]	count	Bronstein & Hoffman 1987
2	1	65	—[f]	count	Howe & Vande Kerckhove 1979

[a] Count types: index = relative index of abundance, e.g., "many," "few"; p/a = presence/absence; count = direct count of fruits.
[b] Also weighed fruits fallen on trail.
[c] Counts conducted on biweekly intervals June 1980 to July 1981 and during 1–3 months in 1979, 1982–1985.
[d] Censused on 2 sequential days, 1–3 times/month.
[e] Censused once prior to fruit maturation.
[f] Censused on 2 sequential days; fruits counted daily on 17 trees.

thington was able to define community boundaries.

Wheelwright (1986a) investigated phenological patterns of 16 common Lauraceae species. He indexed fruit abundance by estimating percentage of canopy area in fruit, but did not obtain an actual count or estimate of fruit production. His research demonstrated the need for long-term studies; even 7 years were too few to represent adequately supra-annual cycles of fruit production displayed by those plants.

Finally, we include under phenological studies those that census fruiting and flowering trees along some set trail or series of trails. (When conducted systematically [i.e., with a set length and width of the sample area] such counts overlap with area-based sampling techniques described later.) Sabatier (1985) and Charles-Dominique et al. (1981), for example, collected and weighed all fallen fruits found along a series of trails. Such a technique is biased since many fruits likely were consumed and others rotted before they were tabulated. Information on general trends in fruit production may be achieved but information on total fruit production will be less reliable. Additional problems associated with sampling fallen fruits and sampling along trails are discussed in the following sections that deal with fruit fall traps and sample plots.

Phenological studies may be useful if the researcher can characterize the diet of the focal bird species (e.g., manakins, Worthington 1982; quetzals, Wheelwright 1983) and thus is able to identify the most important plant species. However, some fruits may be important to birds only in some seasons or years (Loiselle and Blake, this volume), and it may be difficult to determine *a priori* what fruit species should be sampled.

Considerations

Frankie et al. (1974) recommended a minimum of five individuals/species for tropical phenological studies while Wheelwright (1986a) suggested at least 10 individuals/species. However, the rarity of many species may make it difficult to obtain a representative sample, particularly since individuals of many species may show marked variation in phenology (e.g., Wheelwright 1986a). Similarly, unless the researcher knows the relative abundance of species, it will be difficult to estimate community-wide fruit abundance from phenological data.

Researchers should be cautious when using results of phenological studies to interpret results of bird studies conducted in different years. Although phenological patterns may be similar among years, marked annual variation in community-wide fruit production still may occur (Leighton and Leighton 1983, Wheelwright 1986a, Loiselle 1987).

FRUIT FALL TRAPS

Fruit fall traps have been used to estimate canopy fruit production in a variety of lowland trop-

TABLE 2. Selected Studies Using Fruit Fall Traps to Estimate Fruit Production

Study length (months)	Study area	Count interval	No. traps	Trap size (m²)	Total sample (m²)	% of study area	Reference
			Community-oriented studies				
12	±100 ha	2 wk	150	0.08	12	0.0012	Terborgh 1983
18	±100 ha	2 wk	100	0.07	7	0.0007	Janson et al. 1986
12	83 ha	1 wk	312	0.08	26.0	0.0031	Foster 1982a
12	83 ha	2 wk	120	0.08	10.0	0.0012	Foster 1982a
12	10 ha	1 wk	100	0.08	8.3	0.008	Leigh & Windsor 1982
12	10 ha	1 wk	150	0.08	12.5	0.012	Leigh & Windsor 1982
72	10 ha	1 wk	200	0.08	16.7	0.017	Leigh & Windsor 1982
16	10 ha	1 wk	75	2.31	173	0.17	Smythe et al. 1982
			Species-oriented studies				
6	19 trees	1 wk	9 ± 4	1.0	9 ± 4	≥10	Howe 1980
4	17 trees	1 wk	5–18	1.0	5–18	6–23	Howe & Vande Kerckhove 1981
2	7 trees	1–3 d	4[a]	1.0	4	?	Howe 1977
3	0.135 ha[b]	2 d	135	0.20	26.5	1.96	Coates-Estrada & Estrada 1986

[a] Traps were supplemented with belt transects in litter.
[b] Crown area of one tree.

ical areas (Table 2). In general, 75 to 200 traps of from 0.08 m² to 2 m² collecting capacity each are placed throughout the habitats or under specific trees being studied (Table 2). All collected fruits are separated by species and usually weighed to obtain biomass estimates (Smythe et al. 1982, Terborgh 1983). Alternatively, seeds may be counted and then converted to estimates of fruit number and biomass (Foster 1982a, Janson et al. 1986).

Studies using fruit traps vary in focus from a single tree to entire communities. Most studies directed at birds have used fruit traps to estimate production by, or fruit removal from, a single tree or species (Table 2). Some have used traps to examine seasonal and annual patterns of fruit production over considerably larger areas (Table 2), often for studies on mammalian frugivores (e.g., Smythe et al. 1982, Terborgh 1983).

Considerations

Once traps are in place, collection of fruits requires little time and fruits are easily counted. Further, if traps are checked frequently, biomass estimates of fresh material can be calculated (Terborgh 1983). Problems associated with sampling different forest strata in tall lowland rain forest make fruit traps useful in some instances. For example, although the canopy is an area of high fruit production in tropical forests (e.g., Foster 1982a), the great height of lowland forest trees makes estimation of canopy fruit production time consuming and difficult. Direct counts of fruit from the ground are frequently impossible; even if one ascends into the canopy only a few trees can be surveyed effectively (Loiselle, pers. obs.). Highland forests often have comparatively lower canopies than lowland forests, but even here direct enumeration of canopy fruits is difficult. Trees often are shrouded in clouds and the lush growth of epiphytes obscures much of the canopy.

As Terborgh and others have pointed out, fruit fall traps measure "a residual quantity: total fruit production minus amount eaten by arboreal frugivores, including insects" (Terborgh 1983). Further, not all fruiting plants are equally well sampled by fruit fall traps, as we discuss below. Thus, fruit fall data are, at best, an indirect measure of fruit abundance, not an estimate of what is directly available to arboreal frugivores. How patterns of fruit fall reflect patterns of absolute fruit abundance remain undetermined. If, for example, ripe fruits remain on the plant for a long time, they may all be eaten and never recorded in traps, even though they might be an important resource. Similarly, if trees ripen few fruits every day, all fruits may be removed quickly (Howe 1984, Catterall 1985), again preventing collection of fruits in traps. Further, fruit traps can overestimate seasonal variation in fruit production because a larger proportion of ripe fruit is eaten when fruits are scarce than when fruit is abundant (see Terborgh 1983).

A major problem with fruit fall traps is that the area being sampled is usually unknown. Contributions may come from plants not located directly over a particular trap (e.g., Foster 1982a), whose input is difficult or impossible to assess. Similarly, total area of the traps usually is a small fraction of the study area (Table 2). Unless the number of traps is large, estimates of community fruit production can be heavily biased by fruiting of a few individuals. If one uses the data (e.g., biomass or number of fruits/trap or total trap area) to extrapolate fruit production to a much larger total area (e.g., fruit production in kg/ha), substantial errors may occur. Moreover, extrapolated estimates of fruit production will be inaccurate if changes in sampling area occur and are not accounted for. For example, Van Schaik (1986) found that trap area decreased by about 10% in 2 years (about 8% in 1 year) as traps sagged under the weight of water and litter.

Fruit fall traps do not provide comparable estimates of fruit abundance for all types of fruiting plants; understory plants, especially small-seeded shrubs and herbs, are under-represented. In neotropical sites, where up to 98% of understory plants produce animal-dispersed fruits, fruit abundance in the understory may be an important component of community-wide fruit production, particularly for birds that rarely ascend into the canopy. Fruit traps are more likely to sample large-seeded, capsulate, and dry arillate fruits. Small, juicy berries decay rapidly and may become unrecognizable between visits to traps. In such cases, fruits must be identified from seeds remaining in traps and number of fruits must then be estimated from seed counts. Frequent checking also may be necessary if fruits or seeds are removed from the traps by understory frugivores and granivores.

Some fruits, particularly those produced by epiphytes, may not fall to the ground and thus will not be sampled in traps. In highland wet forests, fruit fall traps will provide poor estimates of community-wide fruit production because many fruits of both trees and epiphytes will become lodged in thick vegetation. Underestimation of epiphytic fruits may be a particular problem because epiphytes are important in highland forests (Loiselle 1987; F. G. Stiles, pers. comm.).

If the objective is to estimate community-wide fruit production, placement of traps is important. Most tropical forests support a large variety of fruiting trees, but each trap will sample fruit from only one or two. Consequently, a commensurately large number of randomly placed traps is needed to adequately sample a majority of species. Alternatively, one may place traps on the basis of some stratified design (e.g., based on habitat or tree distribution patterns). If placed in sufficient density, however, use of traps may become time consuming, costly, and unsightly.

Habitats differ in fruit abundance and phenology (Frankie et al. 1974, Opler et al. 1980, Loiselle 1987); thus, extrapolation to a community level may not be warranted unless all habitats are sampled. Unfortunately, habitats such as treefall gaps and early second-growth, which often are rich in fruits (Martin 1985a, Loiselle 1987, Levey 1988), are difficult to sample with traps because of the low, dense vegetation.

Most studies using fruit fall data disregard aborted fruits (which generally are not a resource to frugivores, but are common in traps [Foster 1982a]). Fruit abortion can be high in tropical trees (Stephenson 1981) and the contribution of aborted fruits to fruit abundance measures derived from traps should be discounted.

Studies directed at fruit production by a single tree (Coates-Estrada and Estrada 1986) or species (Howe 1977, 1980; Howe and Vande Kerckhove 1981) suffer from fewer of the sampling problems mentioned above (e.g., area being sampled, input from other species). Placement of traps, number of traps, area sampled, and sampling frequency all can be more specifically tailored to the question being addressed, resulting in less sampling error. However, fruit fall data still represent what is not eaten (except perhaps for capsulate fruits). Without information on rates of fruit removal by frugivores, the total of, temporal variation in, and spatial variation among trees in fruit production may be harder to estimate.

AREA-BASED SAMPLES

Area-based surveys of tropical fruit production have been employed by a number of researchers, primarily to sample understory fruits. Methods include those that sample fruits along long transects, often following trails through a study area (e.g., Wong 1986); and those that rely on circular plots or quadrats situated throughout the area (e.g., Denslow et al. 1986, Loiselle 1987).

Several researchers have sampled fruits along linear transects. In one of the earliest studies that simultaneously monitored fruit production and bird abundance, Davis (1945) made observations at monthly intervals on the presence of fruit growing on trees located within 3 m of a set trail (Table 3), but did not count fruits. Hilty (1980) indexed fruiting activity (0, 10, 50 or 100% of crown area in fruit) of all plants (>3 m tall) within 3 m of a 1000 m trail. He combined this index with an estimate of total crown surface to provide an index of total fruit production. Similarly, Wong (1986) counted all ripe and unripe fruits (or estimated if ≫ 1000 fruits) produced by understory fruits along narrow paths that provided access to mist nets (Table 3).

Location or placement of sample plots varies with study objectives. We have used quadrats placed parallel to mist nets to estimate local fruit production in connection with bird studies (Loiselle 1987, Levey 1988) and circular plots (Denslow et al. 1986) to sample fruits over a wider area (Table 3). Our studies have focused on understory shrubs, treelets, lianas, and epiphytes (<10–20 m above ground) and have not included estimates of canopy fruit production. Leighton

TABLE 3. Selected Studies Using Area-based Samples to Estimate Fruit Abundance

Study length (months)	Study area (ha)	Count interval	Sample no.	Sample area (m²)	Total sample (ha)	% of study area	Count type[a]	Reference
12	?	1 m	100	12	0.12	?	index	Levey 1988
12	2.5	1 m	50	12	0.06	2.4	index	Levey 1988
5	46	1 wk	6	1 ha	6	13	p/a	Gautier-Hion et al. 1981
5	43	1 wk	6	1 ha	6	14	p/a	Gautier-Hion et al. 1981
24	300	1 m	30	5000	15	5	count	Leighton & Leighton 1983
12	300	1 m	25	2500	6.25	2.1	count	Leighton & Leighton 1983
17	?	1 m	50	100	0.5	?	index	Denslow et al. 1986
14	16	1 m	—	—	1.32[b]	4.1	count	Wong 1986
15	4.8	5–6 wk	60	25	0.15	3.1	count	Loiselle 1987
15	4.8	5–6 wk	60	25	0.15	3.1	count	Loiselle 1987
11	4.8	5–6 wk	60	25	0.15	3.1	count	Loiselle 1987
4	8.6	5–6 wk	108	25	0.27	3.1	count	Loiselle 1987
15	?	2 wk	1[c]	3000	0.3	?	index	Hilty 1980
16	40	1 m	1[d]	3100	0.31	0.77	p/a	Davis 1945
16	40	1 m	1[e]	2200	0.22	0.56	p/a	Davis 1945

[a] Count type; see Table 1.
[b] 6.6 km of transects, 2 m wide.
[c] 1 km of transect, 3 m wide.
[d] 1021 m transect, 3 m wide.
[e] 750 m transect, 3 m wide.

and Leighton (1983) used much larger quadrats to sample fruits produced by lianas, epiphytes, and trees (>4 cm diameter) in Borneo (Table 3), where fruiting plants are less abundant than in the neotropics. Gautier-Hion et al. (1981) divided their area into subplots that were sampled on the basis of use by monkey troops (see also Estrada and Coates-Estrada [1986]); unused plots were not sampled while frequently used plots could be sampled every week. A similar, focal animal approach, could be adapted for bird studies.

Considerations

Once quadrats are delineated, it is easy to count fruiting individuals and fruit crops on a regular basis. Quadrat samples (and fruit traps) have the advantage that both spatial and temporal variation in fruit production can be analyzed statistically. Direct comparisons among studies are facilitated as well, although such comparisons necessarily assume that similar foliage strata and life forms were sampled. As with fruit traps, care must be taken in placement of quadrats or transects. Trailside studies are convenient, but understory fruit production is likely to be overestimated if transects are placed along well established trails because of the greater light availability along such trails. Since trails are not randomly distributed, sampling along trails may not produce an accurate assessment of community-wide fruit abundance.

Quadrat size and number will depend on the objectives and scale of study. Estimation of canopy fruits will require larger sample areas than those needed to estimate fruit production of understory shrubs and treelets. For accurate estimates of both canopy and understory fruits, the best area-based method will probably include some combination of large and small quadrats.

DISCUSSION

Quantifying fruits as a resource is not simply a matter of sampling technique, because various attributes influence whether a particular type of fruit is suitable for a particular species or type of bird (Denslow and Moermond 1985, Martin 1985b, Moermond and Denslow 1985). Ideally, a fruit that is never used by a particular bird species should not be included in estimates of fruit available for that bird. From a practical standpoint, this frequently is impossible to achieve, as diets of many neotropical birds are poorly known.

WHAT TO COUNT

Fruit quality

One of the first decisions is whether or not to count unripe as well as ripe fruit. Birds prefer ripe fruits (Moermond et al. 1986), but do feed on unripe fruits, especially during times of fruit scarcity (Foster 1977, pers. obs.). Counts of ripe fruits alone may underestimate fruit production if fruits are removed rapidly as they ripen (Howe 1984, Catterall 1985). Unripe fruits may provide an estimate of future availability of fruits, particularly for species that ripen fruits relatively synchronously (Bronstein and Hoffman 1987).

In addition to ripeness, other factors relating to fruit quality may influence fruit selection by frugivores, including flavor (Sorensen 1981), lipid content (Leighton and Leighton 1983), and sugar content of fruit (Levey 1987a).

Structural attributes

Fruit selection may be limited by a variety of structural characteristics that interact with morphological capabilities of frugivores to determine the bounds of their diet (Janson 1983, Gautier-Hion et al. 1985). Some fruits, especially those enclosed in capsules, are available to few birds (e.g., Leighton and Leighton 1983, Pratt and Stiles 1985) and are unavailable to other species.

Size of fruit also may limit types of fruit that can be consumed (Moermond and Denslow 1985, Wheelwright 1985). This is particularly true for species that swallow fruits whole (e.g., Pipridae); birds that can bite off pieces of fruit (e.g., many Thraupinae) are less limited by fruit size (see also Leighton and Leighton 1983, Foster 1987).

Location of fruits on a plant (e.g., close to a perch, on the tip of a slender twig) can influence choice (Denslow and Moermond 1982, Moermond and Denslow 1983). Accessibility will influence the type of foraging maneuver needed to obtain it and morphological constraints may determine which fruits are accessible to a particular species of bird (Moermond et al. 1986).

The many factors that govern fruit selection will determine its perceived abundance. Ideally, fruit abundance should be weighted by its importance to frugivores. This may be possible if a specific species is being studied, but is difficult for community studies. Information is available on nutrient content of some tropical fruits, but we know too little about the diets of most frugivores, particularly temporal and spatial variation, to determine which fruits to sample. Similarly, too few tropical fruits have been analyzed for energy and nutrient content for complete community analyses. Particularly for community studies, it seems best to sample as thoroughly as possible all fruits that are likely to be eaten by birds. Over time, as more information on diets of specific species becomes available, analyses of bird-fruit interactions may be more precisely addressed.

SAMPLING TECHNIQUE

When sampling fruits, one must decide whether to count all fruits or simply to use an index of relative abundance. We favor direct counts because they likely are more relevant to understanding bird populations. Relative indices of fruit abundance are less likely to be useful for comparative analyses. Quantitative samples of fruit abundance allow one to make direct, statistical comparisons among studies and to make direct correlations with bird populations, either in terms of fruit numbers, biomass, or nutrient content. Direct counts of fruits allow later conversion to a relative scale, but the reverse is not true.

Direct counting of large numbers (i.e., >1000) of fruits can be time consuming and often difficult. When direct counts are not possible, one can count a subsample of fruits (e.g., on one infructescence or branch) and then use those data to estimate total fruit abundance (e.g., Worthington 1982). One must recognize, of course, that such estimates always will involve some level of error, often of unknown magnitude. However, the increase in sample size allowed by the time saved in counting may be substantial. "Knowing that a particular fig bore 32,489 fruits in 1987 is not as valuable as knowing that 10 of 100 trees produced about 20,000 fruits each and the rest produced none" (N. T. Wheelwright, pers. comm.).

Fruit abundance can either be represented in terms of numbers of fruits or in terms of biomass. The latter requires information on weights (including pulp and seed weights) of all species of fruits. Once such data are available, conversion to biomass is easy if quantitative estimates of fruit abundance (numbers) also are available. In her study on reproductive ecology and food selection by two species of manakins, Worthington (1982) measured fresh and dry weight of fruit pulp (minus seeds) and then converted counts of fruit to biomass. Studies that include seed predators (e.g., parrots, many finches and sparrows, pigeons, cracids) would need to modify biomass measures accordingly.

Problems may arise on how to count some fruits, particularly some unripe ones. For example, for arillate fruits enclosed in capsules (e.g., Guttiferae, Malvaceae, Monimiaceae), does one count the capsule as a single unripe fruit or as some number of unripe fruits, dependent on the average number of arillate fruits per capsule? We favor the latter as birds consume fruits separately once they are exposed. Aggregate, spike-like fruits such as *Piper* typically are eaten piecemeal and one could estimate the average number of "bites" available per fruit. Birds vary in amount taken at one time, however, and we favor counting each spike as a single fruit.

WHEN TO SAMPLE

If parallels are to be drawn between fruit and frugivore cycles of abundance, it is necessary that populations be sampled simultaneously. Some studies of birds, for example, have relied on patterns of fruit abundance documented by other researchers at other sites in other years. Because site-to-site and year-to-year variation can be appreciable, this practice may lead to invalid conclusions.

ACKNOWLEDGMENTS

This paper benefitted greatly from the comments of S. Hermann, J. Jehl, T. Martin, K. Smith, J. Verner, and N. Wheelwright. Our work at La Selva Biological Station has benefitted from interactions with many people: D. Brenes, D. A. Clark, D. B. Clark, A. Gomez, R. Marquis, and O. Vargas know the fruits of La Selva particularly well and helped us on numerous occasions. We are grateful to Sr. J. A. Leon for his help in arranging permits, and the Servicio de Parques Nacionales de Costa Rica, especially Srs. F. Cortes S. and J. Dobles Z., for permission to work in Parque Nacional Braulio Carrillo. Financial support has been provided by: National Geographic Society; Jessie Smith Noyes Foundation; University of Wisconsin, Department of Zoology; Douroucouli Foundation; and National Science Foundation.

Quantification of Diets

APPROACHES TO AVIAN DIET ANALYSIS

KENNETH V. ROSENBERG AND ROBERT J. COOPER

Abstract. Direct examination of diets is greatly under-represented in studies of avian biology. Much of our knowledge of food habits of North American birds is still based on the early survey work by "economic ornithologists." Here, we review approaches and techniques of sampling and analysis. For species that cannot be captured alive, collection of stomach or esophageal samples is necessary. Potential biases associated with post-mortem digestion, time spent in nets or traps, and differential passage of food through various parts of the gut are discussed. For species that can be captured alive, flushing the digestive tract or forcing regurgitation with warm water is recommended over use of emetics. Fecal samples and pellets, although more difficult to analyze, also provide accurate estimates of diet. Diets of nestling birds may be sampled with neck ligatures, observed or photographed directly at nests, or examined through the transparent skin of the neck. Aids for the identification of fragmented food samples are discussed, including the use of reference collections, collaboration with specialists, and the conversion of arthropod fragment sizes to total prey length, weight, and energy content. Diet data may be presented as percent occurrence, frequency, volume, or weight, each with its own merits and biases. We recommend presenting at least two kinds of results, as well as the raw data, on a per-stomach basis whenever possible. Finally, we describe two under-used sources of diet information: the U.S. Biological Survey stomach analysis card file at Patuxent Wildlife Research Center and the unanalyzed stomach contents collection at Louisiana State University Museum of Natural Science.

Key Words: Diet analysis; emetics; fecal analysis; ligatures; pellets; stomach contents.

Detailed knowledge of diets is critical to many studies of avian biology and ecology. However, direct measures of diets are rarely attempted. The common use of indirect inferences about diets, based on morphology (e.g., bill shape), behavior, or general food availability, has been questioned in several empirical studies (e.g., Rotenberry 1980a, Rosenberg et al. 1982). The extent to which variation in foraging behavior results in differences in diet (cf. MacArthur 1958, Cody 1974) also remains largely untested. Most recent, frequently cited studies of avian foraging guilds or communities (e.g., Rabenold 1978, Eckhardt 1979, Holmes et al. 1979b, Noon 1981a, Sabo and Holmes 1983, Mountainspring and Scott 1985, Remsen 1985) provide no quantitative measure of local diets, although most make conclusions regarding resource partitioning, optimal foraging, or interspecific competition (for exceptions, see Rotenberry 1980a, Rosenberg et al. 1982, Robinson and Holmes 1982, Sherry 1984). Because we lack clear understanding of the connections between foraging site-selection, food availability, and diet, any assumptions made without further empirical study may be unwarranted.

In a recent symposium on neotropical migrants (Keast and Morton 1980), 15 papers specifically discuss foraging ecology; yet, in only three were diets of individual species described to any extent, and only one study (Morton 1980) provided quantitative data on local diets. In this volume, only one paper (Cooper et al.) specifically addresses the determination of avian diets or provides diet data relevant to the study. In a sample of roughly 200 papers on avian food habits compiled from major ornithological journals since 1960, 68 (34%) concern only waterbirds, 70 (35%) are on raptors, and 13 (7%) deal with gamebirds. Finally, of the 50 papers (25% of total) concerning nongame landbirds, 30 were single-species studies, most from single localities, leaving only a handful that may be useful to community ecologists, ecomorphologists, or other comparative biologists. To date, the only source of diet information for most North American bird species remains the survey data of F. E. L. Beal and W. L. McAtee, summarized in Bent's Life history series, and Martin et al. (1951a). Wheelright's (1986) analysis of the American Robin (*Turdus migratorius*), is the only modern study of geographic or seasonal variation in diet in any North American bird.

Why, then, are avian diets so neglected? We think the reasons are more methodological than philosophical: (1) the variety of alternative approaches and options is not generally appreciated; (2) researchers fear the detail and lack the technical expertise required by the fragmented nature of most diet samples; and (3) data on diets are initially collected, but samples are either not analyzed or the results are not subsequently published. We know the latter to be true in several studies cited above. To the extent that reasons

(1) and (2) are true, we offer this review in the hope of alleviating such fears and stimulating further study.

Our goal is not to provide a handbook of techniques, but rather to lead the reader to appropriate references and provide examples in which each technique has either succeeded or failed. Our biases reflect our own work (primarily with stomach analysis) on temperate and neotropical insectivorous birds, although we have attempted to broaden the scope of our review.

SAMPLING TECHNIQUES

The first major review of avian dietary assessment by Hartley (1948) still applies to most modern studies. Hyslop's (1980) review of methods for analyzing stomach contents of fishes discusses many topics relevant to avian studies and may serve as a basic reference in any dietary investigation. Duffy and Jackson (1986) offer the most recent discussion of sampling and analytical considerations for studies of seabird diets, and most of their discussion applies equally well to terrestrial birds. Ford et al. (1982) review modern, nondestructive methods of sampling gut contents. Other useful discussions of general sampling considerations may be found in Newton (1967), Swanson and Bartonek (1970), and Rundle (1982).

Stomach contents

The most common method of avian diet sampling is direct examination of gut contents. Its primary advantages are (1) adequate samples can usually be obtained relatively easily, and (2) the full contents of a bird's gut are obtained. Disadvantages include the need to kill birds, and the many biases associated with stomach fullness, differential digestion rates, identification of fragmented food items, and presentation of results. These biases, however, are common to all techniques involving gut samples, whether or not the bird is sacrificed.

The techniques used to obtain and analyze gut contents today are similar to those first devised by early researchers attempting to determine the economic importance of North American birds (e.g., MacAtee 1912, 1933). The first consideration is the method and design of specimen collecting. Ideally, only actively foraging individuals will be sampled, controlling for habitat heterogeneity, season, time of day, and the like. These factors are most easily controlled by shooting, and many species (e.g., in the forest canopy or very open habitats) can be sampled only in this way. Duffy and Jackson (1986) criticized the random shooting of birds at sea that may be travelling long distances between foraging sites and thus may have empty or mostly digested gut contents. This problem applies to any species that forages only intermittently at specific sites, including some blackbirds (Gartshore et al. 1979) and shorebirds (Rundle 1982), but probably not to most small landbirds that feed more or less continuously.

Mist-netting or trapping may introduce additional biases. For example, birds caught in nets may not be assignable to a specific habitat or foraging zone (i.e., they may be transients in the area of capture), and age and sex classes may not be sampled equivalently. In addition, birds held alive in nets or holding cages for varying periods of time continue to digest their food and may increase the bias associated with differential digestibility of food items (discussed below).

Sample sizes necessary for any particular study may be difficult to determine a priori, because they depend to a large extent on the variability in diet among individuals. Assessing the adequacy of collected samples is discussed by Sherry (1984), based on the methods of Hurtubia (1973). In general, a cumulative plot of diet composition may be constructed by adding the diets of successive individuals until an asymptote is attained. This point represents the number of stomachs beyond which little additional information about diet composition is obtained. In several studies, 10 or fewer stomachs were adequate for assessing species-specific diets at particular sites within a collecting period (Wiens and Rotenberry 1979; Rosenberg et al. 1982; Sherry 1984; Rosenberg and Cooper, unpubl. data). Larger samples may be necessary for studies of individual, temporal, or geographical variation in diet within species. Sample sizes also influence later statistical procedures, as discussed by Duffy and Jackson (1986); for example, parametric tests usually require larger samples than do nonparametric tests.

Differential digestion rates of dietary items impose the largest potential bias in any study of gut contents and may influence every phase of the study. Koersveld (1950) showed that post-mortem digestion may occur in birds. However, the disappearance of food in birds left at 21°C for 3 days hardly approximates potential problems encountered under normal field conditions. Some researchers have injected formalin (usually 1.0 cc at 10% strength) into the stomach immediately upon death to stop digestion. Dillery (1965) compared 80 stomachs of Savannah Sparrows (*Passerculus sandwichensis*) injected with formalin with 47 (presumably) uninjected samples from the U.S. Fish and Wildlife Service files. More individual arthropods were identifiable in the injected stomachs (13.75/bird vs. 5.13). In addition, soft-bodied Homoptera (e.g., aphids) were under-represented in the uninjected samples (9%

vs. 30% of all items), whereas larval Lepidoptera were over-represented (13% vs. 4%).

Sherry (1984) found an average of 15–30 identifiable arthropods/stomach in a variety of neotropical flycatchers (Tyrannidae). Although these stomachs were not injected with formalin, they were removed immediately and placed in 70% ethanol. An average of 10–13 arthropods/stomach was identified in two species of flycatcher (*Empidonax* spp.) in Louisiana (Rosenberg and E. Robinson, unpubl. data). No attempt was made to stop post-mortem digestion; specimens were usually frozen within 1–2 hours after collection and stomachs were removed to 70% ethanol at the time of specimen preparation. Under similar conditions (but without freezing), an average of 11–14 arthropods/stomach was identifiable in two species of antbird (*Myrmotherula* spp.), and 8–14/stomach in two woodcreepers (*Xiphorhynchus* spp.; Rosenberg and A. Chapman, unpubl. data). Clearly, the necessity for and consequences of not injecting bird stomachs with formalin requires further study.

Differential digestion rates can also bias a sample before a bird is collected. Experiments with bird digestion (Stevenson 1933) showed that wild birds varied greatly in the fullness of their stomachs, and that juveniles of several species contained more food than adults. Stevenson (1933) also determined the time of passage through a bird's gut to average 2.5 hr for a variety of foods including insects, seeds, and fruit pulp. Other studies report much shorter digestion times, with an extreme rate of disappearance of 5 min in the Savannah Sparrow (Dillery 1965). Swanson and Bartonek (1970) found that soft-bodied insects may be gone from gizzards within 5 min, whereas hard seeds may persist for several days. These conflicting results are discussed by Custer and Pitelka (1974), who also provide correction factors for differential digestion rates in the Snow Bunting (*Plectrophenax nivalis*). Similar corrections were made by Coleman (1974) after determining what percentage of various foods persisted in European Starling (*Sturnus vulgaris*) gizzards after 2 hr. A method for determining correction factors for insectivorous birds is given by Mook and Marshall (1965). Following those methods, Cooper (unpubl. data) found that second and third instar gypsy moth (*Lymantria dispar*) larvae were completely digested in less than half the time it took birds to digest fourth and fifth instars. In addition, specialized seed dispersers were shown to have higher gut-passage rates than nonfrugivores of equal body weight (Herrera 1984b). In short, the potential biases associated with rates of digestion are poorly understood and point to a continued need for innovative experimentation with live birds (see also Gartshore et al. 1979, Rundle 1982).

The extent to which stomachs from mist-netted birds may differ from those of shot individuals was addressed for two groups of neotropical species (Rosenberg and A. Chapman, unpubl. data). Among two species of antwren (*Myrmotherula* spp.) and two woodcreepers (*Xiphorhynchus* spp.), the number of identifiable arthropods in shot vs. netted samples was similar (12 vs. 9 and 10 vs. 9, respectively), as was the total number of arthropod orders represented. In the antwrens, more beetles and fewer orthopterans were evident in shot samples of *M. leucophthalma*, whereas the opposite was true in *M. haematonota*. In the woodcreepers, beetles and orthopterans were more prevalent in the netted samples of both species, spiders were proportionally more evident in shot *X. spixii* but not in *X. guttatus*, and Lepidopteran larvae were much more common in shot individuals of both. Thus, these preliminary results do not clearly indicate a consistent bias in netted vs. shot samples, and any potential biases can be lessened by minimizing the time that a bird remains alive in the net.

In species with a well-developed crop, the crop contents are thought to be the most unbiased representation of a bird's diet (Hartley 1948). In larger birds, esophageal contents can be compared with stomach contents (e.g., Goss-Custard 1969, Swanson and Bartonek 1970), with the former usually considered preferable. Rundle (1982) argued strongly in favor of esophageal analysis for studies of shorebird diets, providing examples of marked differences from analyses of gizzard contents alone. Although in most small passerines the esophagus is usually empty and cannot be used to calibrate stomach contents (Custer and Pitelka 1974), careful attention to collecting only actively feeding birds may ensure full gullets. For example, Gartshore et al. (1979) found that most foods persisted for up to 20 min in the gullets of Red-winged Blackbirds (*Agelaius phoeniceus*) feeding under natural conditions. In addition, the gullets of many granivorous species often contain large samples of seeds recently eaten (Newton 1967, Payne 1980, Zann et al. 1974).

In most studies, gut contents are preserved in either formalin or alcohol. In general, formalin is better for preserving flesh (including the stomach itself), but may disolve bone or distort insect or vegetation parts (Duffy and Jackson 1986). Ethyl alcohol (70 to 95%) is preferred by entomologists (Borror et al. 1981) and is probably adequate for most studies of insectivorous birds. Well-preserved gut contents may be stored for long periods. Giuntoli and Mewaldt (1978) successfully examined stomachs of Clark's Nutcrackers (*Nucifraga columbiana*) after storage in formalin for up to 15 years. Thus, samples may be accumulated and stored at central depositories, as discussed below.

Forced regurgitation and flushing

In many cases collecting birds for stomach analysis may be undesirable because of harm to local populations, ethical considerations, or inability to obtain permits. Several approaches allow partial sampling of gut contents via regurgitation or otherwise flushing the digestive tract of live birds. These vary in efficiency and in their effects on individual birds. The various biases associated with differential digestion and sampling concerns, discussed for stomach contents, are equally applicable to any technique involving partially digested or fragmented food samples.

The most common method of forced regurgitation uses chemical emetics. Antimony potassium tartrate (tartar emetic) seems to be the most widely used (Prys-Jones et al. 1974, Zach and Falls 1976a, Robinson and Holmes 1982, Gavett and Wakeley 1986), performing best in comparative trials (Lederer and Crane 1978). Dosages vary but are usually administered orally via a syringe and flexible plastic tubing coated with vaseline. Tomback (1975) found that a 1.5% tartar emetic solution rather than a 1% solution shortened the response period of several species from about 25 min to an average of 10 min, without harming the birds. Other researchers (Prys-Jones et al. 1974, Zach and Falls 1976a, Robinson and Holmes 1982) observed that most insectivores regurgitated samples within 2–3 min using 1% solution. Prys-Jones et al. (1974) found that only 50–64% of the granivores fed tartar emetic regurgitated samples, hypothesizing that more muscular gizzards act as a barrier to regurgitation.

Biases associated with emetics have been examined in several studies. Using captive Ovenbirds (*Seiurus aurocapillus*), Zach and Falls (1976a) found that the action of tartar emetic was independent of the type of prey eaten. Although regurgitation occurred in almost all birds tested, it was not always complete. Thus, no qualitative bias was found, but the material regurgitated did not reflect the quantity of food in the stomach. Gavett and Wakeley (1986) tested the efficiency of emetics in House Sparrows (*Passer domesticus*) by collecting stomachs from a subset of the sampled birds. An average of 58% of the total contents of each stomach was obtained by regurgitation. Although food categories were missing from individual stomachs, the overall emetic sample gave an accurate representation of the total diet in this species.

Mortality caused by emetics can be high and may depend on the species involved, dosage, and stressful effects such as handling. Zach and Falls (1976a) observed 50% mortality in newly caught Ovenbirds fed emetics, and 12.5% mortality in individuals already acclimated to captivity. Successive applications of emetic within a relatively short time invariably resulted in death. Lederer and Crane (1978) observed 20% mortality in wild-caught House Sparrows. Although Prys-Jones et al. (1974) found no difference in survival between treated and control House Sparrows, individuals that regurgitated were more likely to be sighted later than those that did not regurgitate. Emetics also were tried unsuccessfully on Australian honeyeaters (Ford et al. 1982) and various seed-eating species (Zann et al. 1984); in these studies no gut samples were obtained and mortality was often high.

Forced regurgitation also has been used without emetics. Lukewarm water is pumped directly into the stomach using a syringe and thin plastic tube until the stomach and esophageal contents are regurgitated. Brensing (1977) sampled over 2100 migrant passerines of 35 species and reported no loss of weight in birds recaptured after sampling. This technique was also used successfully on 157 Australian passerines (Ford et al. 1982) and on many species of small passerines on migration along the Louisiana Gulf Coast (Franz Bairlein, pers. comm.), with virtually no mortality. Ford et al. (1982) successfully obtained some gut contents from all individuals sampled (13 needed to be flushed twice) and reported equal rates of recapture or resighting of flushed and nonflushed birds.

A variation is flushing the entire digestive tract with warm saline solution (Moody 1970), which was used by Laursen (1978) to study migrant passerines in Europe. Of 396 birds sampled, 14 (3.6%) died during flushing; comparison of the remaining stomach contents with the flushed samples in these individuals indicated an average efficiency of 52%. Jordano and Herrera (1981) used this technique to document the frugivorous diet of the Blackcap (*Sylvia atricapillus*) in Spain. The use of stomach pumps is recommended by Duffy and Jackson (1986) for studies of seabirds and is discussed in relation to dietary studies of fish by Hyslop (1980). Apparently, the efficiency of this technique decreases with the size of the animal sampled. Overall, stomach pumping and flushing hold great promise for many studies and would seem highly preferable to the use of emetics.

Several similar techniques were developed specifically for use on seed-eating birds. Payne (1980) inserted a plastic tube into the crops of of most hummingbirds (Remsen et al. 1986) was not apparent from observations of visitation to various flowers.

Price (1987) observed the seeds eaten by Darwin's Finches (*Geospiza* spp.) and successfully related diet selection to individual morphology and varying ecological conditions. Newton (1967) reported that the foods of cardueline finches that fed above the ground on the seedheads of various plants could be easily quantified, whereas direct observation of the seeds eaten by ground-foragers was not possible.

For insectivorous birds, identification of prey items in the field is much more difficult. Whereas large or common prey may be easy to distinguish, many inconspicuous foods will be overlooked, and such observations by themselves may be highly biased (e.g., Rundle 1982). For example, using direct observation, Cooper (unpubl.) concluded that Scarlet Tanagers (*Piranga olivacea*) preyed almost exclusively on larval and adult Lepidoptera, but stomach contents showed that Lepidoptera comprised only 20–40% of the diet of adult birds. Robinson and Holmes (1982) supplemented gut samples (using emetics) with direct observations of prey captures for 11 species of forest insectivores. Prey were identifiable in from 1.1% (Least Flycatcher, *Empidonax minimus*) to 37.9% (Solitary Vireo, *Vireo solitarius*) of the observed foraging maneuvers. Prey size often may be estimated, even when prey type is unknown, by comparison with bill or head length, although this method has several biases (Bryan 1985, Goss-Custard et al. 1987).

Observations also may be made at nests to determine nestling diets, feeding rates, and other aspects of parental behavior. This technique is most often used for large species such as raptors (e.g., Collopy 1983) but has also been used successfully for passerines (Tinbergen 1960, Sealy 1980, Biermann and Sealy 1982). These observations are often greatly facilitated by the use of blinds, high-powered telescopes, or photography.

Photography

Various photographic devices have been used to record prey brought to nests. A major advantage is that film can be reviewed later, often allowing identification of prey type and size. Probably the most popular apparatus is the 8- or 16-mm movie camera fitted to a nestbox (Royama 1959). Upon entering the nestbox, adult birds trip a switch and a single-frame picture is taken of the bird's head and bill contents. Often a watch and metric ruler are fastened next to the entrance hole, so that the time of feeding and prey size can be determined (Royama 1970, Dahlsten and Copper 1979, Minot 1981). A major advantage is that an observer need not be present, because movie cameras may be operated automatically by a car battery. Because cameras are expensive, the number of nests to be photographed will usually exceed the number of cameras available. This problem can be circumvented by designing nestboxes so that the camera can be fitted to each one (Royama 1970), or by moving nests to a special box fitted with a camera (Dahlsten and Copper 1979).

Video recorders and 35-mm cameras fitted with telephoto lenses have been used to record prey brought to nestlings of open-nesting species. Knapton (1980) and Mcunier and Bedard (1984) placed a stick next to the nest where the adults perched to feed young and were easily photographed. A disadvantage of hand-operated cameras is that an observer must be present, usually in a nearby blind, and must also be a skilled photographer. Video recorders will probably be used with increasing frequency in diet studies, because they record continuously.

DIET ANALYSIS

Diet analysis generally consists of (1) sorting and identifying food items and (2) presenting the results in terms of occurrence, frequency, volumetric, or gravimetric measures (reviewed by Hartley 1948, Hyslop 1980, and Duffy and Jackson 1986). Most researchers recognize the need for presenting diet data in more than one form to minimize biased interpretations (e.g., Otvos and Stark 1985).

Sorting and identification

Little literature exists for sorting and identifying fragmented gut contents, and methods are rarely published in enough detail to be useful to others (but see Calver and Wooller 1982). In general, contents are examined under a dissecting microscope, preferably one with variable power (up to 30×) and fitted with an ocular micrometer. The procedure is more or less a game of matching similar parts and determining the minimum number of prey ingested by counting heads, mandibles, wings, legs, or other parts of known number in the intact state. Seeds are often encountered whole; however, other vegetative matter (e.g., fruit pulp) usually occurs in a form that prevents the enumeration of individual foods. The amount of grit present in a sample may be determined by "ashing" the contents at extremely high temperature (540°F), after identification and weighing (Shoemaker and Rogers 1980).

The ability to detect the full range of dietary items present rests on learning the specific parts, however tiny, that survive digestion. We believe such clues exist for virtually every type of solid food a bird may eat. Ralph et al. (1985) and Tatner (1983) listed commonly encountered fragments representing a variety of arthropod taxa, accompanied by photographs or sketches of the most distinctive parts. Diagnostic structures, at least to the familial level, appear to be remarkably invariate across diverse geographic regions. Accordingly, we found these lists from Hawaiian Islands and Great Britain valuable in identifying stomach samples from the southeastern United States and Amazonian rainforests.

The identification of arthropods and seeds is greatly facilitated by a reference collection of intact food items and of fragmented parts (e.g., mandibles, spider fangs) taken from known, intact specimens and mounted on clear microscope slides for easy comparison. Such reference collections permit determination of original size or weight of ingested foods from identified fragments. Calver and Wooller (1982) provided detailed equations for estimating total length from the size of diagnostic parts (e.g., head width, elytra length) for various families of Diptera, Coleoptera, and Hymenoptera.

Collaboration with entomologists or botanists is also recommended, although even experts may not be familiar with fragmented specimens. In addition, a technician without formal entomological or botanical background may be easily trained to recognize and sort diagnostic parts in fragmented samples (Ralph et al. 1985, Rosenberg pers. obs.). A primer on entomological terms commonly encountered in analysis of bird diets is offered by Calver and Wooller (1982).

In most studies, arthropod remains are identified only to family (sometimes only to order). Levels of prey identification affect the subsequent categories used in dietary comparisons, as discussed by Greene and Jaksic (1983) and Cooper et al. (this volume). In general, more inclusive categories tend to overemphasize the similarities among samples and underestimate diet diversities. Rotenberry (1980a) used the criterion that any taxonomic category represented in at least 5% of his samples would be included in further analyses, with rare taxa not meeting this criterion lumped into the next-most-inclusive category. Prey categories may be combined on the basis of ecological characteristics (e.g., phytophagous or predaceous; Robinson and Holmes 1982), or according to their modes of escape (e.g., flying, jumping, hiding), activity patterns and typical locations (Cooper et al., this volume), or, in the terminology of early diet researchers (e.g., MacAtee 1912), "harmful" vs. "beneficial." Sherry (1984) combined all morphologically identical specimens in his diet samples into "morphospecies" that were assumed to be encountered in patches by the foraging birds. Knowledge of the natural history of the arthropod (and plant) foods, as well as of the birds, will aid in the meaningful assignment of diet categories.

Percent occurrence

Occurrence usually refers to the number of samples in which a particular food type appears, although it is sometimes used as a synonym of frequency. Percent occurrence is the simplest and crudest measure of diet. Its primary advantage is that virtually all food types can be counted, even if individual items ingested cannot be quantified. For example, the presence of certain fruits or wing scales of adult lepidopterans may be detected by a distinctive color, and their occurrence is therefore easily determined. Hyslop (1980) discussed the application of subjective dominance rankings that allow the addition of relative importance values to occurrence measures. In general, species-level comparisons using percent occurrence tend to emphasize similarities among samples, whereas frequency and volume estimates tend to emphasize differences (Hartley 1948).

Frequency

Frequency is usually applied to the enumeration of individual food items. Ideally, the original diet can be "reconstructed" from these identified parts; however, some food types, such as fruit or green vegetation, do not occur in a form that can be counted. Individual samples are often pooled to create a single sample for a particular species, season, or geographic region. In these, the differences between frequency and occurrence measures depend on the patchiness of the foods encountered in nature and, therefore, in the individual diets (Hartley 1948). If the individual samples are considered separately, however, the average frequency per sample with its associated variance would reflect this patchiness. Sherry (1984) discussed the determination of patchiness of food items and its use as an independent characteristic of a species' diet and contrasted the use of pooled vs. individual samples in dietary analyses. In general, we recommend the use of per-sample measures to express frequencies or volume estimates.

The biases associated with differential digestion or passage through the gut are reflected in the differences between frequency and bulk (e.g., volumetric) estimates of diet (Hartley 1948, Hyslop 1980). In general, frequency measures tend to exaggerate the importance of small items and those whose parts persist longest in the digestive tract (MacAtee 1912). For example, a stomach that contains 20 small ants and one large cicada would indicate a diet of mostly ants in a frequency analysis but mostly cicadas in a volumetric analysis. The ants may better reflect the foraging effort and time of the bird, but the cicada may represent the bulk of the energy gained from that collection of food. Correction factors have been applied by Custer and Pitelka (1974) and others to account for these different rates of passage.

Percent volume and weight

The volume of a food type may be estimated as it appears in the sample and then expressed as a percentage of the total volume of the contents or the capacity of the stomach. This procedure allows almost all food types to be considered, including those that cannot be enumerated individually. Therefore, this may be the only way to describe diets of largely vegetarian species. In contrast with frequency measures, volumetric estimates tend to give greater importance to large, mostly undigested food items (Hartley 1948, Duffy and Jackson 1986). MacAtee (1912) considered this the best method to represent the "economic importance" of a bird species (i.e., its potential impact on the range of prey it consumed). He also noted that large samples minimized the potential bias of essentially ignoring the tiny, long-persisting fragments that would be counted in a frequency analysis.

Volumes also can be estimated by reconstructing the original diet based on the frequency and size of various food types (Martin et al. 1946, Hartley 1948). In this way, all food items are counted, but the largest items (at the time of ingestion) are given greatest importance. The determination of original volumes depends on the use of a reference collection of whole specimens or on various correction factors, as discussed by Hyslop (1980) and Calver and Wooller (1982).

Estimates of weight or biomass may be derived in the same ways as for volumes; however, these are often more tedious and time-consuming (Hartley 1948, Duffy and Jackson 1986). The use of wet- vs. dry-weight measures is discussed by Hyslop (1980). Dry weights of arthropods may be estimated from specimens of known or estimated length, using regression equations in Rogers et al. (1976, 1977) and Beaver and Baldwin (1975). Knowledge of original weights is necessary for calorimetric determinations. Estimates of the energy content of various foods are found in Golley (1961), Thompson and Grant (1968), Bryant (1973), Ricklefs (1974a), Norberg (1978), and Bell (this volume). Using these estimates, Calver and Wooller (1982) derived a general equation for determining energy content directly from prey length. Rosenberg et al. (1982) used a similar procedure to estimate the dietary requirements of a bird assemblage preying on cicadas. These measures should be used with caution, however, because of the potential to overestimate the nutritional value of large or long-persisting food types (Hyslop 1980).

DIET INFORMATION SOURCES

Here, we describe two important sources of raw data on the diets of North American and many Neotropical species. The first is the large collection of stomach samples compiled by the U.S. Biological Survey, representing over 250,000 individual birds of over 400 species (see MacAtee 1933). Stomach contents were meticulously identified by expert entomologists and botanists (often to species level). These data appear in various forms in numerous publications by W. L. MacAtee, F. E. Beal, and others and were summarized for most species by Martin et al. (1951a). The raw data are stored on cards filed at the Patuxent Wildlife Research Center of the U.S. Fish and Wildlife Service in Laurel, Maryland. Each card represents a single stomach sample and contains information on the bird's sex, location, habitat, time of day, and date of collection. Contents are listed individually, along with the relative volumes of each food type in relation to the total volume of the contents, and the relative volumes of total plant and animal matter.

This tremendous source of information has barely been exploited by modern ornithologists. Wheelright (1986b) used these data to describe seasonal and geographic variation in the American Robin and urged their wider application. Although the samples for most species are from wide geographic regions and dispersed over many years of collection, precluding many community-level analyses, their potential for studies of ecomorphology, predator–prey relationships, plant–animal interactions, and seasonal variation is great. For example, Hespenheide (1971) reanalyzed the published data for several flycatcher species to test the theoretical relationships between predator and prey sizes.

The second source is the collection of unanalyzed stomach contents at the Louisiana State University Museum of Natural Science (LSUMNS). In most cases these are whole stomachs, taken from birds during routine specimen preparation, and preserved in 70% ethanol. All samples are labeled to correspond with skin or skeleton specimens deposited at LSUMZ and accompanied by complete data on location, habitat, age, sex, reproductive condition, fat, and molt. The ability to measure the morphological features of birds from which diet samples were taken should aid in studies of ecomorphology and individual variation (e.g., Herrera 1978b).

This collection has a strong Neotropical representation, including over 2500 samples from ca. 700 species, mostly from the Andes and low-

TABLE 1. COMPARISON OF COMMON METHODS USED TO OBTAIN AVIAN DIET SAMPLES

Method	Advantages	Disadvantages	Example of use
Direct examination of collected birds	Whole stomachs collected; if shot, then exact bird desired can be obtained.	Birds are killed; multiple samples from one bird impossible.	Rotenberry (1980a), Sherry (1984)
Chemical emetics	Birds not killed directly.	Mortality may still be substantial; multiple samples from one bird often results in mortality; birds must be captured; partial samples obtained; unsuitable for some species.	Zach and Falls (1976a), Robinson and Holmes (1982), Gavett and Wakely (1986)
Stomach pumping	Birds not killed.	Birds must be captured; partial samples obtained.	Moody (1970), Brensing (1977)
Fecal samples	Birds disturbed minimally; samples easily obtained.	Birds usually must be captured; samples highly fragmented; samples must be treated before analysis.	Ralph et al. (1985)
Ligatures	Arthropod prey usually intact; can be effective when combined with direct observation.	Restricted to nestlings; feeding behavior and survival can be affected; estimates of prey size can be biased.	Johnson et al. (1980)
Pellets	Birds not disturbed; samples easily obtained; keys to mammal skulls and hair available.	Restricted to pellet-forming species; may be biased by prey type, size.	Errington (1930)
Direct observation (adult birds)	Birds not disturbed; foraging behaviors that resulted in prey capture are observed.	Difficult for insectivorous birds; observations biased towards large, conspicuous prey.	Robinson and Holmes (1982), Price (1987)
Direct observation (nestlings)	Birds not disturbed; can be effective when used in conjunction with ligatures.	Time consuming, labor intensive; biased as above.	Tinbergen (1960), Johnson et al. (1980)
Photography	Birds not disturbed; automatic movie cameras provide many samples for little effort.	Restricted to nestlings; Equipment relatively expensive; hand operated cameras time consuming, labor intensive.	Royama (1959, 1970), Dahlston and Copper (1979)

land rainforests of Peru and Bolivia. These include many poorly known species for which little basic natural history information exists. Sample sizes for some species are large enough to permit geographic and guild-level analyses. The LSUMNS collection also contains about 1500 stomach samples from common birds in Louisiana, as well as smaller collections from other regions. Research use of any materials is welcomed; inquiries should be directed to: Curator of Birds, Museum of Natural Science, Louisiana State University.

RECOMMENDATIONS AND CONCLUSIONS

With the broad range of techniques now available (Table 1), direct examination of avian diets is possible in nearly any study. For many species that cannot be captured alive, collection of stomach or esophageal contents remains the only means of diet assessment. When collection is necessary, care is needed to maximize sampling efficiency, taking only actively foraging individuals from known habitats or foraging sites, and ensuring adequate sample sizes. When capture is possible, we recommend the use of flushing techniques to force regurgitation of gut contents, avoiding emetics. Fecal samples are probably the easiest to obtain but present added difficulties in analysis and interpretation. When other techniques are unavailable, routine collection of fecal samples will give an adequate representation of many species' diets. For any species that regularly regurgitates pellets, large samples of prey remains can be collected and may give an accurate estimate of diet.

For studies of the diet of nestling birds, several additional techniques are available, including ligatures, photography, and direct observation of the nest. Direct observation of foraging birds may be a sufficient means of assessing diet in some species, particularly in specialized nectarivores or frugivores. Observations of foods eaten can supplement any of the techniques discussed and may aid in the minimization of certain biases associated with highly digested gut contents.

Biases caused by differential rates of passage and digestibility remain poorly documented and understood. Continued experimentation with live birds is needed to determine the advisability or consequences of various collecting, preserving, and analytical procedures. We also urge the publication of additional lists, descriptions, sketches, or photographs that can aid in the identification of fragmented diet samples. Expanded use of reference collections with additional calculations of prey length and weight from fragment size is also recommended.

Each of the several methods of presenting diet data has its advantages and drawbacks. Therefore, more than one method should be presented whenever possible, including at least one that represents occurrence and one that represents frequency or relative volume. Although pooling results may be desirable in cases with small sample sizes or when only population averages are needed, we recommend the use of per-sample measures with their associated variances to characterize species' diets.

We urge that gut contents be routinely preserved from specimens collected for any reason; with limitations placed on present and future collection of birds, the maximization of information from each specimen is highly desirable. We also urge the expanded application of diet analysis techniques to a wide range of ecological pursuits. Our knowledge of avian food habits lags far behind our knowledge of habitat use, foraging behavior, and morphology. In most cases, gathering diet data by any means available is preferable to ignorance. We think that many of the biases and difficulties will be alleviated when more careful attention is paid to sampling design, prey identification, and overall foraging ecology.

ACKNOWLEDGMENTS

C. J. Ralph and T. W. Sherry stimulated us to prepare this review. J. R. Jehl, Jr., J. Verner, C. J. Ralph, M. L. Morrison, J. V. Remsen, and A. Chapman provided helpful comments on various drafts of the manuscript. J. V. Remsen, P. W. Brown, and G. A. Hall helped greatly with the literature review and offered support and advice.

DIETS OF UNDERSTORY FRUIT-EATING BIRDS IN COSTA RICA: SEASONALITY AND RESOURCE ABUNDANCE

BETTE A. LOISELLE AND JOHN G. BLAKE

Abstract. Diets of understory fruit-eating birds were examined in five habitats in northeastern Costa Rica. Diets were quantified by analyzing seeds contained in fecal samples collected from mist-netted birds. We show that neotropical understory frugivores partition fruit resources. Six frugivore guilds were identified by Bray-Curtis ordination. Number of species per guild varied from one to seven. Not all guilds were present at each site (young and old second-growth, and primary forest at 50-m, 500-m, and 1000-m elevations). Guild composition was influenced by morphology, fruit display and type, feeding method, and foraging height. Birds differed in preference or avoidance of fruit species; preferences varied seasonally, annually, and among habitats. Birds were more selective in areas with high fruit abundance (second-growth) than in areas with low fruit abundance (forest).

Key Words: Costa Rica; diet; frugivores; fruits; seasonality.

Plants that produce fleshy fruits and birds that consume fruits are important components of many tropical habitats. From 63 to 77% of understory shrubs and trees produce bird-dispersed fruits in Costa Rican evergreen forests (Stiles 1985a) and fruit-eating birds often constitute a large proportion of tropical avifaunas (Stiles 1985a, Blake et al. in press, Karr et al. in press). The few specific studies have revealed that those birds feed on a wide variety of fruits (Snow 1962a, 1962b, Jenkins 1969, Snow and Snow 1971, Worthington 1982). Even so, diets of most understory fruit-eating birds in neotropical habitats are largely unknown.

Many studies of fruit-frugivore interactions have documented what bird species consume fruit and disperse seeds of a particular species or group of plants. A diverse assemblage of bird species visit trees with abundant fruit crops (e.g., Eisenmann 1961, Willis 1966, Leck 1973, Kantak 1979, Howe 1981). From such studies, some researchers (e.g., McKey 1975, Howe and Estabrook 1977) have proposed a specialist-generalist dichotomy, with small understory frugivores feeding opportunistically and large, canopy frugivores specializing on a limited subset of fruits. However, understory birds also can be highly selective in their choice of fruits (Moermond and Denslow 1985).

Here we present data on the diets of understory frugivores to examine how feeding preferences of birds for common fruiting plants vary in relation to fruit abundance and seasonality. Diet information was obtained from birds in five Costa Rican sites that represent different successional and forest elevational stages. We examine whether understory frugivores are generalists or opportunists, as proposed by previous authors, or selective as suggested by aviary work. Because community-wide fruit production varies among sites, we examine whether and how habitat influences foraging patterns. In habitats with low fruit abundance, understory frugivores likely compete for fruits and partition fruit resources. In areas with high fruit abundance, such as young second-growth, competition for fruits is less. Instead, plants may compete for dispersers. Birds should be more selective in areas with high than low fruit abundance, and if a given fruit species is equally attractive to birds, then feeding preference and diet should overlap broadly.

METHODS

STUDY AREA

The study area was on the Caribbean slope of the Cordillera Central in northeast Costa Rica. Lowland sites were in 5–10 year second-growth, 25–35 year second-growth, and primary (undisturbed) forest at the Estación Biológica La Selva (10°25′N, 84°01′W). We also sampled diets of fruit-eating birds in primary forest at 500-m (10°20′N, 84°04′W) and at 1000-m (10°16′N, 84°05′W) in Parque Nacional Braulio Carrillo, about 15 km and 20 km south of La Selva, respectively.

La Selva receives about 4000 mm rain annually (Hartshorn 1983; Organization for Tropical Studies, unpubl. data). The main dry season lasts from January or February to March or April with a shorter, less pronounced dry season in September and October. Climatological data are not available for the 500-m and 1000-m sites, but annual rainfall probably exceeds 4500 mm at both. During this study, rainfall generally was below the 20-year average and the dry season effectively lasted from January through April. Further descriptions of those sites are in Frankie et al. (1974), Hartshorn (1983), and Pringle et al. (1984).

DATA COLLECTION

Our diet analyses were based primarily on seeds and pulp from feces or regurgitated material (hereafter referred to as "fecal samples") from mist-netted birds. We collected data from January 1985 through April 1986 and from December to mid-April 1987. Samples were collected every 5–6 weeks at each site. Total sampling effort was less at higher sites due to a variety of logistical and weather-related problems. We had not analysed all 1987 data when this paper was written and here include 1987 data from only the youngest site.

We placed all mist-netted birds (except hummingbirds and raptors) in plastic containers lined with filter paper for 5 to 15 min. More than 80% of the birds produced samples; only a few species (e.g., hole nesters such as Wedge-billed Woodcreeper [scientific names of all birds are in Appendix 1]) regularly failed to defecate in containers. We collected 4037 fecal samples; 57% contained fruit pulp, seeds, or both. Using a dissecting microscope, we separated seeds from fecal samples and identified them to species through comparison with a reference collection at La Selva. Some seeds were lumped by genera in our analyses because species could not be distinguished (e.g., *Anthurium, Sabicea, Clusia, Ficus*).

We estimated understory fruit abundance (see also Loiselle [1987]) by counting all fruiting individuals and ripe and unripe fruits in belt transects that paralleled each side of each mist net (50 m²/net). Fruits were sampled at 20 mist nets (1000 m² total sampling area) in each highland area, at 30 mist nets (1500 m² total sampling area) in each second-growth site, and at 54 mist nets (2700 m² total sampling area) in primary forest at La Selva.

DATA ANALYSIS

We used data from all five sites to describe frugivore assemblages in the following analyses. However, because of smaller sample sizes, we did not include our highland sites (500 and 1000-m) in analyses of seasonal or annual variation in fruit use.

Accumulation curves

We plotted cumulative number of fruit species in the diet against sample effort (i.e., number of fecal samples) to construct fruit species accumulation curves. We included all samples (i.e., those containing only insect parts as well as those containing fruit) in the analyses. We fitted accumulation curves to three regression functions: linear (nontransformed), exponential (species/log sample number), and power (log species/log sample number). We used accumulation curves to evaluate sample sizes needed to describe diets of birds and to compare slopes of fruit species accumulation among some bird species.

Multivariate analyses

The original data matrix for each site consisted of the number of times a given fruit species occurred (i.e., at least one seed) in fecal samples for each bird species (bird-species by fruit-species matrix). We simplified the data matrix by combining fruits into 9 to 15 categories defined by fruit presentation, location, and type, and by seed number and size (Appendix 2). Some species or genera (e.g., *Phytolacca rivinoides* and *Passiflora* sp.) did not readily fit into any category and were treated as separate groups. Furthermore, because species composition and representation of fruiting plants in birds' diets varied among habitats, assignment of fruit variables differed among sites. For example, some fruit variables (e.g., aggregate fruits such as *Piper*) were appropriate at one site only, while others (e.g., fruits of aroids and bromeliads) were lumped or divided depending on sample sizes at each site (Appendix 2). This simplification was necessary because of the relative rarity of many fruit species in bird diets. We further simplified the data matrix by excluding birds that rarely ate fruit or that were under-represented among fecal samples.

We relativized the data by rows (birds) (Greig-Smith 1983:248), so that use of a fruit was expressed as a proportion of total fruit used by that species. This "standardization by the norm" eliminates problems arising from unequal sample sizes. This core set of frugivores was ordinated in fruit-species space for each site using a Euclidean distance measure and Bray-Curtis ordination with variance-regression criterion for axis orientation (Beals 1984). Use of a standardized distance with Bray-Curtis construction of axes eliminates the effects of ecologically ambiguous "joint nonuse" of resources that are emphasized by covariance or correlation values used in construction of principal component axes (E. W. Beals, pers. comm.). Fruit variables were correlated with ordination axes. All multivariate analyses were run on PC-ORD (McCune 1987).

Seasonal and annual use of fruit

We divided our samples into four or five (young second-growth) seasons on the basis of rainfall totals to allow evaluation of both seasonal and annual variation in use of fruit by birds. Abundance of ripe fruit at each lowland site was totalled by season for dominant understory fruiting plants. We used an index developed by Jacobs (1974) to evaluate use of a fruiting species in relation to its availability (feeding preference):

$$D_{fr} = \frac{(r - p)}{(r + p - 2rp)}$$

where D_{fr} is an index of fruit use, r is proportion of that fruit species in the diet, and p is proportion of ripe fruit (available) in the habitat accounted for by that species. We followed Morrison (1982) who categorized this index, which ranges from -1 to $+1$, as follows: D_{fr} of 0 to ± 0.15 = no preference, ± 0.16 to 0.40 = slight preference or avoidance, ± 0.41 to 0.80 = moderate preference or avoidance, and ± 0.81 to 1.00 = strong preference or avoidance. Fruit use was evaluated by comparing observed number of fecal samples that contained that fruit species to that expected from availability of ripe fruit (χ^2 analysis, Zar 1984:40–42). We further analysed seasonal and annual use of fruit species by comparing number of occurrences of a particular fruit in the diet in relation to all other fruit species in the diet (χ^2 analysis).

TABLE 1. NUMBER OF SAMPLES CONTAINING FRUIT, NUMBER OF BIRD SPECIES REPRESENTED IN THOSE SAMPLES, AND NUMBER OF FRUIT SPECIES CONTAINED IN THOSE SAMPLES FOR EACH OF FIVE COSTA RICAN SITES. NUMBER OF FRUIT SPECIES WAS UNDERESTIMATED BECAUSE SPECIES OF SOME GENERA WERE LUMPED (SEE TEXT). DATA FROM YOUNG SECOND-GROWTH WERE COLLECTED FROM 1985–1987; DATA FROM ALL OTHER SITES WERE COLLECTED FROM 1985–1986

Site	Number of fecal samples with fruit	Number of bird species	Number of fruit species
Second-growth:			
young (5–7 year)	1119	57	81
old (25–35 year)	339	27	69
Primary forest:			
lowland	366	21	55
500 m	219	21	55
1000 m	252	22	70

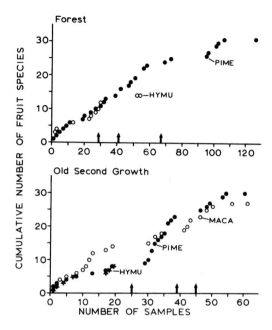

FIGURE 1. Fruit species accumulation curves for representative birds from old second-growth and forest habitats in Costa Rica. PIME = Red-capped Manakin, HYMU = Wood Thrush, and MACA = White-collared Manakin. Arrows along the ordinate axis point to data from a new season for MACA in old second-growth and PIME in forest.

Use of fruit by common fruit-eating birds

We supplemented ordination and feeding preference data by examining diets of some key frugivores in each lowland habitat. We used Kendall's coefficient of concordance (Zar 1984:352–359) to test whether relative use of common fruiting plants found in fecal samples was similar among those birds. We further compared feeding preferences of individual species to those of the entire assemblage.

SAMPLING BIAS

Our index of feeding preference may have overestimated importance of small-seeded fruits. Passage of seeds from such fruits is spread over a longer period than seeds from few- or one-seeded fruits (Levey 1986, 1987b). Nonetheless, within a fruit species or group of small-seeded fruits, seasonal and annual comparisons of this index are valid. Moreover, small-seeded species were not favored by all birds, indicating that potential biases from differences in seed passage time did not affect qualitative interpretations.

Birds that mandibulate fruits ("mashers") often drop seeds (Moermond and Denslow 1985, Levey 1987b). Consequently, large-seeded fruits may be underestimated; we have, however, recorded a wide array of seeds, varying in length from 0.3 mm to about 12 mm, in their diets.

RESULTS

DIET BREADTH

We recorded 226 fruit species in samples from 80 bird species at all five sites combined. Frugivores were most abundant and diverse in the young second-growth site, even after accounting for differences in sample effort (Table 1; Blake et al., in press). By contrast, average number of fruit species in diets of birds overall (total number of fruit species/total number of bird species; Table 1) was lower there than at older sites. Average diversity of fruits in diets was greatest for birds of forest at 1000 m. General trends found in number of frugivores and average dietary diversity among sites were paralleled by trends in abundance and diversity of fruiting plants at each site. Fruit abundance was significantly higher during all seasons in the youngest site than in either of the older lowland sites (Loiselle 1987, see also Levey 1988). Total species richness of fruiting plants, however, was greater in old second-growth and primary forest sites than in the youngest site (Loiselle 1987).

Fruit species still were being added to diets of birds even after 100 fecal samples had been examined (Fig. 1). All three models used to fit accumulation curves produced highly significant ($P < 0.001$) results. The exponential (semi-log) function provided the best fit in only a few cases and few species reached an asymptote with respect to diet diversity. Linear and power functions provided the best fit in an equal number

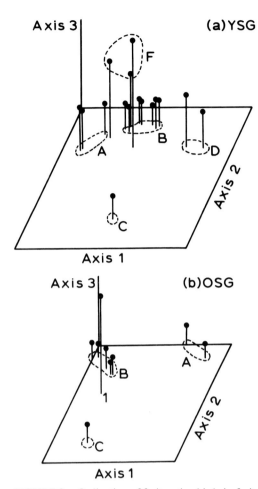

FIGURE 2. Ordination of fruit-eating birds in fruit-species space from (a) young second-growth (YSG) and (b) old second-growth (OSG). Birds are plotted according to their factor scores along first three ordination axes (see Table 2). For young second-growth, Group A = Orange-billed and Black-striped sparrows, and Red-throated Ant-Tanager; Group B = Grey-cheeked, Swainson's, and Wood thrushes, Red-capped and White-collared manakins, Grey-capped Flycatcher, and Grey Catbird; Group C = Ochre-bellied Flycatcher; Group D = Clay-colored and Pale-vented robins; Group F = Scarlet-rumped and Crimson-collared tanagers, and Buff-throated Saltator. For old second-growth, Group A = Red-throated Ant-Tanager and Orange-billed Sparrow; Group B = Dusky-faced Tanager, White-collared and Red-capped manakins, and Swainson's and Wood thrushes; Group C = Ochre-bellied Flycatcher. "1" refers to White-ruffed Manakin (OSG), a bird not readily classified into any group.

of cases. An apparently continuous increase in diet breadth was due partially to differences in plant phenologies; new species were added to the diet as they became available seasonally (Fig. 1).

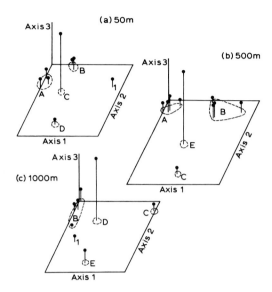

FIGURE 3. Ordination of fruit-eating birds in fruit species space from forest at (a) La Selva, 50-m, (b) 500-m, and (c) 1000-m. Birds are plotted according to their factor scores along the first three ordination axes (see Table 3). For 50-m, Group A = Olive and Tawny-crested tanagers, and White-ruffed Manakin; Group B = Red-capped Manakin and Wood Thrush; Group C = Ochre-bellied Flycatcher; Group D = Pale-vented Robin. For 500-m, Group A = Orange-billed Sparrow, White-ruffed Manakin, and Tawny-created and Olive tanagers; Group B = Red-capped Manakin, Black-faced Solitaire, and *Catharus* and Wood thrushes; Group C = Ochre-bellied Flycatcher; Group E = Tawny-capped Euphonia. For 1000-m, Group B = Slaty-backed Nightingale-Thrush, Swainson's Thrush, White-crowned Manakin, and Black-faced Solitaire; Group C = Olive-striped Flycatcher; Group D = Pale-vented Robin; Group E = Tawny-capped Euphonia. "1" refers to Swainson's Thrush (50-m) or Common Bush-Tanager (1000-m), birds not readily classified into any group.

Despite the continued addition of fruit species with increased sample effort, consistency of guild composition among sites (discussed below) suggests that our sample sizes were adequate to describe the frugivorous bird assemblages through ordination techniques.

If frugivores feed opportunistically, diet diversity should increase more rapidly (i.e., have a higher slope) in habitats that support a wider diversity of fruits (e.g., primary forest understory vs. young second-growth). To test that prediction, we compared slopes of fruit species accumulation curves for four frugivore species that were common in two or three lowland sites (Orange-billed Sparrow, Ochre-bellied Flycatcher,

TABLE 2. CORRELATION (PEARSON'S r) OF FRUIT VARIABLES WITH MAJOR AXES GENERATED BY BRAY-CURTIS ORDINATION OF THE FRUIT-EATING BIRD ASSEMBLAGE IN SECOND-GROWTH HABITATS IN COSTA RICA (SEE TEXT). ONLY VARIABLES WITH SIGNIFICANT CORRELATIONS ($P < 0.05$) ARE SHOWN. DESCRIPTION OF FRUIT VARIABLES IN APPENDIX 2

	Young second-growth				Old second-growth		
Variable	Axis 1	Axis 2	Axis 3	Variable	Axis 1	Axis 2	Axis 3
ARILAT		0.900		ARIL		0.961	
ATTARIL		0.905		LGSDUN			0.940
TERMUN		−0.673		HENOSS			0.740
AXIL	−0.802			CLIBES			
AGGREG			0.944	WITAST			
FICUS	0.708			TERMUN			
TREE1	0.857			FICUS	0.847		0.838
TREES			0.673		0.796		
PASSIF			0.554		−0.838		

Red-capped Manakin, and White-collared Manakin) using equal sample sizes for slope comparisons. Only data for Orange-billed Sparrows supported the hypothesis: fruits were added in the diet at a more rapid rate (higher slope) in older than in younger second-growth ($t = 4.2$, $P < 0.001$). Ochre-bellied Flycatchers actually accumulated fruit species faster in young habitats ($t = 3.6$, $P < 0.001$).

ORDINATION OF FRUIT-EATING
BIRD ASSEMBLAGES

Primary factors separating bird species by diet varied among sites (Tables 2, 3), but a series of distinct groups could be identified (Figs. 2, 3). Interpretation of different groups was based on correlations of fruit variables with major ordination axes for each site. Not all groups were represented at each site and some species fit into different groups at different sites.

Two groups (A and B) were composed of species that fed on different sets of understory fruits. Group A included species that fed on small-seeded axillary or cauliforous fruits. Group B species preferred understory plants with berries displayed on terminal infructescences or with relatively large seeds. Group B was represented by 2 to 7 species at each site, whereas Group A included 2 to 3 species. Group A was not represented among birds at the highest (1000-m) site (Fig. 3c).

Two groups (C and E) were each represented by a single species. Group C species fed principally on arillate fruits and were present at all sites. The Tawny-capped Euphonia fed heavily on fruits of the epiphytic genus *Anthurium* and formed a separate guild (E) at 500 m and 1000 m (Figs. 3b, c). Unlike its lowland counterparts, this euphonia characteristically fed in the understory, most likely because *Anthurium* is more abundant in the understory of highland forests than in lowland forest (Loiselle 1987).

Guild D was composed of birds that fed primarily at subcanopy or canopy levels. It was represented by a single species in lowland and 1000-m forest and was not among common fruit-eating birds captured in the understory of forest at 500 m or in old second-growth.

A final frugivore guild (F) was present only in young second-growth and consisted of two tanagers and a saltator (Fig. 2a). These three species ate a variety of fruits, including *Piper* fruits, whereas most other species only fed rarely on *Piper* or not at all. Since those three species mandibulate fruits, as do members of Group A, their separation into a distinct subset of frugivores, as well as the close alignment along the major axis with fruit-eaters that swallow fruits whole, argues that seed passage rates did not overtly bias the data.

Some species, most notably Swainson's Thrush, did not fit well or consistently into any guild. Swainson's Thrushes primarily are passage migrants through Costa Rica, rarely wintering at La Selva. Their diet thus was restricted to those fruits available during the short time they were present. Similarly, White-ruffed Manakins are altitudinal migrants that descend for two to four months each year to lowland sites at La Selva, where they prefer primary forest. They were present for only a short time in our old second-growth site during January and February 1986 and fed almost exclusively on two species of fruits.

Our sample sizes for most *Tangara* species in forest at 1000 m were too small ($N < 5$) to warrant inclusion into an ordination now. We believe that, once included, they will form a new

TABLE 3. CORRELATION (PEARSON'S r) OF FRUIT VARIABLES WITH MAJOR AXES GENERATED BY BRAY-CURTIS ORDINATION OF THE FRUIT-EATING BIRD ASSEMBLAGE IN COSTA RICAN FOREST HABITATS (SEE TEXT). VARIABLES ARE SIGNIFICANT AT P < 0.05 UNLESS INDICATED BY AN ASTERISK INDICATING P < 0.10. FRUIT VARIABLES ARE DESCRIBED IN APPENDIX 2.

Variable	Lowland forest			Variable	500-m forest			Variable	1000-m forest		
	Axis 1	Axis 2	Axis 3		Axis 1	Axis 2	Axis 3		Axis 1	Axis 2	Axis 3
ARIL	0.848			ARIL		0.937		ARIL	0.881		
LGSDUN			0.980	LGSDUN	0.755			TERMUN	−0.860		
TERMUN		−0.678*		EPISHB			0.803	OSSAEA	−0.770		
HENOSS	−0.834			HENOSS	−0.862			TREE			0.955
AXIL	−0.710			AXIL	−0.710			BROMEL	0.754		
TREE1		0.870		PHYRIV	0.908			ANTHUR		0.789	
TREES	0.898			ANTBRO			−0.608*	UNK112			0.981
				TREE	0.898		0.988				

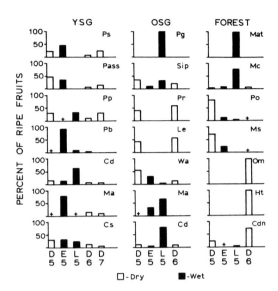

FIGURE 4. Seasonal availability of ripe fruits for common fruiting plants at each of our three lowland sites. YSG = young second-growth, OSG = old second-growth, and FOREST = lowland forest. Percent of ripe fruits represents the number of ripe fruits of a species in one season divided by the total number of ripe fruits of that species over all seasons. D5, E5, L5, D6, and D7 = dry season (January through April) 1985, early wet season (May through August) 1985, late wet season (September through November) 1985, dry season 1986 and 1987, respectively. Ps = *Piper sancti-felicis*; Pas = *Passiflora auriculata*; Pp = *Psychotria pittieri*; Pb = *Psychotria brachiata*; Cd = *Clidemia dentata*; Ma = *Miconia affinis*; Cs = *Conostegia subcrustulata*; Pg = *Psychotria grandis*; Si = *Siparuna* sp.; Pr = *Psychotria racemosa*; Le = *Leandra* sp.; Wa = *Witheringia asterotricha*; Mat = *Miconia "attenuate"*; Mc = *Miconia centrodesma*; Po = *Psychotria officinalis*; Ms = *Miconia simplex*; Om = *Ossaea macrophylla*; Ht = *Henrietella tuberculosa*; Cdn = *Clidemia densiflora*.

subset of frugivores with the Common Bush-Tanager at that site (Fig. 3c). Those tanagers fed primarily in upper levels of forest at 1000 m on fruits of the epiphytic shrubs *Cavendishia* (Ericaceae), *Blakea* (Melastomataceae), and *Topobea* (Melastomataceae), as well as other berries in the family Melastomataceae.

SELECTION OF FRUITS

Overall feeding preferences

Preference for or avoidance of fruits by frugivores was examined for seven common plants at each lowland site (Table 4). Overall indices (absolute values) of fruit use were higher (Mann Whitney U-test; U = 84, P < 0.01) in young

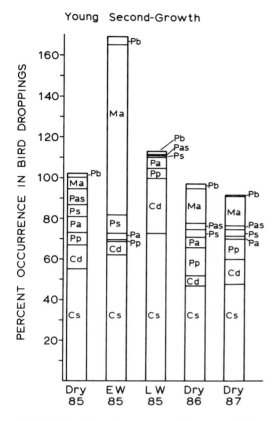

FIGURE 5. Percent occurrence of common fruiting plants in the diets of birds in young second-growth over five seasons. Percent occurrence may exceed 100% because fecal samples often contained more than one seed type. Dry 85 = dry season 1985, EW = early wet season, LW = late wet season, Dry 86 = dry season 1986, Dry 87 = dry season 1987. Pa = *Piper auritum*. Other abbreviations in Figure 4.

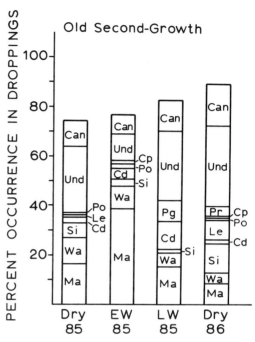

FIGURE 6. Percent occurrence of common fruiting plants in the diets of birds in old second-growth and over four seasons. Percent occurrence may exceed 100% because fecal samples often contained more than one seed type. Can = canopy fruits; Und = understory fruits; Cp = *Clidemia purpureo-violacea*. Other abbreviations in Figure 4.

second-growth than in older sites (Table 4), as would be expected if plants in young second-growth habitats compete more heavily for dispersers. Birds foraging in young second-growth strongly avoided *Psychotria pittieri*. This small shrub (usually <1.5 m tall) produces large crops of "styrofoam" textured fruits, which are low in sugar content (3.4%, from Denslow and Moermond 1982) and consist mostly of epicarp and seeds. With the exception of a moderate avoidance of *Psychotria brachiata*, birds showed a moderate preference for all other fruits tested in the young second-growth.

Birds showed a strong or moderate avoidance of fruits of two or three plant species in lowland forest and old second-growth sites, respectively. Frugivores displayed a moderate preference for *Siparuna* spp. in our old second-growth site (Table 4); other common fruits were eaten roughly in proportion to their availability. In lowland forest, *Ossaea* produces unusually large crops for an understory treelet and fruits ripen quickly. Thus, even though eaten by many birds, the index of fruit use was negative because of the high availability of *Ossaea* over a short period.

Fruit use by common birds

We compared fruit use by three or four common frugivores at each of our lowland sites (Table 5). Relative use of (or preference for) different fruits by those common frugivores was similar in young second-growth (Kendall's coefficient of concordance, $W = 0.54$, df = 7, $P < 0.05$) and in lowland forest ($W = 0.72$, df = 6, $P < 0.05$), but not in old second-growth ($W = 0.32$, df = 6, $P > 0.25$). When use of *Conostegia subcrustulata* was excluded, no significant association of fruit use by common fruit-eating birds existed in young second-growth ($W = 0.34$, df = 6, $P > 0.20$).

Forest birds generally had similar preference indices for fruits, although some differences were

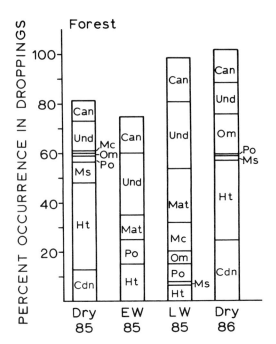

FIGURE 7. Percent occurrence of common fruiting plants in the diets of birds in lowland forest and over four seasons. Percent occurrence may exceed 100% because fecal samples often contained more than one seed type. Abbreviations are in Figures 4, 5, and 6.

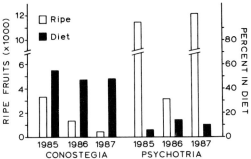

FIGURE 8. Number of ripe fruits of *Conostegia subcrustulata* (Cs) and *Psychotria pittieri* (Pp) and their percent occurrence in the diets of birds over three dry seasons in young second-growth. The number of ripe fruits available are directly comparable among years because the same quadrats and area were sampled in all three years. Note that the peak occurrence of ripe *P. pittieri* fruits was approximately two months earlier in 1986 (see Fig. 4).

noted. Fruit preferences varied more among species in young and old second-growth, as might be expected if fruits were competing for birds at those sites. Red-capped Manakins, for example, avoided *Clidemia dentata*, whereas other species showed weak to strong preferences for it. Scarlet-rumped Tanagers, unlike most other species, preferred *Piper* and *Passiflora* fruits.

SEASONAL AND ANNUAL VARIATION IN FRUIT USE

We analysed seasonal and annual variation in use of fruits produced by seven common fruiting plants at each lowland site. Those plants accounted for 79 to 84% of the total ripe fruit available in the understory (known bird-dispersed plants only) at those sites (Fig. 4).

Seasonal variation

Continuously-fruiting species were more common in the youngest site than in the other two lowland sites, and fruit use by birds was influenced by those phenological patterns. Thus, preference or avoidance of common fruits in young second-growth reflected choice of fruits rather than changes in plant phenologies. *Conostegia subcrustulata* fruited year round at the youngest site (Fig. 4) and was represented in over 40% of fecal samples during all seasons (Fig. 5). In fact, all common fruiting plants, with the exception of *Miconia affinis* in late wet season 1985, appeared in diets of birds during each season in young second-growth. *Miconia affinis*, a fruit relatively rich in sugar content (15.8%, Moermond and Denslow 1983), was unusual in its highly aseasonal production of fruits in young second-growth. When it was available, birds preferred *Miconia* and ate fewer other, generally favored fruits.

Only four species occurred in diets of birds in old second-growth during each season (Fig. 6). *Miconia affinis* was recorded as present in diets of some species, even though we did not record ripe fruits during all seasons (Fig. 4). We often observed birds feeding on unripe or partially ripe berries of that fruit. At the forest site, *Henrietella tuberculosa* and *Psychotria officinalis* occurred in diets of birds during all seasons (Fig. 7), even though our samples failed to detect fruiting of the former species year-round.

Considerable seasonal variation in fruit use occurred at all sites (Table 6), particularly among understory fruits. Use of canopy fruits did not vary in older forests (Table 6), suggesting that birds did not move up into canopy habitats at any one time of the year (see also Loiselle 1988).

Annual variation (dry season samples)

Annual variation in fruit use was pronounced at the young second-growth site (Table 6), but

TABLE 4. INDEX OF FRUIT USE (D_{fr}) BY BIRDS FOR COMMON FRUITING PLANTS AT THREE COSTA RICAN LOWLAND SITES. D_{fr} WAS CALCULATED BY SUMMING DATA FROM ALL SEASONS IN WHICH RIPE FRUIT WAS AVAILABLE. SIGNIFICANCE OF INDICES WAS TESTED BY COMPARING OCCURRENCE OF THE FRUIT IN THE BIRDS' DIETS TO THAT EXPECTED FROM THE AVAILABILITY OF RIPE FRUIT

Fruit species	D_{fr}	P	Fruit species	D_{fr}	P
Young second-growth			Lowland forest		
Conostegia subcrustulata	0.78	***	Clidemia densiflora	−0.24	**
Miconia affinis	0.57	***	Henrietella tuberculosa	0.17	ns
Clidemia dentata	0.66	***	Ossaea macrophylla	−0.46	***
Psychotria brachiata	−0.56	**	Miconia simplex	0.24	ns
Psychotria pittieri	−0.86	***	Psychotria officinalis	−0.84	***
Passiflora auriculata	0.50	***	Miconia centrodesma	−0.25	ns
Piper sancti-felicis	0.52	***	Miconia "attenuate"	−0.07	ns
Old second-growth					
Clidemia dentata	0.28	*			
Miconia affinis	−0.42	***			
Siparuna sp.	0.53	***			
Witheringia asterotricha	−0.48	***			
Leandra sp.	−0.22	ns			
Psychotria grandis	−0.78	***			
Psychotria racemosa	0.32	ns			

* $P < 0.05$, ** $P < 0.01$, *** $P < 0.001$; ns = not significant.

less pronounced in older sites. Because sample sizes from the young second-growth site were larger, a statistically significant χ^2 value was easier to obtain and annual variation in fruit use at that site may be overestimated relative to older lowland sites. Higher annual variation at that site also may have been due to the inclusion of 1987 data, but this is unlikely for two reasons. First, analysis of capture data revealed that 1985 and 1987 were more similar to each other at all sites than to capture data during 1986 (Blake et al. in press). Second, we have observed greater changes in the structure of vegetation through plant mortality and growth at the young site than in either old second-growth or forest sites.

A further illustration of annual variation at the young second-growth site is provided by comparing occurrence of two common fruits, *Conostegia subcrustulata* and *Psychotria pittieri*, in diets (Fig. 8). Abundance of ripe *Conostegia* fruits during the dry season declined steadily from 1985 to 1987, although the proportional representation of this fruit in diets of birds did not change among years ($\chi^2 = 3.7$, df = 2, $P > 0.10$). In contrast, use of *Psychotria pittieri*, an alternative, less preferred fruit (Table 4), increased from 1985 ($\chi^2 = 9.8$, df = 2, $P < 0.01$).

DISCUSSION

FRUGIVORE GUILDS

Previous studies on understory tropical bird communities often recognized a variety of insectivore guilds (e.g., foliage-gleaning, bark, and terrestrial), but only one frugivore guild (e.g., Terborgh and Robinson 1986). Our results show, however, that tropical understory frugivores partition fruit resources. The nonrandomness of the different frugivore guilds was revealed by the consistency of guild composition among sites.

Frugivore guilds, which represented birds that made similar foraging decisions in the field, were separated largely because of differences in morphology and foraging methods. For example, in young second-growth, the Red-throated Ant-Tanager, and Black-striped and Orange-billed sparrows foraged low in the undergrowth or on the ground and rarely ascended into taller shrubs or treelets. All three species have relatively longer tarsi than other emberizids (Loiselle and Blake, unpubl. data) and thus have greater difficulty reaching for fruit from a perch (Moermond and Denslow 1985). Consequently, they characteristically fed on axillary (easily accessible) fruits from low shrubs; terminal (less accessible) fruits were less preferred. In contrast, *Ramphocelus* tanagers and Buff-throated Saltators, with relatively shorter tarsi and, thus, greater perching and reaching ability (Moermond and Denslow 1985), foraged at all heights in young second-growth and fed on a wider range and diversity of fruits than the ant-tanager and sparrows.

Two guilds (C, E) represented birds that specialized on epiphytic (e.g., *Anthurium* sp.) or arillate fruits (e.g., *Clusia* sp.) and consequently, those guilds were defined largely by fruit type.

TABLE 5. Index of Fruit Use (D_{fr}) for Common Fruiting Plants by Three or Four Common Fruit-eating Birds in Each of Three Costa Rican Lowland Sites

Fruit species	Red-capped Manakin	White-collared Manakin	Scarlet-rumped Tanager	Orange-billed Sparrow	Dusky-faced Tanager	Olive Tanager	Wood Thrush
Young second-growth							
Conostegia subcrustulata	0.74	0.78	0.95	0.23			
Miconia affinis	0.44	0.62	0.36	0.00			
Clidemia dentata	−1.00	0.20	0.90	0.93			
Psychotria brachiata	−1.00	0.32	−0.49	−1.00			
Psychotria pittieri	−0.97	−0.78	−0.90	−0.93			
Passiflora auriculata	−1.00	0.00	0.88	−1.00			
Piper sancti-felicis	0.11	−0.42	0.92	−1.00			
Old second-growth							
Clidemia dentata	−1.00	0.41		0.90	0.99		
Miconia affinis	−0.43	−0.22		−1.00	−1.00		
Siparuna sp.	0.23	0.53		−1.00	−1.00		
Witheringia asterotricha	−0.70	0.06		0.62	0.27		
Leandra sp.	0.42	0.23		0.56	0.56		
Psychotria grandis	−0.80	−0.89		−1.00	−1.00		
Psychotria racemosa	0.62	−1.00		−1.00	−1.00		
Lowland forest							
Clidemia densiflora	−0.37					0.46	0.28
Henrietella tuberculosa	0.27					0.72	0.33
Ossaea macrophylla	−0.48					−0.18	−0.68
Miconia simplex	0.61					0.71	0.84
Psychotria officinalis	−0.59					−1.00	−0.48
Miconia centrodesma	−0.05					−1.00	0.10
Miconia "attentuate"	0.50					−1.00	0.50

Most other fruit-eating birds in our analyses often took fruit on the wing by hovering or snatching (see Moermond and Denslow 1985), then swallowed the fruits whole. But several subgroups were identified based largely on foraging height. Feeding decisions, and thus guild composition, were constrained by morphology and influenced by feeding method, fruit type and display, and foraging height.

Does competition explain resource partitioning among frugivores in Costa Rica? This would require that fruit resources be limiting. Fruits may be in short supply in undisturbed forest understory (Foster 1982b), as several lines of evidence suggest. First, few fruits were observed to rot on forest understory plants, suggesting that ripe fruits were taken relatively rapidly. Second, birds have been observed feeding on unripe fruits when fruits were scarce (pers. obs.; also Foster 1977). Third, abundance of frugivores was correlated with abundance of ripe fruits (Blake and Loiselle, unpubl. data; Loiselle 1987; Levey 1988). Fourth, interspecific and intraspecific aggression at and defense of fruit resources has been observed (e.g., Lederer 1977, Martin 1982, Willson 1986).

Ripe fruit was often four-fold more abundant in young second-growth during our study than in forest understory (Loiselle 1987) and may not have been limiting during our study. We often observed fruits rotting on plants and found no correlation between frugivore abundance and ripe fruits. In young second-growth, it appears that fruits may compete for dispersers, rather than the reverse.

Alternatively, partitioning of fruit resources among frugivores may not reflect competition for fruits, but rather may reflect adaptations to exploit other resources, such as insects. Snow and Snow (1971) argued that tanagers and honeycreepers in Trinidad, which overlapped broadly in fruits consumed, coexisted because of their partitioning of insect resources (also Lack 1976a for Jamaican frugivores; but see Moermond and Denslow 1985). Predation also may influence fruit choice and foraging patterns of birds (Howe 1979, Martin 1985b, Snow and Snow 1986). We are not able to evaluate adequately the possible role of competition in structuring frugivore guilds in Costa Rica, but we agree with Fleming (1979) and Willson (1986) that it likely operates in resource partitioning. Particular attention in future

TABLE 6. Significance Values for χ^2 Analyses Testing the Occurrence of Fruits in Diets of Birds Relative to Occurrence of All Other Fruit Species at Three Costa Rican Lowland Sites. "Canopy" Included Fruits of All Known Canopy and Subcanopy Species Recorded in Birds' Diets. Other "Understory" Included All Understory Species in Birds Diets Except Those Tested Separately

Fruit species	Seasonal variation in fruit use			Annual variation in fruit use		
	χ^2	df	P[a]	χ^2	df	P
Young second-growth						
Conostegia subcrustulata	35.5	4	***	3.7	2	0.15
Miconia affinis	379.2	3	***	19.4	2	**
Clidemia dentata	53.5	4	***	12.6	2	**
Psychotria brachiata	6.4	4	0.17	3.2	2	0.21
Psychotria pittieri	29.9	4	***	9.8	2	**
Passiflora auriculata				11.3	2	**
Piper sancti-felicis	9.5	3	*	3.3	2	0.20
Old second-growth						
Clidemia dentata	10.8	3	*	1.0	1	0.75
Miconia affinis	26.3	3	***	2.8	1	0.09
Siparuna sp.	10.1	3	*	2.2	1	0.14
Witheringia asterotricha	3.6	3	0.31	3.0	1	0.08
Leandra sp.				4.5	1	*
"Canopy"	3.9	3	0.27	1.8	1	0.18
Other "Understory"	10.6	3	*	0.8	1	0.38
Lowland forest						
Clidemia densiflora				4.8	1	*
Henrietella tuberculosa	24.4	3	***	0.2	1	0.68
Ossaea macrophylla	6.1	1	*			
Miconia simplex	9.4	2	**	7.0	1	**
Psychotria officinalis	13.2	3	**	1.7	1	0.19
Miconia centrodesma	7.6	1	**			
Miconia "attentuate"	1.4	1	0.23			
"Canopy"	3.4	3	0.33	1.6	1	0.20
Other "Understory"	10.8	3	*	0.1	1	0.85

[a] * P < 0.05, ** P < 0.01, *** P < 0.001.

studies should be given to evaluating alternative hypotheses such as predation, mutualism, and abiotic interactions (Wiens 1977, Brown and Bowers 1984, Martin 1988c).

In contrast to tropical systems, many fewer frugivore guilds, usually two, have been described in temperate forests. In Illinois, frugivore guilds were determined largely by foraging height, but because of annual variation and inconsistency in fruit preference by birds, no single factor explained foraging preference by birds (Katusic-Malmborg and Willson 1988). Sorenson (1981) also was unable to determine reasons for differences in fruit choice among British tits and thrushes.

Seasonal and Annual Variation in Fruit Use

Even in the relatively aseasonal climate of Atlantic slope Costa Rica, fruit abundance varied seasonally (Frankie et al., 1974, Loiselle 1987) among lowland sites. In well lighted areas, more plants produced fruit continuously, whereas in shaded areas, production was highly seasonal (Fig. 4). Consequently, seasonal variation in fruit use by birds was influenced by different factors. In young second-growth, where fruits were more abundant and more species fruited year round, birds were more selective (see Schoener 1971b, Krebs et al. 1977). In contrast, although feeding preferences were observed in lowland forest, seasonal changes in fruit phenology there largely accounted for seasonal variation in fruit use.

The nature of seasonal variation in diet also varied among bird species. Most resident frugivores ate fruit year round and changes in feeding preference or fruit availability accounted for seasonal variation. Some winter residents, such as the Wood Thrush, ate fruit in substantial quantities only during late wet and late dry seasons, times when they were accumulating fat reserves for migration. Wheelwright (1988) demonstrated that even when fruit availability was held constant year round, American Robins showed seasonal variation in fruit use, indicating that physiological needs, and not fruit availability, influenced that seasonal variation.

In spite of large annual variation in fruit abundance and availability, birds of old second-growth and forest showed little annual variation in fruit use. Fruit abundance changed, but phenological patterns (what fruits were available) did not. In contrast, birds of young second-growth showed considerable annual variation, which we attribute to successional changes in vegetation at that site. Our data span only two or three years and interpretation of annual patterns is tentative at best.

Fecal Samples as a Tool for Analysis of Diets

Collection of fecal samples or regurgitated seeds to analyse diets is not new, but only Wheelwright et al. (1984) used it to describe an assemblage of fruit-eating birds. They used a variety of techniques (fecal samples, behavioral observations of birds at fruiting trees, seed traps) and concluded that fecal samples and seed traps placed under display or nest perches generally were the most effective means of obtaining representative diet samples. Clearly, a combination of observational and fecal collection techniques is needed to describe diets in detail, but the difficulty of observing birds in the dark understory of tropical forests often may necessitate use of fecal samples there. Moreover, this method is quick, is not biased by observations at conspicuous plants bearing large fruit crops, and, we suspect, is more likely to include most fruits eaten by birds.

ACKNOWLEDGMENTS

We thank T. C. Moermond for his encouragement and support throughout this study. This paper greatly benefitted from comments by or discussions with E. Beals, R. Holmes, J. Jehl, J. Kitchell, D. Levey, T. Martin, T. Moermond, M. Morrison, K. Smith, J. Verner, and D. Waller. D. Brenes, D. A. and D. B. Clark, J. Denslow, A. Gomez, D. Levey, R. Marquis, and O. Vargas know the fruits of La Selva and helped us often. We are grateful to J. A. Leon for his help in arranging permits, and the Servicio de Parques Nacionales especially F. Cortes S. and J. Dobles Z., for permission to work in Parque Nacional Braulio Carrillo. The following agencies generously provided support for this study: Jessie Smith Noyes Foundation; National Geographic Society; Dept. of Zoology, University of Wisconsin (Guyer Post-doctoral Fellowship [JGB] and Davis Fellowship [BAL]); National Academy of Sciences; Wilson Ornithological Society; and Northeastern Bird Banding Association.

APPENDIX I. ENGLISH AND SCIENTIFIC NAMES OF ALL BIRD SPECIES MENTIONED IN THE TEXT

Wedge-billed Woodcreeper (*Glyphorynchus spirurus*), Olive-striped Flycatcher (*Mionectes olivaceus*), Ochre-bellied Flycatcher (*Mionectes oleagineus*), Grey-capped Flycatcher (*Myiozetetes granadensis*), White-collared Manakin (*Manacus candei*), White-ruffed Manakin (*Corapipo leucorrhoa*), White-crowned Manakin (*Pipra pipra*), Red-capped Manakin (*Pipra mentalis*), Black-faced Solitaire (*Myadestes melanops*), Slaty-backed Nightingale-Thrush (*Catharus fuscater*), Grey-cheeked Thrush (*Catharus minimus*), Swainson's Thrush (*Catharus ustulatus*), Wood Thrush (*Hylocichla mustelina*), Pale-vented Robin (*Turdus obsoletus*), Clay-colored Robin (*Turdus grayi*), American Robin (*Turdus migratorius*), Grey Catbird (*Dumetella carolinensis*), Tawny-capped Euphonia (*Euphonia anneae*), Olive Tanager (*Chlorothraupis carmioli*), Tawny-crested Tanager (*Tachyphonus delatrii*), Red-throated Ant-Tanager (*Habia fuscicauda*), Crimson-collared Tanager (*Ramphocelus sanguinolenta*), Scarlet-rumped Tanager (*Ramphocelus passerinii*), Common Bush-Tanager (*Chlorospingus ophthalmicus*), Dusky-faced Tanager (*Mitrospingus cassinii*), Buff-throated Saltator (*Saltator maximus*), Orange-billed Sparrow (*Arremon aurantiirostris*), Black-striped Sparrow (*Arremonops conirostris*)

APPENDIX II. DESCRIPTION OF FRUIT VARIABLES USED IN ORDINATION OF COSTA RICAN FRUIT-EATING BIRD ASSEMBLAGES (SEE TABLES 2, 3). SEE TEXT FOR DESCRIPTION ON HOW FRUIT VARIABLES WERE DEFINED. SITES AT WHICH FRUIT VARIABLES WERE USED ARE IDENTIFIED. Y = YOUNG SECOND-GROWTH, O = OLD SECOND-GROWTH, L = LOWLAND FOREST AT LA SELVA, M = FOREST AT 500-M, H = FOREST AT 1000-M

Variable	Site	Description
AGGREG	Y	Includes aggregate fruits, e.g., *Cecropia*, *Piper*
ANTHUR	O, H	Fruits of *Anthurium*
ANTBRO	M	Fruits of aroids and bromeliads
ARILAT	Y	Arillate fruits with thin layer of pulp surrounding entire seed, seed usually large, e.g., *Alchornea, Doliocarpus, Dieffenbachia*
ATTARIL	Y	Aril attached to one end of seed only, e.g., *Siparuna, Calathea, Renealmia cernua*
ARIL	O, L, M, H	Includes both ARILAT and ATTARIL
AXIL	Y, L, M, H	Includes juicy berries presented in axils or along stems, e.g., *Clidemia, Besleria, Witheringia, Sabicea,* many seeded
BROMEL	H	Fruits of bromeliads
CLBASP	Y	Fruits of the Compositae: *Clibadium asperum* (Aubl.) DC.
CLIBES	O	A subset of AXIL group, includes juicy fruits of *Clidemia, Besleria,* and *Sabicea*
EPISHB	M, H	Fruits of epiphytic shrubs, e.g., *Cavendishia, Blakea, Topobea*
FICUS	Y, O	Fruits of *Ficus*
HELIC	Y	Fruits of *Heliconia* species
HENOSS	O, L, M	Fruits of some Melastomataceae, e.g., *Henrietella* and *Ossaea*
LGSDUN	all	Large seeded understory fruit from shrubs or small trees, e.g., *Ardisia, Neea, Cestrum*
MICCAN	O, L, M	Canopy and subcanopy trees of *Miconia*
OSSAEA	H	Fruits of *Ossaea* species
PASSIF	Y	Fruits of *Passiflora* species
PHYRIV	Y, M	Fruits of *Phytolacca rivinoides* Kunth & Bouche
STYROF	Y	Includes a selected group of *Psychotria* fruits with a styrofoam rather than juicy texture
TERMUN	O, L, M	Juicy berries of understory shrubs presented on terminal infructescences
TREE1	Y, L	Single or few-seeded subcanopy or canopy trees, e.g., Lauraceae, *Hampea*
TREES	Y, L	Many-seeded subcanopy or canopy trees, e.g., *Dendropanax, Hieronyma, Vismia*
TREE	O, M, H	Incudes TREE1 and TREES
UNK112	H	An unidentified species in the diets of birds at 1000 m, relatively common in some birds
VINE	Y, O, H	Large-seeded vines, e.g., *Cissus, Cissampelos*
WITAST	O, L	A subset of the AXIL group, includes juicy fruits of Solanaceae

DIETARY SIMILARITY AMONG INSECTIVOROUS BIRDS: INFLUENCE OF TAXONOMIC VERSUS ECOLOGICAL CATEGORIZATION OF PREY

ROBERT J. COOPER, PETER J. MARTINAT, AND ROBERT C. WHITMORE

Abstract. In a study of dietary relationships among nine species of insectivorous birds from an eastern deciduous forest, we examined two approaches to prey categorization: (1) taxonomic, using arthropod orders, and (2) subdivisions of orders into ecologically relevant categories. Dietary similarities (correlations) were generally higher within bird species than within period of collection using both categorizations. Similarities using taxonomic categorization generally were higher but were significantly (P < 0.01) less than those using ecological categorization. Using similarity measures and cluster analysis, similarities within bird species and time period that were evident using ecological categorization were not evident using taxonomic categorization. While we cannot specify strict rules concerning appropriate method and level of taxonomic categorization in studies of this sort, we suggest that: (1) prey categories should have sufficient observations to make analysis meaningful and to avoid large numbers of zero counts; (2) prey categories should not be so numerous that procedures such as cluster analysis cannot be readily interpreted; (3) large taxonomic levels (i.e., order) should be subdivided ecologically if subgroups exhibit very different characteristics (e.g., size, location, abundance, behavior); and (4) we encourage input from entomologists in problems of prey categorization.

Key Words: Arthropod prey; diet analysis; dietary similarity; diets; insectivorous birds.

A frequent objective of avian dietary studies is to compare diets among species that feed in similar ways. While some attention has been paid to biases involved in diet analysis, little is known about how the method of prey categorization affects similarity measures. Greene and Jaksic (1983) examined effects of prey identification level in analyses of raptor diets. We know of no similar studies for insectivorous birds, which eat a wide variety of arthropods encompassing many orders and families. Researchers may or may not be able to identify arthropods to the species level, especially if diets are analyzed using highly fragmented stomach contents.

Due to the difficulty of identification of insect parts to species and sometimes family levels, many researchers have compared diets of insectivorous species by categorizing prey at higher taxonomic levels. Arthropod orders are used most often (e.g., Root 1967, Orians and Horn 1969, Robinson and Holmes 1982). Others have used arthropod families or have combined families in some manner (Rotenberry 1980a, Rosenberg et al. 1982, Sherry 1984). Because some studies have involved a limited number of prey types, a few researchers have been able to identify all prey (e.g., caterpillars) to the species level (Tinbergen 1960, Royama 1970). Yet the method by which insect prey are categorized is likely to affect both similarity measures and conclusions drawn from them. Here we address that problem, using dietary data from stomachs of nine foliage-gleaning bird species in an eastern deciduous forest in West Virginia.

METHODS

Cooper (1988) described details of the study area and methods. The study area included 400 ha in Sleepy Creek Public Hunting and Fishing Area, an oak-hickory forest located in Berkeley and Morgan counties, West Virginia. A major feature of this study area is the spring emergence of many larval Lepidoptera, which feed on new foliage of deciduous trees. These caterpillars are a preferred food source eaten by many resident and migrant birds. We collected birds with shotguns from 6 May to 31 July 1985, and from 13 May to 22 July 1986 between 06:00 and 13:00, immediately removing the proventriculus and gizzard and injecting them with formalin to stop digestion. Stomach contents were analyzed in the laboratory under a dissecting microscope. Most prey items could be identified to family.

Several points merit emphasis here. First, intensive sampling of location, abundance, and behavior of canopy arthropods was done by Cooper (1989) simultaneous to collecting. Second, an extensive collection of arthropod voucher specimens was prepared. Third, at least one entomologist was available at all times in the field and laboratory to provide expertise in arthropod identification.

Our unit of measurement was a species-month, pooling all diet samples for a given species in a month (Table 1). Using the Brillouin diversity index (Pielou 1975; also see Sherry 1984), we determined that collection of 3 or 4 individuals/month was adequate to represent the monthly diet of a species. Collections with fewer than four individuals were eliminated from the analysis.

Relative abundances of prey were expressed as percent of total number of dietary items identified. We measured dietary similarity among monthly collections using Spearman's rank-order correlation, which is commonly used as a similarity measure (Clifford and Stephenson 1975). Overall trends of similarity were examined using cluster analysis. Ward's method, which is similar to centroid linkage, was employed using CLUSTER in the Statistical Analysis System (SAS Institute 1985). These analyses used two categorization methods. First, taxonomic categorization used orders as categories with the exception that Lepidoptera were divided into larvae, pupae, and adults (10 total categories). Second, ecological categorization used 15 prey categories based on taxonomy, size, abundance, typical location, and escape behavior of each group (Table 2).

For example, larval Lepidoptera were divided into three categories based on size, substrates occupied, and predator avoidance mechanisms. Smooth-bodied caterpillars typically avoid predation through crypsis, nocturnal feeding, and remaining inactive during the day on the undersurfaces of leaves (see Heinrich 1979c, Heinrich and Collins 1983). They were divided into two groups based on size. A third group, "hairy caterpillars," have long, stiff setae that deter many predators; they commonly forage diurnally in exposed locations. Coleoptera were similarly divided into two categories. One group (primarily Cerambycidae and Elateridae) included individuals that were large (8–16 mm), diurnally active, and found on leaf topsides or bark. The other group (primarily Alleculidae, Chrysomelidae, and Curculionidae) included individuals that were small (5–8 mm), diurnally inactive, and found on leaf undersurfaces.

In this example, we used cluster analysis to examine dietary patterns within and between species and time. If foliage-gleaning species were highly opportunistic, eating the most abundant prey available at any given time, then meaningful clusters should include many species collected at the same time. Conversely, if each species consistently ate unique prey items, meaningful clusters should contain one or a few species regardless of when they were collected.

RESULTS

On average, similarities among collections using taxonomic categorization were greater but were less often significant (P < 0.01, Table 3) than those using ecological categorization (Table 4). Both similarities and significance levels were affected by number of prey categories. Within-species comparisons were correlated more often than within-time-period comparisons using both categorization methods. Several discrepancies between our intuition and results observed using taxonomic categorization were noted. For example, 43% of the May 1986 collections were correlated when prey were categorized ecologically, because many species ate small (<20 mm), smooth-bodied, recently-emerged larval Lepidoptera. However, only 14% of those collections were correlated when prey were categorized taxonomically. Also, no within-species comparisons were significant for Worm-eating Warbler (scientific names appear in Table 1) or Yellow-billed Cuckoo when prey were categorized tax-

TABLE 1. SUMMARIES OF MONTHLY COLLECTIONS MADE OF NINE INSECTIVOROUS BIRD SPECIES DURING 1985–1986

Species	Collection	No. stomachs	No. items identified
Yellow-billed Cuckoo	June 1985	5	93
(*Coccyzus americanus*)	July 1985	7	137
	May 1986	5	217
	June 1986	4	94
Black-capped Chickadee	June 1985	13	282
(*Parus atricapillus*)	July 1985	25	301
	June 1986	4	29
	July 1986	6	89
Tufted Titmouse	June 1985	8	78
(*Parus bicolor*)	July 1985	37	382
	May 1986	5	65
	June 1986	7	47
	July 1986	5	46
Blue-gray Gnatcatcher	June 1985	10	105
(*Polioptila caerulea*)	July 1985	21	257
	May 1986	6	85
	June 1986	8	117
	July 1986	6	59
Red-eyed Vireo	May 1985	9	85
(*Vireo olivaceus*)	June 1985	12	106
	July 1985	17	143
	May 1986	6	81
	June 1986	8	93
	July 1986	5	85
Pine Warbler	June 1985	8	101
(*Dendroica pinus*)	July 1985	18	282
	May 1986	5	44
	June 1986	7	72
Cerulean Warbler	May 1986	4	40
(*Dendroica cerulea*)			
Worm-eating Warbler	June 1985	8	94
(*Helmitheros vermivora*)	May 1986	4	38
	June 1986	5	48
Scarlet Tanager	May 1985	4	29
(*Piranga olivacea*)	June 1985	17	173
	July 1985	22	182
	May 1986	7	91
	June 1986	9	97
	July 1986	6	70

onomically. Cuckoos actually had unique diets, because only they consumed large numbers of gypsy moth (*Lymantria dispar*) larvae. Because gypsy moth larvae were combined with other caterpillars, this trend was hidden.

Cluster analysis using insect orders as prey categories resulted in a dendrogram showing few clear patterns within species or time (Fig. 1). Four major clusters were identified (scree test, Dillon and Goldstein 1984:48–49), each of which contained at least one collection from May, June, and July. Cluster I reflected large percentages of Homoptera in the diet and included collections

TABLE 2. Size, Substrates Used, and Predator Avoidance Mechanisms of Arthropod Categories Used in Ecological Categorization in This Study

Prey category	Taxon	Length (mm)	Substrate	Predator avoidance mechanism
Spiders	Arachnida	2–10	Various	Dropping on thread, crawling
Large, active beetles	Coleoptera			
	Cerambycidae	8–16	Leaf tops, bark	Flying, falling
	Elateridae	8–12	Leaf tops, bark	Falling
Small, inactive beetles	Alleculidae	5–8	Leaf undersides	Falling
	Chrysomelidae	5–8	Leaf undersides	Falling
	Curculionidae	5–8	Leaf undersides	Falling
Large, predatory Hemiptera	Hemiptera			
	Pentatomidae	8–18	Leaf tops	Flying
	Reduviidae	8–18	Leaf tops	Falling
Small, phytophagous Hemiptera	Miridae	5–8	Leaf undersides	Falling
Homoptera	Homoptera			
	Membracidae	5–10	Twigs, branches	Crypsis
	Other	3–10	Foliage	Jumping, flying
Adult Hymenoptera	Hymenoptera			
	Formicidae	3–10	Various	Crawling, flying
	"Wasps"	3–12	Air, foliage	Flying
Orthoptera	Orthoptera			
	Tettigoniidae	>10	Foliage	Crypsis
	Gryllidae	6–18	Foliage	Crypsis
Large "flies"	Mecoptera	10–20	Air, leaf tops	Flying
	Diptera			
	Asilidae	10–20	Air, leaf tops	Flying
	Tipulidae	10–30	Air, foliage	Flying
Small flies	Other	<10	Air, foliage	Flying
Small, smooth-bodied eruciform larvae	Lepidoptera Hymenoptera	8–20	Leaf undersides, rolls or ties	Crypsis
Large, smooth-bodied eruciform larvae	Lepidoptera Hymenoptera	>20	Leaf undersides, rolls or ties, bark	Crypsis
"Hairy" caterpillars	Lepidoptera	>8	Foliage, bark	Unpalatability
Pupae	Lepidoptera	5–20	Foliage, bark	Crypsis
Moths	Lepidoptera adults	3–20	Air, leaf undersides	Flying, crypsis

from four species and all three months of study. All five Blue-gray Gnatcatcher, four of six Red-eyed Vireo, and two of four Pine Warbler collections were included in this cluster. Cluster II reflected diets with a large percentage of Coleoptera. One Worm-eating Warbler and five of six Scarlet Tanager collections were in this cluster. Cluster III reflected a large percentage of larval Lepidoptera in diets. Seven species were represented in this cluster. Cluster IV reflected a moderate percentage (10–20%) of "unusual" prey such as spiders or Orthoptera, and included one representative each of five species.

The dendrogram suggested some dietary similarities within species, especially Blue-gray Gnatcatcher and Scarlet Tanager, but few time patterns, although we strongly suspected their occurrence. For example, the large Cluster III in Figure 1 contained collections with large percentages of Lepidoptera larvae. These included (1) Yellow-billed Cuckoos, which ate many gypsy moth larvae, (2) Tufted Titmice and Black-capped Chickadees, both of which ate numerous longer (>20 mm), smooth-bodied caterpillars in June and July of both years, and (3) a variety of other species that ate smaller, smooth-bodied caterpillars when they were abundant in May and June. These and other patterns might emerge if a more meaningful method of categorization was used.

The cluster analysis that used ecological categorization (Table 2) resulted in a more informative dendrogram (Fig. 2). Five major clusters were identified. Cluster I again reflected a large percentage of Homoptera in diets, including four of five Blue-gray Gnatcatcher collections and both 1985 Pine Warbler collections from June and July. Cluster II contained seven of nine parid

TABLE 3. NUMBER OF COMPARISONS, NUMBER OF SIGNIFICANT ($P < 0.01$) COMPARISONS, AND AVERAGE SIMILARITY IN DIET BETWEEN MONTHLY COLLECTIONS USING TAXONOMIC PREY CATEGORIZATION (ARTHROPOD ORDERS)

Collections compared	Similarity[a]	Number comparisons	Number significant	Percent significant
Within species				
Tufted Titmouse	0.77 ± 0.03	10	4	40
Black-capped Chickadee	0.74 ± 0.04	6	3	50
Worm-eating Warbler	0.67 ± 0.07	3	0	0
Blue-gray Gnatcatcher	0.65 ± 0.05	10	3	30
Pine Warbler	0.64 ± 0.08	6	1	17
Red-eyed Vireo	0.62 ± 0.04	15	3	20
Scarlet Tanager	0.60 ± 0.06	15	4	27
Yellow-billed Cuckoo	0.52 ± 0.07	6	0	0
Total	0.65 ± 0.02	71	18	25
Within collection periods				
May 1985	0.76 ± 0.00	1	0	0
June 1985	0.45 ± 0.04	28	1	4
July 1985	0.39 ± 0.06	21	2	10
May 1986	0.51 ± 0.05	28	4	14
June 1986	0.41 ± 0.05	28	1	4
July 1986	0.50 ± 0.07	10	0	0
Total	0.45 ± 0.02	116	8	7
All other comparisons	0.44 ± 0.01	516	46	9

[a] Values are mean Spearman's rank correlation coefficients among the collections of interest and reported as the mean $\pm$ SE.

collections, all from June and July. Those two species had similar diets, including medium-sized caterpillars (20–30 mm), pupae, and spiders. Cluster III reflected large numbers of smaller, smooth-bodied caterpillars taken during the early part of the breeding season. Seven species were represented in this cluster and all collections except one (YBC 7/85) were from May and June. The Yellow-billed Cuckoo collection was, incidentally, the last single collection to join a cluster and probably reflected a different diet from that in any other collection. Clusters IV and V were both monospecific and reflected the high dependency of Scarlet Tanagers on large beetles at all

TABLE 4. NUMBER OF COMPARISONS, NUMBER OF SIGNIFICANT ($P < 0.01$) COMPARISONS, AND AVERAGE SIMILARITY IN DIET BETWEEN MONTHLY COLLECTIONS USING ECOLOGICAL PREY CATEGORIZATION

Collections compared	Similarity[a]	Number comparisons	Number significant	Percent significant
Within species				
Pine Warbler	0.81 ± 0.05	6	4	67
Black-capped Chickadee	0.74 ± 0.04	6	5	83
Tufted Titmouse	0.70 ± 0.04	10	7	70
Blue-gray Gnatcatcher	0.68 ± 0.03	10	7	70
Worm-eating Warbler	0.62 ± 0.13	3	2	67
Scarlet Tanager	0.58 ± 0.05	15	9	60
Red-eyed Vireo	0.54 ± 0.05	15	5	33
Yellow-billed Cuckoo	0.52 ± 0.09	6	3	50
Total	0.63 ± 0.02	71	42	59
Within collection periods				
May 1985	0.57 ± 0.00	1	0	0
June 1985	0.31 ± 0.06	28	3	11
July 1985	0.36 ± 0.07	21	3	14
May 1986	0.44 ± 0.06	28	12	43
June 1986	0.32 ± 0.06	28	5	18
July 1986	0.55 ± 0.06	10	3	30
Total	0.37 ± 0.03	116	26	22
All other comparisons	0.36 ± 0.01	516	82	16

[a] Values are mean Spearman's rank correlation coefficients among the collections of interest and reported as the mean $\pm$ SE.

FORAGING AND NECTAR USE IN NECTARIVOROUS BIRD COMMUNITIES

BRIAN G. COLLINS, JAMES GREY, AND SHAPELLE MCNEE

Abstract. Nectar-feeding birds, such as honeyeaters, sunbirds, and sugarbirds, usually occupy habitats in which distributions of particular plant species, individual plants, and flowers are patchy. The contribution that each plant species makes to the overall nectar pool is dependent upon plant density, floral abundance, and amount of nectar produced per flower. Nectar availability can be variable: some flowers contain considerable quantities of nectar, youngest flowers usually being most productive, while others are empty. In Australian and southern African habitats, we found interspecific partitioning of nectar resources. The largest species of nectarivore at a given site generally foraged selectively at the most rewarding nectar sources, relying on the most productive plant species and the youngest flowers available. Dominance hierarchies within nectarivore communities helped to sustain partitioning, although incompatibilities between bill and floral morphologies sometimes prevented particular species from utilizing part of the nectar pool. Preliminary observations suggested that intraspecific differences in use of nectar also occurred.

Key Words: Nectarivorous birds; honeyeaters; sunbirds; sugarbirds; foraging; nectar; resource partitioning; community ecology; Australia; Africa.

Nectarivorous birds are abundant in many parts of the world. The most prominent of these are honeyeaters (Meliphagidae) of Australasia, sunbirds (Nectariniidae) and sugarbirds (Promeropidae) of Africa, and hummingbirds (Trochilidae) of northern and neotropical America (Johnsgard 1983, Maclean 1985, Collins and Rebelo 1987, Collins and Paton 1989). Evidence concerning the extent to which these birds use nectar is circumstantial in most cases, although the efficiency and extent of its uptake have been measured precisely for several species (e.g., Wolf et al. 1972, Gill and Wolf 1978, Ford 1979, Collins et al. 1984, Paton and Carpenter 1984). There is a similar dearth of quantitative data regarding the importance of arthropods, fruits, and other potentially useful foods in the diets of nectarivores (e.g., Skead 1967, Johnsgard 1983, Maclean 1985), although a few detailed investigations have confirmed that honeyeaters ingest a variety of materials (e.g., Pyke 1980, Collins and Briffa 1982, Paton 1986). It has been suggested that arthropods are used primarily to provide protein and minerals (e.g., Pyke 1980, Paton 1982). Nectar contains a variety of carbohydrates, as do fruits, and in most instances appears to be the major source of energy for nectarivorous birds (e.g., Hainsworth and Wolf 1976, Baker and Baker 1983, Collins and Paton 1989).

Nectarivorous bird communities in many parts of North America and Africa are simple, often comprising only one or two types of bird that forage for nectar from a small number of plant species at any given time (e.g., Carpenter 1983, Paton and Carpenter 1984). Community organization is considerably more complex in the neotropics, and in most Australian habitats, where numbers of competing nectarivores and potential nectar sources are much greater (e.g., Feinsinger 1976, Ford and Paton 1982, Kodric-Brown et al. 1984, Collins and Newland 1986). Several comprehensive studies have documented the diversity of plants and birds within such habitats, often providing considerable information relating to nectar production and partitioning of nectar between different species of nectarivore (e.g., Wolf et al. 1976; Feinsinger 1978, 1983; Feinsinger and Colwell 1978; Snow and Snow 1980; Ford and Paton 1982; Collins and Briffa 1982; Collins and Newland 1986). Nevertheless, little attention has been paid to intraspecific variations in use of available nectar. Even when such differences have been discussed, small sample sizes have usually been involved, and comparisons limited to territorial male and female birds (e.g., Gill and Wolf 1975b; Wolf 1975; Carpenter 1976; Gass 1978, 1979; Wolf et al. 1976). Almost no data have been supplied for individuals within the same species which differ in age or position within dominance hierarchies (e.g., Gass 1979, Craig 1985, Newland and Wooller 1985).

Most studies of foraging activity by nectarivores other than territorial hummingbirds have produced composite data derived from many observations of (often unmarked) birds, each made over a relatively short period of time (e.g., Collins and Briffa 1983, Collins and Newland 1986). Thus, there has been a tendency for results to be biased in favor of obvious activities, such as insect hawking and foraging at exposed flowers, and birds that are particularly mobile. The purpose of this paper is to demonstrate that collection of data in this manner can conceal inter-

and intraspecific differences in foraging behavior, which are revealed by extended observation, and the use of indirect evidence such as that provided by analysis of facial and fecal smears, for individual birds.

METHODS

STUDY AREAS

Investigations reported here were undertaken at three different sites in southwestern Australia and southern Africa. The African site was located at Betty's Bay, a narrow belt of coastal seepage fynbos (heath), approximately 90 km southeast of Cape Town (B. G. Collins 1983a, b). The two Australian sites occurred within the southwest botanical province of Western Australia. One of these was located in sclerophyllous jarrah forest, 9 km south of Jarrahdale (Collins 1985, Collins and Newland 1986), the other in proteaceous heathland at Fitzgerald River, approximately 25 km northeast of Bremer Bay (Collins et al., unpubl. ms).

The most abundant nectarivorous birds at Betty's Bay were Cape Sugarbirds (*Promerops cafer*) and Orange-breasted Sunbirds (*Nectarinia violacea*). Little Wattlebirds (*Anthochaera chrysoptera*), New Holland Honeyeaters (*Phylidonyris novaehollandiae*), and Western Spinebills (*Acanthorhynchus superciliosus*) were most frequently seen at Jarrahdale. With the exception of Little Wattlebirds, these honeyeaters were also common at Fitzgerald River, where White-cheeked Honeyeaters (*Phylidonyris nigra*), Brown Honeyeaters (*Lichmera indistincta*), and White-naped Honeyeaters (*Melithreptus lunatus*) also were observed.

EXPERIMENTAL DESIGN

Data were gathered during the course of three independent projects. The most recent of these, at Jarrahdale and Fitzgerald River (1985–1987), involved some measurements that were not performed at Betty's Bay, where experimental work was conducted in 1982. In a few instances, the techniques used to obtain comparable information also were slightly different. Notwithstanding these variations, however, two of the major objectives of each study were to document inter- and intraspecific differences regarding the partitioning of available nectar by birds, and to identify possible reasons for the differences.

PLANT DENSITY AND FLORAL ABUNDANCE

At each Australian study site, but not at Betty's Bay, plant densities and floral abundances were measured for species that had been identified previously as major nectar producers (Collins and Newland 1986; Collins et al., ms). Plants that had clearly defined and separate flowers (e.g., *Grevillea wilsonii* at Jarrahdale) had their flowers counted and treated independently. On plants with inflorescences comprising numerous flowers that were tightly packed together (e.g., *Dryandra sessilis* at Jarrahdale or *Banksia nutans* at Fitzgerald River), inflorescences were considered to be the floral units. For convenience, all such units will generally be referred to throughout the remainder of this paper as flowers.

Overall densities of major nectar-producing plant species were estimated, using a plotless, point-centered quarter method with at least 100 points located on a rectangular grid at 10 m centers (Mueller-Dombois and Ellenberg 1974). Only those plants that were judged likely to flower at some time during the year in which investigations occurred were included (for details, see Collins and Newland 1986). Numbers of flowers present were counted on at least 20 randomly chosen plants for each species, during July and/or September. Relative abundances of flowers belonging to different age-classes were scored for selected species (*D. sessilis, G. wilsonii,* and *Mimetes hirtus*) at Jarrahdale and Betty's Bay (for methodology, see B. G. Collins 1983b, 1985; Grey 1985).

NECTAR AVAILABILITY AND PRODUCTION

Fresh flowers were chosen at random for major nectar-producing species on each of 2–3 successive days in July and/or September. These were sprayed with insecticide and "bagged" (i.e., protected from all nectarivores) with perforated fibreglass mesh at dusk. Insect adhesive was wiped around stems supporting the bags and flowers in order to prevent arthropods from reaching flowers via stems (Collins and Newland 1986). Twenty four hours later, the volumes and equivalent sucrose concentrations of nectar in at least 10 bagged flowers were recorded for each species, using techniques described by Collins et al. (1984) and Collins and Newland (1986). Similar measurements were made for separate sets of 10 unbagged flowers at dawn and dusk over the same period of time. The energy equivalent of each nectar sample was estimated as outlined by Collins and Briffa (1983), assuming that 1 mg sucrose yields 16.74 J. Daily (24 hour) nectar productions were calculated by subtracting mean dusk energy values for unbagged flowers from subsequent dusk values for bagged flowers.

In separate experiments at Jarrahdale, approximately 140 flowers, on which anthesis could be induced by a gentle touch to the style(s), were selected at random for each of the two major nectar-producing plant species (*D. sessilis* and *G. wilsonii*). Nectar was collected from subsamples of at least 10 flowers at dusk on day zero, and at dawn and dusk each subsequent day until nectar production ceased. Energy equivalents of samples were calculated as indicated above. Similar experiments were conducted at Betty's Bay, except that standing crops of nectar were measured for unbagged *M. hirtus* inflorescences classified as partly open (some flowers open), and fully-open (all flowers open), rather than for inflorescences whose ages were known more precisely (B. G. Collins 1983a, b). Corresponding data were not obtained for plants at Fitzgerald River.

BIRD MORPHOMETRIC, TIME BUDGET AND ENERGY BUDGET DATA

Honeyeaters present at the Jarrahdale site were captured in mist nets during each of four successive days in July and September. Each bird was weighed using a top-loading electronic balance, color-banded, its bill (exposed culmen) length measured with micrometer calipers, then released. Nectarivores at the other sites were treated in similar fashion, except that honeyeaters at Fitzgerald River were not color-banded.

Time budget data were obtained throughout the day for nectarivores at Betty's Bay and Jarrahdale, using cumulative digital stopwatches. In each instance,

TABLE 1. DENSITIES AND FLORAL AVAILABILITY FOR SOME ORNITHOPHILOUS PLANT SPECIES AT STUDY SITES IN SOUTHWESTERN AUSTRALIA (PARTLY AFTER COLLINS 1985, COLLINS ET AL., UNPUBL. MS)

Location and plant species	Plant density (plants/ha)	Floral abundance (flowers/plant)				Floral density (flowers/ha)
		$\bar{X}$	SD	Range	N	
Jarrahdale (July)						
Adenanthos barbigera	553	2.2	1.1	0–10	30	1217
Calothamnus rupestris	12	0.0	0.0	0–0	30	0
Dryandra sessilis	243	6.9	4.2	0–27	30	1677
Grevillea wilsonii	314	2.2	0.9	0–11	30	691
Jarrahdale (September)						
Adenanthos barbigera	553	4.2	1.8	0–17	30	2323
Calothamnus rupestris	12	115	201	0–815	30	1387
Dryandra sessilis	243	1.9	0.4	0–9	30	1191
Grevillea wilsonii	314	1.8	0.5	0–7	30	565
Fitzgerald River (July)						
Banksia baueri	380	0.8	0.3	0–4	20	304
Banksia coccinea	180	0.7	0.2	0–3	20	126
Dryandra cuneata	230	4.1	1.9	0–12	20	943
Lambertia inermis	310	1.2	0.4	0–7	20	372

amounts of time allocated by a bird to foraging at flowers, gleaning of leaves and bark, hawking, perching ("resting"), hopping between perches and flying were recorded (Collins and Briffa 1983, Collins and Newland 1986). Where data were clearly associated with particular color-banded birds, and had been gathered over intervals of several hours, they were accumulated for the individuals concerned. In most cases, however, birds timed were either unbanded, or were seen only infrequently and for short periods of time. Data for all such birds were pooled according to species and type of activity, thus providing an "overall" indication of the manner in which time was allocated (Collins and Newland 1986). Air temperatures approximately 0.5 m above ground, within vegetation visited by nectarivores, were recorded each hour using shielded thermistors, thus making it possible to construct energy budgets for the birds (see B. G. Collins 1983a, Collins and Briffa 1983).

FORAGING PREFERENCES

Frequencies with which nectarivores visited flowers on various plant species were recorded throughout the day, in conjunction with collection of general time budget data at Jarrahdale and Betty's Bay, and as a separate exercise at Fitzgerald River (Collins 1985). In cases where species had flowers at different ages that could be readily distinguished (e.g., *D. sessilis, G. wilsonii, M. hirtus*), visits to these flowers were scored separately (Collins 1985, Grey 1985).

Supplementary information concerning the types of plants visited was obtained by taking pollen smears from foreheads and throats of birds captured in mist nets and comparing these with type pollen smears from flowers nearby (Wooller et al. 1983, Collins and Newland 1986). Numbers of particular types of pollen grains present in each pair of smears from a given bird were summed and expressed as percentages of total grains counted.

DOMINANCE HIERARCHIES

The outcomes of agonistic encounters between conspecifics and different species while foraging for nectar were recorded opportunistically at Jarrahdale and Betty's Bay, but not at Fitzgerald River. As relatively few encounters occurred between color-banded birds of known age, virtually no data illustrating age-related differences in social status were obtained.

RESULTS

PLANT DENSITY AND FLORAL ABUNDANCE

Nectar-producing species had patchy distributions that tended to overlap one another within all study sites except that at Betty's Bay, where the two principal species (*Mimetes hirtus* and *Erica perspicua*) occurred in fairly discrete, "pure" stands (B. G. Collins 1983b; Collins 1985; Collins et al., ms). Plant densities and numbers of flowers available per plant were not measured at Betty's Bay, although both parameters often differed considerably from species to species at the other two locations (flowers per plant: Jarrahdale [July] $F = 15.5$, $P < 0.001$, [September] $F = 328.1$, $P < 0.001$, Fitzgerald River $F = 11.7$, $P < 0.001$). Variability in floral abundance also was great for individual plants within a given species. Consequently, contributions that particular species or plants made to the total floral pool at a given site often were quite different (Table 1).

FLORAL MORPHOLOGY

All but three plant species involved in this study had gullet-shaped flowers (Table 2). In most cases, individual flowers were arranged in spikes,

TABLE 2. Floral Morphology for Major Plants Visited by Nectarivorous Birds at Study Sites in Southwestern Australia and Southern Africa. N = 30 for All Measurements of Flower Diameter and Stigma–Nectary Distance

Location and plant species	Flower shape[a]	Flower diameter (mm)[b]		Stigma–nectary distance (mm)		Inflorescence type	Flowers/ inflorescence
		$\bar{X}$	SD	$\bar{X}$	SD		
Jarrahdale							
Adenanthos barbigera	gullet	1.8	0.2	28.3	2.7	solitary	1
Calothamnus rupestris	semi-tube	5.1	1.9	35.5	3.1	spike	15–27
Dryandra sessilis	gullet	1.1	0.3	29.8	2.9	capitulum	70–90
Grevillea wilsonii	gullet	2.9	1.3	35.5	3.8	raceme	7–12
Fitzgerald River							
Banksia baueri	gullet	0.9	0.2	31.1	1.7	spike	>5000
Banksia baxteri	gullet	0.9	0.1	29.5	2.1	spike	260–280
Banksia coccinea	gullet	1.0	0.2	24.6	1.9	spike	180–250
Banksia media	gullet	1.1	0.2	27.8	2.2	spike	>5000
Dryandra cuneata	gullet	0.8	0.1	20.2	2.1	capitulum	30–50
Dryandra quercifolia	gullet	0.9	0.1	29.6	2.4	capitulum	40–70
Lambertia inermis	tube	4.9	1.1	31.5	3.3	raceme	7
Betty's Bay							
Erica perspicua	tube	2.8	0.7	19.8	2.4	spike	20–30
Mimetes hirtus	gullet	1.0	0.1	59.4	8.8	capitulum	8–11

[a] Gullet-shaped flowers are categorized by zygomorphic perianth tubes with one or more slits; semi-tubular flowers each comprise four fused staminal bundles which are separate from one another; the only flowers into which at least some of the birds present would have been able to insert their bills were those of *A. barbigera, C. rupestris, G. wilsonii, L. inermis* and *E. perspicua*.
[b] Diameters of individual flowers were measured 10 mm from their bases in all instances, using micrometer calipers.

capitula, or racemes, with more than 5000 small flowers present per inflorescence for species such as *Banksia baueri* and *B. media*. The only species with individual flowers into which bills of at least some nectarivores could be inserted were *Adenanthos barbigera, Calothamnus rupestris, Grevillea wilsonii, Lambertia inermis,* and *E. perspicua*. Birds visiting other species were obliged to use nectar that accumulated between the bases of flowers.

Availability of Nectar

Daily nectar production varied from one plant species to another at each site (Table 3, Jarrahdale [July] F = 37.4, P < 0.001, [September] F = 11.9, P < 0.001, Fitzgerald River F = 1482.1, P < 0.001). Those species with inflorescences comprising numerous, tightly packed, small flowers usually generated the most nectar, regardless of plant density. For instance, production by *Dryandra sessilis* at Jarrahdale in July and September averaged 1614.7 and 663.2 kJ/ha, respectively, compared with 22.8 and 8.5 kJ/ha for *G. wilsonii*. At Fitzgerald River, *B. baueri* produced 2397.0 kJ/ha in July, as opposed to 42.8 kJ/ha by *L. inermis* (estimates made by combining data in Tables 1 and 3). At each site, amounts of nectar produced by individual flowers of a given species also varied considerably.

In general, nectar availability (standing crop) at dawn differed among plant species in much the same way as nectar production (Table 3, Jarrahdale [July] F = 29.6, P < 0.001, Betty's Bay F = 50.96, P < 0.001), some individual flowers containing copious amounts of nectar and others virtually none. Nectar was lost from most flowers during the day, although percentages of dawn standing crops that remained at dusk often varied considerably from species to species. For instance, flowers of *A. barbigera, G. wilsonii,* and *E. perspicua* appeared to retain relatively more nectar than those of *D. sessilis* or *M. hirtus* (Table 3).

A large part of the variability in nectar availability for flowers chosen at random from species such as *D. sessilis, G. wilsonii,* and *M. hirtus* can be attributed to differences associated with floral age. For instance, dawn and dusk standing crops of nectar for all three species varied inversely with floral age, nectar production finally ceasing after approximately 7, 3, and 7 days, respectively (Table 4, at dawn: *D. sessilis* F = 337.3, P < 0.001, *G. wilsonii* F = 5.9, P < 0.001, *M. hirtus* F = 2.4, P < 0.05).

Morphometric and Time Budget Data

At each study site, body masses and bill (exposed culmen) lengths of most nectarivore species differed from one another (Table 5, body mass: Jarrahdale [July] F = 1050.0, P < 0.001, [Sep-

TABLE 3. Nectar Production, Availability and Depletion for Plant Species Visited by Honeyeaters, Sunbirds or Sugarbirds (Partly after B. G. Collins 1983a; Collins and Newland 1986, Collins et al., Unpubl. Ms)

Location and plant species	Nectar production (kJ/24 hour/flower)				Nectar availability at dawn (kJ/flower)				Nectar remaining at dusk (%)[b]
	$\bar{X}$	SD	Range	N	$\bar{X}$	SD	Range	N	
Jarrahdale (July)									
Adenanthos barbigera	0.028	0.013	0.005–0.037	30	0.017	0.009	0.000–0.029	30	42.6
Calothamnus rupestris									
Dryandra sessilis[a]	0.963	0.327	0.375–1.121	30	0.643	0.196	0.179–0.817	30	12.4
Grevillea wilsonii	0.033	0.018	0.017–0.048	30	0.022	0.008	0.005–0.031	30	46.7
Jarrahdale (September)									
Adenanthos barbigera	0.012	0.006	0.003–0.019	30					
Calothamnus rupestris	0.219	0.110	0.112–0.288	30					
Dryandra sessilis[a]	0.557	0.269	0.121–0.783	30					
Grevillea wilsonii	0.015	0.007	0.005–0.037	30					
Fitzgerald River (July)									
Banksia baueri[a]	7.885	3.919	3.899–12.174	10					
Banksia baxteri[a]	0.560	0.221	0.150–0.717	10					
Banksia coccinia[a]	0.125	0.071	0.009–0.184	10					
Banksia media[a]	15.350	8.023	4.632–18.151	10					
Dryandra cuneata[a]	0.500	0.113	0.105–0.788	20					
Dryandra quercifolia[a]	3.111	0.927	1.952–4.923	20					
Lambertia inermis	0.115	0.042	0.034–0.175	40					
Betty's Bay (September)									
Erica perspicua					0.003	0.001	0.000–0.006	100	54.0
Mimetes hirtus[a]					0.081	0.009	0.011–0.116	30	15.6

[a] Species with flowering units that are inflorescences comprising numerous small flowers; other species have widely-spaced flowers.
[b] Nectar present at dusk expressed as percentage of dawn nectar availability.

tember] $F = 1398.7$, $P < 0.001$, Fitzgerald River $F = 407.1$, $P < 0.001$, Betty's Bay $F = 736.7$, $P < 0.001$). For instance, Cape Sugarbirds were considerably larger than Orange-breasted Sunbirds at Betty's Bay, and both New Holland and White-cheeked Honeyeaters larger than other honeyeaters at Fitzgerald River. Intraspecific variability was often quite marked, males generally being larger than females, at least for those species where sexes could be readily distinguished (e.g., Western Spinebills, Cape Sugarbirds, Orange-breasted Sunbirds; body mass: $t = 5.94$, 4.65 and 3.04, respectively, for comparisons of males and females; $P < 0.01$). Body masses and bill lengths were recorded for all birds captured, although the only particular values provided in Table 5 are those for individual birds subsequently involved in extended time budget investigations.

At Jarrahdale, nectarivore body size varied inversely with amounts of time spent foraging, hopping, and flying (Table 6). For instance, relatively large Little Wattlebirds devoted much less time to these activities than either New Holland Honeyeaters, or Western Spinebills in July and September. Similar relationships possibly existed at Betty's Bay, although these are obscured by the fact that hopping and foraging were not measured as separate components of time budgets for either Cape Sugarbirds or Orange-breasted Sunbirds. Differences among time budgets of individual birds and sexes appear not to have been so pronounced as those among species, although male and female Western Spinebills, in particular, may have differed significantly in general activity patterns and foraging behavior.

Energy Budgets and Foraging Efficiencies

At each of the study sites, estimated energy expenditure and requirements vary between different nectarivore species (Table 5). In general, values are greatest for the largest and most active birds. Limited evidence also suggests that expenditures and needs are greater for males than females of a given species, principally because of differences in their body masses.

The efficiencies with which birds extract nectar from flowers can be calculated using data provided in Tables 4 and 5. Extraction efficiency varies according to the species of nectarivore or plant involved, nectar availability and the distances between flowers that are visited. Efficiency is greatest for small birds that visit productive flowers which are close together (Table 7).

FORAGING PREFERENCES

Data (Table 8) have been pooled for individual species of nectarivores, yet reveal some striking differences with regard to types of plant whose flowers were visited by them. For instance, Little Wattlebirds at Jarrahdale visited virtually no flowers other than those of *D. sessilis* in July and *C. rupestris* in September. In contrast, the smaller Western Spinebills supplemented nectar from these species with that from *A. barbigera* and *G. wilsonii*. At Betty's Bay, Cape Sugarbirds relied upon *M. hirtus* flowers; yet, Orange-breasted Sunbirds visited both *M. hirtus* and *E. perspicua*. The situation at Fitzgerald River was more variable. Most honeyeaters at that location appeared to favor *Dryandra cuneata* and *L. inermis*, although two of the smaller species, Brown Honeyeaters and Western Spinebills, visited *Banksia baueri* more frequently than did the larger honeyeaters. Intermediate-sized White-naped Honeyeaters were particularly interesting in that they did not appear to visit *B. baueri* at all, and foraged rarely at *L. inermis*, yet visited *Banksia coccinia*, which is a relatively poor source of nectar.

Analysis of pollen smears provided evidence that generally supported direct observations of the type outlined above (Table 9), although pollen from plant species found only outside the study site was often present in smears obtained at Fitzgerald River (e.g., pollen from *B. media* and *Dryandra quercifolia*). The White-naped Honeyeaters, whose observed foraging preferences at that site differed so markedly from those of other species, had an average of only 14% of the total pollen grains counted that were from plant species listed in Table 9. Marked intraspecific variations in the incidence of pollen types occurred for all nectarivore species. Quite often, particular types were absent from some individual birds of a given species, but present on others (e.g., *G. wilsonii* pollen present on some New Holland Honeyeaters at Jarrahdale, but not on others).

Honeyeaters studied at Jarrahdale, and Cape Sugarbirds at Betty's Bay, all demonstrated clear preferences for flowers of particular ages (Table 10). Invariably, highest preferences were shown for younger flowers that produced the most nectar, although the three Jarrahdale species differed from one another in that they sometimes foraged at flowers whose ages spanned varying ranges (e.g., when visiting *D. sessilis*, Little Wattlebirds visited day 1–2 and day 3–4 flowers only, whereas New Holland Honeyeaters also used day 5–6 flowers).

TABLE 4. AVAILABILITY OF NECTAR AT DAWN AND DUSK ON SUCCESSIVE DAYS DURING THE FLORAL DEVELOPMENT OF UNBAGGED *Dryandra sessilis*, *Grevillea wilsonii* AND *Mimetes hirtus* (PARTLY AFTER B. G. COLLINS 1983A, B). N DENOTES NUMBERS OF FLOWERS TESTED. HORIZONTAL LINES INDICATE RANGES OF FLORAL AGES SPANNED WHEN CALCULATING NECTAR AVAILABILITY FOR *M. hirtus*

Plant species and time of year		Nectar availability (kJ/flower) at dawn on day							Nectar availability (kJ/flower) at dusk on day						
		1	2	3	4	5	6	7	1	2	3	4	5	6	7
Dryandra sessilis (July, N = 10)	X̄	0.603	0.540	0.284	0.060	0.034	0.007	0.005	0.065	0.040	0.018	0.012	0.007	0.004	0.004
	SD	0.031	0.028	0.029	0.018	0.009	0.006	0.003	0.010	0.013	0.008	0.007	0.004	0.004	0.003
Grevillea wilsonii (July, N = 10)	X̄	0.030	0.026	0.011					0.016	0.013	0.004				
	SD	0.011	0.013	0.007					0.007	0.006	0.004				
Mimetes hirtus (September, N = 10)	X̄		0.119				0.021			0.021				0.003	
	SD		0.125				0.039			0.024				0.009	

TABLE 5. MORPHOMETRIC DATA FOR NECTARIVOROUS BIRDS CAPTURED AT STUDY SITES. MEASUREMENTS FOR COLOR-BANDED INDIVIDUALS ARE PROVIDED IF EXTENDED TIME BUDGET DATA WERE OBTAINED FOR THESE BIRDS

Location and bird species	Banding code[a]	Approx. age (years)	N	Mass (g) $\bar{X}$	SD	Range	Exposed culmen length (mm) $\bar{X}$	SD	Range	Photoperiod duration (hours)	Daytime energy expenditure (kJ)[b]	Daytime energy requirement (kJ)[c]
Jarrahdale (July)												
Little Wattlebird (unsexed)	overall		9	61.8	3.1	51.9–67.3	29.6	1.2	27.8–32.0	10.1	74.0	147.8
	LG/R	>1	1	67.3			32.0				76.2	156.6
	MA/R	<1	1	59.3			28.7				67.6	138.4
New Holland Honeyeater (unsexed)	overall		53	20.6	1.3	15.5–23.5	21.3	1.0	19.2–23.7	10.1	39.7	72.8
	R/Y	>1	1	20.7			20.6				39.3	72.5
	R/MA	<1	1	15.5			19.7				27.6	52.5
Western Spinebill (male)	overall		7	11.8	0.9	10.5–14.0	23.7	1.2	21.9–25.9	10.1	21.4	44.5
Western Spinebill (female)	overall		4	9.5	0.8	8.3–10.7	19.5	1.0	18.8–20.6	10.1	21.9	40.5
Jarrahdale (September)												
Little Wattlebird (unsexed)	overall		3	69.1	2.1	66.4–70.1	32.7	2.7	30.1–34.9	11.5	119.7	187.0
New Holland Honeyeater (unsexed)	overall		35	20.1	0.9	16.2–24.4	22.2	0.8	20.1–24.4	11.5	48.9	75.2
Western Spinebill (male)	overall		11	10.7	0.4	10.0–11.2	22.3	0.9	21.0–23.9	11.5	30.8	47.9
Western Spinebill (female)	overall		3	9.5	0.6	9.0–10.4	19.2	0.1	19.1–19.4	11.5	29.6	44.8
Fitzgerald River (July)												
White-cheeked Honeyeater	overall		17	19.1	0.9	17.9–20.7						
White-naped Honeyeater	overall		7	12.8	0.7	11.6–13.2						
Brown Honeyeater	overall		7	10.3	0.6	9.4–11.1						
Western Spinebill (male)	overall		5	11.5	0.7	10.2–11.9	22.9	0.9	20.6–24.4			
Western Spinebill (female)	overall		2	9.6	0.1	9.4–9.8	20.3	1.0	19.3–21.3			
Betty's Bay (September)												
Cape Sugarbird (male)	overall		8	42.6	1.9	38.9–43.7	29.8	0.5	28.4–30.3	12.6	84.9	138.5
	Y/W	>1	1	41.8			29.3					
Cape Sugarbird (female)	overall		5	39.5	1.8	36.2–40.5	28.2	0.3	27.7–28.9	12.6	79.1	129.9
	R/Y	>1	1	40.5			28.5				33.8	52.6
Orange-breasted Sunbird (male)	overall		10	9.9	0.3	9.5–10.6	21.9	0.2	21.6–22.7	12.6		
Orange-breasted Sunbird (female)	overall		4	9.4	0.4	8.9–9.9	21.1	0.3	20.6–21.8			

[a] "Overall" denotes data pooled for several birds; where individual birds have been studied, the code denotes the colors of bands used on left/right legs.
[b] Daytime energy expenditures were calculated using data from Tables 5 and 6, and the methodologies outlined by B. G. Collins (1983a), Collins and Briffa (1983) and Collins and Newland (1986); the equations used for honeyeaters were appropriately scaled versions of those empirically-derived for 10 g Brown Honeyeaters (Collins and Briffa 1983, Collins and Newland 1986).
[c] Daytime energy requirements were calculated by adding daytime expenditures to predicted overnight expenditures that had been divided by 0.875 to allow for 12.5% loss of energy when excess carbohydrate ingested during the day was stored as lipid (B. G. Collins 1983a, Collins and Briffa 1983, Collins and Newland 1986).

TABLE 6. Environmental Temperatures and Time Budgets for Nectarivorous Birds at Study Sites (Partly after B. G. Collins 1983a). Morphometric Measurements for These and Other Birds Are Provided in Table 5, as Are Definitions of Banding Codes. Horizontal Lines and Brackets Denote Range of Activities Included in Time-budget Data for Betty's Bay

Location and bird species	Banding code	Approx. age (years)	Total observ. time (s)	Perching (resting)	Probing flowers	Gleaning	Forward flight	Hopping	Hawking	Mean day/night temp (°C)
Jarrahdale (July)										
Little Wattlebird (unsexed)	overall		16,491	76.6	15.7	0.5	4.6	2.4	0.2	11.5/8.2
	LG/R	>1	4188	74.0	19.7	0.0	4.2	2.1	0.0	
	MA/R	<1	3241	82.0	10.9	0.7	3.3	2.7	0.4	
New Holland Honeyeater (unsexed)	overall		15,095	54.7	30.4	0.6	8.3	4.8	1.2	11.5/8.2
	R/Y	>1	2879	63.7	22.4	0.0	11.2	2.7	0.0	
	R/MA	<1	3325	63.5	24.1	0.2	5.1	5.2	0.2	
Western Spinebill (male)	overall		2564	45.9	43.8	0.0	3.4	6.9	0.0	11.5/8.2
Western Spinebill (female)	overall		2103	54.4	27.2	1.0	4.9	12.5	0.0	11.5/8.2
Jarrahdale (September)										
Little Wattlebird (unsexed)	overall		4369	80.3	7.6	0.2	5.6	5.3	1.0	13.5/9.2
New Holland Honeyeater (unsexed)	overall		14,117	51.1	30.3	0.5	6.6	11.0	0.5	13.5/9.2
Western Spinebill (male)	overall		1413	59.8	17.9	0.5	14.3	7.0	0.5	13.5/9.2
Western Spinebill (female)	overall		1869	31.5	43.0	0.1	8.3	17.1	0.0	13.5/9.2
Betty's Bay (September)										
Cape Sugarbird (male)	Y/W	>1	3600		94.8			5.2		17.7/12.5
Cape Sugarbird (female)	R/Y	>1	28,400		98.4			1.6		17.2/13.3
Orange-breasted Sunbird (male)	overall		3800		89.0			11.0		17.7/12.5
Orange-breasted Sunbird (female)	overall									

DOMINANCE HIERARCHIES

Clearly defined interspecific hierarchies were identified at both study sites where detailed observations of agonistic interactions were recorded (Jarrahdale and Betty's Bay). In each case, larger species (e.g., Little Wattlebirds, Cape Sugarbirds) were consistent winners against smaller species (e.g., Western Spinebills, Orange-breasted Sunbirds) (Table 11). Intraspecific hierarchies were also apparent, with males usually winning out against females. Insufficient data were available to test the hypothesis that the winners of intraspecific encounters are determined on the basis of body size or age.

DISCUSSION

Casual observation of vegetation within plant communities that support nectar-feeding birds usually reveals striking variations in abundance and distribution of different species, although this patchiness has only occasionally been quantified by people interested in resource partitioning (e.g., Wolf et al. 1976, Feinsinger 1978, Collins 1985, Wykes 1985). Data presented in this paper indicate that most nectar-producing species at Jarrahdale and Fitzgerald River have patchy distributions, and are generally supported by independent results for the same areas obtained by Wykes (1985) and Newby (unpubl. data). A similar level of diversity does not occur within the small *Mimetes hirtus-Erica perspicua* community at Betty's Bay, although Boucher (1978) has demonstrated that many additional species occur in adjacent habitats.

For any given plant species, numbers of flowers present on individual plants at a particular time of year also can be quite variable (e.g., Feinsinger 1978, Paton and Ford 1983, Collins 1985, Collins and Newland 1986). Some plants have no flowers, others have many. As a result, differences in flower counts combine with patchy distributions of the plants themselves to present an uneven floral environment to potential visitors.

The contribution that a particular plant species makes to the total nectar pool is clearly related to plant density, floral abundance and the amount of nectar that each flower produces (e.g., Pyke 1983, Collins et al. 1984, Collins and Newland 1986, Paton 1986). We found that standing crops of nectar in flowers that have not been visited by honeyeaters for a considerable time can be quite variable (see also Feinsinger 1978, Carpenter 1983, Gill and Wolf 1977). Genetic and environmental factors are involved (e.g., Cruden

TABLE 7. FORAGING EFFICIENCIES FOR HONEYEATERS, SUGARBIRDS AND SUNBIRDS VISITING FLOWERS AT JARRAHDALE OR BETTY'S BAY

Location and plant species	Age of flower/ time of day	Standing crop energy (J/flower)[a]	Nectarivore species[b]	Foraging efficiency (J)[c]		
				A	B	C
Jarrahdale (July)						
Dryandra sessilis	Day 1/dawn	603	A.s.	597	593	590
			P.n.	594	587	581
			A.c.	589	568	548
	Day 4/dawn	60	A.s.	54	50	47
			P.n.	51	44	38
			A.c.	46	25	5
Grevillea wilsonii	Day 1/dawn	30	A.s.	27	24	24
			P.n.	26 (?)	19 (?)	16 (?)
			A.c.	21 (?)	1 (?)	−20 (?)
	Day 3/dawn	11	A.s.	8	5	1
			P.n.	7 (?)	0 (?)	−7 (?)
			A.c.	2 (?)	−18 (?)	−39 (?)
Betty's Bay (September)						
Mimetes hirtus	Partly-open/dawn	119	N.v.			
			P.c.	107	99	91
	Fully-open/dawn	21	N.v.			
			P.c.	3	−5	−12

[a] Nectar standing crops have been taken from Table 4.
[b] A.s., P.n., A.c., N.v. and P.c. denote Western Spinebills, New Holland Honeyeaters, Little Wattlebirds, Orange-breasted Sunbirds and Cape Sugarbirds, respectively.
[c] Foraging efficiency is the difference between energy intake, assuming 100% ingestion of available nectar when a flower is visited by a bird, and the energy expended in flying to the flower and extracting this nectar; efficiencies have been calculated for situations in which birds had to fly different distances in order to harvest nectar: flight times used were (A) 0.5, (B) 2.0 and (C) 3.5 s; mean extraction times were: *D. sessilis* (A.s. 15.3, P.n. 12.5 and A.c. 6.7 s); *G. wilsonii* (A.s. 4.1, P.n. 3.3(?) and A.c. 1.8(?) s); *M. hirtus* (P.c. male 13.9 and P.c. female 24.7 s, N.v. no data); (?) denotes that extraction times and calculations involving these are estimates, since these birds were not observed visiting such flowers.

et al. 1983), although the ages of flowers are especially significant (this study, see also Gill and Wolf 1977, Grey 1985). For example, *Dryandra sessilis* flowers whose ages range from 1 to 7 days would be expected to have more variable standing crops than a sample of uniform age.

Patchiness of the floral and nectar environments presents nectarivores with a diversity of foraging options. For instance, the most abundant and uniformly distributed plant species at Jarrahdale is *Adenanthos barbigera*; yet, each plant usually bears relatively few flowers, most of which produce small amounts of nectar. In contrast, *Calothamnus rupestris* and *D. sessilis* are less abundant, and often more widely spaced; yet, each has such large numbers of flowers per plant or produces such quantities of nectar per flower that its overall contribution to the nectar pool is much greater. How do nectarivores forage under these conditions? If nectar is harvested selectively, is the choice based on plant density, floral abundance per plant, flower morphology, amount of nectar present per flower, or some combination of these parameters? Even if preferential foraging does occur, variations in size and behavior suggest that species and individual birds may partition nectar resources in different ways.

According to optimal foraging theory (e.g., Pyke et al. 1977, Pyke 1984), nectarivorous birds should maximize their net rates of energy acquisition. This might be achieved by adopting a foraging strategy that maximized energy intake, perhaps by selecting plant species offering the greatest nectar rewards per plant and/or flower, although there would be some energetic sacrifices if the plants were widely spaced (Table 7). Alternatively, birds might select species with the greatest plant and floral densities, at least within certain parts of the habitat. In this situation, the energetic cost of moving between flowers would be minimized, although energy intake would not necessarily be at the highest possible level. Of course, birds could opt for a combination of both strategies (e.g., Gill and Wolf 1977; B. G. Collins 1983a, b; Paton and Ford 1983; Collins 1985; Grey 1985; Collins and Rebelo 1987). For instance, Little Wattlebirds and New Holland Honeyeaters at Jarrahdale foraged selectively at *C. rupestris* or *D. sessilis* flowers, when these were available, but also preferred the densest patches of either species, and individual plants with the most flowers (Collins 1985, Grey 1985). This allowed them to increase their foraging efficiency and satisfy their energy requirements).

Energy expenditures and requirements of larg-

TABLE 8. RELATIVE FREQUENCIES OF VISITS BY NECTARIVOROUS BIRDS TO FLOWERS ON PLANT SPECIES AT STUDY SITES. FIGURES IN PARENTHESES DENOTE RELATIVE ABUNDANCES OF PLANTS VISITED; 5.1% OF THE TOTAL PLANTS PRESENT AT JARRAHDALE BELONGED TO OTHER SPECIES (PARTLY AFTER COLLINS 1985)

Location and bird species	Total visits to flowers	Percentage frequency of visits				
Jarrahdale (July)		*Adenanthos barbigera* (27.7)	*Calothamnus rupestris* (0.7)	*Dryandra sessilis* (16.6)	*Grevillea wilsonii* (18.1)	*Dryandra nivea* (31.8)
Little Wattlebird	40			100.0		
New Holland Honeyeater	411			99.3	0.5	0.2
Western Spinebill	43	37.2		32.6	23.2	7.0
Jarrahdale (September)						
Little Wattlebird	16		100.0			
New Holland Honeyeater	222		64.0	36.0		
Western Spinebill	171	24.6	38.0	26.9	10.5	
Fitzgerald River (July)		*Banksia baueri* (34.5)	*Banksia coccinia* (16.4)	*Dryandra cuneata* (20.9)	*Lambertia inermis* (28.2)	
New Holland Honeyeater	115	7.0	6.1	56.5	30.4	
White-cheeked Honeyeater	107	3.7	2.7	50.6	43.0	
White-naped Honeyeater	15	0.0	20.0	73.3	6.7	
Brown Honeyeater	38	23.7	2.6	26.3	47.4	
Western Spinebill	29	17.2	3.5	27.6	51.7	
Betty's Bay (September)		*Erica perspicua* (?)	*Mimetes hirtus* (?)			
Cape Sugarbird	511	0.0	100.0			
Orange-breasted Sunbird	372	17.5	82.5			

er species, and bigger birds within these, are greater than those for smaller birds, all other things being similar (Table 5). For this reason, one might expect larger birds to be more discerning than others in their choice of nectar resources. As this paper indicates, Little Wattlebirds and New Holland Honeyeaters used young flowers of *C. rupestris* and *D. sessilis* almost exclusively at Jarrahdale, whereas the smaller Western Spinebills also made frequent visits to older flowers of the same species and to flowers of generally less-rewarding plants such as *Grevillea wilsonii* and *A. barbigera* (Table 8). Similarly, Cape Sugarbirds at Betty's Bay preferred partly open inflorescences of *M. hirtus,* and ignored *E. perspicua,* whereas Orange-breasted Sunbirds made considerable use of *E. perspicua.*

Perhaps the most obvious way in which nectar resource partitioning by different species is effected is by the establishment of dominance hierarchies (e.g., Ford and Paton 1982, Craig 1985, Newland and Wooller 1985), and at all three sites studied here larger species often displaced smaller birds from the most rewarding sources of nectar. Differential use of available resources is reflected in the time and energy budgets, larger nectarivores being able to devote less effort to foraging and more to "resting," thereby reducing their energy requirements.

Resource partitioning also occurs because bill lengths and breadths of the birds, and floral morphologies for the plants, are sometimes incompatible. For example, the tubular or gullet-shaped flowers of plants such as *E. perspicua, G. wilsonii* and *A. barbigera* clearly could not be probed by Cape Sugarbirds or Little Wattlebirds, both of which have relatively broad bills (Paton and Collins, unpubl. ms); yet, nectar should have been easily harvested by narrow-billed Orange-breasted Sunbirds or Western Spinebills. We found that the percentage depletion of dawn standing crops of nectar at Betty's Bay and Jarrahdale was not only greatest for plant species which were most productive, but also for those whose nectar was accessible to a wide range of nectarivores (e.g., *D. sessilis, M. hirtus*).

Little is known regarding intraspecific partitioning of nectar resources within bird communities. Rufous Hummingbirds (*Selasphorus rufus*) appear to adjust the sizes of their breeding territories daily (Carpenter et al. 1983, Gass and Lertzman 1980), in a manner that is influenced by their sex and age (Gass 1978, 1979). J. L. Craig (1985) provided some evidence that in-

TABLE 9. RELATIVE OCCURRENCE OF POLLEN GRAINS IN SMEARS TAKEN FROM THE FOREHEADS AND THROATS OF NECTARIVOROUS BIRDS (PARTLY AFTER B. G. COLLINS 1983A). MEANS AND RANGES ARE GIVEN FOR EACH PLANT AND NECTARIVORE SPECIES

Location and bird species	Sex	N	Percentage frequency of pollen[a]				
			Adenanthos barbigera	Calothamnus rupestris	Dryandra sessilis	Grevillea wilsonii	Dryandra nivea
Jarrahdale (July)							
Little Wattlebird	Mixed	4	0.0	0.0	100.0	0.0	0.0
New Holland Honeyeater	Mixed	24	0.0	0.0	95.8 91.4–100.0	4.2 0–7.3	0.0
Western Spinebill	Male	15	1.2 0–3.1	0.0	83.6 79.1–92.7	3.1 0–8.1	12.1 0–15.8
Western Spinebill	Female	3	2.5 0–3.8	0.0	85.4 85.4–89.6	3.8 0–5.1	8.3 0–16.2
Jarrahdale (September)							
Little Wattlebird	Mixed	3	0.0	54.0 36.5–79.3	46.0 33.2–58.4	0.0	
New Holland Honeyeater	Mixed	14	1.0 0–4.2	51.0 32.7–78.8	43.0 29.2–61.5	5.0 0–11.1	0.0
Western Spinebill	Male	5	3.1 0–6.2	47.2 29.3–58.6	48.5 22.5–57.7	1.2 0–4.2	0.0
Western Spinebill	Female	0					

Location and bird species	Sex	N	Banksia baueri	Banksia coccinia	Banksia media	Dryandra cuneata	Dryandra quercifolia
Fitzgerald River (July)[b]							
New Holland Honeyeater	Mixed	7	8.0 0.0–17.0	1.0 0–0.7	26.0 0–31.0	47.0 0–67.6	9.0 0–19.4
White-cheeked Honeyeater	Mixed	7	1.0 0.0–20.0	1.0 0–20.0	28.0 0–39.3	31.0 0–57.0	30.0 0–53.9
White-naped Honeyeater	Mixed	7	0.0	1.0 0–1.3	1.0 0–3.7	11.0 0–17.8	1.0 0–7.1
Brown Honeyeater	Mixed	7	14.0 0.0–36.6	0.0	2.0 0–18.3	54.0 0–97.2	11.0 0–49.3
Western Spinebill	Male	5	23.0 0–29.8	0.0	31.0 0–42.6	31.0 0–58.2	8.0 0–17.2
Western Spinebill	Female	2	17.0 11.0–23.0	0.0	22.3 17.6–27.0	21.0 0–42.0	7.0 0–14.0

Location and bird species	Sex	N	Erica perspicua	Mimetes hirtus
Betty's Bay (September)				
Cape Sugarbird	Male	5	0.0	100.0
Cape Sugarbird	Female	3	0.0	100.0
Orange-breasted Sunbird	Male	7	27.7 13.6–42.6	72.3 39.6–82.1
Orange-breasted Sunbird	Female	1	85.7	14.3

[a] Pollen counts from head and throat of each bird pooled before calculation of percentages.
[b] In cases where frequencies do not add to 100.0, birds also foraged at other plants.

TABLE 10. Preferential Foraging by Honeyeaters or Sugarbirds at Flowers of Different Ages (Partly after B. G. Collins 1983a, 1985)

Plant species and time of year[a]	Age of flowers (days)	Relative abundance of flowers (%)	Foraging preference (%)[b]			
			A.c.	P.n.	A.s.	P.c.
Dryandra sessilis (July, N = 150)	1–2	29.4	76	68	56	
	3–4	39.5	24	28	39	
	5–6	21.3	0	3	5	
	7–?	9.8	0	0	0	
Grevillea wilsonii (July, N = 152)	1–2	53.9			100	
	3–?	46.1			0	
Mimetes hirtus (September, N = ?)	1–4	17.1				99
	5–?	82.9				1

[a] N denotes the number of plants observed for a total time of at least 36,000 s.
[b] A.c., P.n. and A.s. made 91, 208 and 38 visits, respectively, to D. sessilis; A.s. made 30 visits to G. wilsonii; P.c. made 408 visits to M. hirtus. The letters A.c., P.n., A.s. and P.c. denote Little Wattlebirds, New Holland Honeyeaters, Western Spinebills, and Cape Sugarbirds, respectively.

dividual New Zealand honeyeaters partition available nectar, with larger, male birds usually dominating the richest sources. We also found that foraging activity and nectar use by some sunbirds, sugarbirds and Australian honeyeaters varied individually.

Since intraspecific dominance hierarchies exist in honeyeater (Craig 1985, Newland and Wooller 1985) and sunbird-sugarbird (Wooller 1982) communities, there is no obvious reason why larger, dominant birds should not use more rewarding flowers, and spend less time foraging, than subordinates. Although we found this to be true for the larger color-banded New Holland Honeyeaters at Jarrahdale, it was not the case for Little Wattlebirds. Data for Western Spinebills were variable, with males spending less time than females foraging for nectar in July, but more time in September. This discrepancy could have arisen because of the particular (unknown) sizes and positions of individuals sampled in the intraspecific dominance hierarchies at those times.

No direct observational data on intraspecific differences in the types of flowers visited by color-banded honeyeaters, sunbirds, or sugarbirds are available, although smears taken from foreheads and throats of these and other non-banded birds suggest that preferential foraging occurs. However, interpretation of smear data is complicated by the fact that the proportions of various pollen grains present will be biased by the sequence in which plant species are visited and the amounts of pollen that they produce. Sex-related differences were especially obvious at two sites. For instance, all female Western Spinebills at Fitzgerald River carried Banksia baueri and B. media pollen, but some males did not. At Betty's Bay, male Orange-breasted Sunbirds bore significantly more M. hirtus and less E. perspicua pollen than conspecific females.

At best, the evidence currently available merely suggests that intraspecific partitioning of nectar resources occurs. It will only be possible to test this hypothesis satisfactorily if quantitative data are obtained using a variety of techniques, over extended periods of time, for large numbers of individual birds of known age, sex, and social status.

TABLE 11. Outcomes of Aggressive Interactions between Honeyeaters at Jarrahdale, and between Sunbirds and Sugarbirds at Betty's Bay. Figures Denote Either Total Numbers of Interactions or Numbers of Wins/Losses

Winning species	Losing species			
Jarrahdale	A.c.	P.n.	A.s. ♂	A.s. ♀
Little Wattlebird (A.c.)	5	25/0	15/0	10/0
New Holland Honeyeater (P.n.)		52	31/0	10/0
Western Spinebill (A.s. ♂)			9	11/2
Western Spinebill (A.s. ♀)				0
Betty's Bay	P.c. ♂	P.c. ♀	N.v. ♂	N.v. ♀
Cape Sugarbird (P.c. ♂)	4	2/0	45/0	30/0
Cape Sugarbird (P.c. ♀)		0	15/0	12/0
Orange-breasted Sunbird (N.v. ♂)			3	21/5
Orange-breasted Sunbird (N.v. ♀)				0

ACKNOWLEDGMENTS

Our thanks go to the Australian Research Grants Committee, Curtin University of Technology, and the University of Cape Town, all of which have provided financial support for projects that generated data included in this paper. Thanks are also due to G. Maclean and J. Majer, who offered constructive comments on a draft of the paper.

SECTION II

FORAGING BEHAVIOR: DESIGN AND ANALYSIS

Overview

BIOLOGICAL CONSIDERATIONS FOR STUDY DESIGN

MARTIN G. RAPHAEL AND BRIAN A. MAURER

AD HOC AND A PRIORI HYPOTHESIS TESTING

Research on avian foraging is still mostly in the descriptive, empirical stage of development. Most of us, despite intentions to the contrary, simply follow birds and record what they do. Our study designs focus on where we plan to make observations, when we will make them (e.g., time of day, season), how many observations we hope to collect and, perhaps, how we will stratify our observations among groupings of interest (e.g., species, sex, age class, habitat type). We then toss the data into a statistical computing package to test the hypothesis that our dependent variables do not differ among the groupings we defined. Often we reject the null hypothesis, and then we are left searching for biological explanations for the differences we observed. How interesting are conclusions derived from such a process?

Any clever biologist can explain any observation by envisioning a perfectly reasonable series of events that could have led to that observation. One particularly striking example is that of a well-known ornithologist who analyzed bird abundance at a number of sites in relation to the characteristics of vegetation. Based on a multivariate statistical analysis, he developed a very reasonable explanation connecting the patterns of bird abundance to the specific vegetation features; but he then discovered that a keypunching error had caused the data to be shifted by a column. The result was that none of the vegetation data corresponded to the variable names he was using in the analysis. The data were essentially unrelated to the variables he used to explain his results.

The lesson is that retrospective explanations of observed phenomena are not very insightful nor do they lead to strong inferences. Considering all of the sources of error that authors in these proceedings have discussed, we may often be guilty of making biological mountains out of statistical molehills composed of variation attributable to both sampling and measurement error.

There is certainly a place for descriptive studies. After all, strictly empirical observations are the stuff of knowledge, and we are not advocating their abandonment. Rather, we are cautioning that researchers avoid the temptation of going too far in developing ad hoc explanations of descriptive data.

The power of a priori hypotheses, derived from basic biological principles or theory, is much greater than that of ad hoc hypotheses. Real, not illusory, progress is made when such hypotheses are accepted or rejected after analysis of results of a carefully designed and executed study. Such hypotheses are predictions of future outcomes as opposed to explanations of past outcomes. The confirmation of these predictions (which often involve directional or one-tailed hypotheses) is much more difficult to achieve than the usual null hypothesis of no differences. As a result, we are more confident of conclusions derived from results of such hypothesis testing.

OTHER CONSIDERATIONS

STUDY OBJECTIVES

The design of any foraging study must, obviously, be dictated by the objectives of the investigation. Less obvious are the limitations that the design imposes on the legitimate conclusions drawn from the results. It is foolhardy, for the reasons cited above, to draw conclusions about evolutionary fitness from a study designed to gather descriptive data. One can certainly derive evolutionary hypotheses for further testing from such data, but not conclusions. Thus, the objectives of a foraging study should be carefully thought out and explicitly stated.

A wide variety of inferences can be drawn from foraging data if researchers design appropriate studies and collect appropriate data. The objectives of a study are then determined by the level of biological inference that the researcher wishes to achieve. These levels of inference can be ranked based on the amount of information necessary to make specific conclusions (Table 1). Few studies have gone beyond the second level of infer-

Overview

ANALYTICAL CONSIDERATIONS FOR STUDY DESIGN

Barry R. Noon and William M. Block

Studies of the foraging behaviors of birds have been largely descriptive and comparative. One might then expect studies with similar objectives to have similar study designs but that is not the case. Papers in this symposium that focused specifically on study design contain a diversity of biological perspectives. Similarly, there is no accord among statisticians on experimental design and data analysis of multivariable systems. Further, biological and statistical considerations in study design are not always in agreement.

In this paper, we attempt to define the nature of foraging data and to discuss the arbitrary structure of much of the data that are collected. We then touch on the diversity of approaches to study design that appear in this symposium. Finally, we attempt to identify areas of contrasting opinion, offer our own perspectives on controversial issues, and suggest areas in need of further research.

THE NATURE OF FORAGING DATA

Most data on avian foraging are derived from field observations of foraging events that can be classified by one or more nominal attributes. If two or more attributes are recorded for each event, then the data are referred to as cross-classified. Events are now redefined according to each unique combination of attributes assigned to an observation. These classes of events have the property of being mutually exclusive and exhaustive. Given a sample of observations, the final data have the form of counts or frequencies with which certain events were observed. Data with this structure can be portrayed as cross-classified tables with each cell of a table representing the frequency with which a particular event was observed.

Occasionally, event frequencies are estimated across known time intervals, which makes it possible to estimate foraging rates as well as frequencies. If behavioral events are persistent and of sufficient duration, one can construct time budgets. Event-based and time-based approaches are combined when data are collected sequentially and represent a sequence of events. Time intervals can be of fixed or variable length; in the latter case they are dependent upon the cessation of an event. One can estimate event durations and rates from these data but, in addition, one can look specifically at the arrangement of events in the time series and estimate a number of conditional probabilities; for example, given that event A has occurred, what is the probability that it will be followed by event B? The conditional, or transition, probabilities can be arranged in a transition matrix. The event observed at time t is the row variable and the event observed at time $t + 1$, given the event at t, is the column variable. The probability of going from one event to another in a single time step is referred to as a Markov chain.

Regardless of the design of data collection, most foraging studies are event based and the data end up being represented by frequencies. As such, the data are counts of discrete random variables, and relationships among the event categories should be analyzed by discrete multivariate models (cf. Bishop et al. 1975).

The nominal attributes (such as tree species or substrate type) or factors involved in foraging can have many levels. If each event is classified according to bird species, sex, tree species, and foraging substrate, the potential number of mutually exclusive and exhaustive categories is large. A comparative study, for example, of the use of bark versus foliage of four tree species by both sexes of five bird species would result in 80 distinct event categories. Each observed foraging event is classified into the appropriate class for each of the four factors. As such, we can view each observed foraging event as a multinomial trial with a probability of falling in event category i given by p_i, where $i = 1, 2, \ldots, 80$. These probabilities can be estimated from the original frequency data by dividing the frequency of event i by the sum of the frequencies of all events. The data expressed in this form are still discrete, though no longer represented in integer form. If these probabilities are viewed as unbiased estimates of the true multinomial probabilities, assumed constant over the period of study, then the frequencies of each event category can be estimated by multiplying the total number of events (a constant) by the appropriate probability. This exercise will simply reproduce the original data indicating that its basic discrete nature has not been changed.

Ways of Looking at the Same Data: Continuous or Discrete Variables

Viewing the data as continuous random variables

Many authors have analyzed multinomial probabilities rather than event frequencies. That is, they have changed the representation of the data to appear as continuous rather than discrete random variables. Presumably the data have been standardized in this way, because some types of statistical models assume that the input data are continuous. Even so, the data are still discrete.

To analyze data with this structure, most researchers have employed an ordination algorithm such as principal components analysis (PCA) or, less commonly, correspondence analysis (see Miles, this volume). Prior to analysis, the data are arranged in a matrix with each row representing a species and each column a probability associated with a distinct foraging variable. Assuming random sampling, entries in this matrix represent the probability of observing species i engaged in foraging behavior j. To visualize similarities and differences among species, it is useful to think of plotting the rows of this matrix in a j-dimensional space.

A frequent goal of principal components analysis is to plot the rows of the matrix in terms of linear combinations of the column variables. The coefficients defining the linear combination are functions of the eigenvectors estimated from an association matrix of the column variables (usually a correlation or covariance matrix). The scalar product of the jth eigenvector times the ith row of the probability matrix produces the score for the ith individual on the jth principal component. The weights assigned to the foraging variables are estimated so as to maximize the variance of the principal component scores. After the new scores are computed they are plotted according to bird species. The arrangement of species (= points) in this space, viewed in terms of their point-to-point distances, is used to infer similarities and differences among the species. The principal component axes are given biological interpretations in terms of the correlations among the scores and the original columns of the probability matrix.

Correspondence analysis, or reciprocal averaging (RA), is similar to PCA in that it is also based on an eigenanalysis of a two-way matrix (species by probabilities). However, in RA both the rows (species) and columns (foraging behaviors) are analyzed and ordinated simultaneously. The algorithm is referred to as reciprocal because the species ordination scores are averages of the column (foraging variables) ordination scores, and reciprocally, the variable ordination scores are averages of the species ordination scores (Gauch 1982:144). A further difference is that PCA is based on Euclidean distances, provides equal weight to all points, and the ordination is centered at the origin (for mean-corrected data). In contrast, RA is based on chi-square distances, weights are proportional to row and column sums, and the origin is at the center of gravity of the data (Gauch 1982a:147–148). However, the techniques are very similar in their goal of reducing the dimensionality of the original space, and providing some logical ordering of the species that can be given a biological interpretation. One of the most useful aspects of RA is the biplot. In a biplot, both row and column variables of the two-way table are simultaneously plotted with respect to the principal axes (Moser et al., this volume). The biological interpretation of the ordination is based on the relative positions of row and column variables (points) in the plot.

Treating the same data as discrete random variables

It seems somewhat arbitrary to take data that are originally portrayed as a multidimensional, cross-classified matrix and collapse them into two-way matrix of species by foraging variables for analysis by PCA or RA. In doing so we artificially create a series of quasi-independent variables and ignore relationships among the original factors. In light of this concern, RA is to be preferred to either PCA or its variants (e.g., factor analysis). It is possible to use RA complementary to traditional discrete multivariate analyses (van der Heijden and de Leeuw 1985) and to explore both two-way and multidimensional tables based on the original event frequencies (Greenacre 1984; Moser et al., this volume). RA can be used to explore multidimensional contingency tables by the use of dummy variables (Greenacre and Hastie 1987) or by structuring the event frequencies into Burt tables (Greenacre 1984:140–143). A Burt table contains each factor in both rows and columns of the table, thus containing all possible two-way tables (see Moser et al., this volume, for an example).

Since the original data can be arranged as a multiway contingency table, it seems logical to retain this structure for analysis. This is accomplished through the use of log-linear models which explicitly estimate the interdependencies among the factors. For illustration, we return to our previous example of a comparative foraging study of both sexes (S) of five species of birds (B) and their use of bark versus foliage substrates (I) on four species of tree (T). Each of the observed foraging events can be classified by bird species, sex, tree species, and substrate: these are the four factors. The model, presented below, of complete

TABLE 1. HYPOTHETICAL EXAMPLE OF AVIAN FORAGING DATA ILLUSTRATING VARIOUS LOGLINEAR MODELS AND THE INTERPRETATION OF MODEL PARAMETERS

Full model

$\ln f_{ijkl} = u + B_i + S_j + T_k + I_l + BS_{ij} + BT_{ik} + BI_{il} + ST_{jk} + SI_{jl} + TI_{lk} + BST_{ijk} + BSI_{ijl} + BTI_{ikl} + STI_{jkl} + BSTI_{ijkl}$

Parameters:

B_i = bird species $\quad i = 1, 2, \ldots, 5$
S_j = sex (male or female) $\quad j = 1, 2$
T_k = tree species $\quad k = 1, 2, \ldots, 4$
I_l = substrate $\quad l = 1, 2$
f_{ijkl} = cell frequency in the $ijkl$ cell

Model of complete independence

$\ln f_{ijkl} = u + B_i + S_j + T_k + I_l$

Parameter	Interpretation
u	Mean of the logarithms of the expected frequencies
B	One-way term for bird species
S	One-way term for sex
T	One-way term for tree species
I	One-way term for substrate
BS	Sample size effects: the same proportion of males and females were not sampled for all sexes
BT	Not all bird species are utilizing tree species in the same proportions
BI	Not all bird species are utilizing substrates in the same proportions
ST	The two sexes are not using tree species in the same proportions
SI	The two sexes are not using substrates in the same proportions
TI	The proportion of utilized substrates is not the same for all tree species (implicit bird species effect)
BST	The association between sex and tree species depends upon the level of bird species (i.e., males and females differ in the use of tree species according to which species they belong to)
BSI	The association between sex and substrate depends upon the level of bird species
BTI	The association between tree species and utilized substrates is dependent on the level of bird species
STI	The association between tree species and utilized substrates is dependent upon whether the bird is a male or a female
$BSTI$	The association between tree species and utilized substrates is dependent upon whether the bird is a male or a female and this three-way association is in turn dependent upon the level of bird species

association among the factors, would involve all interaction terms of order four or lower plus all individual factors (Table 1):

$\ln f_{ijkl} = u + B_i + S_j + T_k + I_l$
$+ BS_{ij} + BT_{ik} + BI_{il} + ST_{jk}$
$+ SI_{jl} + TI_{lk} + BST_{ijk} + BSI_{ijl}$
$+ BTI_{ikl} + STI_{jkl} + BSTI_{ijkl}.$

In contrast, the model of complete independence of the four factors would contain only the terms for the individual factors (Table 1):

$\ln f_{ijkl} = u + B_i + S_j + T_k + I_l.$

The full model contains 15 classes of parameters: four main effects terms, six two-way interaction terms, four three-way interaction terms, and one four-way term. In all, 80 parameters need to be estimated (5 × 2 × 4 × 2 = 80). However, what we seek is the model with the fewest number of terms that adequately fits the data. By fit we mean that the chi-square statistic, based on the difference between observed and predicted frequencies, is not significant (e.g., P > 0.05). This model will lie somewhere between the model of complete independence and complete dependence. Inclusion of any interaction terms indicates some degree of dependence among the factors. In addition, to make interpretation easier, only hierarchial log-linear models are usually considered. For example, if any three-way interaction term is included in the model, then all two-way interaction terms involving those factors, and the individual factors, are also included in the model.

Model interpretation. Similar to linear models in the analysis of variance, there are alternative ways to block the factors to aid in interpretation. An example would be to define bird species (B) and sex (S) as explanatory or treatment variables and tree species (T) and substrate (I) as response variables. The parameter estimates by factor and interaction, and an interpretation of each parameter, are given in Table 1.

The interaction terms of primary importance are those involving some combination of explanatory and response variables. To illustrate the hierarchical nature of the models, if the highest order term required in the model was BTI, then the terms BT, BI, and TI, and B, T, and I would also be required for an adequate fit of observed and expected frequencies under the hierarchical principle.

Mixtures of continuous and categorical random variables

Foraging studies often involve a mixture of categorical and continuous random variables. For example, Sakai and Noon (this volume) recorded tree species and substrate types (categorical variables) as well as the height and distance from the trunk (continuous variables) of foraging flycatchers. They employed separate analyses, using different statistical models, of the two data types. However, one can use mixtures of variables in some analyses. For example, a PCA of mixed variable data sets is possible because the estimation of eigenvalues and eigenvectors is not dependent upon normality assumptions. Discriminant function analyses (DFA) can also be done with continuous and categorical variables, although logistic regression may be preferred in the two-group case because of its robustness to violations of the normality assumption (Press and Wilson 1978; for a contrasting opinion see Haggstrom 1983).

As an example, consider a multi-species study whose primary data have been arranged in a matrix with the rows partitioned by bird species and the columns representing foraging variables. Each row of this matrix is assumed to represent an independent foraging observation of an individual bird of a particular species. For each observation, bird species, tree species, behavior, substrate, bird height, and distance from the center of the plant are recorded. All but the last two variables are categorical. In general, any factor with k levels can be represented by $k - 1$ dummy (0/1) variables. If there are five possible tree species, then this variable is coded by four dummy, binary variables; four behaviors would be coded by three variables, and so on. (The sum of a set of 0/1 variables has approximately a normal distribution.) The species' groups are to be contrasted on the basis of the foraging variables by DFA.

A problem in discriminant analyses with both continuous and categorical variables is the procedure of selecting variables and thus the biological interpretation of the canonical variates. For example, some continuous variables may supply discrimination only if a particular discrete variable is already in the model (Daudin 1986). Several recent papers discuss the analysis of mixed variable data sets when group discrimination is the goal (Krzanowski 1980, Knoke 1982, Vlachonikolis and Marriott 1982, Daudin 1986) but reach no general consensus. Several authors have argued in favor of the location model approach to DFA, which involves aspects of log-linear analyses and parametric analysis of variance. This requires estimation of a large number of parameters and has not been implemented on any major statistical software package. Analyses of mixed variable data sets with standard statistical packages should be interpreted cautiously.

How are cross-classified categorical data best analyzed

It is possible to take cross-classified data and analyze them as discrete frequencies with log-linear models or to express the data as proportions for analysis by various ordination algorithms (e.g., PCA or factor analysis). But which method provides the clearest insights into the relationships among factors; and do different methods provide complimentary insights?

In the example discussed above of both sexes of five species of birds, a PCA ordination would be based on a matrix whose rows represent bird species-sex combinations (10 distinct categories) and whose columns represent all possible tree species by substrate combinations (8 distinct categories). Entries in this 10×8 matrix would represent the proportion of observations for species-sex combination i observed on tree species-substrate combination j. These entries can also be considered as conditional or multinomial probabilities. For example, entry ij would be interpreted as: given a random observation of species-sex combination i, what is the probability that it is foraging on tree species-substrate combination j. Biological inferences from the ordination of the rows of the matrix are based on distances among the rows plotted as points in the synthetic PC space and from the biological interpretations given to the PC axes. The statistical significance of interactions among the factors (bird species, sex, tree species, and substrate) is not explicitly examined. Rather, these methods of analysis lead to inferences about the similarities or differences among various species-sex combinations in terms of the measured tree species-substrate variables.

In contrast, log-linear analyses explicitly investigate the significance of interactions among the nominal factors and seek the simplest representation of the tabulated frequencies. The factors in these models can be viewed as possessing a treatment-response structure and the significance of any association between factors can be explicitly tested. Relationships among species-sex combinations would be inferred from a comparison of their parameter estimates (the BS_{ij} terms) or by a series of pairwise comparisons of species-sex by tree species and substrate contingency tables (see Raphael, this volume).

Ordination techniques, such as PCA or RA, are not primarily hypothesis testing procedures. Instead, they are most useful for exploring interrelationships among species or foraging variables. In contrast, log-linear models are often explicitly cast in an hypothesis testing context. This suggests that ordination analyses may be more valuable in the initial research into a species' or community's foraging patterns. Log-linear analyses may be used in a subsequent study to explicitly test for significant relationships among some subset of factors implicated by the initial, exploratory analyses.

For a geometric interpretation of factor relationships, ordination analyses are preferred to log-linear analyses. However, if log-linear analyses are done along with RA analyses of combinations of factors, complementary inferences can arise. Van der Heijden and de Leeuw (1985) argue that log-linear analyses yield insights into factor relationships whereas RA analyses provide insights into associations among levels within factors. To illustrate, one could initially analyze the multiway foraging data by log-linear algorithms to estimate the simplest model that adequately fits the observed frequencies. If the model contained significant interaction terms, then these terms could be examined in combination with the treatment factor by correspondence analysis. That is, one or more two-way tables of frequencies, in which the columns of the table represent all possible combinations of levels of factors within a significant interaction term, would be examined for association with the treatment factor and interpreted geometrically. This approach is illustrated by Moser et al. (this volume) and van der Heijden and de Leeuw (1985). A lucid discussion of the geometry of correspondence analysis is presented by Greenacre and Hastie (1987).

We have not seen a comparison of ordination algorithms and log-linear models on the same data set, but suspect that similar inferences about the relationships among factors would be drawn. An explicit comparison of these contrasting methods of analysis is an important area for future investigation. At this time it is not clear if one method is to be preferred over the other and whether more information is extracted from the data by conducting both analyses. However, the complimentary relationship among log-linear and correspondence analyses in the exploration of categorical variables appears most promising at this time.

SEQUENTIAL OR POINT OBSERVATIONS OF FORAGING BEHAVIORS

Two methods of recording foraging events are commonly used. Point samples record the first event observed (or the first recorded after a fixed waiting period to avoid recording only conspicuous behaviors). Sequential samples consist of sequences of events recorded during a fixed or variable time interval. The debate over the use of sequential or point observations focuses, in part, on the issue of statistical independence. Independence of observations is critical for the valid use of most statistical distributions, and thus for tests of hypotheses. Let the events y_1, y_2, and y_3 be mutually exclusive and exhaustive. Define y_1 equal to the event that a bird forages on a leaf, y_2 that it forages on a twig, and y_3 that it forages on bark. Further, let events y_1, y_2, and y_3 occur with probabilities p_1, p_2, p_3, and with the sum $(p_i) = 1.0$. Assuming only first-order correlations, we say that events y_1 and y_3 are statistically independent if the probability of y_3 occurring at time $t + 1$, given that y_1 occurred at time t, is equal to p_3. That is, the conditional probability of an event is equal to its marginal probability. We infer events y_1 and y_3 to be statistically dependent if the probability of observing event y_3 at $t + 1$, given y_1 at t, is not equal to p_3. Tests to examine dependencies in categorical and continuous data are discussed in Hejl et al. (this volume).

When foraging events are recorded in sequence, there is often a tendency for observations close together in either time or space to be more similar than events separated by longer time intervals or distances. Several authors in this volume have addressed issues of temporal dependency, but there has been little discussion of spatial dependency. An exception is Block (this volume), who sampled so as to ensure spatial independence of foraging observations within the same season. Spatial associations may actually be more prevalent, because so many studies are conducted when birds are spatially restricted. For example, subsequent observations of territorial birds, even if separated by long time intervals, may be significantly dependent because territories are likely to encompass different ranges of foraging possibilities and in different proportions. This is an area in need of further research.

Because most statistical models require random and independent observations, many researchers have recorded point observations. Such a sampling design may fulfill the independence assumption, but random sampling is difficult to achieve because the probability of obtaining a foraging observation differs among and within species. An argument, however, in favor of recording sequential foraging acts can be made because most of our data sets are sparse. Maurer et al. (this volume) have estimated that most foraging studies record fewer than 1% of the behaviors occurring during the period of study. Given the size of our sample relative to the sampling frame, we should attempt to collect as much information as possible and to record sequential observations. Such an approach, however, will necessitate recording data so that the temporal sequence of behaviors is documented. This information is needed to estimate the conditional probabilities (given that species i is engaged in foraging act j at time t, what is the likelihood that it will be engaged in act k at time $t + 1$) that form the elements of the first-order transition matrices.

We propose that researchers start with the assumption that sequential behaviors of the same individual are usually dependent (see Hejl et al., this volume). Further, we believe that estimates of the magnitude and direction of these dependencies will yield important insights into a species' foraging ecology and lead to improved predictive models. We support the argument of Raphael (this volume) in favor of Markov analyses, which estimate both the stationary distribution vector of foraging acts (however defined) and model building via log-linear algorithms. The latter analyses allow explicit tests for symmetry (i.e., the likelihood of the transition from behavior j to k equal to that from k to j) as well as comparisons of the transition matrices of different bird species (see Raphael, this volume, for details).

Our suggestion in favor of collecting sequential data is in contrast to that of Hejl et al. (this volume), Bell et al. (this volume), and Recher and Gebski (this volume), who suggested that point observations generally yield more precise parameters for estimating the probabilities of events. If sequences are recorded, then Hejl et al. recommended bootstrap or jackknife methods, because they are less time-consuming than Markov analyses, do not require assumptions about the order of the transitions, and provide estimates with smaller standard errors. However, these studies focused on estimating the mean probabilities of foraging events. We argue, from biological and not statistical grounds, that the transition probabilities themselves are as important in gaining insights to the behavior of foraging birds as are the expected probabilities. We recommend methods that provide both types of estimates.

SAMPLE SIZE REQUIREMENTS

In this symposium approaches to estimate sample sizes range from qualitative interpretations of graphs (Brennan and Morrison) to quantitative calculations of sample sizes based on different target levels of absolute or relative precision (L. Petit et al.). Suggested minimum sample sizes range from 40 to 500 independent observations to an extreme figure of 20,000!

Despite a diversity of approaches, all foraging studies must state what behavioral parameters will be estimated and with what levels of precision. The latter will require at least preliminary knowledge of the species' foraging variability. If the study is comparative, then determining what precision levels can be obtained is essential to estimate the power of any between-species comparisons. For species with variable foraging repertoires, sample size requirements may be so large that the researcher will need to be satisfied with tests of lower power. In this case, only differences among the most disparate species may be detected.

Log-linear analyses

Many papers in this symposium used log-linear models in analyzing categorical foraging data. Recall that the test-statistics for fitting log-linear models are only asymptotically chi-square distributed, and that some minimal sample size is needed for valid statistical inference. For a fixed sample size, the more cell frequencies that are estimated, the more questionable are the probability levels associated with the computed chi-square values. An indication of an inadequate sample size is an excess of small expected cell frequencies. Cochran (1954) suggested that no expected cell frequencies should be <1, and $<20\%$ of the cells should have frequencies <5. A rough guideline is that one should collect about five times as many observations as there are cells in the table (Raphael, this volume). If the table contains one or more rows or columns of all zeroes, the degrees of freedom associated with the test-statistic must be adjusted (Bishop et al. 1975:116).

Surprisingly, an analysis can be affected by too many observations. The result is that most models will fail to fit the data. If too large a sample is taken, any possible model structure will provide a poor fit no matter how minor the discrepancies. This occurs because chi-squares are proportional to the total sample size. If too large a sample is a problem, then the appropriate model may be selected by a stepwise procedure. For example, the magnitude of reduction of the sum of squares of the differences between observed and expected proportions can be computed each time an additional term is added to the model. Terms producing a large decrease in the sum of squares should be considered for inclusion in the final model.

A need to limit the number of factors

A large number of observations is needed to analyze a cross-classified table of even moderate size, because of the number of parameters that need to be estimated. Three factors with four levels each would require the estimation of 64 parameters. In contrast, a multiple regression model with three independent variables and no interaction terms would require, at most, the estimation of seven parameters. Because the number of possible sources of variation in avian for-

TABLE 2. FACTORS AND NUMBERS OF LEVELS CONSIDERED IN A STUDY OF THE FORAGING BEHAVIOR OF THE WESTERN AND HAMMOND'S FLYCATCHERS (FROM SAKAI AND NOON, THIS VOLUME)

Factor	Number of levels
Observers	4
Years	2
Age of forest	3
Stage of breeding cycle	4
Behavior	3
Tree species	6
Substrates	4

Total number of cells = 4 × 2 × 3 × 4 × 3 × 6 × 4 = 6912.

aging behavior is staggering, one cannot estimate all sources of variation, all significant interactions among factors, or investigate all possible factor levels.

For example, Sakai and Noon (this volume) used seven factors (Table 2) in their log-linear model. Considering the levels of all factors there were a total of 6912 cells for each bird species. This value greatly exceeded the total number of data points. The authors had decided *a priori* to pool across forest age because their objective was to estimate foraging patterns across the range of forest types occupied by the species. However, after recognizing the limitations imposed by the size of their data set, they chose to pool across observers and years as well. This probably masked statistically significant interactions and lost information on the joint distribution of some factors. Whether insights into significant biological interactions were lost is unclear.

Our point is that pooling is necessary and justifiable in almost all studies. When possible, interactions among factors that are of minimal biological interest should be controlled in the experimental design and data collection phases, and not in the analysis phase. Our zeal to partition sources of variation as finely as possible needs to be tempered with the recognition that one of our primary objectives is to understand a complex system in terms of a small set of key factors. We are interested in models that can describe and predict the average outcome of samples, not the outcome of individual observations.

MARKOV ANALYSES

We are aware of little published information on sample size requirements for Markov analyses. From unpublished simulation studies conducted by R. M. Fagen (Fagen in Colgan 1978: 107–108), some general guidelines have been proposed. If we let k equal, for example, the number of substrate categories considered, and assuming a first-order Markov model, then a sample of $2k^2$ foraging events is too few, $10k^2$ almost always adequate, and $5k^2$ a borderline value. Thus, if 10 substrate categories are considered, the minimum number of foraging events required is 500.

MULTIVARIATE ANALYSES

Estimates of sample size requirements for multivariate studies are considerably more complex than for univariate studies. We are still concerned with the precision of parameter estimates and the power to reject false null hypotheses, but in addition, one must consider the number of variables, the covariance structure of the data, the number of groups, and the sample size per group. There are "rules of thumb" but few are based on either analytical or simulation studies (e.g., Morrison 1984b). An example of a sample size effect, similar to univariate parameter estimates, is that the confidence interval around a principal component's variance (i.e., its eigenvalue) is a function of the reciprocal of the square root of its sample size (Neff and Marcus 1980: 37). Estimates of confidence intervals, as a function of different sample sizes, can be computed by resampling methods such as the jackknife or bootstrap (Efron 1982; Efron and Gong 1983; Miles, this volume). These computer-intensive methods to variance estimation have considerable application to foraging data.

A clear exception to the lack of information on sample size requirements is the recent study of Williams and Titus (1988). Based on a large scale simulation study, they have developed the following sampling rule: "For discriminant analysis of ecological systems with homogeneous dispersions, choose the total number of samples per group to be at least three times the number of variables to be measured." More guidelines such as these are needed. In their absence, researchers can empirically estimate the variance of many multivariate parameters (i.e., eigenvalues, factor loadings) by the use of jackknife and bootstrap methods. If the resulting confidence intervals on these parameters are too broad for study objectives, then larger sample sizes will be required.

CONCLUSIONS

We believe the papers presented in this symposium represent a significant advancement in the design and analysis of studies of avian foraging behavior. An explicit concern for precise and unbiased parameter estimates, and the necessary sampling design and sample sizes to achieve these goals, should become a regular part of all study designs. In addition, analytical techniques such as log-linear models, Markov processes, and correspondence analysis have become part of the repertoire for the analysis of foraging data. While most of these statistical techniques are not new to the ecological sciences, their application to studies of avian foraging behavior is novel. An additional advancement is the use of computer-intensive methods such as the jackknife and bootstrap. Diversity indices, factor loadings, eigenvalues, discriminant coefficients and other statistics that are regularly computed in foraging studies are usually done without estimates of their variances. Through intensive resampling of the original data, jackknife and bootstrap methods allow estimates of the standard errors of these statistics, yielding better or more appropriate insights into the variability of the systems under study.

Many issues require further work: the variable structure of foraging data and whether it is best analyzed by discrete or continuous multivariate models; the analysis of mixtures of continuous and categorical data; and whether we should sample so as to ensure independent observations or explicitly estimate the dependencies of foraging behaviors. We encourage investigators to address these and related issues in their future research efforts.

Observations, Sample Sizes, and Biases

FOOD EXPLOITATION BY BIRDS: SOME CURRENT PROBLEMS AND FUTURE GOALS

Douglass H. Morse

Abstract. Food exploitation is usually addressed in two major contexts in population and community studies of birds: (1) in consideration of niche relationships (niche theory) and (2) in choice of foods or feeding sites (foraging theory). The two approaches may be, but seldom have been, combined. Studies of niche relationships focus on comparisons of foraging performance. Food-choice or feeding-site (patch) studies compare foragers' performance with an optimum, usually based on maximizing resource intake. Both niche and foraging studies typically assume that resources (food) are limiting, but this assumption is seldom verified. Failure to test for resource limitation weakens most foraging studies, but this failure will be difficult to rectify. Few studies have concentrated on periods during which food limitation is likely to be most serious.

Foraging studies must determine how resources should be defined, when and how often foraging activities should be measured, which members of a population should be studied, how to compare foraging results with resource availability, and what effect other species' densities will have on foraging. Foraging theory evaluates efficiency in resource use. Failure of birds to realize foraging predictions may point to the mechanisms that shape foraging behavior. Studies combining niche and foraging theory should advance understanding of how communities develop structure. All of the studies discussed need to be evaluated in terms of fitness considerations. It is not sufficient merely to assume that selection exists for foraging variables independent of other life-history variables.

Key Words: Competition; fitness; food limitation; food exploitation; foraging; niche theory; optimal foraging theory.

The study of food exploitation, including foraging (searching and selecting) has been a major preoccupation of avian ecologists and behaviorists over the last 35 years. One might thus think that little work remains to be done, but a closer look will quickly change that impression. I will focus here on an evaluation of past and current work, and suggestions for a future agenda.

I will discuss four major areas: (1) food limitation and related competition; (2) some major foraging variables that often do not receive adequate attention; (3) the hiatus between optimal foraging theory and niche theory; and (4) fitness considerations. I have worked extensively in all of these areas, for the first three in studies of paruline warblers and mixed-species foraging flocks (reviewed in Morse 1980a), and for the first and fourth in current studies on other animals, primarily crab spiders (reviewed in Morse and Fritz 1987).

These topics deal with three distinct hierarchical levels: individual, population, and community. Food exploitation involves many different variables that interact to predispose a bird to forage where, when, and how it does. As such, it is a complex topic to study.

Much of the early work on foraging attempted to establish how species' ecologies differed and how these differences were related to coexistence. Often these studies compared the foraging patterns of coexisting species and used the resulting data to infer ecological relationships among the participants. In my opinion, these studies are unlikely to provide much further insight into basic understanding of community ecology. More recently, interest has shifted to optimal foraging, a subject largely concerned with how individuals can enhance or retain their efficiency in gathering food. This work focuses on the individual, rather than the community hierarchical level. As a result it has usually been treated as an issue distinct from niche partitioning studies; however, it is important to link these two bodies of study.

FOOD LIMITATION AND COMPETITION

Studies of niche partitioning deal with problems in which competition, often taken to be for limited food resources (or places to hunt for it), is assumed to be a driving force in niche differentiation. Few studies, however, have directly addressed the problem of food limitation.

The obvious way to test for food limitation is by manipulating the food supply. A few workers have attempted this technique with passerine birds (e.g., Krebs 1971), and in a non-controlled way we do so when we set up a feeding station. Most food supplementation studies have been done during the winter, perhaps for two reasons: because the investigator can readily manipulate the food (usually seeds), and because it is believed by many (e.g., Lack 1954, Pulliam and Millikan 1982) that northern residents are limited during this season. Unfortunately, foraging studies have not accompanied most food supplementation studies (but see Grubb [1987] for data on flock foraging). A logistical problem in studying food limitation may be that periods of limitation ("crunches") are infrequent events (Wiens 1977, Dunham 1980), so even carefully designed experiments on food limitation may not accomplish what the investigator intended.

Winter is not the only time that food limitation may occur, however. It is probably easiest to view this possibility from the perspective of total lifetime fitness. Foraging studies have focused on the short-term survival of adults, especially for species provided with winter food supplements. Yet, winter survival constitutes only part of the birds' problem; another major factor is fecundity (reproduction), which may not be directly related to winter food considerations. Reproduction in the vast majority of passerine birds, at least temperate-zone species, occurs during the spring and summer. At the period most advantageous for breeding it is just as important that conditions permit birds to accumulate the additional resources required for breeding as it is for birds to survive the winter. The ultimate result of failure to reproduce during the summer or to survive the contingencies of the winter is a net fitness of zero. Failure of an iteroparous individual to breed successfully during a given summer clearly is not equivalent to failure of the same individual to survive a winter, in the former instance it can try again. However, since many passerine species have high mortality rates, the mean number of seasons to breed may not greatly exceed one, so that the importance of breeding and winter contingencies may not differ greatly. Depending on the relative importance of winter or summer limitation, pressure on foraging efficiency may differ. Experimental tests of problems such as these would prove daunting in the field.

The crisis can thus occur at either season. Consecutive, catastrophic, breeding seasons at my Maine coast study areas (1972 and 1973) were associated with a population decline of up to 50% for some warbler species in certain study areas (Morse 1976a). I interpreted this poor level of success to the parents' inability to feed young during extended periods of stormy weather, with resultant high juvenile mortality (Morse 1971a, 1976a). Thus, although breeding contingencies may not be food-based, they can be. Sorting out these relationships requires that more attention be paid to these problems.

Thus, the problem of limitation is complex, and it probably differs among species, within species, spatially, and from one time to the next. Apparent niche shifts do not qualify as strong evidence for competition, notwithstanding the extensive pleading to parsimony that sometimes occurs (for example, see Diamond [1978:327]). Equally inappropriate are statements that since few satisfactory experimental demonstrations of competition exist, we may assume that competition is not an important structuring factor in communities (see Connell 1975). What *is* clear is that tests of food limitation or competition are not easy to perform in the field, especially for certain groups of animals, unfortunately including birds. Nevertheless, Connell's arguments, as well as those of Simberloff and his colleagues (e.g., Connor and Simberloff 1979), have had the salutary effect of encouraging workers to address these problems seriously. It is encouraging to see Schoener's (1983) report of some 150-odd studies in which he concluded that competition was adequately demonstrated experimentally in the field, although only seven of them came from birds (probably partly because of the extreme difficulty of performing the appropriate studies, and partly because investigators have not been in the habit of attempting to do so).

This difficulty should encourage us to look for indirect evidence. For instance, more food probably exists during insect outbreaks, such as spruce budworm (*Choristoneura fumiferana*) infestations, than birds can eat; perhaps that could form the basis for comparison with situations in which such a visible outbreak does not occur. Similar assumptions may be valid in instances of temporal ecological release; that is, when the exploitation patterns of an individual change from moment to moment with the presence or absence of another one or more individuals (see Morse 1967a, 1970, 1980a). These observations provide stronger evidence than Diamond's (1978) putative niche shifts, in that the same birds can be observed both in the presence and absence of other individuals.

Two other observations have to be made here. First, behavior normally associated with limiting situations, such as aggressive behavior, either intra- or interspecific, may occur even if resources themselves are not directly limiting. For instance, one can observe hostile interactions among spruce-woods warblers during major insect outbreaks, and that behavior may affect foraging patterns. This seemingly inappropriate behavior could be a consequence of these birds existing under limiting conditions at other times, with behavioral repertoires that function effectively then. During periods of superabundant food, the seemingly inappropriate venting of aggressive behavior may not exact a significant decrease in foraging efficiency. The important point is that interpreting aggressive behavior uncritically as evidence for resource limitation may lead to error. It is important, however, to ask whether one can assume the existence of unlimited resources if hostile encounters lower success. Numbers of some species of spruce-woods warblers may actually decline during spruce budworm outbreaks (e.g., Morris et al. 1958). These declines could be either a response to declines in the numbers of alternative prey or to the warblers themselves.

The second point is that competition takes place at an individual level, as do its consequences, even though ecologists have generally considered it as a population, or community, level phenomenon. This perception is largely a consequence of interest in population densities or species diversity. However, Martin (1986) has stressed that the concern to an individual is related to the pressures it personally experiences. If it is territorial, those concerns are pressures on its territory, rather than events going on in other parts of its population. This consideration assumes major importance if a strong gradient of food or habitat acceptability exists (the ideal free space of Fretwell and Lucas 1970). It should assume less importance in an ideal (hypothetical) homogeneous habitat, where the obvious solution is to space out. However, given that habitats are heterogeneous and individuals exhibit preferences for a part of the habitat, the ideal homogeneous habitat seems unlikely. The significance of competition as a between-individual phenomenon in the community remains to be worked out. Assuming that resource limitation exists in places, populations and communities can be divided into two categories of individuals, those that exist under varying levels of competitive stress and those that do not. This difference is a potential selective force, even if its consequences at population and community levels are not clear. For our present purposes, it may mean that we should separate these two groups of individuals for studies of food or foraging. To what degree are these individuals otherwise randomly distributed within a population, as in their food choices and foraging repertoires? If habitat or resource gradients are worth contesting, dominant and submissive individuals may experience secondary selective pressures for somewhat different patterns of resource exploitation.

Food limitation thus probably affects foraging strongly, but that result has seldom been directly demonstrated in bird populations. Bird populations exhibit a variety of characteristics, such as apparent niche shifts, which can be interpreted as evidence for competition, often food-based competition, and the tendency has been to accept as sufficient, far weaker evidence than I feel is appropriate. The constancy of the studies cited supports the importance of food-limitation and competition, but most of the individual studies in themselves provide only weak backing for this explanation.

SOME COMMENTS ON FORAGING VARIABLES

Potentially ecologically distinct categories have often been lumped in foraging studies. This procedure may produce erroneous conclusions, but probably more often, equivocal, no-difference results that may obscure major variables upon which natural selection may act. If these studies are to have an evolutionarily relevant context, it is important to identify and concentrate on such variables. If they are difficult to study, that is a serious problem, but if they are to be explored in a bird system, there may be no alternative to hard work. I will consider several foraging variables; the following list is not complete, but it should suffice to make my point. They can be broken down into two basic categories: (1) problems of scale and (2) problems of individual variability.

FORAGING CATEGORIES

The most basic sampling problem in studies of food exploitation is the investigator's selection of foraging categories (foraging sites, foraging motions, etc.). This problem is one of scaling: dividing the habitat into either too many or too few components will misrepresent the way in which foraging birds respond to it. Prior to MacArthur's (1958) study, ecologists seriously entertained the possibility that the spruce-woods warblers provided an important counterexample to the competitive exclusion principle, since these birds coexisted, in high diversity and large numbers, in seemingly homogeneous spruce forests. However, MacArthur quickly established that this conception grossly misinterpreted the warblers' space allocation patterns, for they do not respond to the forest as a single homogeneous entity, but as a highly divisible one. That conclusion indicated the necessity of using a scale similar to those used by the birds themselves. I will largely confine discussion to within-habitat divisions, but between-habitat distinctions may be important as well.

To select foraging categories from the viewpoint of adaptive or fitness considerations, the investigator should assume the perspective of the foraging bird. Detailed pilot studies may help to resolve the problem of which foraging categories to adopt, although they may greatly increase the effort necessary. In their calculations of foliage-height diversity, MacArthur and his colleagues (MacArthur and MacArthur 1961, MacArthur et al. 1962) attempted to discover what features were important for the presence of different species. Although their initial results were promising, attempts to extrapolate from some eastern forests and their inhabitants to other communities did not prove to be very successful. These were ambitious attempts to discover simplification and generality, and they may have foundered on those points. Their techniques assigned birds to habitats on the basis of a very few kinds of information, and now knowing that extensive within-habitat partitioning takes place in some groups (but not others), it is not surprising that they did not generally succeed. Their efforts were nevertheless important because they explored new methodologies. In contrast, if one selects too many categories in a mechanistic quest to establish whether quantifiable (although questionably biologically-based) differences occur among species, the data sets necessary may become prohibitively unwieldy, and the likelihood of finding spurious correlations increases.

Two apparent alternatives exist. The first is to divide foraging sites into what appear to be biologically meaningful subdivisions (e.g., crown, understory, ground, and perhaps with within-layer categories like trunk and large limb). The second may be not to attempt such biological divisions at this point, but to separate the habitat into arbitrary categories of such a size range that the members of the community as a whole will use all so-designated parts with reasonable frequency. Height intervals could be used (e.g., 3-m heights), and horizontal (within-layer) separation might be by distance, or by dividing the range of available substrate sizes into several categories. Both methods have their advantages and disadvantages; the first may be botanically relevant, but partition the habitat in a way that the bird never would; the second may avoid any unwarranted assumptions about a species' biology, but at the possible expense of creating biological redundancy and biologically irrelevant categories. The latter technique has the redeeming feature of presenting results that have not incorporated major, and possibly fallacious, biological assumptions into the data gathering at this early stage. Pilot studies that initially record data on a small scale may help to resolve this difficulty.

A related sampling problem concerns how foraging data are gathered. Many workers have gathered substantial numbers of observations from single individuals, an expedient way to obtain the large data sets needed for quantitative analysis. A positive feature of this technique is that it minimizes bias associated with the different visibility of individuals in different parts of the habitat. If an individual is more easily discovered in some parts of its habitat than in others, the larger number of subsequent observations gathered on it, the less the data should reflect the bias of initial observation (the "spotting" bias). But, these are not independent data points. The problem of independence of foraging data points is usually ignored (but see Morrison 1984a; Hejl et al., this volume), so that such studies present artificially (and incorrectly) inflated n's, and the specter of pseudoreplication. Ideally such difficulties can be redressed with analyses that compare bouts of foraging among individuals, but that has often not been the approach.

FOLIAGE SAMPLING AND BIRD FOOD CHOICE

A question of central interest to students of insect-gleaning birds is, "Do these birds specialize on certain types of food, and if they do, how?" Many and varied efforts have been made to sample the food supply in order to answer these questions. They differ in accuracy, difficulty, and human effort. Even if they sample the foliage accurately, that does not mean that the birds sample it in the same way, however (see Hutto, this volume). For instance, in some of my work in which I used exhaustive methods of foliage analysis that appeared to be very accurate (Morse 1976a, 1977), I found that Black-throated Green (*Dendroica virens*) and Yellow-rumped (*D. coronata*) warblers specialized strongly on large caterpillars that they gathered on spruce foliage, even though these caterpillars sometimes appeared in very low frequencies in the foliage samples. These same studies showed that large numbers of insects less than 2 mm in length, mostly psocids, regularly occurred in the samples, but seldom in the stomach contents. This absence might simply result from their not being visible in the stomach remains, but more likely, judging from their behavior, the birds did not perceive these insects because of their small size, or they eschewed them. If they were not profitable prey, not seeing them might actually improve the birds' foraging efficiency, in terms of energy gain per unit time. More time-efficient, but less accurate, estimates of insect standing crop raise additional questions and debate over how far one can extrapolate; for instance, what data from sticky traps can tell us, since these traps take highly biased samples (Southwood 1978). Thus, it is important not to interpret bird food intake from foliage studies alone, even accurate ones.

THE EFFECT OF ABUNDANCE OF SPECIES ON THEIR COMMUNITY IMPACT

The foraging impact of a species on its own members and on other species will differ with its abundance. This factor may assume considerable importance at the community level, but is often ignored, although it enjoyed considerable attention in the theoretical literature under the term of "diffuse competition" (MacArthur 1972). Thus, an abundant species that overlaps another species slightly may have a considerably heavier impact on it than will a third species that is relatively uncommon but overlaps it heavily. Ulfstrand (1976) has emphasized the importance of this role for the Willow Warbler (*Phylloscopus trochilus*), an abundant species in many parts of Europe. Since the density of this migratory species may greatly exceed that of any resident species, its impact upon them is likely to be major. In a collective sense the same relationship may exist between the spruce-woods *Dendroica* warblers, whose numbers may make up 70% or more of the total summer bird fauna, and the permanent residents. In both Europe and North America, some of the residents exhibit habitat shifts between seasons that strongly suggest competitive displacement.

TEMPORAL VARIATION

Temporal variation may also compromise the precision of foraging studies. It may occur at several time scales. Short-term studies run a high risk of presenting misleading results, for they may record only part of the variation inherent in a system, and possibly a very atypical part at that. The scales in question may range over several time frames: part of a day vs. an entire day, part of a season vs. an entire season, part of a year vs. an entire year, or one year vs. more than one year.

Some atypical periods, or the intervals between them, considerably exceed one year. They include both the "crunches" (periods of shortage) to which Wiens (1977) refers and periods of temporary superabundance. Representing the greatest inflections from a long-term mean, these two kinds of fluctuations are of great overall importance to the birds.

Extreme droughts may have a devastating effect on foraging opportunities. Grant (1986:191) found that drought affected the foods available to Darwin's finches, with many of the foods normally taken becoming unavailable, necessitating concentration on certain others. A severe population decline followed, with accompanying selection for individuals best able to exploit the remaining food types.

In my study of mixed-species foraging flocks I observed a major shift in foraging associated with a periodic mast crop of longleaf pine seeds (Morse 1967a). Foraging by Brown-headed Nuthatches (*Sitta pusilla*) changed markedly with this gradation; they shifted from a primarily insectivorous diet to one of over 80% seeds. To get the pine seeds they worked farther out into the foliage than they did at other times. There they came into frequent contact with the abundant and aggressive Pine Warbler (*D. pinus*). Simultaneously, fights between these two species increased markedly. The consequences for the nuthatches may not have been significant, because of the abundant source of food available to them; however, the consequences to the Pine Warblers, in terms of energy and time expenditure, may have been more severe. The warblers did not feed heavily on seeds, and thus probably profited marginally if at all from them. One might be somewhat at a loss to explain this strong hostile response, which obviously detrimentally affected the warblers' foraging efficiency, if one had conducted the study only during the one winter of the three that I devoted to this system. Mast years of longleaf pine occur every six years or so (Wahlenberg 1946).

Gradations over a somewhat longer time scale, or with highly mobile species, may result in striking population changes, which in turn are bound to affect interactions, and consequent food choice and foraging patterns as well. Sustained spruce budworm outbreaks, sometimes lasting a few years, produce marked shifts in the abundance of their predators. Three-fold increases in numbers of warblers and other insectivorous species may occur during budworm years, as revealed by comparing Kendeigh's (1947) censuses during a budworm outbreak at Lake Nipigon, Ontario, with those of Snyder (1928) and Sanders (1970) when few budworms were present. Aggressive behavior does not disappear during an outbreak, even among the budworm specialists, the Baybreasted (*D. castanea*) and Cape May (*D. tigrina*) warblers. These interactions might even be responsible for the declines in numbers of some species at this time, such as Blackburnian Warblers (*D. fusca*).

Foraging shifts may occur over shorter periods, also. Foraging may change during the course of a "normal" breeding season under equilibrium conditions, as in the activities of female spruce-woods warblers during the incubation period and at other times. At incubation time, the females forage at an unprecedentedly rapid rate, which probably affects the types of substrates used, their efficiency of using them, and the abundance of resources required for success. This contingency comes about because the females perform all of the incubation and also must hunt for themselves, resulting in an intensity of foraging unmatched at other times (Morse 1968).

Major changes may even occur over the period of a day. Holmes et al. (1978) found that the American Redstart (*Setophaga ruticilla*) changed its frequency of flycatching strikingly over the day, a shift correlated with the activity of its insect prey. Early in the day, while it was cold and the number of flying insects low, redstarts remained relatively inactive and did little flycatching; as it became warmer and the day drew on, flycatching became the prevalent technique. This type of shift in foraging behavior may be widespread in flycatching species. I have observed a similar pattern in Yellow-throated Warblers (*D. dominica*) (Ficken et al. 1968), which concentrated their activities on insects hiding in old pine cones during the cold of the early morning, but reverted to flycatching as the air warmed on early spring mornings and insects became active.

The social environment may affect the foraging patterns of these birds as well. Members of mixed-species foraging flocks exhibit this relationship especially clearly (Morse 1970). Members of socially subordinate species shift their foraging patterns in the presence of dominant species, and this change should affect the resources available to them. Even more important for many species are intraspecific dominance patterns (e.g., Black-capped Chickadee [*Parus atricapillus*], Glase [1969]). The effects of both interspecific and intraspecific flock relationships often shift over a period of minutes, and the results that one obtains inside and out of flocks may also differ markedly.

The tendency to participate in a mixed-species foraging flock may itself depend on food considerations, or perhaps predator avoidance is of primary importance. Social groups may also shift in character as a consequence of changes in climatic conditions. For instance, on warm winter days members may leave the groups, usually to take up a territory. The largest species, presumably the least vulnerable to surface-volume ratios of heat loss, quit the flocks first during warm stretches of winter weather (Morse 1978a).

Weather can strongly affect foraging patterns in other ways. Wet foliage may be one of the most serious factors for foliage gleaners, and the conclusions that one draws from observing foraging on wet and dry foliage may differ markedly. Carolina Chickadees (*P. carolinensis*) (Morse 1970) shifted from foliage-gleaning to large-limb hunting during rainy periods in the winter, thereby sparing their plumage from the wet foliage. Since the temperatures during these observations were near freezing and were preceded by freezing temperatures, these foraging shifts are unlikely to result from insect movements.

The problem of wet foliage assumes fundamental importance during the stormy weather that sometimes occurs while spruce woods warblers are incubating or feeding nestlings. They are extremely vulnerable to the loss of nestlings at this time (Morse 1971a, 1976a, 1977), and they, too, concentrate their foraging away from wet foliage, using areas such as the inner parts of branches, where most of them seldom forage at other times. This shift may also affect the one species that normally uses these areas most frequently and might therefore appear least vulnerable, the Yellow-rumped Warbler. Being the most subordinate of the *Dendroica* warblers in these communities, one might expect the added interactions to affect them adversely. More work needs to be done on the wet-foliage problem (also see Morrison et al. 1987a), but it is often difficult to gather these observations. Students of foraging tend to gather foraging data only on good days, and even if those data accurately portray the usual foraging patterns, they probably do not adequately represent the "crunch" situations. Birds may forage most efficiently when the foliage is dry and may even lack strong adaptations for the wet conditions, which could be so severe as to obviate the possibility of feeding young, anyway. If so, the birds are playing a game of chance during the breeding season, in which the odds favor escaping these extremely inclement conditions in any given breeding season.

Variability among the Members of a Population

Members of bird populations are not homogeneous in their characteristics, which leads to predictions of differences in foraging patterns and possibly in food secured. This must be considered in any study program. Size varies profoundly within species of many animal groups; for instance, as foragers, most fishes or salamanders vary over several orders of magnitude of mass during a lifetime (Werner 1977, Fraser 1976). The case of metamorphosing anurans, which shift from herbivorous to carnivorous existences at metamorphosis, is even more dramatic.

In contrast, within-population differences of birds are modest; indeed, with few exceptions passerine birds do not become foragers until they have reached full size. Even so, a number of ecologically significant differences in foraging occur regularly within bird populations, and they may turn out to be commonplace. I will consider two, male-female differences and adult-immature differences. Note, too, that dominance-related differences often have a size- or age-related element.

Male-female foraging differences may seem most likely to occur in association with marked sexual dimorphism. Differences in foraging repertoire between male and female Hispaniolan Woodpeckers (*Centurus striatus*), in which males and females differ in beak length by over 20% and tongue length by nearly 35% (Selander 1966), are thus not surprising. However, marked between-sex foraging differences are not confined to strikingly dimorphic species. They occur among monomorphic male and female spruce-woods warblers during the breeding season (Morse 1968), with males foraging higher in the vegetation than their females. The male heights matched their display heights more closely than those of the nest sites. Female foraging heights, in contrast, resembled nest heights more closely than male display heights. Only the females incubate. Others have subsequently found similar differences in several warblers (e.g., Sherry and Holmes 1985) and in vireos (Williamson 1971). Perhaps the most interesting example is that male Black-throated Blue Warblers (*D. caerulescens*) forage lower than their females, a difference that is associated with a tendency to display in open areas below the canopy (Sherry and Holmes 1985). Thus, an adequate display site, rather than height alone, seems to be the governing variable in this partitioning.

Adults and immatures may differ in foraging success, a likely consequence of the difficulty of learning how to forage. Such differences may not be apparent in most species; however, if they involve particularly difficult foraging repertoires, significant differences in success rates as well as in foraging patterns, foraging time, or items caught may exist. These differences have been reported for various seabirds (Ashmole and Tovar 1968, Orians 1969a) and wading birds (Recher and Recher 1969). I do not know of similar examples among territorial species of small birds. However, passerine fledglings learn by trial-and-error and narrow their foraging repertoires in the process (e.g., Davies and Green 1976). Further, heavy mortality often occurs at this time (e.g., Lack 1966), probably largely due to the inefficiency of foraging by these birds as they become completely independent. Consequently, although brief, this period may be one of fundamental importance and involve some of the most critical foraging decisions of a lifetime.

Thus, a diverse range of variables may affect the foraging patterns of birds. Not all will be of concern to each individual or at all times. Part of the challenge involves determining when such variables constrain success and when they do not. Knowledge of them and when they apply can provide insight to major fitness considerations.

OPTIMAL FORAGING THEORY

These niche-related studies differ from those of basic food-choice and foraging strategies. The latter type of work follows from the recent popularity of optimal foraging theory, the proposition that animals forage in a way that optimizes their success. In practice, workers usually substitute "maximize" for optimize, and energy gain per unit time for success, and implicitly use foraging success as an estimate of fitness. This work operates at the level of the individual, albeit with strong population implications.

Most of these studies are really not tests of optimal foraging theory (Krebs et al. 1983, Pyke 1984). Rather, they state whether their results are consistent with the predictions of a particular model, even though they may claim to do more. Nevertheless, these studies are of importance here, because my main concern is food choice and foraging behavior at the individual and population levels, rather than testing theory. These studies, as well as the more direct tests, reveal a variety of complications at the individual level, also. The first optimal foraging theory models were simple ones with no constraints and, depending on whose interpretation one accepts, were either quite successful (e.g., Pyke et al. 1977) or not very successful (Gray 1987). Quantitative predictions often were only approximate, suggesting complications. These deviations from theoretical predictions are generally attributed to constraints not built into the models, including inadequate memory, predator-avoidance, competition, dietary constraints, morphological constraints, and risk-minimizing. The nonconformities should not be surprising, but are of interest because they provide possible insight into the food and foraging problems discussed above, and their resolution may help to predict which species can prosper in different situations. Here I will put these studies into the context of food exploitation and suggest how to relate them to niche theory studies.

For these purposes one may divide optimal foraging theory studies into those concerned with diet-choice and those concerned with patch-choice. Patches deal directly with the use of space, which equates them somewhat with the niche relationships I have already discussed. Diet studies deal directly with food acquisition, rather than substrates exploited. Foods are ranked according to their energy value to the foragers, and foragers are expected to take only those items that will improve their overall energy balance (reviewed in Pyke et al. 1977). This general pattern often holds, although foragers frequently take items relatively low in value more often than predicted (e.g., Krebs et al. 1977). Krebs et al. attributed this deviation to the birds sampling the environment in a way that favored a long-term strategy; that is, obtaining information on food characteristics for possible future use when conditions have changed, such that these items might assume high positions on the birds' list of preferences. This simple model does not take into account such problems as memory; knowledge of the intricate detail necessary to make perfect choices; the problem that items are often discovered sequentially, rather than simultaneously, in many sorts of foraging situations; or the substantial hunting times required to find cryptic organisms, which will enhance the probability that most cryptic items, once found, will be taken despite an otherwise low value. Cryptic organisms are less likely than others to require substantial handling times or special physical abilities to exploit, which would further favor eating them once discovered.

This simple approach thus brings several problems with it. However, my purpose here is not to critique simple optimal foraging theory models, but to show that difficulties in fitting results to models indicate the existence of variables of basic importance. This work exposes a deficiency of understanding about foraging and related factors. Problems of learning and manipulation have received considerably more attention, primarily from the psychologists. It is in this area between experimental psychology and behavioral ecology that the lacuna exists (Kamil and Sargent 1981).

Patch-choice studies make similar predictions, but relate to aggregation of food items in space and strategies necessary to exploit them with maximum efficiency. Distances between patches, sizes of patches, and the like will have major effects upon decisions to move. If the forager has incomplete information on the alternatives available, this deficiency will complicate the result. Many of the same variables as those associated with diet choice will affect the patch-choice decisions made.

Several other optimality problems, such as optimal flock participation, are related in varying degrees to food choice and foraging. However, they tend to incorporate parts of the diet and patch-choice considerations, or play off food-patch contingencies against other demands such as reproductive considerations and social relationships, and therefore I will not discuss them.

Food availability differs considerably in its predictability, which confounds the probabilities of accomplishing feeding or foraging "goals." Birds require a high minimum energy input, and it may be necessary for them to adopt foraging strategies that incorporate this constraint. The alternatives are often referred to as risk-prone and risk-adverse. When food is the critical variable, starvation is the crisis that they must avoid. Life cycles will be heavily influenced by the patchiness of the environment as well as the abundance of resources. Risk-prone and risk-averse strategies assume major importance with high temporal variation in foraging conditions. If predictability of finding food is low, but overall resource availability is adequate to support the individuals present, individuals should adopt a risk-averse pattern; that is, they should use techniques that minimize the probability of starving because of an inability to locate food within the habitat. Strategies might include flocking, in which many eyes search for the occasional large reward that might feed all of the members of the flock. (If an individual cannot defend such an item, one need not invoke group-selective advantages for this system to operate.) However, if the average food availability is inadequate to feed an individual, it pays to play a risk-prone game (if one cannot leave the area). If one adopted a risk-averse strategy "successfully," the inevitable results would be starvation. A risk-prone strategy gives it a chance to survive and should be adopted by individuals in imminent danger of starvation. Indeed, the behavior of some individuals suggests that this is the case; at least, traits such as predator avoidance may largely disappear at this time. Whether their disappearance constitutes more than a physiological consequence of poor body condition is not always clear.

Risk assessment is a relatively new area of interest in foraging, and has not been developed extensively for ecological problems. However, Caraco (1981b), Clark and Mangel (1984), and others have studied it from the viewpoint of winter flock participation. Foragers may also adopt similar strategies in comparable, if not so extreme, situations. Moore and Simm (1986) reported that migrating Yellow-rumped Warblers adopted a risk-prone strategy when rapidly fattening, choosing variable rewards over constant ones of the same average abundance, consuming more items in a foraging bout, handling them more rapidly, and selecting especially profitable ones. However, upon attaining maximum body mass, they shifted to a risk-averse strategy, selecting predictable rewards rather than unpredictable ones of the same average abundance.

Diet-choice, patch-choice, and predictability thus all appear likely to play a major role in determining the food-choice and foraging strategies adopted by birds. The general guidelines of optimal foraging theory may provide a good framework from which to start, recognizing that the goal is not to test optimal foraging theory, but to use it as a tool to generate testable hypotheses about food and foraging choices.

These optimal foraging theory studies thus help to identify the spatial and temporal patterns and mechanisms by which animals obtain food. In turn, they should provide insight into how populations and communities are composed if food is a limiting resource.

FITNESS

Optimal foraging theory rests on the assumption that animals foraging as predicted maximize their fitness. Foraging animals may satisfy this assumption, but it is short-sighted to treat energy gain, or some other measure of foraging success, as an adequate or sufficient estimate of fitness, because it is an extremely indirect estimate. The behavior in question is often separated from eventual fitness payoffs by the better part of a life cycle. Considerable question exists, for example, about whether the strongest selective pressures occur in the winter or summer. Several workers have argued that winter is the critical time for some permanent residents (Lack 1966, Fretwell 1972), and this may hold for some species that remain at high latitudes. But, the question of payoffs will only be resolved in terms of reproductive success, and since the breeding season is remote from the time at which winter crunches occur, the matter may receive little attention. Lifetime fitness is what matters. If individuals survive several breeding and winter seasons, these periods all have to be taken into consideration, which makes isolated bouts of foraging behavior difficult to evaluate. It seems impossible to demand such information routinely, but the central nature of this assumption must be recognized. Lifetime fitness information, especially that put in the context of foraging repertoires or success, is virtually lacking; in fact, only a few studies of lifetime fitness have been made (Clutton-Brock et al. 1981, Arnold and Wade 1984). Birds, as iteroparous, supra-annual, highly mobile and often migratory animals, present especially difficult problems, but even partial tests, such as comparing the relationships of high foraging success at certain times with reproductive success, would advance foraging studies to a more critical level. If foraging considerations were not correlated with reproductive success, assumptions made in optimal foraging theory studies, and in niche-level studies as well, would have to be re-evaluated.

Grant (1986) has found that factors associated with feeding play a dominating role intermittently in the survival and consequent fitness of Darwin finches during serious drought periods. He also demonstrated that finch populations underwent strong directional selection at this time. However, if directional selection occurred then, one wonders what other factors normally act to produce a population not maximally adapted to these drought conditions in the first place. Other forces may dominate through the rest of their lifetimes, and possibly in ways that counter this feeding-related advantage of certain individuals (large beaks that facilitate feeding on seeds that smaller beaks cannot crack). If different forces really do act at different times, one should be very careful in interpreting optimal foraging theory results.

INTEGRATING FORAGING THEORY AND NICHE THEORY

Little effort has been made to integrate niche theory and foraging theory. My suggestions are largely based on Werner (1977), who used optimal foraging theory techniques to derive predictions about niche relationships and coexistence among three centrarchid fishes (bluegill [*Lepomis macrochirus*], green sunfish [*L. cyanellus*], and largemouth [*Micropterus salmoides*]).

Werner constructed cost curves of prey under controlled laboratory conditions, using data from the pursuit time, handling time, and capture efficiency of several food items by different-sized individuals of the three predatory species. Using estimates of resource distribution and abundance from the field and calculating cost-benefit ratios from the prey capture—handling data and caloric estimates of these prey—he set boundaries on the predators' predicted niche dimensions (food-size axes). The fishes' shapes and sizes affected the results strongly. These results can be matched against predictions from species-packing theory (niche overlap), thereby facilitating the introducing of food-exploitation patterns into predictions of community structure. Species-packing theory (MacArthur and Levins 1967, MacArthur 1972, May 1973) addresses the problem of how closely species can be fit into a community if sustained on one principal resource axis.

Werner's technique allows insight into mechanisms that drive community-level organization, and provides a link among morphology, efficiency of resource use, and overlap in resource use. Werner's food-size axes predicted the presence and abundance patterns of these species well; typically small lakes supported two of the three species, the smallmouth bluegill and the largemouthed bass; the intermediate species' (green sunfish) absence was usually predicted. The latter species was uncommon and coexisted only by habitat segregation. The bluegill and bass totally overlapped in habitat, but were complementary along the food-size axis; the green sunfish strongly overlapped the other two species on the food-size axis, but where it coexisted it was largely confined to a shallow fringe of habitat along the shore that was seldom used by the other species.

I will not discuss Werner's procedures in detail, because they are unlikely to be useful for food studies of birds. Gleaning species of birds may expend considerable time and effort in finding individual food items, so that once they are found, they will probably be taken. Selectivity of discovered prey items should thus not be as high as for consumers with relatively low searching costs, like sunfishes, although specialized hunting procedures could lead to a food intake unrepresentative of the standing crop. Nevertheless, it may be profitable to adopt an approach analogous to Werner's, especially to explore patterns of coexistence among closely related bird species, or for ecologically similar members of a community. Comparisons of species groups exhibiting high (spruce-woods warblers) and low (*Empidonax* flycatchers) levels of coexistence would assume particular interest.

Since birds do not share the complications of the tremendous intraspecific size variation seen in fishes and in many other animal groups, they have an important compensating advantage for studies conducted at a population or community level. If birds concentrate on a relatively few food types, as my Yellow-rumped and Black-throated Green warblers did (Morse 1976a), the problem of modeling efficiency may be tractable. Approximate energetic costs of different activities are known for several birds (e.g., King 1974), and can be readily estimated. Holmes and his colleagues (Holmes and Sawyer 1975, Holmes et al. 1979a) have estimated energy expenditure of several northern passerines. Their results suggest that it would be feasible to concentrate on the foraging strategies of different co-occurring species and to generate cost curves for exploiting the various stations recognized in bird foraging studies. These curves would be based on food availability and foraging efficiency (the major problem) in different sites and, similar to Werner's curves, could be used for predicting the presence or absence of species. Measuring food intake would constitute the most difficult aspect of such a study, but the procedure nevertheless warrants serious attention.

If one can establish the conditions under which species coexist, it should be possible to focus on which situations are the limiting ones and how they act in limitation. This approach should also provide insight into the conditions that permit the insinuation of non-equilibrium species (Bay-breasted and Cape May warblers among the spruce-woods *Dendroica*), as well as why some equilibrium species decline at these times.

Werner recognized foraging generalists and specialists, and habitat generalists and specialists, in his fish community. Members of bird communities also clearly differ in this way (Morse 1971a, 1977, 1980a). Some bird species may even differ in their tendencies to specialize or generalize along different foraging axes, thereby presenting potentials for segregation (Cody 1974, Ulfstrand 1977, Morse 1978a). For instance, the participants of English mixed-species foraging flocks that I studied (Morse 1978a) varied in relative specialization and separation from each other along dimensions of foraging substrate (e.g., limb, twig), height, and tree species. In contrast, species-poor North American flocks did not clearly separate along a tree-species gradient (Morse 1970). Thus, bird communities offer many opportunities for disentangling problems of niche complementarity and coexistence.

SYNTHESIS

Integration of work done at different organizational levels (community, population, individual) is needed to maximize advance in the understanding of food exploitation. Studies of niche-partitioning, as it relates to foraging, are well developed in their basics, although often suspect in light of questions about resource limitation or competition. They require considerable attention, however, to accommodate a wide range of variables in ways that focus attention on foragers at the level of the individual, in this way reflecting the action of selection. In that sense, a substantial part of the work needed might be considered corrective. In particular, this work needs to be focused toward periods of unusual demands or want, the "crunches" of Wiens (1977).

Although optimal foraging theory itself is not concerned with the mechanisms by which foragers make choices, it addresses foraging problems at a level that draws attention to these matters. An understanding of these mechanisms seems vital to comprehending fully the decisions that determine resource exploitation patterns and why some apparent options are exercised and others not (morphogenetic and phylogenetic constraint). Optimal foraging theory also addresses questions at a level that permits one to relate the behavior to fitness, a subject in great need of attention, both as it relates to foraging and to other problems. By doing so, it may be possible to start piecing together the events and interactions taking place in a community in a way that will reflect the action of initial selective pressures, adjustments to them, and possibly, evolutionary change.

ACKNOWLEDGMENTS

I thank W. M. Block, M. L. Morrison, S. D. Oppenheimer, and J. C. Robinson for commenting on the manuscript. The National Science Foundation has supported much of my work on warbler foraging and mixed-species foraging flocks.

A CLASSIFICATION SCHEME FOR FORAGING BEHAVIOR OF BIRDS IN TERRESTRIAL HABITATS

J. V. REMSEN, JR. AND SCOTT K. ROBINSON

Abstract. Studies of avian foraging behavior in terrestrial habitats suffer from a lack of standardization in the kinds of data gathered and in the terminology used to classify different activities. These inconsistencies partially reflect the variety of questions asked about foraging. If a standard terminology were used, then data on foraging behavior could be included among the standard data (e.g., clutch size, body weight, and mating system) routinely recorded for the biology of a bird species.

We propose a system for gathering foraging data for landbirds in which the five basic, sequential components of foraging (search, attack, foraging site, food, and food handling) are quantified separately. Data on searching behavior involve measuring continuous variables and are particularly critical for studies of energetics. The "attack" component is most in need of standardized terminology. The system that we propose separates the attack perch from the attack maneuver, and further subdivides the maneuvers into near-perch, subsurface, and aerial maneuvers. Each of these general categories is further subdivided according to details of attack movements and ways in which substrates are manipulated. Data on attack methods are primarily useful for studies of ecomorphology, but may also be important in bioenergetic and community-level studies.

Quantifying the foraging site involves measuring the following variables: general habitat (location in a study area), vertical position, foliage density, and substrate. Although identification and quantification of foods taken in the field is difficult, it can provide valuable information on food size (and taxon for larger items). Dietary data from stomach samples are useful for studies of resource partitioning when they show dramatic differences, but overlapping diets do not necessarily indicate that two birds forage in the same way. Food-handling behavior is seldom measured in the field, but is valuable in studies of optimal foraging behavior and ecomorphology.

Intercorrelations between each of these aspects of foraging can be determined from standard multivariate analyses. How finely to subdivide categories depends upon the kinds of questions being asked.

Key Words: Foraging behavior classification; foraging maneuvers; search; attack; foraging site; diet; food-handling; glean; sally; probe; manipulate.

There are almost as many ways of classifying and quantifying foraging behavior as there are papers on the subject. In part, this variety reflects the fact that no two species or groups of species forage in exactly the same way, and that no two habitats present exactly the same foraging opportunities. It is difficult, for example, to quantify the foraging methods of bark-foragers in the same way that one quantifies the foraging of foliage-foragers (Jackson 1979). Another factor contributing to the lack of standardization is that different kinds of questions often require different kinds of data. Many studies that focus on resource partitioning record only the details of foraging site selection and omit data on search and attack movements (e.g., Hertz et al. 1976). In contrast, studies of ecomorphology emphasize prey-attack methods (e.g., Osterhaus 1962; Fitzpatrick 1980, 1985), whereas bioenergetic and optimal foraging studies emphasize searching movements as well as prey handling (e.g., Sherry and McDade 1982, Robinson 1986). Even studies addressing the same questions in ecologically similar birds do not always measure the same variables.

This lack of standardization, however, reflects fundamental inconsistencies in the importance attached to the individual variables used to quantify foraging. For example, the "hawk" category of Sherry (1979) and Holmes et al. (1978, 1979b) includes attacks on prey animals that were flying when first seen, attacks on prey flushed from foliage by the foraging activities of the bird, and chases after prey that the bird attacked and missed. Later papers by Robinson and Holmes (1982, 1984) and Sherry (1983, 1984), however, showed that differences among these aerial maneuvers have important implications for diet. Remsen (1985) showed that the fine details of substrate use (e.g., dead leaves, moss) that are often ignored in many community studies were particularly important in resource partitioning in a tropical bird community. Rosenberg (this volume) further showed that even within a group as specialized as dead-leaf foragers, species differ in the kinds of suspended dead leaves that they search. The classification of foraging behavior, therefore, is more than a semantic problem: one can reach different conclusions simply by classifying foraging methods differently.

In this paper, we propose a system for measuring and classifying foraging behavior for non-raptorial landbirds. Our goal is to standardize data-gathering methods and terminology to permit among-site and among-species comparisons that are currently handicapped by the absence of a common vocabulary. If some standard system and terminology were used by those studying foraging behavior, then we could ask questions concerning the frequencies of various behaviors among communities. Such comparisons may provide important insights into community organization: in a field such as foraging behavior, in which community-wide experimental manipulations are difficult or limited, a comparative approach among species and communities may be the only method available for testing many hypotheses.

The system presented here separates five sequential components of foraging behavior, each of which is quantified separately when data are gathered in the field: (1) search (movements leading up to sighting of food or food-concealing substrates); (2) attack (movements directed at food item or the substrate that conceals it once sighted); (3) foraging site (including general location and specific substrate); (4) food (including type and size); and (5) food handling (after food item obtained). We recognize that many of these components are not necessarily independent, but we prefer to quantify each separately and to allow subsequent analyses to show intercorrelations.

At the outset, we recognize that any classification system inflicts typology on what may be gradients of behavior, but we see no other practical solution for organizing foraging information. Our goal is to distinguish among what we subjectively perceive to be functionally different categories. By using standardized terminology, data on foraging behavior can be included among the standard biological data reported for bird species. At present, reference works on birds, which typically include detailed, quantitative data on variables such as clutch size, body weight, molt and migration schedule, and mating system, tend to omit, or describe in superficial, qualitative ways, all aspects of foraging behavior except perhaps diet. Foraging data should be included in such reference works, because foraging behavior is an integral part of a species' biology, and because it relates to time-activity budgets, morphology, habitat selection, and social system. The foraging behavior of many species may also be as "typical" of that species as any other aspect of its biology. A standard vocabulary will eventually allow a more sophisticated review of the prevalence of various foraging behaviors in birds; such a review will be able to replace the often anecdotal, "who-does-what" approach with quantitative comparisons between taxa and regions.

The heart of this paper is the section on the attack, which is the phase of foraging that is most in need of a standard terminology based on functionally different categories (cf. Moermond, this volume). We discuss briefly data on bird diets, and also propose a standard terminology for food-handling techniques. The final section of our paper deals with some of the ways in which data can be analyzed to address questions of resource-partitioning, bioenergetics, and ecomorphology.

SEARCHING BEHAVIOR

Searching behavior includes those movements used to search for food or substrates that hide food; under our definition of searching behavior, "search" ends once food or food-hiding substrates are sighted and attacked. Searching methods have usually been quantified by recording the lengths and rates of movements between perches and the time intervals between movements (e.g., MacArthur 1958, Cody 1968, Williamson 1971, Fitzpatrick 1981, Robinson and Holmes 1982). Other variables are: (1) distance covered per unit time; (2) number of stops and time-spent-stopped per unit time; (3) number of attacks (and % successful) per unit time; (4) direction of movement between stops, in three dimensions if appropriate; (5) sequential distribution of locations of stops (to calculate return rates to previous stops). Between-foraging-site movements can be categorized as: (1) walk, (2) hop, (3) jump (leg-powered leaps that cover more space than the typical hop), (4) run, (5) climb (with notations on whether or not the tail is used as a brace), (6) glide, (7) flutter, and (8) fly. Robinson and Holmes (1984) further distinguished between hops in which American Redstarts (*Setophaga ruticilla*) fanned their tails and lowered their wings, and hops in which there were no extra movements. Some birds also use their wings for support when hopping on thin, weak perches, a movement that could be called a "flutter-hop" (Robinson, unpubl. data). Also of interest are postures during searching that are seldom qualitatively described (for exceptions, see any E. O. Willis reference) or quantified but that may have subtle morphological correlates. Postures are particularly difficult to categorize because most species move and change postures frequently, and because head, wing, and tail orientations are all simultaneously involved. Perhaps the advent of telephoto video-cameras will permit such analyses; in this paper we deal with postures only peripherally. More amenable to quantification are changes of orientation while searching from a perch, either with head or body movements. For example, many species have characteristic side-to-side movements, whereas others maintain a straight-ahead orientation. Many species have characteristic wing-flicking or tail-wagging movements that accompany foraging, the significance of which is not often understood; the frequency of such actions could also be quantified. There is also a parallel literature on the searching behavior of various lizards (Moermond 1979, Huey and Pianka 1981).

In general, searching movements of birds form a continuum from "active" to "passive" modes (Eckhardt 1979). Active foragers change perches at a high rate, including many hops (or steps in species that walk), and most flights are short. Passive foragers seldom change perches, but fly long distances when they do move. The subset of birds that Eckhardt (1979) chose from the community that he studied fit into active (primarily wood-warblers) and passive (primarily tyrannids) foragers. A subsequent study of a different forest bird community (Robinson and Holmes 1982), however, found that many species did not fit cleanly into either category. Tanagers and vireos, for example, were intermediate in their rates of hopping and flying and in the lengths of their flights. Another species, the American Redstart appeared to add many movements of its wings and tail designed to flush prey to its already active foraging behavior. Furthermore, the searching movements of the Black-capped Chickadee (*Parus atricapillus*) depended upon the distribution of whatever substrate (e.g., dead leaves) an individual of this species was searching at the time. Searching movements, therefore, cannot necessarily be categorized typologically in the same way as attack maneuvers (see below). We agree with the approach of most authors who quantify and present data on searching movements separately (e.g., Robinson and Holmes 1982, Landres and MacMahon 1983, Holmes and Recher 1986b).

Many birds use food-searching methods that are similar to attack methods (see below). Robinson (1986), for example, quantified the rate at which Yellow-rumped Caciques (*Cacicus cela*) used the following search tactics: "probe-searches," in which an individual searched a dense cluster of leaves; "hover-searches," in which an individual searched the tips of foliage while hovering under them in stationary flight; and "hang-searches," in which an individual searched a nearby substrate while suspended below it. If any of these searching tactics led to a prey capture, then they were quantified separately as prey attack maneuvers. The only difference between a "probe-search" and a "probe-attack" (see below) was in whether or not prey was located. This illustrates a problem in distinguishing between searching and attack methods in species that search for concealed food: any classification system inevitably includes data on both searching and attacking behavior. Woodpeckers, for example, both search for and attack prey by removing an outer layer of bark. We have chosen to classify these methods as attacks (see below) on food-concealing substrates. Distinguishing between search and attack phases, however, represents an unresolved issue in some parts of our classification scheme, especially for those species that peer closely and unambiguously at particular substrates, such as curled leaves, leaf tops, and crevices. Although these species search a particular substrate in a manner that is analogous to any of the substrate-manipulation maneuvers outlined in the attack maneuver section (see below), our scheme does not include "peering" or "scanning" as an attack maneuver. In the field, we record substrates that are unambiguously searched with the notation "visual search," but we are uncertain as to how to include such data in analyses. Certainly, such visual searches are important in analyses of substrate use.

Several other kinds of search behavior are sufficiently distinct to warrant separate treatment. Many birds follow disturbances that expose prey; such disturbances include fires, other animals (particularly other bird species, ungulates, monkeys, and army ants), and humans and their machines. Many recent studies place disturbance-followers in a separate guild (e.g., Karr 1980, Terborgh 1980a, Terborgh and Robinson 1986).

Many birds steal food from other birds, and some species rely on this tactic (kleptoparasitism) for locating and capturing food (see Brockman and Barnard [1979] for review). Finally, some landbirds form mutualistic food-searching associations either with conspecifics (e.g., Kilham 1979, Mindell and Black 1984) or other bird species (Jackson 1985), especially within mixed-species flocks (e.g., Munn 1986). These sorts of associations should always be recorded.

ATTACK BEHAVIOR

We define the "attack" phase as that portion of foraging behavior from the moment when a food item, or food-concealing substrate, is sighted to the moment when a capture attempt is made. Thus, we include within "attack" phase those behaviors aimed at dislodging or revealing food before it is sighted, such as various kinds of substrate manipulation (e.g., flaking, hammering, gaping). We further subdivide the attack phase into (1) perch and (2) maneuver.

Few studies have quantified parameters concerning the characteristics of the perch from which an attack is launched. The numerous studies by E. O. Willis (see references) have shown that a species' presence in a particular habitat or microhabitat may be determined in part by availability of suitable perches. Certain species may also specialize on perch types not used by other species. For example, two small tanagers (*Hemispingus xanthophthalmus* and *H. verticalis*) that characteristically search the uppersides of leaves in dense clusters do so by perching on the clusters themselves (Parker and O'Neill 1980; Parker et al. 1980, 1985). Several species in the vireonid genus *Hylophilus* characteristically grasp the margins of leaves for perches to reach the undersides of these leaves (T. A. Parker and JVR, unpubl. data). Furthermore, studies such as those by Partridge (1976a; cf. Leisler and Winkler 1985) have revealed important morphological adaptations associated with particular perch types.

Perch type can be quantified using the same variables as those used in our scheme for "substrate" (see below). In practice, most measurements taken for the substrate will be the same as those for the perch except for the details of perch angle and diameter; therefore, the increase in volume of data to be recorded in the field is minimal. Those species that search while moving do not really have a "perch" *per se*; for instance, some species search and attack while in continuous flight ("screening," see below) or while hovering (e.g., Say's Phoebe [*Sayornis saya*], Grinnell and Miller 1944; bluebirds [*Sialia* spp.], Power 1980; and Restless Flycatcher [*Myiagra inquieta*], Ford et al. 1986).

Our classification of attack maneuvers categorizes them with respect to the complexity and required agility of each behavior. For example, we assume that aerial maneuvers and substrate manipulation require greater agility and more energy than those maneuvers in which a food item is removed from a substrate next to the bird's perch. Our classification also attempts to remove where possible the influence of substrate; thus, foraging motions that appear similar, but are directed at different substrates, are grouped together. Such similarities may be superficial, and foraging motions are certainly influenced by the types of substrates at which they are directed. Nevertheless, we prefer to group to-

TABLE 1. PROPOSED CLASSIFICATION SCHEME FOR ATTACK METHODS OF THE FORAGING BEHAVIOR OF NON-RAPTORIAL LANDBIRDS, WITH SYNONYMS OR PRESUMED SYNONYMS FROM OTHER STUDIES (SEE TEXT)

I. Near-perch maneuvers
   A. Surface maneuvers
      1. Glean (= pluck, perch-glean, pick)
      2. Reach (= stretch)
         a. Reach-up (= crane)
         b. Reach-out
         c. Reach-down (= lean, duck-under)
      3. Hang (= hang-glean)
         a. Hang-up (= hang vertical, hang head-up, vertical clinging)
         b. Hang-down (= hang head-down)
         c. Hang-sideways (= hang-side, vertical clinging)
         d. Hang-upsidedown (= hang horizontal)
      4. Lunge (= dart, rush)
   B. Subsurface maneuvers: no substrate manipulation
      1. Probe
   C. Subsurface maneuvers: substrate manipulation
      1. Gape
      2. Peck (= tap)
      3. Hammer (= drill)
      4. Chisel
      5. Flake (= bill-sweep, toss)
      6. Pry
      7. Pull
      8. Scratch
II. Aerial maneuvers
   A. Leg-powered maneuvers
      1. Leap (= jump-glean, jump). Include leap-distance and leap-angle.
   B. Wing-powered maneuvers
      1. Sally (= hawk, snatch, flycatch, hover-glean, hover). Include sally-distance and sally-angle.
         a. Sally-strike (= outward strike, upward strike, snatch)
         b. Sally-glide
         c. Sally-stall (= hover, hover-glean)
         d. Sally-hover (= hover, hover-glean)
         e. Sally-pounce (= land-and-glean, pounce, dive-glean)
      2. Flutter-chase
      3. Flush-pursue
      4. Screen (= hawk)

In general, studies of attack behavior in landbirds have distinguished the conspicuous aerial maneuvers from other attack behaviors, but the nonaerial attack maneuvers have often been lumped in one category, usually "glean." Such merging of nonaerial maneuvers into one or a few categories might obscure important behavioral differences among species that have implications for adaptive morphology (e.g., Richards and Bock 1973, Partridge 1976a, Leisler and Winkler 1985), search tactics, niche overlap, and food selection.

Our classification (Table 1) does not include certain maneuvers that appear to be rare, such as digging in ground by using a strongly curved bill in a hoe-like motion (Engels 1940), using spines and twigs as tools for probing (reviewed by Boswall 1977), using the head as a brace to provide leverage for foot scraping (DeBenedictis 1966, Kushlan 1983) or as a buttress to move or dislodge substrate (Keast 1968), vibrating feet to startle prey in leaf-litter (Hobbs 1954, Wall 1982 as cited by Edington 1983), rustling leaf-litter to startle prey (Potter and Davis 1974), and crushing twigs with the bill to expose prey therein (Mountainspring 1987). The system in Table 1 can be expanded as needed to include any such rare behaviors.

Many studies of foraging behavior that make interspecific comparisons, or intraspecific comparisons among seasons or habitats, have presented their data in the form of a diversity index and have not included the original data with percentage of observations in each foraging category. Other studies have identified the number of species in a community associated with various foraging categories, but have neglected to identify which species belong in each category. We think that the original data themselves should be presented to facilitate comparisons with other studies; they should at least be published as appendices.

An outline of the categories with definitions of each attack behavior follows. Some categories are not mutually exclusive. For example, a bird that "sally-hovers" might also "probe" while hovering. Therefore, many attack maneuvers can have compound names, such as "sally-hover-probe" or "reach-out-gape."

Each maneuver category is accompanied by some examples from the literature. Our literature survey is intended to be illustrative rather than encyclopedic. We tend to cite examples from recent, quantitative studies, rather than older, more qualitative material. Although the descriptive sections of the latter are often superior, much of the older material is contained within more general life-history studies and is therefore more difficult to locate.

In choosing a standard vocabulary, we have attempted to use simple, descriptive terms, which, if possible, are already frequently used in studies of foraging behavior; we have not hesitated to "synonymize" many favorite terms, including many of our own.

I. Near-perch maneuvers (target food item can be reached from bird's perch)
   A. Surface maneuvers
      1. *Glean:* to pick food items from a nearby substrate, including the ground, that can be reached without full extension of legs or neck; no acrobatic movements are involved. Emlen's (1977) and Mountainspring's (1987)

gether similar-appearing behaviors to alert morphologists to these potential similarities, rather than to allow the substrate category to separate automatically such behaviors. Because the system presented here also requires that the substrate also be recorded, no information is lost by excluding substrates from the behavior categories.

"pluck," Fitzpatrick's (1980) "perch-glean," and Moermond and Denslow's (1985) and Remsen's (1985) "pick" are synonyms. Perhaps the majority of maneuvers performed by most foliage- and ground-searching birds are "gleans." For example, 51% of the forest species studied in the Andes by Remsen (1985) and 53% of the forest species studied in Australia by Ford et al. (1986) used glean as their principal foraging maneuver. Because gleaning is presumably the least costly maneuver in terms of energy expenditure, it is not surprising that it is used so frequently (Remsen 1985; Moermond, this volume).

2. *Reach:* to extend completely the legs or neck upwards, outwards, or downwards to reach food (after Moermond and Denslow 1985, Remsen 1985). Because most studies seldom distinguish "glean" from "reach," the frequency with which "reach" maneuvers are used is not generally known. Strong interspecific differences among congeners in ability to reach are associated with morphological differences (Snow and Snow 1971, Moermond and Denslow 1985). Some frugivores, especially toucans and some tanagers, obtain their food by reaching (Moermond and Denslow 1985). Morse (1967b), who distinguished "stretching"—which is probably equivalent to our "reaching"—from gleaning, found that it was used seldomly (0–5% of all maneuvers) in the six species of wood-warblers studied. Three further subdivisions may be made with respect to direction:

   a. *Reach-up:* to reach above the bird. This is synonymous with the "crane" of Greenberg (1987b). This maneuver is used especially frequently to pick prey from undersides of leaves. The Pale-legged Warbler (*Basileuterus signatus*) uses this motion, along with the next, more frequently than any other maneuver (Remsen 1985).
   b. *Reach-out:* to reach lateral to the bird. A maneuver used especially frequently to pick prey from nearby leaves and branches.
   c. *Reach-down:* to reach below the plane of the feet. This is synonymous with the "lean" of Greenberg (1987b) and probably the "ducking-under" of Rabenold (1980). This maneuver is used by many tanagers, especially *Tangara,* when foraging on branches (Snow and Snow 1971; Skutch 1981; Parker and Parker 1982; Remsen 1984, 1985; Hilty and Brown 1986; Isler and Isler 1987); tanagers often reach-down alternately on opposite sides of a branch as they move along the branch, as does the wren *Odontorchilus branickii* (Parker et al. 1980). A bird-of-paradise (*Parotia carolae*) apparently uses a similar maneuver when searching branches (Forshaw and Cooper 1979). At least one hummingbird (*Metallura tyrianthina*) uses reach-down maneuvers to reach more than a third of its flowers (Remsen 1985).

3. *Hang:* to use legs and toes to suspend the body below the feet to reach food that cannot be reached from any other perched position. "Hang-glean" of Recher et al. (1985) and Robinson (1986) is a synonym. Differences in frequency of use of "hang" among similar species may have subtle consequences for morphology (Partridge 1976a, Leisler and Thaler 1982). Parrots use "hang" frequently (Forshaw 1973 and references therein). Chickadees and titmice (Paridae), bushtits (Aegithalidae), and some thornbills (*Acanthiza*) frequently "hang" to reach undersides of branches and leaf tips (e.g., Gibb 1954; Root 1964, 1967; Grant 1966; Sturman 1968; Partridge 1976b; Rabenold 1978; Moreno 1981; Alatalo 1982; Bell 1985b; Recher et al. 1985, 1987; Laurent 1986). The Palm Tanager (*Thraupis palmarum*) in Trinidad "hangs" almost exclusively when searching for insects in foliage (Snow and Snow 1971). The Blue-backed Conebill (*Conirostrum sitticolor*; Thraupinae) also uses this maneuver as its primary means of attack (Remsen 1985). Other insectivores that use "hang" regularly include: Rufous-browed Wren (*Troglodytes rufociliatus*; Skutch 1960), some wood-warblers (Root 1967, Ficken and Ficken 1968, Elliott 1969, Andrle and Andrle 1976, Rabenold 1980), Speckled Tanager (*Tangara guttata*; Snow and Snow 1971); some white-eyes (*Zosterops*; Gill 1971, Earlé 1983); Ruby-crowned Kinglet (*Regulus satrapa*; Rabenold 1978); the furnariid *Siptornis striaticollis* (Eley et al. 1979); and Sharpbill (*Oxyruncus cristatus*; De L. Brooke et al. 1983, Stiles and Whitney 1983). Some vireos "hang" when grasping the margins of leaves to reach food that cannot be reached from branches (*Vireo griseus,* Nolan and Wooldridge 1962; *V. huttoni* and *V. gilvus,* Root 1967); several tropical vireos (*Hylophilus*) use this maneuver frequently if not predominately (Greenberg 1984a; T. A. Parker and JVR, unpubl. data). Many species that extract prey from hanging dead leaves "hang" (and "reach") to investigate isolated dead leaves (Skutch 1969 for *Automolus ochrolaemus*; Greenberg 1987b; K. V. Rosenberg, unpubl. data; JVR, unpubl. data). Among frugivores that "hang" to reach fruit are *Euphonia violacea* (Snow and Snow 1971) and two species of woodpeckers (Moermond and Denslow 1985). Several hummingbirds "hang" to reach flowers (Parker and O'Neill 1980, Parker and Parker 1982, Parker et al. 1985, Remsen 1985). Four types of "hang" maneuvers (Fig. 1) should probably be distinguished (modified after Partridge 1976b,

Rabenold 1980, Alatalo 1982, Earlé 1983, and Greenberg 1987b):
a. *Hang-up:* to hang, head-up.
b. *Hang-down:* to hang, head-down. This differs from reach-down only in that the bird is clinging to a vertical surface or side of a horizontal surface, rather than perching on the upperside of a surface in reach-down. There is probably a continuum between the two maneuvers, and in fact, Moermond and Denslow's (1985) "reach" would include maneuvers here considered to be "hang-down."
c. *Hang-sideways:* to hang on the side of a substrate with body axis parallel to the ground and with the bird's side oriented upwards.
d. *Hang-upsidedown:* to hang, belly-up, on underside of horizontal or diagonal surface.

These same four categories may also be applied to the foraging behavior of all specialized bark-foraging birds (e.g., woodpeckers, dendrocolaptids, certhiids) that characteristically hang while searching and attacking; in bark-foraging birds, these maneuvers are probably best considered postures rather than maneuvers.

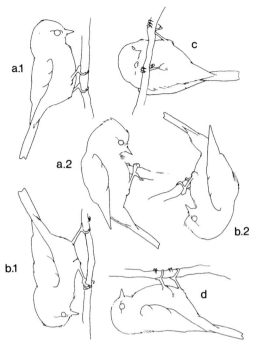

FIGURE 1. "Hang" maneuvers: (a.1) = "hang-up" on vertical perch; (a.2) = "hang-up" on horizontal perch; (b.1) = "hang-down" on vertical perch; (b.2) = "hang-down" on horizontal perch; (c) = "hang-sideways"; (d) = "hang-upsidedown." Drawing by Donna L. Dittmann.

4. *Lunge:* those maneuvers in which the food item is beyond the range of "reach," but rapid leg movements rather than flight are used to approach and capture the prey. This is synonymous with the "lunge" of Greenberg (1984a), except that Greenberg's lunge would include movements that we call "reach-out." Root's (1967) "rush" is a combination of our "sally-pounce" (see below) followed by our "lunge." Some studies have used "dart" for foliage-gleaning birds and "rush" for ground-foraging birds as presumed equivalents. Several ground-foraging birds, particularly thrushes (Heppner 1965; Smith 1973; Tye 1981; Willis 1985a, 1986) and ground-cuckoos (*Neomorphus*: Willis 1982a), and also some bulbuls (*Bleda*; Willis 1983a), tyrannids (*Muscisaxicola*, Smith and Vuilleumier 1971; *Corythopis torquata*, Willis 1983b), and antbirds (*Gymnopithys*, Willis 1968; *Grallaricula nana*, Parker et al. 1985) use the lunge maneuver regularly. Arboreal birds that also regularly lunge include: Red-crowned Ant-Tanager (*Habia rubica*; Willis 1960), Plain-brown Woodcreeper (*Dendrocincla fuliginosa*; Willis 1972), Chestnut-crowned Gnateater (*Conopophaga castaneiceps*; Hilty 1975); Black-headed Grosbeak (*Pheucticus melanocephalus*; Airola and Barrett 1985), the tiny, canopy antwrens of the *Terenura callinota* superspecies (Remsen et al. 1982; Stiles 1983; T. A. Parker, unpubl. data), White-shouldered Tanager (*Tachyphonus luctuosus*; Greenberg 1984a; JVR, unpubl. data), and an undescribed species of *Cercomacra* (Formicariidae; Parker and Remsen 1987).

B. Subsurface maneuvers (bird penetrates or manipulates the substrate rather than removing food from its surface; the attack is directed at food that cannot be seen from the surface without substrate manipulation).
1. *Probe:* to insert the bill into cracks or holes in firm substrate or directly into softer substrates such as moss or mud to capture hidden food. This tactic is often associated with specialized morphologies adapted for specific substrates. Most probers have long, slender, decurved bills for reaching deep into crevices, tubes, holes, and soft substrates such as mud or moss. Those that probe bark often have specialized hindlimb morphology and tail structure for climbing on and bracing against branches (Richardson 1942, Bock and Miller 1959, Feduccia 1973, Norberg 1979). Several unrelated groups have converged on similar morphology associated with bark probing: the creepers (Certhiidae), some woodcreepers (Dendrocolaptidae), and the Australian treecreepers (*Climacteris* spp.). The scythebills (*Campylorhamphus* spp.) and the Long-billed Woodcreeper (*Nasica longirostris*), with some of the longest bills relative to body size of any passerines, use their bills for probing deep into holes in tree trunks and

bamboo stems and into large bromeliad clusters (Pierpont 1986; T. A. Parker, unpubl. data; JVR, unpubl. data). Some woodpeckers do more probing (and gleaning) than the more "typical" woodpecker maneuvers, such as "peck" and "hammer" (e.g., Bock 1970; Short 1973, 1978; Cruz 1977; Alatalo 1978; Cruz and Johnston 1979, 1984; Stacey 1981; Askins 1983; Pettersson 1983; Kattan 1988). Bark-foraging birds that also probe epiphytic vegetation include the Brown Creeper (*Certhia familiaris*; Stiles 1978), many woodcreepers (Willis 1983c, d), and some woodpeckers (Kilham 1972, Short 1973, Cruz 1977, Askins 1983), and certain furnariids (Skutch 1969, Eley et al. 1979, Parker et al. 1985, Remsen 1985). Some species without obvious morphological adaptations for climbing also frequently probe bark or epiphytes on branches. Although examples include continental birds, such as the Sharpbill (Stiles and Whitney 1983), some wrens (Root 1964, Parker 1986a), and the Red Warbler (*Ergaticus ruber*; Elliott 1969), they are particularly frequent on islands or in regions such as New Guinea and Australia where specialized bark-searching taxa are rare or absent (Keast 1968, Zusi 1969, Cruz 1978). Examples include: a pachycephalid (*Colluricincla harmonica*), a scrub-wren (*Sericornis magnus*), and some meliphagid honeyeaters (Keast 1968, Recher et al. 1985); a mimid (*Cinclocerthia ruficauda*; Zusi 1969); some Hawaiian honeycreepers (Richards and Bock 1973); some icterids (*Nesopsar nigrerrimus* and *Icterus leucopteryx*; Cruz 1978); and several birds-of-paradise (*Ptiloris* spp., *Astrapia mayeri*, *Pteridophora alberti*, *Diphyllodes magnificus*; Forshaw and Cooper 1979). Many species that search hanging dead leaves for hidden arthropods probe into these curled leaves (Remsen and Parker 1984 and references therein; K. V. Rosenberg, unpubl. data; T. A. Parker, unpubl. data). Similarly, some species of small tanagers (*Dacnis* spp. and *Cyanerpes* spp.; Snow and Snow 1971, Isler and Isler 1987) use their slender bills to probe inside curled living leaves. Some populations of Yellow-throated Warbler (*Dendroica dominica*) probe pine cones (Ficken et al. 1968, Emlen 1977) or dense clusters of pine needles or small leaves (Lack and Lack 1972). Some ground-foraging birds probe in soil, mud, or deep leaf-litter; examples include thrashers (*Toxostoma*, Mimidae; Fischer 1981), White's Thrush (*Zoothera dauma*; Edington 1983), Rook (*Corvus frugilegus*; Waite 1984b), Whitewinged Chough (*Corcorax melanorhamphos*; Ford et al. 1986), and the woodcocks (*Scolopax* spp.; Sheldon 1971). The furnariid *Cinclodes excelsior* probes moss and lichens on rocks and the ground (Fjeldså et al. 1987). Hundreds of species of nectar-feeding birds around the world probe flowers, especially in the Trochilidae, Nectariniidae, and Meliphagidae. Woodpeckers and hummingbirds also extend their tongues to probe crevices, holes, and flowers; such probing could be labelled "tongue-probing."

C. Subsurface maneuvers with *Substrate Manipulation* (maneuvers in which the substrate is manipulated beyond insertion of a probe).

1. *Gape:* to insert the bill into the substrate as in a probe, but the bill is opened to widen the opening. This maneuver is characteristic of many starlings and American blackbirds (Icteridae), which have bills and jaw musculature adapted for gaping (Beecher 1951, 1978; Orians 1985b). Various icterids use their bills to open holes in curled living and dead leaves (e.g., orioles [*Icterus* spp.]), dead branches and stems, moss, bromeliad clusters, seed clusters, leaf-litter, soil (*Sturnella* spp.), clumps of grass, flowers, and large fruits (Cruz 1978; Orians 1985b; Robinson 1985, 1986, 1988); they also use "gape" to turn over stones, twigs, dung, and other objects that might conceal prey on the ground (Orians 1985b). Several species of wood-warblers, including several *Vermivora* spp. (Ficken and Ficken 1968), the Swainson's Warbler (*Limnothlypis swainsonii*; Meanley 1970), and the Worm-eating Warbler (*Helmitheros vermivorus*; Greenberg 1987b), use the gape maneuver for probing buds, dead leaves, and flowers. The Sharpbill "gapes" to open tightly rolled young leaves and dead leaves (Stiles and Whitney 1983), as does the woodhoopoe *Phoeniculus bollei* to open crevices in loose bark (Löhrl 1972). Instances of gaping are occasionally reported in other taxa, such as Meliphagidae (Keast 1968) and Dendrocolaptidae (Willis 1983c).

2. *Peck:* to drive the bill against the substrate to remove some of the exterior of the substrate. This maneuver is characteristic of many woodpeckers (Picidae) that excavate holes in bark or wood to expose prey. "Peck" is synonymous with the "tap" maneuver of some studies of woodpeckers; we recommend restricting "tap" to those motions that are probably exploratory pecks for detecting wood-borer tunnels or movements, as described by Davis (1965) and Kilham (1972). Many parids and at least one icterid (*Nesopsar nigrerrimus*; Cruz 1978) also peck to excavate holes in rotted wood. Ground-foraging birds use this maneuver in combination with "flake" (see below) to dig small holes to reach food in the ground (e.g., thrashers [*Toxostoma*], Engels 1940, Fischer 1981; and some thrushes, Tye 1981). Some frugivorous birds use "peck" to break the outer skin of large fruit (Snow and Snow 1971). The Bananaquit (*Coereba flaveola*; Gross 1958), some hummingbirds (Colwell 1973, Stiles 1985c), some white-eyes (*Zos-*

terops; e.g., Gill 1971), and icterids (Robinson, unpubl. data) may use this maneuver, usually described as "piercing," to make a hole in the base of flower corollas for "stealing" nectar, but the actual maneuver used to make the hole is uncertain. The flowerpiercers (*Diglossa*) hold the flower with their hooked upper mandible and pierce with their sharp, upturned lower mandible (Skutch 1954).

3. *Hammer:* to deliver a series of pecks without pausing between pecks. This maneuver is mainly restricted to certain woodpeckers that use it for excavation of deep holes to reach bark- or wood-dwelling insects or sap. The twig-foraging furnariid *Xenops minutus* also uses this maneuver frequently (Skutch 1969). Some chickadees and titmice (Paridae) may use this maneuver occasionally to open acorns, galls, seeds, and fruits (e.g., *Parus inornatus* and *P. rufescens*; Root 1964, 1967), but the pecks are not delivered as rapidly as in woodpeckers. The distinction between hammer and peck, which rests on whether there is a pause between pecks, may be vague. Counting the number of pecks per unit time, and thereby eliminating the "hammer" category, is an alternative treatment.

4. *Chisel:* like "peck," but rather than the bill being pounded almost perpendicularly into the substrate, it is aimed more obliquely at the substrate—usually bark or dead stems— and the bill is used as a chisel or lever to dislodge portions of the substrate. The direction of head movements is forward and upwards. Slightly to strongly upturned lower mandibles that give the bill a somewhat chisel shape are often associated with species specialized on chiseling. Species that seem to have converged on this foraging behavior and morphology are some *Xenops* spp. (Furnariidae; Skutch 1969), the dendrocolaptid *Glyphorynchus spirurus* (Skutch 1969), the furnariid *Simoxenops ucayali* (JVR and T. A. Parker, unpubl. data), and the antbird *Neoctantes niger* (Hilty and Brown 1986); and to a lesser degree, nuthatches (*Sitta* spp.) and sitellas (*Sitella* spp.; Holmes and Recher 1986a, b). We invented this category to match our expectations of how chisel-shaped bills are used rather than on any data on movements used by these species. Although some brief descriptions (e.g., *Glyphorynchus*; Skutch 1969) fulfill our expectations, the reality of our "chisel" maneuver remains unclear.

5. *Flake:* to brush aside loose substrate with sideways, sweeping motions of the bill. Not as much force is required as in chisel or pry because the substrate dislodged is already loose or unattached. This category combines two types of motions that are often difficult to distinguish in the field: the closed bill tip is used to brush aside the substrate, and the substrate is grasped briefly between the mandibles (which can be called "toss" when the distinction can be made). "Flake" is synonymous with "bill-sweeping" (Clark 1971) except that it applies to substrates other than leaf-litter. "Flake" is also apparently synonymous with R. J. Craig's (1984) "leaf-pull." Many bark-foraging woodpeckers "flake" to dislodge loose sections of bark (Tanner 1942; Kilham 1965, 1983; Conner 1981). The term "scaling" used in many studies of woodpeckers to describe removal of loose bark presumably refers to a combination of our "pecking," "flaking," and "prying." Some dendrocolaptids (Willis 1983c, Pierpont 1986), furnariids (JVR, unpubl. data), and a meliphagid (*Melithreptus brevirostris*; Keast 1968) use this maneuver to search through debris clusters and loose bark. Ground-foraging birds that "flake" leaf-litter include some thrushes (*Turdus* [Skutch 1960, 1981; Clark 1971; Tye 1981]; *Hylocichla* [Clark 1971; Holmes and Robinson 1988]; *Alethe* [Willis 1986]), antbirds (*Formicarius*, Skutch 1969, Willis 1985b; *Rhopornis*, Willis 1981a), leaftossers (*Sclerurus*, Furnariidae; Skutch 1969, Hilty and Brown 1986), thrashers (*Toxostoma*; Clark 1971, Fischer 1981), bulbuls (*Bleda*; Willis 1983a), the waterthrushes (*Seiurus*; R. J. Craig 1984), and horneros (*Furnarius*; Robinson, unpubl. data). The Dune Lark (*Mirafra erythrochlamys*) uses "flake" to dislodge sand to excavate small craters to expose hidden seeds (Cox 1983). The furnariid *Cinclodes excelsior* "flakes" moss and lichens from rocks (Fjeldså et al. 1987).

6. *Pry:* to insert the bill into a substrate and use it as a lever to lift up portions of the substrate. This differs from "flake" in that the sides of the bill, rather than the tip, accomplish the movement of the substrate while the tip remains relatively stationary. Substrates for which "pry" is needed are generally more firmly attached than those dislodged when a bird "flakes." "Pry" differs from "chisel" in that the tip of the bill is stationary, instead of moved forward and upward as in chisel. Examples of species that use "pry" are: Band-backed Wren (*Campylorhynchus zonatus*; Skutch 1960), some species of dendrocolaptids (Skutch 1945; Willis 1983c, d), a meliphagid (*Melithreptus validirostris*; Keast 1968), many woodpeckers (e.g., Short 1973), and a bird-of-paradise (*Astrapia mayeri*; Forshaw and Cooper 1979), all of which pry up sections of loose bark; and Sharpbill, which pries moss from branches (Stiles and Whitney 1983).

7. *Pull:* to grasp, pull, or tear, and thereby remove or dislodge sections of the substrate with the bill. Pulling differs from "flaking" in that the target substrate is grasped in the bill because extra force is needed to dislodge

more firmly attached potions of substrate. Birds that pull off loose bark or lichens to attack hidden insects include Band-backed Wren (Skutch 1960), Plain Titmouse (*Parus inornatus*; Root 1967), Crested Shrike-Tit (*Falcunculus frontatus*; Recher et al. 1985, Ford et al. 1986), a bird-of-paradise (*Macgregoria pulchra*; Forshaw and Cooper 1979), some orthonychids (Holmes and Recher 1986a), some dendrocolaptids (Willis 1983c, d) and Giant Cowbird (*Scaphidura oryzivora*; Robinson 1988). The Plain Titmouse also pulls apart leaf galls, flowers, lichens, and curled dead leaves (Root 1967). *Thripadectes rufobrunneus* (Skutch 1969) and several other furnariids (T. A. Parker, unpubl. data) pull leaves from bromeliads to expose prey. Most New World barbets (*Capito, Eubucco*) also pull open large dead leaves, twig galls, and sections of rotting wood to search for prey (Remsen and Parker 1984; T. A. Parker, unpubl. data; SKR, pers. obs.). The Plush-capped Finch (*Catamblyrhynchus*) pulls the leaf whorls at the nodes on bamboo stems, presumably to reveal insects (Hilty et al. 1979, Remsen 1985). The ground-foraging Song Thrush (*Turdus philomelos*) uses "pull" in its foraging repertoire (Henty 1976). Many parrots use "pull" for opening fruits, seeds, flowers, and rotting wood (Forshaw 1973 and references therein).

8. *Scratch:* to dislodge section of substrate with foot movements. This maneuver is used by many ground-foraging birds around the world; examples include: some orthonychids (Zusi 1978, Frith 1984), Australian lyrebird (Menuridae; Recher et al. 1985, Holmes and Recher 1986b), and some megapodes (e.g., *Alectura lathami*; Frith 1984). Although most species scratch using one foot at a time, many emberizid sparrows (Davis 1957, C. J. O. Harrison 1967, Hailman 1973, Greenlaw 1976 and references therein) and occasionally some thrushes (*Turdus*; Clark 1983) and icterids (Greenlaw 1976) move both feet simultaneously to expose food under leaf-litter or snow.

II. Aerial maneuvers (bird must leave substrate to reach food)
  A. Leg-powered maneuvers
    1. *Leap:* to launch into the air to reach a food item too far for a "reach" but too close for a "sally." This differs from "sally" in that the upward thrust seems to come mostly from leg movements rather than wing movements (Davies and Green 1976); it is equivalent to the "jump-glean" of Holmes and Robinson (1988) and presumably the "jump" of Hutto (1981b). Distinguishing "leap" from short sallies is often difficult. Davies and Green (1976) found that "leap" was the most frequent maneuver used by Reed Warblers (*Acrocephalus scirpaceus*), and Greenberg (1984a) found that it comprised 25% of the maneuvers of Chestnut-sided Warblers (*Dendroica pensylvanica*) in winter. Holmes and Robinson (1988) found that about one-fifth of all maneuvers used by Ovenbirds (*Seiurus aurocapillus*) and Dark-eyed Juncos (*Junco hyemalis*) were leaps. Greenberg and Gradwohl (1980) considered leaping (from ground to foliage) to be the primary foraging maneuver of Kentucky Warblers (*Oporornis formosus*) and Chestnut-backed Antbirds (*Myrmeciza exsul*). The Chestnut-crowned Gnateater (*Conopophaga castaneiceps*) leaps to nearby perches to attack prey (Hilty 1975). Many species that follow army-ant swarms probably "leap-down" from low perches above the ants to capture flushed insects (e.g., *Gymnopithys,* Willis 1968; *Rhegmatorhina,* Willis 1969; *Phlegopsis,* Willis 1981b; *Dendrocincla,* Willis 1972, 1979). Some seed-eating species apparently leap onto stems to pull seed heads to the ground (Emlen 1977). The direction and distance of the leap should be recorded, just as it is for "sally" (see next account), particularly because a "leap" downward (i.e., dropping) probably requires only a fraction of the energy than does an upward or outward leap against gravity.
  B. Wing-powered maneuvers
    1. *Sally:* to fly from a perch to attack a food item (and then return to a perch). Most authors have used separate terms to distinguish sallies directed at aerial prey from those aimed at nonflying prey. We do not, because the foraging site (i.e., air vs. anything else) will automatically be recorded more appropriately in our scheme under the "substrate" category (see below); and the maneuver itself appears to us to be very similar whether directed at air or hard substrate. Although we acknowledge that the movements directed at flying vs. nonflying food may be different, we prefer to remove the substrate-bias from terminology as much as possible. Another difference between our system and others is that the term "hawk" has been used frequently to describe what we here call "sally" (e.g., Holmes et al. 1979b). We use "sally" rather than "hawk" because: the dictionary definition of "sally" is closer to this behavior than is "hawk," and hawks rarely if ever fly from a lookout perch to attack flying prey. Similarly, the term "flycatch" has been used frequently for sallies after flying prey, but most "flycatchers," whether tyrannids or muscicapids, do not "flycatch" *per se,* but instead glean or sally to substrates (e.g., Fitzpatrick 1980). Greenberg (1984a) distinguished sallies in which a bird returned to the perch from those in which the bird continues in the same direction by calling the latter "darts." There is probably more among-author variability in terms used to describe aerial maneuvers (e.g., hawk, hover, hover-glean, snatch, sally, flycatch) than

in any other broad category of foraging behavior.

Many species have characteristic directions or distances associated with their sallies that provide an index of the average search radius (Fitzpatrick 1981, Robinson and Holmes 1982), and these are important to record (after Fitzpatrick 1980):

a. *sally-distance* (distance of the sally from perch to food item).
b. *sally-angle* (the qualitative divisions "up," "down," "horizontal," "diagonal-up," and "diagonal-down" probably represent maximum possible resolution under most field conditions). Certain species or species groups may have characteristic sally angles. Willis (1984), for example, noted that most manakins (Pipridae) typically sally only at a horizontal angle, and Holmes and Recher (1986a) found that two species of thornbills differed in the angles of their sallies.

Sally-distance and sally-angle should refer to the initial attack attempt only; subsequent pursuit of a missed target should be recorded separately. We distinguish five types of sallies based on the bird's foraging motion at the end of the sally:

a. *Sally-strike:* to attack in a fluid movement without gliding, hovering, or landing (after the "outward striking" and "upward striking" of Fitzpatrick [1980] and the "snatch" of Moermond and Denslow [1985]). The "sally-strike," whether aimed at flying prey or stationary substrates, is the characteristic attack behavior of many Tyrannidae (Hespenheide 1971; Fitzpatrick 1980, 1985; Sherry 1984), Muscicapinae and other Old World "flycatchers" (e.g., Croxall 1977, Davies 1977b, Fraser 1983, Moreno 1984), Pipridae (Skutch 1969), Bucconidae (e.g., Skutch 1948; Willis 1982b, c), Galbulidae (Hilty and Brown 1986), Meropidae (Fry 1984), Momotidae (e.g., Skutch 1947; Willis 1981c), Alcedinidae (Fry 1980), and Conopophagidae (Willis 1985b). Numerous species in other families use the sally-strike maneuver to varying degrees, accompanied by morphological adaptations that parallel those seen in more typically sally-striking groups (Partridge 1976b, Norberg 1979, Schulenberg 1983). Most species that use this maneuver are sit-and-wait predators that watch for prey while sitting motionless on an elevated perch, although others search more actively (e.g., tree-climbing dendrocolaptids [Willis 1972, 1982d; Pierpont 1986] and some vireos [Robinson and Holmes 1982]). Ground-foraging birds that "sally-strike" to capture insects on foliage above them include the tyrannid *Corythopis torquata* (Fitzpatrick 1980, Willis 1983b) and *Catharus* thrushes (Paszkowski 1984, Holmes and Robinson 1988). Other ground-foraging birds "sally-strike" to catch flying insects. Examples include ground-tyrants (*Muscisaxicola* spp. [Smith and Vuilleumier 1971; Fitzpatrick 1980, 1985]), *Rhipidura leucophrys* (Ford et al. 1986), and wheatears (*Oenanthe* spp.; Leisler and Seinbenrock 1983). Some species also use this maneuver to obtain fruit (Skutch 1969; Fitzpatrick 1980, 1985). Sally-striking species often have wide, scoop-like bills and wide gapes that presumably facilitate prey capture in flight (Fitzpatrick 1985).

b. *Sally-glide:* like sally-strike except the final approach at the target is a glide (vs. continuous flapping in sally-strike). Moermond and Denslow (1985) pointed out that many sally-strikers do not use continuous, flapping flight in their approach, and they made a convincing case for distinguishing those species that used a brief glide from those that did not. It is likely that some or many of the examples of sally-strikers above are actually sally-gliders. Other than Moermond and Denslow's (1985) data on frugivores, the prevalence of sally-gliding (which they called "sally-scooping") vs. sally-striking will be revealed only by careful observations.

c. *Sally-stall:* to stall in front of the target briefly with fluttering motions at the end of the sally. Moermond and Denslow (1985) noted that many species usually considered to sally-hover (see below) do not engage in true hovering (flying in place), but rather flutter awkwardly in a stalling motion after a steep attack angle at the final approach of the sally. Such species, mainly trogons and some cotingas, use different flight motions and have different morphological adaptations from those that hover. We suspect that many of the examples of "sally-hover" noted below may actually be "sally-stalling." As with sally-gliding, only careful observations (or high-speed photography?) will reveal its true prevalence among sallying birds.

d. *Sally-hover:* like other sallies except that the bird hovers at the target substrate at the end of the sally. This is synonymous with Fitzpatrick's (1980) "hover-glean." Most studies do not distinguish between sally-strike and sally-hover (much less sally-glide and sally-stall), and many other studies appear to label all sallies to foliage as "hovering" (e.g., Holmes et al. 1979b), even though few of these maneuvers actually involve hovering flight. Unless these maneuvers are distinguished, the possibility that they require different morphological adaptations, as found for frugivores by Moermond and Denslow (1985), cannot be addressed. Some tyrannids use the sally-hover maneuver regularly (Fitzpatrick 1980, 1985), as do kinglets (*Regulus*; Rabenold 1978, Moreno 1981, Franzreb 1984), the Blue-gray Gnatcatcher (*Polioptila caerulea*; Root 1967), some sylviid warblers (*Phylloscopus*; Gaston 1974), some wood-warblers (*Ergaticus ruber*, Elliott 1969; *Dendroica*, Rabenold 1978, 1980; Greenberg 1984a); an acanthizid (*Sericornis magnirostris*; Frith 1984), the Restless Flycatcher (*Myiagra inquieta*; Ford et al. 1986), and some puffbirds (Sherry and McDade 1982; Willis 1982c, e). Bell (1984) found that at a forest site in New Guinea, 5 of 83 bird species studied in detail used this maneuver in 18–24% of his foraging observations: two monarch-flycatchers (*Monarcha* and *Arses*), a cracticid (*Peltops*), a meliphagid (*Melilestes*), and a drongo. Similarly, Remsen (1985) found that at a forest site in the Andes, 4 (all tyrannids) of 33 species studied in detail used this maneuver in 16–33% of their foraging observations. In contrast, hovering accounted for only 1% of all prey attacks observed in an Australian eucalypt forest where 41 species were studied in detail (Recher et al. 1985). Many species use this maneuver when taking fruit. Examples include many tyrannid flycatchers, manakins, and some tanagers (Fitzpatrick 1980, Willis 1984, Moermond and Denslow 1985). Some species, including kinglets (Leisler and Thaler 1982, Franzreb 1984), some wood-warblers (Morton 1980a), and the Yellow-rumped Cacique (Robinson 1986), occasionally hover under surfaces to search for food that cannot be seen from a perch. Hummingbirds, of course, use this maneuver extensively when feeding at flowers or searching foliage and branches; for nectar-feeding, however, the parameters "sally-distance" and "sally-angle" are usually irrelevant.

e. *Sally-pounce:* to land briefly at the end of the sally to take food from substrate. Although the bird is perched when it takes the food item, we classify this maneuver as a "sally" because it involves a flight after food is spotted at a distance from the lookout perch. It is probably synonymous with Fitzpatrick's (1980) "landing-and-gleaning," Recher et al.'s (1985) "pounce," and Holmes and Robinson's (1988) "dive-glean." Examples of birds that use this maneuver are: many open-country tyrannids and muscicapids (Fitzpatrick 1980, 1985; Fraser 1983), bluebirds (Power 1980), Australian robins (*Petroica, Eopsaltria*; Recher et al. 1985, Ford et al. 1986, Holmes and Recher 1986b), and Fan-tailed Cuckoo (*Cuculus pyrrhophanus,* Recher et al. 1985), some *Catharus* thrushes (Dilger 1956, Paszkowski 1984), some puffbirds (Willis 1982b, c), and the Field (*Spizella pusilla*) and Chipping (*S. passerina*) sparrows when foraging for insects (Allaire and Fisher 1975). Some vireos (Vireonidae) use this maneuver when attacking prey on branches (James 1976, Robinson and Holmes 1982). Some tropical vireos (*Hylophilus*) characteristically use this maneuver followed immediately by hanging on leaf margins when attacking undersides of leaves (T. A. Parker and JVR, unpubl. data). A special kind of sally-pounce is used by some seed-eating birds that sally to a grass stem, grasp the stem in their feet, and then allow their weight to pull the stem to the ground, where seeds can be removed more effectively (Allaire and Fisher 1975).

2. *Flutter-chase:* to flush or dislodge prey from a substrate and to then chase the prey. This maneuver is used regularly by foliage-gleaning birds that flutter after a falling or flying prey item that has escaped their normal attack behavior and is often preceded by a lunge. Root's (1967) "tumble" is synonymous (because "tumble" refers to out-of-control, sommersaulting movements, we have chosen a new term). Root (1967) found that Blue-gray Gnatcatchers (*Polioptila caerulea*) used this maneuver in 23% of all sallies directed at insects in the air; however, Root suspected that the frequent tail-flashing of this species may function to startle insects, therefore making these "flutter-chases" into "flush-pursuits" (see below) in our scheme. Morse (1968) found that four wood-warblers (*Dendroica*) used this maneuver in about 5% of their foraging motions. We see this maneuver most frequently in foliage-gleaning birds in mixed-species flocks in the canopy of tropical forests; apparently, the escape behavior of many of their arthropod prey involves falling from the substrate at the approach of a bird predator. In particular, the White-shouldered Tanager (*Tachyphonus luctuosus*) uses the flutter-chase maneuver frequently (Snow and Snow 1971; JVR, unpubl. data). We use this term mainly for species that are not typically salliers. We recomend recording the distance and angle of the chase, just as in the sally maneuvers.

3. *Flush-pursue:* similar to "flutter-chase" except that species that use this maneuver deliberately (vs. accidentally) flush prey from hiding places and then pursue the flying or falling prey. This maneuver tends to be prominent in the foraging repertoire of species that use it, most of which have conspicuous wing or tail spots or stripes that are flashed to startle hidden prey. Distin-

guishing this maneuver from "flutter-pursuit" may be difficult, but because each involves fundamentally different tactics, we believe that to do so where possible is valuable. Among North American species, the American Redstart (*Setophaga ruticilla*; Robinson and Holmes 1982) and, on the ground, the Northern Mockingbird (*Mimus polyglottos*; Hailman 1960) most frequently use this maneuver. Other examples include: *Dendrocincla* woodcreepers (Willis 1972, 1979), fantails (*Rhipidura*; Recher et al. 1985, C. J. O. Harrison 1976, Holmes and Recher 1986a), *Monarcha* flycatchers (Pearson 1977b), *Myiobius* tyrannids (Fitzpatrick 1980), Ruddy-tailed Flycatcher (*Terenotriccus erythrurus*, Sherry 1984), and the *Myioborus* redstarts (Parulinae; Remsen 1985).

4. *Screen:* to attack in continous flight (after Emlen 1977, Fitzpatrick 1980). (Note that this is a searching behavior as well as an attack maneuver.) This is synonymous with "hawk" as used by Remsen (1985) and others for birds that feed in flight. Swallows, swifts, and nighthawks (*Chordeiles*, Caprimulgidae) use this maneuver almost exclusively. Other birds that may use this maneuver occasionally include European Starling (*Sturnus neglectus*; Cayonette 1947), Golden-naped Woodpecker (*Melanerpes chrysauchen*; Skutch 1969), Lewis' Woodpecker (*M. lewis*; Bock 1970), some tyrannids (Fitzpatrick 1980), and probably the puffbird *Chelidoptera tenebrosa* (Burton 1976).

FORAGING SITE

We suggest recording the following parameters with respect to the foraging site used by a foraging bird: (1) general habitat, (2) vertical position, (3) horizontal position, (4) foliage density, and (5) the precise substrate from which the food was taken. We discuss each catgory briefly.

I. *Habitat:* Many study areas contain more than one habitat or microhabitat. Each foraging record should be assigned to one of the investigator's general habitat or microhabitat categories to permit examination of the influence of habitat on foraging behavior (e.g., Bilcke et al. 1986). Classification of habitats, a complex and critical topic, is beyond the scope of this paper.

II. *Vertical position:* It has been recognized for decades that important differences in vertical position separate the foraging activities of many closely related birds. Furthermore, foraging behavior may change with changes in height above ground. Therefore, every foraging record should be assigned two values to allow its position to be plotted: (1) height-above-ground and (2) distance-to-canopy (above bird). We have also found a third parameter to be of interest: (3) height of the individual plant in which the bird was foraging. This allows us to distinguish species that frequently use small trees or saplings within the foliage column from those that use the lower foliage of canopy trees at the same height as the small trees. Provided that only one observer records the data, a visual estimate of height (vs. precise measurements) may be the only practical way to obtain such data. Not only does the time required to make precise measurements reduce the volume of data that can be collected, but it seems unlikely that the birds recognize vertical subdivisions sufficiently precisely to warrant such a time investment. Heterogeneity in canopy height, light penetration, and foliage distribution obliterate such precise boundaries. However, differences among observers in the accuracy of such visual estimates (Block et al. 1987) reveal the unreliability of such visual estimates and provide support for use of objective measures of height.

III. *Horizontal position:* Many researchers have recorded the "horizontal" position (e.g., "inner," "middle," "outer") of the bird in the tree or bush. Many species of foliage- and branch-gleaning birds characteristically favor one of these foraging zones (e.g., MacArthur 1958 and numerous other studies). Whether birds select such zones *per se*, or are keying on differences in foliage density (next category) is unknown. It is possible that "horizontal position" and "foliage density" measures are largely redundant. However, Greenberg and Gradwohl (1980) and Holmes and Robinson (1981) showed the importance of branch and leaf arrangement around the bird in determining which surfaces can be attacked effectively. Greenberg (1984a) used a system for "horizontal" position designed specifically to place the foraging bird in categories with respect to foliage and branch geometry.

IV. *Foliage density:* Foliage density at the point of foraging observation can be recorded using a qualitative scale. For example, the system that we have found to be useful (e.g., Remsen 1985; modified from Wiley 1971) is a scale from "0" to "5" of increasing foliage density within a one-meter radius around the bird: "0" = no vegetation within the imaginary 1-m sphere; "1" = very low vegetation density within the sphere, 95–99% of all light passes through sphere); "2" = low density, 75–95% of light passes; "3" = moderate density, 25–75% of all light passes; "4" = high density, only 5–25% of light passes; and "5" = extremely dense, 0–5% of light passes.

V. Substrate. We have found the following substrate categories to be useful:
A. Living Foliage
1. *Plant species* or "type" (species, genus, or family when possible; otherwise "broadleaf tree," "vine," "palm," "grass," "bamboo," "fern," "cactus," and the like; note if epiphytic). Many studies (e.g., Hartley 1953; Gibb 1954; Willson 1970; Reller 1972; Holmes and Robinson 1981; Woinarski and Rounsevell 1983; Robinson and Holmes 1984; Franzreb 1984; Bell 1985b; Morrison et al. 1985, 1987b) have emphasized the importance of distinguishing plant species. In the tropics, many bird species specialize on distinctive plant types such

as bromeliads, bamboo (Parker 1982, Remsen 1985), and palms.

2. *Leaf size* (visual estimate of length and width of leaf searched). This is probably necessary mainly in areas where complexity of plant communities prevents quick taxonomic identification of plant species (and therefore subsequent, more accurate assessment of leaf size). Leaf buds should also be distinguished, although these can be "food" as well as substrate.

3. *Top* or *Bottom*. See Greenberg and Gradwohl (1980) and Greenberg (1984a) for the importance of distinguishing leaf tops from leaf bottoms. Greenberg and Gradwohl (1980) also found that a foliage-gleaning tanager (*Dacnis cayana*) may inspect brown, insect-damaged areas on leaves; therefore, observers should be careful to record when such leaf sections are investigated.

B. Dead foliage. See Gradwohl and Greenberg (1982b), Remsen and Parker (1984), and Rosenberg (this volume) for the importance of distinguishing live from dead leaves. Size of leaf should also be recorded, as well as condition (curled, tattered, or entire; see Rosenberg, this volume) and general type (e.g., palm, broadleaf, bamboo).

C. Bark or stem surfaces. Observers should note that careful observations often reveal that many species generally thought to be foliage-searchers direct considerable proportions of their attacks at branches and stems, such as some species of vireos (Nolan and Wooldridge 1962; Root 1967; James 1976, 1979; Robinson and Holmes 1982; Airola and Barrett 1985), tanagers (Snow and Snow 1971, Isler and Isler 1987), wood-warblers (Morse 1967a, b, 1968; Lack and Lack 1972; Emlen 1977; Greenberg 1984a), sylviids (Earlé 1983), Hawaiian honeycreepers (Richards and Bock 1973), shrikes (Earlé 1983), chats (Frith 1984), Old World sallying flycatchers and drongos (Bell 1984), honeyeaters, whistlers, and babblers (Keast 1968, Thomas 1980, Wooller and Calver 1981), and thornbills (*Acanthiza*; Bell 1985b, Recher et al. 1987). When recording use of this substrate category, the observer can record:

1. *Diameter* (visual estimate)
2. *"Angle"* of branch (i.e., vertical, horizontal, or diagonal).
3. *Upper* or *Lower* side (for horizontal or diagonal branches). Some species may characteristically forage on the undersides of limbs, such as the woodcreeper *Xiphorhynchus lachrymosus* (Willis 1983c).
4. *Plant species,* when possible, or plant type (see A.1. above). See Jackson (1979) and Morrison and With (1987) for examples of the importance of tree species for woodpecker feeding-site selection.
5. *Surface type* and *texture* (especially critical where identification of plant species is not possible). Examples include: (a) smooth-green; (b) smooth bark; (c) rough bark (with perhaps a qualitative scale to indicate degree of corrugation); (d) seam between two closely growing branches or between vine and supporting trunk (such seams appear to be particularly favored foraging sites for some dendrocolaptids; e.g., *Hylexetastes perrotii* [Willis 1982f]); (e) lichen- or moss-covered (mossy branches are favored sites for furnarids, dendrocolaptids, several birds-of-paradise, and tanagers [Skutch 1969, 1981; Forshaw and Cooper 1979; Parker and O'Neill 1980; Remsen 1984; Parker et al. 1985; Remsen 1985]); (f) hard, dead wood with bark removed; (g) soft, rotted dead wood (see Alatalo [1978], Cruz and Johnston [1979], Pettersson [1983], and Morrison et al. [1987b] for examples of the importance of distinguishing live from dead branches in bark-foraging birds; the furnariid *Xenops minutus* seems to be specialized on dead branches, especially those that have fallen but are caught up in the canopy [Skutch 1969; T. A. Parker and JVR, unpubl. data]); and (h) holes (favored foraging sites for some dendrocolaptids [Willis 1982d, f]).

D. Ground
1. *Surface type* (e.g., mud, bare soil, leaf-litter, moss, gravel).
2. *Distance* to *nearest cover.*
3. *Slope* (e.g., flat, moderate slope, steep slope).

E. Rock
1. *Size.*
2. *Surface type* (e.g., smooth, rough, crevice).
3. *Surface "angle"* (top, bottom, side; vertical or diagonal slope).

F. Air

G. Flower (when identification of plant unknown); as noted by Emlen (1977), it is often difficult to distinguish whether some species use flowers as sources of food (nectar feeding) or as substrates for searching for arthropods.
1. *Corolla length.*
2. *Color.*
3. *Flower density* (estimate no. flowers/unit area; e.g., per 0.5 m^2).

G. Miscellaneous. Almost every habitat will have some substrates that do not fit into the above scheme. For example, some species of birds search pine cones (Morse 1967a, Ficken and Ficken 1968, Emlen 1977, Moreno 1981), termite nests (Bell 1984), wasp nests (Willis 1982f), spider webs (Young 1971, Burtt et al. 1977, Douglass 1977, Waide and Hailman 1977, Bell 1984, Brooks 1986, Tiebout 1986, Parrish 1988, Petit and Petit 1988), dung (Anderson and Merritt 1977), and even the skin of other vertebrates (Rice and Mockford 1954, Orians 1983, Isenhart and DeSante 1985 and references therein, Robinson 1988). For fruit-eating birds, we do not record a substrate *per se,* but note certain characteristics of the fruit under "food taken" (see next section).

Although the number of parameters to be recorded in this classification of foraging maneuvers and sub-

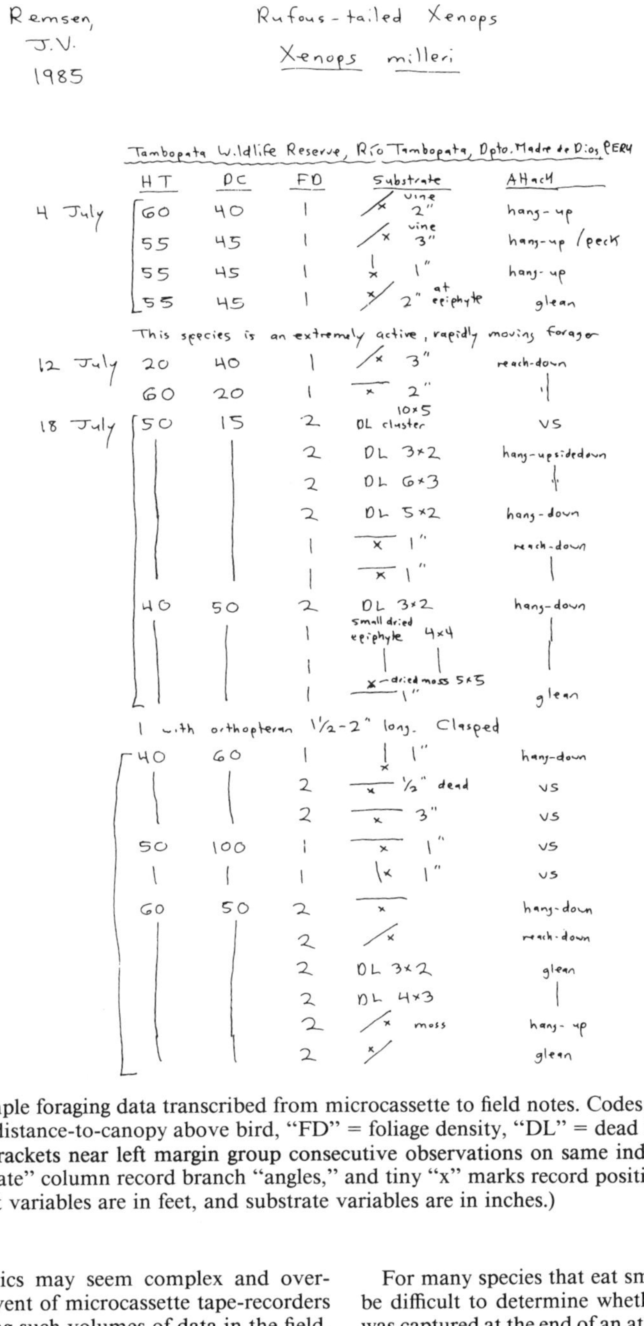

FIGURE 2. Sample foraging data transcribed from microcassette to field notes. Codes: "HT" = height above ground, "DC" = distance-to-canopy above bird, "FD" = foliage density, "DL" = dead leaf, and "vs" = visual search. Vertical brackets near left margin group consecutive observations on same individual. The thin lines under the "Substrate" column record branch "angles," and tiny "x" marks record position of bird with respect to branch. (Height variables are in feet, and substrate variables are in inches.)

strate characteristics may seem complex and overwhelming, the advent of microcassette tape-recorders facilitates recording such volumes of data in the field. Also, transcription of the data can be simplified by using codes and symbols (Fig. 2).

FOOD TAKEN

Data on diets are useful for virtually every kind of foraging study. Differences in food taken may provide information on niches, morphology (principally of the bill), and energetics. Unfortunately, dietary data are usually difficult to obtain in the field, especially for insectivores.

For many species that eat small insects, it can even be difficult to determine whether or not a prey item was captured at the end of an attack. For these reasons, most field studies of insectivores include only limited data on prey. Variables measured include prey size (usually in relation to bill length, but see Bayer [1985] and Goss-Custard et al. [1987] for cautions) and prey type (for large prey items such as caterpillars and orthopterans). Some authors (e.g., Greenberg 1984a) recorded each time that a bird wiped its bill after a prey attack as an index of success. Reasonably accurate estimates of capture rates can be obtained for large prey, such as orthopterans that require extensive handling before they are eaten (Robinson 1986). Many neotropical insectivores evidently obtain most of their energy from large katydids (Orthoptera: Tettigoniidae) and have bills adapted specifically to handle them (Greenberg 1981a). Most temperate-zone insectivores, on the other hand, have smaller bills, presumably adapted for the smaller arthropods or less agile larvae available during the breeding season. Because large food items have more biomass than small items, we think that food size should always be recorded where feasible.

For frugivores, the most important variable is the plant species. Secondary variables include the color (as a measure of ripeness), size (especially if the plant species is unknown), and shape of the fruit. For nectarivores, the plant species is again the primary variable of interest. If this is unknown, then color, shape, and corolla length should be recorded.

Data obtained from stomach samples are discussed elsewhere in this volume (Rosenberg and Cooper). Here we wish only to emphasize that stomach samples can be very useful when they reveal major ordinal levels of dietary differences among species being compared. Sherry (1984), for example, showed that species that are generally similar in size and foraging behavior can differ strikingly in their diets. Dietary analyses of Least Flycatchers (*Empidonax minimus*) and American Redstarts, which are strikingly similar in many aspects of their foraging behavior and foraging-site selection (Sherry 1979), revealed surprisingly little overlap (Robinson and Holmes 1982). In this case, knowledge of diet from stomach samples (redstarts catch many Heteropteran leafhoppers) provided information on the functional significance of the "flush-chase" attack maneuver described previously.

Data from stomach samples should, however, be treated with caution. Because prey items in stomach samples can usually only be identified to the level of order or family, the categories are crude. It is quite possible that two species that eat the same orders, families, or even genera of insects could overlap very little in other aspects of their foraging behavior, particularly substrate use. Information on diet in the absence of data on other components of foraging (e.g., Wiens and Rotenberry 1979) therefore could be misleading.

FOOD-HANDLING TECHNIQUES

Once food is "captured," it may be eaten, delivered to offspring or mate, stored (cached), or rejected. We here consider only the techniques associated with the first of these options. The way that food is handled is important because (1) food-handling time must be considered in the cost : benefit ratio of any food type (e.g., Sherry and McDade 1982), (2) it is a factor in studies of adaptive morphology (e.g., Sherry and McDade 1982, Moermond and Denslow 1985, Foster 1987), and (3) it has important implications for the study of plant-frugivore interactions (Howe and Smallwood 1982, Moermond and Denslow 1983, Levey 1987b). Food-handling techniques, however, have been largely ignored in studies of arthropod-foraging behavior (for exception, see Sherry and McDade 1982). Fortunately, the detailed descriptions by some observers (e.g., E. O. Willis and A. F. Skutch) have revealed the distinctive behaviors associated with handling of various food types. The lack of data on food-handling techniques, particularly in insectivores, prevents an evaluation of their relative frequencies of use. In addition to quantifying the time taken to manipulate food before swallowing, we recommend the following terms to describe techniques that we feel are appropriate for field observations of landbirds:

1. *Engulf:* to capture and swallow in one continuous motion, without being held by the bill.
2. *Gulp* (after Moermond and Denslow 1985): to swallow upon capture without any noticeable manipulation other than being held briefly by the bill.
3. *Snap:* to pinch momentarily, usually between tips of mandibles and usually to kill prey before further handling.
4. *Mash* (after Moermond and Denslow 1985): to squeeze or move around between the mandibles before swallowing (apparently to kill prey or remove undesirable portions, such as wings, legs, shells, and husks); sometimes, juices or pulp are squeezed out of the food and solid portions discarded (Moermond and Denslow 1985, Foster 1987). This category almost certainly lumps distinct types of mandibulation that could be revealed by analysis of high-speed photography of food-handling.
5. *Shake:* to shake food item violently (to remove undesirable portions).
6. *Beat:* to beat food item against hard substrate (as in above, to kill or remove undesirable portions). Many small insectivorous birds typically beat insects against branches in a diagonally downward-facing position (e.g., Root 1967).
7. *Rub:* to rub food along substrate (usually to remove distasteful substances or undesirable portions such as hairs and stingers [Sherry and McDade 1982]).
8. *Jab:* to peck food item with bill tip (to kill it or open it), usually while clasped with feet.
9. *Tear:* to eviscerate or dissect food item into smaller pieces, usually while the food is clasped by one or both feet.
10. *Bite:* to bite and remove a section of food item (after Foster 1987). This technique applies as far as we know only to frugivores that take bites from fruit too large to swallow whole.
11. *Juggle:* to reposition food item, sometimes by tossing into air and catching it (to allow or facilitate swallowing; many species juggle prey to maneuver it into a head-first position before swallowing).
12. *Clasp:* to hold food item with feet.
13. *Anchor:* to immobilize food item by fixing it to substrate, such as by impaling with sharp objects or by wedging food item into crack.
14. *Drink:* intake of liquid food, such as fruit juices and nectar.

In practice, we have found that in the field, we have time to note only those food-handling behaviors that are not "gulping," which seems to be the predominant food-handling technique in most insectivorous and frugivorous birds, with the notation that all "blank" records refer to gulping. Our scheme leaves out certain techniques that are presumably very rare, such as scraping (to remove fruit pulp in snake-like jaw motion; Schaeffer 1953), dropping (to break open), soaking, and drowning.

ANALYSES OF FORAGING DATA

This classification system contains many finely subdivided categories. Although too many can create problems (e.g., small or empty data cells) for statistical analyses, we think that fine subdivisions are preferable during the data-gathering stage. Their retention allows maximum data resolution, which in turn, even if sample sizes are too small for statistical analysis, might generate insights that can be developed to answer specific questions in subsequent studies. Here we provide examples of how categories might be combined or subdivided.

I. *Ecomorphological studies.* Fine subdivisions of attack methods, foraging substrate, and searching behavior are most likely to be useful in studies of adaptive morphology. Fitzpatrick (1985), for example, showed that many aspects of bill and wing shape were strongly associated with the details of aerial attack methods (see Table 1) and substrate in tyrannids, whereas leg morphology was more closely related to searching movements and perch types. Fitzpatrick's (1985) classification system of foraging methods, therefore, combined searching movements, perch types, substrate type, and attack method in an attempt to include all of the variables that affect flycatcher ecomorphology. The bill morphologies of bark-foraging birds are also affected by the methods used to manipulate the substrate to attack concealed food. The finer subdivisions of near-perch maneuvers (see Table 1) also may be related to leg and foot morphology (Partridge 1976a, Leisler and Winkler 1985). The bill shapes of frugivores and some insectivores may also be associated with particular kinds of food (Greenberg 1981, Moermond and Denslow 1985, Foster 1987).

II. *Community-level studies.* Community-level studies probably require the least finely subdivided categories. Communities in wooded habitats are likely to include birds that use most of the attack methods described in Table 1. If each method were to be broken down by substrate, the resulting data matrix would be prohibitively large and would contain many zero values. For this reason, most studies that seek to identify guilds use only a few general attack categories (e.g., Holmes et al. 1979b) or use only data on foraging site (Anderson and Shugart 1974). Holmes et al. (1979b), for example, divided the attack methods of birds in a northern hardwoods forest into "gleans" (lumping all "near-perch" maneuvers in Table 1), "hovers" (all sallies to substrates other than air in Table 1), "probes" (including all subsurface maneuvers in Table 1), and "hawks" (all sallies directed at flying prey in Table 1). Each of these attack methods was then combined with a foraging site. The resulting analysis showed that such variables as substrate and tree species were more important in assigning species to guilds than attack methods. By contrast, in a similar analysis of an Australian bird community, which added categories for flush-chase and manipulative prey-attacks, Holmes and Recher (1986a) found that attack methods were also important. The different guild structures in the two areas may have been influenced, therefore, by their differing classification systems. In general, we recommend that manipulative attack-methods be distinguished from methods in which food is simply plucked from surfaces or the air in studies of entire communities.

III. *Single-guild studies* (taxonomic guilds, *sensu* Terborgh and Robinson 1986). Studies that focus on ecologically similar species should benefit from fine subdivisions of substrates and attack methods. The members of a guild are only likely to use a subset of the attack methods shown in Table 1, which should simplify the matrices and allow finer subdivisions. Rosenberg (this volume), for example, included data on the size and shape of dead leaves searched, and Greenberg's (1987a, b) study of a dead-leaf forager included data on searching postures similar to the subdivisions of near-perch attacks shown in Table 1. Conner's (1980, 1981) studies of bark foragers showed the importance of different methods of manipulating substrates in distinguishing among species. Fitzpatrick (1980, 1981) showed the different ways that syntopic tyrannids differ in the subtle details of how they sally to catch prey.

IV. *Foraging modes* (*sensu* Huey and Pianka 1981) or adaptive syndromes (*sensu* Eckhardt 1979). Studies of foraging modes seek to identify suites of intercorrelated foraging variables. Many researchers have shown that the rates and lengths of searching movements are associated with the lengths and kinds of attack methods (e.g., Williamson 1971; Eckhardt 1979; Fitzpatrick 1981; Robinson and Holmes 1982; Holmes and Recher 1986b; Holmes and Robinson 1988; see also Moermond 1979a and Huey and Pianka 1981 for similar analyses of foraging in lizards). In general, birds that move short distances between perches also obtain food on nearby substrates. Similarly, species that fly long distances between perches also search and attack over long distances. Studies of adaptive syndromes therefore include detailed data on searching movements (including rates and lengths), attack tactics (including lengths of attacks), and the use of special foraging tactics such as tail-fanning. Table 2 gives examples of adaptive syndromes or foraging modes that have been identified in New World insectivorous birds (modified from Eckhardt [1979], Fitzpatrick [1981], and Robinson and

TABLE 2. ADAPTIVE SYNDROMES OR FORAGING MODES OF INSECTIVOROUS BIRDS

Foraging mode	Search movements	Associated prey-attacking maneuvers
Open perch or passive searching	Infrequent, long flights	Long sallies
Medium-distance searching	Frequent medium-length flights and bouts of hopping	Sallies and near-perch gleans
Near-surface searching	Frequent hops and short flights	Near-perch maneuvers, probes
Flush-Chasing	Conspicuous, frequent flights and hops, wing and tail flicking	Flush-chases
Manipulative	Short periods of movement between long periods at the substrate	Flake, peck, tear, hammer, scratch, chisel

Holmes [1982]). Whether these relationships have global generality remains to be determined.

V. *Energetics and optimal foraging.* Studies of energetics or optimal foraging primarily use data on time intervals between movements and food-capture rates. Robinson (1986), for example, measured intervals between flights of at least one meter as an index of foraging speed and prey-capture rate, and prey size as an index of foraging success. Energetic studies therefore require long, timed sequences on individual birds in which the length and kinds of every movement are recorded. As already noted, food-handling time is a critical variable in studies of optimal diet selection (e.g., Sherry and McDade 1982).

CLOSING REMARKS

Although portions of our classification scheme have been used by us or other researchers for many years, other portions were novelties generated by rethinking the underlying logic of earlier versions or by incorporating suggestions from other researchers. We regard this scheme as a first step towards standardization of the organization and vocabulary of studies of foraging behavior of birds. We anticipate that it will be modified as it is tested and refined by us and, we hope, other researchers.

ACKNOWLEDGMENTS

We thank C. L. Cummins, R. Greenberg, R. T. Holmes, T. C. Moermond, M. L. Morrison, T. A. Parker, K. V. Rosenberg, and T. S. Schulenberg for valuable comments on the manuscript or portions thereof, and Donna L. Dittmann for preparation of Figure 1. The authors' fieldwork, critical for the formulation of the ideas and proposed scheme presented here, has been supported generously by (for JVR) John S. McIlhenny, H. Irving and Laura Schweppe, Babette Odom, and the National Geographic Society, and (for SKR) the Chapman Fund of the American Museum of Natural History, the National Science Foundation, and the Society of Sigma Xi.

PROPORTIONAL USE OF SUBSTRATES BY FORAGING BIRDS: MODEL CONSIDERATIONS ON FIRST SIGHTINGS AND SUBSEQUENT OBSERVATIONS

Graydon W. Bell, Sallie J. Hejl, and Jared Verner

Abstract. This study presents a mathematical approach to comparing results from initial observations of foraging birds to sequential observations of repeated foraging maneuvers by the same individuals. We consider the case in which the objective is to compare the proportions of use of each of several substrates by a single species. Results suggest that only initial observations should be used, and that subsequent observations do not carry information about the question of proportional use. Generalizations are given for a wide class of probability distributions and also to the problem of comparing proportional use by two bird species.

Key Words: Birds; foraging behavior; initial observations; sequential observations; mathematical models; substrate comparisons.

Avian ecologists use two basic approaches when collecting data on foraging behavior. In the first, the observer records only one event from each bird observed. In the second, the observer records each event in a sequence of events by each bird for as long as it can be observed. Modifications of the second approach have included time-based and location-based constraints on data collection, as well as various criteria for truncating sequences (see Hejl et al., this volume, for examples). Although sequential observations of this sort generate longer sample sizes than if only one event were recorded, the samples are flawed for certain kinds of analyses by a lack of independence. Studies about behavioral transitions of foraging birds must, of course, record sequential events. However, when using foraging observations to characterize the proportional use of different substrates, sites, maneuvers, or other categorical measures, observations should be independent or some adjustment should be made for dependency among observations.

Application of a chi-square goodness-of-fit test to test the null hypothesis of equal proportions among substrates used by birds, for example, assumes independent events, which is a problem when using sequential observations. One way to use sequential observations and to be reasonably assured of independence among units is to treat all foraging attacks of a single bird as a unit, as done by Airola and Barrett (1985). (Note, however, that different record lengths among individual birds may create problems of unequal weighing.) Another approach is to use Markov chain analyses or bootstrapping to assess the effects of dependency among observations on results (e.g., Hejl et al., this volume; Raphael, this volume). Tests of independence can be applied to sequential data but should consider the advice about power given by Swihart and Slade (1986).

Studies that assume independence among sequential observations when data are analyzed also assume that each event in a sequence adds to our knowledge of proportional use of categorical measures. Our primary objective here is to test that assumption mathematically. We describe possible mathematical models, giving specific assumptions, resulting probability distributions, and some of the parameters of those distributions. We further describe likelihood-ratio tests of the hypothesis of equal proportions. Although we use the substrate at which a bird is observed directing an apparent foraging attack as the measure for consideration, results would be the same for whatever categorical measure we might have selected.

MODELS AND ASSUMPTIONS

Let there be k substrates in all. Assume that birds are detected singly and forage from one of the substrates, with the detections following a Poisson process. The number of birds to be observed is not fixed in advance and this is a case of Poisson sampling (Fienberg 1980:15). The intensity of the process (the mean of the Poisson) will be denoted as λ_i in the ith substrate. If the means for all substrates are equal, the proportions are equal. The λ_i values may depend on a variety of factors, including: (1) the quantity of the resources available, (2) the nutritional and energetic values of the different resources, (3) weather conditions, (4) the apparent safety from predators, and (5) the effects of interference from other individuals of the same or different species.

Once a bird has selected a substrate and made a foraging strike, it is counted for that substrate. The total number of birds for the whole sampling period will be denoted by X_i for the ith substrate. The random variables $X_1, X_2, \ldots, X_k$ are assumed to be independent, making their sum,

TABLE 1. Field Counts of Initial Observations (X), Subsequent Observations (Y), and Total Observations (T) of Bushtits

Sub-strate	Period 1			Period 2		
	X	Y	T	X	Y	T
1	11	4	15	10	6	16
2	42	15	57	40	39	79
3	8	1	9	31	24	55
4	38	24	62	35	17	52
5	7	11	18	3	0	3
6	5	0	5	10	20	30

TABLE 2. Some Moments of the Random Variables—First Sightings, Subsequent Observations, and Total, per Substrate (Subscript Suppressed)

Moment	X	Y (conditional)	Y (unconditional)	T
Mean	λ	$x\mu$	$\lambda\mu$	$\lambda(1 + \mu)$
Variance	λ	$x\mu$	$\lambda\mu(1 + \mu)$	$\lambda(1 + 3\mu + \mu^2)$
Correlation $(X, Y) = \mu/[\mu(1 + \mu)]^{0.5}$				

ΣX_i, a Poisson variate with parameter $\Sigma \lambda_i$ (Hogg and Craig 1978:131). After the initial foraging strike, the bird may make additional strikes on the same substrate (perhaps interspersed with other activities) or leave the area. (Other possibilities exist: [1] The bird may disappear from view but still be on the same substrate, and then reappear to be counted again. The frequency of such events cannot be known and is ignored here. [2] The bird may exhibit a transition to a different substrate. We do not address such events here, as transition to a new substrate by the same bird cannot be treated as the beginning of an independent sequence of observations.) Additional strikes by the bird on the substrate are assumed to follow another Poisson process, with intensity μ_i in the ith substrate. Counts of the number of subsequent strikes by different birds are assumed to be independent, thus their sum is Poisson, this one denoted by Y_i in the ith substrate. (Note that we do not adopt a notation for the number of strikes made by a single bird, only for the total made by all birds on that substrate.) The sum that yields Y_i has x_i terms, once $X_i = x_i$ is observed, hence the Y_i rate is $x_i\mu_i$. Thus the Y mean depends on the number of individual birds seen, as does Y itself. To summarize, Y is a Poisson random variable with parameter $x\mu$ conditional on the number of birds seen.

A logical trap exists at this point. The X data and their associated parameters are of primary interest for comparing proportional use of substrates. The Y data (number of subsequent foraging attacks) might be expected to carry additional information about the X parameters, because the Y's depend directly on the X's. Some observers may combine the two counts, letting $X_i + Y_i = T_i$ denote the total in the ith substrate. This is not implausible, because T_i is the total number of foraging strikes seen. On the other hand, T is a total with mixed units, individual birds and foraging strikes, which helps focus attention on the issue addressed in this study.

Additional random variables exist in this setting. The unconditional distribution of Y_i, obtained by averaging over all possible values of X_i, is that of a Neyman Type A random variable and T is known as a Thomas variable. These are two of the well-known "contagious" distributions used for modeling clumped or clustered data (references in Johnson and Kotz 1969:213–215, 236–237; Pielou 1977:118–123).

Foraging data collected by these methods on Bushtits (*Psaltriparus minimus*) (Table 1) can be used to clarify the notation. For example, consider Substrate 5, Period 1. Before observations began, we expected to obtain values for three random variables, X_5, Y_5, and T_5. The observed counts were $x_5 = 7$ birds sighted, $y_5 = 11$ additional strikes made by those 7 birds, and $t_5 = 18$ total foraging strikes seen. The latter number, by itself, conceals important details about the distribution of observations. They might have resulted from single observations of 18 different birds, from 18 observations of a single bird, or from some intermediate combination. Also, 7 is an observed value of a Poisson variable with parameter λ_5 and, conditional on $x_5 = 7$, 11 is an observation on a Poisson variable with parameter $7\mu_5$.

The theoretical or expected performance of these random variables may be summarized by their means, variances, correlations, or other moments. These may be found in Johnson and Kotz (1969:209, 218); some are shown in Table 2. Two columns are needed for Y, as it may be treated conditionally or unconditionally. Note the equality of the mean and variance for X and Y (conditional) but not for Y (unconditional) or for T. The λ factor in some of the Y moments suggests that the subsequent observations can be used in a chi-square test of equal proportions across substrates. The absence of the λ factor in the correlation suggests that the correlative information available does not refer to the λ's.

HYPOTHESES AND TESTS

Two main possibilities are considered in this section. A test may be based on X, Y, or T, or on some combination of these variables. These are addressed as univariate tests or bivariate tests, respectively. In the following paragraphs, log refers to natural logarithm; alternative hypotheses are logical alternatives of the null hypothesis, and approximate chi-square test statistics are denoted by χ^2.

UNIVARIATE TESTS

The null hypothesis is that the rates are equal, $\lambda_1 = \lambda_2 = \ldots = \lambda_k$. Once the total of the X_i is known, the set of substrate counts is a multinomial random variable, with proportion $\lambda_i/\Sigma \lambda_j$ for the ith substrate (Johnson and Kotz 1969: 93). Thus a chi-square test of equal proportions or a G-test (e.g., see Sokal and Rohlf 1981:705–708) may be used, provided expected counts are not too small. For the first sighting data from Period 1, $\chi^2 = 76.40$, with 5 degrees of freedom ($P < 0.001$). This test should not be run on the Y (or T) data alone, because both Y and T depend on two parameters per substrate, and we cannot estimate two parameters from a single observation. Formally, this is a problem of identifiability (Ferguson 1967:144). Intuitively, a decision based on the T data, for example, cannot be attributed to differential values of the λ's or differential values of the μ's. While univariate tests must be restricted to the X data, it seems possible that in a bivariate test the Y data can be used to supplement the information from the X's.

BIVARIATE TESTS

We now consider hypotheses based on the joint distribution of X and Y. This discussion is based on the likelihood ratio (e.g., see Morrison 1976: 17–22), a test principle that leads to G-tests or other approximate chi-square tests. The likelihood function is essentially the product of the density function of the random variable, the product extending over the sample. After the data are obtained, the likelihood function depends only on the parameters. Parameters are estimated to maximize this function twice, once under the constraints of the null hypothesis, H, and then with no constraints. If the ratio of the maximum of the likelihood function constrained by H to the unconstrained maximum is denoted by L, then $-2 \log L$ is an approximate chi-square variate.

Consider the composite hypothesis that the substrates are equally used while the within-substrate foraging rates are unconstrained.

$$H_{01}: \lambda_1 = \lambda_2 = \ldots = \lambda_k;$$
$$\mu_1, \mu_2, \ldots, \mu_k \text{ are unspecified.}$$

For this hypothesis

$$\chi^2 = 2(\Sigma x_i \log x_i - \Sigma x_i \log \bar{x}).$$

Note the absence of y's in this expression. The test based on the joint distribution of the X's and Y's uses only the data on the X's. It is the same test found using the distribution of the X's only. (It is not exactly the same as the chi-square test usually applied; it is more similar to the G-test; see Kendall and Stuart 1967:421.)

Consider next a hypothesis that does constrain the μ's:

$$H_{02}: \lambda_1 = \lambda_2 = \ldots = \lambda_k, \text{ and}$$
$$\mu_1 = \mu_2 = \ldots = \mu_k.$$

The approximate chi-square for testing this hypothesis is

$$\chi^2 = 2(\Sigma x_i \log x_i - \Sigma x_i \log \bar{x})$$
$$+ 2[\Sigma y_i \log(y_i/x_i)$$
$$- \Sigma y_i \log(\Sigma y_i/\Sigma x_i)].$$

The first line of this expression is the χ^2 of the previous hypothesis, so

$$\chi^2(H_{02}) = \chi^2(H_{01}) + \text{other terms}.$$

The "other terms" in this expression can be shown to be those obtained to test

$$H_{03}: \mu_1 = \mu_2 = \ldots = \mu_k,$$

with no constraints on the λ's. Evidently the X's carry information about the Y's, but not conversely. This is consistent with the observation made about the correlation.

We consider only one further hypothesis; this time the two parameter sets are related proportionately.

$$H_{04}: \lambda_1 = \lambda_2 = \ldots = \lambda_k;$$
$$\mu_i = c_i \lambda_i, c_i \text{ unspecified,}$$
$$i = 1, 2, \ldots, k.$$

It can be shown that the approximate chi-square statistic is now exactly that for H_{01}. The likelihood ratio essentially ignores the subsequent observations.

OTHER RESULTS

We have generalized the problem in several ways, but do not include the details here. We have proven that the overall results hold when comparing the substrate distribution for two species and also for comparing two sampling periods. We have also extended the results by replacing the Poisson distribution of X by any single parameter-discrete random variable and Y by any discrete variable whose parameter depends on the observed value of X. The test statistics are different, but conclusions remain unchanged.

QUALITATIVE ASSESSMENT OF ASSUMPTIONS

POISSON ASSUMPTIONS

Consider first the Poisson assumptions. From the previous paragraph, it is clear that the results are virtually independent of these assumptions. Almost any pair of discrete random variables will lead to the same conclusions.

INDEPENDENCE ASSUMPTIONS

These are critical, and probably least amenable to verification. The first is the requirement that observations be of birds foraging singly. Our modeling did not address the problem of species that forage in flocks, although results may still apply if observation is limited to a lead bird. Independence between substrates is easier to accept, because data from additional substrates must come from sightings of different individual birds. Finally, we assumed that birds within a substrate act independently. This may require that we have only one bird in sight at a time.

THE CASE WHEN X IS UNKNOWN

We may wish to assume that X is unknown (per substrate), or that we are unsure of how many distinct birds have contributed to our counts. Then we treat Y in its unconditional distribution and Y must be taken to carry all information about both the numbers of birds and extent of their foraging. In the k substrate problem, we have $2k$ parameters, but only k data values. Additional data must be obtained to carry out any useful test on the substrate proportions. Additional data can perhaps be collected by another observer in a different area, or by means of shorter, repeated, observation periods.

Another method for handling this case would be to simply record all foraging strikes, making no attempt to separate sightings from subsequent observations. These data, from Thomas distributions, again depend on two parameters, and some device must be employed to replicate the sampling.

ILLUSTRATIONS WITH FIELD DATA

The following analysis is based on data in Hejl et al. (this volume), recorded at the San Joaquin Experimental Range, in Madera County, California, during March through May 1980. Field observations were made on a 19.8 ha (300 × 660 m) plot, gridded into quadrats 30 m on each side. To gather foraging information, observers walked back and forth along alternate gridded lines on the study area. The lines walked and the direction of travel were selected to ensure even coverage of all segments of the grid during daylight hours. When a bird was detected, one that had not obviously been disturbed, it was selected for observation. To reduce dependence of the data between individual birds, information was recorded only for the first bird detected in a flock or pair of birds and only if that bird species had not been seen in the last 30 m or for the last 10 min. The activity of the bird was noted at the count of "5". If it was foraging, then sequential observations were recorded for each apparently successful foraging strike that was noted up to 11 observations. Counts were made for several categorical variables including foraging substrate. Foraging substrate as used in Table 1 included plant species, the ground, and the air. In the modeling discussion, "substrate" could represent either foraging substrate or any other categorical variable.

An inconsistency between our assumptions and the study as done was the fact that sequences were truncated at 11 observations, but no adjustment was made for this. Truncation was rarely needed, however, because birds could seldom be followed for that many consecutive foraging strikes.

The data on foraging Bushtits (Table 1) can be used to test the Poisson assumptions for X_i and Y_i, provided we assume that the means did not change between periods. Poisson variables have a variance-to-mean ratio of 1.0. The average variance-to-mean ratio for X between periods across substrates was 3.40, but dropped to 1.36 on deletion of Substrate 3. Using the results of Ratcliffe (1964), these gave (approximate) chi-square values of 20.38 and 6.82, with 6 and 5 degrees of freedom, respectively. The apparent shift in mean for Substrate 3 caused the large value; the remaining data did not contradict the Poisson assumption. For Y the mean ratio was 10.73, with a chi-square of 64.42, far too large to confirm Poisson variation with constant means.

The field objective of substrate comparisons should be addressed by only the data on first sightings. The chi-square values were 76.40 and 56.81 for the separate periods, indicating that some substrates were used more frequently than others. When the same computations were done on total foraging strikes, the values were 114.01 and 100.33, biased upwards in this case by likely differences in the Y rates. By studying the conditional distributions of the subsequent observations, one could test the equality of the within-substrate foraging rates, but this lies outside the scope of this paper. Finally, consider the T data again. Substrate 2, across periods, furnished a good example of the risks inherent in this problem. Virtually the same numbers of birds gave quite different values of t.

CONCLUSIONS

The objective at the outset was to consider the information furnished about one process by data from another. The data on the discovery process seemed straightforward, but the status of the data on subsequent observations was less clear. The two extremes of data analysis are to use only numbers of distinct birds or to use counts of all observed foraging acts. A reasonable compromise was to model the two main aspects of the problem as related processes.

Of the many possible ways to model the joint distribution of initial detections and subsequent events, we have dealt with only one. We focused on the ultimate totals of birds and subsequent events per substrate, since that seemed the natural way to summarize the data. As a result, our modeling of the actions of a single bird may seem artificial; the reproductive property of Poisson variables (totals of Poissons are Poissons) had some influence on our choice of model since it makes the mathematics tractable. However, reproductivity is not really necessary. The total of subsequent strikes need not follow the same distributional form as the variables in the sum.

We have also limited the scope of this discussion by insisting that the question is to discover what subsequent observations tell about proportional use of substrates. The broader question of what can be done with those observations has not been addressed; questions that are within-substrate in content seem more approachable by these data. Hejl et al. (this volume) apply and discuss some methods appropriate for analysis of the subsequent observations.

The use of subsequent observations in the present problem is clearly a case of pseudoreplication (Hurlbert 1984). It is similar to the use of multiple readings per experimental unit in a treatment design. One can know more about the experimental unit by subsampling, but gains no degrees of freedom to compare the treatments. In the same way, subsequent observations tell more about the individual birds that forage on a substrate, but give no advice about the comparison of proportions.

ACKNOWLEDGMENTS

Jeff Brawn, Dave Siemens, and Bob Woodman critically discussed these ideas with us. Robert Campbell, John Marzluff, and Hildy Reiser reviewed earlier drafts of the manuscript. These people, and our consulting editors, have made important contributions.

SEQUENTIAL VERSUS INITIAL OBSERVATIONS IN STUDIES OF AVIAN FORAGING

SALLIE J. HEJL, JARED VERNER, AND GRAYDON W. BELL

Abstract. During the breeding season, we compared sequential and initial observations of the foraging locations of five species of permanent residents in an oak-pine woodland of the western Sierra Nevada. Sequential observations were more dependent—that is, the conditional probabilities of occurrence of any locations were greater when from a sequence—than were initial observations. No visibility biases were associated with either method. Using bootstrap simulations, standard errors calculated for all observations (initial + sequential), without adjustment for dependency, underestimated the true standard error in 68% of the cases, with no difference in 32%. For common foraging locations, the mean proportions of used foraging sites and foraging substrates were similar with both methods, but initial observations gave more precise estimates of foraging locations than did all observations. The two methods differed in their estimates of means and standard errors for uncommon foraging locations. We also created a model using Markov chain analysis to investigate a larger population of sequential observations. Both Markov chain and bootstrap analyses resulted in similar implications. We prefer the use of initial observations in statistical tests that assume independence between observations and the use of statistical techniques that adjust for dependency with dependent, sequential observations. Suggestions for appropriate statistical analyses of sequential observations are given.

Key Words: Foraging; dependent observations; independent observations; statistical analysis; bootstrap; Markov chains.

Martin and Bateson (1986) emphasized that a common error in behavioral research is to treat repeated measures of an individual as though they were independent. One problem likely to result from analyses of such data is underestimation of sample variance. Although the problem of dependence is acknowledged by some students of avian foraging behavior, most have nonetheless used repeated observations from the same individual during the same period without testing for independence between observations from a single individual (but see Holmes et al. 1979b, Porter et al. 1985).

Researchers have used all sequential observations that they could obtain from an individual (Holmes et al. 1979b, Holmes and Robinson 1981, Sabo and Holmes 1983, Keeler-Wolf 1986), or have allowed sequential records of the same individual only after elapse of a specified period of time (e.g., Landres and MacMahon 1980, Wagner 1981a, Morrison 1984a, Porter et al. 1985) or after the bird moved to a new location (e.g., Hartley 1953, Root 1967, Peters and Grubb 1983). Hartley (1953) recorded the first observation on each separate plant while following the same bird. Root (1967) recorded up to three observations from the same individual, always separated by at least 2 min, and they were recorded only if the bird moved to a new substrate between records. Peters and Grubb (1983) recorded up to four observations of a given bird, but only after it moved to a new location for each record.

In addition to obtaining larger samples, many researchers prefer using all observations (initial + sequential observations) because they believe that initial observations are biased toward birds in conspicuous locations (e.g., Sturman 1968, Wiens et al. 1970, Austin and Smith 1972, Hertz et al. 1976). Wagner (1981a) and Morrison (1984a) both compared the results from initial observations with those from all observations. Wagner concluded that the method of data collection had an effect on her results but that different visibility biases were associated with each method. Morrison (1984a) concluded that similar results were obtained by the two methods for most measures, but he preferred sequential sampling because more rare behaviors were observed in his sequential data set. Bradley (1985) compared methods for biases in time-budget studies, concluding that counting only initial contacts was the least satisfactory of the four methods and was especially prone to discovery bias.

We studied the foraging behaviors of five species of birds—Scrub Jay (*Aphelocoma coerulescens*), Plain Titmouse (*Parus inornatus*), Bushtit (*Psaltriparus minimus*), Bewick's Wren (*Thryomanes bewickii*), and Brown Towhee (*Pipilo fuscus*)—in an oak-pine woodland in the foothills of the western Sierra Nevada. Our objectives were: (1) to test for independence among sequential observations of foraging sites and foraging substrates used by the birds, (2) to explore whether all observations gave the same information about foraging locations as did initial observations, and (3) to consider various analytical procedures that can be used to make appropriate adjustments in variance derived from sequential observations.

STUDY AREA AND METHODS

This study was done at the San Joaquin Experimental Range in March, April, and May 1980, and May 1982, during the breeding season. The Range is located approximately 32 km north of Fresno in Madera Co., California. Vegetation was characterized by intermixed patches of blue oak (*Quercus douglasii*, 5.4% cover) interior live oak (*Q. wislizenii*, 7.2% cover), gray pine (*Pinus sabiniana*, 12.5% cover), chaparral, mainly buckbrush (*Ceanothus cuneatus*, 18.6% cover), and annual grassland. Combined cover of the nine remaining tree and shrub species was 4.5% (J. Verner, unpubl. data). The climate is characterized by hot, dry summers and cool, wet winters.

Field observations were made on a 19.8 ha plot (300 × 660 m) gridded into 30-m quadrats, located in approximately 32 ha of foothill woodland that has not been grazed by livestock, burned, or otherwise disturbed since 1934.

BIRD OBSERVATIONS

Three observers recorded data in 1980 and five in 1982; two were the same observers in both years. Observers walked along alternate, numbered lines in the long dimension of the grid. Lines walked and the direction of travel were selected to ensure even coverage of all segments of the grid. Walking and stationary searches for birds were alternated approximately every 15 min. We attempted to obtain an equal number of observations of each species during each quarter of the daylight period, from sunrise to sunset, although sample sizes were smaller during the early afternoon quarter than during other quarters.

Only certain birds were selected for observation. Observers did not search out singing birds, as this would have biased our sample toward singing birds, although most birds sang or called during the period that they were observed. Only the first bird detected in a flock or pair was used as a subject, as locations of flock or pair members might not be independent. Further, a new individual of a given species was chosen as a subject only if the observer had traveled at least 30 m or unless 10 min had elapsed, since the last record of that species.

When a bird was accepted as a subject, we recorded its species and several aspects of its behavior and location. From the time of first detection, the observer counted slowly to 5 (approximately 5 s), allowing time to assess the bird's activity. Its activity at the count of "5" was recorded as an instantaneous sample. (We distinguish between the "state" of foraging, as being in the process of searching for and/or procuring food, and the "event" as actually procuring or attempting to procure a food item; see Altmann 1974, Martin and Bateson 1986.) If the bird was looking for food (in the state of foraging, but not the event of foraging) when the instantaneous sample was taken, but it did not appear to procure or attempt to procure a food item at that instant, the observer followed it visually until it appeared to procure or attempt to procure food. All subsequent locations of food procurement (sequential observations) were recorded, to a limit of 11 in 1980 and without limit in 1982.

In this paper we analyzed two measures of the location where a bird appeared to procure a food item, based on data obtained only in 1980: (1) foraging site (gray pine, blue oak, interior live oak, buckbrush, ground, or "other"); and (2) foraging substrate, the part of the plant or environs toward which a foraging maneuver was directed (air, twig [<5 mm in diameter], small branch [5 mm to 10 cm in diameter], large branch [>10 cm in diameter], flower bud, flower, catkin, cone, staminate cone, forb, fruit, ground, leaf bud, leaf, and trunk).

The effect of concealing cover on the detectability of a bird and the time between its subsequent foraging maneuvers were recorded only in 1982. Concealing cover for each observation was described as (1) little (the bird was completely in view), (2) moderate (vegetation obscured some of the bird), and (3) much (vegetation nearly obscured the bird). Observations ceased when the observer could no longer see the foraging behavior of the bird.

STATISTICAL ANALYSES

We used an alpha level of 0.05 for tests of significance.

Dependency among sequential observations

We created transition matrices and corresponding Pearson's contingency coefficients (Conover 1971:177) for the sequential observations to compare with matrices and coefficients for the initial observations used as a standard, assuming that initial observations were independent. These were then used to investigate dependency between sequential observations with zero, one, two, and three intervening observations, to compare values for sequential observations to those obtained from initial observations and to evaluate the effects of repetitive foraging habits on the Pearson's value that would be obtained from independent samples. The chi-square distribution provides a test of the significance of Pearson's contingency coefficients. To examine observations separated by one intervening observation, we compared the first observation to the third, the second to the fourth, and so on. A similar approach was used to compare observations separated by two and three intervening observations. For example, to examine observations separated by two intervening observations, we compared the first observation to the fourth, the second to the fifth, and so on. Pearson's contingency coefficients were corrected by dividing each coefficient by the maximum value possible for each contingency table.

Visibility bias

Places where birds were first observed may have been biased toward locations where they were most conspicuous. We tested this in two ways. First, we tested whether initial observations in certain sites or substrates more often resulted in records of subsequent observations. We used chi-square analysis to test whether the frequency of first observations differed by record length as a function of site or substrate at the initial location. Bonferroni adjustments (Miller 1981: 67) compensated for multiple comparisons.

Second, we used McNemar's test (Conover 1971: 127) to compare first and second observations from sequences. This test adjusted for dependency between observations, and Bonferroni adjustments compensated for multiple comparisons. Our sample size of matched pairs from third or later observations in sequences was too small for this test. Initial observations as a group could not be compared statistically to subsequent observations as a group, because not all records included the same number of observations (i.e., weighting problems and many unmatched initial observations; some initial observations did not have subsequent observations and some had many). Further, comparison of initial observations as a group with all observations as a group (as many researchers have done) is inappropriate, because the initial observations are a subset of all observations and often comprise a substantial proportion thereof, and unequal record lengths result in weighting and matching problems.

Estimating means and standard errors

Because they adjust for dependency within samples, bootstrap simulations (Efron and Gong 1983) were used to compare means and standard errors (precision) of the sample proportions of each class of site and substrate by initial observations and by all observations (initial + sequential observations). Five hundred random samples were drawn, with replacement, from the observed data.

To see whether large numbers of sequential samples would provide additional information, we used Markov chain analyses (Bishop et al. 1975:257–267, Isaacson and Madsen 1976) to compare differences in results based on initial observations and all observations. Assuming that our initial observations were independent of each other and that they approximated true proportions, our Markov chain model had characteristics similar to our data. We further assumed that the generation of successive observations in a sequence occurred according to a first-order Markov process. Transition matrices from sequential observations in the 1980 data set were estimated to approximate the true probability of change from one foraging site to the next and one foraging substrate to the next. Probabilities of the length of each sequential record were also estimated from our sample. Simulations of foraging records were then created from 500 runs for each species, drawing the same sample size as in the original data set for each species, and weighting each record length according to its proportion in the original data. Means and standard errors for initial observations as a group and all observations as a group were then computed for each simulation.

Both bootstrap and Markov chain analyses were also used to examine standard errors of all observations with and without adjustment for dependency among sequential observations. We compared the bootstrap estimate of standard error to the usual standard error created when assuming that all sequential observations were independent. From the Markov chain analyses, we compared the standard errors generated from each of the 500 simulations with the measure of standard error calculated from the mean estimate of proportions from all 500 simulations.

Because the means and standard errors generated by Markov chain analyses approximate the true values, based on the assumptions given, statistical comparisons are unwarranted. As a conservative criterion, we assumed that any difference between initial observations and all observations was biologically meaningful if the absolute difference exceeded 0.2 times the value from the initial observations. We used the same criterion to interpret bootstrap results.

RESULTS

The primary data set (1980) used in this study contained 1070 records of foraging events; 66% of those were of Plain Titmice and Bushtits, the two most commonly detected species on the plot. Sixty-five percent of all observations consisted of single records of foraging birds. We were seldom able to follow the same individual long enough to observe five consecutive foraging maneuvers, and records of eight or more consecutive behaviors were rare (Table 1).

In 1982, the only year we timed foraging sequences, the duration of a record was highly variable. For example, collective results from the five species gave a mean of 36 s (SD = 63; N = 173; range = 5 s to 6 min 38 s) to complete five consecutive maneuvers.

Our ability to record sequential observations differed among the bird species (Table 1). Sequential observations were obtained in 59% of the records of Brown Towhees but in only 31–34% of the records of the four other species. Data on the percent of sequences with 10 or 11 observations indicated that, if Scrub Jays, Plain Titmice, and Brown Towhees could be followed at all, they could be followed up to our self-imposed limit 7%, 7%, and 20% of the time, respectively. Bewick's Wrens changed foraging sites during 29% of the sequential observations, but the other species did so in only 6–10% of them. Thus we were more likely to get new information from sequential observations of Bewick's Wrens than from any other species.

DEPENDENCY AMONG SEQUENTIAL OBSERVATIONS

All analyses showed that sequential observations were highly dependent, with all values exceeding 0.64 and all but 4 of 40 values exceeding 0.81 (Table 2). For comparison, the Pearson's contingency coefficients that we created as standards using initial observations of foraging sites were 0.42 (Scrub Jay), 0.35 (Plain Titmouse), 0.38 (Bushtit), 0.57 (Bewick's Wren), and 0.52 (Brown Towhee); and of foraging substrates were 0.38 (Scrub Jay), 0.52 (Plain Titmouse), 0.59 (Bushtit), 0.42 (Bewick's Wren), and 0.39 (Brown Towhee). The transition matrix for foraging sites of Scrub Jays—for sequential observations with no intervening observation—had the highest Pearson's contingency coefficient (1.00) (Table

TABLE 1. Distribution of Records According to the Number of Foraging Maneuvers by an Observed Bird During a Continuous Observation (Proportions Shown below the Number of Records) in the 1980 Data Set

Species	Number of foraging maneuvers										
	1	2	3	4	5	6	7	8	9	10	11
Scrub Jay	117	13	15	4	6	1	1	1	0	12	0
	0.69	0.08	0.09	0.02	0.04	0.01	0.01	0.01	0.00	0.07	0.00
Plain Titmouse	230	40	19	15	5	6	2	1	2	23	0
	0.67	0.12	0.06	0.04	0.02	0.02	0.01	0.00	0.01	0.07	0.00
Bushtit	244	44	36	20	7	1	0	1	0	6	0
	0.68	0.12	0.10	0.06	0.02	0.00	0.00	0.00	0.00	0.02	0.00
Bewick's Wren	65	9	11	3	6	1	2	2	0	0	0
	0.66	0.09	0.11	0.03	0.06	0.01	0.02	0.02	0.00	0.00	0.00
Brown Towhee	41	10	9	6	9	3	1	0	0	16	4
	0.41	0.10	0.09	0.06	0.09	0.03	0.01	0.00	0.00	0.16	0.04
Totals	697	116	90	48	33	12	6	5	2	57	4
	0.65	0.11	0.08	0.05	0.03	0.01	0.01	0.01	0.00	0.05	0.00

3). The transition matrix for foraging sites of Bewick's Wrens—for sequential observations with three intervening observations—had the lowest coefficient for sequential data (0.65) (Table 4). Transition matrices created from initial observations for Scrub Jays on foraging sites (Table 3) and for Bewick's Wrens on foraging sites (Table 4) showed much less emphasis on transitions between the same foraging sites (visually depicted in the matrix as a high proportion of numbers on the diagonal from the upper left corner to the lower right).

Visibility Bias

The concealing cover of a bird when initially located apparently had no effect on whether it could be followed for subsequent observations. For example, a similar proportion of initial observations led to subsequent observations as did not, irrespective of the initial foraging site or foraging substrate. Only one of 105 comparisons had a significant chi-square value.

Percentages of observations in 1982 that were in little, moderate, and much concealing cover showed that first and subsequent observations were made in similarly difficult-to-see locations. For initial observations (N = 130), 19% were in little, 54% in moderate, and 27% in much cover. For subsequent observations (N = 403), 13% were in little, 61% in moderate, and 26% in much cover. No statistically significant differences appeared in any of the comparisons of the proportions of foraging sites and substrates that were used in the first and second maneuvers in a sequence. To convince ourselves that there were no differences, we set a standard for differences in proportion equal to 0.10 for the half-width of the 95% confidence interval and 21 of the 105 comparisons were inconclusive. We cannot reject the null hypothesis of no differences for these comparisons, but we cannot view it as confirmed either because of the large width of the confidence interval. Twelve of these 21 comparisons were for Bewick's Wrens.

Estimating Means

All observations sometimes gave markedly different estimates of means than did initial observations, particularly in the case of uncommon foraging locations (defined here as representing 10% or less of the observations). Forty-two of 81 bootstrap comparisons met our criterion of a meaningful biological difference (Table 5). Thirty-seven of the 42 differences were on uncommonly used sites and substrates. When compared to initial observations, Markov chain analyses indicated that all observations overestimated the mean in 3% and underestimated it in 13% of the comparisons of foraging sites; all of these were on uncommonly used sites. All observations overestimated the mean in 25% and underestimated it in 31% of 51 comparisons of foraging substrates. Seventy-one percent of all comparisons of uncommon substrates satisfied our criterion of a meaningful biological difference, but only 15% of all comparisons of common substrates did so.

Estimating Standard Errors: All Observations vs. Initial Observations

Bootstrap and Markov chain analyses differed slightly in their estimates of standard errors (Table 5). In bootstrap comparisons, initial observations estimated common foraging locations

TABLE 2. TESTS OF INDEPENDENCE OF SEQUENTIAL OBSERVATIONS OF FORAGING SITES AND SUBSTRATES WITH CORRECTED PEARSON'S CONTINGENCY COEFFICIENTS. A COEFFICIENT OF 1.00 IS THE HIGHEST POSSIBLE INDEX OF AUTOCORRELATION

Species	Number of intervening observations	Pearson's contingency coefficients[a]	
		Foraging sites	Foraging substrates
Scrub Jay	0	1.00	0.98
	1	0.93	0.97
	2	0.94	0.97
	3	0.99	0.97
Plain Titmouse	0	0.99	0.94
	1	0.90	0.88
	2	0.99	0.90
	3	0.99	0.84
Bushtit	0	0.97	0.92
	1	0.90	0.87
	2	0.99	0.91
	3	0.99	0.96
Bewick's Wren	0	0.93	0.79
	1	0.90	0.68
	2	0.92	0.78
	3	0.65	0.88
Brown Towhee	0	0.98	0.92
	1	0.93	0.87
	2	0.92	0.85
	3	0.83	0.82

[a] All values statistically significant at $P < 0.05$.

TABLE 3. TRANSITION MATRICES FOR FORAGING SITES OF SCRUB JAYS BASED ON SEQUENTIAL OBSERVATIONS WITH NO INTERVENING OBSERVATIONS (TOP) AND BASED ON INITIAL OBSERVATIONS ONLY (BOTTOM) (PROPORTIONS OTHER THAN 0 IN PARENTHESES)

Initial foraging site	Subsequent foraging site				
	Blue oak	Gray pine	Live oak	Ground	Other
Sequential observations					
Blue oak	80 (1.00)	0	0	0	0
Gray pine	1 (0.02)	60 (0.98)	0	0	0
Live oak	0	0	7 (0.88)	1 (0.13)	0
Ground	1 (0.02)	0	0	50 (0.94)	2 (0.04)
Other	0	0	0	1 (0.02)	50 (0.98)
Initial observations					
Blue oak	29 (0.46)	11 (0.17)	5 (0.08)	13 (0.21)	5 (0.08)
Gray pine	10 (0.43)	1 (0.04)	4 (0.17)	7 (0.30)	1 (0.04)
Live oak	4 (0.33)	2 (0.17)	1 (0.08)	3 (0.25)	2 (0.17)
Ground	14 (0.25)	7 (0.13)	1 (0.02)	25 (0.45)	8 (0.15)
Other	6 (0.38)	3 (0.19)	1 (0.06)	6 (0.38)	0

more precisely than did all observations in 59% of all cases, but less precisely in only 7%. Conversely, all observations estimated uncommon locations more precisely than initial observations in 46% of all cases and less precisely in 26%. In the Markov chain analyses, estimates of standard error from all observations differed from estimates from initial observations in 79% of the comparisons of common foraging sites and in 75% of the uncommon foraging sites; the estimates of standard error from all observations differed from those from initial observations in 46% of the comparisons of common foraging substrates and 82% of the uncommon foraging substrates.

ESTIMATING STANDARD ERRORS:
ADJUSTED VS. UNADJUSTED DEPENDENCY IN ALL OBSERVATIONS

Both bootstrap and Markov chain procedures generally showed that standard errors estimated from all observations in the usual (unadjusted) way, assuming them all to be independent records, were smaller than true standard errors after adjustment for dependency. Using bootstrap, the usual standard error underestimated the adjusted standard error in 68% of all cases, using our criterion of a meaningful biological difference. The two estimates were similar in 32% of the cases, and in no case did the usual procedure overestimate standard error. Markov chain analyses showed that the usual procedure underestimated true standard error for foraging site by a mean of 45%, and 28 of 30 comparisons were underestimated. For foraging substrate, the usual procedure underestimated true standard error by a mean of 34%, and 42 of 53 comparisons were underestimated. The mean underestimate differed among species, but it was not significantly correlated with sample size (either for initial observations or for all observations).

DISCUSSION

Our results suggest that using dependent sequential observations is inadvisable for the estimation of proportions of foraging locations unless appropriate statistical analyses are used to adjust for autocorrelation. We were not able to obtain sequential records that were far enough apart in time to appear independent. We were seldom able to follow an individual long enough to obtain more than five sequential records of its foraging, and all analyses showed that the fifth observation in a sequence was dependent on the

TABLE 4. Transition Matrices for Foraging Sites of Bewick's Wrens Based on Sequential Observations with Three Intervening Observations (Top) and Based on Initial Observations Only (Bottom) (Proportions Other Than 0 in Parentheses)

Initial foraging site	Subsequent foraging site			
	Live oak	Buckbrush	Ground	Other
Sequential observations				
Live oak	0	1 (0.25)	2 (0.50)	1 (0.25)
Buckbrush	1 (0.09)	7 (0.64)	0	3 (0.27)
Ground	0	2 (0.20)	8 (0.80)	0
Other	0	1 (0.33)	0	2 (0.67)
Initial observations				
Live oak	8 (0.38)	6 (0.29)	0	7 (0.33)
Buckbrush	8 (0.21)	19 (0.50)	1 (0.03)	10 (0.26)
Ground	1 (0.11)	2 (0.22)	2 (0.22)	4 (0.44)
Other	5 (0.17)	10 (0.33)	6 (0.20)	9 (0.30)

TABLE 5. Bootstrap and Markov Chain Analyses for Common and Uncommon (10% or Less of Observations) Foraging Sites and Substrates. AO = All Observations (Initial + Subsequent Observations) and IO = Initial Observations Only. Described Differences in the Means and Standard Errors Are Those for Which the Absolute Value of the Difference between AO and IO Was Greater Than 0.2 IO

Differences	Bootstrap		Markov chain	
	Site	Substrate	Site	Substrate
Means				
Common locations				
AO > IO	1	2	0	1
No difference	13	9	14	11
IO > AO	0	2	0	1
Means				
Uncommon locations				
AO > IO	3	9	1	12
No difference	5	12	11	11
IO > AO	8	17	4	15
Standard errors				
Common locations				
AO > IO	10	6	11	4
No difference	4	5	3	7
IO > AO	0	2	0	2
Standard errors				
Uncommon locations				
AO > IO	4	10	8	14
No difference	5	10	4	7
IO > AO	7	18	4	17

first as indicated by a higher value than those created for initial observations. However, sequential observations of some species approached an equivalent level of independence to that obtained by the use of initial observations. For example, one of the contingency coefficients for sequential observations (0.65, Table 2) of Bewick's Wrens was nearly as small as that obtained from initial observations (0.57).

Dependency between observations in a sequence leads to inaccurate estimates of variance. Unadjusted standard errors from all observations were consistently less than those adjusted for dependency. One is thus more likely to conclude erroneously that two sample means are different with unadjusted standard errors that are artificially small due to the lack of adjustment for dependency.

The use of initial observations is preferable for estimating common foraging locations, but we are not sure which method is better for estimating uncommon foraging locations. As shown by bootstrap and Markov chain analyses, estimates of means of common foraging locations were similar with both methods, and initial observations more precisely estimated common foraging locations. However, the estimates of means and standard errors from uncommon foraging locations differed between the two methods, and we do not know which method estimates the true population parameters more accurately and precisely.

We had no conclusive evidence of a visibility bias in our habitat; however Recher and Gebski (this volume) found some evidence of a tendency for first-recorded prey attacks to be of particularly conspicuous individuals in their study in an open eucalypt woodland in Australia. We may not have detected any biases because we waited 5 s before recording any observations. Recher and Gebski concluded that the problem of over-representation of conspicuous behaviors or individuals might be minimized by rejecting initial observations. Rejecting initial observations may have the same effect as our 5-s waiting period. However, this solution may not be tenable in habitats other than eucalypt woodland. For example, we would not want to reject initial observations in our study, because we were unable to follow birds for sequential observations for 41–69% of our cases.

At least three solutions can be used to deal with problems of autocorrelation in sequential records. First, observers could record only initial

TABLE 6. Some Appropriate and Inappropriate Statistical Tests for Sequential Data

Suggested analyses to examine or adjust for dependency among sequential records			Inappropriate analyses with dependent, sequential records
To compare proportions between initial and subsequent observations	To estimate variance of proportions for sequential records	To examine dependency among sequential records	(e.g., to compare proportions among foraging locations, to examine dependency among sequential records)
Categorial data (e.g., site, substrate)			
McNemar's test (Fleiss 1981:113–119)	Bootstrap Efron and Gong 1983)	Pearson's contingency coefficient (Conover 1971:177)	G-test (Bishop et al. 1975:125–130),
Cochran's Q (Fleiss 1981:126–133)	Jackknife (Efron and Gong 1983)	Runs test (Conover 1971:349–356)	Chi-square (Steel and Torrie 1960:346–387)
			Two-sample t-test (Steel and Torrie 1960:73–78, 82–83)
To compare means between initial and subsequent observations	To estimate variance of means for sequential records	To examine dependency among sequential records	(e.g., to compare mean values of foraging locations)
Continuous data (e.g., dbh, height)			
Paired t-test (Steel and Torrie 1960:78–80)	Bootstrap Jackknife	Durbin-Watson D (Durbin and Watson 1951)	Two-sample t-test Analysis of variance (Steel and Torrie 1960:99–160, 194–276)

observations from each bird (e.g., Gibb 1954, Morse 1970, Lewke 1982, Franzreb 1985), or only second observations as Recher and Gebski (this volume) have suggested, and use a study design that ensures that all such records are independent. This method may not always be easy or practical for answering certain biological questions. For example, we designed this study so that we would rarely observe the same individual bird more often than once each day. Even if the design succeeded with this objective, however, the same individual was likely observed in the same territory repeatedly over a period of several days. Our primary objective was to study the changes in foraging behavior of a particular population over time. The extent to which obtaining foraging information from the same individuals over time may have biased results of the present analysis is unknown. The most obvious way to obtain completely independent records is to select new areas with new individuals for each observation, but many questions that students of avian foraging behavior choose to answer would not be compatible with this design.

Second, one could make sequential observations for extended periods in a pilot study, analyze for autocorrelations, and select a time interval between observations to ensure independence. Others have used intervals of 10 s (Wagner 1981a), 15 s (Landres and MacMahon 1980), and 60 s (Morrison 1984a). Because our average interval between the first and fifth sequential records was 36 s, we consider the 10-s and 15-s intervals probably insufficient to ensure independence. Porter et al. (1985) followed six individually marked Red-cockaded Woodpeckers (*Picoides borealis*) for extended periods and concluded that records separated by 10-min intervals were independent. But few species are so amenable to study; we could not have followed many individuals for 10 min, and most studies without individually marked birds would likely have the same difficulty. Further, the effects of within-season, seasonal, and annual variation on avian foraging should be considered when establishing appropriate intervals.

Third, as in this study, one could record all possible sequential observations from each individual and analyze the data with procedures capable of adjusting for autocorrelation. We recommend bootstrap or jackknife procedures, both of which can be used with sequential records of unequal length. However, the discrepancies in mean proportions for uncommon foraging locations found for all observations and initial observations in this study show that the two methods may give different estimates of proportions, and we do not know which method would produce a more accurate estimate of true proportions.

Airola and Barrett (1985) used sequential observations but treated each *sequence* as an equally-weighted independent sample. Each measure was expressed as a proportion of the total for that measure in the sequence, so each individual's record was weighted as one in the total sample. We question the validity of giving equal weight to records of unequal length, although the problem may be significant only for relatively short sequences where biases are high (J. T. Rotenberry, pers. commun.). A solution is to use only records greater than a standard length, for example, a 3-min minimum, although this would require rejection of all records shorter than the standard, and longer records may be biased toward more visible locations or more visible bird species or individuals.

Assumptions of statistical analyses have rarely been achieved in studies of avian foraging behavior. First, most errors in the application of statistics result from assumptions of independence among sequential records (see Table 6). Probably the most common example of such errors is the use of G-tests or chi-square tests (that assume independence between records) to examine differences in proportions of behavioral measures using sequential records without first establishing that the records within each sequence are independent. Further, when comparing initial observations with sequential observations, the two data sets must be perfectly matched, the sequential observations must be weighted equally, and a test that deals with matching must be used. For example, G-tests are commonly used incorrectly to compare initial observations with all observations to decide whether sequential data may be used.

Finally, sequential observations are useful, even essential, for certain ethological studies of foraging, such as transitions among various behaviors. They also allow one to include time as a measure to estimate rates at which birds make foraging strikes, move from substrate to substrate, and move from one tree or shrub to another. They may also help to correct for visibility bias, because birds in relatively concealed locations may not be detected as often by initial observations. Although our data did not provide evidence of such a bias, it is probably a valid concern in some habitats.

ACKNOWLEDGMENTS

C. L. DiGiorgio, P. M. Gunther, C. Hansen, K. L. Purcell, and L. V. Ritter assisted with field observations. J. A. Baldwin provided statistical advice and assistance with the analyses. G. W. Salt gave constant support and advice. W. M. Block, L. A. Brennan, C. Kellner, K. A. Milne, M. L. Morrison, and J. T. Rotenberry reviewed an earlier draft. To all we express our sincere appreciation.

ANALYSIS OF THE FORAGING ECOLOGY OF EUCALYPT FOREST BIRDS: SEQUENTIAL VERSUS SINGLE-POINT OBSERVATIONS

HARRY F. RECHER AND VAL GEBSKI

Abstract. Up to five consecutive prey attacks were recorded for each individual encountered of five species of Australian warblers (Acanthizidae) foraging in eucalypt woodlands near Sydney, New South Wales. A comparison of the first (single-point observations) against all subsequent prey attacks (sequential observations) revealed no significant differences in the use of plant species or foraging heights for the species studied. First observations were biased towards birds foraging in foliage, but the differences between first and subsequent observations were not significant. For all species active prey-attack behaviors (snatch, hover, hawk) were recorded more often on the first than on subsequent observations. However, only a few of these differences were significant: White-throated Warblers (*Gerygone olivacea*) snatched more often ($P < 0.005$), Little Thornbills (*Acanthiza nana*) gleaned less ($P < 0.02$) and hawked more often ($P < 0.02$), and Weebills (*Smicrornis brevirostris*) hovered more often ($P = 0.054$) on the first than the second observations. Differences between the first and subsequent observations were greatest for the more active species [White-throated Warbler, Weebill and Buff-rumped Thornbill (*A. requloides*)] and least for the less active Striated (*A. lineata*) and Little Thornbills. The differences between first and subsequent prey attacks were insufficient to affect interpretations of resource use or of possible interactions between species. Other than for foraging height, where samples of 110–120 individuals were necessary, observations of 60–70 individuals were required to stabilize sample variances of the foraging behaviors of all species, irrespective of the number of consecutive prey attacks recorded. At least for open habitats this suggests that it is necessary to record only one prey attack for each individual encountered. These estimates of minimum sample sizes generally fell within the range required for 90–95% confidence intervals. Greater precision requires much larger samples.

Key Words: Sequential observations; single-point observations; foraging ecology; sample size; prey-attack behavior.

Studies of the foraging ecology of birds usually employ one of two methods: single-point or sequential observations. With single-point observations only one set of data, usually obtained at the first sighting of the bird or whenever it first performs the behavior being studied, is recorded for each individual encountered (e.g., Hartley 1953, Morse 1970). Sequential observations require the bird to be followed and data recorded continuously (e.g., Hertz et al. 1976) or at intervals (e.g., Morrison 1984a). Most observers employing sequential sampling procedures have well-defined rules for stopping and starting which specify minimum and maximum periods of observation (e.g., Morrison 1984a, Recher et al. 1985).

A decision as to which method to use may largely depend on the hypotheses being tested and the ease of studying the birds in question (Bradley 1985). It is also necessary to have information on the extent to which observations may be biased by conspicuous behaviors, the importance of inconspicuous or uncommon events, and the minimum sample sizes required for an acceptable level of precision (Wagner 1981a, Morrison 1984a). Few studies have presented data comparing the two methods (Wagner 1981a, Franzreb 1984, Morrison 1984a) and only Morrison (1984a) has suggested a minimum sample size. These studies are from North America and compare closely related or ecologically similar species.

As part of a study of the foraging ecology of Australian warblers (Acanthizidae) in eucalypt woodland near Sydney, New South Wales (Recher 1989 b), data were recorded for up to five consecutive prey-attacks for each individual encountered. Data were obtained for five species of three genera which differed in their use of substrates, foraging height distribution, and prey-attack behavior (Recher 1989 b). In this paper we compare interpretations of the behaviors of these species based on the first recorded observation (single-point method) to interpretations based on all and subsequent observations (sequential method). Minimum sample sizes required for analysis are also examined.

METHODS

STUDY AREA

The foraging ecology of Australian warblers was studied during 1984 on a 25 ha plot located within a large block (ca. 400 ha) of regrowth eucalypt forest at Scheyville, 40 km northwest of Sydney, New South Wales. The study site was dominated by narrow-leaved ironbark (*Eucalyptus crebra*) (42% of eucalypt foliage) and grey box (*E. molluncana*) (50% of eucalypt foliage). Other plant genera were absent from the canopy and understory. Blackthorn (*Bursaria spinosa*) (>98% of shrub foliage) dominated the shrub layer. Ground vegetation was dominated by exotic grasses and herbs. Total tree canopy cover was 40–45% with the tallest trees emerging to 25 m from an average canopy height of 14–18 m. Patches of dense sapling regrowth of both eucalypts occurred throughout the plot.

The study area was flat and forms part of the Cumberland Plain, an extensive area of low to undulating terrain west of Sydney. Soils in this area were primarily derived from shale formations. The area receives about 775 mm of rain annually with a tendency for spring (August–October) to be drier and for summer (December–March) to be wetter than other months. Summers are hot (January mean maximum 30°C) and winters are mild (July mean minimum 3°C). Recher (1989 b) provided additional details of the plot.

Data were obtained for five species of birds: Little Thornbill (*Acanthiza nana*), Striated Thornbill (*A. lineata*), Buff-rumped Thornbill (*A. reguloides*), Weebill (*Smicrornis brevirostris*), and White-throated Warbler (*Gerygone olivacea*). All foraging data were collected by H. Recher. He recorded up to five prey-attacks for each bird encountered. Most birds were located visually.

Where birds occurred in either single- or mixed-species flocks, data were recorded for as many individuals as possible without repeating observations on the same birds. Generally this meant that fewer than half the birds present in the flock were recorded. Although it is likely that some of the same individuals were observed on more than one occasion, observations were made on different parts of the study site on successive days to reduce the duplication of observations on the same individuals.

As it was not always possible to determine success, all prey-attacks were recorded irrespective of whether or not they were successful. Bird species, type of prey-attack behavior (e.g., glean, snatch, hawk), foraging height, substrate of prey, and plant species, where appropriate, were recorded for each observation. These are the same procedures used by Recher et al. (1985). Prey-attack heights were estimated to the nearest meter and later grouped into height categories (0–0.1 m, 0.1–2 m, 2.1–8 m, >8 m) corresponding to ground, shrub, understory, and canopy vegetation layers.

Observations were made during spring (September–November), summer (January–February), autumn (April–May) and winter (July–August). With the exception of spring, when observations were made over a 6-week period, seasonal data were collected during a 2-week period with most data obtained on 4–6 mornings of fieldwork (20–30 hours). Observations generally began within an hour of sunrise and ceased at 11:00–12:00 EST. Additional details are in Recher (1989 b).

DATA ANALYSIS

We compared the proportions of different foraging behaviors for the first recorded prey-attacks to the proportions for all foraging sequences (i.e., the first through the fifth prey-attack) and also to those calculated from foraging sequences where the first observation was deleted (i.e., the second through the fifth prey attack).

The comparison was performed as follows. Suppose prey-attack heights for 0–0.1 m are being considered. Then, for each bird of each species, we calculate the proportion of times the bird is present in the 0–0.1 m height range. An overall or average proportion of birds for the species in this height range may then be obtained. We next calculate the proportion of birds present in the 0–0.1 m height range using the first observation only. If there is no bias in the observation, then this proportion should be substantially similar to that calculated using all the observations. A formal statistical test such as the chi-squared test may then be employed. When the data compared are in terms of proportions, the chi-squared test is identical to a two sample t-test. Each behavior or foraging category was tested separately.

As the inclusion of observations where only a single prey-attack was recorded may influence the results towards conspicuous behaviors, the data were re-analyzed using only sequences where two or more prey-attacks were recorded and the proportions of foraging behaviors recorded for the first observation tested against proportions recorded for the second.

Small changes in the standard error (SE) of the population mean can be used as a simple estimate of minimum sample sizes beyond which further observations provide little additional information relative to the "cost" of obtaining more data. To estimate the SE's of different sized samples (n) (at increments of 5), we assumed the proportion (P) of each foraging parameter for the total sample approximated the proportion (p) for all sample sizes (i.e., 5, 10, 15, ... n). This is justified by the large sample sizes available for each species. The SE of p was then calculated from

$$\text{SE}(p) = \sqrt{\frac{PQ}{n}}$$

when n is large (i.e., $nP > 5$, $nQ > 5$), $Q = (1 - P)$: "When P is the underlying proportion, the sample p is approximately normally distributed with mean P and standard error" (Fleiss 1981:13).

As in other analyses, the first recorded observation was used to estimate sample sizes for single-point observations. For sequential observations the mean value for each foraging category was calculated for each foraging sequence and these values used to estimate the proportion (P) for each foraging category. In this instance, P is a weighted average of the proportion of each foraging parameter. A weighted average is preferred as sequential observations are not independent. Thus, for example, we could not observe one bird five times, another three times, and a third once and say we had nine individuals. All observations were used including individuals for which only a single prey attack was recorded. Only sequential data were used in calculating SE's for foraging height data. In this instance SE's were calculated progressively from the field data.

RESULTS

SINGLE-POINT VERSUS SEQUENTIAL SAMPLES

There were seasonal differences in foraging behavior (Recher, 1989 b) and the proportions of behaviors in each foraging category for single-

TABLE 1. COMPARISON OF FORAGING DATA, OBTAINED BY SINGLE-POINT (A) AND CONTINUOUS OBSERVATIONS (2–5) (B) OF AUSTRALIAN WARBLERS (ACANTHIZIDAE)

Species	Little Thornbill		Striated Thornbill		White-throated Warbler		Buff-rumped Thornbill		Weebill	
Method	A	B	A	B	A	B	A	B	A	B
No. individuals	324	200	421	209	84	39	160	110	252	168
No. prey attacks	324	758	421	790	84	207	160	450	252	653
Prey-attack behavior (%)										
Glean	61.7	66.5	40.1	46.3	34.1	43.5	63.7	75.6	28.9	33.7
Hang-glean	2.5	2.4	36.7	33.5	0	0	0	0	4.4	9.8
Hover	11.1	10.9	12.3	12.0	7.1	12.1	16.2	13.3	53.6	42.3
Snatch	19.4	15.3	9.0	6.3	51.8	37.5	13.7	3.6	11.1	12.2
Hawk	5.2	4.9	1.8	1.8	7.1	6.9	6.3	7.6	2.0	2.0
Substrate (%)										
Foliage	80.8	76.5	90.3	87.2	89.5	86.6	53.1	41.8	94.8	92.6
Bark	13.9	18.0	7.9	11.0	3.9	2.6	21.3	25.1	2.4	3.5
Ground	0	0	0	0	0	0	17.5	24.0	0	0
Aerial	5.3	5.5	1.8	1.8	7.0	10.8	8.1	9.0	2.8	3.9
Plant species (%)										
Ironbark	63.9	65.2	62.8	66.4	48.8	51.7	36.7	33.9	87.1	88.8
Box	23.3	19.9	29.3	27.4	39.0	38.6	27.3	26.9	9.5	8.3
Other eucalypts	7.5	7.3	7.9	6.2	12.2	9.7	5.1	4.2	3.3	2.8
Blackthorn	5.2	7.7	2.2	2.0	0	0	30.8	35.0	0.1	0.1
Height intervals (m) (%)										
0	0	0	0	0	0	0	16.4	14.4	0	0
0,1–2	5.0	7.6	5.0	4.0	3.3	2.2	42.1	50.1	2.0	1.7
2–8	32.1	31.5	62.7	62.5	63.7	51.3	37.7	32.8	39.6	42.4
>8	62.9	60.9	32.3	33.5	33.0	46.5	3.8	2.7	58.4	55.9

point and sequential samples were first tested for seasonal effects. As seasonal differences did not affect the proportions of observations recorded in any of the foraging categories for the first observation (single-point method) compared to the total data set or to the second plus subsequent observations (sequential method) (G-tests, P's > 0.05), seasonal data were combined in subsequent analyses.

The proportions of foraging behaviors for single-point observations were similar to those recorded for sequential observations (Table 1). This was the same whether single-point data were compared to sequential observations with the first prey-attack deleted (Table 1) or to the total data set. None of the species studied had rare or unusual behaviors that required prolonged study (i.e., >20–25 observations) to observe or that affected interpretations of their use of resources and interactions with other individuals (see Recher 1989 b, for details).

There were no significant differences in the use of foraging substrates, plant species or height intervals (P's > 0.05) between the first and subsequent observations. However, there were some consistent, although not significant, differences in the proportions of foraging behaviors recorded for the first and subsequent prey-attacks. For most species foliage was over-represented whereas bark and aerial foraging were under-represented on the first observation (Table 1). The exception was the White-throated Warbler, for which the proportion of bark foraging decreased with subsequent observations.

Apart from Buff-rumped Thornbills, the proportion of prey-attacks by birds foraging in ironbarks increased after the first observation, whereas the proportion in grey box and other eucalypts decreased (Table 1). Ironbark has smaller leaves and denser foliage than grey box and the other eucalypts on the plot, which made the detection of birds in ironbark more difficult. Buff-rumped Thornbills were the only birds to forage extensively in blackthorn and there was an increase in the use of blackthorn and a decrease in the use of ironbark and grey box subsequent to the first observation (Table 1). The foliage of blackthorn is much denser than that of the eucalypts and the detection of birds foraging in blackthorn more difficult. The increased use of the shrub layer by Little and Buff-rumped thornbills with the second and subsequent observations (Table 1) reflects their use of blackthorn.

Active prey-attack behaviors (i.e., snatch, hover, hawk) were recorded more often and less active behaviors (i.e., glean, hang-glean) less often on the first compared with subsequent observations (Table 1); these differences were significant for two species. Buff-rumped Thornbills snatched and hovered more often and gleaned less often on the first than subsequent observations (P < 0.001). Weebills gleaned and hang-gleaned less often and hovered more often on the first than subsequent observations (P < 0.025).

Testing the first against second prey-attacks, there were no significant differences for any species in the proportions of plant species, foraging heights, or substrates recorded for the first and second prey-attacks (P's > 0.1). However, there was a tendency for active behaviors to be recorded more often and less active behaviors to be recorded less often on the first than on the second prey-attack. Little Thornbills gleaned less often (P < 0.02), but hawked (P < 0.02) and snatched more often (P = 0.1) on the first than second observation. White-throated Warblers snatched more often (P < 0.005) and gleaned less often (P = 0.1) on the first than second observation. Weebills hovered (P = 0.054) more often and hang-gleaned less often (P < 0.004) on the first than second observation. Buff-rumped Thornbills snatched more often on the first than the second observation (P = 0.07). Other differences were not significant (P's > 0.1).

ESTIMATE OF SAMPLE SIZE

The standard error of the mean for different sized samples stabilized (i.e., a small change in value with increasing sample size) at about ±0.05 for single-point and sequential methods for all foraging categories and all species (Figs. 1–3). This value can therefore be used to estimate sample sizes beyond which additional observations add little information on the proportions of different foraging behaviors. Although sample sizes differed between species, generally observations of 60–70 individuals were needed before standard errors stabilized (Figs. 1–3). For the proportional data reported here, samples of 60–70 individuals fall between the sample sizes estimated for 90 and 95% confidence intervals (15–365 individuals) (Snedecor and Cochran 1980: 441–443). For a 99% confidence interval, samples exceeding 5900 individuals are required.

Foliage and bark were the two most commonly used foraging substrates (Recher 1989 b). For Little and Striated thornbills, which took 70–80% of their prey from foliage, standard errors

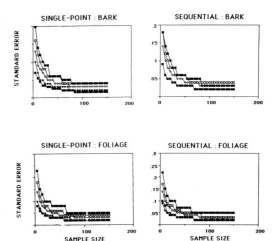

FIGURE 1. The standard error of the mean for the proportions of foraging substrates used by Australian warblers at Scheyville is plotted against sample size for foliage and bark.

stabilized at a maximum of 50–60 individuals (Fig. 1). Smaller sample sizes (30–40 individuals) were required for Weebill and White-throated Warbler which took more than 85% of their prey from foliage. The largest sample sizes (65–70 individuals) were required for Buff-rumped Thornbills which used the greatest diversity of substrates and often foraged on the ground and among debris as well as taking prey from foliage and bark.

Snatch, glean, and hover were the most common foraging behaviors used by Australian warblers at Scheyville (Table 1; see also Recher 1989 b). Gleaning was the most frequently used prey-attack behavior (35–70% of observations). Standard errors for the proportion of gleaning stabilized for all species at 60–70 individuals (Fig. 2). Hovering by Weebills and snatching by White-throated Warblers were the most common behaviors (40–50% of prey-attacks) used by these two species. Standard errors for these behaviors stabilized at 65–70 individuals for Weebills and White-throated Warblers and for the other species at 45–50 individuals (Fig. 2).

Ironbark and grey box dominated the study site and accounted for >90% of foraging by Australian warblers on eucalypts at Scheyville (Recher 1989 b.) Weebills foraged almost exclusively on ironbark (>90% of observations). For Weebills foraging on ironbark single-point observations stabilized at 55–60 individuals and sequential observations at 65–70 individuals (Fig. 3). For all other species standard errors for the use of ironbark as a foraging substrate stabilized at 60–70 individuals.

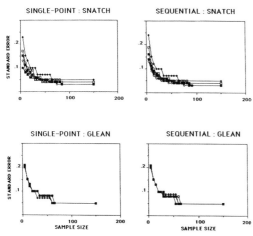

FIGURE 2. A plot of standard error against sample size for the three most commonly used prey-attack behaviors: snatch, glean, and hover.

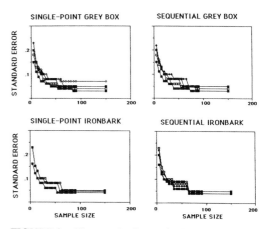

FIGURE 3. The standard error in the use of the two dominant eucalypts at Scheyville by Australian warblers is plotted against sample size.

Apart from White-throated Warblers, which used grey box as a foraging substrate more frequently (39% of observations) and Weebills which used it less often (9% of observations) than other species (Table 1), standard errors for single-point and sequential observations for grey box stabilized at 55–60 individuals. For White-throated Warblers 70–75 individuals were required for sequential observations and 65–70 individuals for single-point observations (Fig. 3). Standard errors for the proportion of foraging on grey box by Weebills stabilized with observations of only 40–45 individuals for both single-point and sequential observations.

The most variable foraging parameter measured was mean foraging height. All species foraged from the shrub layer into the canopy and the Buff-rumped Thornbill foraged extensively on the ground (Table 1). Relative to other foraging categories, large samples were required to stabilize standard errors. After weighting for the number of observations per individual (see Methods), sequential data were used to calculate the standard error of mean foraging height with increasing sample size (Fig. 4).

For all species the rate of change in foraging height standard error decreased markedly after 70–80 observations with standard errors between ±0.2 m for Buff-rumped Thornbills and ±0.4 m for Little Thornbill. Standard errors stabilized between ±0.2–0.3 m for all species after 110 observations. Estimates of the minimum required sample sizes (Snedecor and Cochran 1980: 53) for an 80% confidence interval about the mean with standard errors between 0.2 and 0.3 m range from 110 to 140 individuals. For a confidence interval of 95% the required sample size is 440 and for a 99% interval it is 725.

DISCUSSION

Sequential observations of the same individual are not independent, posing problems for the statistical analysis of the data (Wagner 1981a, Morrison 1984a, Bradley 1985). In addition, results may be biased towards individuals or behaviors that are easy to follow (Franzreb 1984, Bradley 1985). Single-point observations have the advantage of statistical independence, but may be biased towards particularly conspicuous individuals (e.g., singing males) or behaviors (e.g., hawking) (Wiens 1969, Wagner 1981a). Single point observations are also useful in that details of the substrate (e.g., plant species, substrate height, prey concentrations) can be recorded without the necessity of following the bird and losing track of the foraging stations that had been used. Sequential observations have the advantage that a large amount of data can be collected for each bird, and uncommon or inconspicuous behaviors are more likely to be recorded (Hertz et al. 1976, Sturman 1968, Austin and Smith 1972). Thus it is tempting to use sequential recording techniques when little is known of a species' behavior or when individuals are difficult to locate. For these reasons sequential observations have generally when preferred (e.g., Hertz et al. 1976, Wagner 1981a, Morrison 1984a, Recher et al. 1985), but with the caveat that large sample sizes may be necessary to overcome problems of the lack of statistical independence (Morrison 1984a) or that special methods are needed to analyze the data (Bradley 1985).

Data collected over 12 months for five species of Australian warblers suggests that there may be a tendency for the first recorded prey-attack to be of particularly conspicuous individuals. Despite the openness of the habitat in which observations were made, birds that foraged in foliage were more readily detected than those foraging on bark. Probably this is because eucalypt foliage tends to be clumped and clustered towards the ends of branches. Foliage gleaners are seldom concealed by leaves and the terminal position of the foliage makes them easy to detect. Similarly, conspicuous foraging behaviors such as snatching and hovering were over-represented on the first recorded prey-attack. The reduced frequency of aerial foraging (an active behavior usually associated with hawking and/or hovering) on first compared to subsequent observations may have resulted from a tendency by the observer to avoid recording particularly conspicuous behaviors when birds were first sighted. For Weebills the greater frequency of hovering on first observations and the increased incidence of gleaning and hang-gleaning with sequential observations results from hovering being an exploratory as well as a prey-attack behavior, with hovering birds landing to feed after locating prey.

With the exception of the White-throated Warbler, none of the species studied was sexually dimorphic. Male White-throated Warblers were the only birds studied that sang and which were located by sound. Although there was a tendency for singing males to forage in the upper canopy (Recher 1989 b), first observations tended to be biased towards individuals foraging in lower vegetation (Table 1). Thus, there is no indication that the detection of some birds by song affected results. Probably this is because of the small numbers of males located while they were singing. All other birds were located visually. This probably contributed to the tendency to first see birds that were in the outer foliage of trees or that were foraging actively. The greater proportion of first observations of birds in grey box than in ironbark may result from the more open foliage and larger leaves of grey box than ironbark, where birds were more easily concealed.

Despite the tendency to locate individuals that were conspicuous, there were few significant differences between the proportions of the various foraging parameters recorded on the first prey-attack (single-point method) versus subsequent behavior (sequential method). Such differences did not affect any of the conclusions relating to the use of resources by these birds or their interactions with each other. At least in the open eucalypt habitats where this work was done, problems of conspicuous behavior or individuals might be minimized by rejecting the first prey-attack observed for each bird encountered or having a set waiting period before recording the

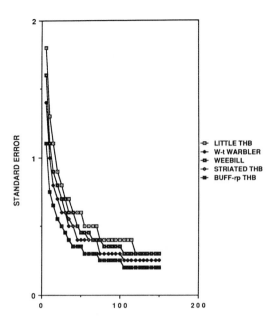

FIGURE 4. A plot of standard error against sample size for mean foraging height. In this plot standard errors were calculated for all prey attacks as the data were collected in the field.

first observation. Either procedure could be used without greatly increasing the effort required to obtain adequate sample sizes.

Regardless of the sampling procedure a minimum of 60–70 individuals was required to stabilize sample variances for most foraging parameters by both single-point and sequential methods. This estimate of minimum sample size assumes that the proportion of each foraging behavior recorded for the total sample approximates the underlying proportion for the population (Fleiss 1981). As such, the estimate of sample size is independent of the time period over which the sample is obtained. Where there are significant temporal or spatial changes in the proportions of foraging behaviors within a population, similar sized (i.e., 60–70 individuals) samples are required for each time period or area.

The estimates of minimum sample size presented here are greater than Morrison's (1984a) estimate of a minimum of 30 individuals or 150 sequential observations. Inspection of Morrison's data suggests a sample size of 30–40 individuals is required for single-point observations and 60–180 observations is required for sequential sampling, although more than 200 observations may be needed to ensure that some rare behaviors are sampled. Both our estimates of minimum sample sizes and those of Morrison (1984a) fall within the range required for 90-95% confidence intervals. Greater precision requires much larger samples.

Although Morrison (1984a), Wagner (1981a), and Hertz et al. (1976) advocated sequential sampling, unless the objective of the study was to record series of events (e.g., rates of movement, search and quitting times), there appears to be little justification for these procedures in habitats where it is easy to locate birds. Similar numbers of individuals are required for both procedures and sequential recording failed to detect rare and/or unusual behaviors that might affect interpretations regarding the use of resources or the ways in which species interacted with each other.

The large sample sizes needed to stabilize sample variances for mean foraging height can be used to establish an upper limit for data recording, which is easily calculated progressively in the field. The time saved by recording only a single prey-attack for each individual located can be used to obtain other habitat data (e.g., details of substrate) or to reduce the time taken to obtain a sample, thereby reducing effects of weather, time of day or seasonal changes in food resources on avian behavior.

ACKNOWLEDGMENTS

William Block, Hugh Ford, Michael Morrison and Andrew Smith were constructively critical of early drafts of this paper. Their assistance in improving the manuscript is appreciated. The fieldwork at Scheyville was conducted while HFR was on the staff of the Australian Museum, Sydney. Grey Gowing assisted with data tabulation and analysis. Analysis of the data was made possible by a grant from Harris-Daishowa Pty. Ltd.

USE OF RADIOTRACKING TO STUDY FORAGING IN SMALL TERRESTRIAL BIRDS

Pamela L. Williams

Abstract. Radiotracking can be used to study foraging of small birds (approximately 30 g and larger), often allowing a more accurate description of behavior than can be obtained by visual observation. I describe methods used to study foraging of Northern Orioles (*Icterus galbula bullockii*) during the breeding season and compare them with methods used in other radiotracking studies of small terrestrial birds. Transmitters revealed that nesting orioles foraged as far as 1 km from their nests, returning repeatedly to foraging sites 200–850 m away. Individuals from different nests within the same valley used foraging sites within the same general area and in some cases were found within the same patch of trees, sometimes simultaneously. These distant foraging sites, and this consistent overlap in foraging activity, were not discovered until transmitters were attached to birds.

Key Words: Radiotracking; Northern Oriole; *Icterus galbula bullockii*; foraging behavior.

Radiotracking is useful for studying the spatial and temporal distributions of the activities of individual animals because: (1) it allows studies of animals where detection might be difficult; (2) it can locate foraging sites that are distant from a central place (nest vicinity or roost); (3) it allows continuous observation of an individual to determine its use of different parts of its home range. Thus, it allows calibration of the amount of observational time at different sites so that it is proportional to the actual use at that site, and thus is less biased than observational methods.

Sampling of behaviors, as well as locations, may be improved by radiotracking. The ability to continuously follow and identify an individual avoids biasing observations toward conspicuous individuals or behaviors, a common problem (Altmann 1974). With radiotracking, an observer can detect with higher confidence differences between sex and age classes in foraging sites, substrates, and distances (see Grubb and Woodrey this volume), or follow the behavior of nonterritorial as well as the more obvious territorial individuals. The option to stay farther away from an individual also allows testing of the observer's effects on behavior and site use at different distances. Radiotracking can also directly detect simple changes in behavior. If a bird is not moving, the signal transmitted is constant, whereas when the bird moves the signal varies. Additional activities and orientations can be monitored by using simple radio circuits with variable resistors (Kenward 1987:39–43).

Reduction in the size and weight of the electronic components of transmitters and batteries in the last 20 years has allowed radiotracking of birds weighing as little as 29 g (e.g., Great Tits [*Parus major*; East and Hofer 1986], *Catharus* thrushes [Cochran et al. 1967, Cochran and Kjos 1985], and Brown-headed Cowbirds [*Molothrus ater*; Raim 1978, Dufty 1982, Rothstein et al. 1984, Teather and Robertson 1985]). The greater horseshoe bat (*Rhinolophus ferrumequinum*), which ranges in weight from 17–27 g, is one of the smallest species that has been radiotracked (Stebbings 1982).

I radiotracked the foraging activities of nesting Northern Orioles (*Icterus galbula bullockii*) at Hastings Reservation, Monterey Co., California. While many authors have described Northern Orioles as nesting and feeding on all-purpose territories (Grinnell and Storer 1924, Miller 1931, Bent 1958), I observed considerable overlap in space use among individuals of different pairs. For example, in a case where two pairs nested in the same tree, I observed at least nine individuals perching, foraging, and even singing there, although the nonresidents were not usually present at the same time as the resident pairs. Spacing of nests within a 100-m radius circle varied from solitary pairs with no neighbors to clusters of up to 13 pairs (Williams 1988), and I suspected that although the former pairs might have all-purpose territories, the latter did not. I used radiotracking to compare distances of foraging trips from the nest and the amount of overlap in foraging areas, if any, in relation to the density of nesting conspecifics. The technique was used because I could not otherwise locate an individual's foraging areas or determine if individuals overlapped on foraging sites. Using my data and a brief literature review I report here on data obtained by radiotracking that could not have been discovered by traditional observational methods.

METHODS

Equipment

The transmitter package was a Cochran design (Cochran et al. 1967, Wilkinson and Bradbury 1988) with a single-stage transmitter, battery (zinc-air, mercury 312, or silver oxide), and a stainless steel fishing-trace whip antenna. Transmitters were supplied by Biotrack, Wareham, Dorset, England BH20 5AJ and AVM Instrument Co. Ltd., Dublin, CA 94566. Dental acrylic was used as potting to seal out moisture. A thin piece of cloth with a finished edge (seam binding) extending 2–3 mm beyond the transmitter was attached. This created a larger surface area for attachment of the transmitter to the bird. The joint between the antenna and the transmitter was covered by the manufacturer with heat-shrunk tubing, and I constructed a cone of silicone glue around the joint to further protect it from breaking. The transmitters came with 15-cm antennae, which I trimmed to extend 2–3 cm beyond the tail. Weight at attachment for transmitters from Kenward averaged 2.1 g (N = 9, range = 1.8–2.3 g); those from AVM averaged 2.7 g (N = 3, range = 2.4–3.0 g). Weights of the birds before transmitters were attached averaged 34.8 g (N = 12, range = 29–39 g), and the transmitters averaged 6% of body weight (range = 4.9–7.7%).

After I soldered the battery lead to the transmitter and completed potting over the solder joint, I located the frequency of the strongest signal and any weaker signals from the transmitter on each receiver. (Weaker signals may result from problems in transmitter construction; knowledge of them may be useful later in locating the signal if it shifts with time or temperature.) When possible, batteries were activated 24 hours before needed because early battery failures often occurred within that time and because shifts in signal frequency sometimes occurred soon after activation. I used three receivers of model CE-12 from Custom Electronics (the same as the LA-12 model from AVM). I located the signal on all receivers, because slight differences in fine tuning occurred between receivers.

Attachment

I attached the transmitter while an assistant restrained the bird; placing the toe of a baby's sock over the bird's head calmed most individuals. I weighed the bird and the transmitter before attachment to closely monitor the effects of the relationship between transmitter weight and individual behavior. The transmitter was then placed anterior to the articulation of the humeri, as high on the back as possible without interfering with the movement of the head (see illustrations in Cochran et al. 1967, Raim 1978, Perry et al. 1981). Transmitters were attached to six birds with contact cement and to 11 birds with cyanoacrylate glue. Feathers in an area slightly larger than the transmitter were trimmed to a length of 1–2 mm and the area was cleaned with acetone or alcohol. Trimming the feathers rather than removing them prevented stimulation of the growth of new feathers that would push the transmitter off. Before releasing the bird I again located the signal on the receiver to confirm that the frequency had not shifted during attachment.

Following the bird

Immediately after release, many newly radioed Northern Orioles flew to a nearby hillside and foraged there for several hours. All birds had resumed normal behavior patterns after 3–4 hours and showed no difficulty in flying or other activities. Although I usually followed the birds immediately after release, only data collected at least three hours after release were analyzed. By that time I was aware of no differences in behavior due to the transmitter. I followed individuals on foot, carrying a receiver and three-element Yagi antenna (see Mech 1983 for details on methods of following animals).

RESULTS

SUCCESS OF METHOD FOR NORTHERN ORIOLES

The five radio-tagged, nesting females I followed in 1984 returned in 1985 and four of these again nested on the study area. I recaptured two about a week after they had lost their transmitters. They had lost the feather quills where the transmitter was attached but the skin appeared healthy. I recaptured one of these birds in 1985, and she showed no evidence of the previous year's transmitter attachment. Four of the five nesting females tracked in 1984 and six of nine nesting females tracked in 1985 successfully fledged young, while the average nest success in these years for the study population was 62% (N = 42) and 68% (N = 34), respectively (Williams 1988). I did not monitor the return of individuals in 1986, but I believe these results indicate that the transmitters did not adversely affect survival and reproduction.

I placed transmitters on 17 females and gathered sufficient data to analyze movement patterns of 13. I was able to follow birds an average of 9 days (range = 3–15 days; SD = 4 days) before either the battery failed or the transmitter fell off.

The average life of batteries active for more than 24 hours was 11.9 days (N = 13, SD = 8.0 days). One transmitter retrieved after 13 days was monitored until the battery failed after 35 days. The zinc-air batteries had a higher failure rate than the mercury batteries within the first 24 hours after being activated.

It was not always possible to tell if a female was still carrying a transmitter after it stopped working, because it was preened into the feathers, with only the antenna remaining visible. Five transmitters attached with contact cement remained attached for 14 ± 8 days SD, whereas 10 attached with cyanoacrylate glue stayed attached for 16 ± 18 days SD. Two (one attached with each type of glue) that fell off after two days were recovered and re-attached to the same individuals for 13 and 14 days. In most cases the attachment lasted longer than the battery. This was especially true using cyanoacrylate glue, with two females carrying their transmitters a minimum of 42 and 55 days.

Using hand-held equipment, I was able to detect line-of-sight distances up to 1 km. The signal from a bird on the ground could be detected from about 300 m.

DISTRIBUTION OF FORAGING SITES

Assuming that the movements of nesting females were primarily influenced by food availability, and that females would minimize the distance traveled from their nests, I compared the spatial distribution of foraging sites of females to the density of conspecifics near their nest site. Observations alone yielded little information. I located dispersed pairs readily when they were near the nest, but only rarely after they left the area. Where nests were clustered, it was easier to locate a foraging individual, but it was not possible to follow a particular individual or relocate it on enough occasions to adequately describe its foraging area. Soon after departing the nest individuals usually disappeared into dense foliage or over a hill, occasionally flying directly out of sight. Given that I was often unable to locate birds foraging, I could not know whether they were present and camouflaged or had left the nest area. Although I observed birds from different pairs foraging sequentially in the same tree, and sometimes even simultaneously with a minimum of aggressive interactions, it was not possible to determine whether these were rare or common occurrences.

Using transmitters I discovered that individuals sometimes foraged undetected in the canopy of a tree for as long as an hour and that they could enter or leave a tree undetected. They sometimes appeared to move only to the next tree or over a small hill but were located next at sites up to 1 km from their nests. I found no consistent association between the direction they departed from the nest and the direction of their destination. In the first month after their arrival in the spring, I discovered that the orioles abandoned their nesting areas during cold or rainy weather and spent whole days on nearby hillsides, sometimes with other individuals in the same tree, as well as occasionally making trips of several hours duration to sites at least as far as 1 km from their nests. Only by using transmitters was I able to determine the proportion of time females spent foraging at different sites, the distance traveled from the nest to foraging sites, or whether there was overlap in foraging areas among different females either sequentially or simultaneously.

During incubation I followed seven females, two in 1984 and five in 1985, for varying numbers of days. I used three 3-hour samples from different days to compare foraging by these females. Because of considerable individual variation in movement patterns, even among females nesting at the same density, I have presented data for each female separately (Fig. 1). Each female spent on average 2 hours of a 3-hour watch in her nest tree ($\bar{X}$ = 128 min, SD = 13 min). Females foraged farther than 200 m from their nests between 10% and 92% of the time. Four of the seven females spent more than 50% of their foraging time at these distant sites.

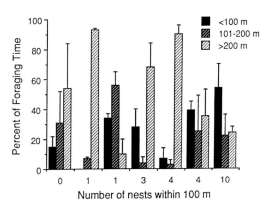

FIGURE 1. The relationship between nest density and proportion of foraging time spent at increasing distances from the nest for incubating female Northern Orioles at Hastings Reservation during 1984–1985. Each set of three bars represents the mean and standard error from three 3-hour observations of one female, with the density of nests within a 100 m radius around her nest on the X axis below the data for each female. The first of two females nesting with one nest within 100 m is represented by only two bars because she was not observed foraging less than 100 m from her nest.

Of the three females that spent less than 50% of their time at distant sites, one spent 58% foraging 100–200 m from her nest, whereas the other two did almost half of their foraging (40% and 54%) within 100 m of the nest. The two females with only one other pair nesting within 100 m spent 92% and 10% of their time foraging more than 200 m from their nests. A similar contrast was noted between the two females with four neighboring pairs, who spent 90% and 35% of their time at distances more than 200 m from the nest. The variation in distance to foraging sites between females nesting at the same density, and the lack of correlation between foraging distance and nest density, suggest that density of conspecifics near the nest was not an important determinant of foraging patterns. Although this conclusion is only tentative because of the small sample size, the fact remains that I would not have known about foraging sites beyond 100 m from the nest without the use of telemetry. This would have eliminated more than 53%, on average, of the foraging time of these females.

Between 5 and 23 May 1985, I tracked six females, each for a varying number of days. Four were incubating, one nest building, and one laying. This revealed extensive overlap in foraging sites among five females nesting in a valley within 0.5 km of each other, but solitary foraging by the sixth female nesting on a ridge over 0.5 km from the nearest nest in the valley (Fig. 2). This female, nesting 230 m from her nearest neighbor, did more than two-thirds of her foraging 200–

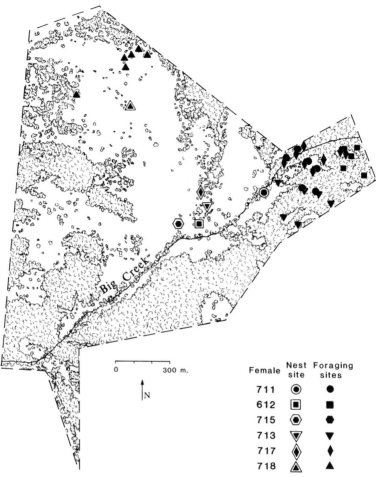

FIGURE 2. The study area at the Hastings Reservation showing nest sites and foraging sites of six female Northern Orioles tracked for different numbers of days each between 5 and 23 May 1985. Nest sites of additional orioles present in the same season are not shown. Big Creek runs through a lowland area with terrain rising both to south and north as well as along the creek east of nest site 711.

320 m from her nest, by herself or with her mate. In contrast, the females nesting in the valley overlapped considerably in foraging areas. They did the majority of their foraging at sites on Buckeye Hill and in neighboring ravines at the north end of the valley, flying 200–850 m from their nests to these sites. While these five birds overlapped in general foraging area, they also sometimes overlapped in exact foraging sites. Thus, females 711 and 713 foraged near each other in neighboring trees on one occasion, and female 612 left a foraging site just before female 717 arrived at that site. It was more common to detect sequential overlap in foraging sites, as evidenced by other observations of these same females. Additionally, female 717 was observed overlapping sequentially with females 715 and 711. While at these foraging sites, I could usually see or hear a number of other Northern Orioles, either on the sites or flying overhead up and down the hillside or ravine. In a few cases I could identify banded individuals in addition to the birds with transmitters. The only way I could monitor the females at this time was by following their signal and seeing them fly in and out of an area. They left their nests in a variety of directions, giving no visual cues of their final destinations. While foraging they were hidden from view in the canopy. Without the use of radiotracking I would never have discovered this considerable overlap in foraging areas.

DISCUSSION

Contrary to the prevailing view regarding spacing in breeding populations of orioles (e.g., Lowther 1975, Orians 1985a), Northern Orioles in this study were not consistently territorial. This is shown by clustered nests and by radiotracking data, which showed (1) significant overlap in foraging areas of breeding birds, and (2) recurrent use by the same individuals from one breeding area of a localized foraging area. The latter suggests some form of communication among these individuals. Their nesting dispersion and feeding overlap remind one of other icterids such as the Brewer's Blackbird (*Euphagus cyanocephalus,* Horn 1968).

The occurrence of dispersed as well as clustered nesting in the Northern Oriole may vary geographically (see Pleasants 1979, Williams 1988). A contributing factor in central California is the seasonal summer drying and local uncertainties in insect food levels. Thus, overlap in foraging is not surprising.

OTHER SPECIES

Radiotracking has been used mainly to determine home-range size and to follow social behavior (e.g., Bradbury 1977, MacDonald 1978, Marquiss and Newton 1981, Pruett-Jones 1985, Wilkinson 1985, Wood 1986), but is now being used increasingly for foraging studies to obtain information that is not available by observation alone.

East and Hofer (1986) found that Great Tits foraged intensively at small patches interspersed among similar-sized areas of low use. This confirms laboratory studies showing that Great Tits concentrate foraging in areas with high food density while continuing to appraise food availability elsewhere. The two territorial males they followed ranged over substantial areas outside their territorial boundaries, foraging on the territories of other males. The single nonterritorial bird also ranged over a large area. "Radio signals suggested that Great Tits spent a large percentage of their time during the late morning and afternoon foraging near the ground in dense vegetation, explaining why Great Tits are so difficult to observe after an active period following dawn" (East and Hofer 1986).

The Woodcock (*Scolopax rusticola*) is an elusive and secretive species. Using radiotracking, Hirons and Owen (1982) established that in winter and early spring birds foraged mainly in pastures at night, returning to woodlands during the day. As nights got shorter, the birds switched to feeding in woodland during the day and roosting at night. As with Great Tits, individual Woodcocks used intensively only small patches within preferred habitat, and these were areas where earthworm densities were highest. Hirons and Johnson (1987) found no evidence that Woodcocks preferred swampy patches, as described by other authors, e.g., Cramp and Simmons (1982).

Nesbitt et al. (1978) found a consistent pattern of foraging movements for three groups of Red-cockaded Woodpeckers. By placing a transmitter on one bird they followed the daily movements of all clan members along a 1.9-km foraging path. Each clan began moving and feeding soon after leaving the roost hole in the morning and moved quickly until late morning or early afternoon, reaching the farthest distance from the roost, 0.72 km on average, early in the afternoon; they returned in the late afternoon, sometimes in one direct flight.

Radiotracking of two species of brood parasites, Brown-headed Cowbirds and the Common (or European) Cuckoo (*Cuculus canorus*) supported qualitative information that these birds have separate breeding and feeding ranges. Wyllie (1981:96) found that cuckoos moved 4 km between breeding areas in reed beds and feeding areas in orchards and scrublands. Several males and females used the same feeding areas, although foraging was usually solitary. Rothstein et al. (1984) found that Brown-headed Cowbirds spent the early mornings on breeding areas, and in late mornings and afternoons flocked at favored feeding areas. Females visited fewer feeding sites, traveled shorter distances between sites, and spent more time at feeding sites than males. Some males commuted between disjunct breeding and feeding sites; others stayed at feeding sites all day.

Common Grackles (*Quiscalus quiscula*) at three roosts in Oklahoma foraged on successive days at sites an average of 11.9 km apart, and did not always return to the same roost (Bray et al. 1979). European Starlings (*Sturnus vulgaris*) wintering in Oregon also foraged at different sites each day, although they returned to the same roost each night. The average distance between sites used on succeeding days was 4.8 km (Bray et al. 1975). In contrast, in New Jersey this species used several roosts, with individuals using up to five during the 4-month study, while each bird returned regularly to the same diurnal activity center (Morrison and Caccamise 1985). Multiple roost sites may have been used to exploit rich sources of supplemental food near those roosts, while maintaining foraging territories in areas of persistent food abundance.

CONCLUSIONS

Radiotracking allows the gathering of important qualitative and quantitative information on the foraging activities of individuals that could not be discovered otherwise. Large amounts of data can be accumulated, albeit on a small number of individuals. However, the procedure is both expensive and time-intensive, and equipment failures are not uncommon. Using automatic monitoring equipment can save considerable time, but at great initial expense and loss of direct observations of behavioral details, and the procedure is not appropriate for all studies. Despite these problems, radiotracking is an important component of thorough modern studies of resource use in avian populations.

ACKNOWLEDGMENTS

I am indebted to J. Davis, J. Griffin, and the Museum of Vertebrate Zoology for use of facilities at Hastings Reservation. F. Pitelka, my graduate advisor, contributed invaluable guidance throughout the study, and this paper is dedicated to him in honor of his 70th birthday. M. and S. Pruett-Jones provided expert guidance in the techniques of radiotracking. R. Etemad, A. Kieserman, D. McNiven, and S. Ostby helped in the field. The manuscript was improved by the comments of W. Carmen, H. Greene, P. Hooge, J. Jehl, W. Koenig, S. Laymon, M. Morrison, F. Pitelka, M. Reynolds, T. Smith, and J. Verner. K. Klitz drew the map. Field work was supported by the Museum of Vertebrate Zoology, the George D. Harris Foundation, and Sigma Xi. Financial assistance was provided by the Betty S. Davis Memorial Fellowship made possible through the generosity of Fanny Hastings Arnold. I especially wish to acknowledge the generosity of Mrs. Arnold for her ongoing support of Hastings Reservation.

INFLUENCE OF SAMPLE SIZE ON INTERPRETATIONS OF FORAGING PATTERNS BY CHESTNUT-BACKED CHICKADEES

LEONARD A. BRENNAN AND MICHAEL L. MORRISON

Abstract. We used sequential sampling techniques and statistical estimation of sample size to analyze the influence of sample size on interpretations of seasonal patterns of foraging by a resident population of Chestnut-backed Chickadees (*Parus rufescens*). We found that estimates of central tendency and dispersion for use of tree species, use of foraging substrate, and foraging behavior stabilized when 40 or more samples were used and that 30–50 samples were usually required for 95% confidence that an estimated mean would be within 10% of the mean of the entire sample. Although seasonal patterns obtained from two month and one month sampling periods were similar, the one month period provided greater information on changes in foraging patterns.

Key Words: Sample size analysis; seasonal foraging patterns; use of tree species; use of substrates; foraging behavior; Chestnut-backed Chickadee.

Variations in sample size can have a strong and potentially confounding influence on observed patterns of behavior (Kerlinger 1986:109); yet, little attention has been paid to the influence of sample size on analyses of avian foraging behavior. There are techniques for determining the minimum number of samples needed to see whether an estimate of a parameter falls within a selected confidence interval (see Cochran 1977, Scheaffer et al. 1986, and references therein). Until recently, however, ornithologists have generally neglected the use of statistical and graphical procedures for assessing factors that influence analyses of foraging behavior and habitat use (but see Wagner 1981a; Morrison 1984a, b; Block et al. 1987). Typically, most investigators collect as many samples as possible and then base their analysis on all samples collected, without regard to the adequacy of their sample size. This study was designed to expand upon Morrison (1984a) by extending the assessment of the influence of sample size to include seasonal changes in foraging behavior. Using the Chestnut-backed Chickadee (*Parus rufescens*) as an example, our objectives were to (1) determine the number of samples required for obtaining precise (based on the stability of means and variances) estimates of foraging behavior during different times of the year, and (2) evaluate how different time scales affect the outcome of patterns of seasonal changes in the use of tree species, use of foraging substrates, and foraging behaviors.

METHODS

STUDY AREA

We studied the foraging behavior and habitat use of Chestnut-backed Chickadees in the mixed-conifer forest zone of the western Sierra Nevada approximately 8 km east of Georgetown in El Dorado County from May 1986 through April 1987. Data were collected on and around the Blodgett Forest Research Station, University of California, at approximately 1100 meters elevation. This area is a mature mixed-conifer second-growth forest dominated by Douglas-fir (*Pseudotsuga menziesii*), white fir (*Abies concolor*), incense cedar (*Calocedrus decurrens*), ponderosa pine (*Pinus ponderosa*), sugar pine (*P. lambertiana*) and California black oak (*Quercus kelloggii*). See Morrison et al. (1986) for a description of the study area.

DATA COLLECTION

The data used in this study were collected as part of an ongoing study of seasonal variation in foraging and habitat use by chickadees in the western Sierra Nevada. Observers walked random transects through the forest and recorded timed (8–30 s) observations of foraging chickadees. The observer waited a minimum of 10 s after seeing the bird, and then recorded a series of variables which corresponded to the tree species, substrate, and mode of foraging. We used the focal animal technique described by Altmann (1974) and Martin and Bateson (1986). Each recorded observation consisted of between two and ten records, or lines of data. Each time a bird changed tree species, substrate, foraging mode, or foraging height, a new record, or line of data, was added to the observation until the bird was lost from sight. Thus, each observation consisted of 1–9 sequential records of foraging observations. Each sequential series of 1–9 foraging records was treated as a single (N = 1) sample (see Data Analysis section below).

When flocks were encountered, we allowed at least 10 min to elapse between recording foraging observations. At Blodgett, chickadees forage in flocks from July until late March or April, and as solitary birds or pairs during nest building and breeding (mid to late April through early July; Brennan, pers. obs.). Thus, the detectability of foraging chickadees varied during the annual cycle. During the breeding season, most foraging observations were of breeding birds near (within 100 m) nests. Foraging observations of family groups (parents and fledglings) make up a major part of the July and August observations. Family groups of chickadees and mixed species-flocks were treated in a similar manner when foraging observations were made. Mixed flocks of Chestnut-backs and other species (e.g., Mountain Chickadee [*P. gambeli*], Red-breasted Nuthatch [*Sitta canadensis*], Golden-crowned Kinglet [*Re-*

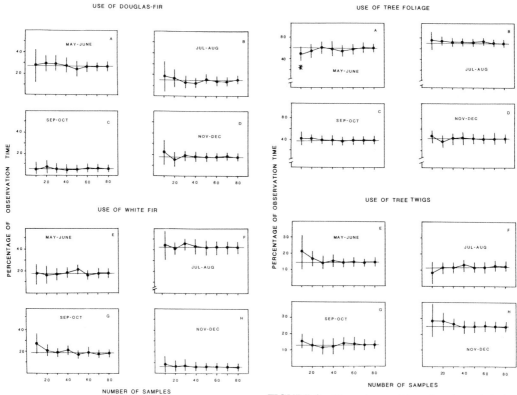

FIGURE 1. Percent use of two tree species (Douglas-fir and white fir) by Chestnut-backed Chickadees during four different sampling periods at Blodgett Forest Research Station, 1986–1987. Solid dots represent mean values at sample sizes ranging from 10 to 80 observations, vertical bars represent one standard deviation. Horizontal lines represent means calculated from all 80 samples.

FIGURE 2. Percent use of two foraging substrates (tree foliage and tree twigs) by Chestnut-backed Chickadees during four different sampling periods at Blodgett forest Research Station, 1986–1987. Symbols as in Figure 1. Asterisks denote means that were statistically significant from the remaining homogeneous subset (P < 0.05, SNK-ANOVA).

gulus satrapa]) also foraged on the study area for much of the year.

Observations were made during all daylight hours and under the range of climatic conditions of the western Sierra Nevada (30°C during summer to freezing rain and snow in winter). Data were collected by four different people. Interpretations of observations were standardized during training exercises every time an observer had not continuously collected data during the previous three week period.

DATA ANALYSIS

We selected variables that represent three important aspects of foraging by Chestnut-backs: (1) use of tree species, (2) foraging substrate, and (3) foraging mode. Chestnut-backs spent nearly 99% of the time foraging in six species of trees, using four different substrates and eight foraging modes (Brennan and Morrison, unpubl. data). For this study we used data that illustrate the variability of foraging by Chestnut-backs on two species of trees (Douglas-fir and white fir), in two substrates (tree foliage and tree twigs), and using two foraging modes (gleaning and hanging). We selected these variables because they represent aspects of foraging that are used in varying amounts during different seasons.

The raw data from each foraging observation were transformed into a matrix of percentages of the total time Chestnut-backs used each tree species, substrate, and foraging behavior. Transforming the data from a discrete (e.g., frequency of tree species use) to a continuous form (percent of observation time), by mathematically combining the frequency data with corresponding seconds of observation time, allowed us to analyze the data using standard one-way analysis of variance and associated tests for homogeneity of means and variances (see below). It also served to standardize the data because of the variation in observation time (8–30 s). Furthermore, this method allowed us to calculate confidence intervals around mean values. Incorporating sequential records of foraging behaviors into a single sample allowed us to circumvent problems of dependency that arise when each sequential record is treated as an individual sample.

We selected two-month intervals for our sample size analyses for several reasons. First, we needed sufficient samples to insure stability of means and variances. We considered estimates of means and variances to be stable when they converged with the estimates obtained from all (N = 80) samples used within a sampling period. The sample size of 80 was selected because this represented the largest number of samples collected during sampling periods in the fall and winter. Second, a two month period can be aligned with significant biological events during the chickadees' annual cycle: May through June is typically the core of the breeding period; family groups frequently forage as flocks during July and August; the onset of fall rains and leaf abscission for deciduous trees (most notably *Q. kelloggii*) occurs during September and October; the onset of winter and the first snows begin in the western Sierras during November and December; January and February are typically the coldest months; pre-breeding events (pair bonds and nest building) begin in March and April.

During each two month sampling period, we randomly subsampled (with replacement) each data set ten times, using sample size increments of ten. We used Student-Newman-Keuls multiple comparisons with one-way analysis of variance (Zar 1974:151) to test for differences in means of each different sample size for each variable.

For the statistical estimation of sample size, we used Stein's two-stage technique (Steel and Torrie 1960:86), which employs the following equation:

$$n = (t^2)(s^2)/(d^2)$$

where t is the t-value for the desired confidence interval with $n - 1$ degrees of freedom for the sample used, s is the standard deviation, and d is the half-width of the desired confidence interval. To be 95% confident that the mean of a given variable would be within 10% of the mean from all 80 samples from a particular sampling period, we sequentially calculated the standard deviations from 10, 20, 30 ... n samples until the estimated sample size converged with the sample size of the subset being used. To analyze the effect of the length of sampling period on seasonal patterns of foraging we compared one month and two month sampling periods. This allowed us to examine seasonal patterns in relation to 6 and 12 intervals, each of which represents a different portion of the annual cycle.

RESULTS

INFLUENCE OF SAMPLE SIZE

Our data indicated that the size of the sample significantly affected the outcome of the analysis. At sample sizes >30, the estimated means appeared to converge with the mean value of a particular variable for the entire sampling period. Along with convergence of mean values, the standard deviations of the estimates also stabilized when 40 or more samples were used (Figs. 1–3).

Although the mean values varied widely between some sampling periods (see, for example

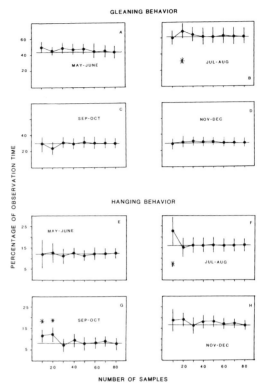

FIGURE 3. Percent time spent gleaning and hanging by Chestnut-backed Chickadees during four different sampling periods at Blodgett Forest Research Station, 1986-1987. Solid dots represent mean values at sample sizes ranging from 10 to 80 observations, vertical bars represent one standard deviation. Horizontal lines represent means calculated from all 80 samples. Asterisks denote means that were statistically different from the remaining homogeneous subset (P < 0.05 SNK-ANOVA).

the use of white fir [Figs. 1E,F], or the use of twigs [Figs. 2E,F]), time of year did not appear to affect the number of samples required for a stable estimate of means and variances.

In all cases involving variables and sampling periods, variances were not equal with different sample sizes (Bartlett's test for homogeneity of variances, P < 0.001). In four instances the mean values of the subsample estimates did not equal the other means from the subsamples of each variable (P < 0.05 Student-Newman-Keuls one-way analysis of variance [SNK-ANOVA]). These were: N = 10 for the May-June analysis of tree foliage use (Fig. 2A); N = 20 for the July–August analysis of gleaning behavior (Fig. 3B); N = 10 for the July–August analysis of hanging behavior (Fig. 3F) and N = 10–20 for the September–October analysis of hanging behavior (Fig. 3G).

TABLE 1. SAMPLE SIZES REQUIRED FOR 95% CONFIDENCE THAT THE ESTIMATED MEAN IS WITHIN 10% OF THE MEAN VALUE, CALCULATED FROM THE ENTIRE GROUP OF 80 SAMPLES FOR EACH FORAGING BEHAVIOR VARIABLE USING CHESTNUT-BACKED CHICKADEE FORAGING DATA COLLECTED AT BLODGETT FOREST, MAY–DECEMBER 1986. MEAN VALUES AND STANDARD DEVIATIONS USED FOR THE SAMPLE SIZE CALCULATIONS ARE GIVEN IN FIGURES 1–3

Sampling period	Variable	Size of sample used for calculation	Number of samples required[a]
May–June	Use of Douglas-fir	10	97
		20	30
	Use of white fir	10	91
		20	105
		30	36
	Use of tree foliage	10	17
	Use of tree twigs	10	203
		20	54
		30	30
	Gleaning behavior	10	10
	Hanging behavior	10	135
		20	59
		30	33
July–August	Use of Douglas-fir	10	204
		20	79
		30	70
		40	40
	Use of white fir	10	156
		20	25
	Use of tree foliage	10	85
		20	20
	Use of tree twigs	10	153
		20	25
	Gleaning behavior	10	10
	Hanging behavior	10	117
		20	33
September–October	Use of Douglas-fir	10	148
		20	126
		30	92
		40	73
		50	50
	Use of white fir	10	112
		20	48
	Use of tree foliage	10	22
	Use of tree twigs	10	40
		20	22
	Gleaning behavior	10	43
		20	20
	Hanging behavior	10	21
November–December	Use of Douglas fir	10	140
		20	41
	Use of white fir	10	305
		20	198
		30	57
		40	41
	Use of tree foliage	10	27
	Use of tree twigs	10	61
		20	21
	Gleaning behavior	10	22
		20	16

[a] Based on Stein's two-stage technique, see text for equation.

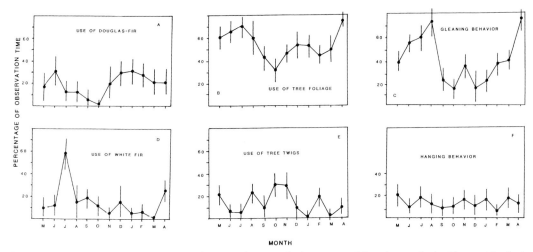

FIGURE 4. Seasonal variation in use of tree species, use of substrates, and foraging modes by Chestnut-backed Chickadees at Blodgett, using a one month interval. Dots represent mean values, vertical bars represent one standard deviation.

Otherwise, the means derived from subsampling 10–80 samples represented homogeneous groups of estimates that were not statistically different ($P < 0.05$; SNK-ANOVA).

SAMPLE SIZE ESTIMATION

The number of samples required to be within 10% of an estimated mean 95% of the time varied widely (Table 1). For example, common foraging behaviors, such as percent time foraging on foliage, or percent time gleaning from all substrates generally required 10–20 samples, whereas uncommon or highly variable behaviors such as use of white fir, use of Douglas-fir, or use of tree twigs required 30–50 samples (Table 1). In only one case were more than 40 samples required for estimating a variable: the use of Douglas-fir in September–October (Table 1).

LENGTH OF SAMPLING PERIOD

We found similar patterns for both the one month and two month sampling periods (Figs. 4 and 5). The use of tree species, substrates, and foraging modes varied dramatically across the year in both analyses. For example, use of Douglas-fir decreased during the summer and then rose during late fall and early winter. Use of white fir increased dramatically during July and August, but was low during the rest of the year. The use of twigs increased and the use of foliage decreased during the fall (Figs. 4 and 5). Gleaning peaked during late summer, whereas time spent hanging from terminal buds, twigs, and foliage varied widely (Figs. 4 and 5).

DISCUSSION

SAMPLE SIZE ANALYSES

The number of samples required to obtain reliable estimates of the relative amounts of time chickadees spend foraging was variable. Common behaviors typically required 10–20 samples for estimates of central tendency and dispersion, whereas less common behaviors required up to 40 (and in one case 50) samples. These results generally support Morrison's (1984a) findings that confidence intervals and mean values remained virtually unchanged at sample sizes ≥40 or larger; he concluded that samples from at least 30 individuals were required for a reliable estimate. We found, however, that some estimates based on 20 or fewer samples differed from the overall (all 80 samples) mean. These differences may be related to the different species studied: Morrison studied two species of migrant *Dendroica*, whereas we used a resident parid. Morrison collected data from April to July; thus, his results are most comparable with ours from May–June. None of the mean values calculated for the different sample sizes in our analyses was statistically different from the overall means for each variable during the May–June sampling period; perhaps there is less variation in behavior of foliage-gleaning birds during the breeding season than at other times of the year, and this accounted for the lack of statistical differences in the means for this sampling period.

Our estimates of the number of samples required for a reliable estimate of foraging behavior were considerably lower than those calculated by Petit et al. (this volume), who found that sev-

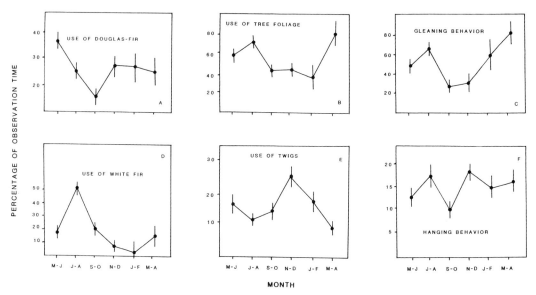

FIGURE 5. Seasonal variation in use of tree species, use of substrates, and foraging modes by Chestnut-backed Chickadees at Blodgett, using a two month interval. Dots represent mean values, vertical bars represent one standard deviation.

eral hundred samples were generally needed. The differences are most likely a function of analytical approaches. We used individual variables, whereas Petit et al. considered sets of foraging behavior categories simultaneously. As a result, behaviors used less than 5% of the time strongly influenced their calculations of sample sizes.

INFLUENCE OF SAMPLING TIME SCALE

Although the one month and two month sampling periods showed similar seasonal foraging patterns, much detail was lost as the length of sampling period increased. Whether this is important depends on the questions being asked. For example, an assessment of interactions between a population of birds and changes in food availability would require numerous, short sampling periods, whereas a general assessment of foraging behavior could be done using longer (2–3 month) sampling intervals. The inherent variability and shifts in foraging behavior are "smoothed out" as the time interval is increased.

Chickadees are, in many respects, generalists with a wide repertoire of foraging behaviors. Our results indicated that reliable estimates of their foraging behavior require at least 40–50 behavior samples per sampling period. Year-round analysis would require a minimum of 240–480 samples, depending on sampling interval (two months vs. one month). For a year-round investigation of an assemblage of, say, ten species, a minimum of 2400 behavior samples would be required, depending on the behavioral variability of individual species. Species with less varied behavior would probably require fewer samples. There is no sound biological or statistical justification for attempting such community-level analyses if adequate numbers of samples cannot be collected; even cursory survey work would be suspect. Thus, researchers would be advised to restrict their sampling to the number of species for which adequate samples—and thus meaningful results—can be obtained. In all cases sample size analysis is essential.

ACKNOWLEDGMENTS

A. Franklin helped collect data in 1986; W. Block also assisted with field work. I. Timossi wrote the Pascal computer program that converted the raw frequency data to the percent time matrix format; J. Dunne provided computer software and other important help. This manuscript benefitted greatly from reviews by P. Beier, W. Block, and J. Buchanan. R. Heald and S. Holmen provided logistic help at Blodgett Forest. D. Dahlsten, K. Hobson, and J. Gurulé provided interesting discussions on the meaning and interpretation of the seasonal foraging patterns.

PRECISION, CONFIDENCE, AND SAMPLE SIZE IN THE QUANTIFICATION OF AVIAN FORAGING BEHAVIOR

LISA J. PETIT, DANIEL R. PETIT, AND KIMBERLY G. SMITH

Abstract. We used equations presented by Tortora (1978) to estimate minimum sample sizes for avian foraging data. Calculations using absolute precision provided considerably lower estimates of sample size than those using relative precision. When sample sizes were estimated using absolute precision more observations were required to accurately represent foraging behavior of a generalist than of a specialist, but, for a precision of ±5% with $k = 3$ categories, no more than 572 observations were ever required. The opposite trend was observed with relative precision, such that, for extreme specialists, with $k = 3$ categories, >100,000 observations were needed to achieve relative precision of 5% around extremely rare behaviors. Because foraging studies typically focus on common behaviors, absolute precision is usually adequate for estimating sample size. Estimates of sample size acquired using Tortora's (1978) equations are dependent upon desired levels of confidence and precision. The estimation method can also be used *a posteriori* to determine precision associated with a sample.

Key Words: Sample size; avian foraging behavior; generalist; specialist; precision.

The increased use of statistics over the last two decades to analyze avian foraging behavior has heightened awareness of the problem of obtaining enough observations for proper analysis. Sample size clearly has a considerable effect on one's ability to make statistical inferences; yet, few attempts have been made to determine the number of observations needed to quantify avian foraging behavior. It would appear that most researchers simply gather the greatest number of observations possible, without much regard for which sample sizes may be appropriate for their analyses. Thus, a great variation in sample sizes of foraging behavior has been reported in the literature, ranging from 20–30 (e.g., Eckhardt 1979, Tramer and Kemp 1980, Maurer and Whitmore 1981) to >1000 (e.g., Holmes et al. 1979b, Sabo 1980, Landres and MacMahon 1983) single point and sequential foraging observations on individual species. Data collected in two or more field seasons are often combined to increase sample sizes, but that practice may not be appropriate because of between-year differences (e.g., Landres and MacMahon 1983).

Only Morrison (1984a) has directly assessed influence of sample size. Based on stabilization of means and narrowing of confidence intervals with increasing sample size, he suggested that a minimum of 30 independent observations (i.e., individual birds) were necessary to quantify foraging behavior of two species of warblers. The point at which confidence intervals are sufficiently narrowed, however, may be difficult to ascertain through simple inspection. In addition, because avian foraging behavior data often are made up of multiple variables dissected into many categories (e.g., "glean," "hover," and "hawk" within the variable, "foraging mode"), Morrison's (1984a) method involved calculating confidence intervals for each category of observations separately, such that minimum sample sizes in his study varied among different categories within the same variable. Further, it is not clear whether Morrison's estimate of sample size is readily generalizable to other passerine species.

Another factor that may influence sample size is variation of behavioral repertoires among species. For example, for a species with a fairly limited repertoire, with most observations falling into one or very few categories (i.e., a specialist; Morse 1971a), adequate sample sizes might be smaller relative to those required to quantify the more diverse repertoire of a foraging generalist. On the other hand, Tacha et al. (1985) indicated that large sample sizes were needed to capture rare behavioral events. If so, more observations will be needed to characterize a specialist's behavior compared to that of a generalist because of difficulty associated with quantification of rare events.

To maximize efficiency in collecting foraging data, some criteria are needed to determine a minimum sample size necessary to quantify such behaviors. Goodman (1965) introduced a procedure based on calculation of simultaneous confidence intervals for a multinomial population. Tortora (1978) modified that procedure for application to the situation in which a random sample of observations (i.e., independent and unbiased observations) are classified into k mutually-exclusive categories, and the proportions in those categories sum to one. (While we acknowledge that there are difficulties associated with obtaining a truly random sample of behaviors in avian foraging studies [e.g., Altmann 1974, Wagner 1981a, Morrison 1984a, Tacha et al. 1985], this is an assumption of all sample size estimation techniques [e.g., Cochran 1977, Steel

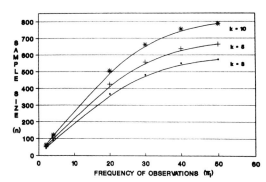

FIGURE 1. Estimation of sample sizes with absolute precision (b_i) of 5% as a function of the frequency of observations in one of 3, 5, or 10 mutually-exclusive categories (k). Confidence level (α) for these estimations is 0.05. See text for further explanation.

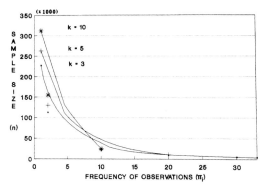

FIGURE 2. Estimation of sample sizes with relative precision (b_i') of 5% as a function of the minimum frequency of observations in one of 3, 5, or 10 mutually-exclusive categories (k). Confidence level (α) for these estimations is 0.05. See text for further explanation.

and Torrie 1980], and it is our intention only to present one of these techniques rather than to discuss the related but separate question of how foraging data are obtained.) In contrast to the methods used by Morrison (1984a), Tortora's (1978) procedure considers all categories simultaneously and allows for estimation of the sample size needed to achieve a specified level of confidence (α-level) such that percentages in all k categories are within some specified range (precision) of the true population values.

Our objectives were to: (1) determine a minimum sample of independent observations necessary to quantify foraging behavior, and (2) determine whether minimum sample sizes are different for specialist and generalist species.

METHODS

Tortora (1978) presented equations for calculating sample sizes based on either absolute or relative precision. (Precision is a measure of variance around the true population mean. Therefore, for these equations we assume that the true population mean is known [i.e., representation of the true mean is accurate]. We discuss below what can be done when the true mean is not known.) Absolute precision refers to the situation in which the acceptable variation around a small proportion is relatively greater than that around a larger proportion. This means that we are more interested in the ability to quantify the most common behavior at the expense of the precision associated with the rarest behaviors. For example, assume that gleans, hovers, and hawks occur with frequencies of 96%, 2%, and 2%, respectively, for a hypothetical foliage-gleaning bird. If we specify an absolute precision of 5%, we would accept foraging behavior estimates of 91–100% (96% ± 5%) for glean and 0–7% (2% ± 5%) for both hover and hawk. The equation given by Tortora (1978) for calculating sample size (n_a; the subscript refers to the type of precision used) with absolute precision is:

$$n_a = B\Pi_i(1 - \Pi_i)/b_i,$$

where B is the critical value of a χ^2 with 1 degree of freedom at a probability level of α/k (k = number of categories), b_i is a specified absolute precision (i.e., acceptable deviation from the true value) for each category i, and Π_i is the proportion of observations in the ith category. Sample sizes (n_a) increase to a maximum as Π_i approaches 0.50 (see Results). Thus, if $b_i = b$ for all categories, one calculates n_a using the Π_i closest to 0.50. If that frequency is >50%, its complementary frequency (i.e., 1 − percent frequency) is used. If $b_i = b$ for all categories, the largest n_a is chosen as the minimum sample size, and if the true population mean is unknown, one can calculate a "worst case" sample size by using $\Pi_i = 0.50$ (see also Discussion).

Relative precision refers to when the acceptable relative variation around the smallest proportion is the same as around the largest proportion. For the example mentioned above, we would accept estimates between 91.2–100% for glean for a relative precision of 5% (i.e., ±5% of 96%), but we would now only accept estimates between 1.9–2.1% for hover and hawk (i.e., ±5% of 2.0%). Here, sample sizes will be greatly influenced by attempting to quantify precisely the rarest foraging event. Tortora's (1978) equation for calculating sample sizes (n_r) with relative precision is:

$$n_r = B(1 - \Pi_i)/\Pi_i b_i'^2,$$

where $b_i'^2 = b_i/\Pi_i$, and, if $b_i' = b'$ for all categories, Π_i is the minimum proportion of the k observed proportions (e.g., 2% in the example above). As with absolute precision, if $b_i' = b'$ for all k, choose the largest n_r calculated for the sample size.

RESULTS

APPLICATION OF EQUATIONS

We calculated sample sizes necessary to represent with absolute precision means for six different frequency combinations, for $k = 3$, 5, and 10 categories (Fig. 1). The relationship between Π_i and sample sizes with absolute precision (n_a)

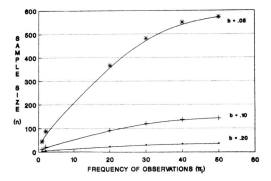

FIGURE 3. Effect of variation in absolute precision (b_i) on estimation of sample size for different frequencies of observations at $\alpha = 0.05$ and for $k = 3$ categories.

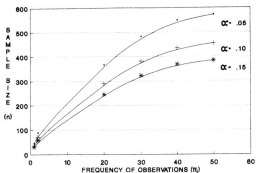

FIGURE 4. Effect of variation in confidence level (α) on estimation of sample size for different frequencies of observations, with precision (b_i) of 5% and $k = 3$ categories.

is such that, as any one categorical frequency approaches 50%, sample size increases for a given α and b_i. Thus, the curve in Figure 1 is symmetrical around $\Pi_i = 0.50$. Consider the situation in which $k = 3$ categories, $\alpha = 0.05$, $B = 5.724$ (χ^2 critical value for $P = 0.05/3 = 0.0167$), and $b = 0.05$ (absolute precision of 5%). If 98% of the observations are in one category and 1% are in each of the two remaining categories, about 45 independent observations would be necessary to have 95% confidence that the observed (sample) mean is within 5% of the true population mean (Fig. 1). Based on this approach, no more than 572 independent observations would ever be needed to quantify a variable with $k = 3$ categories (e.g., glean, hover, and hawk) at our specified levels of b (=0.05) and α (=0.05). Note, however, that n_a increases as number of categories (k) increases, particularly as Π_i approaches 0.50 (Fig. 1). Assuming those frequency combinations are representative of specialist or generalist species, the results suggest that: (1) minimum sample size is smaller for a species that is specialized in its foraging behavior (i.e., frequency in any category diverges substantially from 50%); and (2) influence of k on minimum sample size is greater for a generalist than for a specialist (Fig. 1).

A potential problem with an absolute precision of 0.05 is that, for example, in the extreme specialist case (98%, 1%, 1%), an acceptable mean would range from 93–100% for the first category and 0–6% for the others, which produces an acceptable range of 600% around the means for the two "rare event" categories. This problem can be remedied by calculating n_r with a relative precision (b_i') for each category. Unfortunately, this results in a large increase in minimum sample sizes (Fig. 2). Those data show that, contrary to estimations using absolute precision, sample sizes estimated with relative precision increase substantially as a species becomes more specialized (i.e., min [$\Pi_i, \ldots, \Pi_k$] approaches 0). Thus, in the case of an extreme specialist with a repertoire of three foraging modes with percent frequencies of 98%, 1%, and 1%, the minimum required sample size (with $b_i' = 0.05$) is 226,670 independent observations. Again, as with absolute precision, sample sizes calculated with relative precision increase as number of categories (k) increases (Fig. 2). Increases in both specified α and b_i levels cause decreases in sample size estimates with the greatest influence being exerted by changes in b_i (Figs. 3 and 4).

We applied the equations above to foraging data (Table 1) to determine how precisely sample sizes have allowed estimations of "true" population values. Note that all but Morrison's (1984a) are based upon sequential observations. Thus, the assumption of independence of observations for Tortora's equations may be violated, such that precisions we report probably are lower (i.e., better) than the actual precisions associated with those data sets (Tacha et al. 1985).

Table 1 shows that, for example, Morrison (1984a) reported that Hermit Warblers gleaned 78.8% of the time, hover-gleaned 11.5%, fly-caught 3.8%, and performed some other maneuver 5.8% of the time. Assuming those are the true proportions for the population then, based on a sample of 60 independent observations, with $k = 4$ and $B = 6.239$ (for $\alpha/k = 0.0125$), we calculated an absolute precision of 0.1319, or 13.2% (Table 1), meaning that one can expect to estimate within 13.2% of the true values for that distribution of proportions using 60 observations. To achieve 5% absolute precision, Morrison would have needed approximately 417 independent observations (n_a). To achieve relative precision of 5%, he would have required 63,178 independent observations (n_r)!

TABLE 1. REPORTED SAMPLE SIZES (N) WITH ESTIMATED SAMPLE SIZES (n_a AND n_r) BASED ON EQUATIONS PRESENTED IN TORTORA [1978]) FOR THE VARIABLE, FORAGING MODE, FOR SELECTED "SPECIALIST" AND "GENERALIST" SPECIES IN STUDIES OF AVIAN FORAGING BEHAVIOR. PRECISIONS ASSOCIATED WITH REPORTED SAMPLE SIZES ARE REPRESENTED BY b_i (ABSOLUTE) AND b_i' (RELATIVE). CALCULATIONS OF n_a AND n_r ARE BASED ON $\alpha = 0.05$ AND b_i AND $b_i' = 0.05$; k IS THE NUMBER OF BEHAVIORAL CATEGORIES WITHIN THE VARIABLE FORAGING MODE

Species	k	Greatest percent	Smallest percent	n	n_a	b_i	n_r	b_i'	Study
Hermit Warbler *Dendroica occidentalis*	4	78.8	3.8	60	417	0.13	63,178	1.62	Morrison (1984a)
Bushtit *Psaltriparus minimus*	3	96.0	2.0	270	88	0.03	112,190	1.02	Landres and MacMahon (1983)
White-breasted Nuthatch *Sitta carolinensis*	3	51.0	3.0	1430	572	0.03	74,030	0.36	Landres and MacMahon (1983)
Solitary Vireo *Vireo solitarius*	3	66.6	2.6	114	506	0.11	85,772	1.37	Holmes et al. (1979b)
Least Flycatcher *Empidonax minimus*	3	75.4	3.0	609	425	0.04	74,030	0.55	Holmes et al. (1979b)
Western Wood-Pewee *Contopus sordidulus*	3	94.0[a]	1.0[a]	419	129	0.03	226,670	1.16	Eckhardt (1979)
Wilson's Warbler *Wilsonia pusilla*	3	75.0[a]	8.0[a]	427	429	0.05	26,330	0.39	Eckhardt (1979)
Red-eyed Vireo *Vireo olivaceous*	3	61.0	11.5	150	545	0.09	17,620	0.54	James (1976)
White-eyed Vireo *Vireo griseus*	3	59.0	10.0	132	553	0.10	20,606	0.62	James (1976)
Black-throated Blue Warbler *Dendroica virens*	3	43.3	26.7	30	562	0.22	6,286	0.76	Maurer and Whitmore (1981)
Acadian Flycatcher *Empidonax virescens*	3	59.7	1.4	72	551	0.14	161,253	2.37	Maurer and Whitmore (1981)
Prothonotary Warbler *Protonotaria citrea* (males)									Petit et al. (1990)
Pre-nestling period	3	91.0	0.4	1393	188	0.02	570,110	1.01	
Nestling period	3	77.8	3.5	630	396	0.04	63,128	0.50	

[a] Proportions estimated from histograms.

Absolute precisions (b_i) associated with reported sample sizes (n) in Table 1 ranged from 0.02 (Petit et al., unpubl. data) to 0.22 (Maurer and Whitmore 1981) and, in general, most observed sample sizes corresponded to absolute precisions within 10% of the true proportions (with 95% confidence) in each category of foraging mode (Table 1). It is perhaps not surprising that none of the observed sample sizes (n) provided acceptable relative precisions.

DISCUSSION

Tortora's (1978) equations provide a useful and straightforward method for estimating sample sizes for quantifying foraging behavior. However, such dramatic differences between sample sizes calculated using relative and absolute precision prompts the question: How much precision is necessary? A minimum necessary sample size of 600 is infinitely more attractive (and attainable) for field researchers than is one of 50,000. Although some attention has been paid to methods that quantify rare events (e.g., Wagner 1981a, Morrison 1984a, Tacha et al. 1985), most studies have focused only on common behaviors, because extremely rare behaviors (e.g., 1–5% of all maneuvers) are usually relatively unimportant in characterizing the general foraging behavior. Thus, for most studies, it may be sufficient to calculate sample size based on absolute precision, provided that the acceptable confidence interval is relatively small. The decision of what constitutes an acceptable absolute precision or confidence level may depend on the objectives of the study in question and is always at the discretion of the investigator. We chose $\alpha = 0.05$ and $b_i = 0.05$ based on standard statistical criteria (i.e., α-level of significance [α/k is similar to calculating an experimental error rate]). However, these specifications may be unnecessarily stringent. Several recent papers (e.g., Thompson 1987; Angers 1979, 1984) have criticized Tortora's method for being too conservative (i.e., estimating larger sample sizes than necessary), and proposed variations in the estimation technique, making it more liberal (i.e., lowering estimated sample sizes). The technique proposed by Angers (1979, 1984), however, involves tedious calculations. Moreover, the methods proposed by both Thompson (1987) and Angers (1979, 1984) do not improve greatly on the applicability of Tortora's original modification of the estimation technique, and thus, do not decrease its validity.

Given the conservative nature of Tortora's method, one may be justified in relaxing levels of confidence or precision or both when using the equations. It is reasonable to set $\alpha/k = 0.05$ and/or to accept a precision of 10% or even 15%, either of which will lower the minimum number of samples needed (Figs. 3 and 4).

An implicit assumption in using Tortora's equations is that the theoretical frequency to be observed in each category does not change through time. This is difficult to meet in foraging studies because a species' behavior can differ between sexes (e.g., Morse 1968), within a season (Morse 1968, Sherry 1979), and between years (Landres 1980). To meet that assumption, sample sizes would have to be estimated for each category depending on the temporal or spatial scale at which the research is conducted and the objectives of that research. Using the equations presented in this paper, researchers can estimate a required sample size at any required confidence level (α) or precision.

Although sample sizes calculated using absolute precision are considerably lower than those using relative precision, it still may be difficult for researchers to obtain even 100 independent observations (depending on how one achieves that independence; e.g., single point observations) for a population. The estimation method presented here allows researchers to assign a precision, *a posteriori*, to any sample of independent observations, thereby getting an idea of the "power" of their sample and attaining a certain level of confidence in their data.

SUGGESTED SAMPLING PROTOCOL

To estimate sample size using techniques described above, one must have some *a priori* idea of the number of categories (k) and the proportions of observations that will be found in each category. Because those proportions usually are not known, one may consider using the "worst case" (e.g., using $\Pi_i = 0.50$ in the equation for absolute precision above) sample size in order to ensure an adequate sample. While this approach is justifiable, it could lead to gross oversampling. One might also rely on published data to gain an idea of the proportions for a particular species, provided that those data are accurate representations of behaviors exhibited by the species. However, many species exhibit highly plastic foraging behaviors (Petit, Petit, and Petit, this volume), such that predicting foraging behaviors for one population based on previous studies conducted at other locations, or even at the same location using different methods or observers, may be tenuous.

A more reasonable approach would be to collect a preliminary sample of observations (say N = 100; these would not necessarily have to be independent observations) to estimate the proportions $\Pi_1, \ldots, \Pi_k$. For each estimate of Π_i, decide the acceptable absolute precision, b_i, and confidence (α) levels (see above) for Π_i and calculate the estimated sample size (n_a) using the formula above, realizing that it will be necessary to then collect $n_i - N$ additional observations (if N is made up of independent observations). As for the formula above, if $b_i = b$ for all categories, the Π_i closest to 0.50 should be used. Because the required sample size will increase with an increase in number of categories within a variable, researchers perhaps should calculate a required sample size based on the minimum n_i required for the variable with the most k categories.

ACKNOWLEDGMENTS

We thank W. M. Block, J. E. Dunn, M. L. Morrison, and especially C. A. O'Cinneide and J. T. Rotenberry for comments which greatly improved the quality of this manuscript. K. G. Smith was supported by National Science Foundation Grant BSR 84-08090 during preparation of this paper. This work is dedicated to the memory of Richard E. Petit.

INTEROBSERVER DIFFERENCES IN RECORDING FORAGING BEHAVIOR OF FUSCOUS HONEYEATERS

HUGH A. FORD, LYNDA BRIDGES, AND SUSAN NOSKE

Abstract. We independently recorded foraging of the Fuscous Honeyeater (*Lichenostomus fuscus*), a small, generalized insectivore-nectarivore, at the same site in northern New South Wales, SN from January to July 1981, LB from August 1981 to January 1982, and HF throughout this period. Single observations were recorded for each bird at each encounter, with behavior being classified by method and substrate. All observers recorded leaf-gleaning as the most frequent activity (47–59%) with probing flowers second (12–27%). Hawking, hovering at foliage, gleaning and probing at bark, and ground foraging were less frequent. Significant differences were noted in the use of some categories by HF and the other two observers for the common time periods. HF apparently overestimated feeding at flowers, perhaps because he was attracted to flowering trees. All three observers differed in the incidence of aerial foraging, probing into bark for insects, and hovering they recorded. Nevertheless, all three observers presented the same general pattern of foraging. Interobserver overlaps were high (73–83%), despite the latter two observers recording data at different times. Differences in the foraging behavior of the species between the two periods were not great, as HF's data overlapped 91% between the two periods.

Key Words: Foraging behavior; observer bias; honeyeaters; eucalypt woodlands; Australia.

Quantifying an animal's behavior in the field is difficult. Species, individuals, and activities differ in their conspicuousness. In addition, because field recording is a skill requiring many hours of practice, it is usually impossible to employ naive recorders, as can be done in the laboratory (Balph and Romesburg 1986). Observers will probably bias their results compared with the true behavior, and bias may differ among observers. For instance observers may differ in experience, which will not only result in different levels of skill but also different expectations. They could also differ in visual or aural acuity and in classification of behaviors.

This paper describes differences among three observers in their observations of foraging behavior of the Fuscous Honeyeater (*Lichenostomus fuscus*). We sought significant differences in foraging methods or substrates. If these occurred, using the same method in the same area, they would indicate caution when comparing observations between different observers in different areas or years.

METHODS

Most data were collected in about 30 ha of Eastwood State Forest, 10 km SE of Armidale (30°35'S, 151°44'E), with a few collected at Hillgrove Creek State Forest, 12 km E of Armidale (<10% for each observer). Both sites have been described in detail elsewhere (Ford et al. 1985). They were both in eucalypt woodland with 345–415 trees/ha and a canopy cover of 16–32%. The habitat was open with good visibility into the canopy. As eucalypts are evergreen, the conspicuousness of birds in the canopy varied little through the year. Fuscous Honeyeaters are small (18 g), active, vocal, and aggressive throughout the year. They were also the commonest bird in eucalypt woodland near Armidale (3–5 birds/ha at Eastwood) at the time of the study. SN collected data from January to July 1981, LB from August 1981 to January 1982, and HF throughout this period. We compared data between HF and SN and between HF and LB (same sites and periods in both cases, and between SN and LB (same sites, different periods). In a separate study, Fuscous Honeyeaters showed seasonal changes in foraging (Ford, Huddy, and Bell, this volume), though these were not substantial.

Foraging observations were recorded by walking slowly through the habitat until a bird was sighted. It was then observed until it foraged, when a single record was taken. For birds that were already foraging when sighted, the next foraging move was recorded to reduce the bias in favor of conspicuous activities. No particular effort was made to seek Fuscous Honeyeaters, because we collected data on all species. Although the sites were not homogeneous, we made an effort to cover different sub-habitats in the proportion in which they occurred. Data were analyzed and observers did not discuss their results until after field work was completed.

The overall foraging behavior of Fuscous Honeyeaters has been discussed previously along with that of 39 other species (Ford et al. 1986). Here we concentrate on foraging substrates and methods. Substrates were: flowers, leaves, bark (twigs, branches and trunks), ground, and air. Methods were: gleaning, probing, hovering (includes snatching), and hawking.

Observers were compared using a 2 × N contingency test in which N = 5 substrates and 4 methods. If a significant difference was found, cells were examined to identify the factors that contributed to this difference.

RESULTS

Fuscous Honeyeaters spent about half of their foraging time gleaning from leaves (Fig. 1). They also hovered to take insects from leaves, and

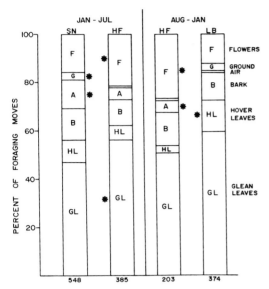

FIGURE 1. Percentage of foraging moves for each observer in the two time periods (using a method-substrate classification). An * next to the column designates an activity recorded significantly more frequently by one observer than the other in the same time period (based on χ^2 value in individual cells in contingency tests). Sample sizes at base of each column.

took insects from bark, from the air, and rarely from the ground. Many of these foraging moves were directed at items such as manna (exudate from damaged leaves), honeydew, lerp (sugary coats of psyllids), as well as at arthropods. Fuscous Honeyeaters also visited flowers of eucalypts and mistletoes (*Amyema*) for nectar.

Results of SN and LB both differed significantly from those of HF for the common periods for substrates (χ^2 = 24.1 and 38.4, df = 4), and for methods (χ^2 = 28.8 and 47.3, df = 3); $P <$ 0.01 in all cases. SN and LB also differed for substrates (χ^2 = 51.4, df = 4) and for methods (χ^2 = 48.1, df = 3), $P <$ 0.01 in both cases. In the case of SN and LB, observed differences may include seasonal effects. HF's observations did not differ significantly between periods, either for substrate (χ^2 = 4.23, df = 4, $P >$ 0.30) or method (χ^2 = 0.96, df = 3, $P >$ 0.80). As method and substrate were not independent (e.g., all hawking was in the air and all flowers were probed), we have shown interobserver differences in Figure 1 by six substrate-method categories. These differences were evident in most categories, HF recorded more foraging on flowers than both SN and LB, SN recorded the most aerial feeding, and LB the most foraging at leaves.

The magnitude of differences was not great, however, ranging up to 14.5% of total observations for a category. Overlaps ($100[1 - \Sigma |P_{ij} - P_{ik}|]$, where P_{ij} and P_{ik} were proportions of observations in category i for observers j and k) between observers were also high: SN × HF = 80% (common period), LB × HF = 73% (common period), and SN × LB = 83% (different periods). Overlap was highest (91%) between data from the two periods for HF.

A few significant differences also occurred among some of the lesser categories that were not represented in Figure 1. Twigs (a subset of bark) were recorded significantly more frequently by SN than HF (χ^2 = 6.56, df = 2, $P <$ 0.05), but significantly less often by LB than HF (χ^2 = 15.7, df = 2, $P <$ 0.001) when comparing twigs, branches, and trunks within the bark category, between observers. Within the bark-foraging categories, HF recorded significantly more probing than SN (χ^2 = 34.4, df = 2, $P <$ 0.01) and less gleaning than LB (χ^2 = 11.2, df = 2, $P <$ 0.01).

DISCUSSION

The size and number of statistical differences between data collected by the observers indicate that such differences are not due to sampling error. However, observations were collected by each observer on a small number of days, and usually on different days. If differences among days in weather, for instance, influence behavior of the birds, then apparent differences between observers may have been accentuated. The facts that Fuscous Honeyeaters displayed only small seasonal changes in foraging (Ford et al., this volume), and that these data for the two periods collected by HF were very similar, argue against day-to-day differences causing interobserver differences.

The observers' levels of experience differed, perhaps influencing perception and expectation. For instance, HF's greater experience with honeyeaters may have caused him to be attracted to flowering trees, thus overestimating feeding at flowers. Classification of less frequent activities may have been imprecise (e.g., twigs could be classified as leaves [petioles] or branches).

In any event, comparisons between the same species in different areas or years, recorded by different observers, need to be treated cautiously, especially when observers have not previously agreed on standard methods of observation, or classification of terms. Adoption of a universal classification for foraging methods and substrates would reduce, but probably not eliminate, interobserver variability. Indeed it may be unrealistic to attempt to differentiate between some categories. As implied above, experience may reduce or increase bias.

Perhaps the most important result from this study was the basic similarity in the results from the three observers. We should emphasize similar patterns in comparative studies rather than seek too carefully to demonstrate statistical differences that may not have much biological significance, as they may represent idiosyncrasies of individual birds or observers.

ACKNOWLEDGMENTS

We thank Richard Noske, Gillian Dunkerley, Harry Recher, and Jared Verner for comments on the paper and Sandy Higgins and Viola Watt for typing the manuscript.

Interspecific, Spatial, and Temporal Variation

WITHIN-SEASON AND YEARLY VARIATIONS IN AVIAN FORAGING LOCATIONS

SALLIE J. HEJL AND JARED VERNER

Abstract. We studied monthly and yearly differences in the foraging sites and substrates of Plain Titmice (*Parus inornatus*) and Bushtits (*Psaltriparus minimus*) in a foothill oak-pine woodland in the central Sierra Nevada during the breeding seasons of 1979 and 1980. The greatest intraspecific differences observed for both species were monthly changes in the use of foraging sites (primarily plant species) and substrates (plant part to which the foraging maneuver was directed) and yearly differences in foraging substrates. The main interspecific differences were in foraging sites used overall and in monthly usages of substrates. Several patterns of resource use paralleled phenological changes in the plant species upon which the birds foraged. For example, both species foraged more on buckbrush (*Ceanothus cuneatus*) during the flowering stage, and Plain Titmice foraged more on blue oak (*Quercus douglasii*) as new leaves reached full growth. Pooling data across months in the same breeding season would have hidden these variations. Furthermore, ignoring site-substrate interactions makes it difficult to interpret patterns in avian foraging.

Key Words: Foraging; within-season variation; yearly variation; Plain Titmouse; Bushtit; oak-pine woodlands; California.

Researchers have commonly pooled observations of avian foraging behaviors within seasons and across years (James 1976, Holmes et al. 1979b, Holmes 1980, Conner 1981, Holmes and Robinson 1981, Morrison 1981, Lewke 1982, Franzreb 1983a, Airola and Barrett 1985, Morrison et al. 1985). Seasonal differences in foraging behavior have often been acknowledged (Conner 1981, Lewke 1982, Morrison et al. 1985), but within-season and yearly differences usually have not, in spite of the fact that such differences are reflected in diets (Holmes 1966, Root 1967, Busby and Sealy 1979, Rotenberry 1980a) and behaviors of birds (Holmes 1966; Root 1967; Busby and Sealy 1979; Alatalo 1980; Wagner 1981b; Ford, Huddy, and Bell, this volume; Sakai and Noon, this volume; Szaro et al., this volume). Pooling heterogeneous data sets in this manner could obscure important short- and long-term differences in avian foraging and lead to incorrect interpretations of ecological relationships.

Within-season and yearly differences in diets and foraging behaviors have been demonstrated in many habitats. In five seasons near Barrow, Alaska, Holmes (1966) documented within-season and yearly changes in prey availability and in the associated foraging behavior and diet of Dunlins (*Calidris alpina*). Root (1967) recorded seasonal and yearly differences in prey availability and in the associated diet of Blue-gray Gnatcatchers (*Polioptila caerulea*) in a coastal oak woodland in California. Both the sandpiper and the gnatcatcher also selected certain prey types. On the other hand, although Busby and Sealy (1979) found monthly and yearly differences in the foraging behavior and diet of Yellow Warblers (*Dendroica petechia*) in Manitoba, the warblers consumed prey in proportion to their availability. Alatalo (1980) studied the foraging behaviors of five bird species in coniferous forests in Finland throughout 1 year and for 3 months of another year, observing within- and between-season shifts in their foraging behaviors. Similarly, Rotenberry (1980a) found within-season, between-season, and yearly differences in diets of three ground-foraging passerines in shrubsteppe habitats of southeastern Washington during two breeding seasons and one complete year. Wagner (1981b) documented seasonal and yearly differences in foraging behavior of a foliage- and bark-gleaning guild in a California oak woodland.

We studied the foraging locations of Plain Titmice (*Parus inornatus*) and Bushtits (*Psaltriparus minimus*) in a foothill oak-pine woodland to: (1) discern possible intraspecific variations in foraging locations between years or from month to month in the same year, (2) assess the similarities and differences in foraging locations of the two species during the same time periods, and (3) learn whether monthly and yearly differences in foraging locations of either species reflected observed changes in plant phenology.

STUDY AREA AND METHODS

Study area. The study was done during the breeding season of both species at the San Joaquin Experimental Range in March, April, and May during 1979 and 1980. The Range is located approximately 32 km north of Fresno, in Madera Co., California. Elevation ranges from 215 to 520 m. The climate is one of hot, dry summers and cool, wet winters.

Field observations were made on a 19.8 ha (300 × 660 m) plot gridded at 30-m intervals and situated within approximately 32 ha of foothill woodlands that had not been grazed by cattle or managed in any other significant way since 1934. Vegetation on the plot was mainly oak-pine woodland, with some small patches of blue oak (*Quercus douglasii*) savanna, chaparral, and annual grasslands. Buckbrush (*Ceanothus cuneatus*), with 18.6% crown cover, was the most abundant shrub on the plot. Among the trees, gray pine (*Pinus sabiniana*) had a crown cover of 12.5%, interior live oak (*Q. wislizenii*) had 7.2%, and blue oak had 5.4%. The nine remaining tree and shrub species contributed only 4.5% crown cover.

Bird observations. One observer recorded data in 1979 and three did so in 1980; the observer in 1979 also observed in 1980. Observers walked along alternate, numbered lines in the long dimension of the grid. Lines walked and the direction of travel were regularly selected to ensure even coverage of all segments of the grid. Walking and stationary search for birds were alternated approximately every 15 min. Observations were made from sunrise to sunset.

Only certain individuals were selected for observation. To avoid bias toward singing birds, observers did not hunt out singing birds. However, most birds sang or called during the observation period. Only the first bird detected in a flock or pair was used as a subject, as locations of flock or pair members would not be expected to be independent. A new individual was chosen as a subject only if the observer had traveled at least 30 m or at least 10 min had elapsed since the last record of a given species. This constraint was imposed in an attempt to increase independence among samples.

From the time a bird was selected, the observer counted slowly to 5 (approximately 5 s) to give time to assess the bird's activity. Its activity at the count of "5" was recorded as an instantaneous sample. If the bird was obviously searching for food at that instant, observations continued until it executed a distinct foraging maneuver (assumed to indicate an attempt to secure food). Two aspects of the location of the foraging maneuver will be examined in this paper as follows: (1) site (gray pine, blue oak, interior live oak, buckbrush, and other, including all other plants, air and ground); and (2) substrate, the exact part of the plant or environs toward which a foraging maneuver was directed (twig [<5 mm in diameter], small branch [5 mm–10 cm in diameter], large branch [>10 cm in diameter], flower bud, flower, catkin, fruit, leaf bud, leaf, trunk, air, and ground).

Plant phenology. Phenology of the major woody plant species was sampled weekly during both years and summarized by 2-week periods. Trees sampled were gray pine, blue oak, interior live oak, and California buckeye (*Aesculus californica*). Shrubs sampled were buckbrush, redberry (*Rhamnus crocea*), California coffeeberry (*R. californica*), mariposa manzanita (*Arctostaphylos mariposa*), bush lupine (*Lupinus albifrons*), poison oak (*Toxicodendron diversilobum*), bush penstemon (*Keckiella breviflora glabrisepala*), and blue elderberry (*Sambucus mexicana*).

Random samples of 10 shrubs and trees of each species were selected, except for species with fewer than 10 individuals on the plot, in which case all individuals were sampled. Eight branches (two each on the north, east, south, and west sides) were selected on each plant, at approximately breast height, and labeled with small, numbered, metal tags. The phenology of each branch was recorded weekly during both growing seasons. Some branches were grazed during the course of the study; these were replaced with the nearest neighbor. All phenological stages present on a given branch were noted. Vegetative growth was recorded as budding, swollen buds, elongated buds, new leaves, stem elongation, and full-sized leaves. Reproductive phenological states included initial budding, swelling of the bud, opening of the bud, full flowers present, fruits set, fruits developing, fruits developed, catkins emerged, and pollen released when evident.

Statistical analyses. Because log-linear models can be used to describe data from a multiway contingency table (Fienberg 1970, 1977; Bishop et al. 1975), we searched for log-linear models that best fit our data. We would have preferred to analyze our data in one comprehensive analysis, since we know that important interactions between foraging site and substrate exist. However, data on foraging sites were analyzed separately from foraging substrates, because our data set was too small to classify each record by site and substrate as well as by year, month, and bird species in a multiway contingency table. (Too many sampling zeros would have occurred. According to our statistical consultant, the total number of observations should be at least four times the number of cells in the contingency table; J. A. Baldwin, pers. comm.) Because birds may use a hierarchical decision-making scheme in which they first choose a site and then a substrate within that site (an extension of the habitat selection ideas of Hutto [1985a]), we thought it reasonable to analyze site and substrate separately.

To find the best model for foraging site, we categorized each record into four variables: (1) bird species, (2) year, (3) month (= March [the first two phenological periods], April [the second two phenological periods], or May [the last two phenological periods]), and (4) site. The result was a 2 × 2 × 3 × 5 contingency table. To find the best model for foraging substrate, we pooled across foraging sites. We categorized each record by bird species, year, month, and foraging substrate for the second model. The month variables were defined as above. Foraging substrate included four categories: (1) bark surface (= twig, small branch, large branch, or trunk), (2) foliage (= leaf or leaf bud), (3) reproductive parts (= flower bud, flower, catkin, or fruit), and (4) other (= air or ground). The result was a 2 × 2 × 3 × 4 contingency table. Foraging site and foraging substrate were treated as response variables in the chosen models. The biological relevance of the interactions entering the models, which included foraging site and foraging substrate, are discussed later. Other interactions that entered models indicated sampling differences; these interactions are discussed in less detail.

We chose a model based on three criteria. Initially, we determined which models had P-values that were close to but greater than 0.05. From those models, we then chose the simplest ones (those with fewer and

TABLE 1. The Chosen Log-linear Models for Foraging Site and Foraging Substrate. Sample Sizes for Plain Titmice Were 35 in March 1979, 36 in April 1979, 84 in May 1979, 63 in March 1980, 86 in April 1980, and 204 in May 1980. Sample Sizes for Bushtits Were 93 in March 1979, 87 in April 1979, 76 in May 1979, 114 in March 1980, 110 in April 1980, and 140 in May 1980

Model I: foraging site
 A. $\ln x_{ijkl} = u + B_i + Y_j + M_k + I_l + BM_{ik} + BI_{il} + MI_{kl} + BY_{ij}$
    Chi-square = 43.62, df = 36, P = 0.18
 B. $\ln x_{ijkl} = u + B_i + Y_j + M_k + I_l + BM_{ik} + BI_{il} + MI_{kl} + YM_{jk}$
    Chi-square = 48.00, df = 35, P = 0.07

Model II: foraging substrate
 $\ln x_{ijkm} = u + B_i + Y_j + M_k + S_m + BY_{ij} + BM_{ik} + BS_{im} + YM_{jk} + MS_{km} + YS_{jm} + BMS_{ikm}$
    Chi-square = 27.07, df = 17, P = 0.06

Parameters
B_i = bird species $i = 1, 2$
Y_j = year $j = 1, 2$
M_k = month $k = 1, 2, 3$
I_l = foraging site $l = 1, 2, 3, 4, 5$
S_m = foraging substrate $m = 1, 2, 3, 4$
x_{ijkl} = cell frequencies in the (x_{ijkl}) cell
x_{ijkm} = cell frequencies in the (x_{ijkm}) cell

lower-order interaction terms). To choose the best model from similarly simple models with similarly low P-values, we used four assessment techniques. These were comparisons of: (1) the linear predictors with fitted and observed responses, (2) the nonstandardized residuals with expected responses, (3) the standardized residuals with expected responses, and (4) the standardized residuals with the linear predictors.

Initially we used BMDP4F (Dixon 1983) to determine which level of interaction terms should be included in the final model. These choices ranged from the saturated model (the four-factor interaction and all of those below it) to complete independence of all variables (the four main effects and no interaction terms). Log-linear models are hierarchical: if a three-way interaction is included in the model, then all two-way interactions between the variables in the three-way interaction are included in the model. Inclusion of interaction terms indicates dependence between the variables in the interaction. For example, a model including the two-way interaction between foraging site and month indicates that foraging sites differed among the 3 months. A model with four main effects (foraging site, month, year, and bird species) and no interaction terms would indicate that foraging site did not differ among months, years, or bird species.

TABLE 2. Estimated Parameter Values for the Significant Interactions Involving Foraging Site from the Two Log-linear Models Chosen to Describe Plant-species Use (See Methods)

	Site				
	Gray pine	Live oak	Blue oak	Buckbrush	Other
Model A					
Bird					
Plain Titmouse	0.167	−0.277	0.673	−0.446	−0.116
Bushtit	−0.167	0.277	−0.673	0.446	0.116
Month					
March	−0.392	0.485	−0.535	0.670	−0.229
April	−0.260	−0.048	0.385	−0.064	−0.013
May	0.651	−0.437	0.150	−0.606	0.242
Model B					
Bird					
Plain Titmouse	0.166	−0.279	0.678	−0.449	−0.116
Bushtit	−0.166	0.279	−0.678	0.449	0.116
Month					
March	−0.391	0.487	−0.543	0.672	−0.226
April	−0.252	−0.058	0.400	−0.077	−0.013
May	0.643	−0.429	0.143	−0.595	0.239

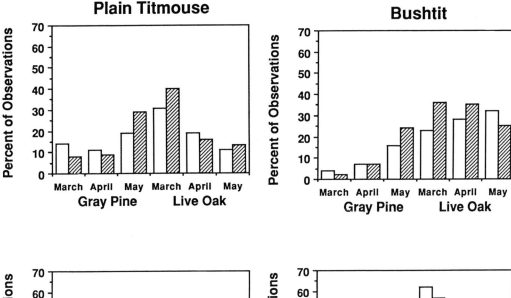

FIGURE 1. Percent of all foraging maneuvers by Plain Titmice on each of the four major plant species in each month in 1979 (open bars) and 1980 (shaded bars). March = phenological intervals 1 and 2, April = phenological intervals 3 and 4, May = phenological intervals 5 and 6. Sample sizes were 35 in March 1979, 36 in April 1979, 84 in May 1979, 63 in March 1980, 86 in April 1980, and 204 in May 1980.

FIGURE 2. Percent of all foraging maneuvers by Bushtits on each of the four major plant species in each month (see Fig. 1 for definition of phenological intervals) in 1979 (open bars) and 1980 (shaded bars). Sample sizes were 93 in March 1979, 87 in April 1979, 76 in May 1979, 114 in March 1980, 110 in April 1980, and 140 in May 1980.

Backward and forward selection procedures from BMDP were examined to select several models that were similarly good, based on their P-values for the log-likelihood test statistic that approximates the chi-square statistic for larger sample sizes. We sought the simplest model that would adequately explain our data (P > 0.05). Use of the General Linear Interactive Model, GLIM (Royal Statistical Society 1986), further refined our choice. We could add or delete terms easily and quickly on GLIM and compare linear predictors, fitted, observed and expected responses, and nonstandardized and standardized residuals. The procedures led to two similarly simple models for foraging site and one model for foraging substrate. We next employed the four assessment techniques to choose between the two competing models for foraging site. Based on the assessment techniques, neither model for foraging site seemed better. Therefore, we present results from both models. Judgments were made on complete models. All terms in the chosen models are significant and their biological meanings are discussed.

For the chosen log-linear model, parameters were estimated to assess the sign and magnitude of each component of each variable in each interaction term. Bishop et al. (1975:62) refer to estimates of parameter values as u-terms. The estimates sum to zero across categories. The magnitude reflects the importance of the component, and the sign indicates the direction of the effect. Bishop et al. (1975) give a mathematical description of log-linear models and parameter estimates. A good biological example of the use of parameter estimates is in Page et al. (1985); Schoener (1970),

TABLE 3. ESTIMATED PARAMETER VALUES FOR THE SIGNIFICANT INTERACTIONS INVOLVING FORAGING SUBSTRATE FROM THE LOG-LINEAR MODEL CHOSEN TO DESCRIBE SUBSTRATE USE (SEE METHODS)

		Substrate			
		Bark	Foliage	Reproductive parts	Other
Year					
1979		0.042	−0.000	−0.169	0.127
1980		−0.042	0.000	0.169	−0.127
			Bird species		
Substrate	Month	Plain Titmouse	Bushtit		
Bark	March	0.065	−0.065		
	April	−0.007	0.007		
	May	−0.058	0.058		
Foliage	March	−0.043	0.043		
	April	0.251	−0.251		
	May	−0.208	0.208		
Reproductive parts	March	−0.240	0.240		
	April	−0.050	0.050		
	May	0.290	−0.290		
Other	March	0.218	−0.218		
	April	−0.194	0.194		
	May	−0.024	0.024		

Jenkins (1975), and Harris (1984) provide other biological examples using log-linear models.

RESULTS

MODEL I: FORAGING SITE

Based on P-values, simplicity, and the four assessment techniques, GLIM showed that four of the two-way interactions alone created two different but equally satisfactory models (Table 1).

Within-season changes. The two species foraged differently among the five sites in the 3 months, as indicated by the significant interactions between months and sites in both models (Table 1). These changes were parallel in the two species. In general, the greater use of live oak and buckbrush and the concomitant lesser use of blue oak by both species in March, the increased use of blue oak in April, and the increased use of gray pine and other sites in May were indicated by the size and sign of the estimated parameter values for the site-by-month interactions (Table 2 and Figs. 1 and 2).

Yearly differences. The relative number of observations among plant species was the same in both years (both models excluded the year-by-site interaction).

Foraging differences between Plain Titmice and Bushtits. Although parallel changes in site use occurred in the two species, the overall use of plant species was significantly different between the two species, as reflected by the inclusion of the bird-by-site interaction in both models (Table 1). Overall, Plain Titmice foraged more often on blue oak and Bushtits foraged more often on buckbrush (Table 2 and Figs. 1 and 2).

MODEL II: FORAGING SUBSTRATE

According to the criteria described above, one simple, satisfactory model was the best for foraging substrate (Table 1).

Within-season differences and foraging differences between Plain Titmice and Bushtits. The two species foraged from the four substrates differently across the 3 months, as evidenced by the inclusion of the bird-by-month-by-substrate interaction in the chosen model (Table 1). Plain Titmice emphasized other substrates in March, foliage in April, and reproductive parts in May, while Bushtits foraged from reproductive parts in March, from other substrates in April, and from foliage in May (Table 3).

Yearly differences. The use of foraging substrates by the two species differed significantly between years, as indicated by the inclusion of the year-by-substrate interaction in the chosen model (Table 1). However, the relatively small sizes of the estimated parameter values suggested that the weight of this interaction in this model was small (Table 3).

DIFFERENCES IN FORAGING SITES, SUBSTRATES, AND PLANT PHENOLOGY

Several monthly differences in emphasis of foraging substrates on certain foraging sites par-

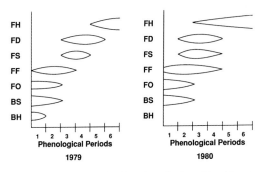
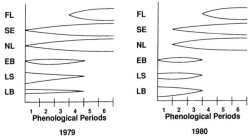

FIGURE 3. Reproductive phenologies of buckbrush in 1979 and 1980. BH = flower buds hard, BS = flower buds swollen, FO = flowers open, FF = full flowers present, FS = fruits set, FD = fruits developing, FH = fruits hard. Dates for phenological periods as follows: 1 = 12 to 23 March 1979, and 10 to 21 March 1980; 2 = 26 March to 6 April 1979, and 24 March to 4 April 1980; 3 = 9 to 20 April 1979, and 7 to 18 April 1980; 4 = 23 April to 4 May 1979, and 21 April to 2 May 1980; 5 = 7 to 18 May 1979, and 5 to 16 May 1980; 6 = 21 May to 1 June 1979 and 19 to 30 May 1980. Widths of symbols are based on relative percentages of total branches in each phenological state. Curves were drawn by hand to connect the points from each 2-week sample.

FIGURE 4. Vegetative phenologies of blue oak in 1979 and 1980. LB = leaf buds present, LS = leaf buds swollen, EB = elongated buds, NL = new leaves, SE = stem elongation, FL = full-sized leaves. Dates for phenological periods as in Figure 3. Widths of symbols are based on relative percentages of total branches in each phenological state. Curves were drawn by hand to connect the points from each 2-week sample.

alleled changes in plant phenology. For example, the peak period of flowering by buckbrush occurred in March and April in both years (Fig. 3), and fruit replaced flowers by the end of April each year. Blue oaks began leafing out in March, and stem elongation and the surge of new leaves occurred by mid-April in both years (Fig. 4). Concomitantly, Bushtits so emphasized buckbrush flowers as a substrate in March of both years that they comprised nearly 50% of all substrates on all foraging sites (Table 4). On buckbrush alone, flowers comprised 71% of the substrates in 1979 and 84% in 1980. Plain Titmice exhibited a similar pattern in March of both years (Table 4), although buckbrush leaves comprised a larger proportion of their foraging substrates than flowers in 1979. Use of buckbrush flowers by both species dropped markedly in April and did not occur at all in May, but both species increased their use of blue oak leaves as a foraging substrate in April and May of both years (Table 5).

DISCUSSION

Our results of significant within-season differences are like those of many other investigators (Holmes 1966, Busby and Sealy 1979, Alatalo 1980, Rotenberry 1980a). Our yearly differences were not pronounced and did not seem as great as those found by Holmes (1966), Root (1967), Busby and Sealy (1979), Rotenberry (1980a), Wagner (1981b), and Szaro et al. (this volume).

Several researchers have found that within-season trends in the foraging behavior and diet of one species often parallel those of other species in the same habitat (Morse 1970, Alatalo 1980, Rotenberry 1980a, this study), probably because prey availability changed (e.g., Holmes and Pitelka 1968). Seasonal changes in the foraging behavior of gnatcatchers, and the availability of their prey in a California oak woodland, corresponded with plant phenology (Root 1967). We believe that within-season shifts in the foraging behavior of the Plain Titmice and Bushtits in this study also resulted from changing prey availability in relation to different stages of plant phenology.

Observer differences cannot be ruled out as contributing to some of the yearly differences observed in this study, although we do not believe they had a major effect. For example, the patterns of shifting foraging substrates on certain sites with plant phenology were similar in both years, even though only one observer sampled in 1979 but three observers sampled in 1980. Our ability to detect yearly differences may have been increased by the disparity in sample sizes between years. However, each of our monthly sample periods included more than 30 observations of each species, thus exceeding the minimum sample size recommended by Morrison (1984a) for studies of avian foraging behavior (but see Brennan and Morrison, this volume), and the bird-by-month and bird-by-year interactions in the models for foraging sites and substrates act as blocking factors for sample size differences (M. F. Bryan, pers. comm.).

TABLE 4. PERCENT OF FORAGING ON EACH SUBSTRATE ON BUCKBRUSH (IN RELATION TO FORAGING ON ALL SITES AND SUBSTRATES) BY PLAIN TITMICE AND BUSHTITS DURING THE SPRINGS OF 1979 AND 1980

Substrate	1979			1980		
	March	April	May	March	April	May
Plain Titmouse						
Branch	0	3	1	2	4	3
Flower	11	6	0	22	2	0
Flower bud	0	0	0	2	0	0
Fruit	0	0	0	0	0	1
Leaf	17	11	6	3	13	8
Twig	6	6	0	3	4	3
Bushtit						
Branch	1	2	3	1	1	3
Flower	44	9	0	48	4	0
Flower bud	0	0	0	0	0	0
Fruit	0	0	0	0	9	0
Leaf	13	23	16	6	19	26
Twig	4	10	8	2	4	5

Our results caution against generalizations made from data gathered during one month, one year, or for differing numbers of months and years. Ignoring either short- or long-term variations in foraging behavior can lead to oversimplifications and even obscure ecologically significant patterns. We also think this can happen when researchers uncouple components of foraging behavior for ease of analysis. For example, important interactions between foraging substrates and foraging sites were missed in this study when we created models for foraging sites independent from foraging substrates. The birds shifted their emphasis on substrates from buckbrush flowers to blue oak leaves in a similar pattern in both years. Because of our relatively small sample sizes, important relationships between these site-substrate combinations could be shown only in tables and figures. We suggest that researchers with larger data sets include foraging site and substrate in the same multiway contingency table for analysis. Structural zeros (cells in the contingency table that necessarily contain zeros; for example, the cell for buckbrush flowers in May contained a zero because buckbrush does not flower in May) will inevitably occur with

TABLE 5. PERCENT OF FORAGING ON EACH SUBSTRATE ON BLUE OAK (IN RELATION TO FORAGING ON ALL SITES AND SUBSTRATES) BY PLAIN TITMICE AND BUSHTITS DURING THE SPRINGS OF 1979 AND 1980

Substrate	1979			1980		
	March	April	May	March	April	May
Plain Titmouse						
Branch	8	5	17	5	1	9
Catkin	0	0	0	0	1	0
Large branch	0	5	0	3	0	<1
Leaf	0	25	25	5	38	20
Leaf bud	3	0	0	3	0	0
Trunk	0	0	1	0	0	<1
Twig	0	3	5	2	7	4
Bushtit						
Branch	0	0	1	0	1	2
Catkin	0	1	0	0	2	0
Large branch	1	1	0	1	0	0
Leaf	0	10	8	0	6	4
Leaf bud	1	0	0	1	0	0
Trunk	0	0	0	1	0	0
Twig	2	1	1	0	1	3

changes in plant phenology. However, the BMDP program can be instructed to deal with them and to adjust the degrees of freedom appropriately. Sampling zeros are problematic only when the marginals (row or column totals) are zeros. A large proportion of the literature on avian foraging behavior includes data pooled across months and years and data analyzed separately for foraging site and substrate, but the extent to which such procedures may have biased conclusions is unknown.

ACKNOWLEDGMENTS

K. L. Purcell and L. V. Ritter collected field observations and shared ideas on avian foraging behavior and plant phenology. R. M. Caldwell helped collect phenological data, keypunch data, and offered support and advice throughout the study. D. A. Duncan, J. R. Larson, D. L. Neal, and R. D. Ratliff helped formulate appropriate phenological methods. G. W. Salt provided support, advice, and criticism throughout the study. J. A. Baldwin, M. F. Bryan, S. R. Mori, G. W. Bell, and W. R. Rice offered statistical help. C. J. Evans, J. H. Harris, S. A. Laymon, B. A. Maurer, G. W. Page, and L. E. Stenzel helped clarify the use of log-linear models for us. R. L. Hutto, B. A. Maurer, K. L. Purcell, C. J. Ralph, and M. G. Raphael reviewed an earlier version of the manuscript. We appreciate the help of all.

THE IMPORTANCE AND CONSEQUENCES OF TEMPORAL VARIATION IN AVIAN FORAGING BEHAVIOR

DONALD B. MILES

Abstract. Monthly and yearly differences in foraging technique, substrate use, and tree species use were examined for three bird species from desert scrub and oak woodland habitats of southeastern Arizona from 1981 to 1983; all species displayed significant variation. Birds in desert scrub showed a strong temporal variation in their foraging behaviors, whereas species from the oak woodland appeared to be relatively temporally invariant. The strongest shifts in behavior appeared to be for choice of foraging site within a plant (substrate) and the differential use of particular plants within a season. Such variation within a season and among years suggests that caution is necessary in drawing inferences about species interactions, community organization, and resource partitioning that are based on data from a single year or data pooled over several years.

Key Words: Foraging behavior; temporal variation; seasonality; yearly variation; desert scrub; oak woodland.

Recent insights into community processes, whether determined directly by dietary variation or indirectly by differences in foraging behaviors, are commonly derived from single sample surveys or a pooled sample from long term studies (e.g., Holmes et al. 1979b; Alatalo 1982; Robinson and Holmes 1982, 1984; Airola and Barrett 1985; Holmes and Recher 1986a). Unfortunately, few studies have examined temporal variation in avian foraging behavior (Ulfstrand 1976, 1977; Smith et al. 1978; Alatalo 1980; Saether 1982; Morrison and With 1987), so that its importance has been uncertain. Furthermore, few data are available regarding the changes in community parameters (e.g., niche overlap and niche breadth) as a consequence of variation in foraging behavior within and among years as well as among seasons. The presence of seasonal or yearly variation may have a significant effect on the confidence we place in the estimates of habitat use, niche breadth and overlap, and consequently on the subsequent inferences drawn about species interactions.

In this study, I examined temporal variation in foraging behaviors of foliage-gleaning passerine birds from two habitats during three breeding seasons. Several questions are pertinent to quantifying avian foraging behavior and to more general problems in avian ecology. First, at what level are temporal fluctuations likely to be detected? Second, are there habitat-related differences in the seasonal patterns of resource use? As a corollary, are these patterns related to fluctuations in abiotic environmental factors that birds use as proximate cues? Third, are temporal fluctuations in foraging behaviors more prominent in particular guilds? Fourth, which aspects of a bird's foraging repertoire are most likely to exhibit seasonal variation? That is, do birds alter the locations that they forage rather than the particular feeding technique?

STUDY AREA AND METHODS

Study area. The foraging patterns of foliage-gleaning birds were studied at two locations in southeastern Arizona: Saguaro National Monument (Tucson Mountain Unit) and Madera Canyon, Santa Rita Mountains. Data were recorded during the breeding season (May–August) for 3 years (1981–1983). The vegetation of the former site consisted primarily of the scrub trees, ironwood (*Olneya tesota*), and foothill paloverde (*Cercidium microphyllum*) with a variety of cacti (principally the saguaro [*Carnegia gigantea*] and several species of cholla [*Opuntia* spp.]) and bursage (*Ambrosia deltoidea*). The Madera Canyon site was characterized by open oak woodland, which contained an admixture of Mexican blue oak (*Quercus oblongifolia*), emory oak (*Q. emoryi*), Arizona white oak (*Q. arizonica*) and alligator juniper (*Juniperus deppeana*).

Bird observations. Three 1000-m-long transects were established in each locality; each had ten sampling stations spaced approximately 100 m apart. Observations were recorded throughout the morning until midday. The statistical analyses of foraging behavior data tend to exhibit serial dependencies among the observations (see Morrison 1984a; Hejl et al., this volume; Raphael, this volume). Therefore I employed the following sampling protocol. Upon sighting a foraging bird, I dropped the first foraging observation and recorded the second and subsequent observations every 15 s for 2 min. I visited each sampling station once per day and remained for 15–20 min to avoid repeated sampling of individuals. For every bird that was observed to capture or attack a prey item, I quantified foraging technique, substrates, and the plant species. I collected data for all passerine species and classified foraging behavior as glean, hover, probe, or hawk. I also recognized six foraging substrates: leaves, flowers and fruits, twigs, branches, trunks and ground, including litter. I analyzed seasonal shifts in only the most commonly used species of plant.

I compiled monthly and yearly estimates of the foraging repertoire of the species from each habitat. Following the recommendations of Morrison (1984) I retained only those bird species that had at least 40 observations/month/year. Three species from each habitat fulfilled this criterion: Verdin (*Auriparus flaviceps*), Cactus Wren (*Campylorhynchus brunneicapillus*) and Black-tailed Gnatcatcher (*Polioptila melanura*) from the desert habitat; and Bridled Titmouse (*Parus wollweberi*), Bewick's Wren (*Thryomanes bewickii*), and Black-throated Gray Warbler (*Dendroica nigrescens*) from the oak woodland habitat.

Statistical analyses. The categorical behavior data were tabulated into frequencies for use in subsequent analyses. I determined whether species in the desert habitats exhibited significant temporal variation in their foraging behavior through a log-linear analysis of multiway contingency tables (Bishop et al. 1975; Fienberg 1977; Hejl and Verner, this volume). I tested for monthly and yearly heterogeneity in the use of each category of foraging variable. I used PROC CATMOD (SAS 1985) to calculate the main effects and interaction (month, year) terms for the models pertaining to each foraging variable.

Two foraging variables, technique and substrate, were crosstabulated and the month and year estimates of each category for all six bird species were used to calculate a similarity matrix. Out of a total of 24 possible combinations among the foraging categories, only 14 were used by the species. I estimated the similarity in foraging behavior between and within-species using the percent similarity coefficient:

$$PS_{ij} = 1 - 0.5 \sum |p_{ik} - p_{jk}|,$$

where p_{ij} represents the percent use of the jth foraging category for the ith species. The similarity matrix was used as a basis to cluster species using the unweighted pair group average (UPGMA) method. Because both habitats exhibited similar patterns of seasonality, I present the results from the oak woodland habitat.

I calculated estimates of niche overlap and breadth for each species. Niche breadth was estimated using the Shannon-Wiener index:

$$H' = - \sum p_i \ln p_i,$$

where p_i is the proportionate use of the ith category. The change in niche breadth between months and years was individually calculated for each foraging category. Estimates of the variance of H' were calculated using formulae given in Poole (1974:393). Niche overlap with other species combined was derived from the percent similarity index described above.

RESULTS

SEASONALITY OF DESERT HABITATS

Total yearly rainfall is bimodally destributed with the majority of precipitation occurring in late summer (Fig. 1). Two distinct seasons are evident: an early period of drought, lasting from late April through late June to early July, followed by the summer rains, which occur largely as convective thunderstorms. Their onset and amount is temporally and spatially variable. This

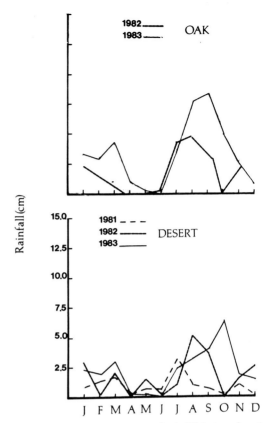

FIGURE 1. Yearly patterns of rainfall in two desert study sites. Data are from 1981–1983 for the desert scrub habitat, and 1982–1983 for the oak woodland habitat.

pattern of rainfall has a profound effect on the vegetation structure and resource base of each habitat. Several species of plants flower throughout the period of drought especially in the desert scrub habitat, where species such as ironwood, foothill paloverde, saguaro, and mesquite (*Prosopis juliflora*) produce flowers from late April through early June. Associated with this period of flowering is an increase in the abundance of arthropods.

After the summer rains begin there is a period of vegetative growth, which is mainly attributable to the emergence of annual plants, particularly grasses (Cable 1975; Maurer 1985b; Miles, unpubl. data). This is correlated with an increase in the number of arthropods, which is mainly manifested in a high density of lepidopteran larvae (pers. obs.). Thus, the foraging behaviors of birds in desert habitats, if tracking resource flushes, should vary between the drought and rainy seasons.

TABLE 1. SIGNIFICANCE OF LOG-LINEAR ANALYSIS OF SEASONAL VARIATION IN FORAGING TECHNIQUE, FORAGING SUBSTRATE AND PLANT SPECIES. DATA WERE ANALYZED FOR THREE SPECIES FROM THE DESERT SCRUB AND OAK WOODLAND STUDY LOCALITIES. FORAGING DATA FROM THE BREEDING SEASON FROM THREE YEARS, 1981, 1982 AND 1983, WERE ANALYZED

Species	Temporal component	Foraging[a]		
		Technique	Substrate	Plant species
Desert Scrub				
Black-tailed Gnatcatcher	Month	ns	$P < 0.05$	ns
	Year	ns	ns	ns
	Month × Year	ns	ns	ns
Cactus Wren	Month	$P < 0.005$	$P < 0.03$	$P < 0.001$
	Year	$P < 0.02$	$P < 0.05$	ns
	Month × Year	ns	ns	ns
Verdin	Month	$P < 0.003$	$P < 0.001$	$P < 0.001$
	Year	$P < 0.001$	$P < 0.001$	$P < 0.001$
	Month × Year	ns	ns	ns
Oak Woodland				
Black-throated Gray Warbler	Month	ns	$P < 0.05$	ns
	Year	ns	ns	ns
	Month × Year	ns	ns	ns
Bridled Titmouse	Month	ns	ns	$P < 0.001$
	Year	$P < 0.01$	ns	ns
	Month × Year	ns	ns	ns
Bewick's Wren	Month	ns	$P < 0.05$	ns
	Year	ns	ns	$P < 0.01$
	Month × Year	ns	ns	ns

[a] ns = not significant.

LOG-LINEAR ANALYSIS OF TEMPORAL VARIATION IN AVIAN FORAGING BEHAVIORS: TESTS OF MONTHLY AND YEARLY EFFECTS

The log-linear analysis of the variation in foraging behavior within a species between months and years and the interaction between month and year yielded heterogeneous results (Table 1). In spite of the divergence in temporal responses, several trends were evident. First, most of the temporal variation seemed attributable to monthly changes in the foraging repertoire. Second, the species rarely showed statistically significant patterns of variation among years in foraging, except for the Verdin and Cactus Wren. Both of these species were generalized in their foraging repertoire (Table 2). Species from the oak woodland habitat showed greater temporal heterogeneity in substrate choice and plant species preference rather than vary their foraging techniques (Table 1). The species showed a complex pattern of changing their choice of foraging substrate as the breeding season progressed as well as showing significant yearly variation in substrate and plant species use. Lastly, the species from the desert scrub habitat exhibited a far larger magnitude in their temporal shifts than the oak woodland species. This may be attributable to the larger effect summer rains have on desert scrub vegetation than in the oak woodlands.

CLUSTER ANALYSIS OF AVIAN FORAGING BEHAVIORS

I present the results for the oak woodland species, although a similar pattern was found for the desert scrub species. Four clusters are apparent from the dendrogram (Fig. 2). The first cluster consisted of observations on the Black-throated Gray Warbler and Bridled Titmouse, mainly during May and June, the period of low rainfall. At this time the birds foraged primarily by gleaning leaves (Table 3). A second cluster includes the same species, but the samples were taken in July–August, after the beginning of the summer rains. At this time the birds spent more time gleaning from small and large twigs (Table 4). The third and fourth clusters describe monthly variation in Bewick's Wren, which presumably reflects a response to changes in the vegetation related to rainfall. The former cluster represents the foraging behavior of the wren during the early summer (May and June) and the latter represents the late summer (July). This species foraged in the lower strata of the vegetation and often would glean or probe at leaves on the ground (Tables

TABLE 2. CHANGES IN THE PROPORTION OF FORAGING TECHNIQUES EMPLOYED BY BIRDS OF THE DESERT HABITAT SITE BY MONTH IN 1981, 1982, AND 1983

Tech-nique	1981			1982			1983		
	June	July	August	May	June	July	May	June	July
				Black-tailed Gnatcatcher					
Hawk	0.0	2.0	2.7	1.5	0.0	0.0	3.7	0.0	0.0
Glean	86.7	91.8	83.8	87.9	95.8	98.1	81.5	80.0	88.9
Hover	0.0	2.0	8.1	6.1	4.2	0.0	14.8	10.0	11.1
Probe	0.0	2.0	2.7	3.0	0.0	0.0	0.0	0.0	0.0
Other	13.3	2.0	2.7	1.5	0.0	1.9	0.0	10.0	0.0
				Cactus Wren					
Hawk	0.0	0.0	0.0	0.0	0.0	0.0	0.0	0.0	0.0
Glean	100.0	38.7	17.9	41.2	20.8	25.4	35.0	14.3	13.3
Hover	0.0	0.0	7.1	0.0	0.0	0.0	0.0	0.0	0.0
Probe	0.0	48.4	53.6	47.1	79.2	73.0	65.0	85.7	86.7
Other	0.0	12.9	21.4	11.8	0.0	1.6	0.0	0.0	0.0
				Verdin					
Hawk	0.0	0.9	0.0	0.6	0.0	0.0	0.0	0.0	0.0
Glean	93.8	96.4	97.7	96.2	73.0	68.0	95.4	88.5	97.9
Hover	0.0	2.7	0.0	1.9	0.0	0.4	0.9	1.6	2.1
Probe	6.3	0.0	2.3	0.6	27.0	30.2	3.7	9.8	0.0
Other	0.0	0.0	0.0	0.6	0.0	1.5	0.0	0.0	0.0

3, 4). Thus, it is likely that the Bewick's Wren would show a greater response to the flush of annual plants that follows the summer rains.

CONSEQUENCES OF TEMPORAL VARIATION IN FORAGING BEHAVIOR

Changes in niche breadth

As shown for the Black-tailed Gnatcatcher, the diversity of foraging techniques changes both between the dry and rainy seasons and between years (Fig. 3). Prior to the onset of summer rains the species displays a high niche breadth, utilizing a number of foraging maneuvers. During the rainy season the gnatcatcher tends to specialize on gleaning at foliage. For example, during 1981, which was characterized by a relatively dry summer, the range of foraging techniques increased during the latter part of the breeding season. However, the difference was not statistically significant. Estimates of niche breadth were not significantly different between years for the months of May or June. There were significant differences between years when comparing the niche breadth estimates for July (1981–1982, t = 2.05, $P < 0.05$, df = 46; 1982–1983, t = 2.16, $P < 0.05$, df = 68). Thus the gnatcatcher had a significantly narrower niche breadth during July 1982 than in either 1981 or 1983. This was also the driest July among the three years, suggesting that the gnatcatcher was specializing on gleaning maneuvers, mainly at leaves and fine substrates (Table 2). A comparison of the estimates of niche breadth for each month during the 1982 breeding season showed that June and July significantly differed from May (t = 2.24, $P < 0.05$, df = 77; and t = 2.87, $P < 0.05$, df = 91). Estimates for June and July were not significantly different. There were no differences in the estimates of niche breadth between months for 1981 and 1983.

Similarly, the range of substrates used within a season and between years was characterized by a certain amount of variation (Fig. 4). In 1981 and 1982 the gnatcatcher showed a decrease in the breadth of substrates used over the course of the breeding season. Yet in 1983, which was characterized by the lowest amount of rainfall through the drought months and a late arrival of the summer rains, the gnatcatcher exhibited a contraction of the number of substrates used and specialized on small twigs and leafy substrates entirely. This decrease in niche breadth was statistically significant only during June (1981–1983, t = 2.06, $P < 0.05$, df = 23; 1982–1983, t = 2.38, $P < 0.05$, df = 33). There were no significant differences between years for the May or July estimates of niche breadth. There were no differences among months within a year for 1981 and 1982. But significant differences among months were evident for 1983 (May–June, t = 3.42, $P < 0.05$, df = 18; June–July, t = 2.83, $P < 0.05$, df = 41; May–July, $P > 0.05$). Thus, only the 1983 estimates of niche breadth showed a

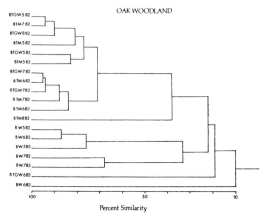

FIGURE 2. Average Linkage Cluster Analysis of the oak woodland species. The analysis was based on the percent similarity coefficient calculated from 14 cross-classified foraging categories. Species codes are BTGW = Black-throated Gray Warbler, BTM = Bridled Titmouse, and BW = Bewick's Wren. The numbers following each code represent the sample month and year: 5 82 = May 1982, 6 82 = June 1982, 7 82 = July 1982, 8 82 = August 1982, 5 83 = May 1983, 6 83 = June 1983, 7 83 = July 1983.

TABLE 3. CHANGES IN THE PROPORTION OF FORAGING TECHNIQUES EMPLOYED BY BIRDS OF THE OAK WOODLAND SITE BY MONTH IN 1982 AND 1983

Tech-	1982			1983		
nique	May	June	July	May	June	July
Bewick's Wren						
Hawk	3.7	0.0	0.0	5.6	0.0	0.0
Glean	18.5	18.2	54.3	16.7	18.2	64.7
Hover	0.0	4.5	0.0	77.8	0.0	0.0
Probe	77.8	77.3	45.7	0.0	81.8	35.3
Other	0.0	0.0	0.0	0.0	0.0	0.0
Bridled Titmouse						
Hawk	0.0	0.0	0.0	0.0	0.0	0.0
Glean	93.8	92.9	76.4	100.0	82.2	79.2
Hover	4.2	3.6	0.0	0.0	2.2	16.7
Probe	2.1	3.6	0.6	0.0	15.6	4.2
Other	0.0	0.0	0.0	0.0	0.0	0.0
Black-throated Gray Warbler						
Hawk	0.0	0.0	0.0	0.0	0.0	3.7
Glean	95.7	90.9	100.0	89.5	100.0	92.6
Hover	4.3	9.1	0.0	5.3	0.0	3.7
Probe	0.0	0.0	0.0	0.0	0.0	0.0
Other	0.0	0.0	0.0	5.3	0.0	0.0

significant decrease within the breeding season suggesting an increase in specialization of substrate use.

Unlike the previous two variables, the temporal change in breadth of use of plant species exhibited a complex pattern (Fig. 5). In two of the years (1981 and 1982), gnatcatchers tended to specialize on a few plant species early in the breeding season, primarily foothill paloverde and ironwood. Later in the breeding season the gnatcatcher broadened the number of plant species it would search for food, primarily by searching saguaro cactus and white-thorn acacia. Comparing months within years supported this pattern and revealed several statistically significant differences for 1981 (June–August, t = 1.98, P < 0.05, df = 48) and 1982 (May–June, t = 2.24, P < 0.05; June–July, t = 2.61, P < 0.05, df = 76). The decrease between May and June during 1983, which was attributable to relatively high usage of foothill paloverde and ironwood, was not statistically significant. Furthermore, the change in niche breadth was not significant between years.

CHANGES IN NICHE OVERLAP

Mean overlap, which was calculated as the average overlap with each of the other species, fluctuated among months and between years (Fig. 6). Two trends were evident for all species. Low overlap values (i.e., dissimilar foraging behaviors) tended to occur during the dry summer months and high overlap values in the wetter summer months. The two foliage-gleaning species—Black-tailed Gnatcatcher and Verdin—showed almost coincident patterns of niche overlap. On the other hand, the Cactus Wren was characterized by low overlap values; peak overlap occurred at the onset of summer rains, but low overlap values were found both in early summer and late summer.

DISCUSSION

I found significant differences in bird foraging behaviors between early and late summer months and among years. Several generalizations emerged from these results. First, the response to environmental fluctuations appeared to be greater in the desert scrub. This result follows the pattern found by Smith et al. (1978) for *Geospiza* finches in the Galapagos. Significant differences were found in the foraging behavior and diet of the finches between the wet season and dry season. Second, species in both habitats responded similarly to the environmental variation, mainly by moving to new foraging locations and plant species, rather than adjusting their foraging behaviors. The seasonal shifts in foraging behavior tended to be consistent among years. Third, the cluster analysis of foraging behaviors failed to reveal a strong within-species grouping, which is similar to the findings of Rotenberry (1980a). Thus, foliage-gleaning birds exhibited similar be-

TABLE 4. PERCENT USE OF THE 14 CROSS-TABULATED FORAGING CATEGORIES BY THE BIRDS OF THE OAK WOODLAND STUDY AREA. DATA ARE PRESENTED FOR THE BREEDING SEASON IN 1982 AND 1983. SPECIES CODES ARE: BTGW = BLACK-THROATED GRAY WARBLER, BTM = BRIDLED TITMOUSE, BW = BEWICK'S WREN. CODES FOR FORAGING CATEGORIES ARE: GLST = GLEAN AT SMALL TWIGS (5 CM < IN DIAMETER), GLBT = GLEAN AT BIG TWIGS (5 CM >), GLBR = GLEAN AT BRANCHES, GLTR = GLEAN AT TRUNK, GLGR = GLEAN AT GROUND, GLLF = GLEAN LEAVES, HVBR = HOVER AT BRANCH, HVLF = HOVER AT LEAVES, PRBT = PROBE BIG TWIGS, PRBR = PROBE BRANCH, PRTR = PROBE TRUNK, PRGR = PROBE GROUND, PRFDWD = PROBE FALLEN DEADWOOD.

Species		Hawk	GLST	GLBT	GLBR	GLTR	GLGR	GLLF	HVBR	HVLF	PRBT	PRBR	PRTR	PRGR	PRFDWD
BTGW	5/82	0.0	13.0	4.3	4.3	0.0	0.0	73.9	0.0	4.3	0.0	0.0	0.0	0.0	0.0
	6/82	0.0	0.0	0.0	0.0	0.0	0.0	90.9	0.0	9.1	0.0	0.0	0.0	0.0	0.0
	7/82	0.0	20.0	0.0	0.0	0.0	0.0	80.0	0.0	0.0	0.0	0.0	0.0	0.0	0.0
BTM	5/82	0.0	0.0	0.0	4.2	0.0	0.0	89.6	0.0	4.2	0.0	2.1	0.0	0.0	0.0
	6/82	0.0	7.1	14.3	0.0	0.0	0.0	71.4	0.0	3.6	3.6	0.0	0.0	0.0	0.0
	7/82	0.6	15.9	4.0	2.3	0.0	0.0	76.7	0.0	0.0	0.0	0.6	0.0	0.0	0.0
	8/82	0.9	25.5	0.0	0.9	0.0	0.0	50.0	0.0	0.9	17.3	0.0	0.0	0.0	0.0
BW	5/82	0.0	0.0	18.5	0.0	0.0	0.0	0.0	0.0	0.0	0.0	33.3	33.3	3.7	3.4
	6/82	0.0	0.0	0.0	13.6	0.0	0.0	4.5	4.5	0.0	0.0	45.5	27.3	4.5	0.0
	7/82	0.0	2.9	2.9	11.4	5.7	25.7	5.7	0.0	0.0	2.9	28.6	5.7	8.6	0.0
BTGW	5/83	5.3	0.0	0.0	15.8	0.0	0.0	73.7	5.3	0.0	0.0	0.0	0.0	0.0	0.0
	6/83	0.0	0.0	0.0	85.7	0.0	0.0	14.3	0.0	0.0	0.0	0.0	0.0	0.0	0.0
	7/83	3.7	0.0	11.1	0.0	0.0	3.7	77.8	0.0	3.7	0.0	0.0	0.0	0.0	0.0
BTM	5/83	0.0	11.1	0.0	22.3	0.0	0.0	66.7	0.0	0.0	0.0	0.0	0.0	0.0	0.0
	6/83	0.0	2.2	20.0	8.9	0.0	0.0	51.1	0.0	2.2	0.0	15.6	0.0	0.0	0.0
	7/83	0.0	0.0	20.8	0.0	0.0	4.2	54.2	0.0	16.7	4.2	0.0	0.0	0.0	0.0
BW	5/83	5.6	0.0	0.0	11.1	5.6	0.0	0.0	0.0	0.0	5.6	27.8	38.9	0.0	5.6
	6/83	0.0	4.5	0.0	9.1	0.0	0.0	4.5	0.0	0.0	0.0	31.8	0.0	50.0	0.0
	7/83	0.0	0.0	5.9	5.9	0.0	47.1	5.9	0.0	0.0	0.0	23.5	5.9	0.0	0.0

SEASONAL DIFFERENCES IN FORAGING HABITAT OF CAVITY-NESTING BIRDS IN THE SOUTHERN WASHINGTON CASCADES

RICHARD W. LUNDQUIST AND DAVID A. MANUWAL

Abstract. For each of four cavity-nesting bird species we compared winter and spring foraging habitat in second-growth (42–190 yrs) and old-growth (>210 yrs) stands in the western hemlock (*Tsuga heterophylla*) zone of the southern Washington Cascades. We measured the availability of live trees and snags and observed foraging birds in 48 stands during the breeding seasons of 1983 to 1986 and during the winters of 1983–1984 and 1984–1985. Although most species fed in large diameter (>50 cm dbh) trees more than expected in both seasons, the foraging methods as well as the tree portions used differed among species. In winter, Red-breasted Nuthatches (*Sitta canadensis*) shifted foraging activities inward to the trunk and to lower relative postions in trees. Brown Creepers (*Certhia americana*) and Hairy Woodpeckers (*Picoides villosus*) showed more subtle shifts in foraging location. Chestnut-backed Chickadees (*Parus rufescens*) differed from the other species in remaining in the outer branches and high in the crown profile of trees while feeding. Most species selected Douglas-fir (*Pseudotsuga menziesii*) trees in both winter and spring. Chickadees selected western hemlocks disproportionately in winter, but in spring they used tree species about as available. Relative use of dbh classes and tree species also differed between forest age classes for most species. The importance of large Douglas-firs to foraging birds appears to be related to abundance and diversity of prey species inhabiting its fissured bark.

Key Words: Seasonal differences; foraging; cavity-nesting birds; Cascade Mountains.

Seasonality is an important aspect of natural variation in temperate ecosystems that affects community structure and habitat use of birds (Fretwell 1972). For winter survival, permanent residents must be able to respond to changes in the distribution and abundance of food resources brought on by climatic changes (Gordon et al. 1968). Many authors have confirmed seasonal changes in patterns of habitat use and foraging activities in several bird species in other regions of North America (e.g., Stallcup 1968, Willson 1970, Austin 1976, Travis 1977, Conner 1981, Lewke 1982, Morrison et al. 1985, Morrison and With 1987). No study, however, has examined seasonal changes in foraging behavior in the productive Douglas-fir (*Pseudotsuga menziesii*)/ western hemlock (*Tsuga heterophylla*) forests of the Washington Cascades. Characterization of seasonal change is important not only in theoretical studies of niche overlap (or segregation) and community structure (Alatalo 1980), but also in forest management, because managers may have to provide for a different set of habitats for the needs of each species in the nonbreeding vs. the breeding season (Conner 1981). To the extent that intensive timber management changes the species composition and structure of forest stands, it may also affect the winter survival of resident birds.

Of particular concern are cavity-nesting birds, which typically nest in standing dead trees, or "snags," because snags are usually removed during timber harvesting. Birds may focus foraging activities on different species and sizes of trees from those used for nesting, and foraging activities may change seasonally, so characteristics of foraging habitats should not be overlooked (Conner 1980). Our objective in this study was to compare the foraging activities of cavity-nesting birds during winter and spring (breeding season) in old-growth and second-growth forests. Specifically, we examined seasonal changes in foraging behavior and location (both horizontal and vertical), as well as selection of different tree species, sizes, and conditions (in relation to availability).

STUDY AREA AND METHODS

The area studied was the southern Washington Cascades in the Gifford Pinchot National Forest (GPNF) and in Mt. Rainier National Park (MRNP). Forty-eight forest stands (25–30 ha each) representing second-growth (42–190 years old) and old-growth (200+ years old) forest age classes were selected as part of the vertebrate community studies of the USDA Forest Service's Old-Growth Wildlife Habitat Program (OGWHP) (Ruggiero and Carey 1984). All stands were within the Western Hemlock Vegetation Zone (Franklin and Dyrness 1973) and ranged in average elevation from 404 to 1218 m. Western hemlock was the most abundant tree species in old-growth, followed by Pacific silver fir (*Abies amabilis*), Douglas-fir, and western redcedar (*Thuja plicata*). Douglas-fir structurally dominated old-growth stands, however, as most of the largest trees (>100 cm dbh) were of this species. Douglas-fir was the most abundant species (in all size classes) in second-growth stands, followed by western hemlock and western redcedar.

Because winter access to many of the stands was limited, we selected a subset of eight stands in the southern part of the study area near the Columbia River Gorge for winter study (December through early March) in 1983–1984. The winter study was expanded in 1984–1985 to include eight additional stands in the northern portion of the study area. A more detailed description of the stands included in this study is found in Manuwal and Huff (1987) and Lundquist (1988).

FORAGING OBSERVATIONS

We observed foraging birds while conducting OGWHP studies during the winters of 1983–1984 and 1984–1985 and the springs (late April through June) of 1983 through 1986. The species analyzed, all permanent residents, included the following: Brown Creeper (*Certhia americana*), Chestnut-backed Chickadee (*Parus rufescens*), Hairy Woodpecker (*Picoides villosus*), and Red-breasted Nuthatch (*Sitta canadensis*). The observed foraging activities (≥ 2 s duration) of an individual bird on a single "host" (e.g., a tree, shrub, or log) comprised one foraging observation. Each observation ended when the bird flew to a new "host" or a time limit of 99 seconds was reached. In the springs of 1985 and 1986, up to five sequential observations were also taken on individual birds. Because of questions concerning independence (e.g., Morrison 1984a; Hejl et al., this volume; Bell et al., this volume), all but the initial observations were excluded from analysis. By attempting to monitor a bird's foraging activities on a single host for the maximum duration, and by establishing a minimum observation time of 2 s, we have attempted to minimize discovery (or visibility) bias, which may affect estimates of resource use (Bradley 1985). Loss bias may also be a problem (Wagner 1981a), but it may not be possible to avoid both biases simultaneously with one sampling method (Bradley 1985).

We recorded the following information on each foraging bird: species; sex and age class (where discernible); primary feeding behavior (e.g., gleaning, probing); horizontal part of tree or snag (i.e., trunk, or base, middle, or ends of branches); and vertical zone of the tree (e.g., upper, middle, lower crown, below crown), if applicable. Recorded attributes of the "host" included species, diameter breast height (dbh) class (10-cm intervals), condition (dead or alive, top condition), and position relative to the forest canopy (above, co-canopy, lower canopy, or understory). One exception to the above was during the first winter (1983–1984), when the dbh class of trees was recorded in 20-cm intervals. As a result, when analyzing use patterns in relation to tree availability (see below), we had to exclude observations in trees of dbh classes (e.g., 1–20 cm, 41–60 cm) that could not be placed in dbh categories to match those of the vegetation data.

VEGETATION SAMPLING

In analyses of resource selection, described below, we used vegetation data collected in 12 nested circular plots (0.05 ha, 0.2 ha) systematically located on each study stand. In the 0.05-ha plots, all live trees ≤ 100 cm dbh and all snags 10-19 cm dbh were tallied by tree species and dbh class. Live trees <10 cm dbh were tallied as well, but snags of this size were not, so we excluded this size from the analyses. Stem counts in the 0.2-ha plots included live trees >100 cm dbh and snags ≥ 20 cm dbh by species and dbh class. We summarized the data in 11–50 and >50 cm dbh classes to obtain overal frequency distributions of trees (live and dead combined) in each forest age class (old-growth and second-growth).

DATA ANALYSIS

Various aspects of foraging behavior of birds have been shown to differ by sex and age class (Ligon 1968a, Jackson 1970, Austin 1976, Morrison and With 1987), among years (Root 1967; Grant and Grant 1980; Wagner 1981b; Szaro et al., this volume), and even within a season (Holmes 1966; Busby and Sealy 1979; Alatalo 1980; Hejl and Verner, this volume; Sakai and Noon, this volume). Unfortunately, our data samples were too small to analyze data comprehensively in multiway contingency tables (too many empty cells would have resulted) and to search for interactions among all these factors (e.g., by development of log-linear models, as in Hejl and Verner, this volume). Thus, we combined data for the two winters and four breeding seasons in analyzing seasonal changes in foraging patterns. In addition, the sexes of most species could not be distinguished in the field; this, together with limited data sets, prevented us from including intersexual comparisons in the analyses. Rather, we focused on the degree to which attributes of winter foraging by each species differed from foraging during the breeding season.

These analyses of seasonal shifts by each bird species were done separately for each attribute (i.e., behavior, horizontal location, vertical location) by means of two-way log-likelihood contingency tests of independence (G-tests). Log-likelihood G-tests are analogous to, and often preferred over, the Chi-square statistic (Sokal and Rohlf 1981:704, Zar 1984:52–53). We employed the Williams (1976) correction to the G-statistic to obtain a better approximation to the Chi-square distribution, even in cases with only one degree of freedom. This correction appears to be superior to the Yates correction for continuity in such cases (Sokal and Rohlf 1981).

Where sample sizes permitted, we also statistically evaluated the use (i.e., selection) of tree conditions (live or dead), size (dbh class), and tree species by each bird species in winter and spring separately by means of single-dimension log-likelihood G-tests. Expected frequencies for these analyses were calculated from tallies of trees and snags on the stands on which the foraging observations were made. Because the frequency distributions of size classes and species of trees differed between old-growth and second-growth, we evaluated use of trees by foraging birds separately in each forest age class. Low sample sizes for some bird species (see Results) prompted us to group some of the rarer tree species together for statistical analysis. Vegetation was summarized using the SPSSX computer package (SPSS 1986); log-likelihood G-tests were run using modifications of programs developed for the Hewlett-Packard HP-41CX hand calculator (Hewlett-Packard 1984.)

Estimates of minimum sample sizes required for statistical evaluation may vary considerably with the level of precision or confidence required (Sokal and Rohlf 1981; Petit, Petit, and Smith, this volume; Recher and

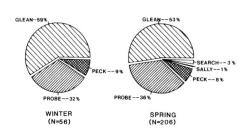
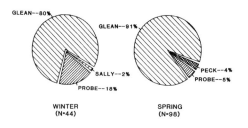
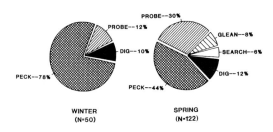
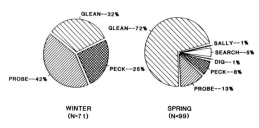

FIGURE 1. Primary foraging behaviors of four cavity-nesting bird species during winter and spring.

Gebski, this volume), as well as with the species and habitats studied (Morrison 1988). Although different rules have been suggested for goodness-of-fit tests, we followed the general rule commonly used in Chi-square tests that no expected frequency should be less than 1.0 and no more than 20% of the expected frequencies should be less than 5.0 in any test (Cochran 1954). In addition, in most cases our sample sizes for each species, season, and univariate attribute of foraging were above the minimum of 30 recommended by Morrison (1984a) for analysis of avian foraging behavior (but see Brennan and Morrison, this volume; Petit, Petit, and Smith, this volume). Where samples were near or below this minimum, the results were viewed as suspect and interpreted with caution.

RESULTS

PRIMARY BEHAVIORS

Brown Creepers, primarily bark gleaners, showed no seasonal shift in behavior (Fig. 1A). Chestnut-backed Chickadees, also gleaners of insects, but from foliage, probed more frequently in winter than in spring ($P < 0.05$) (Fig. 1B). Hairy Woodpeckers (Fig. 1C) and Red-breasted Nuthatches (Fig. 1D) shifted behaviors more substantially than the other two species. Both species pecked for food items more frequently in winter and nuthatches also probed more frequently in winter than in spring.

FORAGING LOCATION: HORIZONTAL AND VERTICAL

Creepers and chickadees showed no substantial horizontal or vertical shifts in foraging location between seasons, though creepers fed primarily in different locations in trees (Figs. 2A, 3A) than chickadees (Figs. 2B, 3B). The apparent relative decrease in trunk foraging by creepers ($P < 0.001$) and the increase in outer limb foraging by chickadees ($P < 0.005$) in winter, while statistically significant, could have been due to the great disparity in sample sizes between the seasons for each species. Hairy Woodpeckers foraged on the same portion of the trees (trunks) during both seasons (Fig. 2C), but they fed less frequently in the crown zones of trees and more frequently in snags without branches during winter than during spring ($P < 0.0001$) (Fig. 3C). Hairies rarely fed on logs in either season. Nuthatches shifted foraging locations most substantially between the sesons: they fed significantly more frequently further inward (Fig. 2D) and downward (Fig. 3D) in tree profiles during winter compared with spring ($P < 0.0001$ in both tests).

USE OF TREES IN RELATION TO AVAILABILITY

Tree condition. None of the four species shifted significantly their relative use of live or dead trees

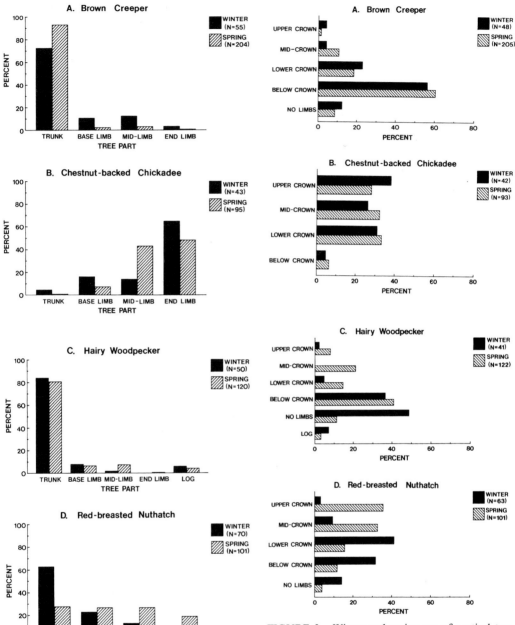

FIGURE 2. Winter and spring use of horizontal tree parts by foraging birds.

FIGURE 3. Winter and spring use of vertical tree zones by foraging birds. Live and dead trees were pooled: "no limbs" category represents snags without limbs; "logs" are fallen dead trees ≥10 cm diameter.

between seasons in old-growth (log-likelihood contingency analysis, df = 1, P > 0.05) (Fig. 4A). Sample sizes were generally too small in second-growth (Fig. 4B) to analyze seasonal shifts in resource use, but relative use of live and dead trees was similar to that in old-growth. Likewise, samples were too small to analyze resource selection for the winter data in second-growth (for all bird species).

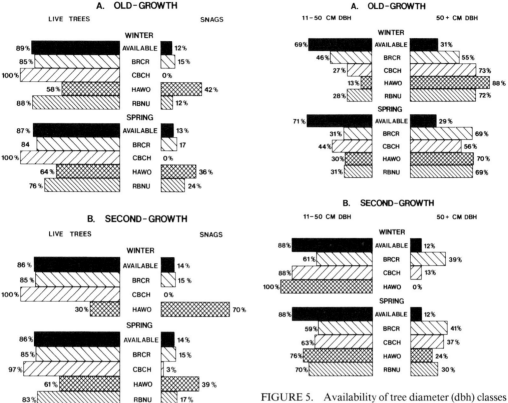

FIGURE 4. Availability of live and dead trees (snags) and their use by foraging birds during winter and spring in both (A) old-growth and (B) second-growth stands. Live and dead trees <10 cm dbh, as well as logs, were excluded from the analysis. Bird species codes (Klimkiewicz and Robbins 1978) are as follows: BRCR, Brown Creeper; CBCH, Chestnut-backed Chickadee; HAWO, Hairy Woodpecker; RBNU, Red-breasted Nuthatch.

FIGURE 5. Availability of tree diameter (dbh) classes and their use by foraging birds during winter and spring in (A) old-growth and (B) second-growth stands. Trees <10 cm dbh, as well as logs, were excluded from the analysis. Bird species codes are as in Figure 4.

Brown Creepers used live and dead trees in proportion to their availability in both forest age classes during both seasons ($P > 0.05$, all tests). Chickadees, which fed almost exclusively in live trees, appeared to select live trees over snags in all cases tested (G-tests, $P < 0.005$ in old-growth; $P < 0.05$ in second-growth [spring]). Hairy Woodpeckers, on the other hand, selected snags disproportionately in all cases tested ($P < 0.001$). Red-breasted Nuthatches used live and dead trees about as available in old-growth during the winter and in second-growth during the breeding season ($P > 0.05$). However, in old-growth during spring, nuthatches apparently selected snags over live trees as foraging substrates ($P < 0.01$) (Fig. 4A), despite the fact that no significant shift was detected between winter and spring in the contingency analysis.

Diameter. In old-growth, no significant changes in relative use of tree dbh classes were noted for any of the bird species (Fig. 5A). All species fed in large trees (>50 cm dbh) significantly more than expected during both seasons ($P < 0.01$ for creepers, $P < 0.005$ for chickadees, and $P < 0.001$ for the others). While no seasonal comparisons could be made in second-growth, all bird species except Hairy Woodpeckers again selected large diameter trees disproportionately as foraging substrates ($P < 0.01$ for nuthatches, $P < 0.001$ for creepers and chickadees, and $0.05 < P < 0.10$ for Hairies) (Fig. 5B). In contrast to old-growth, however, all bird species were observed primarily in smaller diameter trees (11–50 cm dbh) during both seasons in these stands.

Tree species. The Chestnut-backed Chickadee was the only bird species that significantly shifted relative use of tree species in old-growth stands between seasons ($P < 0.05$) (Fig. 6B). During the

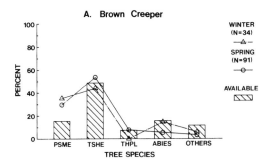
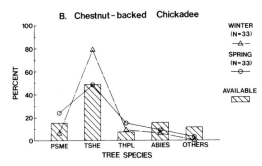
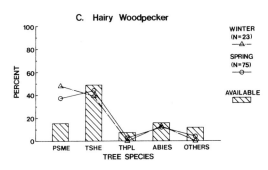
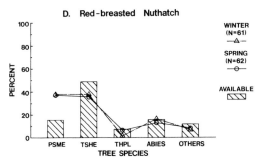

FIGURE 6. Availability of tree species and their use by foraging birds in old-growth stands during winter and spring. Trees <10 cm dbh, as well as logs, were excluded from the analysis. Tree species codes are: PSME, Douglas-fir; TSHE, western hemlock; THPL, western redcedar; ABIES, true firs, including Pacific silver fir, noble fir (*Abies procera*), and grand fir (*A. grandis*); OTHERS, including conifers such as Pacific

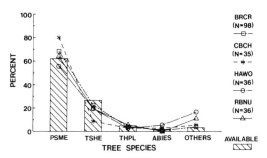

FIGURE 7. Availability of tree species and their spring use by foraging birds in second-growth stands. Trees <10 cm dbh were excluded from the analysis. Tree species codings are as in Figure 6.

breeding season they used tree species in proportion to availability, but during the winter they selected western hemlock significantly more than expected ($P < 0.001$). The other bird species all selected Douglas-fir disproportionately in old-growth during both seasons ($P < 0.001$ for Hairy Woodpeckers, and $P < 0.01$ for the others) (Figs. 6A,C,D). Again, no seasonal comparisons of tree species use could be made for the second-growth data, because of small winter samples. Interestingly, in contrast to old-growth, all bird species fed primarily in Douglas-fir during spring in second-growth (Fig. 7), but only chickadees appeared to select this species significantly more than expected ($P < 0.05$).

DISCUSSION

Our results generally confirm seasonal changes in foraging activities, as other investigators have observed in other regions. Not surprisingly, analysis of foraging data pooled across seasons may then mask significant variation. Some of the shifts we noted may reflect differences in prey distribution on different tree parts in winter and spring. During spring, insects are constantly appearing and are readily available on all parts of trees. In winter, small branches, which have thinner bark and are more exposed to harsh weather conditions, provide fewer places for insects to survive (Jackson 1970, Travis 1977). Thus, we might expect resident birds to concentrate winter foraging activities on the tree bole or under the

yew (*Taxus brevifolia*) and western white pine (*Pinus monticola*), and hardwoods such as vine maple (*Acer circinatum*), bigleaf maple (*A. macrophyllum*), red alder (*Alnus rubra*), and black cottonwood (*Populus trichocarpa*).

wood surface, and perhaps lower in tree profiles, than during spring.

However, the nature and degree of seasonal changes differed by species, depending upon the attribute in question. These differences may partly reflect evolved morphological differences among species and thus their relative abilities to extract prey items, which in turn determine which types of prey are exploitable (Kisiel 1972; Conner 1980, 1981). Of the species we studied, Hairy Woodpeckers were the most capable of finding prey beneath bark and bare wood surfaces. Their increased use of branchless snags in winter, and the increase in their pecking activities, probably reflected a shift toward prey items under the bark.

Nuthatches, which have smaller bills and are less able to extract subsurface prey, nevertheless can adequately chip bark pieces from tree trunks. In spring, they can exploit abundant insect populations in a variety of locations without resorting to more energetically-demanding means. In winter, these easily attainable foods were not available, so nuthatches concentrated activities on tree trunks and lower in the tree profile, where they pecked and probed more frequently. In Colorado pine forests, Stallcup (1968) noted similar seasonal changes by White-breasted Nuthatches (*Sitta carolinensis*) in winter.

Curiously, we found no shift inward and downward by Chestnut-backed Chickadees like that observed for nuthatches. Chickadees, adapted to foliage-gleaning, are less able to extract prey from bark or under wood surfaces than the bark-foraging species and probably focus on different food items. While the other species selected Douglas-fir in winter (and spring), chickadees markedly increased their use of western hemlocks. The specific benefits of western hemlock to chickadees are unclear, but their seeds (Manuwal and Huff 1987) might provide a reliable winter food for chickadees, which cannot compete with the other species for bark- and wood-dwelling prey. The need for quantification of potential food resources in both seasons is obvious.

Creepers, on the other hand, are bark specialists, highly adapted for removing prey items from crevices on tree trunks, a relatively more seasonally uniform source of food than other parts of trees (Jackson 1970). Thus, no substantial seasonal changes in foraging methods or location would be expected. The creepers' concentration on the lower bole then may have been due to visibility bias, even though our procedures should have minimized this problem. Other researchers (e.g., Willson 1970, Morrison et al. 1987b) have found that creepers concentrate activities on trunks and at lower relative heights than other bark-foraging species, particularly in winter.

Factors other than prey abundance may also have influenced seasonal shifts in foraging behavior and location. Grubb (1975, 1977, 1978) found that birds in deciduous woods foraged relatively lower in cold, windy periods, which mainly affected species using small outer branches. This may help explain the shifts that we observed in nuthatches. Hairy Woodpeckers and Brown Creepers, which already concentrated activities on trunks and foraged lower in trees than nuthatches in spring, may have been less affected in winter. Why chickadees remained in the outer branches is still unclear. Grubb (1975) suggested that birds may benefit from solar warming by foraging slightly higher when the sun is shining than during overcast conditions, even if air temperatures are lower with clear (but calm) skies. Because we observed birds during calm conditions and avoided severe weather in both seasons, we may not have witnessed its full impact.

The differences we observed strengthen the argument against treating all species within the same nesting or foraging guild together. The species we studied are all cavity or crevice nesters, and all but chickadees are bark-foragers. Analyzing data pooled over members of the same guild not only may lead to misleading conclusions with respect to resource selection by individual species (Mannan et al. 1984), but also may mask seasonal changes. While species may respond similarly to changes in food abundance or distribution within a season (e.g., Morse 1970; Hejl and Verner, this volume; but see Sakai and Noon, this volume), this is not consistently the case across different seasons (e.g., Conner 1981, this study). Management schemes based on the requirements of a single "indicator species" (e.g., Graul et al. 1976, Severinghaus 1981) or upon data pooled over all species in a guild (*sensu* Verner 1984) or over different seasons may therefore be inadequate.

The importance of large-diameter Douglas-fir to bark-foraging birds in winter (as well as spring) is probably due, in part, to its thick bark with deep furrows. Such trees may provide important places for insect larvae and pupae to overwinter (MacLellan 1959). Furthermore, Nicolai (1986) found that smooth-barked tree species in central Europe were dominated by a single arthropod species, whereas species with fissured bark had a higher density and diversity of arthropods, particularly spiders. Although we have no data on prey abundance during the winter season, Mariani and Manuwal (this volume) found that the relative abundance of bark-dwelling spiders and large, soft-bodied insects (several families) was highly correlated with bark furrow depth in Douglas-firs on our study sites during spring. Moreover, spiders were an important and consistent component in the diet of Brown Creepers (Mariani and Manuwal, this volume).

Similarly, Morrison et al. (1985) attributed increases in winter bird use of incense cedar (*Calocedrus decurrens*) in California to the presence of an abundant prey clearly associated with its bark characteristics (relative to other tree species). They also noted use of significantly larger Douglas-fir, and Red-breasted Nuthatches increased relative use, albeit slightly, of Douglas-fir (all sizes pooled), in winter.

Our results with regard to forest age class, though incomplete because of inadequate winter samples, further caution against pooling data across sites differing in physiognomy, even within the same forest type (see also Szaro et al., this volume). Although all bird species appeared to select similar dbh classes and tree species in old-growth and second-growth in relation to availability, the proportions used differed with changes in the proportions of trees in the different categories. Because birds exhibited some plasticity in resource use, conclusions regarding resource selection based on data from any particular forest age class, or from pooled data, may be misleading.

We did not take into account variability among individual stands, which can be quite marked (Manuwal, unpubl. data). Also, frequency distributions, or densities, of trees may not be the most appropriate measure of resource availability. Measures such as total canopy volume, basal area, or bark surface area (Jackson 1979; Mariani and Manuwal, this volume) may be more representative. Nevertheless, our data revealed not only seasonal changes in relative use of resources, but also differences among the species, and at least the potential for selection of different kinds of trees by foraging birds in winter and spring. Future investigators should consider such factors when designing studies or formulating management plans.

ACKNOWLEDGMENTS

We express our appreciation to M. Q. Affolter, B. Booth, J. B. Buchanan, C. B. Chappell, M. Emers, K. W. Hegstad, A. Hetherington, A. B. Humphrey, D. A. Leversee, J. M. Mariani, B. R. North, M. J. Reed, B. A. Schrader, and L. W. Willimont for assistance with the field work. R. N. Conner, R. W. Mannan, M. G. Raphael, and J. Verner provided valuable comments which greatly improved earlier drafts of the manuscript. We thank USDA Forest Service personnel at the Gifford Pinchot National Forest and at Mt. Rainier National Park for various assistance. Computer funds were made available through the Academic Computer Center at the University of Washington. We gratefully acknowledge the support of L. F. Ruggiero, A. B. Carey, and F. B. Samson at the USDA Forest Service, Forestry Sciences Laboratory, Olympia. This study was funded by USDA Forest Service contracts PNW-83-219, PNW-84-227, and PNW-86-244 and is Contribution No. 34 of the Old-Growth Wildlife Habitat Research Program.

YEARLY VARIATION IN RESOURCE-USE BEHAVIOR BY PONDEROSA PINE FOREST BIRDS

ROBERT C. SZARO, JEFFREY D. BRAWN, AND RUSSELL P. BALDA

Abstract. Foraging patterns of breeding birds in a ponderosa pine (*Pinus ponderosa*) forest of northern Arizona were studied from 1973 to 1975. Significant yearly differences occurred for many bird species in activity patterns, foraging mode, tree species selection, substrate use, foraging posture, horizontal tree positioning, and vertical tree positioning. Relationships determined over a single year or using data pooled across years can lead to misinterpretations about community organization, competitive interactions, and foraging ecology of single species, guilds, or entire communities.

Key Words: Foraging ecology; annual variation; ponderosa pine.

Foraging patterns and resource partitioning are popular areas of investigation by avian ecologists. Many studies emphasize differences in foraging technique (Airola and Barrett 1985), food selection (Kuban and Neill 1980), substrate or vegetation preferences (Holmes and Robinson 1981; Parker 1986b; Morrison et al. 1986, 1987b), vegetation structure (Maurer and Whitmore 1981, Robinson and Holmes 1984, Morrison and With 1987), search tactics (Robinson and Holmes 1982, Holmes and Recher 1986a), resource availability (E. P. Smith 1982), foraging efficiency (Pulliam 1985, Rogers 1985), foraging height (Szaro and Balda 1979, Alatalo 1981), and feeding posture or position (Alatalo 1982, Saether 1982) when investigating foraging relationships in bird communities. Yet, few studies examine annual variation in resource use and foraging (Grant and Grant 1980; Saether 1982; Ford et al., this volume; Hejl and Verner, this volume).

Most studies have not attempted to examine annual changes in foraging behavior even when substantial differences in bird density, species composition, and weather patterns could affect the availability of food items and/or territory selection and establishment (Grubb 1975, 1977, 1978; K. G. Smith 1982; Szaro and Balda 1986; Szaro 1986). This paper examines annual changes in foraging patterns and resource use by birds in ponderosa pine (*Pinus ponderosa*) forests during three breeding seasons, in Arizona.

STUDY AREAS AND METHODS

Four 15-ha plots within a 21-km radius in the Coconino National Forest, 43–63 km southeast of Flagstaff, Arizona were studied from 1973 to 1975. Ponderosa pine, gambel oak (*Quercus gambelii*), and alligator juniper (*Juniperus deppeana*) were the only tree species present, with ponderosa pine dominant on all sites. A wide spectrum of silvicultural treatments were represented by four study plots (Brown et al. 1974). Large trees (>25 cm diameter at breast height) and small dense thickets were selectively removed from the lightly cut area, resulting in a density of 263 trees/ha and a crown volume of 1.70×10^3 m^3/ha. The moderately cut area was thinned in a pattern of strips of trees 36 m wide alternating with cleared areas 18 m wide, resulting in a density of 181 trees/ha and a crown volume of 0.65×10^3 m^3/ha. The heavily cut plot was severely thinned to 69 trees/ha and a crown volume of 0.40×10^3 m^3/ha, with slash piled in windrows spaced at regular intervals. The uncut area had 646 trees/ha and a crown volume of 1.94×10^3 m^3/ha. Yearly precipitation, mostly in the form of snow, was 135 cm in 1973, 40 cm in 1974, and 64 cm in 1975. Annual mean temperature was 4.8°C in 1973, 6.7°C in 1974, and 4.9°C in 1975. For a more complete description of the study sites, see Szaro and Balda (1979).

The spot-mapping method (Robbins 1970) was used to estimate breeding bird densities. Eight visits were made annually to each plot between May and July from 1973 to 1975. Counts began within 15 min after sunrise and continued for 3 hours. Starting points differed for each count to minimize temporal bias. Sampling was done beyond plot boundaries where bird territories extended beyond the study area to provide a better estimation of territory size. All field data were collected by the senior author in all years.

Plots were systematically traversed and data were recorded for the first observation of each bird encountered. Data recorded at each sighting included bird species, activity pattern (singing, foraging, or resting), foraging mode (pick and glean, hover and hawk, tear and peck, or ground probing), substrate (trunk, branch, twig, foliage, ground, or air), posture (upright, hanging, head up, or head down), tree species (ponderosa pine, gambel oak, or alligator juniper), horizontal position in tree (outer foliage-twigs or inner trunks-branches), and height of bird. All bird species were recorded and observed but only those species in each foraging guild with the greatest number of observations were used in further analyses: pick and glean—Grace's Warbler (*Dendroica graciae*), Pygmy Nuthatch (*Sitta pygmaea*), Solitary Vireo (*Vireo solitarius*); hover and hawk—Western Bluebird (*Sialia mexicana*), Western Wood-Pewee (*Contopus sordidulus*); tear and peck—Northern Flicker (*Colaptes auratus*), White-breasted Nuthatch (*Sitta carolinensis*); and ground probing and walking—Dark-eyed Junco (*Junco hyemalis*) and Chipping Sparrow (*Spizella passerina*). Observations on the ground were excluded from analyses of horizontal tree position and tree species selection. We compared vertical tree

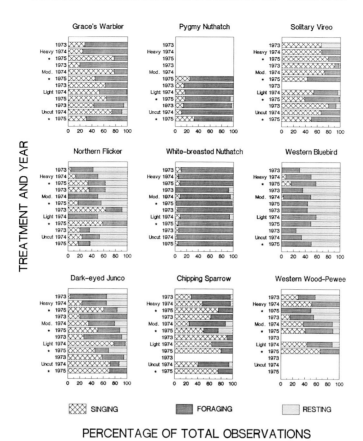

FIGURE 1. Activity patterns of nine ponderosa pine forest bird species by treatment and year. An asterisk under treatment indicates that the association among activity pattern and years was significant (G-test, P ≤ 0.05).

profiles for Grace's Warblers and Yellow-rumped Warblers (*Dendroica coronata*), because this was the only observed occurrence of one species potentially responding to the presence of another in the same guild. Yellow-rumped Warblers were not used in other analyses of resource-use behaviors; yearly comparisons were not possible because these warblers were not observed on any site for all 3 years.

Analysis of frequencies for each behavioral attribute was initially attempted with three-way tables (i.e., treatment by year by behavioral attribute) using log-linear models (Sokal and Rohlf 1981:747). The occurrence of significant interaction terms in all cases precluded fits to any simpler models. All further analyses were based on separate two-way tests of independence within each level of treatment (i.e., we examined the association between a behavioral attribute and years). We specifically asked the question, "Are activity, foraging method, substrate use, posture, tree species use, or position in the tree independent of year?" All significant yearly differences in proportions of a given behavior were determined with a goodness-of-fit test (G-test) at P ≤ 0.05 (Sokal and Rohlf 1981).

RESULTS

On an individual species basis, significant yearly differences in resource-use behaviors occurred for Pygmy Nuthatches in 92% (N = 12) of all cases, as contrasted with only 42% (N = 24) for Chipping Sparrows. The changes were basically conservative, however, and none would result in classifying a species in different guilds in different years. No marked yearly differences were noted between treatments, as 69%, 63%, 52%, and 63% of resource-use behaviors by all species on the heavily cut, moderately cut, lightly cut, and uncut plots, respectively, were significant.

Examination of yearly differences for each behavioral attribute (33 possible comparisons) revealed that activity patterns varied significantly among years for all species on at least one treatment (79% of all cases; Fig. 1). For most species, shifts were from singing-calling to foraging or

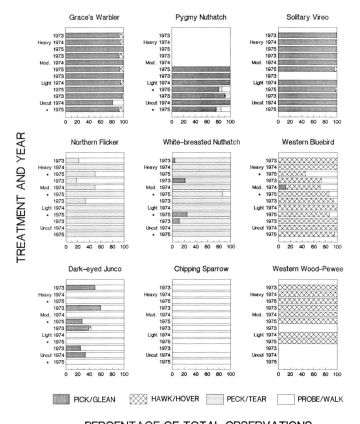

FIGURE 2. Foraging modes of nine ponderosa pine forest bird species by treatment and year. An asterisk under treatment indicates that the association among foraging mode and years was significant (*G*-test, P ≤ 0.05).

vice versa. But for aerial feeders such as the Western Bluebird and Western Wood-Pewee the shifts in activity pattern occurred between foraging and resting/preening. Significant yearly differences for all treatments were found only for Pygmy Nuthatches, Solitary Vireos, Dark-eyed Juncos, and Western Wood-Pewees.

Significant differences in foraging mode occurred on at least one treatment for six of the nine species (Grace's Warbler, Pygmy Nuthatch, Northern Flicker, White-breasted Nuthatch, Western Bluebird, and Dark-eyed Junco; Fig. 2). Overall, foraging mode varied significantly between years in 46% of the cases (N = 15). The Dark-eyed Junco was the only species whose foraging mode varied significantly between years on all plots. No changes were found for Solitary Vireos, Chipping Sparrows, and Western Wood-Pewees.

Substrate selection by individual species varied significantly on at least one plot among years for all species and in most cases (82%; Fig. 3). Substrate selection by four species—Northern Flicker, White-breasted Nuthatch, Western Bluebird, and Dark-eyed Junco—varied significantly between years on all plots. In contrast, substrate selection by the Solitary Vireo differed between years only on the uncut plot, where it was observed on branches 47% of the time in 1973, but only 16% and 21% of the time in 1974 and 1975, respectively.

Posture significantly differed between years in only 30% of all cases (Fig. 4). In fact, postures of only Pygmy Nuthatches, Northern Flickers, and White-breasted Nuthatches differed significantly between years on more than one plot. Pygmy Nuthatches significantly decreased their use of hanging, head up, and head down postures in 1975, compared to either 1973 or 1974. The shift in upright posture by White-breasted Nuthatches from 23% and 26% of all observations in 1973 and 1974 to 7% in 1975, to a greater proportion

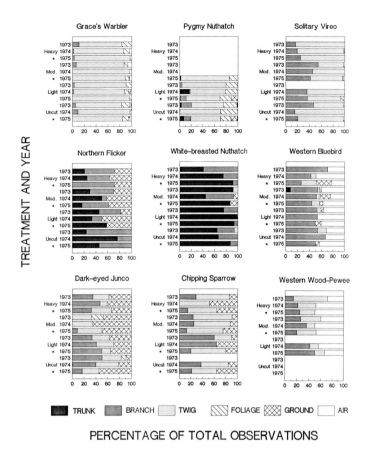

FIGURE 3. Substrate use of nine ponderosa pine forest bird species by treatment and year. An asterisk under treatment indicates that the association among substrate use and years was significant (G-test, $P \leq 0.05$).

of time in the head up or down positions, corresponded with an increase in the use of the trunk as a substrate.

Horizontal tree position varied significantly between years on at least two plots for all species (Fig. 5). Similar to the situation for substrate use, 82% of the comparisons for horizontal tree position varied significantly between years. For most species, the amount of time spent in the inner tree (i.e., inner trunk and branch area not associated with foliage) increased from 1973 to 1974. For example, the Solitary Vireo increased its use of the inner trunks and branches from 0–11% in 1973 to 14–52% in 1974.

Tree species selection varied significantly between years in 17 of 33 cases (51%; Fig. 6). However, no bird species' use of tree species was significantly different between years on all four plots. Solitary Vireos and White-breasted Nuthatches had most significant differences between years (3 of 4), whereas Pygmy Nuthatches, Western Bluebirds, and Chipping Sparrows had the fewest.

Significant annual differences ($P \leq 0.05$) in vertical tree positions were observed for all species (see Szaro [1976] and Szaro and Balda [1979] for foliage-use profiles by year and treatment). We found only one case, between Grace's and Yellow-rumped warblers, of an apparent interspecific interaction (Fig. 7). When Yellow-rumped Warblers were absent on the heavily cut, lightly cut, and control plots, Grace's Warblers were observed from 6 to 14, 6 to 14, and 0 to 18 m in 1973, and from 6 to 14 m on the heavily cut plot in 1974, and from 0 to 26 m on the untreated plot in 1975. In contrast, when Yellow-rumped Warblers were present and using foliage <18 m, Grace's Warblers shifted higher in the foliage, ranging from 0 to 34 and 0 to 30 m on the lightly cut plot in 1974 and 1975, and from 0 to 28 m on the untreated plot in 1974. The differences in

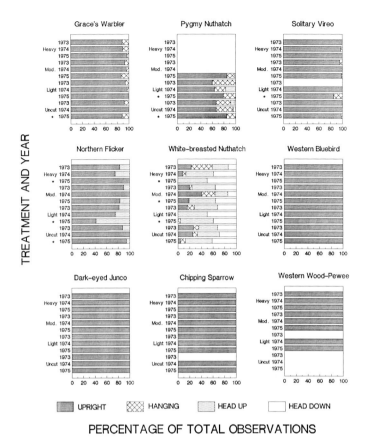

FIGURE 4. Postures of nine ponderosa pine forest bird species by treatment and year. An asterisk under treatment indicates that the association among postures and years was significant (*G*-test, P ≤ 0.05).

the amount of shifting higher in the trees might be related to differences in densities of the two species on the two plots. There were 15 and 9 breeding pairs/40 ha of Yellow-rumped Warblers on the lightly cut plot in 1974 and 1975, respectively, but only 3 breeding pairs/40 ha on the untreated plot in 1974 and none in 1975. At the same time, 18.7 and 19.5 breeding pairs/40 ha of Grace's Warblers were on the lightly cut plot in 1974 and 1975, respectively, and only 12 and 6 breeding pairs/40 ha in 1974 and 1975, respectively, were on the untreated plot.

DISCUSSION

Avian foraging studies frequently examine seasonal, intersexual, and overall variation in resource and habitat use, and foraging behavior. Our review of more than 150 papers that dealt with attributes of avian foraging behavior over the past 10 years found that only seven examined yearly differences. This seems peculiar, especially because so much has been written about the necessity of long-term studies for determining relationships between population densities and habitat (Wiens 1984, Raphael et al. 1987). The same reasoning should apply to annual variation in foraging behavior. Wiens et at. (1987b) made a strong case for multiple-year studies to determine the dynamics of variation in behavior.

Modifications in foraging patterns by bird species in response to changing environmental conditions and resource availability should be expected. Inter- and intraspecific foraging patterns between years are not static. Resource-use behavior tends to be plastic and varies considerably among years and study sites, particularly in our case, in which alterations in habitat physiognomy resulted in 73% of all behavioral attributes varying significantly with treatment (Szaro and Balda 1979). Moreover, in 62% of all cases (N = 198), the six behavioral attributes examined varied significantly between years.

Annual variation in resource-use behaviors is considerable, since birds tend to be opportunistic

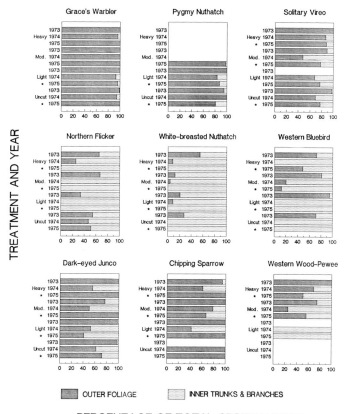

FIGURE 5. Horizontal tree position of nine ponderosa pine forest bird species by treatment and year. An asterisk under treatment indicates that the association among horizontal tree position and years was significant (G-test, $P \leq 0.05$).

foragers and often make ready use of superabundant food resources. For example, substantial changes were noted in the percent of total foraging on seeds and other foods by two species of Galapagos finches (*Geospiza fortis* and *G. scandens*) from foraging primarily on soft seeds in March 1976 to *Opuntia* buds, flowers, and nectar in March 1977 and to insects in March 1978 (Boag and Grant 1984). Ford et al. (this volume) observed significant annual differences in bark versus flower foraging by Fuscous Honeyeaters (*Lichenostomus fuscus*) and ground foraging versus leaf-gleaning by Buff-rumped Thornbills (*Acanthiza reguloides*). Robinson (1981) concluded that if he had studied Red-eyed Vireos (*Vireo olivaceous*) and Philadelphia Vireos (*V. philadelphicus*) in 1978 alone, he would have found no interaction between the two species. Yet, the frequency and intensity of aggressive encounters between them varied considerably between years and seasons as a function of changes in spatial overlap while foraging. In this study, the significant shifts along several niche dimensions for all species among years emphasize the need for both longer-term studies and the examination of annual variation in any foraging study. Moreover, the most biologically meaningful annual changes may be those in which the potential for differences in prey density and dispersion should have the most impact. More species showed differences in substrate use and horizontal tree position than in foraging mode.

Inter- and intraspecific shifts in resource-use patterns in response to changing densities are hard to document without experimentation to separate confounding factors such as changes in food availability and weather conditions. We were unable to discern any density related patterns in resource-use behaviors. Significant differences in resource-use behaviors were just as likely by Northern Flickers, which had almost equal densities on all study sites and years, as by White-

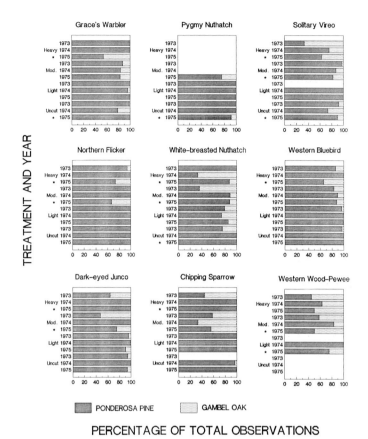

FIGURE 6. Tree species selection of nine ponderosa pine forest bird species by treatment and year. An asterisk under treatment indicates that the association among tree species selection and years was significant (G-test, P ≤ 0.05).

breasted Nuthatches, which had the most variable density patterns (Table 1). Although, Grace's Warblers apparently foraged higher on plots and during years when Yellow-rumped Warblers were present than on plots and during years when they were absent. At the Hubbard Brook Experimental Forest in New Hampshire, Philadelphia Vireos were observed foraging in the upper canopy and Red-eyed Vireos in the lower canopy, but Robinson (1981) was unable to determine if the Philadelphia Vireos were displaced by the Red-eyed Vireos or merely preadapted to foraging conditions of the upper strata. Relative population densities of both species, the stage of the nesting cycle, and food abundance may affect whether these two vireos segregate vertically, horizontally, or at all. Moreover, Wiens (1977) suggested that little parallelism may exist between increasing niche overlap and intensified competition, and that overlap may be greatest when competition is least, particularly when resources are superabundant and ecological overlap carries no selective penalties.

Brawn et al. (1987) found the variation in bird species densities in picking-gleaning and aerial feeding guilds in ponderosa pine forests was independent of changes in densities of other species within the guild over an 8-yr period, on the same heavily cut and lightly cut plots used in this study. They were unable to demostrate any indication of competitive interactions even with controlled manipulation experiments to increase bird densities. This implies that competition for food is unimportant to the structure of breeding bird communities in ponderosa pine forests of north-central Arizona.

Year-to-year changes in foraging may be either simply the consequence of incomplete sampling of the population or may result from the idiosyncracies of the birds that happen to end up in

TABLE 1. Density, Number of Resource-use Observations, Number of Possible Comparisons (Based on Six Resource-use Behaviors), and Number of Significant Comparisons between Resource-use Behaviors and Years

Species Study site Year	Density (pairs/40 ha)	N	Possible annual comparisons	Significant annual comparisons[a]
Grace's Warbler			24	12 (50.0%)
Heavily cut				
1973	3.8	36		
1974	6.0	50		
1975	7.5	97		
Lightly cut				
1973	7.5	47		
1974	18.7	72		
1975	9.8	84		
Moderately cut				
1973	11.2	92		
1974	18.7	111		
1975	19.5	82		
Uncut				
1973	7.5	143		
1974	12.0	96		
1975	6.0	92		
Pygmy Nuthatch			12	11 (91.7%)
Moderately cut				
1973	7.5	50		
1974	15.0	72		
1975	18.0	72		
Uncut				
1973	13.5	122		
1974	15.0	156		
1975	13.5	120		
Solitary Vireo			24	12 (50.0%)
Heavily cut				
1973	3.8	48		
1974	6.0	143		
1975	6.0	72		
Lightly cut				
1973	6.0	44		
1974	12.0	132		
1975	6.0	84		
Moderately cut				
1973				
1974	6.0	89		
1975	6.0	60		
Uncut				
1973	1.5	127		
1974	3.0	73		
1975	3.0	80		
Northern Flicker			24	18 (75.0%)
Heavily cut				
1973	3.0	48		
1974	3.0	32		
1975	3.0	48		

TABLE 1. CONTINUED

Species Study site Year	Density (pairs/40 ha)	N	Possible annual comparisons	Significant annual comparisons[a]
Lightly cut				
1973	2.3	40		
1974	3.8	36		
1975	3.0	48		
Moderately cut				
1973	3.0	33		
1974	3.0	36		
1975	3.0	36		
Uncut				
1973	3.0	31		
1974	3.0	34		
1975	3.0	35		
White-breasted Nuthatch			24	19 (79.2%)
Heavily cut				
1973	5.2	22		
1974	9.0	36		
1975	6.0	24		
Lightly cut				
1973	4.5	25		
1974	9.0	118		
1975	12.0	53		
Moderately cut				
1973	3.0	25		
1974	7.5	46		
1975	15.0	48		
Uncut				
1973	3.0	39		
1974	10.5	37		
1975	3.0	37		
Western Bluebird			24	13 (54.2%)
Heavily cut				
1973	6.0	30		
1974	8.3	85		
1975	3.0	57		
Lightly cut				
1973	6.7	62		
1974	12.0	96		
1975	15.0	48		
Moderately cut				
1973	5.2	88		
1974	8.3	36		
1975	7.5	59		
Uncut				
1973	4.5	41		
1974	6.0	48		
1975	3.0	48		
Dark-eyed Junco			24	18 (75.0%)
Heavily cut				
1973	9.8	26		
1974	6.7	60		
1975	6.0	53		

TABLE 1. CONTINUED

Species Study site Year	Density (pairs/40 ha)	N	Possible annual comparisons	Significant annual comparisons[a]
Lightly cut				
1973	6.0	25		
1974	10.5	52		
1975	12.0	48		
Moderately cut				
1973	12.7	78		
1974	22.5	130		
1975	15.0	48		
Uncut				
1973	9.0	97		
1974	18.0	60		
1975	12.0	102		
Chipping Sparrow			24	10 (41.7%)
Heavily cut				
1973	6.0	21		
1974	6.0	23		
1975	3.0	24		
Lightly cut				
1973	4.5	21		
1974	12.0	48		
1975	6.0	24		
Moderately cut				
1973	3.0	20		
1974	7.5	36		
1975	4.5	30		
Uncut				
1973				
1974	1.5	32		
1975	3.0	24		
Western Wood-Pewee			18	9 (50.0%)
Heavily cut				
1973	3.0	28		
1974	3.0	37		
1975	3.0	24		
Lightly cut				
1973	8.2	80		
1974	9.0	124		
1975	9.0	104		
Moderately cut				
1973	2.3	0		
1974	3.0	48		
1975	1.5	24		
Total			198	122 (61.6%)

[a] Goodness-of-fit test (G-statistic) at $P \leq 0.05$.

the areas studied. While observed differences could be due to sampling errors or biases resulting from small sample sizes for some species in this study, the substanial proportion of resource-use behaviors (62%) that varied significantly between years strongly indicates the potential for significant annual variation in resource-use behavior. Thus, studies attempting to determine the presence or absence of competition between bird species by examining shifts in

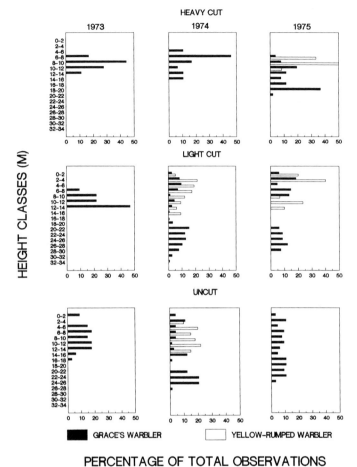

FIGURE 7. Vertical tree profiles of Yellow-rumped and Grace's warblers on the heavily cut, lightly cut, and uncut plots for 1973 to 1975 in a ponderosa pine forest of northern Arizona.

niche dimensions should be done over several years and seasons, and obtain information on resource conditions, to adequately determine patterns in species overlap. Relationships determined over a single year or pooling data across years are insufficient. The ideal study would be based on marked individuals followed from year to year. Studies should be specifically designed to identify the proximate mechanisms that "drive" this variation in foraging behavior. Some possibilities are resource availability, weather, predation, and phenology.

ACKNOWLEDGMENTS

Special thanks go to Richard L. Hutto, Fritz L. Knopf, Brian A. Maurer, Martin G. Raphael, Jared Verner, and Kimberly A. With for their helpful reviews of this manuscript.

VARIATION IN THE FORAGING BEHAVIORS OF TWO FLYCATCHERS: ASSOCIATIONS WITH STAGE OF THE BREEDING CYCLE

HOWARD F. SAKAI AND BARRY R. NOON

Abstract. The foraging characteristics of Hammond's and Western flycatchers in northwestern California varied with different stages of the breeding cycle during the breeding seasons (early April–mid August) in 1984 and 1985. The species' behaviors did not always vary in parallel nor were all foraging behaviors distributed equally during the breeding cycle. For example, the direction of aerial foraging movements for both species did not differ between stages. In contrast, the predominant type of foraging activity (either hover-glean or flycatch) differed by stage of the breeding cycle for Western Flycatchers but not for Hammond's Flycatchers. Both birds differed in their use of foraging substrates and plant species among breeding stages. Western Flycatchers did not differ in position (height of foraging bird or distance to the canopy edge) among stages of the breeding cycle, but Hammond's Flycatchers did. Both species foraged in trees with different structural characteristics (diameter-at-breast height, tree height, and bole height) during different stages of the breeding cycle. For both species, differences in foraging patterns within specific stages of the breeding cycle were apparent when compared with data pooled across the breeding stages. Failure to partition the data by stage of the breeding cycle may mask significant sources of variation and preclude important insights into a species' breeding biology.

Key Words: Hammond's Flycatcher; Western Flycatcher; breeding cycle; foraging behavior; northwestern California.

Most studies of avian foraging behavior have estimated foraging patterns by pooling observations within a season even though a species' foraging behaviors may change seasonally. Pooling data may thus mask significant variation, as noted by several authors (Busby and Sealy 1979, Sherry 1979).

Our study of Hammond's (*Empidonax hammondii*) and Western (*E. difficilis*) flycatchers allowed us to test whether tree species selection, forage substrate characteristics, and the overall distribution of foraging behaviors were associated with specific stages of the breeding cycle. Because both Western and Hammond's flycatchers are sexually monomorphic, we were unable to test for intersexual effects which may also provide a significant source of variation. Our objectives are to: (1) test the hypothesis of no difference in the distribution of foraging behaviors between stages of the breeding cycle separately by species; (2) compare our estimates of foraging patterns based on specific stages of breeding cycle with data pooled across the breeding cycle; (3) discuss the insights that arise from information on the within-season variation in foraging pattern; and (4) compare our results with other studies that have ignored sources of variation associated with stage of the breeding cycle.

METHODS

STUDY STANDS

Nine stands, selected to provide three replicates of each combination of three forest development stages (young, mature, and old-growth), were located in Humboldt and Trinity counties of northwestern California (refer to Sakai 1987 for specific details). A young stand was defined as 30–90 years, mature 91–199 years, and old-growth >200 years. Stand age was determined from increment cores of 4–6 dominant Douglas-firs (*Pseudotsuga menziesii*) or by counting annual rings of Douglas-fir stumps found in adjacent clearcuts. The stands were dominated by Douglas-fir and tanoak (*Lithocarpus densiflora*). Pacific madrone (*Arbutus menziesii*) and canyon liveoak (*Quercus chrysophylla*) were the associated hardwoods and incense-cedar (*Calocedrus decurrens*), sugar pine (*Pinus lambertiana*), and whitefir (*Abies concolor*) the associated softwoods.

Study plots ranged in size from 12 to 20 ha and in elevation from 710 to 1235 m. The 12 ha stand contained one transect. The 20 ha plots were rectangular and contained two transects. Located along each transect were six evenly spaced bird census sampling points. These points, located 150 m apart, defined the center of circular plots, which subsequently became the focus for the vegatation and foraging sampling.

FORAGING SAMPLES

To compare variation in foraging behaviors associated with each stage of the breeding cycle across the entire range of habitats occupied by the species in the forests of northwestern California, we pooled data for each species across all study plots. Sakai (1987) discussed, in detail, the association between variation in stand age and vegetation with variation in species' foraging behaviors. In general, he found that variation in foraging behaviors paralleled changes in vegetation structure and floristics associated with stands of varying ages.

Data were collected during the breeding seasons (early April–mid August) in 1984 and 1985. Four observers (HFS plus three others) and two observers (HFS plus one other) were involved in data collection in 1984

TABLE 1. FREQUENCY OF FORAGING BEHAVIORS OBSERVED FOR HAMMOND'S AND WESTERN FLYCATCHERS BY STAGE OF THE BREEDING CYCLE FOR THE THREE STAND AGE GROUPS, NORTHWESTERN CALIFORNIA

Stage of the breeding cycle	Stand age		
	Young	Mature	Old-growth
Western Flycatcher			
Pre-incubation	27	47	71
Incubation	54	88	116
Brooding	61	99	131
Post-brooding	54	118	119
Hammond's Flycatcher			
Pre-incubation	a	23	19
Incubation		45	96
Brooding		43	97
Post-brooding		50	57

[a] Did not occur.

and 1985, respectively. In this analysis we pooled data across observers and years as well as study stands. We acknowledge that these factors may contribute additional variation. However, partitioning our data by these additional factors would have greatly reduced the power of our analyses and, for the log-linear analyses, produced more cells than data points. We believe that pooling our data across years was justified because the environmental conditions both years were very similar. This is exemplified by almost identical arrival times for the birds and consistent timing of the breeding stages (Sakai 1988). Pooling across observers was justified on the basis of rigorous training as well as frequent monitoring of observers throughout the period of data collection by the senior author.

Study stands were sampled equally, in terms of visits to each stand, along the bird census transects out to 30 m on either side, in an attempt to obtain 35 foraging birds/flycatcher species/stand/sampling period. Sampling periods for both species were divided into pre-incubation (10 April to 15 May), incubation (16 May to 15 June), brooding (16 June to 15 July), and post-brooding (16 July to 15 August). Despite some individual differences in the timing of the nesting cycle, the populations' nesting behaviors were highly synchronous (Sakai 1988). The dates bounding the periods were chosen such that the majority of the nests were at the same stage of the breeding cycle. Given the degree of synchrony, we feel justified in partitioning the foraging observations by the stage of the species' breeding cycles. Sample sizes for each species by stage of the breeding cycle and stand age are given in Table 1. Hammond's Flycatchers were not found in the younger stands.

Foraging behaviors were recorded from sunrise to late afternoon. The behavior of each flycatcher was recorded from its initial contact for 10–100 s. Once a foraging bird was located, information was taken on its behavior, position in the habitat, and characteristics of the forage substrate (see Table 2). When a bird foraged at more than one location within 100 s, we analyzed only the initial observation. Usually only one observation per individual per day was obtained, but sometimes two were taken on the same individual after 10 min had elapsed. Estimates of specific foraging variables as well as distance and direction of aerial flight movements (Table 2) were collected at those points where a prey was captured. Samples used in individual analyses varied because some data were collected on non-foraging birds (Table 3).

STATISTICAL ANALYSES

The foraging observations of each species could be classified by: stage of the breeding cycle (pre-incubation, incubation, brooding, post-brooding), behavior (flycatch, hover-glean, glean), aerial flight movement (up, down, horizontal), tree species (Douglas-fir, tanoak, Pacific madrone, and other broad-leaf deciduous trees), and substrate (leaf, twig and small branch, medium and large branch, and trunk). The result is a $4 \times 3 \times 3 \times 4 \times 4$ contingency table with 586 cells. Because (1) this number of cells exceeded our sample size, (2) the expected values within a cell should be >1, and (3) no more than 20% of the cells should have expected values <5 (Cochran 1954), the size of our contingency table had to be reduced to 3-way tables of breeding cycle by tree species by substrate. We used log-linear analyses to examine the interactions among these variables (Bishop et al. 1975). We viewed breeding stage as an explanatory variable and tree species and substrate as response variables. The simplest models that fit the observed data and chi-square test statistics were estimated by algorithms in BMDP program 4F (Dixon et al. 1985).

Tests of the null hypothesis between stage of the breeding cycle and the variables behavior and aerial flight movement were tested by 2-way contingency tables (Sokal and Rohlf 1981:731). By conducting these tests separately from the log-linear analyses we were unable to test for significant interactions between these variables and plant species and substrate. Chi-square values were considered significant at $P < 0.05$. Graphic starplots (Gower and Digby 1981) were used for visual comparisons, by stage of the breeding cycle, of the direction and distance flown by foraging birds that successfully captured prey.

The structural characterisitics of the tree in which the bird was foraging and the bird's position (Table 2) were analyzed separately using MANOVA computed using BMDP program 7M (Dixon et al. 1985), with stage of the breeding cycle as the grouping variable. Each MANOVA tested the null hypothesis of equality of the breeding stage centroids (i.e., multivariate means). The relative contributions of the original variables to separation of the stages were based on the magnitude of structure coefficients, which are simple bivariate correlations between the original variables and the canonical variates. Along a canonical variate axis or in a ≥ 2-dimensional canonical space, the origin represents the multivariate mean (centroid) of the pooled sample. To determine whether the sample partitioned by stage of the breeding cycle differs from the pooled sample, one simply needs to determine whether the 95% confidence ellipses about stage centroids overlap a similar ellipse surrounding the origin. If a significant MANOVA resulted, all possible pairwise combinations of stage specific centroids were tested for equality. These a posteriori comparisons were adjusted to maintain an overall experimentwise error rate of $P \leq 0.05$.

TABLE 2. FORAGING VARIABLES RECORDED FOR HAMMOND'S AND WESTERN FLYCATCHERS IN NORTHWESTERN CALIFORNIA DURING THE BREEDING SEASONS IN 1984 AND 1985

Variable	Explanation
Tree species	Tree species in which bird was foraging
Height of foraging bird	Estimate to nearest 1 m. Clinometer used to check estimates
Bird location on forage branch	Estimate to the nearest $\frac{1}{10}$ m of birds' location from the canopy edge
Diameter-at-breast height (dbh) of foraged tree	Measured diameter in cm at 1.1 m height from tree base
Tree height	Estimate to nearest 1 m from ground. Clinometer used to check estimates
Bole height	Estimate to nearest 1 m from ground of first live branch
Types of foraged substrates	Items to which birds direct attention: twigs, <1 cm diameter; small branches, 1–5 cm diameter; medium branches, 5–15 cm diameter; large branches, >15 cm diameter; trunks
Distance to prey	Estimate to $\frac{1}{10}$ m from perched bird to prey capture
Foraging behavior	Behaviors such as: flycatch (pursuit of aerial prey); hover-glean (removal of stationary prey while in flight); glean (removal of prey from substrate while perched)
Aerial flight movements	Direction of initial flight from perch (down, up, and horizontal)

RESULTS

PREY CAPTURE AND FORAGING ACTIVITY

The direction of aerial flight movements made in pursuit of prey differed between stages of the breeding cycle in Hammond's ($\chi^2 = 15.3$, df = 6, P = 0.018; Fig. 1a) and Western flycatchers ($\chi^2 = 16.1$, df = 6, P = 0.013). Aerial attack movements of the two species within each stage of the breeding season suggested that both species had almost identical distributions (Fig. 1a). In addition, a comparison of starplots suggested that horizontal attack flights by both species were favored during the pre-incubation and incubation periods, but both birds used vertical attack flights more frequently later in the breeding cycle. There was also an inverse relationship between the proportion of attacks or aerial flight movements in a particular direction and the distance traveled in that same direction to obtain prey (Fig. 1b). We found a significant correlation for Western Flycatchers (r = −0.61, df = 10, P = 0.035), and a marginally significant correlation for Hammond's Flycatchers (r = −0.55, df = 10, P = 0.064). Collectively, the foraging movements of both species suggest that the shortest distance to prey was generally the favored aerial flight direction in all breeding stages (Fig. 1).

Both species gleaned insects from leaves and woody substrates, but too rarely (<2% of the observations) to be included in the contingency analysis. Use of a particular foraging maneuver (either hover-glean or flycatch) by Western Flycatchers differed by stage of the breeding cycle (Table 4). Western Flycatchers hover-gleaned more than expected during the pre-incubation and incubation periods, but flycatched appreciably more than expected during periods with young in nests ($\chi^2 = 19.9$, df = 3, P < 0.01). Hammond's Flycatchers did not differ in use of hover-glean and flycatch activities between the different stages of the breeding cycle (Table 4).

VARIATION IN PLANT SPECIES AND SUBSTRATE USE

For both the Western Flycatcher and the Hammond's Flycatcher, the only log-linear model that

TABLE 3. RANGE OF SAMPLE SIZES USED IN ANALYSES OF WESTERN AND HAMMOND'S FLYCATCHERS FORAGING BEHAVIOR BY STAGE OF THE BREEDING CYCLE, NORTHWESTERN CALIFORNIA

Species	Stage of breeding cycle			
	Pre-incubation	Incubation	Brooding	Post-brooding
Western Flycatcher	120–140	226–255	219–284	228–281
Hammond's Flycatcher	32–40	110–133	95–133	78–100

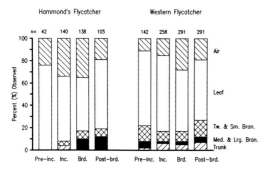

FIGURE 2. Percent of substrates used by foraging Western and Hammond's flycatchers for each of the four stages of the breeding cycle in northwestern California. Breeding cycle codes are: Pre-inc. = pre-incubation period, Inc. = incubation period, Brd. = brooding period, and Post-brd. = post-brooding period.

Hammond's Flycatcher forage tree characteristics by breeding stage.

COMPARISONS WITH POOLED BREEDING CYCLE DATA

Differences in foraging pattern for each species were apparent when patterns based on specific stages of the breeding cycle were compared with data pooled across the breeding stages. For the categorical data these comparisons are indirect. From the log-linear analyses we found that the variables tree species and substrate were significantly associated with stage of the breeding cycle. Also, removal of the variable categorizing breeding stage cycle caused a significant lack of fit of observed to expected values for both species. In addition, the two-way contingency analyses detected significant associations between stage of the breeding cycle and the distribution of other aspects of foraging behavior for almost all analyses.

For the continuous variables, comparisons with the pooled sample can be illustrated graphically. The mean foraging position within trees for Hammond's Flycatchers differed significantly among pre-incubation, incubation, and post-brooding periods. All stages, except for the brooding period, differed significantly from the

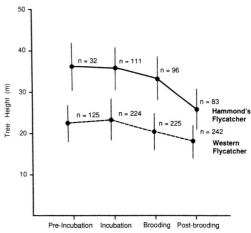

FIGURE 4. Mean tree height used by foraging Western and Hammond's flycatchers for each of the four stages of the breeding cycle in northwestern California. Ninety-five percent confidence intervals are shown.

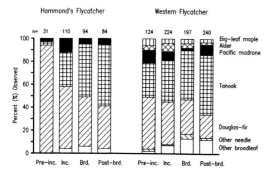

FIGURE 3. Percent of plants used by foraging Western and Hammond's flycatchers for each of the four stages of the breeding cycle in northwestern California. Breeding cycle codes are: Pre-inc. = pre-incubation period, Inc. = incubation period, Brd. = brooding period, and Post-brd. = post-brooding period.

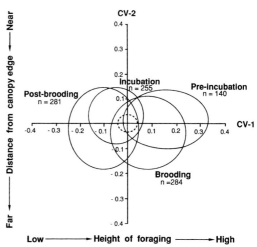

FIGURE 5. Mean canonical variate scores characterizing Western Flycatchers' position in the forage trees for each of the four stages of the breeding cycle in northwestern California. The canonical variate represents variation in the height of the foraging bird. Ninety-five percent confidence intervals are shown.

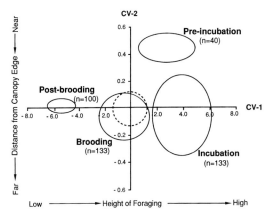

FIGURE 6. Mean canonical variate scores characterizing Hammond's Flycatchers' position in the forage trees for each of the four stages of the breeding cycle in northwestern California. CV-1 represents variation in the height of the foraging bird, CV-2 variation in distance from the canopy edge. Ninety-five percent confidence ellipses are shown.

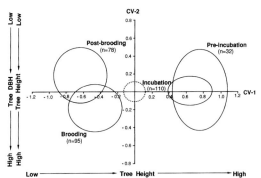

FIGURE 8. Mean canonical variate scores characterizing the structural characteristics of the forage trees used by Hammond's Flycatchers during four stages of the breeding cycle in northwestern California. CV-1 represents variation in tree height, CV-2 variation in tree dbh and tree height. Ninety-five percent confidence ellipses are shown.

centroid of the pooled sample (Fig. 6). In contrast, the mean foraging position of Western Flycatchers within trees did not differ among stages of the breeding cycle, nor did these means differ from data pooled across the breeding stages (confidence intervals around canonical variate scores all overlapped with the confidence ellipse around the origin, Fig. 5).

The structural characteristics of trees used by Hammond's Flycatchers differed significantly from early to late stages and all stages differed significantly from the pooled sample centroid (Fig. 8). Structural features of trees used by Western

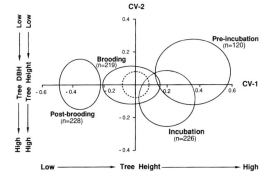

FIGURE 7. Mean canonical variate scores characterizing the structural characteristics of the forage trees used by Western Flycatchers during four stages of the breeding cycle in northwestern California. CV-1 represents variation in tree height, CV-2 variation in tree dbh and tree height. Ninety-five percent confidence ellipses are shown.

Flycatchers were significantly distinct in the post-brooding stage. In addition, the positions of the pre-incubation, incubation, and post-brooding periods differed significantly from the pooled sample centroid (Fig. 7).

DISCUSSION

EFFECTS OF POOLING DATA ACROSS THE BREEDING CYCLE

Because foraging behaviors vary significantly between stages of the breeding cycle, pooling data across the breeding stages may mask significant variation in foraging behavior. Intraspecific variation in foraging behaviors between sexes or between seasons is well known. And our research has shown that variation associated with stage of the breeding cycle may also be pronounced (see also Brennan and Morrison, this volume). As a result, partitioning of a species' foraging niche by sex or season is essential to increase our understanding of its life history.

Pooling data may be justified for some variables. For example, Hammond's and Western flycatchers direction of flight movements by stage of the life cycle did not vary significantly between breeding stages; therefore, pooling the data would not have changed our inferences. In contrast, comparison of bird position (height in tree and distance to the canopy edge) showed no difference between pooled data and stages of the life cycle for Western Flycatchers but did for Hammond's Flycatchers. Further, both species showed evidence of significant changes in the use of tree structural characteristics when the data were compared by stage of the life cycle. We conclude that whenever sample size is adequate, analysis

HETEROGENEITY OF FORAGING BEHAVIORS WITHIN THE BREEDING CYCLE

For both Western and Hammond's flycatchers, direction of foraging movements while pursuing prey was not related to stage of the breeding cycle. However, for both species the shortest distance to prey was generally the favored flight movement direction in all breeding stages. Because nearby prey are easier to detect and require less energy to capture, on average, this is not surprising and explains the inverse relationship between attack frequency and attack distance.

Note that this consistent relationship between aerial flight movement and distance occurred in the context of an otherwise variable foraging repertoire, with both species changing aspects of their distribution of foraging behaviors (positions within the forage trees, frequency of use of different tree species, tendency to forage lower) as the breeding cycle progressed. We speculate that these changes were due to changes in prey availability, as reflected in the inverse relationship of vector movement and distance to prey.

Western and Hammond's flycatchers differed intraspecifically in foraging activity and substrate use throughout their breeding cycles, but both species essentially used the same substrates during the same stages of the breeding cycle. Overall, the variation in forage activity, substrate use, and vertical distribution by the flycatchers suggests differences in their food resources throughout the breeding cycle. Both species hover-gleaned off leaves more often during the early breeding stages and switched later to flycatching insects from the air or gleaning off woody substrates.

Hammond's Flycatchers consistently selected taller trees and foraged higher in the canopy and subcanopy than Western Flycatchers. However, both species were similar in that they used Douglas-fir more in the early stages of the breeding cycle and tanoak and Pacific madrone in the later stages. Douglas-fir, tanoak, and Pacific madrone, the most common tree species in the study areas, had a high insect density (Sakai 1987). Even assuming a strong relationship between plant species and their associated arthropods, we can not determine if shifts in utilized plant species were caused by within-season shifts in prey availability or by necessary dietary changes.

Changes in use of tree species for foraging in our study area could also be associated with differences in the tree structural characteristics as Robinson and Holmes (1982, 1984) found in New Hampshire, or the amount of air-space available for flycatching (Sakai 1987), since these factors ultimately influence the foraging opportunities and the bird's position in the vegetation. The primary causes for the stage-specific changes in foraging cannot be determined from our data set. However, the simplest explanation is that the observed differences occurred as a result of within-season changes in prey availability coupled with a need to maintain high foraging efficiency.

ACKNOWLEDGMENTS

We thank Scott Edwards, Holly Hutcheson, Michael Schroeder, and John Sterling for assistance in the field. Special thanks to Sallie Hejl, Martin Raphael, Michael Morrison, and Nancy Tilghman for their constructive comments on the manuscript.

SEASONAL CHANGES IN FORAGING BEHAVIOR OF THREE PASSERINES IN AUSTRALIAN EUCALYPTUS WOODLAND

HUGH A. FORD, LEONIE HUDDY, AND HARRY BELL

Abstract. The foraging behavior of the Fuscous Honeyeater (*Lichenostomus fuscus*), Scarlet Robin (*Petroica multicolor*), and Buff-rumped Thornbill (*Acanthiza reguloides*) was compared in different seasons in evergreen eucalypt woodland. Fuscous Honeyeaters mostly gleaned from leaves (52–85% of observations), an activity that tended to increase slightly in autumn and winter. Flowers of eucalypts and mistletoes were visited for nectar when available. Bark and aerial foraging did not change consistently between seasons, but all categories changed between years. Scarlet Robins foraged by hawking, snatching and pouncing, and changed from leaf, bark, and aerial foraging to mostly ground foraging in winter. This change reflected the abundance of arthropods on the different substrates and perhaps the influence of temperature on the activity of arthropods and birds. Buff-rumped Thornbills also fed on the ground most in winter, moving onto foliage and bark in spring and summer. They also occurred in larger groups in winter. Seasonal changes in foraging may also be influenced by vulnerability to predation.

Key Words: Foraging; seasonal changes; eucalypt woodland; Fuscous Honeyeater; *Lichenostomus fuscus*; Scarlet Robin; *Petroica multicolor*; Buff-rumped Thornbill; *Acanthiza reguloides*.

There have been numerous studies of the foraging behavior of birds, principally during the breeding season. Several studies have compared behavior in different seasons, in particular foraging of birds in winter with that during the breeding season (e.g., Ulfstrand 1976, Morrison and With 1987). Most of these studies have been in seasonal habitats in the Northern Hemisphere. Birds in deciduous woodland frequently change their behavior through the year, because a major foraging substrate, foliage, is absent or scarce for part of the year (Hartley 1953, Gibb 1954, Willson 1970). Birds in coniferous forests also may show marked seasonal changes (Ulfstrand 1976, Alatalo 1980), although foliage is not shed in winter, as days are short, substrates may be snow-covered, and food is scarce. In addition, many potentially competing species have left the area. Migratory species themselves may alter their foraging behavior between breeding, migratory and winter areas (Greenberg 1987b; Martin and Karr, this volume). In addition, birds may show marked changes in foraging behavior among years (Szaro et al., this volume).

Fewer studies have been conducted at low latitudes, where changes in the weather and bird community may be less drastic than in cool temperate regions. A notable exception is the series of studies on Darwin's finches (*Geospiza* spp.) on the Galapagos Islands (Smith et al. 1978, summarized in Grant 1986). Here, a dry season, severe in some years, led to food shortage and a divergence in the diet of congeners.

Our objective is to contrast seasonal changes in the foraging behavior of three species of passerines in eucalypt woodland in Australia. We also examine changes in foraging between years in two of the species. This habitat is evergreen and not strongly seasonal in climate or food abundance (Bell 1985a); therefore, we might not expect to find substantial seasonal changes in foraging. Alternatively, such seasonal changes as do occur may still affect the foraging behavior of birds, or it may be influenced by other factors such as predation and competition.

The Fuscous Honeyeater (*Lichenostomus fuscus,* Meliphagidae) is a dull-colored, small (18 g), insectivore-nectarivore. It is active, loosely social, and aggressive. The Scarlet Robin (*Petroica multicolor,* Eopsaltriidae) is sexually dimorphic with a brilliant red, black, and white male and pink, brown, and gray female. It weighs 13 grams and resembles in appearance and behavior the small Palearctic muscicapines. The Buff-rumped Thornbill (*Acanthiza reguloides,* Acanthizidae) is a tiny (7 g) brown, actively gleaning Australian warbler. All three species, as well as most other common species, are sedentary.

STUDY SITES AND METHODS

Fuscous Honeyeaters were studied at Eastwood State Forest, 12 km southeast of Armidale (30°30′S, 151°30′E), whereas the other two species were studied at Wollomombi Falls Reserve, 40 km east of Armidale. Both sites were at about 1000 m elevation in the Northern Tablelands of New South Wales and have already been described in detail (Noske 1979, Ford et al. 1985, Bell 1985a, Bell and Ford 1986). The habitat is eucalypt woodland with trees fairly well spaced and a deep canopy up to about 20 m. Virtually all trees belong to the genus *Eucalyptus* with four common species at each site (*E. caliginosa* and *E. melliodora* at both sites,

E. laevopinea and *E. conica* at Wollomombi, and *E. viminalis* and *E. blakelyi* at Eastwood). These eucalypts differ principally in their type of bark. Eucalypts have sclerophyllous, evergreen leaves that tend to hang down and show less variation between species than do those of trees in northern deciduous forests. A tall shrub layer of *Acacia* was present along with smaller shrubs of *Olearia, Exocarpos, Cassinia,* and *Jacksonia*. These were more common at Wollomombi. Armidale has a cool winter (mean minimum 1°C, maximum 12°C in July) and warm summer (mean minimum 12°C, maximum 26°C in January) with an annual rainfall of 790 mm peaking in summer. The period from early 1980 to middle 1982 was very dry with some defoliation and death of trees and shrubs and a marked scarcity of insects (Bell 1985a).

INVERTEBRATE SAMPLING

Arthropods were sampled from foliage and the ground at Wollomombi from September 1978 and March 1979, respectively, to August 1981 (details in Bell 1985a). Samples were taken at 09:00, mid-monthly on calm, sunny days, from six plant genera (*Eucalyptus, Acacia, Olearia, Jacksonia, Exocarpos* and *Cassinia*). From 1–2 kg of leafy branches were collected from at least 20 plants of each genus. Where there was more than one species per genus, each was sampled in proportion to its relative abundance. Foliage was placed into plastic bags before removal and sprayed with Baygon household insecticide. It was then vigorously beaten to dislodge invertebrates, which were later identified, mostly to order, and measured. Measurements were converted to dry weight using the formulae of Zug and Zug (1979).

One hundred sweep-net samples were made each month at 11:00 over the ground vegetation. Arthropods were killed, identified, counted, and measured as for the foliage samples.

Also at Wollomombi, litter samples were taken from February to July 1979 (Huddy 1979). Six 30 cm × 30 cm samples of plant and loose material were collected each month; arthropods were sorted from it and identified. In addition, 120 sweeps each covering 5 m were made just above ground level each month. Arthropods were killed and later identified. Arthropods on bark were estimated by counting them on 50 cm × 50 cm grids on 32 trees each month (method described in Noske 1985).

AVIAN FORAGING

As the studies were independent, different methods were used for each species, and several methods were used for the Fuscous Honeyeater. Fuscous Honeyeaters were studied from January 1981 to December 1982 and from February 1984 to March 1986. In the first period a single foraging observation was recorded for each individual bird at each encounter. The foraging method, substrate, plant species, and height were recorded for the first attempt at prey capture after the bird was seen (details in Ford et al. 1986). In this paper only six categories have been analyzed: aerial hawking, bark foraging, ground foraging, probing flowers, gleaning leaves, and hovering at leaves. For the second period individual birds were followed for at least 20 s and up to 10 min, during which each activity was timed and accumulated into different categories. Major activities were perched (inactive but including preening and incubating), flying, and foraging. Foraging behavior was subdivided into six categories as for the first period of study. Discrete records were taken whenever an individual changed method or substrate or moved to a new tree. The comparisons between the two periods included methods as well as years and in the second period the methods of time-budgeting and discrete sampling were compared.

Scarlet Robins were studied from February to July 1979. Foraging was recorded in the same way as for Fuscous Honeyeaters in the first period except that sequential records were taken. The four categories of ground, bark, leaves, and air were compared by month. Most individuals were color-banded and were resident and territorial in the area (Huddy 1979).

For Buff-rumped Thornbills five consecutive foraging moves were recorded for each individual encountered from September 1978 to August 1981. Again most birds were color-banded and consisted of a resident clan made up of several breeding pairs or groups (Bell and Ford 1986).

Observations between different periods were compared using contingency tests. Comparisons made and sample sizes are outlined in Appendix 1. Where tests indicated significant differences, cells were examined to identify categories and periods with higher values. In some cases rows or columns were combined and further tests applied. For all except honeyeaters in the first period there is a risk of non-independence due to sequential sampling. This is alleviated by the large sample sizes and setting significance levels at $P < 0.01$. For thornbills all frequencies were divided by 5 to make the contingency test more rigorous (virtually all sequences were of five observations). Niche breadth was calculated from the formula $B = -\Sigma P_i \log_e P_i$, where p_i = proportion of observations in the *i*th category. Ten categories were used for niche breadth in Fuscous Honeyeaters (hawking, hover and gleaning leaves, gleaning twig, branch, trunk and ground, probing branch, and probing flowers of mistletoe and eucalypt). Five substrate categories were used for the other two species.

RESULTS

ARTHROPODS

The abundance of arthropods on foliage from all plant genera combined in the proportion to which they contribute to the foliage are shown in Bell and Ford (this volume:Fig. 1). The first year showed a marked spring and summer peak in abundance, with lower numbers in autumn and winter. Spring and summer peaks were much less noticeable in the last two years, which were exceptionally dry. Indeed, numbers throughout the last 20 months were similar to those in winter 1979. As trees had lost many of their leaves by mid-1980, this indicates a period of relative and probably absolute shortage of food for leaf-gleaning birds. Arthropods from the ground showed a summer peak in 1979 and 1980, with less marked peaks in spring 1980 and summer 1981 (Bell and Ford [this volume:Fig. 1]). Levels in

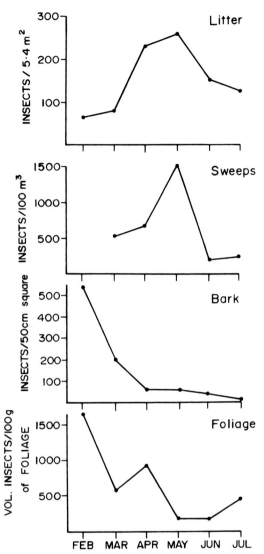

FIGURE 1. Estimated abundance of arthropods from litter, air, bark and eucalypt foliage at Wollomombi (from Huddy 1979, Noske 1982, Bell 1985a).

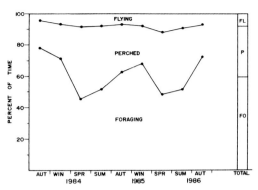

FIGURE 2. Seasonal changes in proportion of time spent foraging, perched, and flying by Fuscous Honeyeaters, at Eastwood, based on time budgets. (AUT = autumn, WIN = winter, SPR = spring, SUM = summer.)

winter were very low. Arthropods in litter and the air peaked in autumn; those from bark and foliage declined from summer to winter (Fig. 1).

FUSCOUS HONEYEATER

Fuscous Honeyeaters spent a greater proportion of the day foraging in autumn and winter than in spring and summer, though this was never above 80% (Fig. 2). The proportion of time perched and flying peaked in spring when territorial vigilance and defense were highest and activities such as nest-building and incubation were observed. The breeding season lasts from September to January (Dunkerley, unpubl. data).

Use of the six foraging categories differed significantly between seasons in both 1981 and 1982 ($\chi^2 = 76.6$, df = 15 for 1981; $\chi^2 = 147.9$, df = 20 for 1982; both $P < 0.001$; Fig. 3). The categories that contributed most to these values differed between years, however. In 1981, hawking was more frequent in autumn yet less frequent in winter, and hovering at leaves was less frequent in winter and more frequent in spring. Leaf gleaning was higher than expected in winter. In 1982 seasonal changes in leaf gleaning and flower probing contributed most to the significant value, with the former high in winter and low in both summers and the latter showing the reverse pattern. Bark feeding was high in spring.

Data from 1981 and 1982 also differed ($\chi^2 = 28.1$, df = 5, $P < 0.001$; Fig. 4), mostly because hovering at leaves was more frequent in 1981 than in 1982. There were also differences between discrete data from 1984 and 1985 ($\chi^2 = 24.5$, df = 5, $P < 0.001$; Fig. 4), with less bark foraging but more flower foraging in 1984 than 1985. The first and second periods also differed ($\chi^2 = 197$, df = 5, $P < 0.001$), with high residuals in all but bark foraging. The different recording methods employed in the two periods probably contributed to these substantial differences.

Time-budget data were used to calculate expected values for the discrete observations. This indicated that the two methods, although carried out simultaneously, yielded different results ($\chi^2 = 372$, df = 5, $P < 0.001$; Fig. 4). In particular, time-budgeting underestimates aerial feeding compared with discrete data. The time-budget data indicate that flower probing and leaf gleaning changed seasonally in 1984–1985, but this was not tested statistically (Fig. 3).

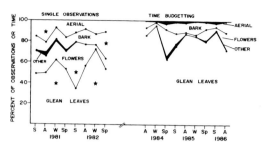

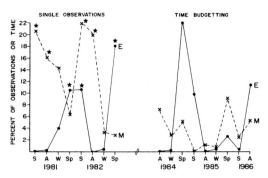

FIGURE 3. Percentage of observations (1981–1982) and percentage of time (1984–1986) spent foraging in each category in each season for Fuscous Honeyeaters at Eastwood. A star indicates significantly more foraging in this category in this season than average (S = summer, A = autumn, W = winter, SP = spring.)

FIGURE 5. Percentage of observations or time spent foraging on flowers of mistletoes (M, *Amyema*) and eucalypts (E) each season by Fuscous Honeyeaters at Eastwood. A star indicates significantly ($P < 0.05$) more foraging on the genus in a season within each year for 1981 and 1982 (seasons as in Fig. 3).

There were marked seasonal changes in the incidence of foraging on the two main genera producing nectar ($\chi^2 = 161$, df = 8, $P < 0.001$ for 1981–1982; Fig. 5). Mistletoes (*Amyema pendulum* and *A. miquelii*) tended to flower regularly from late summer to winter, at which time honeyeaters visited them. This pattern was less clear for the second than the first period (Fig. 5). *Eucalyptus melliodora* flowered fairly regularly in spring and *E. blakelyi* irregularly in summer. Feeding on eucalypt flowers tended to peak in spring or summer, but there were differences between years.

The foraging niche breadth tended to be lowest in winter (Fig. 6), the season when more time was spent foraging and leaf gleaning. Niche breadth tended to be highest in spring and summer, though summer 1985–1986 was an exception. Niche breadths were consistently lower in 1984–1986 than in 1981–1982.

SCARLET ROBIN

Scarlet Robins are sit-and-wait foragers, which typically snatch insects from bark, pounce on them on the ground, or hawk for them in the air (Huddy 1979, Recher et al. 1985, Ford et al. 1986). However, they showed marked seasonal changes in method and particularly substrate ($\chi^2 = 2360$, df = 15, $P < 0.001$). Hawking and snatching from bark and leaves declined from summer through to winter ($r_s = -0.83$, -1.0, -0.94 for air, bark, and leaves, respectively, against month; all $P < 0.05$; Fig. 7). Ground foraging, mostly by pouncing, progressively increased during this period ($r_s = 1.0$, $P < 0.005$).

Niche breadths declined from autumn to winter. Robins also showed changes in foraging within the day ($\chi^2 > 26.2$, df = 12, $P < 0.01$ for all months except May), which were similar in all months (Fig. 8). Aerial foraging peaked around

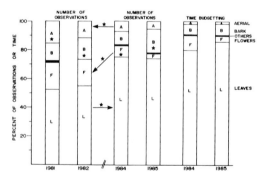

FIGURE 4. Percentage of observations by foraging substrate by Fuscous Honeyeaters in each of 4 years, and percentage of time in each category in 1984 and 1985. A star indicates significantly ($P < 0.05$) more foraging in this category than in other years in the same period. Arrows indicate significantly ($P < 0.05$) more foraging in the category in one 2-year period than the other.

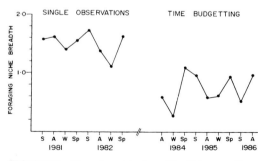

FIGURE 6. Foraging niche breadth of Fuscous Honeyeaters in each season at Eastwood (seasons as in Fig. 3).

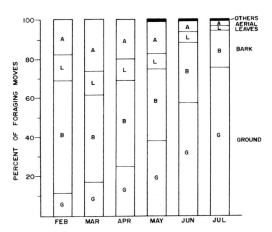

FIGURE 7. Percentage of observations of foraging on four substrates each month by Scarlet Robins at Wollomombi in 1979.

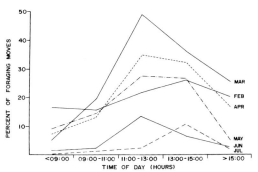

FIGURE 8. Changes in the proportion of aerial foraging in each period through the day by Scarlet Robins at Wollomombi in 1979.

midday and early afternoon, while ground foraging was lowest during these periods. Bark and foliage feeding did not change consistently during the day.

Data were separated according to the maximum temperature on the day that they were collected. Overall, ground foraging was the predominant activity at low temperatures, whereas bark and aerial foraging were more common at higher temperatures ($\chi^2 = 1660$, df = 9, $P < 0.001$). Leaf foraging increased slightly with an increase in temperature (Table 1). Differences in foraging at different temperatures were shown in data collected within the months of April, May, and June, but not in February and March (Table 1). There were data from only one temperature range in July, so no within-month between-temperature comparisons were possible. Within the lower temperature ranges there were increases in ground foraging over successive months, whereas at the highest temperature range there were declines in bark foraging and increases in aerial foraging from February to April.

BUFF-RUMPED THORNBILL

Buff-rumped Thornbills are typically gleaners (Bell 1985b). Their use of different substrates varied consistently through the year ($\chi^2 > 108$, df = 9, $P < 0.001$ each year; Fig. 9). Leaf gleaning reached a peak in spring, at which time bark foraging predominated. From late summer to late winter use of the gound increased so that in winter it was the most frequently used substrate.

Superimposed on the regular seasonal changes in foraging were some changes between years ($\chi^2 = 40.5$, df = 6, $P < 0.001$). Ground feeding was more frequent in autumn 1980 than in 1979 or 1981. At this time leaf gleaning was particularly infrequent. The drought was severe in autumn 1980, when many trees and shrubs were defoliated.

Niche breadth fluctuated sharply between months but there was a tendency to decline towards the winter of 1980, when the drought was at its worst, and rise thereafter.

DISCUSSION

The small, though consistent, changes shown by the Fuscous Honeyeater are perhaps not surprising in a leaf-gleaner that lives in an evergreen habitat where the abundance of insects does not vary greatly between summer and winter (Woinarski and Cullen 1984; Bell 1985a; Lowman, unpubl. data). The ratios of high to low abundance estimates of arthropod biomass on leaves are 10:1 or less, even in years with a marked spring or summer peak (Bell and Ford, this volume:Fig. 1; see also Woinarski and Cullen 1984). Deciduous and coniferous forests of the Northern Hemisphere display much more marked seasonal changes in abundance of arthropods (e.g., Gibb 1950, Perrins 1979:Fig. 58). Even in the latter habitats some species do not show marked seasonal changes in foraging. For instance the rather generalized Willow Tit (*Parus montanus*) did not show marked seasonal changes in foraging in coniferous forests in Scandinavia (Ulfstrand 1976, Alatalo 1980). Titmice (Paridae), whose seasonal changes in foraging have been extensively studied in Europe, may show broader (Ulfstrand 1976) or narrower (Alatalo 1980) niches in winter. In deciduous woodland, Gibb (1954) found that niches were broadest in autumn and early winter, but narrowest in early summer. We found that our species showed a narrower foraging niche when food was least abundant.

There were no major seasonal changes in the abundance of leaf-foraging birds at Eastwood

TABLE 1. PERCENTAGE OF FORAGING MOVES BY SCARLET ROBIN BETWEEN 11:00 AND 15:00 ON EACH SUBSTRATE AT DIFFERENT DAILY MAXIMUM TEMPERATURES EACH MONTH. NOTE THAT SOME TEMPERATURE RANGES WERE NOT ENCOUNTERED IN SOME MONTHS

Temp.	Substrate	February	March	April	May	June	July	N[a]	P
0–15°C	Ground				49.7	54.4	69.9	781	*
	Bark				30.4	37.8	23.2	341	
	Air				16.1	4.4	5.8	84	*
	Leaves				3.7	2.0	1.0	20	
15–20°C	Ground	0		21.5	22.5	45.5		508	*
	Bark	69.0		42.2	49.6	34.6		615	
	Air	28.6		29.1	20.8	11.6		274	*
	Leaves	2.4		7.3	5.8	6.6		97	
20–25°C	Ground	13.5	6.5	25.6	11.1			110	
	Bark	51.4	41.8	34.4	42.9			362	
	Air	25.7	44.0	30.0	37.5			326	
	Leaves	9.5	7.7	10.0	8.4			75	
>25°C	Ground	0	8.1	3.1				11	
	Bark	86.5	45.0	38.5				120	*
	Air	11.5	41.4	50.8				85	*
	Leaves	1.9	5.4	7.7				12	

[a] N = number of foraging moves on each substrate at different temperatures.
[b] Significant (P < 0.01, based on cell values from contingency tests) trend in foraging site at same temperature level over months.

(Ford et al. 1985), so the potential for competition between Fuscous Honeyeaters and other species probably does not change seasonally. Indeed, Fuscous Honeyeaters aggressively dominate all other foliage gleaners and may drive them from the most productive areas (Dunkerley, pers. comm.). This is in contrast to the situation in Scandinavia where, for instance, an abundant summer visitor, the Willow Warbler (*Phylloscopus trochilus*), may exclude resident species from deciduous trees (Ulfstrand 1976).

Flowers provided a resource whose abundance and use showed some seasonal patterns, but where abundance can be very high when, for instance, some eucalypts flower prolifically. At such times larger honeyeaters like Red Wattlebirds (*Anthochaera carunculata*) and Noisy Friarbirds (*Philemon corniculatus*) and lorikeets (*Glossopsitta*) feed on nectar (Ford et al. 1986). Fuscous Honeyeaters may be aggressively excluded from flowering trees or their foraging efficiency may be reduced by exploitation of nectar by other species. Smaller honeyeaters (Eastern Spinebill [*Acanthorhynchus tenuirostris*] and Scarlet Honeyeater [*Myzomela sanguinolenta*]) also feed on the less productive but more regular mistletoe flowers, but these can be driven away by Fuscous Honeyeaters (Ford et al. 1986; Ford, pers. obs.; Dunkerley, pers. comm.).

Differences in foraging behavior and niche breadth between the two periods and within the second period between discrete observations and time budgets indicate that the observational method can greatly influence the results. Time budgets relative to discrete observations apparently overestimated activities that continue for long periods, but underestimated those that were brief, though more conspicuous, such as hawking.

Scarlet Robins showed a change from leaf, bark, and aerial foraging to ground feeding from midsummer to mid-winter (Fig. 7). This was partly because food did not decline seasonally as much on the ground as on bark and foliage (Fig. 1). Indeed, arthropods may be more common on the ground surface in winter than in summer. It is also partly because aerial and bark foraging increased with increasing temperature, whereas ground feeding declined (Table 1). This may be because insect activity on most substrates, and hence conspicuousness, is greater when it is warmer (e.g., Taylor 1963). In addition, when ground temperature becomes elevated, insects may leave the ground surface to travel to other substrates or burrow deep into the litter.

A similar change was shown by the distantly related European Robin (*Erithacus rubecula*), which changed from ground gleaning to pouncing at higher temperatures (East 1980). Bark feeding (in females), leaf gleaning and aerial foraging (in both sexes) also increased, though these activities were still fairly infrequent. This was perhaps because the temperature range in that study was only 0–10°C. Grubb (1975) noted temperature-related changes in foraging behavior of North American birds. Birds tended to move lower and more to the center of trees at low temperatures. This was most obvious below 0°C, lower than is usually experienced in daytime in our study sites.

Sit-and-wait predators among passerines in the Northern Hemisphere (e.g., Muscicapidae, Tyrannidae, Laniidae) are almost without exception migratory, as indeed are some Australian species, so there is little scope for studying seasonal changes in foraging within a site. Several studies have found temperature-related changes in foraging during the day (e.g., Spotted Flycatcher [*Muscicapa striata*], Davies 1977b; American Redstart [*Setophaga ruticilla*], Holmes et al. 1978; Eastern Kingbird [*Tyrannus tyrannus*], Murphy 1987). Also, Sakai and Noon (this volume) found changes in foraging behavior in *Empidonax* flycatchers through the breeding season. The seasonal change in foraging by Scarlet Robins is not due to changes in interspecific competition, as many aerial foragers leave Wollomombi in winter (Bell, unpubl.). Also, all the data that we collected were outside the breeding season.

Buff-rumped Thornbills showed a marked and consistent change in foraging substrate from bark and leaves to the ground from late summer onwards. This change to ground foraging was early in 1980 when insects on foliage were scarce. A change to ground feeding in autumn and winter was shown in European titmice (Gibb 1954, Alatalo 1980) and some finches (Newton 1975). These birds changed from feeding predominantly on arthropods in the breeding season to taking more seeds in autumn and winter. Thornbills fed mostly on arthropods throughout the year, though some seeds were taken in winter (Bell 1983).

The ground may be a more profitable place to forage than bark or leaves in winter. As shown in Figure 1, arthropods may actually be more common on the ground in winter than in summer, although Bell (1985a) showed a decline in arthropods on surface vegetation from summer to winter (see Bell and Ford, this volume: Fig. 1).

Buff-rumped Thornbills show a marked seasonal change in social behavior (Bell and Ford 1986). Pairs or trios defend small territories in spring to which any young reared are added. During the summer, breeding pairs or groups combine so that by January or February most birds occur in a clan of 7–15 birds. In August the clan breaks up and breeding pairs or groups are formed. Thus the high incidence of ground feeding coincides with large group size. Indeed, at such times other species join Buff-rumped Thornbills, which are nuclear species in mixed species feeding flocks (Bell 1980). These flocks probably provide benefits such as increased vigilance to predators (Morse 1980a). Some ground predators, such as snakes, are also less active in

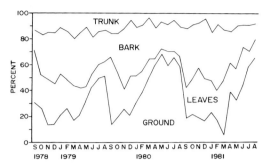

FIGURE 9. Changes in use of different substrates by Buff-rumped Thornbills each month from September 1978 to August 1981 at Wollomombi.

winter. Individuals or pairs in the breeding season should avoid conspicuous foraging not only for their own safety, but to avoid disclosing the position of their eggs and young.

To summarize, the three species we studied displayed three patterns of change. Fuscous Honeyeaters displayed some seasonal changes which were to some extent consistent from year to year. Scarlet Robins showed seasonal changes in foraging that resulted in part from seasonal changes in the availability of arthropods, and at least partly from short-term changes in weather. Buff-rumped Thornbills showed substantial and consistent seasonal changes in substrate, along with a change in social behaviour.

Our studies have indicated that it is easy to find statistically significant seasonal changes in foraging behavior of birds, provided that sample sizes are large. This suggests that birds usually show seasonal changes in foraging (see also Sakai and Noon, this volume; Hejl and Verner, this volume; Martin and Karr, this volume). However, the methods employed in any study may influence the results. Comparisons of different methods for collecting data on Fuscous Honeyeaters show that a consistent method must be adopted within a study and indeed between studies carried out by different observers or at different sites. Time budgets or many sequential observations may overestimate common activities, whereas single discrete observations may overestimate less common but conspicuous activities. A corollary to this is that even a rigidly consistent method may yield spurious results where the conspicuousness of an activity varies seasonally. For instance, in deciduous woodland gleaning on twigs will be more conspicuous in winter than in summer.

Sequential observations and even single observations on the same bird on different occasions suffer from the problem of nonindependence. This may mean that impressive statistical differences do not reflect real biological changes. Single observations per bird at each encounter are preferable and the more individuals that are sampled the better. Seasonal changes in foraging should be consistent over several years (true for Buff-rumped Thornbills but not always for Fuscous Honeyeaters) or at least should be substantial (as in Scarlet Robin) before they should be accepted as biologically significant.

Birds may show day-to-day changes in behavior, based for example on temperature changes; they may also show changes within a day. Our data on Scarlet Robins indicate that these changes may be substantial. These changes are interesting in themselves, but may mask or exaggerate apparent seasonal changes in behavior. Studies investigating seasonal changes should cover many days each month or season, and either restrict data to one type of weather or sample all types of weather. Also, data should be collected at the same time of day in different periods, or at intervals evenly spread through the day.

We separated our data into seasons or months, which are rather arbitrary divisions. The stage of the breeding cycle (Sakai and Noon, this volume), time in relation to migration (Martin and Karr, this volume) and phenological stage of vegetation (Hejl and Verner, this volume) are more biologically realistic separators. Our species though had long, rather asynchronous breeding seasons, were residents and lived in woodland where trees may show fresh foliage from September through to May. However, comparisons of the behavior of individually marked birds at different stages of breeding could be attempted.

If there are problems in collecting data to indicate meaningful seasonal changes in foraging behavior, there are even more in explaining these changes. Our species were principally dependent on arthropods that they obtained from a variety of sites. Seasonal changes in the abundance of arthropods can be estimated in a variety of ways. Most are tedious and time-consuming, and data are hard to standardize because arthropods are heterogeneous in form and often patchily distributed. Indeed, data on arthropod abundance are far easier to criticize than to collect and interpret. Despite this, general patterns between changes in food abundance and foraging behavior can be found, but it is unlikely that these will be close.

Factors other than food influence foraging behavior. Species frequently join flocks in the nonbreeding season (Morse 1980a), which may allow them more safety from predators, particularly on exposed substrates. Predator avoidance and choosing areas where food is abundant will interact in their influence on foraging behavior and it may be hard to identify the proximate factor. For instance, in our study Buff-rumped Thornbills might feed on the ground in winter because food is available there. Because the ground is exposed to predators they consequently join flocks. Alternatively, because they only join flocks in the nonbreeding season, this may be the only time when it is safe to venture onto the ground, which could always be the more productive substrate. Identification of seasonal changes in foraging behavior of birds is clearly only the first step in opening up a whole series of interesting questions on the factors that influence a bird's behavior.

ACKNOWLEDGMENTS

We are grateful to Richard and Susan Noske, Harry Recher, Michael Morrison, and Barry Noon for commenting on the manuscript and to Susan Noske, Gillian Dunkerley and Lynda Bridges for assistance in the field. HAF was supported by research grants from the University of New England and HB received an Australian Commonwealth Postgraduate scholarship. Vi Watt and Sandy Higgins typed the manuscript.

APPENDIX 1. A SUMMARY OF THE MAIN COMPARISONS MADE IN THIS PAPER WITH SAMPLE SIZES FOR EACH TIME PERIOD AND DETAILS OF CONTINGENCY TABLE USED TO TEST FOR SIGNIFICANT DIFFERENCES. THE PROPORTION OF TIME IN EACH CATEGORY FROM TIME BUDGETS WAS USED TO CALCULATE EXPECTED VALUES FOR COMPARISON WITH SEQUENTIAL OBSERVATIONS

	Sample sizes	Contingency table (time period by foraging category)
FUSCOUS HONEYEATER (6 categories)		
Single foraging observations		
Between seasons—1981	237–465	4 × 6
Between seasons—1982		
(includes 2 summers, Jan. + Feb. and Dec.)	75–181	5 × 6
Between years 1981 vs. 1982	775, 1425	2 × 6
Sequential observations		
Between years 1984 vs. 1985	555, 611	2 × 6
Different periods and methods		
1981 + 1982 vs. 1984 + 1985	1166, 2200	2 × 6
Different methods, same period 1984–1985		
Time-budgeting vs. sequential observations	1166, 74,714 s	1 × 6
Feeding on mistletoes vs. eucalypts		
Between seasons 1981–1982; flower-probing data only used	9–68	8 × 2
SCARLET ROBIN (4 substrates)		
Sequential observations		
Between months February–July	856–2525	4 × 6
Between months, same temperature	225–1494	3 × 4 or 4 × 4
Within day, different time	94–1934	5 × 4
Between temperatures, all months	724–4655	4 × 4
BUFF-RUMPED THORNBILL (4 substrates)		
5 sequential observations		
Between seasons, each of 3 years	1235–2445	4 × 4
Between 3 years	5115–9536	3 × 4

GEOGRAPHIC VARIATION IN FORAGING ECOLOGY OF NORTH AMERICAN INSECTIVOROUS BIRDS

Daniel R. Petit, Kenneth E. Petit, and Lisa J. Petit

Abstract. There is little information on geographic variation in foraging ecology of North American insectivorous birds during the breeding season. We summarized foraging data for 22 species of arboreal Passeriformes. Four to 11 ($\bar{X}$ = 5.6) populations per species were compared using foraging technique (i.e., glean, hover, and hawk) and prey location (i.e., branch, trunk, leaf, ground, and air) to characterize foraging niches. Detrended correspondence analysis and an index of ecological overlap were employed to quantify interpopulational foraging plasticity (variability). Of 11 species that had data for both foraging technique and prey location, the Blue-gray Gnatcatcher, Ash-throated Flycatcher, and Warbling Vireo had the highest levels of plasticity, whereas the Yellow-rumped Warbler, White-breasted Nuthatch, and Red-eyed Vireo were relatively stereotyped. The Solitary Vireo, Black-throated Green Warbler, Acadian Flycatcher, and Yellow Warbler exhibited high degrees of foraging plasticity. In contrast, the Brown Creeper, Pine Warbler, White-breasted Nuthatch, Red-breasted Nuthatch, and Mountain Chickadee revealed substantial stereotypy in foraging techniques. Bark gleaners showed less geographic variation than leaf gleaners and leaf hoverers. Those differences may be related to the differential accessibility of arthropods on the two types of substrates. We suggest that behavioral plasticity exhibited by many species is due to simple functional responses associated with local environmental conditions (e.g., vegetation structure).

Key Words: Detrended correspondence analysis; foraging behavior; foraging niche; foraging plasticity; geographic variation; guilds.

Studies of foraging behavior have provided important insights into many aspects of avian ecology, including intersexual relationships (e.g., Kilham 1965, Morse 1968, Franzreb 1983a), temporal variation in behavior (e.g., Kessel 1976, Holmes et al. 1978, Hutto 1981b), guild structure (e.g., Willson 1974, Pearson 1977a, Holmes et al. 1979b), morphological constraints imposed on species (e.g., Selander 1966, Ricklefs and Cox 1977, Miles and Ricklefs 1984), and factors affecting bird community structure and composition (e.g., Morse 1968, Holmes et al. 1979b, Holmes and Recher 1986a). Most of these studies recorded species' foraging behavior over one to several years and drew conclusions pertaining to the ecology of the species. Although numerous authors (e.g., Sabo 1980, Sabo and Holmes 1983, Petit et al. 1985, Emlen et al. 1986) have assumed that individual species occupy similar foraging niches across study areas, this assumption remains largely untested.

Several studies have compared the foraging niches of species inhabiting two distinct habitats (e.g., James 1979, Maurer and Whitmore 1981, Sabo and Holmes 1983), tree species (e.g., Franzreb 1978, Szaro and Balda 1979), or communities (e.g., Crowell 1962, Morse 1971a, Rabenold 1978). Most authors found significant differences, showing that many species were capable of responding to changes in the external environment. Sabo and Holmes (1983) called for study of avian foraging niches across multiple sites, thereby allowing for more definitive examination of niche theory. With the exception of Morse (1971a, 1973), we know of no published studies that have quantitatively compared foraging niches among more than two populations of a species in North America.

The profusion of studies of foraging behavior and the concomitant development of foraging theory have spawned predictions about how individuals should forage under certain prescribed conditions. In addition, terminology has been introduced that categorizes the behavorial and temporal aspects of species' niche shapes. Morse (1971a, b) defined specialists as individuals, populations, or species of birds that exploit a narrow range of available resources. Resource utilization commonly is used in reference to food, foraging area, or habitat preference (Morse 1971b, Pianka 1983). Conversely, generalists are birds that use many of the resources available to them.

Morse (1971a, b) also introduced terminology describing the temporal or interpopulational consistency of niche shape. (Niche shape is used here in the sense of Hutchinson's [1957] "hypervolume.") Stereotypy refers to an individual, population, or species that uses a certain subset of resources with high predictability. Alternatively, birds that exhibit plasticity use resource types with little regularity, varying their use of prey types, behavior, or habitat in response to environmental stimuli.

Based on those concepts, ecological theoreticians have postulated a number of hypotheses regarding how animals should alter their niche shape when they encounter various combinations of resource availabilities and habitat types (e.g., MacArthur 1965, Emlen 1966, MacArthur and Pianka 1966, Charnov 1976a). However, despite more than 200 published reports on the foraging ecology of North American insectivorous birds, no attempt has been made to integrate those results into a comprehensive analysis of geographic variation of foraging behavior. Compounding the difficulties of such an investigation are substantial differences among investigators in describing, quantifying, and analyzing foraging behavior. Considering the volume of literature, we believe that such an investigation is long overdue, and may provide insight into factors that influence shapes of foraging niches. The objectives of this paper are to assess the degree of interpopulational variability (plasticity) of foraging behavior in some insectivorous birds and relate any plasticity to the natural history of the species.

METHODS

THE DATA SET

Data were taken from 27 published and unpublished scientific papers and dissertations (see below) which met the following criteria: (1) observations of foraging behavior were gathered during the breeding season, i.e., the period between the time of arrival on breeding territories and the end of nesting. To minimize temporal variation, when possible we restricted use of data to those collected during incubation and nestling stages of the breeding cycle; (2) the foraging behavior documented could be classified into three "technique" or five "prey location" categories (see below); (3) species were observed in forests, woodlands, or second growth woodlands (in the Temperate Zone) with canopies >4 m tall (most were >8 m tall); (4) species were passerines that typically did not forage from the ground and that devoted (as a species) >33% of their foraging maneuvers to techniques other than hawking (= flycatching; see below). This criterion emphasized species that frequently had direct foraging interaction with vegetation; and (5) data on ≥4 populations were available for each species. Some studies (e.g., Rabenold 1978, James 1979, Landres 1980) provided data on more than one population per species. We subjectively chose four as the minimum number of studies needed to judge a species' behavioral variability.

To determine if differences in behavioral plasticity existed among groups with distinct foraging modes, each species was placed into a trophic group or guild (*sensu* Root 1967) based upon the predominant foraging behavior of the populations we surveyed: (1) glean—leaf, (2) glean—bark, (3) hover—leaf, and (4) hawk.

FORAGING BEHAVIOR

A variety of methods and terminology permeates the foraging ecology literature and, therefore, a synthesis of studies necessarily will be ambiguous unless data are standardized. Documentation of foraging behavior in most studies, including those used in this paper, followed one of four techniques: (1) one observation was made per bird, usually taken when the individual was first sighted (e.g., Franzreb 1983a, 1984); (2) multiple, consecutive records were taken on each bird sighted, and there may (e.g., Williamson 1971, Rabenold 1978) or may not (e.g., James 1976, Eckhardt 1979, Holmes et al. 1979b) have been a limit placed on total number of foraging maneuvers recorded for an individual or total time an individual was watched on any given day; (3) multiple observations were recorded at given time intervals, usually with a maximum number allowed per bird (e.g., Landres 1980, Morrison 1984a); and (4) a stopwatch was used to measure time devoted to a given foraging behavior (e.g., MacArthur 1958; Morse 1967b, 1968). Because data used in this study were collected under such varied manners, it was not possible to categorically describe how data were recorded and we refer the reader to the original papers. Also, although there may be statistical biases associated with some of those techniques of gathering data (e.g., see Wagner 1981a, Morrison 1984a), we assumed that this potential problem was minimal and each investigator accurately quantified behavior of the population(s) under study.

The schemes into which behaviors were classified also varied among studies. In our analyses, we were concerned mainly with two measures of passerine foraging ecology, the technique used to attack prey and the location of attacked prey. Although other behaviors (e.g., foraging rates, distances travelled, height) may be important in quantifying species' niche characteristics, they often are not recorded by researchers or are peripheral to the scope of this study. Terminology used to describe passerine foraging behavior is often ambiguous and often designed so as to accentuate species' differences in studies of guild-community ecology. We used the simplest divisions that we deemed adequate to describe foraging ecology of arboreal passerines. Our definitions are taken largely from James (1976), Eckhardt (1979), and Holmes et al. (1979b).

Foraging technique was partitioned into three mutually exclusive categories: (1) glean, a maneuver directed toward a prey item on a substrate (or, rarely, in the air) while the bird was perched or hopping, and included such maneuvers as probe (Holmes et al. 1979b, Landres 1980, Franzreb 1983a), peck (Williamson 1971, Sabo 1980), pounce (Eckhardt 1979), and hang (Morse 1968, Rabenold 1978, Greenberg 1987b); (2) hover, a maneuver in which prey located on a substrate is attacked by a nonperching (i.e., hovering or flying) bird, which some authors (e.g., Rabenold 1978, Landres 1980, Sabo 1980) have termed sally and hawk; and (3) hawk, a behavior in which both insect and bird are in flight, which is sometimes termed flycatching (e.g., Sabo 1980), sallying (e.g., Eckhardt 1979, Hutto 1981b), and chase (Morse 1967b).

Prey location (i.e., the location of the arthropod prey when a bird made an attempt to procure it) was apportioned into five mutually exclusive categories: (1) branch, which included all surfaces covered by bark, except trunk; (2) trunk; (3) leaf, including petioles and flowers; (4) ground; and (5) air.

Most data could be adapted to our classification scheme. However, in several instances, frequencies within categories in the original paper did not equal 100%, or an extra division (e.g., "other") was given that did not conform to our categories. In the former case, we changed the values relative to one another, so that the total equalled approximately 100%. For the latter, we distributed the anomalous observations equally across those categories in which there was a possibility that they belonged. The error we introduced into estimates of foraging behavior was negligible using this method because we manipulated percentages only when the unassigned observations were ≤10% of all records gathered for that study.

ANALYSES

Species were divided into two groups based upon the amount of information that was available: (1) Group A, species for which observations had been made in all eight (i.e., technique and prey location) foraging categories; and (2) Group B, species that were represented by ≥4 studies for the technique variable only. Groups were not mutually exclusive (e.g., all Group A species were also included in Group B analyses), but were necessary due to the varied amounts of data that were available from individual studies. These studies were: Airola and Barrett 1985 (Groups A and B), Bennett 1980 (AB), Eckhardt 1979 (B), Ficken et al. 1968 (B), Franzreb 1983a (AB), Franzreb 1984 (B), Holmes et al. 1979b (AB), Hutto 1981b (B), James 1976 (AB), James 1979 (AB), Landres 1980 (AB), MacArthur 1958 (AB), Maurer and Whitmore 1981 (AB), McEllin 1979 (AB), Morrison et al. 1987b (B), M. L. Morrison et al., unpubl. data (AB), Morse 1967b (AB), Morse 1968 (AB), Morse 1973 (AB), Morse 1974a (B), D. R. Petit et al., unpubl. data (AB), Rabenold 1978 (AB), Rogers 1985 (B), Root 1967 (AB), Sabo 1980 (AB), Sherry 1979 (AB), and Williamson 1971 (AB).

We used three techniques to assess the degree of behavioral plasticity-stereotypy exhibited by different arboreal passerines: (1) detrended correspondence analysis (DCA)-interval method, (2) DCA-standard deviation method, and (3) overlap method. Because all three types of analyses have minor biases associated with them (when applied to quantifying niche breadths), we developed a scheme to rank species' behavioral plasticity based on a combination of the three methods.

DCA-interval method

Detrended correspondence analysis (Hill 1979, Hill and Gauch 1980) was used to evaluate the degree of behavioral plasticity both within and between species. DCA is an improved version of reciprocal averaging and may be superior to other ordination procedures (e.g., principal components analysis) in characterizing relationships in ecological data sets (e.g., Sabo 1980, Gauch 1982a). For both groups A and B, species from each study (a "species-sample") were ordinated as separate samples along with all other species-samples from that group. Scores of the species-samples on each DCA axis were used to describe quantitatively each species-sample's position on the derived "foraging behavior" gradient and its relationship (distance) to all other samples of a given species. Ecological interpretation of axes was determined from correlations between axis scores and original variables. Following Johnson (1977) and Rotenberry and Wiens (1980b), we divided each DCA axis into four divisions of equal length. Next, the distribution of behavior along the derived resource gradients was estimated by counting the proportion of samples for each species that fell into each interval. Species behavioral variability (plasticity) was defined using the niche breadth equation of Levins (1968):

$$B = 1/\sum p_i^2,$$

where p_i is the proportion of samples that were contained in the ith interval. Niche breadth values (B) ranged from 1, if all samples fell within one interval, to 4, the number of intervals available.

Several biases are inherent in this method. One shortcoming of the niche breadth measure is that the maximum B (B_{max}) for any data set depends on the number of samples in that set, especially with small (e.g., <10) sample sizes. To correct for this bias, we divided all niche breadth values (B) by their maximum possible values (B_{max}) to produce a relative measure of variability. B and B_{max} were highly correlated (Pearson's r's > 0.90) for all axes in all analyses.

Another source of potential error with using this technique is that adjacent divisions along a multivariate axis are usually more similar in "ecological space" than are intervals separated by some distance (e.g., Gauch 1982a). Therefore, in disjunct distributions along these axes, the "space" (distance) is not acknowledged and the distribution is treated as continuous (see, for example, Rotenberry and Wiens 1980b:Fig. 5). Because of small sample sizes, the discontinuous distribution of several species on axes in our analyses may be artifactual. Therefore, both the distribution of species-samples across intervals and the relationships among intervals may be important in describing species' behavioral niche breadths. Our second technique (DCA-standard deviation) took into account these concerns (see below).

To assess species' overall niche breadths, we calculated species' responses across all niche dimensions simultaneously. The orthogonality of DCA axes allowed us to use the product of the first two axes as a measure of overall niche variability because the axes are independent (May 1975). Because the niche breadth values were scaled relative to the maximum possible niche breadth for that species, potential overall niche breadth ranged from 0.06 (0.25 × 0.25) to 1.0 (1.0 × 1.0). We also performed ordinations on the 11 Group A species based on separate analyses of prey location and foraging technique. The purpose of this was to determine if one variable showed greater variability along the derived gradients than the other.

DCA-standard deviation method

Species-samples' scores on the DCA axes were used to calculate a standard deviation for each species on each axis. Large standard deviations indicated a wide range of foraging behaviors between studies, while small standard deviations represented relatively stereotyped behavior. The standard deviation of sample scores is proportional to the projection of a confidence ellipse onto that axis (Noon 1981a). Thus, our technique is mathematically and biologically comparable to plotting the commonly used confidence ellipses (e.g., Green 1974, Smith 1977).

TABLE 1. Species Used in This Study, Along with Their Predominant Foraging Mode

Taxon	Foraging mode
Family Tyrannidae	
Acadian Flycatcher, *Empidonax virescens*	Hover—leaf
Least Flycatcher, *E. minimus*	Hover—leaf
Dusky Flycatcher, *E. oberholseri*	Hawk
Ash-throated Flycatcher, *Myiarchus cinerascens*	Hawk
Family Paridae	
Mountain Chickadee, *Parus gambeli*	Glean—leaf
Family Sittidae	
White-breasted Nuthatch, *Sitta carolinensis*	Glean—bark
Red-breasted Nuthatch, *Sitta canadensis*	Glean—bark
Family Certhiidae	
Brown Creeper, *Certhia americana*	Glean—bark
Family Muscicapidae	
Golden-crowned Kinglet, *Regulus satrapa*	Glean—leaf
Blue-gray Gnatcatcher, *Polioptila caerulea*	Glean—leaf
Family Vireonidae	
Solitary Vireo, *Vireo solitarius*	Glean—bark
Yellow-throated Vireo, *V. flavifrons*	Glean—bark
Red-eyed Vireo, *V. olivaceus*	Hover—leaf
Warbling Vireo, *V. gilvus*	Glean—leaf
Family Emberizidae	
Black-throated Green Warbler, *Dendroica virens*	Glean—leaf
Yellow Warbler, *D. petechia*	Glean—leaf
Yellow-rumped Warbler, *D. coronata*	Glean—leaf
Magnolia Warbler, *D. magnolia*	Glean—leaf
Blackburnian Warbler, *D. fusca*	Glean—leaf
Pine Warbler, *D. pinus*	Glean—bark
Northern Parula, *Parula americana*	Glean—leaf
American Redstart, *Setophaga ruticilla*	Hover—leaf

Overlap method

Because the positions of species-samples along derived gradients are determined, to some extent, by their relationships to other species-samples in the data set, we used a simple index of overlap to evaluate within-species variability in foraging behavior. This measure of overlap (O), based on Lotka-Volterra principles (MacArthur 1972, Hurlbert 1978), was formulated by Pianka (1973):

$$O = \frac{\sum (x_i y_i)}{[\sum x_i^2 \cdot \sum y_i^2]^{1/2}},$$

where x_i and y_i are proportions of behavioral observations for populations x and y in the ith resource category. For our analysis, eight resource states, those of technique (3) and prey location (5), were recognized. This symmetrical measure of overlap ranges from 0 (no overlap) to 1 (total overlap) and was computed for all pairwise combinations of species-samples within a species. Mean overlap values were used to assess degree of behavioral plasticity within a species. Comparatively large overlap values were interpreted as stereotyped behavior, while small overlap values depicted species with high degrees of behavioral plasticity.

RESULTS

We located 123 species-samples of foraging behavior representing 22 species of arboreal passerines (Table 1). More than one-third were wood-warblers (Parulinae); Tyrannidae, Vireonidae, Muscicapidae, Sittidae, Paridae, and Certhiidae were also represented.

GROUP A SPECIES

DCA-interval method

Only 11 of the 22 species had ≥4 samples for all eight foraging categories. The 59 species-samples were ordinated using DCA and the distribution of each species was plotted across four equally-spaced divisions along the first two DCA axes (Fig. 1). Only axes I (eigenvalue = 0.43) and II (0.10) were used because of their disproportionately large eigenvalues as compared to axes

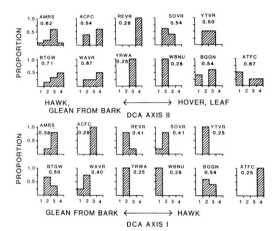

FIGURE 1. Distribution of (Group A) species' sample scores across four equally spaced intervals on DCA (detrended correspondence analysis) axes I and II. The ordination is based on three technique and five prey location categories for each sample (i.e., Group A species). The number in each graph is a measure of niche breadth (B/B_{max}; see text for description) along that axis. See Table 2 for species' acronyms.

III (0.05) and IV (0.02) (see Hill 1979). Axis I represented a gradient from gleaning from branches and trunks to hawking flying insects. Axis II separated the hawkers and branch gleaners from the species that hover and take prey from leaves. Most species exhibited stereotyped use of one or both foraging niche axes (Fig. 1). On axis I, only the Black-throated Green Warbler and Blue-gray Gnatcatcher had niche breadth measures >0.50, while 8 of the 11 species exceeded this value on axis II. The Red-eyed Vireo, Yellow-rumped Warbler, and White-breasted Nuthatch were highly stereotyped on both axes.

The White-breasted Nuthatch and Yellow-rumped Warbler showed the lowest overall foraging variability (i.e., product of axes I and II) with the DCA-interval method. In contrast, the Black-throated Green Warbler, Blue-gray Gnatcatcher, and Warbling Vireo demonstrated relatively high plasticity (Table 2). Most species, however, demonstrated a moderate amount of restriction in their foraging niches.

Based on separate ordinations, the magnitude of the overall niche breadth scores for technique ($\bar{X} = 0.28$, median = 0.25) was not significantly different (Mann-Whitney U-test, $P > 0.20$) from that of prey location ($\bar{X} = 0.22$, median = 0.22).

DCA-standard deviation method

The DCA-standard deviation method produced results somewhat similar (Spearman's $r_s = 0.55$, $P < 0.10$) to those of the previous technique. However, the DCA-standard deviation method recognized the discontinuous distributions of the Blue-gray Gnatcatcher, Ash-throated Flycatcher, and Acadian Flycatcher on DCA axis II and ranked those species as the three most behaviorally diverse (Table 2). Clearly, the Red-eyed Vireo, White-breasted Nuthatch, and Yellow-rumped Warbler still were the most stereotyped species in Group A.

In separate ordinations, neither technique ($\bar{X}$ overall niche breadth = 672, median = 588) nor prey location ($\bar{X} = 811$, median = 365) was consistently larger than the other (Mann-Whitney U-test, $P > 0.20$).

Overlap method

The overlap method produced results comparable to the DCA-interval method ($r_s = 0.59$, $P < 0.10$), but diverged more from the DCA-standard deviation results ($r_s = 0.36$, $P > 0.10$). The overlap method seemed to rank species as a compromise between the other two indices (Table 2). The Blue-gray Gnatcatcher and Ash-throated Flycatcher were (as for the standard deviation method) the two most behaviorally plastic species, and the Warbling Vireo showed the third highest level of diversity (as for the interval method). As before, the Red-eyed Vireo, White-breasted Nuthatch, and Yellow-rumped Warbler showed the least geographic variability. In concurrence with previous results, technique ($\bar{X}$ overlap = 0.85, median = 0.75) was no more variable than prey location ($\bar{X}$ overlap = 0.84, median = 0.86) when separate analyses were performed (Mann-Whitney U-test, $P > 0.20$).

We determined an overall rank of behavioral plasticity for each species by averaging its ranks from the three analyses. We concluded that, for Group A species, the Blue-gray Gnatcatcher, Ash-throated Flycatcher, and Warbling Vireo showed the most geographic variability in foraging behavior, while the Red-eyed Vireo, White-breasted Nuthatch, and Yellow-rumped Warbler had relatively narrow foraging niches (Table 2). We used this same averaging procedure for both prey location and technique separately, and found that the Yellow-throated Vireo, Ash-throated Flycatcher, and Acadian Flycatcher took prey from a variety of substrates, whereas the Solitary Vireo, Acadian Flycatcher, and Blue-gray Gnatcatcher used a diversity of techniques. For both variables, the White-breasted Nuthatch, Red-eyed Vireo, and Yellow-rumped Warbler were highly stereotyped. Both measures were positively correlated with overall niche plasticity (technique: $r_s = 0.70$, $P < 0.05$; prey location: $r_s = 0.59$, $P < 0.10$) and between themselves ($r_s = 0.59$, $P < 0.10$).

TABLE 2. COMPARISON OF THREE METHODS USED TO EVALUATE BEHAVIORAL PLASTICITY OF SPECIES WITH ≥4 SAMPLES OF BOTH FORAGING TECHNIQUE AND PREY LOCATION (GROUP A). DCA REFERS TO DETRENDED CORRESPONDENCE ANALYSIS. SEE TEXT FOR DETAILS OF THE DIFFERENT METHODS

Species	Acronym	Method			
		DCA-interval	DCA-standard deviation	Overlap	Overall average rank
Blue-gray Gnatcatcher	BGGN	0.36 (1)[a]	2240 (1)	0.717 (1)	1.3 (1)
Ash-throated Flycatcher	ATFC	0.17 (6)	1464 (2)	0.760 (2)	3.3 (2)
Warbling Vireo	WAVR	0.27 (3)	395 (7)	0.773 (3)	4.3 (3)
Acadian Flycatcher	ACFC	0.15 (7)	827 (3)	0.775 (4)	4.7 (4)
Black-throated Green Warbler	BTGW	0.36 (1)	559 (6)	0.914 (8)	5.0 (5)
American Redstart	AMRS	0.24 (4)	748 (5)	0.876 (7)	5.3 (6)
Solitary Vireo	SOVR	0.22 (5)	288 (8)	0.781 (5)	6.0 (7)[b]
Yellow-throated Vireo	YTVR	0.13 (8)	793 (4)	0.787 (6)	6.0 (7)[b]
Red-eyed Vireo	REVR	0.12 (9)	113 (9)	0.971 (10)	9.3 (9)
White-breasted Nuthatch	WBNU	0.08 (10)	105 (10)	0.949 (9)	9.7 (10)
Yellow-rumped Warbler	YRWA	0.06 (11)	102 (11)	0.973 (11)	11.0 (11)

[a] Rank; 1 = most variable foraging behavior (i.e., plastic); 11 = least variable foraging behavior (i.e., stereotyped).
[b] Tied with ≥1 other species.

Behavioral plasticity and guild membership

Most Group A species took prey from leaves (leaf gleaners, N = 4; leaf hoverers, N = 3), but there were three bark gleaners and one flycatcher (Table 1). Although bark gleaners tended to be more stereotyped than leaf gleaners, the difference was not significant (Mann-Whitney U-test, P > 0.20). However, when average percent use of bark for each species was correlated with overall foraging plasticity, the use of branches and trunks was positively related to foraging stereotypy ($r_s = 0.68$, $P < 0.05$, Fig. 2).

GROUP B

Group B was comprised of 22 species that were represented by ≥4 samples of the technique variable. The 123 species-samples were analyzed in a way comparable to that of Group A species.

DCA-interval method

DCA axis I (eigenvalue = 0.528) separated samples dominated by gleaning from those characterized by high percentages of hawking (Fig. 3). The Black-throated Green Warbler, Blue-gray Gnatcatcher, Solitary Vireo, and American Redstart showed wide distributions along axis I, while 10 species were found in only one interval. Axis II (eigenvalue = 0.096) placed the hoverers and the hawkers at opposite ends of the gradient. The Black-throated Green Warbler, Solitary Vireo, Acadian Flycatcher, Yellow Warbler, and Golden-crowned Kinglet demonstrated high variability along this axis, whereas many others were highly stereotyped. Overall, the Solitary Vireo, Black-throated Green Warbler, and Acadian Flycatcher showed substantial geographic variation in capture methods, whereas the Brown Creeper, Mountain Chickadee, Pine Warbler, Blackburnian Warbler, White-breasted Nuthatch, and Red-breasted Nuthatch did not (Table 3).

DCA-standard deviation method

This technique produced species' niche breadths that were highly correlated with those of the interval method ($r_s = 0.89$, $P < 0.01$). Considering both axes, the Solitary Vireo, Black-throated Green Warbler, Ash-throated Flycatcher, Acadian Flycatcher, and Yellow Warbler were highly diverse in their foraging repertoires, while the Mountain Chickadee, Red-breasted Nuthatch, White-breasted Nuthatch, Brown Creeper, and Pine Warbler were not (Table 3).

Overlap method

Average species' overlap values corroborated results obtained for species in Group B, as they were highly correlated with both the DCA-standard deviation ($r_s = 0.89$, $P < 0.01$) and DCA-interval ($r_s = 0.87$, $P < 0.01$) methods. The only major difference was that the Blue-gray Gnatcatcher showed the least overlap (i.e., greatest variation) between different populations, when it ranked no better than eighth previously (Table 3).

Consolidating results from the three analyses showed that foraging behavior of the Solitary Vireo, Black-throated Green Warbler, and Acadian Flycatcher varied the most from area to area. In contrast, the Brown Creeper, Pine Warbler, White-breasted Nuthatch, Red-breasted Nuthatch, and Mountain Chickadee were highly predictable (Table 3).

Behavioral plasticity and guild membership

Group B was comprised of 6 bark gleaners, 10 leaf gleaners, 4 leaf hoverers, and 2 species that

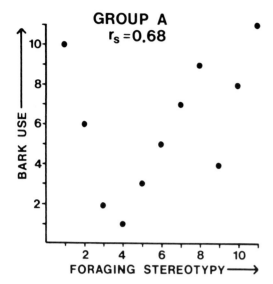

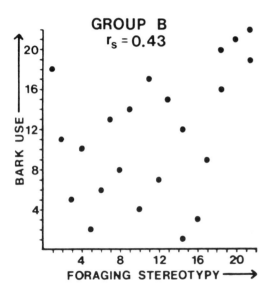

FIGURE 2. Relationship between foraging plasticity and use of bark (twig, branch, trunk) for both Group A (foraging technique and prey location) and Group B (foraging technique only) species. Numbers on all axes represent ranks for individual species.

hawk insects (Table 1). There were no significant differences in foraging plasticity among bark gleaners, leaf gleaners, and hoverers (Kruskal-Wallis one-way analysis of variance, $P > 0.10$). However, as a group, species that take prey from leaves exhibited significantly more behavioral plasticity than did bark gleaners (Mann-Whitney U-test, $P < 0.10$). In support of this claim, there was a significant positive relationship between use of bark and foraging stereotypy ($r_s = 0.43$, $P < 0.05$, Fig. 2).

DISCUSSION

Although each species occupied a recognizable foraging niche, there was considerable variation among populations of many species. Bark gleaners appeared to be more stereotyped than birds that take prey from foliage, in that bark foragers almost always gleaned prey. This trend may have been created by our consolidation of foraging techniques, such that the variety of maneuvers by bark gleaners was lumped under gleaning for our analysis. However, we grouped several foraging modes under each of the three main foraging techniques, so a bias towards the bark gleaners seems unlikely. Foliage insects often were taken by both gleaning and hovering. The stereotyped behavior of many bark gleaners may be due to the types of arthropods found on bark, which may be generally less mobile and thus more accessible, than those inhabiting foliage. Jackson (1979) found that ants, spiders, and hemipterans were the most commonly found arthropods on tree trunks in Mississippi. More importantly, though, may be the differences in accessibility of arthropods on bark vs. leaf surfaces. Species that glean from bark can usually perch on the same substrate (e.g., branch or twig) as their prey items. In contrast, arthropods on leaves (especially leaves with relatively long petioles) are less easily gleaned because they are farther from the bird's perch (Holmes and Robinson 1981). In those cases, techniques such as hovering must be employed.

Several previously held beliefs on species foraging diversity and plasticity were not supported by our analyses. Morse (1967b, 1971a) and Sabo and Holmes (1983) considered the Black-throated Green Warbler to be behaviorally stereotyped; however, our examination of 11 populations demonstrated high levels of plasticity. This was surprising because most populations were studied in similar geographical locations and habitat types. In fact, even comparing records from the same study area, but in different years, revealed substantial plasticity (e.g., compare Holmes et al. 1979b with Bennett 1980), suggesting that Black-throated Green Warblers may not only respond to changes in vegetation structure and bird community composition, but also to annual variation in resource availability and distribution.

Although American Redstarts have been reported to be highly plastic (e.g., Holmes et al. 1978, Sherry 1979, Maurer and Whitmore 1981), we found them no more opportunistic, on an interpopulational scale, than many other species. This does not preclude, however, the possibility

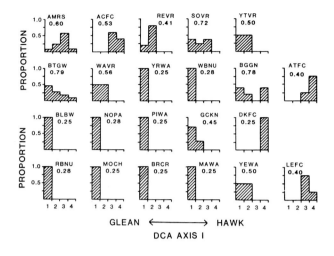

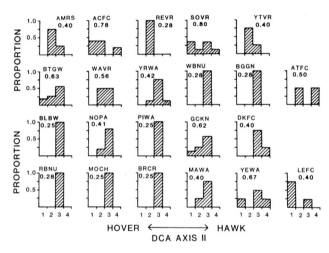

FIGURE 3. Distribution of (Group B) species' sample scores across four equally spaced intervals on DCA (detrended correspondence analysis) axes I and II. The ordination is based on three technique categories for each sample (i.e., Group B species). The number in each graph is a measure of niche breadth (B/B_{max}; see text for description) along that axis. See Table 3 for species' acronyms.

that within-site plasticity is different than the foraging variation measured among populations. Because the terms plasticity and stereotypy are relative, researchers should use this terminology only in comparing with specific populations or species.

The substantial intraspecific behavioral plasticity and interspecific differences in the degree of plasticity detected in this study may be due to a multitude of factors, such as: (1) differences in methods, abilities, and biases of the various data gatherers; (2) nonrandom samples of some (or all) species' populations; and (3) genuine ecological responses by species to the proximate and ultimate constraints imposed on them.

Whether the diverse array of techniques that have been used in the past to quantify passerine foraging ecology biases our results is not known. We believe that the various methods and observers did not obscure the general picture because sample sizes in each study (when reported) were sufficiently large to overcome the error introduced by different sampling techniques (e.g., Petit et al., this volume). We cannot assess observer-expectancy bias (Balph and Balph 1983, Balph and Romesburg 1986), but assume that it was equal across studies. In addition, we safeguarded against the possibility of one aberrant study greatly influencing results by considering a minimum of four populations of any species.

TABLE 3. COMPARISON OF THREE METHODS USED TO EVALUATE BEHAVIORAL PLASTICITY OF SPECIES WITH ≥4 SAMPLES OF FORAGING TECHNIQUE (GROUP B). DCA REFERS TO DETRENDED CORRESPONDENCE ANALYSIS. SEE TEXT FOR DETAILS OF THE DIFFERENT METHODS

		Method			
Species	Acronym	DCA-interval	DCA-standard deviation	Overlap	Overall average rank
Solitary Vireo	SOVR	0.58 (1)[a]	1780 (1)	0.652 (2)	1
Black-throated Green Warbler	BTGW	0.50 (2)	1298 (3)	0.725 (3)	2
Acadian Flycatcher	ACFC	0.42 (3)	1090 (5)	0.751 (4)	3
Yellow Warbler	YEWA	0.33 (4)	1255 (4)	0.778 (5)	4
Ash-throated Flycatcher	ATFC	0.20 (9)[b]	1339 (2)	0.797 (7)	5
Blue-gray Gnatcatcher	BGGN	0.22 (8)	725 (10)	0.626 (1)	6
Golden-crowned Kinglet	GCKN	0.28 (6)	906 (6)	0.810 (8)	7
American Redstart	AMRS	0.24 (7)	880 (8)	0.781 (6)	8
Warbling Vireo	WAVR	0.31 (5)	763 (9)	0.837 (9)	9
Least Flycatcher	LEFC	0.16 (11)	890 (7)	0.878 (10)	10
Yellow-throated Vireo	YTVR	0.20 (9)[b]	286 (11)	0.981 (16)	11
Red-eyed Vireo	REVR	0.12 (12)[b]	210 (15)	0.939 (12)	12
Yellow-rumped Warbler	YRWA	0.11 (14)	245 (12)	0.973 (15)	13
Dusky Flycatcher	DKFC	0.10 (15)[b]	173 (16)	0.929 (11)	14[b]
Magnolia Warbler	MAWA	0.10 (15)[b]	225 (14)	0.967 (13)	14[b]
Northern Parula	NOPA	0.12 (12)[b]	115 (17)	0.984 (17)	16
Blackburnian Warbler	BLBW	0.06 (19)[b]	232 (13)	0.969 (14)	17
Mountain Chickadee	MOCH	0.06 (19)[b]	42 (18)	0.992 (18)	18[b]
Red-breasted Nuthatch	RBNU	0.08 (17)	24 (19)	0.998 (19)	18[b]
White-breasted Nuthatch	WBNU	0.08 (18)	20 (20)	0.998 (20)	20
Pine Warbler	PIWA	0.06 (19)[b]	1 (22)	0.999 (21)	21[b]
Brown Creeper	BRCR	0.06 (19)[b]	2 (21)	0.999 (22)	21[b]

[a] Rank: 1 = most variable foraging behavior (i.e., plastic); 21 = least variable foraging behavior (i.e., stereotyped).
[b] Tied with ≥1 other species.

Examination of DCA ordination plots revealed few outliers and there was no relationship (r_s = 0.05, P > 0.80) between behavioral plasticity and the number of populations used in analyses of Group A species. Nevertheless, because of these concerns, we encourage more interpopulational comparisons conducted by the same research team using one technique to quantify behavior.

This study has demonstrated that data recorded on 4–12 groups of individuals cannot be considered to depict precisely the behavior of each population of that species. If we used species-samples for analysis that were in a restricted geographic area, such that they did not represent a species' full range of behaviors, but used other species' data from a diverse set of areas, then our interspecific comparisons may have erred because of nonrandom sampling of those populations. We can address this indirectly by comparing behavioral plasticity and average distances between study plots for each species. If behavioral variability was positively related to geographic distances between populations, then measures of behavioral plasticity may be incomplete for some species. We measured the distances between all study sites for all 11 Group A species and then averaged the distances for each species. We found no significant relationship between behavioral plasticity and average distance between study sites (r_s = −0.04, P > 0.80). These results suggest that our measures of species' behavioral plasticity were not noticeably biased due to the populations sampled.

We suggest that the observed trends of behavioral flexibility are real—an outcome of past and present selective pressures imposed on each species—and represent simple functional responses to their environments. Holmes and Robinson (1981), Robinson and Holmes (1982), and D. R. Petit (unpubl.) have demonstrated that foliage-gleaning birds alter their foraging behavior in apparent response to variation in vegetation structure and composition. The ability to modify foraging behavior may allow species to occupy a diverse array of habitat types that are geographically distant (e.g., Morse 1971b, Cody 1974, S. L. Collins 1983).

That insectivorous passerines exhibit extensive interpopulational variation in foraging behavior has several ramifications. First, researchers should restrict conclusions to study sites and species on which the data were gathered. Because many species vary their behavior from area to area, we do not accept the concept "adaptive syndromes" (Eckhardt 1979), which attempts to categorize insectivorous birds into groups that exhibit very specific behavioral foraging responses to their environment. Strict delineation of foraging niches at the species (vs. population) level can be misleading and counterproductive to elucidation of ecological trends.

Second, in the design of some ecological studies, species are divided into guilds or trophic groups based on published reports or the general knowledge of researchers. As a result, conclusions are determined to some extent by the classification of each species. Blake (1983), for example, categorized the trophic characteristics of birds breeding in forest tracts in Illinois, then conducted analyses based on those groups. Of the eight species that are common to both Blake (1983) and our study, three were classified differently. Blake (1983) even used a cautious approach by only considering broad-based trophic groups. Likewise, James and Boecklen (1983) divided a Maryland bird community into foraging guilds and then tested very specific ecological theories based on those classifications. One-third (2 of 6) of the species used in both James and Boecklen (1983) and our study were categorized differently. Thus, avian ecologists should consider the extensive variation that exists among different populations of the same species and the consequences of assuming that foraging niches remain constant geographically. Similarly, use of the guild concept in environmental impact assessment (e.g., Severinghaus 1981) will have limited value if species do not occupy similar foraging niches in different geographic regions or habitats.

ACKNOWLEDGMENTS

We gratefully acknowledge the editorial comments of M. L. Morrison, J. R. Jehl, and B. A. Maurer. K. G. Smith read a preliminary draft of this manuscript and made numerous suggestions for its improvement. This paper is dedicated to our father, Richard E. Petit, who passed away while we were preparing this manuscript. His financial and enthusiastic moral support of our research for the past 10 years has allowed us to surpass that level of achievement constrained by our abilities alone.

GEOGRAPHIC VARIATION IN FORAGING ECOLOGIES OF BREEDING AND NONBREEDING BIRDS IN OAK WOODLANDS

WILLIAM M. BLOCK

Abstract. I studied geographic variation in the foraging ecology of four breeding and four nonbreeding species in three oak (*Quercus* spp.) woodlands of California. Variations were evident for all species. Variations in tree-species use, foraging tactics, substrates, and behaviors were species-specific. For example, White-breasted Nuthatches (*Sitta carolinensis*) used tree species with different frequencies at each study area, although they specialized in where and how they foraged within a tree. The foraging behavior of Yellow-rumped Warblers (*Dendroica coronata*) varied little among study areas, although they were generalists in their use of trees and in their foraging locations. Because foraging ecologies of birds can be highly site-specific, studies should be conducted at different locations. To preserve site-specific characteristics of a species' foraging ecology, researchers should not pool samples from different geographic locations.

Key Words: Geographic variation; oak woodlands; foraging ecology; tree-species use; foraging location; foraging behavior.

Because foraging ecologies are strongly influenced by the types and abundances of resources available to birds (e.g., Gibb 1960, Karr 1976, Morse 1980b), different patterns of foraging may occur at different seasons or locations (Morse 1971b, Laurent 1986).

Interpretations of all aspects of avian foraging ecology probably depend on the scale of observation (Allen and Starr 1982, Hurlbert 1984, Wiens et al. 1986b). Results of studies that are restricted to a single location or a single season might apply only to the place or time of study (Wiens et al. 1987b); studies across space and time include greater variation in patterns of resource use and may provide different interpretations. Many studies have related seasonal variations in resource abundance to foraging behaviors (e.g., Austin 1976, Travis 1977, Alatalo 1980, Conner 1981, Hutto 1981b, Wagner 1981b); Arnold (1981) found no studies of geographic variation of foraging by birds, a topic that has been addressed only in recent years (Maurer and Whitmore 1981; Sabo and Holmes 1983; Wiens et al. 1987b; Petit, Petit, and Petit, this volume).

Foraging by birds encompasses three primary aspects of resource use. These are: (1) general characteristics of where foraging occurred (Holmes and Robinson 1981), such as species of plant as well as its size, shape, and vigor; (2) specific characteristics of where the bird foraged in relation to the plant used for foraging (MacArthur 1958, Hutto 1981b), including the relative position of the bird within the canopy, the foraging perch, and the height of the bird; and (3) the behavior of the bird: foraging maneuver (e.g., glean, peck, probe), rate of foraging, and foraging substrate (Root 1967, Morrison 1982).

In this paper, I explore geographic variation in the foraging ecology of several breeding and nonbreeding species found in oak and oak-pine woodlands of California. The study was done along a latitudinal gradient of 580 km. My specific objectives were to compare characteristics of the species of plant where foraging occurred, the location of the bird in relation to the plant, and the foraging behavior of the bird.

STUDY AREAS AND METHODS

All study areas were in oak or oak-pine woodlands; each differed from the others in topography and in the structure and composition of the vegetation.

Sierra Foothill Range Field Station (SF), Yuba County, is in the foothills of the Sierra Nevada, about 25 km northeast of Marysville. Elevation ranged from 200 to 700 m on a general west-northwest facing slope. Blue oak (*Quercus douglasii*), interior live oak (*Q. wislizenii*), and digger pine (*Pinus sabiniana*) were the major species of trees with lesser amounts of California black oak (*Q. kelloggii*), valley oak (*Q. lobata*), California buckeye (*Aesculus californicus*), and ponderosa pine (*Pinus ponderosa*). Most stands consisted of mixtures of blue oak, interior live oak, and digger pine, although relatively pure stands of blue or interior live oak were not uncommon. The composition and structure of the canopy and understory had been modified by historic land-use practices. With the exception of 60 ha of fenced areas, the remaining 2500 ha of the Station have been grazed by cattle which has reduced much of the shrub and herbaceous understory. In addition, trees in many stands had been removed, either selectively or completely, as part of range management practices.

San Joaquin Experimental Range (SJ), Madera County, is in the foothills of the Sierra Nevada, about

40 km north of Fresno. Elevation ranged from 200 to 500 m on a generally southwest facing slope. Blue oak, interior live oak, and digger pine were the major tree species and they generally occurred in mixed-species stands. Stands of pure blue oak and blue oak savanna were not uncommon. About 20 ha of SJ were fenced to exclude cattle grazing. Because of shallow soils and a southerly exposure, SJ supported a sparser shrub understory than that found at SF. This understory was further reduced by cattle grazing, resulting in widely scattered stands of mature shrubs.

Tejon Ranch (TR), Kern County, is about 50 km south of Bakersfield in the Tehachapi Mountains. Elevation ranged from 1100 to 1700 m; aspect included all cardinal directions. Major trees found on the ranch included blue oak, valley oak, California black oak, interior live oak, canyon live oak (*Quercus chrysolepis*), Brewer's oak (*Q. garryana* var. *breweri*), and California buckeye. These trees generally occurred in pure stands of single species, with mixes found along narrow ecotones. The terrain of TR was steep and rugged. Consequently, aspect and vegetation changes were distinct, creating a mosaic landscape of various vegetation types. Cattle grazing modified the composition and structure of the shrub and herbaceous layers. Trees were selectively removed as part of a commercial firewood enterprise, which altered the density and size structure of some stands of oaks. Most stands, however, never had any trees removed.

FIELD METHODS

I established about 30 km of transects each at SF and TR. At SJ, I used 40 km of transects previously established by J. Verner for a different study. During the breeding season (April–June 1987) I studied four species of birds that used different modes of foraging: Plain Titmouse (*Parus inornatus*), a foliage- and bark-gleaning generalist; White-breasted Nuthatch (*Sitta carolinensis*), a bark forager; Ash-throated Flycatcher (*Myiarchus cinerascens*), a flycatcher and foliage gleaner; and Western Bluebird (*Sialia mexicana*), a ground forager. During the nonbreeding season (September 1986–February 1987) I studied: Plain Titmouse; White-breasted Nuthatch; Yellow-rumped Warbler (*Dendroica coronata*), a foliage gleaner; and Dark-eyed Junco (*Junco hyemalis*), a ground forager. Three observers collected data during the breeding season, and four in the nonbreeding season. All observers participated in training exercises to standardize the way data were collected. Observers followed the general direction of a transect, staying within 100 m of either side. Once a bird was encountered the observer watched it for 10 s but recorded no data. This allowed the bird to resume "normal" activities after being disturbed and minimized the observer's likelihood of recording only conspicuous activities. During the following 10 s the observer recorded the species, height, diameter, and crown radius of the plant (tree generally) where foraging was observed; the height of the bird and its relative position from the center to the edge of the canopy; foraging and perch substrates (twig [≤1 cm diameter], small branch [1–10 cm diameter], medium branch [10–30 cm diameter], large branch [>30 cm diameter], leaf, ground, air); and the foraging maneuver (search, glean, peck, probe, pluck, flycatch, fly-glean, hover-glean). The location of each foraging sample was marked with plastic flagging and mapped to try to ensure spatial independence of foraging observations for individual species within the same season.

At each study area I used the point-center quarter method (Cottam and Curtis 1956) to sample the relative frequencies, heights, and diameters of the trees occurring there. One hundred points were established at TR and SF using a systematic-random sampling design (Cochran 1977; see Block and Morrison 1987 for the procedures used in point placement). At SJ, I selected 100 points from bird counting stations previously established by Verner. Although I recorded data on the closest tree or shrub within each quarter, I report here only data pertaining to trees.

DATA ANALYSIS

Because some of the species used in this analysis were monochromatic, I pooled sexes of all species. Similarly, ages of many of these birds were difficult to distinguish in the field (i.e., adults, first-year hatching birds), so I pooled data from all ages.

Prior to analysis I developed new variables from the original data using simple arithmetic transformations. Relative height was calculated as the activity height of the bird relative to the total height of the plant. Foraging and search rates were defined as the number of foraging or searching motions per unit of time. Foraging and search times were the total number of seconds during each 10 s sample spent foraging and searching, respectively. Foraging and search speeds were the distances moved per unit of time while foraging or searching.

Variables were analyzed differently depending on whether they were continuous or categorical. I used one-way analysis of variance (Sokal and Rohlf 1969: 204) to test for among-area differences of each species' foraging ecology for diameter, height, and crown radius of the foraging plant; relative position of the bird from the crown center and activity height; foraging rate, time, and speed; and search rate, time, and speed. I used log-linear analyses (Fienberg 1980:13) to compare plant species, foraging activities, and perch and foraging substrates of each species at the three study areas.

I also used log-linear analyses to compare relative frequencies of the tree species used by each bird with the relative frequencies of trees occurring at each study area. I used Mann-Whitney U-tests (Conover 1980) to compare the heights and diameters of the trees used by each species with those of the trees present at each area.

RESULTS

BREEDING BIRDS

Tree use

All birds used tree species with different frequencies in comparisons among and within study areas, except Plain Titmice at SF (Fig. 1; likelihood ratio chi-squares, P < 0.05). White-breasted Nuthatches and Western Bluebirds, however, appeared to use blue oaks more frequently than they occurred at each study area (Fig. 1). Both the nuthatch and titmouse exhibited significant

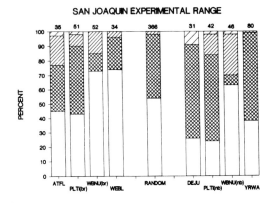

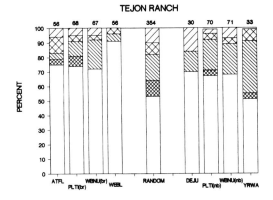

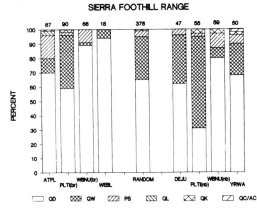

FIGURE 1. Relative frequencies of tree species used by and available to Ash-throated Flycatchers (ATFL), Plain Titmice (PLTI), White-breasted Nuthatches (WBNU), Western Bluebirds (WEBL), Dark-eyed Juncos (DEJU), and Yellow-rumped Warblers (YRWA) in oak woodland habitats in California from September 1986 to February 1987 and March to June 1987. Tree codes are blue oak (QD), interior live oak (QW), digger pine (PS), valley oak (QL), California black oak (QK), and other (OT). Not all bars total 100% because rare categories were not included. Sample sizes are presented on the top of the bars.

TABLE 1. SUMMARY OF COMPARISONS AMONG STUDY AREAS FOR CONTINUOUS VARIABLES OF THE FORAGING ECOLOGIES OF FOUR SPECIES OF BREEDING BIRDS FOUND IN OAK WOODLANDS OF CALIFORNIA FROM MARCH–JUNE 1987

Variable	Ash-throated Flycatcher (58, 42, 71)[a]	Plain Titmouse (69, 55, 92)	White-breasted Nuthatch (67, 52, 66)	Western Bluebird (56, 35, 18)
Tree characteristics				
Diameter		**[b]	**	*
Height	*	**	**	
Crown		*	**	
Foraging location				
Crown position		**	**	
Activity height	*		**	**
Relative height		**	**	**
Foraging behavior				
Foraging rate	**			**
Foraging time	*	**	*	**
Foraging speed	*			**
Search rate		**		
Search time		**		
Search speed				

[a] Number of foraging samples from TR, SJ, and SF, respectively.
[b] One-way analysis of variance: * $P < 0.05$; ** $P < 0.01$.

geographic differences for the heights, diameters, and crowns of the trees (Table 1). In contrast, Ash-throated Flycatchers were consistent in their use of diameters and crowns, and Western Bluebirds consistently used trees of a similar height and crown. There were no significant differences in the diameters of the trees used by Plain Titmice at TR compared to the size of the trees present, nor were there any significant differences in the heights of trees used by titmice at TR and SJ (Mann-Whitney U-tests, $P > 0.05$). The heights of trees used by Ash-throated Flycatchers and Western Bluebirds were similar to their occurrences at TR and SJ, and also for the White-breasted Nuthatch at TR (Mann-Whitney U-tests, $P > 0.05$).

Activity location

Plain Titmice and White-breasted Nuthatches exhibited geographic variation in foraging locations within the tree canopy. The foraging height of the Plain Titmouse was consistent among study areas, but its relative foraging height varied, as did that of the nuthatch and bluebird (Table 1). Geographic variation in the use of foraging perches occurred for all four species (Fig. 2; likelihood ratio chi-squares, $P < 0.05$).

Foraging behavior

Foraging activities and foraging substrates differed among areas for all four species (Fig. 2;

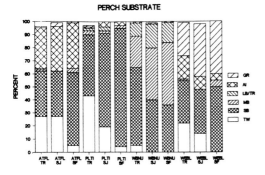

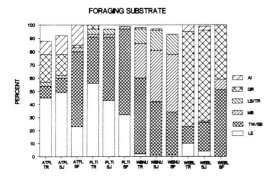

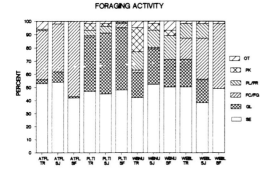

FIGURE 2. Relative frequencies of breeding bird use of perch substrates, foraging activities, and foraging substrates by four species at three oak woodland habitats in California from March to June 1987. Species codes are given in Figure 1, area codes are given in the text. Codes for perch and foraging substrates are leaf (LE), twig (TW), small branch (SB), medium branch (MB), large branch (LB), trunk (TR), ground (GR), and air (AI). Codes for foraging activities are search (SE), glean (GL), flycatch (FL), fly-glean (FG), pluck (PL), probe (PR), peck (PK) and other (OT). Not all bars total 100% because rare categories were not included. Sample sizes are given in Table 1.

likelihood ratio chi-squares, P < 0.05). White-breasted Nuthatches, however, foraged consistently on bark, regardless of location. Depending on the study area, Ash-throated Flycatchers and

TABLE 2. SUMMARY OF COMPARISONS AMONG STUDY AREAS FOR CONTINUOUS VARIABLES OF THE FORAGING ECOLOGIES OF FOUR SPECIES OF NONBREEDING BIRDS FOUND IN OAK WOODLANDS OF CALIFORNIA FROM SEPTEMBER 1986–FEBRUARY 1987

Variable	Dark-eyed Junco (30, 43, 49)[a]	Plain Titmouse (71, 46, 60)	White-breasted Nuthatch (71, 46, 59)	Yellow-rumped Warbler (33, 85, 51)
Tree characteristics				
Diameter	*[b]	**		
Height	**	**	**	*
Crown	*	*		
Foraging location				
Crown position	**	**		**
Activity height				*
Relative height	*	**		**
Foraging behavior				
Foraging rate			**	
Foraging time	**	**	**	**
Foraging speed	**	*		
Search rate	**			
Search time	**			
Search speed	*	**		*

[a] Number of foraging samples from TR, SJ, and SF, respectively.
[b] One-way analysis of variance: * $P < 0.05$; ** $P < 0.01$.

Western Bluebirds foraged at different rates, speeds, and durations. Foraging time differed among areas for both the titmouse and nuthatch, and the titmouse searched at different rates and for different periods of time at each study area.

NONBREEDING BIRDS

Tree use

All birds generally used tree species with different frequencies among and within study areas (Fig. 1; likelihood ratio chi-squares, P < 0.05). Plain Titmice used trees in proportion to their occurrence at TR and Dark-eyed Juncos did so at TR and SJ (likelihood ratio chi-squares, P > 0.05). The trees used by Plain Titmice and Dark-eyed Juncos also differed among areas in diameter, height, and crown radius (Table 2). The heights of trees used by White-breasted Nuthatches and Yellow-rumped Warblers differed among study areas (Table 2). The diameters of trees used by each bird differed from those present at each study area in all comparisons, except for titmice at TR (Mann-Whitney U-tests, P < 0.05). The heights of trees used by titmice, juncos, and warblers were similar to those present at TR and SJ, and similarly for nuthatches at TR (Mann-Whitney U-tests, P > 0.05).

Activity location

The use of foraging perches differed among study areas for all birds (Fig. 3; likelihood ratio

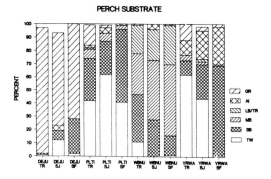

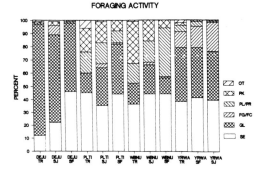

FIGURE 3. Relative frequencies of nonbreeding bird use of perch substrates, foraging activities, and foraging substrates by four species at three oak woodland habitats in California from September 1986 to February 1987. Species codes are given in Figure 1, area codes are given in the text. Substrate and activity codes are given in Figure 2. Not all bars total 100% because rare categories were not included. Sample sizes are given in Table 2.

chi-squares, $P < 0.05$). Dark-eyed Juncos, Plain Titmice, and Yellow-rumped Warblers foraged at different positions in the canopy, and at different relative heights. Activity heights for the warbler differed among study areas (Table 2).

Foraging behavior

All birds but the Yellow-rumped Warbler showed geographic variation in foraging activities (Fig. 3; likelihood ratio chi-squares, $P < 0.05$). Foraging substrates used by warblers, however, differed among study areas, as did those used by juncos, nuthatches, and titmice (Fig 3; likelihood ratio chi-squares, $P < 0.05$). Dark-eyed Juncos foraged at a different speed and for a different duration, and they searched at different rates, speeds, and durations at each study area. Foraging times also differed for the Plain Titmouse, White-breasted Nuthatch, and the Yellow-rumped Warbler. Titmice searched and foraged, and warblers searched, at different speeds at the study areas. Nuthatches foraged at different rates among areas (Table 2).

DISCUSSION

I found geographic variation in all aspects of the foraging ecologies of both breeding and nonbreeding birds in California oak woodlands. The types and magnitudes of this variation appeared to be species-specific, as species exhibited variation differently for tree use, foraging location, and foraging behavior. This agrees with and extends studies during the breeding season (e.g., Maurer and Whitmore 1981, Sabo and Holmes 1983, Wiens et al. 1987b) to the nonbreeding season.

Wiens et al. (1987b) attributed variations in foraging behaviors to differences in local environmental conditions. The vegetation at each of my study areas was unique in structure and composition. Consequently, the types and dispersion of suitable foraging patches, biotic pressures, such as competition and predation (Wiens 1977, Ekman 1986), and weather (e.g., see Grubb 1975) all probably interacted to influence local patterns of foraging.

Studies restricted to one location probably provide little insight into the type or magnitude of a species' plasticity. By contrast, studies that pool data across areas may include too much variation and, subsequently, the results may conceal site-specific behaviors. Investigators should not extend results from one location to different locations or base analyses on data pooled from geographically distinct areas.

ACKNOWLEDGMENTS

This paper benefitted from the comments of J. Bartolome, P. Beier, L. A. Brennan, B. Maurer, M. L. Morrison, and J. Verner. B. Griffin, J. C. Slaymaker, and M. L. Morrison assisted in the field. J. M. Connor, D. A. Duncan, and D. A. Geivet facilitated study at Sierra Foothill Range Field Station, San Joaquin Experimental Range, and Tejon Ranch, respectively. I thank I. Timossi for invaluable computer advice and

G. Jongejan for assistance in preparation of the figures. The University of California, Division of Agriculture and Natural Resources, Integrated Hardwood Program, and United States Forest Service, Pacific Southwest Forest and Range Experiment Station provided funding for this project. The Department of Forestry and Resource Management, University of California, provided computer time and resources.

SEX, AGE, INTRASPECIFIC DOMINANCE STATUS, AND THE USE OF FOOD BY BIRDS WINTERING IN TEMPERATE-DECIDUOUS AND COLD-CONIFEROUS WOODLANDS: A REVIEW

Thomas C. Grubb, Jr. and Mark S. Woodrey

Abstract. Most reports addressing the importance of food resources for the biology of wintering forest birds do not distinguish among sex, age, and intraspecific social dominance categories of the individuals studied, even though these can have widespread effects on the relationships between birds and their food. Here we review a selection of recent findings, emphasizing birds wintering in bark-foraging guilds in temperate-deciduous and cold-coniferous forests. We first examine how birds of different sex, age, and dominance differ in where, when, and how they look for food, and in the kinds of food they eat. We include varying tendencies to forage socially and to locate food by copying the behavior of others. We then consider how birds of different sex, age, or dominance status may store energy acquired during feeding, either externally (e.g., caching) or internally (storage of subcutaneous fat). The third section examines how stored energy is consumed by birds differing in dominance status and considers diurnal existence metabolism, nocturnal hypothermia, and roosting. This is followed by an overview of sex, age, and social dominance status and food use.

Key Words: Age; cold-coniferous forest; dominance; food acquisition; food consumption; food storage; forest birds; sex; temperate-deciduous forest.

Woodland birds have been prime material for population and community ecologists. Since the pioneering work of Hartley (1953), Gibb (1954, 1960), Betts (1955), and MacArthur (1958), dozens of reports have focused on the dimensions of species-specific niches and how such niches might overlap enough to control population sizes. Largely, it seems, because of the theoretical framework constructed by Hutchinson (1957), research on foraging niches and food has concentrated at the level of the population. Although the situation has changed enough recently to provide sufficient material for this review, the bulk of work on the foraging, feeding and physiology of terrestrial birds still lumps results for all birds of a species, regardless of sex, age or intraspecific social dominance status. Such a procedure rests on the often unappreciated assumption that niche differences among birds of different sex, age and dominance status are so minor that any population of a given species may be characterized by one realized niche. The major aim of our review will be to demonstrate that this assumption is incorrect.

We emphasize studies on bark-foraging birds wintering in temperate-deciduous and cold-coniferous forests. In practice, most of the available literature on members of this guild concerns either woodpeckers (Family Picidae) or true tits, titmice and chickadees (Family Paridae). Lesser amounts of material exist for nuthatches (Family Sittidae), creepers (Family Certhiidae), Bushtit and Long-tailed Tit (*Psaltriparus minimus* and *Aegithalos caudatus* respectively, Family Aegithalidae) and for the Goldcrest (*Regulus regulus*), Firecrest (*R. ignicapillus*) and the kinglets (Subfamily Sylviinae, Family Muscicapidae).

We begin with examples of how sex, age, and dominance status are correlated with where, when, and how such birds look for food and what they eat. We include differing tendencies to forage in the company of conspecifics and heterospecifics and to copy the food-finding activities of other foragers. In this section we first encounter a difficulty that recurs throughout the review; sex, age, or dominance status are usually autocorrelated. This makes it difficult to assign differences among birds in their relations to the food supply to any one of these variables. For example, among the parids, most of which winter in social groups larger than two, social dominance rank is highest in adult males, all males (both adult and juvenile) usually dominate all females (both adult and juvenile), and juvenile females have the lowest dominance rank (e.g., Smith 1967; Saitou 1979; Ekman 1987; Hogstad 1987a; Grubb and Waite, unpubl.). Nevertheless, we will often treat the relationships between sex, age, or dominance status and foraging separately, partly because many authors have studied only one or two of these factors.

From food acquisition, we examine how bark foragers of differing status may store energy for future use. We consider both the external storage involved with caching or hoarding food items within the home range and the internal storage of energy in the form of subcutaneous fat, as well as variations in existence metabolism during daytime, nocturnal hypothermia, and roosting.

TABLE 1. Sex-specificity of Foraging by Woodpeckers along Two Niche Axes

Species	Location	Tree type[a]	Sex higher in tree	Sex on smaller diameter substrate	Reference
Ladder-backed Woodpecker (*Picoides scalaris*)	Arizona	EB	F	F	Austin (1976)
Nuttall's Woodpecker (*Picoides nuttalli*)	California	DB EB	F	F	Jenkins (1979)
White-headed Woodpecker (*Picoides albolarvatus*)	California	C	M	M	Koch et al. (1970)
	California	C	M	M	Morrison and With (1987)
Downy Woodpecker (*Picoides pubescens*)	New Hampshire	DB	M	M	Kilham (1970)
	Kansas	DB	F[b]	M	Jackson (1970)
	New Jersey	DB	M	M	Grubb (1975)
	Illinois	DB	F	M	Willson (1970)
	Ohio	DB	M	M	Peters and Grubb (1983)
	Ohio (in lab)	DB		Neither	Pierce and Grubb (1981)
	Virginia	DB	M	M	Conner (1977)
Hairy Woodpecker (*Picoides villosus*)	Virginia	DB	F	F	Conner (1977)
	California	C	M	M	Morrison and With (1987)
Three-toed Woodpecker (*Picoides tridactylus*)	Norway	C	F	F	Hogstad (1976)

[a] EB = Evergreen broadleaf; DB = deciduous broadleaf; C = conifer.
[b] On live trees.

FOOD ACQUISITION: FORAGING AND FOOD

Sex

Because of noticeable differences in bill or body size between the sexes (e.g., Selander 1966) and because free-ranging birds can usually be sexed by plumage differences, sex-specific foraging in woodpeckers has been well documented. A sample of sex differences in five species of similar size, taken from studies since Selander's (1966) review (Table 1), reveals some cases in which the male forages higher in trees on thinner diameter substrates and other cases in which the opposite result holds. So far as we know, males are socially dominant to females in all the species.

Two experimental studies suggest that males may choose the more productive portion of the forest and exclude females from such sites. Pierce and Grubb (1981) showed that when isolated Downy Woodpeckers were given a choice of branch sizes under controlled laboratory conditions, both males and females foraged on 5-cm-diameter substrates most often, 2.5-, 10.0-, and 20.0-cm-diameter branches about equally, and ignored 0.5- and 1.0-cm-diameter twigs. This was approximately the range of substrates selected by free-ranging male Downy Woodpeckers in the studies cited in Table 1. Furthermore, in one study in which males were removed experimentally, females became male-like in their foraging-substrate selection, but males did not change their substrate selection after females were removed from another site (Peters and Grubb 1983). Observations by O. Hogstad (pers. comm.) suggest that sex-specific foraging is also a function of male dominance in Three-toed Woodpeckers. In a subalpine, mixed-deciduous, broadleaf, and coniferous forest in central Norway, he found that males and females often maintained separate winter territories. When in separate territories, the two sexes foraged mostly low (<5 m) on dead birch trunks with diameters >15 cm. However, when the sexes were together in a territory, females foraged mainly higher and on thinner trunks.

During the winter, those woodpeckers that are primarily inscctivorous subsist chiefly on beetles and ants and, overall, the sexes do not differ markedly in their diets in any consistent fashion (Table 2). One interesting exception has been noted in Downy Woodpeckers. The old-field plant, Canadian goldenrod (*Solidago canadensis*), is host to several gall-inducing insects. Where old fields abut woodland habitat, wintering Downy Woodpeckers excavate galls and extract the larval insects (Confer and Paicos 1985, Confer et al. 1986), but the birds involved seem always to be males (J. L. Confer, pers. comm.; TCG, pers. obs.). Whether this difference occurs elsewhere and what might be the mechanism for segregation is unknown.

The reliance of certain melanerpine woodpeckers on stored mast crops during the winter will be dealt with in the section concerning food storing. It is worth noting here, though, the striking lack of sex-specific plumage in those solitary species that store food in defended "larder

TABLE 2. SEX-SPECIFICITY OF DIETS IN WINTERING WOODPECKERS

Species	Location	Principal foods	Differences between sexes	References
Grey-headed Green Woodpecker (*Picus canus*)	Japan	Ants, berries	Little	Matsuoka and Kojima (1985)
Black Woodpecker (*Dryocopus martius*)	Japan	Ants	Little	Kojima and Matsuoka (1985)
White-headed Woodpecker (*Picoides albolarvatus*)	California	Coleoptera, Homoptera, Hymenoptera, pine seeds	Males eat more pine seeds, fewer homopterans	Koch et al. (1970) Otvos and Stark (1985)
Hairy Woodpecker (*Picoides villosus*)	California	Coleoptera, Hymenoptera	Males eat more beetles	Otvos and Stark (1985)
Downy Woodpecker (*Picoides pubescens*)	California	Coleoptera, Hymenoptera	Little	Otvos and Stark (1985)
	Illinois	Coleoptera, Hymenoptera	Males eat more ants, berries and seeds, fewer homopterans and spiders	Williams and Batzli (1979)
Three-toed Woodpecker (*Picoides tridactylus*)	Norway	Coleoptera	Little	Hogstad (1970)

hoards" rather than spreading it out through the home range in nondefended "scatter hoards," perhaps because females must defend their larder hoards against intruding males (Kilham 1978).

Much less is known about sex-specific foraging and food supplies in other types of bark-foraging birds, partially because sexual status may be difficult to ascertain. In one study of Great Tits (*Parus major*) in southern England, fewer males than females were actively looking for food at the time they were sighted (Grubb 1987). Also, males tended to forage more while solitary and less in mixed-species flocks than did females. These differences are consistent with the idea that males monopolized the supply of hazel nuts (*Corylus avellana*) by supplanting any feeding females. Accordingly, the females turned to foraging in flocks of heterospecifics, where they presumably could reduce the proportion of time committed to vigilance for predators (Pulliam 1973).

Grubb and Waite (unpubl.) have recently examined the extent to which birds in mixed-species flocks imitate the food-finding behavior of other flock members. They made a cryptic source of mealworms available to free-ranging birds wintering in Ohio and observed how the information circulated among the members of a resident mixed-species flock. In comparing the use of this resource by males and females of four species using the same feeding site, regardless of the dominance status of birds within each sex (Fig. 1), they found that male Carolina Chickadees (*Parus carolinensis*) copied by local enhancement (*sensu* Thorpe 1963) significantly sooner than did females; such comparisons were nonsignificant for Tufted Titmice (*P. bicolor*), White-breasted Nuthatches (*Sitta carolinensis*) and Downy Woodpeckers. Also, the average female titmouse and woodpecker took significantly more food items than did their male counterparts; the sexes of chickadee and nuthatch did not differ. After the food source had been artificially "depleted," males and females of all four species inspected the location where the food had been about two or three times each before leaving the vicinity; none of the within-species male-female differences was significant.

Among White-breasted Nuthatches, differences in male and female foraging behavior have been observed in Colorado and Ohio. In a Colorado pine forest, McEllin (1979) found that females foraged significantly higher and on significantly smaller branches. In the deciduous forests of central Ohio in winter, males and females were statistically indistinguishable (Grubb 1982a); furthermore, no differences were detected when Ohio nuthatches were brought into the laboratory and tested in isolation (Pierce and Grubb

1981). Whether the degree of sex specificity of foraging in White-breasted Nuthatches is a response to food distribution, prevailing climate, or the interspecific social environment remains to be determined.

AGE

Age and dominance status are so tightly correlated in bark-foragers that we deferred most evidence concerning age effects to the category of dominance status. However, one finding that warrants attention here concerns differences in the abilities of adult and juvenile tits to imitate food-finding activities (Sasvari 1985). In a laboratory study, adult Great Tits copied the food-locating behavior of a "teacher" slightly sooner than did juveniles regardless of whether the teacher was a conspecific or heterospecific. Conversely, juvenile Blue Tits (*Parus caeruleus*) responded faster than did adults. Unfortunately, the results of this study are difficult to interpret because appropriate control groups were lacking. Also, every experimental group was arranged so that the "teacher" was socially subordinate to the "learner." Commonly, a potential learner stole the food item retrieved by the teacher, so that the apparent copying of food finding could have been confounded by the effects of kleptoparasitism, rather than being an unambiguous example of local enhancement.

Many woodpeckers defend territories and exist more or less independently of conspecifics during the winter. Relationships between age and foraging behavior may be more clearly disentangled from dominance status in this group, but we have not found attempts to do so.

DOMINANCE

Social dominance is a mechanism of interference competition in which one animal uses its dominant status to secure priority of use of some resource. In practice, dominance ranking among members of a social group is determined using several techniques (Dixon 1965, Smith 1976, Baker et al. 1981, Brawn and Samson 1983, DeLaet 1984, Schneider 1984). Among the woodpeckers that subsist largely on (uncached) invertebrates, males dominate females socially and apparently use their dominant status to reserve certain portions of the habitat for their own food searching. In those melanerpine woodpeckers that store mast in larder hoards, both adults and juveniles may defend small territories containing their caches. Although differences in this behavior have apparently not been examined systematically, it appears that in Ohio, beech-mast storing adult Red-headed Woodpeckers (*Melanerpes erythrocephalus*) may defend larger,

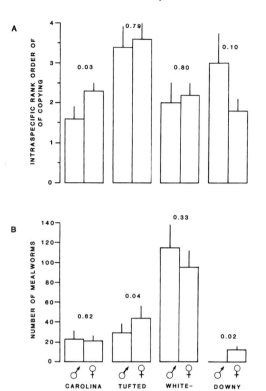

FIGURE 1. Relationships between two measures of copying by local enhancement and sex in four species of a deciduous-forest bark-foraging guild. A. Mean rank order of copying by males and females of each species. B. Mean number of mealworms taken from the hidden supply over the course of 9 hours by males and females of each species. The vertical bars represent standard errors of the mean. All sample sizes are five, the number of different flocks studied. Numbers associated with each pair of bars are probabilities of significance derived from paired t-tests on the means of the sexes (from Grubb and Waite, unpubl.).

better-quality territories than juveniles (TCG, unpubl. data).

Among bark-foraging guilds, the relationships between social dominance status, food-related activity and their consequences for survival have been best studied in the Family Paridae. In virtually all species studied, birds overwinter in small parties consisting of the adult male and female that bred in the area and two or more juveniles (birds of the year). Usually juveniles disperse from their natal sites in late summer and early fall (Greenwood and Harvey 1982), so that the members in any one group are seldom parents and offspring. The number of males and females is usually the same within any given flock, and even during the early winter flock members appear to

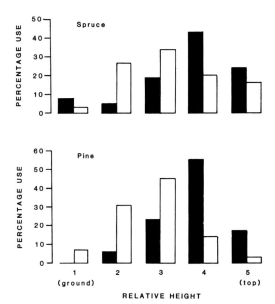

FIGURE 2. Relative heights at which dominant (solid bars) and subordinate (open bars) Willow Tits foraged in spruce and pine trees in Sweden during December (after Ekman 1987).

be organized into male-female pairs. By midwinter, the strongest attachments are between males and females that will form breeding pairs the following spring (Ficken et al. 1981). Thus, although all males in a flock dominate all females in "one-on-one" encounters, females mated to high-ranking males attain some additional social status, which, for instance, may confer more ready access to a food supply (Dixon 1963, 1965; Ritchison 1979; Hogstad 1987a; also see below). In addition to birds integrated into sedentary nonbreeding flocks, some individuals remain as "floaters." Most are apparently juveniles of very low rank (Ekman 1979); some incorporate themselves into flocks over the course of the winter, as group members die (Ekman et al. 1981).

Ekman and his colleagues have performed an impressive investigation of the relationship between social dominance status and foraging activity in Willow Tits (*Parus montanus*) wintering in southwest Sweden; and Hogstad (1987c) has conducted similar work in central Norway. In both locations, the birds live in flocks that defend group territories against conspecifics throughout the winter. In Sweden, a flock usually begins in the fall with two dominant adults and two subordinate juveniles, while in Norway the two dominant adults are joined by four juveniles.

The members of the flock gain in feeding efficiency by pooling their vigilance time. However, the gain is not symmetrical, because in Sweden the dominant birds force subordinates to forage lower in the more open parts of conifers (Ekman 1987; Fig. 2), where they are more susceptible to predation (Ekman 1987, Ekman et al. 1981). The subordinates compensate by raising their level of vigilance beyond that of the more protected adults, but not to the level shown by birds foraging only in pairs. Thus, among juveniles, foraging with dominant conspecifics is favored over leaving their company to forage as a pair.

Hogstad (1987c) performed a removal experiment with Willow Tits that supported the hypothesis that social dominance was responsible for age-specific foraging niches. When adults were removed from flocks, the juveniles moved significantly higher in pine trees to forage in adult-like locations. In a similar experiment in Norway, Hogstad (1988) positioned feeders in a treeless bog 1, 3, 5, 10, and 20 m from the edge of a forest. Dominant tits tended to visit only those close to the woodland, forcing subordinates to feeders out in the bog. When the feeders 10 m and 20 m from cover were the only ones baited, only juveniles visited them. In general, the level of vigilance increased as Willow Tits used feeders farther from the forest, a finding consistent with the notion that predation risk increased with distance to cover. In juvenile females, however, the most subordinate category of bird, birds spent more time scanning the environment from a feeder 1 m from the woodland than from feeders 3 m and 5 m out, perhaps because they were trying to avoid supplanting attacks by dominants that preferred the feeder closest to the forest.

Waite (1987a, b) proposed that some of the vigilance time of subordinate birds is used to keep track of dominant flock mates. In laboratory experiments involving both Tufted Titmice and White-breasted Nuthatches, the dominant member of a pair scanned its surroundings more when kept in isolation than when housed with the subordinate member. However, the subordinate bird of a pair was more vigilant in the company of the dominant conspecific than when alone.

The study on imitative foraging in mixed-species flocks (Grubb and Waite, unpubl.) that we cited above provided insight into how intrasexual dominance status can affect food finding. In five different flocks tested in five different woodlots in central Ohio, a strong tendency existed for the dominant male or dominant female Carolina Chickadee, Tufted Titmouse, or White-breasted Nuthatch to take the first food item from a cryptic source. In Carolina Chickadees the dominant male or female took a food item after seeing the fewest similar items taken by other birds. In male titmice and female nuthatches, the dominant individual took significantly more food items over the course of the experiment than did individuals of lower social rank. The same trend occurred for male and female chickadees, female titmice and male nuthatches, but was not significant.

Earlier we mentioned that the pair bond maintained in winter flocks appears to be stronger in the dominant than in more subordinate birds. This may influence the way birds find food (Fig. 3). In both Carolina Chickadees and Tufted Titmice, the two birds of the alpha pair first used the hidden food supply more quickly than did the male and female of subordinate pairs.

EXTERNAL FOOD STORAGE

Caching or hoarding behavior has been shown in 12 of 170 bird families (D. F. Sherry 1985). Two major patterns have been distinguished. Larder-hoarding, the storage of food items in large central caches in an animal's home-range, is best demonstrated by the Acorn Woodpecker (*Melanerpes formicivorous*; MacRoberts and MacRoberts 1976). Among scatter-hoarders, those animals that store food items at dispersed sites, the most studied in North America is the Clark's Nutcracker (*Nucifraga columbiana*; Tomback 1982). Several studies of scatter-hoarders have found nonoverlapping distributions of caches for individual birds but have not associated caching behavior with age, sex and/or dominance status (Cowie et al. 1981, Sherry et al. 1982, Clarkson et al. 1986).

SEX AND DOMINANCE

Because nuthatches (Family Sittidae) are typically sexually dimorphic in plumage, this group lends itself to the study of sex-specific caching behavior. Because sex, age and dominance are autocorrelated, we have combined consideration of sex and dominance. In the European Nuthatch (*Sitta europaea*), males dominate females, and adults dominate juveniles of the same sex (B. Enoksson, unpubl.). Within mated pairs of White-breasted Nuthatches in North America, males are dominant over females (Grubb 1982a).

Moreno et al. (1981) studied the hoarding behavior of one pair of European Nuthatches during autumn and winter in central Sweden. They found that the male stored food items at the periphery of the pair's communally held territory significantly more frequently than did the female, and that the female placed significantly more of her caches under lichen than did the male. The female was more generalized in all niche dimensions of hoarding, perhaps indicating the female's subordinate status, causing her to broaden the use of storage sites due to interference competition from her mate. On five occasions, they

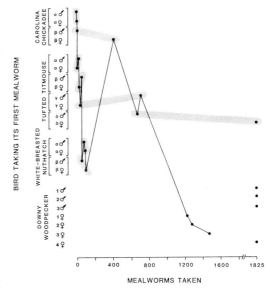

FIGURE 3. Copying by local enhancement in 23 deciduous-forest, bark-foraging birds recorded as having been in attendance during an experimental trial at one woodland site. The independent variable was the number of mealworms that other birds had removed from a cryptic supply before each bird took its first mealworm from the same place. Birds connected by stippling were mated pairs. The gamma female Tufted Titmouse and the four Downy Woodpeckers denoted by the closed circles at the far right of the figure never copied (from Grubb and Waite, unpubl.).

observed the male apparently watching the female store a food item. Each time, the male removed it and then cached it at a different site. Together, the observation of cache covering by the female and the instances of stolen caches suggest that the members of the pair did not share their caches.

Sexual differences in caching behavior of European Nuthatches have also been found by B. Enoksson (unpubl.) who found that males made longer visits to a feeder and also took more seeds per visit. This pattern was not influenced by age.

In a study of diurnal caching rhythms in free-ranging White-breasted Nuthatches, Woodrey and Waite (unpubl. ms) found that caching intensity for both the dominant male and female nuthatches at a feeder was negatively correlated with time of day. Within each alpha pair, the female cached a higher proportion of kcal than did the male (Fig. 4). In another context, Hogstad (1987c) observed five instances of adult male Willow Tits kleptoparasitizing seeds newly cached by juvenile conspecifics. Four of the seeds were eaten immediately, while the fifth was removed and cached elsewhere.

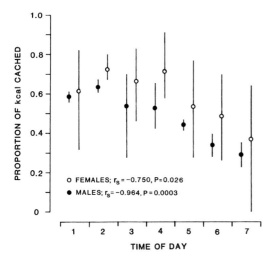

FIGURE 4. The relationship between time of day and the proportion of kcal cached by dominant male and female White-breasted Nuthatches. The circles represent means for a given septile and the vertical bars represent ranges. The P-values are for one-tailed Spearman's rank correlation tests (after Woodrey and Waite, unpubl.).

studies have distinguished among sex, age or dominance categories.

SEX

From studies of passerines in Finland, Lehikoinen (1986, 1987) proposed that the increase in minimum morning weights, a measure of the degree of winter fattening, was determined by seasonally variable risk of starvation, and that temperature and photoperiod were proximate factors that mainly regulated the daily weight amplitude. The least winter fattening was found in the Certhiidae, Sittidae and Passeridae, and in those parids with predictable food availability as a result of their hoarding behavior. Although age and sex were unrelated to fat levels in most species, Lehikoinen did find sexual differences in the Greenfinch and the Blue Tit (Table 3). In both species males showed a significantly greater increase in winter weight than did females.

AGE

Crested and Willow tits wintering in Finland showed age related differences in fat loading (Table 3). In both species, juveniles gained significantly more weight than adults. Because adults are dominant (Ekman and Askenmo 1984), the age differences in winter fattening cannot be explained by social constraints (Lehikoinen 1986), but may be connected with their habit of caching food (Haftorn 1956). Caches are presumably a predictable food resource for adults and decrease the benefit of fat deposition.

Lima (1986) modelled the body mass of small wintering birds as a trade-off between the risks of predation and starvation. Because of age differences in dominance, juveniles could be forced into areas of increased predation risk (e.g., Ekman and Askenmo 1984, Hogstad 1987c), and attain lower fat levels than adults. Lehikoinen's (1986) finding of heavier weights in juveniles seems to be inconsistent with this prediction.

AGE

Few studies have addressed age-related differences in caching behavior. Enoksson (unpubl.) found that adult European Nuthatches collected more seeds per visit to a feeder than did juveniles during summer. Between summer and autumn, however, the juveniles increased both the number of seeds collected per unit time and the total number taken per visit. Changes in dominance status did not fully explain the observed changes because not all juveniles were subordinate to all adults. Enoksson concluded that juveniles became more proficient foragers later in the year, due both to increasing experience and to enhanced ability to handle seeds, the latter possibly associated with growth of young birds' bills.

Haftorn (1956) documented the development of scatter-hoarding in young tits. In July, juvenile Willow Tits (*Parus montanus*) and Crested Tits (*P. cristatus*) were not yet proficient in caching. Of 15 attempts by birds of both species combined, 47% were unsuccessful, largely because the young birds tended to remove a newly cached item immediately and transplant it to another site. By August, however, only 14% of attempts were unsuccessful and only 24% of newly cached seeds were transplanted.

INTERNAL FOOD STORAGE

Winter fattening is a common phenomenon in small birds in the temperate zone (King 1972; Blem 1976; Lehikoinen 1986, 1987). Although the literature is extensive (e.g., Blem 1976), few

FOOD CONSUMPTION: BIOENERGETICS AND ROOSTING

Food stored externally in caches or internally as subcutaneous fat is consumed by birds during the diurnal and roosting phases of their diel cycles. In light of the differences in food acquisition and storage among sexes, ages and dominance categories described above, comparison of intraspecific categories at the physiological level could prove worthwhile. Unfortunately, such comparisons have not been made, so we combine discussion of the three category types.

DIURNAL METABOLISM

Roskaft et al. (1986) studied the daytime metabolic rates of Great Tits captured in winter near Trondheim, Norway, and kept for one week in

TABLE 3. Increase in the Winter Weight and the Seasonal Minimum Morning Weight in Passerine Species Retrapped in Southwest Finland. Weight Is Defined as the Mean of the Lowest Individual Morning Weights in March through April (after Lehikoinen 1986)

Species	Group	Weight (g)	Increase of winter weight			Increase (%)	P[a]
			$\bar{X}$	SE	N		
Great Tit (*Parus major*)	Male	18.4	0.88	0.073	125	4.8	***
	Female	17.3	0.86	0.114	85	5.0	***
Blue Tit (*P. caeruleus*)	Male	11.0	0.98	0.089	46	8.9	***
	Female	10.5	0.48	0.154	14	4.6	**
Coal Tit (*P. ater*)	All	8.9	0.70	0.210	3	7.8	0
Crested Tit (*P. cristatus*)	Adult	11.3	0.08	0.147	8	0.7	ns
	Juv.	10.9	0.54	0.014	8	5.0	**
Willow Tit (*P. montanus*)	Adult	11.1	0.45	0.136	11	4.0	**
	Juv.	10.8	0.72	0.093	32	6.7	***
Greenfinch (*Carduelis chloris*)	Male	26.5	3.33	0.472	21	12.6	***
	Female	28.3	3.09	0.798	10	10.9	**
Bullfinch (*Pyrrhula pyrrhula*)	Male	32.3	4.45	1.242	9	13.8	**
	Female	28.5	6.92	1.331	5	24.3	**
Yellowhammer (*Emberiza citrinella*)	Male	28.9	1.50	0.216	29	5.2	***
	Female	28.7	1.47	0.276	22	5.1	***

[a] P values taken from paired *t*-tests are denoted as ns if not significant, 0 if <0.1, ** if <0.01, and *** if <0.001.

an indoor aviary prior to testing. They found a positive correlation between metabolic rate and the width of a tit's breast-stripe. After noting that social dominance is correlated with and, indeed, may be indicated to other birds by the width of a Great Tit's breast-stripe, Roskaft et al. concluded that dominant birds in nature appear to have a higher energy requirement than subordinate birds. However, because the breast stripe is wider in males than in females and wider in adult males than in juvenile males (A. G. Gosler, pers. comm.; TCG, pers. obs.), Roskaft and his colleagues may have actually demonstrated a sex-related rather than dominance-related effect.

Similar results from a study of Willow Tits in central Norway (Hogstad 1987b) are more convincing, because the compositions of test flocks were natural and because the design included an experimental manipulation of dominance status. Metabolic rates were determined for Willow Tits newly captured from six different flocks. Each flock consisted of six birds in the usual arrangement for parids, with the numbers of males and females being equal, in this instance divided into one adult pair and two juvenile pairs. All males dominated all females and within a sex the adults dominated the juveniles. This order of dominance was almost perfectly correlated with the order of metabolic rates, which were highest in the most dominant male and lowest in the most subordinate female. Interestingly, when the dominant male was removed from a flock, the formerly beta male assumed the dominant position and his metabolic rate increased to that of the previous dominant. No other bird in the hierarchy changed its metabolic rate. Furthermore, when a female was removed, no other bird, female or male, changed its metabolic rate (Fig. 5). Within a flock of wintering Willow Tits (and, apparently, in most other parids as well; e.g., Dixon 1963) the dominant adult male is most involved with defending the group's territorial boundaries and in most cases he will breed in the area in which his flock overwintered. Hogstad interprets his results as suggesting that the alpha bird of a flock requires a high metabolic rate to support the level of aggressive behavior required in boundary defense.

Related to physiological differences among sex, age, and dominance categories is the study of Silverin et al. (1984) on the hormonal levels of Willow Tits wintering in Ekman's study area east of Gothenburg. Within each flock of four birds (adult male and female and juvenile male and female), juvenile females were found to have higher levels of testosterone and corticosterone than adult birds and a higher level of dihydrotestosterone than juvenile males and adult birds, perhaps indicating that juvenile females—the most subordinate of the four age-sex categories—were the most stressed (O. Hogstad, pers. comm.).

NOCTURNAL METABOLISM AND HYPOTHERMIA

Much is known about nocturnal hypothermia in bark-foraging birds, but little of this information bears on differences among sex, age and dominance classes. We draw the following major conclusions from Reinertsen's (1983) review. (1) Nocturnal hypothermia involves the controlled reduction in nighttime body temperatures to values <10°C below body temperature during the day. (2) Such a reduction results in a 10% re-

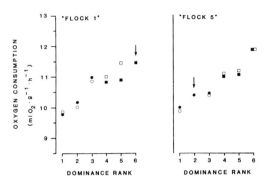

FIGURE 5. Oxygen-consumption rates of males (squares) and females (circles) in two Willow Tit flocks in Norway. Solid and open symbols, respectively, denote oxygen consumption rates before and after the bird designated by the arrow was removed from each flock (redrawn from Hogstad 1987b).

duction in consumption of stored energy. (3) Among members of the bark-foraging guild, nocturnal hypothermia is well-known among parids, having been documented in the Black-capped Chickadee and in Siberian Tits (*Parus cinctus*), Great Tits, and Willow Tits, as well as the Carolina Chickadee (Munzinger 1974). (4) No other taxa of the bark-foraging guild of temperate-deciduous and cold-coniferous woodlands appear to have been tested for nocturnal hypothermia; a search among the Picidae, Sittidae, Sylviinae and Certhidae could be successful. During midwinter, one of us (TCG) has encountered roosting Downy Woodpeckers that appeared to be in a torpid state. (5) Among the parids in which hypothermia has been shown, no published information exists for sex, age and dominance classes about either the tendency to enter hypothermia or the degree to which the body temperature is allowed to drop. However, Hogstad reports (pers. comm.) that although dominant male Willow Tits had the highest metabolic rate during the day, the metabolic rates of dominant males and subordinate females were the same while they roosted at night.

Certain results with Willow Tits suggest that the extent of nocturnal hypothermia may vary with sex, age or dominance status. The body temperature of a roosting bird is the result of an interaction among the ambient temperature, the bird's insulative properties, and its metabolic rate. In Willow Tits implanted with temperature-reading radio transmitters, it was found that the reduced resting metabolic rate and, therefore, the extent of steady-state hypothermia were apparently maintained at constant values from shortly after roosting time until arousal the following morning. Furthermore, the extent of reduction in metabolic rate was related to a bird's fat store at roosting time. The leaner the bird, the lower the resting metabolic rate and the more profound the hypothermia (Reinertsen and Haftorn 1983). Thus, if birds of different sex, age or dominance rank go to roost with characteristically different energy reserves, as outlined above, the extent of hypothermia could vary according to their status.

ROOSTING

Wintering bark-foraging birds usually roost under some sort of protective cover. Birds wintering in northern forests generally roost in tree cavities (Kendeigh 1961) and some may burrow through snow cover to roost in underground rodent burrows (e.g., Willow Tit; Zonov 1967). Cavity roosting confers a considerable savings in metabolic expenditure because of the reduction in radiative and convective heat loss (e.g., Askins 1981). Where nighttime temperatures are warmer, birds of "cavity roosting species" and other members of the bark-foraging guild may sometimes roost outside of cavities (Red-bellied Woodpecker [*Melanerpes carolinus*], Saul and Wassmer [1983]; Red-cockaded Woodpecker [*Picoides borealis*], Hooper and Lennartz [1983]). Downy Woodpeckers wintering in Ohio dug cavities in artificial snags made of polystyrene, a highly insulative material, and roosted there throughout the winter (Grubb 1982b, Peterson and Grubb 1983), whereas only one bird of the same species wintering in the considerably milder conditions of east Texas did so (R. N. Conner, pers. comm.).

Whether in cavities or in the cover of heavy foliage or snow, most bark foragers roost in isolation, but huddling in clumps is not unknown, particularly among birds of small body size in which surface-to-volume ratios, and consequently metabolic costs, are high (e.g., Long-tailed Tit [*Aegithalos caudatus*], Lack and Lack [1958]; Bushtit [*Psaltriparus minimus*], Smith [1972]). A flock of 29 Bushtits wintering on the University of Washington campus normally maintained an individual distance of about 5 cm while roosting. However, in a 2-week period one January, during which nighttime temperature fell below freezing—an unusual occurrence in Seattle—virtually all of the birds packed tightly against one another along a perch (Smith 1972).

Little evidence points directly to differences among sex, age and dominance classes with respect to roosting. An exception is an analysis of nestbox use by Great Tits in the Netherlands. During the time of leaf-fall in late October and early November Kluyver (1957) found that Great Tits stopped roosting in the crowns of trees and began to compete for nest boxes. Sixty-nine percent of adult males, 63% of adult females, 54% of juvenile males, but only 31% of juvenile females were found roosting there, a ranking that parallels dominance status. Winkel and Winkel (1980) reported similar asymmetries among Great Tits roosting in nest boxes in Germany. Kluyver (1957) concluded that competing to roost in nest boxes was adaptive because, although the overall overwinter survivorship of males (50.0%) was significantly greater than that for females (45.5%), enhanced survivorship for both sexes (67%) occurred in males and females that roosted there.

In both of these studies Great Tits were constrained to accept or reject a roost site positioned by the experimenter. It could be instructive to compare sites chosen by sex, age and dominance classes when a variety of potential sites is available. For example, Downy Woodpeckers in central Ohio were provided with trios of artificial trees or snags made from polystyrene cylinders (Grubb 1982b). Each trio was comprised of cylinders 1.21, 2.42, and 3.63 m in length, positioned vertically in a woodlot and arranged in an equilateral triangle 3 m on a side. Cylinders were checked at daily intervals until it was determined that a complete cavity had been excavated in one of each trio. While in 10 of 16 cases birds dug cavities in the snag of intermediate (2.42 m) height, there was some indication that the sexes differed in their snag preferences, with males tending to roost higher. Although the significance of the differences in cavity-site choice between the sexes of this woodpecker remains obscure, the results of this one controlled test suggest that the sexes may segregate along a niche dimension for cavity height.

CONCLUSIONS

Population and community ecologists are interested in determining why only a certain number of species or individuals occur in any given area. As the abundance and distribution of animals are at least partially the consequences of the animals' behavior, the behavioral ecologist must seek the behavioral mechanisms underlying ecological patterns, recognizing that these are affected by sex, age and social dominance. The meager literature supporting our survey prompts the conclusion that analysis has only begun.

The substantial return to be gained by studying the food-related behavior of sex, age and dominance categories is demonstrated by the pioneering studies of Ekman and his colleagues on Willow Tits and the subsequent manipulative experiments of Hogstad on the same species. It seems that population size in wintering Willow Tits, before it is pruned by predation, is set by some sort of cost-benefit analysis performed by juveniles on whether to remain in the company of a dominant pair. At least some of the factors in the "analysis" done by these young birds appear to be the harshness of the climate, presence of heterospecifics, and density of the food supply. In no other taxon or geographical locale does such a depth of analysis exist; the prospects for comparative research seem bright.

ACKNOWLEDGEMENTS

The comments of R. N. Conner, B. Enoksson, O. Hogstad, and T. A. Waite improved the manuscript. Preparation of this review was supported by NSF grants BSR-8313521 and BSR-8717114 to TCG.

EFFECTS OF UNKNOWN SEX IN ANALYSES OF FORAGING BEHAVIOR

JoAnn M. Hanowski and Gerald J. Niemi

Abstract. Foraging data were collected on Yellow-rumped (*Dendroica coronata*) and Palm (*D. palmarum*) warblers, both of which nest in spruce bogs of northern Minnesota. Male and female Palm Warblers cannot be distinguished, whereas sexes of the Yellow-rumped Warbler are dichromatic. We used our data to examine effects of unknown sex in analyses of foraging data. Male Yellow-rumped Warblers foraged higher than females. When data from sexes were combined, the mean was closer to the value for males because the sample size was larger for that sex. It follows that foraging data for sexually monochromatic species may not be representative for either sex. In addition, interspecific comparisons of foraging behavior may not be appropriate when sexes are unknown. Appropriate statistical tests may compensate for unknown sexes. Analysis of frequencies (chi-square or G-test) may eliminate this bias because means are not compared. However, this test does not compensate for the unknown proportion of male and female observations, which obviously skews the frequency distribution.

Key Words: Warbler; foraging behavior; date analysis

MacArthur's (1958) study of foraging behavior in spruce-woods warblers initiated a voluminous literature on avian foraging behavior. Nevertheless, little attempt had been made to standardize collection or analysis of such data. One topic that is rarely considered is how observations of birds of unknown sex affect the results of foraging analyses. Our objectives were to: (1) quantify foraging height and tree height used by two spruce-woods warblers, the monochromatic Palm Warbler (*Dendroica palmarum*) and the sexually dichromatic Yellow-rumped Warbler (*D. coronata*); (2) determine the consequences of unknown sex in interspecific comparisons of foraging and tree heights; and (3) examine statistical tests that best quantify interspecific differences or similarities in foraging behavior when sexes are unknown.

STUDY AREA AND METHODS

We studied birds in a black spruce (*Picea mariana*) bog in 1980 and 1981 (Hanowski 1982). The study area, located within the Red Lake Peatland in northwest Minnesota, is relatively homogeneous, with 95% of the trees black spruce and 5% tamarack (*Larix laricina*) (Niemi and Hanowski 1984). Mean tree height was 5.7 m (Hanowski 1982) and the stand age was 150–200 yr (Heinselman 1963). Labrador tea (*Ledum groenlandicum*) and leatherleaf (*Chamaedaphne calyculata*) were common in the understory and the ground was covered with mosses (*Sphagnum* spp., *Dicranum* spp., and *Polytrichum* spp.).

We collected foraging data three times weekly from early May through early July in two 17.5-ha study areas (500 × 350 m) (Niemi and Hanowski 1984), primarily in the morning (04:30–12:00 CDT). Data for several foraging variables (e.g., method, substrate size) also were collected, but here we concentrate on data for height of tree and height of foraging bird in 1-m intervals. The relatively short stature of trees allowed us to record heights accurately. We recorded data at 30-s intervals for up to five observations/sighting. Height means (t-test; $P < 0.05$) or frequencies (chi-square; $P < 0.05$) for the first observation in a series were not different from subsequent observations in a series for any bird group. We used only the initial observation in all statistical analyses.

Foraging data were analyzed by comparing means (parametric tests) and frequencies (nonparametric tests). Mean bird and tree heights were first compared with a three-group (male Yellow-rumped, female Yellow-rumped, and Palm warblers) analysis of variance (ANOVA). If a significant ($P < 0.05$) difference was found, Scheffe's test (Sokal and Rohlf 1981) was used to compare individual group means. Frequencies of bird and tree heights were compared with a 3 × 12 chi-square contingency-table test. Paired comparisons were also computed with chi-square tests if the initial three-group test was significant ($P < 0.05$).

RESULTS

The three-group ANOVA indicated that these two species selected trees of different heights and also foraged at different heights ($P < 0.05$; Table 1). Scheffe's multiple comparisons showed that the male Yellow-rumped Warblers foraged higher than females ($P < 0.01$) but did not select taller trees (Table 2, Fig. 1). The Palm Warbler (Fig. 2) foraged lower and selected shorter trees than the yellow-rumps ($P < 0.01$; Table 2). However, no difference ($P > 0.05$) existed between foraging height or height of trees selected by the Palm and female Yellow-rumped warblers (Table 2). By contrast, when we combined data for male and female yellow-rumps, we found that yellow-rumps selected taller trees (Fig. 3, Table 2) but that the height of the foraging bird did not differ between species.

The three-group chi-square test, like the t-test, showed that the male and female yellow-rumps and the Palm Warbler differed from each other in their use of tree and foraging heights ($P <$

TABLE 1. Mean, Variance, Sample Size and Results of ANOVA and Chi-square Tests for Bird and Tree Heights Selected by Male and Female Yellow-rumped and Palm Warblers (Sexes Combined) while Foraging

Species	N	Bird height		Tree height	
		$\bar{X}$	s^2	$\bar{X}$	s^2
Male Yellow-rumped Warbler	46	5.6	2.9	7.5	3.5
Female Yellow-rumped Warbler	22	3.3	2.8	6.8	1.9
Palm Warbler	82	4.1	4.9	6.2	4.1
Male and female Yellow-rumped Warbler[a]	68	4.8	3.6	7.2	2.9
Overall ANOVA		$P < 0.01$		$P < 0.04$	
Overall chi-square		$P < .0.02$		$P < 0.01$	

[a] Not included in overall ANOVA or chi-square test.

0.05; Table 1). Two of eight paired chi-square tests comparing foraging data had different results from Scheffe's test. In both situations, the chi-square test was significant when the Scheffe's test was not (Table 2). The two tests that were not in agreement concerned tree height between female yellow-rumps and the other two groups (male yellow-rumps and Palm Warblers).

DISCUSSION

If the goal of foraging analyses is to assess resource use by a species, distinguishing sexes may not be critical. Frequency distributions of a species' use of trees and its height when foraging will provide an adequate indication, although the space used by males will probably be over-represented due to their conspicuousness. In contrast, if the objective is to define and separate microhabitat use for species that are close, both morphologically and in the habitat that they occupy, ignorance of sexual differences may obscure ecologically important differences between species.

The inability to distinguish between sexes is a problem because approximately 50% of North American passerine species are sexually monochromatic. Furthermore, when sexes have been distinguished, studies have found that males and females forage at different heights and select different heights of trees for foraging (Morse 1968, 1971a, 1973, 1980b; Morrison 1982; Franzreb 1983b; Holmes 1986; Grubb and Woodrey, this volume). For example, our data for yellow-rumps agree with a previous study by Franzreb (1983b), who showed that males foraged higher than females. Nevertheless, in other studies male and female data have been combined with little consideration for the consequences of this procedure.

We have shown that neither a comparison of mean heights (t-test) or frequency of heights (chi-square or G-test) may be appropriate when sexes cannot be distinguished because: (1) interspecific and intraspecific sexual differences will not be detected, and (2) means compared will be skewed toward the sex with the larger sample size. For example, t-tests are not appropriate because the mean is skewed toward the sex with the larger sample. Combining data from Yellow-rumped Warbler sexes obscured the actual difference between female yellow-rumps and Palm Warblers. If the ratio of male to female observations in the sample is not equal for both species, interspecific comparisons will be inaccurate.

The alternative test is nonparametric comparisons of frequency distributions. Where sexes are unknown, these comparisons should be more appropriate because means are not compared. Indeed, most investigators have used them (Sturman 1968, Balda 1969, Hertz et al. 1976, Morse 1979). Our nonparametric tests detected more differences between species groups than the ANOVAs or Scheffe's test. Chi-square tests indicated that male yellow-rumps selected taller trees

TABLE 2. Results of Scheffe's (P-values) and Paired Chi-square (χ^2) Tests between Male and Female Yellow-rumped Warblers and Palm Warblers for Bird and Tree Heights Selected while Foraging

Comparison	Bird height		Tree height	
	Scheffe's	χ^2	Scheffe's	χ^2
Male versus female Yellow-rumped Warbler	0.01	0.01	0.18	0.03
Male Yellow-rumped versus Palm Warbler	0.01	0.01	0.01	0.01
Female Yellow-rumped versus Palm Warbler	0.08	0.01	0.16	0.01
Yellow-rumped (sexes combined) versus Palm Warbler (sexes combined)	0.29	0.39	0.02	0.01

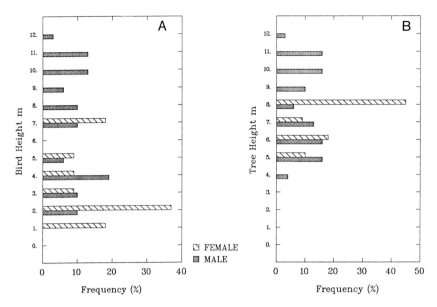

FIGURE 1. Distribution (relative percent) of (A) bird height (m) and (B) tree height (m) selected for foraging by male (N = 46) and female (N = 22) Yellow-rumped Warblers.

than females and that females selected taller trees than Palm Warblers. Neither of these differences was detected with Scheffe's test. However, as in Scheffe's test, combining the yellow-rumped sexes masked the difference in foraging height between species.

Chi-square tests are not an adequate solution to the problem because the proportion of male and female observations in the sample is still unknown. The frequency of height observations in data that identified sexes had two unimodal normal distributions (Wilk's-Shapiro test; Sha-

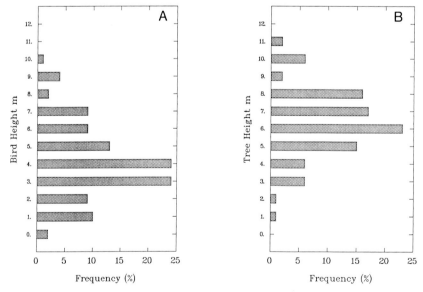

FIGURE 2. Distribution (relative percent) of (A) bird height (m) and (B) tree height (m) selected for foraging by Palm Warblers (N = 82).

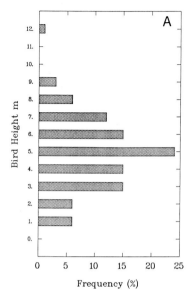 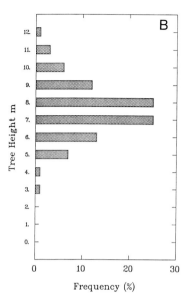

FIGURE 3. Distribution (relative percent) of (A) bird height (m) and (B) tree height (m) selected for foraging by Yellow-rumped Warblers sexes combined (N = 68).

piro and Wilk [1965]), which is expected if males forage higher than females. However, when we combined data between sexes, the frequency distribution remained unimodal and normal (Fig. 3), due to the overlap of observations in the center of the distribution. Similarly, we would not expect the frequency distribution of Palm Warbler foraging heights to show two unimodal distributions, because data were a combination of heights for both sexes. Therefore, nonparametric tests are not a remedy if the goal is to identify interspecific differences.

One cannot assume that random foraging observations for sexually monochromatic species will contain similar proportions of males and females because males are usually more conspicuous than females (pers. obs.; Holmes 1986). This assumption is not a solution to the problem of unknown sex.

ACKNOWLEDGMENTS

We thank J. G. Blake, R. F. Green, and an anonymous reviewer for their helpful suggestions on an earlier draft of the manuscript. This work was completed by the senior author in partial fulfillment of a Master of Science degree at the University of Minnesota. Monetary support was provided by Northern States Power, Minneapolis, Minnesota. This is contribution number 22 of the Center for Water and the Environment.

DIFFERENCES IN THE FORAGING BEHAVIOR OF INDIVIDUAL GRAY-BREASTED JAY FLOCK MEMBERS

LAURIE M. MCKEAN

Abstract. Among Gray-breasted Jays (*Aphelocoma ultramarina*) subordinates are less successful in certain foraging situations than dominants, the foraging preferences of young birds are not as defined as those of adults, and subadults change their foraging behavior after observing the feeding of other flock members. By averaging behavioral data without regard for variables such as age, status, or social context, information may be lost, and emerging patterns may not be representative of any individual in the study population.

Key Words: Individual variation; age-related differences; cooperative breeding; Corvidae; *Aphelocoma ultramarina*; cultural transmission of information.

In the past, emphasis on detecting patterns in biological systems has minimized appreciation of individual deviations in behavior. Models such as optimal foraging models often assume homogeneity in the foraging abilities of their subjects (Charnov 1976a, Pyke et al. 1977) and may be confounded by the presence of much individual variation; yet, this variation may be important in uncovering the mechanisms producing phenomena of larger scale (Sibly and Smith 1985, Hassell and May 1985), such as cooperative breeding.

I studied individual variation in foraging success and patterns related to differences in the age or dominance status of Gray-breasted Jays by using field observations and experimental manipulations of certain behavioral parameters. My objectives were to examine age, dominance status, and social context as sources of individual behavioral variation that must be considered when designing sampling programs for foraging studies.

Gray-breasted Jays are cooperative breeders that live in groups of 5 to 20 individuals (Gross 1949, Brown 1987). Birds up to three years old (subadults) can be distinguished by pale patches in their bills; by age three (adults) most birds have completely black bills (Pitelka 1961, Brown 1963). Jays at the Southwestern Research Station were color-banded by Jerram and Esther Brown and at Santa Catalina by J. B. Dunning and me. The experiment in the Chiricahua Mountains involved three flocks, including fourteen subordinates and ten dominants. Parts of this study done in Bear Canyon involved one flock with three dominants and seven subordinates; of the subordinates in this flock, three were subadults and four were adults.

The study was done at the Bear Canyon Recreation Area in the Santa Catalina Mountains near Tucson, AZ, and at the Southwestern Research Station of the American Museum of Natural History, in the Chiricahua Mountains, Cochise County, AZ.

METHODS

Dominance trials

Dominance status was determined by calculating the individual binomial probabilities of winning an aggressive dyadic encounter with a particular flock member at a localized food source (McKean 1988). An individual was categorized as the winner of a bout when it: (1) continued eating while a new arrival (the loser) waited at the food source; (2) displaced an individual already at the food source by displaying, pecking, or merely by approaching the feeding area; or (3) chased the loser away. The technique and most of the criteria were discussed by Barkan et al. (1986). The most dominant bird in a flock was the individual with a significant binomial probability of winning an encounter with all other birds in the flock; the second most dominant individual had a significant probability of winning a bout with every flock member except the most dominant; and so on. I designated at least the top third of the individuals of a flock as dominants or until the binomial probability of a bird's winning dropped below 0.025.

In several cases, I observed ≤5 encounters between particular individuals. In these cases I looked for evidence of avoidance behaviors by looking at (1) whether one waited in the trees adjacent to the feeding area, and (2) which individuals used the food source at the same time. When an individual waited for a specific bird to leave the feeder in more than two instances, it was counted as a loser (McKean 1988).

Feeder experiment

This experiment tested (1) the ability of subordinates and dominants to solve a novel foraging problem, and (2) compared the success of individuals when alone to that after observing another flock member at the feeder. To test for effects of learning on success, I divided the trials into early and late periods, with approximately the same numbers of observations in each.

The Santa Catalina experiments were carried out in nine trials of about three hr each between 9–22 October 1984. The Station flock was the subject of 10 trials of up to 2.5 hr each from 28 January to 23 February 1985.

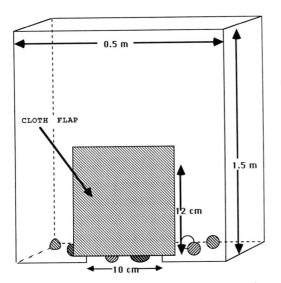

FIGURE 1. Diagram of feeder apparatus.

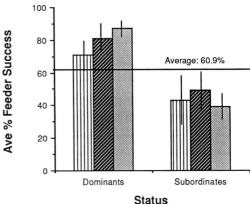

FIGURE 2. Mean percent feeder success for subordinates (N = 18) and dominants (N = 13) comparing doughball selections made alone (vertically striped bars) with those made after observing another subordinate (cross-hatched bars) or dominant (stippled bars) individual. Each individual made at least 10 attempts.

Trials ended when the flock left the area. I also performed 15 similar trials with the other two Chiricahua flocks. The trials were rotated between the periods of 06:30–09:00, 09:30–12:00, and 13:00–15:30 and tested for time of day (which had no significant effect).

I constructed two plexiglass feeders (Fig. 1), which I operated simultaneously to minimize competition between feeding individuals and other birds or squirrels. Peanuts and sunflower seeds, used as bait, were visible through the plexiglass sides of the feeder. In each feeder, rocks or cardboard slats were used to keep the peanuts near the door where the birds could reach them. The opening to the feeder was covered with a blue cloth flap, which was attached at the top of the opening so that the birds could lift it with their bill or feet and enter the feeder. This feeder was designed to be opened with a sweeping bill movement that the jays use in natural foraging.

I recorded approaches to the feeders, proximity of other individuals, and whether or not the individual was successful, i.e. in removing food. I calculated percent success for each individual (successful attempts/approaches), and I performed arcsine transformations to normalize percentages before executing an ANOVA.

Doughball color choice: color preferences

This experiment tested (1) color preferences of subadult versus older individuals, and (2) cultural transmission of feeding information, which I defined as the spread of information without direct experience with a phenomenon through observation of experienced others (Cavalli-Sforza and Feldman 1981; Fisher and Hinde 1949; Giraldeau 1984; Klopfer 1958, 1961).

I studied the jays' color preferences (25 March to 4 April 1984) for peanut butter doughballs (natural brown, red, and green). Each 1.5-cm-diameter ball was made of a mixture of peanut butter, flour, and egg. Vegetable food coloring was used to color the red and green balls. I placed 20 doughballs of each color on a neutral background on top of a table within the flock's home territory, arranged so that access by birds was not biased. After approximately half of any particular color had been removed, I replenished the supply to 20 of each color.

I recorded the order of the birds' appearances, color of bait removed, and the identities of other birds present. I used arcsine transformations of all percentages to normalize the distribution (Sokal and Rohlf 1981), and compared colors chosen by each individual within an age-dominance class (dominants, older subordinates and younger subordinates) with those in the other classes using MANOVA (SAS 1985), I also compared color choices made alone with those made after observing another individual.

RESULTS

Dominants (N = 10) were significantly more successful than the subordinates (N = 14) (Fig. 2) (two-way ANOVA, $F = 8.978$, $df = 24,48$, $P < 0.0001$), but success rates were not significantly different between birds that had or had not observed another bird (two-way ANOVA, $F = 1.044$, $df = 2,48$, $P > 0.25$).

Brown was the preferred color overall, but less so by birds that had not observed others foraging (Fig. 3). When the selections of older birds made when alone were compared with those of subadults, they differed significantly (MANOVA, $F = 6.7$, $df = 1,9$, $P < 0.03$). After observing others, the differences between the selections of birds in the two age classes were not significantly different (MANOVA, $F = 2.7$, $df = 1,9$, $P > 0.14$).

When alone, older birds exhibited a preference for the brown doughballs and their selections were not significantly altered after watching the choices made by flock mates (paired t-test = 3.328, $df = 6$, $P > 0.1$). However, when young

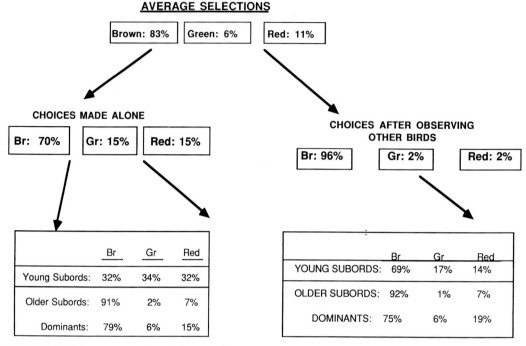

FIGURE 3. Mean percent color choice partitioned into three levels: (1) overall average percent of each color chosen, (2) choices made when feeding alone versus feeding after observing others, and (3) choices made when partitioned by status-age class. Sample sizes were three young subordinates, four older subordinates, and three older dominants, a minimum of 10 selections per individual.

individuals observed the choices of others their selections changed significantly, approximating those of the adult (paired t-test = -15.03, df = 2, $P < 0.005$) (Fig. 3). After observing another flock member, the young birds chose at least 25% more brown doughballs than when alone.

DISCUSSION

It is becoming increasingly apparent to students of behavior, including many in this symposium (e.g., Grubb and Woodrey), that selection affects individuals differently and is influenced by age, sex, dominance status and specific social context. In order to more accurately describe patterns of foraging (or other) behavior, it may be useful to categorize observations of individuals in a species by these criteria. In a population polymodal for a behavior, a model using averaged responses may not accurately describe the behavior of any individual in the population.

My research suggests that age, dominance rank, and social context contribute to the variability in the feeding behavior of individual Gray-breasted Jays. In this study, I observed that: (1) subordinates were less successful in certain foraging situations than dominants; (2) the foraging preferences of young birds, as defined by color choice, were not as specific as those of adults; and (3) subadults changed their foraging behavior after observing the feeding of other flock members.

Reduced foraging efficiency in young animals is a well known phenomenon. In species in which foraging behavior is relatively complex (Orians 1969a, Blus and Keahey 1978, Morrison et al. 1978, MacLean 1986, Sasvari 1985, Goss-Custard and Durrell 1987), prolonged subadult periods may reflect the time necessary to acquire foraging and other skills, and depending upon individual abilities may affect reproductive success and other functions, at least short term. Researchers studying other corvid species (Lawton and Guindon 1981, Reese and Kadlec 1985, Hochachka and Boag 1987) have also found age-related differences in abilities, including some foraging skills, affecting reproductive timing and overall success. Even in species such as the Yellow-eyed Junco (*Junco phaenotus*), in which foraging behavior appears relatively simple, age-related variation in foraging efficiency may have profound effects on juvenile mortality patterns (Sullivan 1988).

The variation in feeding patterns, illustrated in this study by disparate selections of colored doughballs by jays of different ages, suggests the ontogeny of a feeding behavior. Young individuals exhibited less defined preferences than older birds, suggesting that the older individuals had developed a foraging rule narrowing their initial color selections.

Cultural transmission of information may offer an efficient means of acquiring foraging information in a social context (Clark and Mangel 1984, Giraldeau 1984), with less skilled individuals benefiting by following experienced flock members and imitating their behavior (Lawton and Guindon 1981). In my study, the foraging behavior (doughball color choice) of young jays changed after they observed older birds. By averaging the color choices of individual birds without regard for social context, evidence of such interactions would have been blurred.

In this study, as in others (e.g., Ekman and Askenmo 1984, Baker et al. 1981), dominants had significantly more success than subordinates at acquiring food from the feeders, even when the subordinates approached the feeders alone. If the experimental results reflect natural patterns, increased access to food may provide selective advantages to dominants and their offspring.

In order to accurately model behavior, variation resulting from differences in age and dominance classes must be considered, as these may be important in the understanding of behavioral mechanisms fundamental to community structure and demography. Particularly for social species, understanding of dispersal, territoriality, and patch choice, all of which are instrumental in producing large scale phenomena, may hinge upon predicting individual variation in behavior.

ACKNOWLEDGMENTS

The James Silliman Memorial Award provided funding for room and board at the Southwestern Research Station of the American Museum of Natural History. I would also like to thank J. Michael Reed, Dr. John B. Dunning, Dr. Steve Russell and Dr. Jeffrey Walters for helpful discussions and aid in editing this manuscript.

Analytical Methods

USE OF MARKOV CHAINS IN ANALYSES OF FORAGING BEHAVIOR

MARTIN G. RAPHAEL

Abstract. For logistical reasons, observers often record sequential movements of birds among foraging resources. An appropriate method of analysis involves Markov chains, which summarize the frequency of movement from one resource to another. Such data are summarized into a transition matrix, where numbers of observations of movement of a bird from one habitat are tallied into all categories to which the bird subsequently moves. When such data are gathered for several species, or other groupings, tests of homogeneity can be performed using log-linear models. These data can also be used to generate tables of transition probabilities, and these in turn can be reduced, through eigenanalyses, to steady-state vectors that give the probability of use (over the long run) of each habitat. These vectors can be compared (through goodness-of-fit tests or tests of independence) to measures of habitat availability and measures of habitat selectivity can be calculated. Analyses are described for use with popular statistical computer packages.

Key Words: Birds; environmental grain; foraging behavior; log-linear models; Markov chain; transition probability.

For mostly logistical reasons, ornithologists usually record foraging behaviors of birds as a sequential series of observations. The reason data are gathered sequentially is valid: birds can be difficult to find and it is more efficient to follow a bird, once it is found, than to abandon it after one or two observations and search for another bird. However, analyses of such data using traditional chi-square or other similar techniques may not be valid because sequential observations are not necessarily independent, and independence of observations is a critical assumption of most statistical tests. Other methods of analysis are available that take advantage of the sequential structure of such data (e.g., time-series analyses). This paper describes one of these methods involving Markov chains and log-linear modeling.

First, a few definitions (following Vandermeer 1972) may be useful. *Operational habitat* denotes an identifiable habitat unit, for example, each of the s tree species in a study site. *Environment* denotes a specified set of operational habitats. *Environmental grain* denotes the way in which a particular species moves from one operational habitat to the next during a specified time interval that is short relative to the lifespan of the species (MacArthur and Levins 1964, Levins 1968). For example, one might observe two species of birds. Individuals of one species stay in one operational habitat for a long time (the birds forage mainly in one tree species), whereas individuals of another species forage in trees of all species at random. The environment is coarse grained for the first species and fine grained for the second. *Markov chain* denotes a series of operational habitats, and the probability of passing to a new one by some defined process (Keller 1978). To illustrate, suppose we have a system that moves from habitat i at time t to habitat j at time $t + 1$. At each time interval, the system can be in any one of s habitats. We define $p_{i,j}^{(1)}$ as the frequency (probability) of moving from habitat i at time t to habitat j at time $t + 1$, where the superscript (1) indicates a transition occurring in one, discrete time interval. For example, let the habitats be tree species in a forest. We observe frequencies of birds flying from one tree species to the next. Next, we observe the bird flying at two time intervals; from habitat i at time t to habitat k at time $t + 2$, with the resulting probability $p_{i,k}^{(2)}$. Thus:

$$p_{i,k}^{(2)} = p_{i,1}^{(1)}p_{1,k}^{(1)} + p_{i,2}^{(1)}p_{2,k}^{(1)} \\ + \ldots + p_{i,j}^{(1)}p_{j,k}^{(1)} \\ + \ldots + p_{i,s}^{(1)}p_{s,k}^{(1)}$$

or

$$p_{i,k}^{(2)} = \sum_{j=1}^{s} p_{i,j}^{(1)}p_{j,k}^{(1)}.$$

This is the sum of all the different pathway-probabilities between habitat i and habitat k, each passing through exactly one intermediate habitat.

We can extend the last result to n consecutive time intervals to obtain $p_{i,j}^{(n)}$, and in general

$$p_{i,k}^{(m+n)} = \sum_{j} p_{i,j}^{(m)}p_{j,k}^{(n)},$$

where $p_{i,j}^{(m)}$ is the m-step transition probability from habitat i to habitat j, and $p_{j,k}^{(n)}$ is the n-step transition probability from habitat j to habitat k.

Now let

$$\lim_{n \to \infty} p_{i,j}^{(n)} = u_j.$$

It can be shown (Chiang 1980:123) that if the limit u_j exists, then u_j is independent of the initial state i, and the vector **u** is called a stationary distribution, with the sum of the vector elements equal to 1.0. Formally, if

$$u_j = \sum_i u_i p_{i,j}^{(1)},$$

u is a stationary distribution. As before, suppose $p_{i,j}^{(1)}$ defines the probability (frequency) of a bird of one particular species flying from tree species i to tree species j, during one time interval, in an environment that contains s tree species. The transition probabilities $p_{i,j}^{(1)}$ ($i, j = 1, 2, \ldots, s$) define a one-step matrix $\mathbf{G}^{(1)}$, which is called the grain matrix (Vandermeer 1972, Colwell 1973); that is:

$$\mathbf{G}^{(1)} = \begin{matrix} p_{1,1}^{(1)} & p_{1,2}^{(1)} & \cdots & p_{1,s}^{(1)} \\ p_{2,1}^{(1)} & p_{2,2}^{(1)} & \cdots & p_{2,s}^{(1)} \\ \cdot & \cdot & & \cdot \\ \cdot & \cdot & & \cdot \\ \cdot & \cdot & & \cdot \\ p_{s,1}^{(1)} & p_{s,2}^{(1)} & \cdots & p_{s,s}^{(1)} \end{matrix}$$

Given $\mathbf{G}^{(1)}$, one can obtain the grain matrix of the stationary distribution $\mathbf{G}^{(\infty)}$ that contains s identical row vectors composed of s elements:

$$\mathbf{G}^{(\infty)} = \begin{matrix} u_1 & u_2 & \cdots & u_s \\ u_1 & u_2 & \cdots & u_s \\ \cdot & \cdot & & \cdot \\ \cdot & \cdot & & \cdot \\ \cdot & \cdot & & \cdot \\ u_1 & u_2 & \cdots & u_s \end{matrix}$$

Note that u_i is also the reciprocal of the mean return time to habitat i (Hoel et al. 1972:60). Thus, a large u_i also indicates a relatively small number of steps before a bird returns to u_i after having left.

If we denote the frequency of the jth operational habitat (e.g., relative frequency of tree species j in the study site) by e_j, we can then define the environmental matrix (**E**) composed of s identical row vectors, each with s elements:

$$\mathbf{E} = \begin{matrix} e_1 & e_2 & \cdots & e_s \\ e_1 & e_2 & \cdots & e_s \\ \cdot & \cdot & & \cdot \\ \cdot & \cdot & & \cdot \\ \cdot & \cdot & & \cdot \\ e_1 & e_2 & \cdots & e_s \end{matrix}$$

As a scaler u_j approaches unity, the environmental grain becomes coarser; and, as u_j tends to e_j, the environment becomes fine grained. Therefore, we have upper and lower limits to grain coarseness. An index of grain coarseness (C) can be calculated as

$$C = \sum_{j=1}^{s} |u_j - e_j|,$$

where (Colwell 1973),

$$C_{\max} = \sum_{j=1}^{s} [1 - e_j + (s - 1)e_j]$$
$$= 2(s - 1).$$

Thus, C can vary between 0 and $2(s - 1)$. By dividing C by $C_{\max}$ one can calculate a relative index that is independent of s. One can also compare the vector **u** to a row vector from **E**, in which case $C_{\max} = 2.0$ (e.g., if $u_1 = 1$, then all other u_i are 0 since $\Sigma\ u_i = 1.0$; if $e_2 = 1.0$, then all other $e_i = 0$ and $\Sigma\ |u_j - e_j| = 2.0$).

AN APPLICATION

Study Site and Methods

Field work was conducted within a 20-km radius of the University of California Sagehen Creek Field Station near Truckee, California. Birds were observed June and August of 1976 and 1977 at elevations varying from 1800 to 2300 m. The basin is dominated by Jeffrey pine (*Pinus jeffreyi*) and white fir (*Abies concolor*). Meadow stands with lodgepole pine (*Pinus contorta* var *murrayana*) and aspen (*Populus tremuloides*) occur in the moist areas near springs and streams. Red fir (*Abies magnifica*) and mountain hemlock (*Tsuga mertensiana*) dominate at higher elevations.

Bird observations

For this analysis, two types of data were collected: first, an index of relative abundance of operational habitat units (in this case relative frequency of tree species); and second, a record of sequential moves by individual birds between tree species. A single observation started when a bird left one tree and ended when it landed on the next. An individual bird was sometimes followed as it flew from one tree to the next for up to 10 moves.

I tabulated the movements of four species of woodpeckers and three species of nuthatches among the four most common tree species (operational habitats) (Table 1). For this example, I eliminated all observations of birds landing on or departing from rarer tree species because sample sizes in these species were too small for analysis.

I also estimated the availability of stems of each tree species from a randomly located sample of 100, 0.04-ha circular plots. All stems >8 cm dbh were tallied among the four tree species.

Data analysis

Environmental (**E**) and grain (**G**) matrices were obtained from the data in Table 1 by dividing each cell

TABLE 1. One-step Transition Frequencies for Movements among Tree Species by Foraging Woodpeckers, Sagehen Creek, California

Species	Tree species[a] at t_i	Tree species[a] at t_{i+1}				
		LP	JP	WF	RF	Totals
Williamson's Sapsucker	LP	16	2	1	0	19
	JP	3	12	4	0	19
	WF	1	5	14	2	22
	RF	1	2	1	11	15
Red-breasted Sapsucker	LP	36	0	1	0	37
	JP	0	8	2	1	11
	WF	0	3	11	1	15
	RF	1	1	2	3	7
Hairy Woodpecker	LP	38	1	2	1	42
	JP	0	29	15	0	44
	WF	2	13	38	4	57
	RF	0	0	2	18	20
White-headed Woodpecker	LP	3	1	1	0	5
	JP	1	47	5	1	54
	WF	1	5	24	2	32
	RF	0	1	1	13	15
Pygmy Nuthatch	LP	0	0	0	0	0
	JP	0	33	3	0	36
	WF	0	3	5	1	9
	RF	0	1	0	3	4
Red-breasted Nuthatch	LP	20	1	2	1	24
	JP	1	53	5	1	60
	WF	3	4	10	2	19
	RF	0	1	2	40	43
White-breasted Nuthatch	LP	0	0	1	0	1
	JP	0	87	11	1	99
	WF	0	10	10	1	21
	RF	1	2	0	3	6
Random sample[b]		142	95	139	97	473

[a] LP = lodgepole pine, JP = Jeffrey pine, WF = white fir, RF = red fir.
[b] Frequency of each tree species counted on 100, 0.04 ha circular plots, randomly located on the study area.

value by its corresponding row total (Table 2). Stationary grain vectors **u** were calculated for each bird species so that $\mathbf{uG} = \mathbf{u}$. The row vector **u** is the eigenvector associated with the dominant eigenvalue of the transposed grain matrix, which in this case is always equal to unity (Vandermeer 1972:115). This calculation was accomplished using a FORTRAN program (available from the author on request) incorporating the EIGRF subroutine of the IMSL library (IMSL 1982). The eigenvector was normalized so that all values summed to 1.0. Harlow (1986a, b) provided a BASIC program that could also be used for the eigenanalysis.

Statistical inferences regarding the similarity of transition frequencies among bird species, and between each species and the randomly sampled trees (environmental matrix), were tested using log-linear models (Bishop et al. 1975) and chi-square tests (Neu et al. 1974, Riley 1986). Because these analyses assume a one-step, stationary, Markov process (i.e., the habitat unit occupied by a bird at time t depends only on its habitat occupied at time $t - 1$, and probabilities do not change over time), Bishop et al. (1975:265) discuss a goodness-of-fit approach for testing the assumption of one-step stationarity. I tested the grain matrices for symmetry ($G_{ij} = G_{ji}$ for all species) prior to computing among-species comparisons (Bishop et al. 1975:282).

To compare grain matrices of each species (**G**) to the environmental matrix (**E**), I used a chi-square test of independence based on the row frequencies of **G** and a row vector from matrix **E**. Interspecific comparisons were computed using a log-linear model that included main effects (row, column, species) and the interaction of row and column. All computations were performed using the HILOGLINEAR module of the SPSS/PC+ statistical program (Norusis 1986). Full descriptions of statistical inference tests are provided by Bishop et al. (1975), Basawa and Prakasa Rao (1980), and Chatfield (1973).

RESULTS

I recorded a total of 736 foraging transitions of seven bird species (Table 1). Birds were most likely to move to another tree of the same species rather than to another tree species in all cases except White-breasted Nuthatches using white fir. In the latter case, White-breasted Nuthatches were equally likely to switch to Jeffrey pine.

TABLE 2. COMPUTATION OF TRANSITION PROBABILITIES FROM TRANSITION FREQUENCIES FOR WILLIAMSON'S SAPSUCKER (TABLE 1)

Tree species at t_i	Tree species at t_{i+1}			
	Lodgepole pine	Jeffrey pine	White fir	Red fir
Lodgepole pine	16/19 = 0.84	2/19 = 0.11	1/19 = 0.05	0/19 = 0.00
Jeffrey pine	3/19 = 0.16	12/19 = 0.63	4/19 = 0.21	0/19 = 0.00
White fir	1/22 = 0.05	5/22 = 0.23	14/22 = 0.64	2/22 = 0.09
Red fir	1/15 = 0.07	2/15 = 0.13	1/15 = 0.07	11/15 = 0.73

The overall test of symmetry, based on the entire 4 rows × 4 columns × 7 species contingency table (Table 1), was not significant (χ^2 = 11.03, df = 36, P > 0.50), indicating that birds were equally likely to move from tree species i to species j as from species j to i.

Comparisons of tree-species use by each bird species with tree availability estimated from randomly sampled plots showed that all birds, except Williamson's Sapsucker, departed significantly from expected frequencies of use (Table 3). This was evident both from direct comparisons of the steady-state vectors **u** with the environmental vector e_j (assessed using the index of grain coarseness [Table 3]), and by chi-square tests of independence between the marginal row frequencies of each bird and the numbers of randomly sampled trees of each species (Table 1). White-breasted and Pygmy nuthatches differed most from the random sample; Williamson's Sapsucker differed least.

Interspecific comparisons, based on tests of homogeneity (Table 4), revealed significant differences among all pairs of species except Williamson's versus Red-breasted sapsuckers, Williamson's Sapsucker versus Hairy Woodpecker, White-headed Woodpecker versus Pygmy Nuthatch, and White-breasted versus Pygmy nuthatches. Significant differences indicated that birds differed in their probabilities of moving to a particular tree species at time t_i, given the tree species they used at time t_{i-1}.

DISCUSSION

DESIGN CONSIDERATIONS

Analyzing sequences of behavior using Markov chains appears to be a useful technique, primarily because such chains allow explicit recognition of the potential interdependence of sequential observations. The technique can be applied to any type of behavior—including spatial distribution—that can be categorized into discrete units. For example, Colwell (1973) used the method to analyze visit frequencies of hummingbirds to flower species and used the results to predict the relative abundance of phoretic mites in the various flowers. Cane (1978) used Markov chains to examine grooming behavior of a blowfly (*Calliphora erythrocephala*) in which sequences of 10 different types of behavior were analyzed, and to analyze 11 social behaviors (aggregated from 123 original categories) of rhesus monkeys (*Macaca rhesus*). Raphael and White (1984) used Markov chains to compare the use of snags, living trees, and other substrates among

TABLE 3. STEADY-STATE VECTORS OF TREE SPECIES USE DERIVED FROM EIGENANALYSES OF GRAIN MATRICES (TABLE 1), AND INDEX VALUES OF DEPARTURE FROM FREQUENCIES OF AVAILABLE TREE SPECIES

Bird species	Steady-state vector (**u**) of relative use of tree species				Index of grain coarseness[a]	Significance[b]	
	Lodgepole pine	Jeffrey pine	White fir	Red fir		χ^2	P
White-breasted Nuthatch	0.005	0.785	0.177	0.033	0.584	160.32	0.000
Red-breasted Nuthatch	0.155	0.354	0.125	0.367	0.315	43.01	0.000
Pygmy Nuthatch	0.000	0.787	0.147	0.065	0.587	70.53	0.000
Williamson's Sapsucker	0.391	0.288	0.239	0.081	0.062	1.35	>0.500
Hairy Woodpecker	0.129	0.244	0.351	0.276	0.171	9.10	0.027
Red-breasted Sapsucker	0.389	0.250	0.288	0.074	0.138	15.12	0.002
White-headed Woodpecker	0.019	0.499	0.268	0.214	0.205	55.71	0.000
Random sample[c]	0.300	0.201	0.294	0.205			

[a] ($\frac{1}{2}\sum_{j=1}^{s} |u_j - e_j|$). Values can vary from 0 to 1, with greater values indicating greater departure from the fine-grained limit (random use of habitat units).

[b] Significance of chi-square test of independence based on data in Table 1 comparing row frequencies (marginal totals) of each bird species to random frequencies (df = 3).

[c] Proportional abundance of each tree species estimated from 100, randomly selected, 0.04 ha plots (e_j).

foraging cavity-nesting birds. Following techniques of Colwell (1973), they computed steady-state vectors of substrate use but did not conduct any tests of statistical significance of patterns; they simply described the values obtained. Investigators have used Markov analyses to examine sequences of song phrases in wood pewees (*Contopus* sp.) and cardinals (*Paroaria* sp.) (Chatfield and Lemon 1970), and to compare foraging-substrate use between male and female Emerald Tocanets (*Aulacorhynchus prasinus*) (Riley 1986). Mangel and Clark (1986) based their development of a unified foraging theory on what they call "Markovian decision processes," which are analyzed using Markov models.

Most of these analyses were based upon first-order or one-step chains, but analyses of higher order processes are also possible. Suppose that the following are five successive observations of habitat units (or behavior), A, B, and C:

A B A C B C.

In this sequence there are five pairs of first-order observations (A–B, B–A, A–C, C–B, B–C) and four second-order triplets (A–B–A, B–A–C, A–C–B, C–B–C). The data could be arranged in a $3 \times 3 \times 3$ table that contains the frequencies of each unit (A, B, or C) at time t_i that lead to each of the next two possible combinations at times t_{i+1} and t_{i+2}:

		t_{i+2}		
t_i	t_{i+1}	A	B	C
A	A	x_{111}	x_{112}	x_{113}
A	B	x_{121}	x_{122}	x_{123}
A	C	x_{131}	x_{132}	x_{133}
B	A	·	·	·
B	B	·	·	·
B	C	·	·	·
C	A	·	·	·
C	B	·	·	·
C	C	·	·	x_{333}

Chi-square tests of goodness-of-fit of such higher-order models compared to lower-order alternatives can be assessed using log-linear analyses described by Bishop et al. (1975:269).

Another important assumption of these analyses, especially important in interpreting the steady-state vector **u**, is that transition probabilities are stationary. In reality, these probabilities may shift during different times of day, times of year, across different years, or among subgroups within the animal population (e.g., age groups, sexes, demes), as described by other authors in

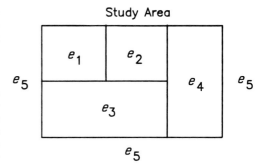

Transition Matrix

	e_1	e_2	e_3	e_4	e_5
e_1	0	1/4	1/4	0	1/2
e_2	1/4	0	1/4	1/4	1/4
e_3	1/6	1/6	0	1/6	3/6
e_4	0	1/6	1/6	0	4/6
e_5	2/10	1/10	3/10	4/10	0

Steady–state Vector
u = 0.133 0.133 0.200 0.200 0.333

FIGURE 1. Hypothetical study area composed of four habitat patches (e_j) and surrounded by a fifth (e_5). The probability that an animal will move from e_i to e_j from time t_i to t_{i+1} is the ratio of the perimeter that abuts against e_j and the total perimeter of e_i. The matrix **E** represents the matrix of probabilities of transition from any patch e_j along a row to any other patch, and the vector **u** represents the long-term probability (after an infinite number of transitions) of an organism being observed in each of the five patch-types.

this symposium. This temporal heterogeneity of resource use or behavior is not a unique concern in Markov analyses; indeed, any behavioral study must consider these effects and must restrict conclusions to the appropriate season or time period. Under a Markov analysis, one could collect observations within each relevant time unit and then compute and compare transition matrices between units to explicitly test for differences. If none is found, the units may be grouped for further analyses. If they do differ, all subsequent analyses must be restricted to comparisons across animal groups within the same time unit.

Sample size is another important issue in these analyses. Although I am aware of no explicit treatment of sample size requirements for Markov analyses, the considerations appropriate for contingency table analyses probably apply. In general, the study should be designed so that none of the expected values of cells in the table

TABLE 4. Comparisons[a] of Transition Matrices (Table 1) among All Bird Species. Chi-square Values Are Given with Significance in Parentheses

Bird species	WISA[b]	RBSA	HAWO	WHWO	PYNU	RBNU
White-breasted Nuthatch (WBNU)	83.47	106.50	104.33	34.93	3.82	73.95
	(0.00)	(0.00)	(0.00)	(0.00)	(<0.50)	(0.00)
Red-breasted Nuthatch (RBNU)	30.34	59.36	50.14	29.62	29.07	
	(0.01)	(0.00)	(0.00)	(0.01)	(0.02)	
Pygmy Nuthatch (PYNU)	41.31	56.76	53.52	12.17		
	(0.00)	(0.00)	(0.00)	(>0.50)		
White-headed Woodpecker (WHWO)	31.86	68.62	42.15			
	(0.01)	(0.00)	(0.00)			
Hairy Woodpecker (HAWO)	17.57	30.65				
	(0.29)	(0.01)				
Red-breasted Sapsucker (RBSA)	22.44					
	(0.10)					

[a] Chi-square tests of homogeneity, df = 15.
[b] Williamson's Sapsucker.

is <1 and no more than 20% of the cells should be <5 (Cochran 1954). Thus, a rough guide is that one should collect at least 5 times the number of cells in the analysis. For the bird data I used to illustrate the technique, I used a 4 × 4 × 7 table (=112 cells), which would require a sample size of at least 5 × 112 = 560 observations. This is a minimum estimate; greater numbers of observations (up to some asymptotic sample size) will lead to more robust results.

As in any study of animal behavior, an observer's actions must not influence the behavior of the observed animal. Because one is most interested in the movement among habitat units, it is critical that the observer does not disturb the animal, forcing it to move to a new location that it might not otherwise have chosen.

REFINEMENTS AND OTHER APPLICATIONS

The methods described here do not take into account the time spent in each habitat unit before moving to the next unit. It is certainly realistic to believe that an organism might spend more time in some habitat units (or behavior) than in others. For example, Raphael and White (1984: 38) reported that foraging time on a tree increased from averages of 30–73 s as tree diameter increased. Cane (1959) described methods to incorporate time effects into what she calls "semi-Markov" chains.

Another important improvement on the technique I have described involves a better sample-design and analysis of the distribution of available habitat units or environmental matrix (E). Most applications I have described assume a homogenous distribution of habitat units so that, at any time t_i, the choices available at time t_{i+1} are estimated from the habitat units that were randomly sampled over the entire environment (study area). However, if habitat units are patchily distributed, then the choices presented to the organism differ from one time to the next. Suppose, for example, that a study area contained only 10% lodgepole pine, occurring in one patch. If a woodpecker flew into the patch of lodgepole pine, its next choice of tree would probably be another lodgepole pine. The grain matrix for this bird could show a strong tendency to remain in lodgepole pine, even though the bird's actual behavior may have been random with respect to tree species when the environmental matrix was estimated from the overall study area. There are two solutions to this problem. First, one could estimate the total area occupied by each habitat-patch unit and then record the transitions between patch types and the transitions between units within patches. Colwell (1973) encountered a similar situation where hummingbirds foraged in patches of flowers; his techniques should be followed where resources are patchily distributed. If there is "preference" for one or more patch-types the observed transition probabilities will differ from expected transition probabilities.

A second approach would involve resampling the available habitat units at each successive location. From a bird's perspective, the available habitat probably lies in some radius (average distance flown between habitat units at time t_i and t_{i+1}) around its current location. Thus, selection should really focus on the units in this immediate environment rather than the whole study area. An observer could follow the birds from point to point, mark the successive locations (without disturbing the bird), and then estimate frequency of available habitat units in an area bounded by, say, a circle of radius r, which could be determined from pilot studies of movement distances. A grain matrix could be calculated as usual from the observed transition data, but the environmental matrix would be calculated from the sample of available units recorded at each foraging stop. Such an approach should provide a reasonable picture of the bird's selection of habitat units.

Markov analyses might also be useful in analyses of an organism's spatial distribution among geographically defined patches of habitat. To illustrate such an analysis, consider a hypothetical study area (Fig. 1). Each habitat unit e_j is a recognizable patch, such as a timber type or any mapped area. The question to be addressed is "What is the probability that an animal will be found in any unit e_j after n trials?" Note that a trial consists of a move from one unit to another. The probability that an animal will move from, say, patch e_1 to e_2, might be estimated from the proportion of the perimeter of e_1 that abuts against e_2 (in this case $1/4 = 0.250$). Similar values can be computed for each combination of units (Fig. 1). Over the long run, the expected distribution of animals in each habitat unit can be calculated using the eigenanalysis described above. In this example, the steady-state vector **u** equals 0.133, 0.133, 0.200, 0.200, and 0.333 for e_1 to e_5, respectively. One could then compare the observed distribution of animals to the steady-state vector using the chi-square goodness-of-fit or log-linear analyses described earlier. Such an approach could be used in radio-telemetry studies or any other studies where the spatial distribution of mobile organisms is investigated over time.

ACKNOWLEDGMENTS

I thank Yosef Cohen, Marc A. Evans, and Rudy King for helpful discussion of these analyses; the Department of Forestry and Resource Management of University of California, Berkeley, for support of field studies; and Marc A. Evans, Lyman L. McDonald, Michael L. Morrison, Barry R. Noon, and Pham X. Quang for comments on an earlier draft of the manuscript.

A COMPARISON OF THREE MULTIVARIATE STATISTICAL TECHNIQUES FOR THE ANALYSIS OF AVIAN FORAGING DATA

DONALD B. MILES

Abstract. This study discusses the complexities of analyzing foraging data and compares the performance of three multivariate statistical techniques, correspondence analysis (CA), principal component analysis (PCA), and factor analysis (FA) using five sample data sets that differ both in numbers of species and variables. Correspondence analysis consistently extracted more variation from the data sets (measured per eigenvalue or cumulatively) than either PCA or FA. Percent variance associated with the first axis and cumulative variance associated with the first five axes were negatively correlated with sample size, although the trend was stronger with PCA. There was also a significant positive relationship between percent variance and number of variables for PCA. CA showed a similar but nonsignificant trend. All three methods exhibited the "arch" effect or curvilinearity of the data when the positions of species were plotted along the first two derived axes. This suggests that the curvature trend in foraging data may represent a characteristic of the data rather than be solely an artifact of data reduction. Consistency in the biological interpretation of the derived foraging axes was determined using an analysis of concordance. Of the three methods, PCA and CA showed a high level of consistency in magnitude and sign of the coefficients from the first three eigenvectors. The concordance of the results from a factor analysis with the other two methods was low. Further, jackknife and bootstrap analyses revealed relatively stable estimates of the eigenvectors for only CA and PCA. Overall the analysis indicates that CA is a preferred method for analyzing foraging data.

Key Words: Foraging behavior; multivariate analysis; correspondence analysis; principal component analysis; factor analysis; jackknife; bootstrap.

Many analyses of avian ecology, particularly community oriented studies, rely on data representing the foraging behavior of coexisting species to address questions pertaining to guild structure, resource partitioning, community organization, habitat use and competition (e.g., Holmes et al. 1979b, Landres and MacMahon 1983, Sabo 1980, Sabo and Holmes 1983, Miles and Ricklefs 1984, Morrison et al. 1987b). Because most community studies assume that the manner in which a species exploits food resources represents an important niche dimension, a primary goal is to describe such resource axes indirectly through the measurement of foraging behavior. Thus, these studies attempt to estimate an unknown and underlying gradient of foraging behavior. Having determined this gradient, species may be positioned relative to one another along a foraging axis and inferences drawn about the ecological determinants of resource partitioning, guild structure, or community organization.

The resulting data set from a behavioral study of avian foraging usually consists of many variables measured on several species. Consequently, the investigators may choose to extract the key relationships embedded in the multidimensional data through a multivariate analysis. Several methods have been used to derive resource (niche) axes or foraging gradients from foraging behavior data. One approach adopted by avian ecologists for analyzing foraging data has been cluster analysis, based on various distance or similarity metrics (e.g., Landres and MacMahon 1980, Airola and Barrett 1985, Holmes and Recher 1986a). However, many investigators have turned to more advanced multivariate techniques, namely ordination methods, for deriving ecological patterns in multidimensional data. The prevalent ordination methods used in avian foraging studies include principal component analysis (Landres and MacMahon 1983, Leisler and Winkler 1985), factor analysis (Holmes et al. 1979b, Holmes and Recher 1986a) and correspondence analysis (Sabo 1980, Miles et al. 1987).

While several studies have compared the performance of multivariate methods in relation to vegetational gradients (e.g., Fasham 1977, Gauch et al. 1977), few attempts have used foraging data (Sabo 1980, Austin 1985). This paper assesses the "best" method for analyzing foraging data and tests the degree to which the unique characteristics of such data, in particular the "constant sum constraint," affect the results from a principal component analysis and factor analysis. Data from five avian studies spanning four habitat types (Sub-Alpine Forest, Deciduous Forest, Desert Scrub and Evergreen Oak Woodland) were analyzed using principal component analysis (PCA), factor analysis with Varimax rotation (FA), and correspondence analysis (CA). The criterion employed to determine efficacy of analysis was the percent variance summarized by the first four axes. Because many significance tests of multivariate methods require large sample sizes and most data infrequently meet this assumption, I generated standard errors and confidence limits for the coefficients and eigenvalues associated with each multivariate technique by jackknife and bootstrap procedures (Mosteller and Tukey 1977, Efron 1982).

CHARACTERISTICS OF FORAGING BEHAVIOR DATA

Investigations of avian foraging behaviors often depend on data gathered by observational methods. In such studies, *a priori* decisions are made about the types of foraging categories to recognize; these include the distinctiveness of various foraging substrates and the characterization of the foraging repertoire. Hence the range of categories included is determined by the ecological perceptions and subjective biological judgment of the investigator; the inclusion or definition of a category is largely arbitrary. Further, the nonindependent nature of most foraging observations, which is affected by the particular design of the study, presents an additional complication in the analysis of resource exploitation. The latter point may be addressed by using an appropriate sampling design when collecting the foraging observations. Accordingly, the choice of statistical technique for analyzing foraging data is constrained by these inter-relationships among the variables.

The analysis of foraging data presents two major difficulties; one involves a biological dilemma, and the second is one of statistical assumptions. Data collected on the foraging behavior of species may be envisaged to consist of observations apportioned among various cells in a multidimensional contingency table (see Miles and Ricklefs 1984). Such a contingency table represents a classification of foraging techniques by the type of substrate. A frequent method of analyzing such data is to treat each category as a separate, independent variable and use PCA or FA on the correlation matrix. However, such a procedure ignores the underlying relationships and biological interdependencies among the foraging variables and arbitrarily adds dimensions to the ecological space. That is, certain combinations of maneuvers and substrates are more likely to be employed because of energetic or biomechanical factors. Yet, other combinations may be physically unavailable to a species. For example, techniques such as gleaning, hovering, and probing may represent intermediate points along an underlying continuum. Similarly, foraging substrates may be intuitively ordered in some unknown manner, such as from coarse substrates, trunk and branches, to finer substrates, such as leaves. Overall, we may imagine that gleaning and hovering at leafy substrates lie at one end of an axis, and probing or pecking at ground substrates fall on the opposite end. Therefore, we may be justified in the assumption that the cross-tabulated foraging categories are discrete estimates of a continuous ecological axis that is to be estimated.

A second characteristic of foraging data is that the measurements are frequencies rather than continuous variables. This presents difficulties in the use of ordination techniques such as principal components analysis. Two main problems emerge by transforming the data from raw counts to proportions. First, frequency data exhibit marked curvature (Aitchison 1983). Second, as has been recognized in geological analyses, correlations among proportions may be subject to misinterpretation. When a vector of raw counts for p observations $(x_1, \ldots, x_p)$ is normalized, that is $y_i = x_i / \Sigma_i^p (x_1, \ldots, x_p)$ it becomes a vector of proportions (or compositional data) that are correlated. This property of frequency data has been termed the "constant sum constraint" by Aitchison (1981, 1983) because the terms in each vector must sum to unity. This constraint restricts the estimates of the correlation structure of the variables and results in a bias towards negative correlations. The statistical problem involves the recognition of this artifact, that is, how can the correlations that are artificially negative be detected. Thus, a principal component analysis of a categorical matrix may result in a biologically uninterpretable space. Such a conclusion leads to the question "how can foraging data be analyzed?" Further, can we develop confidence limits for our estimates? A comparison of the analysis of frequency data using several multivariate techniques may yield important insights into their behavior and biases.

EVALUATION OF MULTIVARIATE TECHNIQUES USED IN FORAGING ANALYSES

A chief goal of most investigations of avian foraging behavior is to summarize a cross-tabulated matrix of maneuver by substrates in a few axes that accurately represent the interrelationships of the species. Thus, we wish to position species along a foraging continuum that may be used later for interpreting those factors responsible for separating species in the ecological space; that is, we may look for clumping or clustering of species, which would suggest possible guilds. Further, we may be interested in discovering those foraging variables that contributed most to determining the inferred guild structure. Because the multivariate methods are used both to reduce a complex multidimensional data set to a lower number of uncorrelated variates or axes, and to position species along these derived gradients, one must examine the assumptions and properties of the three commonly employed multivariate techniques as well as the biological interpretability of these techniques.

Principal component analysis

The most prevalent technique used for analyzing foraging variables is principal component analysis (PCA). It is a variance-maximizing procedure, based on a Euclidean distance metric. PCA derives a small number of independent axes that extract the maximum amount of variance from the original data (Dillon and Goldstein 1984, Pielou 1984). No assumptions are necessary about the distribution of the data used by the method, although the data are assumed to be linearly or at least monotonically distributed. However, to perform significance tests of the eigenvalues one must assume that the data are approximately multivariate normally distributed. Apart from calculating the covariance or correlation matrix, PCA does not estimate parameters that fit an underlying statistical model. PCA is not scale invariant; variables that differ in units of measurement or vary in magnitude will affect the results. Because PCA attempts to maximize the total variation in a reduced number of axes, those variables with the highest variance will tend to contribute more to the derived axes. Many studies avoid the problems of scale in PCA by standardizing the variables by their corresponding standard deviation. This procedure concomitantly distorts the distances between points. Consequently the derived principal axes are unique to the particular data set and preclude generalizations from one study to another.

PCA transforms the original data matrix, composed of many presumably intercorrelated variables, into a reduced set of uncorrelated linear combinations that account for most of the variance present in the original variables. The first principal component (PC 1) is the linear combination that accounts for the greatest amount of variation relative to the total variation in the data. The second principal component (PC 2) extracts the largest amount of remaining variation, subject to the condition that it is uncorrelated (orthogonal) to the first. Similarly, PC 3 is calculated as the linear combination of original variables with the largest amount variance, but it is uncorrelated to the second and first PC axes.

Interpretation of the principal axes is arrived at by inspection of the coefficients of the eigenvectors and the correlations of the original variables with the principal component or component loadings. Because all principal components are linear combinations of the original data, the orientation of the axis projected through the cloud of points that maximizes the explained variation is determined by the coefficients of each eigenvector. The contribution of a variable to the principal component axis is determined by an examination of sign and magnitude of the component loadings (Dillon and Goldstein 1984).

Factor analysis

Whereas PCA is concerned with maximizing the total variation in a reduced number of axes to arrive at a more parsimonious representation of the data, FA is a technique for determining the intercorrelation structure among the variables (Dillon and Goldstein 1984). That is, FA attempts to portray the interrelationships among the variables in a reduced number of axes that maximize the variance common to the original variables. Implicit in this definition of a FA model is the assumption that a variable may be partitioned into two components, a unique factor and a common factor. As the terms suggest, the common factor represents an hypothetical and unobserved variable that jointly shares a fraction of the variation among all variables; the unique factor is an unobserved, hypothetical variable in which the variation is fixed and distinct to one variable. A second assumption made in FA is that the unique fractions are uncorrelated both with one another and with the common fraction. Thus, the factor analytic model is an analysis of the common variation among the variables (Dillon and Goldstein 1984). FA may be summarized by the model

$$X = \Lambda f + e,$$

where X is the matrix of observations, f is a matrix of the unknown and hypothetical common factors, e is a matrix of unique factors, and Λ is a matrix of unknown factor loadings. Simply stated, FA seeks to describe the complex relationships that characterize the observed variables in terms of a few, unknown, unobservable quantities known as factors. These factors allow one to determine the structure of the data and to derive common axes that unite the variables. However, few ecologists have critically examined the extent to which the factor model is relevant for their analytical goals. Because of the complex nature of the factor model and the assumptions made about the nature of the variation associated with the variables, ecologists must be keenly aware of the differences between FA from PCA before deciding on an analytical technique. Direct solution of the complex factor model is difficult, because of the presence of several hypothetical and unknown quantities (Dillon and Goldstein 1984). A common *approximate* solution is given by a principal component analysis of the reduced correlation matrix (i.e., a correlation matrix that has had the unique variation removed). In this solution an estimate of the unknown matrix of factor loadings is derived by multiplying each element of the eigenvectors by the square root of the corresponding eigenvalue. The "meaning" of each factor axis, in terms of identifying the underlying pattern of variation that is common to the variables, usually proceeds by the examination of the magnitudes of all loadings. A variable is retained for interpretation if it exceeds a critical threshold, which may be defined either arbitrarily, as in a loading exceeding a certain minimum value, or by the statistical significance of the loading.

Orthogonal rotation of the factor axes often follows the extraction of the components as an aid to interpretation of the extracted factor pattern. The justification for rotating the axes is, in most instances, that the factor pattern may be difficult to interpret; one or two variables might have high loadings, but most may be of similar magnitude. This additional transformation of the factor axes is coupled with the goal of restricting the interpretation of each axis to as few of the variables as possible. The most commonly used method, Varimax rotation, seeks to maximize the square of the factor loadings. The end result is an exaggeration of the magnitude of the loadings: the larger loadings are made larger and the smaller loadings are diminished (Dillon and Goldstein 1984). Most examples of FA in the ecological literature simply employ a Varimax rotation of the derived PCA axes. Several disadvantages accompany the use of FA. First, the solution to the factor model is unique to the particular study. That is, it is very difficult to generalize the results of one study to another. Second, the rotation of the axes distorts the distance relationships among the observations, which precludes comparing the positions of species in the ecological space from one study to another.

Correspondence analysis

Correspondence analysis, also known as reciprocal averaging analysis (Hill 1974, Miles and Ricklefs 1984, Moser et al., this volume) is a dual ordination procedure. Both species and foraging categories are analyzed simultaneously on separate but complementary axes. The dispersion of species is accomplished by means of the distributions across foraging categories. Conversely, the categories are ordinated according to the patterns of their use by each species. The technique reveals the presence of underlying ecological and phenotypic variables pertinent to the manner in which birds forage (Sabo 1980, Miles and Ricklefs 1984).

Correspondence analysis uses an eigenvector algorithm similar to that of PCA (Hill 1973, 1974; Gauch et al. 1977; Pielou 1984). However, it differs from PCA in three principal qualities: (1) the use of chi-square distances rather than Euclidean; (2) a double standardization of the data; and (3) an additional division step (Gauch et al. 1977). This first quality is useful, for it allows confidence intervals to be placed about points in the reduced space. Axes are computed that maximize the correspondence between species and the foraging categories. As in PCA and FA, the number of CA axes required to explain most of the variation in the data set is fewer than the number of categories in the original matrix. One advantage of CA is its resistance to distortion when analyzing curvilinear or nonmonotonic data (Gauch et al. 1977, Lebart et al. 1984, Moser et al., this volume).

I specifically did not include detrended correspondence analysis in this study (Sabo 1980) because of its use of an arbitrary, ad hoc standardization of the second and successive axes based upon the assumption of a single dominant axis. It further employs a rescaling of the data as an aid to interpret intersample distances (Miles and Ricklefs 1984, Pielou 1984). In a study comparing four ordination methods, Wartenberg et al. (1987) showed that detrended correspondence analysis and CA arrived at a similar ordering of species along a single gradient. For a detailed discussion of the weaknesses of detrended correspondence analysis see Wartenberg et al. (1987).

MATERIAL AND METHODS

I analyzed five sets of data (Table 1) that had the following dimensions: 20 species by 14 variables, 19 species by 14 variables, 11 species by 16 variables, 14 species by 15 variables, and 12 species by 15 variables. Because the data consisted of proportions, I used the arcsine-square root transformation before performing the PCA or FA.

Each data set was subjected to analysis by CA, PCA, and FA. The last two techniques had as input the correlation matrices generated from the foraging data. To make comparisons among studies I followed the methods of previous studies, and used the principal factor method to derive a reduced set of factor axes in the FA. All factor axes whose associated eigenvalues exceeded one were used in subsequent analyses. Next, I performed a Varimax orthogonal rotation of factor axes to further reduce the structure of the data to a few combinations of original variables. In this study, PCA and FA extracted eigenvalues using a similar algorithm and generally arrived at common solutions, therefore I only analyzed the PCA eigenvalues for patterns in explained variance. Unlike the previous two analyses, CA was performed using the untransformed proportions. Interpretation of the results was accomplished by a simultaneous plotting of the foraging category coordinates and the species (sample) coordinates. The magnitude and sign of the coordinate indicates its contribution to the structure of the data. Previous evaluations of CA considered it to lack rigorous statistical tests for the eigenvalues and eigenvectors. However,

TABLE 1. SOURCES OF FORAGING DATA ON PASSERINES USED IN THIS STUDY

Location	Habitat type	Number of species	Number of variables	Source
Mt. Moosilauke, New Hampshire	Sub-alpine forest	20	14	Sabo (1980)
Hubbard Brook, New Hampshire	Deciduous forest	19	14	Holmes et al. (1979b)
Purica, Mexico	Evergreen oak woodland	11	16	Landres and MacMahon (1980)
Santa Rita Mtns., Arizona	Encinal	14	15	Miles (unpubl. data)
Saguaro National Monument, Arizona	Desert scrub	12	15	Miles (unpubl. data)

the unique distributional qualities of chi-square distances allow for several significance tests (see Lebart et al. 1984).

All three multivariate techniques share two common goals: (1) the determination of common themes of covariation among a strongly correlated group of variables and (2) the reduction of a high-dimensional data set into a few derived axes that preserve as much of the original variation as possible. Therefore, I based my evaluation of the performance of these procedures on the percent variation extracted per axis. This criterion allows a direct comparison of PCA and CA whose eigenvalues are not interchangeable. I examined (1) the number of axes necessary to explain at least 90% of the variation and (2) the proportion of variation associated with the first axis. The multivariate technique that consistently explained a larger fraction of the original variation in the least number of axes and resulted in easily interpretable axes should be preferred. This also has direct bearing on the number of axes to retain for subsequent analyses and interpretation. Because most studies that use multivariate techniques depend on the loadings for interpreting the results, I compared the three procedures for consistency in the direction and magnitude of the axis loadings.

Jackknife and bootstrap estimation of variability

Several common problems plague ecological investigations that employ multivariate methods. The first is how many axes should be interpreted, or kept for further analyses. The second involves which of the coefficients in the eigenvector may be used to interpret the patterns suggested by a PCA or CA. Because of the small sample sizes, unknown sampling distribution, and the large number of categories that characterize foraging studies, formal statistical testing of eigenvalues is impossible. Consequently, predominant solutions to the above dilemmas are actually ad hoc guidelines. Computation of PCA by using the correlation matrix further complicates hypothesis testing, for most of the statistical tests are based on the variance-covariance matrix.

However, bootstrap and jackknife resampling techniques can replace the arbitrary and ad hoc procedures. Both are receiving increased use in ecological studies (e.g., Gibson et al. 1984, Stauffer et al. 1985). Their use provides an estimate of a statistic as well as a measure of variance associated with the estimate. These methods are particularly crucial for deriving confidence limits about a complex statistic that lacks an analytical sampling distribution.

The premise of the jackknife is to determine the effect of each sample on a statistic by iteratively removing successive samples and recalculating the statistic (Mosteller and Tukey 1977, Efron 1982, Efron and Gong 1983). The jackknife analysis begins by computing the desired statistic for all the data. A single observation is then removed from the data and the statistic is recalculated using the remaining $n - 1$ observations. Let y_{all} represent the statistic calculated for the full sample. Define a pseudovalue to equal

$$y^* = ny_{all} - (n - 1)y_{(j)}, j = 1, 2, \ldots, n,$$

where n is the sample size. The jackknifed estimate of the statistic is defined as the mean of the pseudovalues

$$y^* = 1/n \sum y^*_j,$$

and the variance of the jackknifed statistic is given by

$$s^{2*} = [(y^*_j - y^*)^2/n(n - 1)]^{1/2},$$

where s^2 is the variance of the pseudovalues. One can use the jackknife estimate of variance to calculate confidence intervals based on the t distribution (Mosteller and Tukey 1977).

I used the jackknife method of variance estimation for the principal component analysis, factor analysis, and correspondence analysis of foraging data from all five data sets. Two statistics were subjected to this resampling plan. Upon deleting a single observation from the original data set and recalculating the three multivariate procedures, I derived the pseudovalues for the first four eigenvalues and the elements of the first three eigenvectors. This procedure resulted in the calculation of jackknife estimates of the statistics and a measure of their variability. Following Mosteller and Tukey (1977), I also computed the jackknife error ratio, which is simply the jackknife estimate divided by its standard error. The ratio may be viewed as a t statistic with $(n - 1)$ degrees of freedom. Because the results of the jackknife method were similar for all data sets, in this paper, I present only the results for the Santa Rita data set.

The bootstrap is a conceptually simple, but computer-intensive, nonparametric method for determining the statistical error and variability of a statistical estimate. The premise of the bootstrap is that, through resampling of the original data, confidence intervals may be constructed based on the repeated recalculation of the statistic under investigation. An assumption made

TABLE 2. PERCENT VARIANCE EXPLAINED BY THE FIRST FIVE EIGENVALUES FROM PRINCIPAL COMPONENT ANALYSIS AND CORRESPONDENCE ANALYSIS

	Axis									
	I		II		III		IV		V	
Sample	PCA	CA	PCA	CA	PCA	CA	PCA	CA	PCA	CA
Mt. Moosilauke	27.3	36.7	23.4	20.1	16.2	18.4	9.2	9.7	6.6	6.2
Hubbard Brook	27.5	32.4	23.1	28.6	14.7	12.1	12.3	10.4	7.2	9.0
Purica	37.1	41.9	19.4	28.7	13.7	12.7	9.2	8.7	7.9	5.1
Santa Rita	31.6	33.5	21.4	30.0	15.8	13.9	11.3	10.7	6.4	4.5
Saguaro	39.4	36.1	19.7	31.1	12.4	12.2	8.2	8.2	6.3	5.2

by the bootstrap is that the data follow an unknown but independent and identical distribution.

To begin the bootstrap procedure, the following steps were executed. First, I pooled the original data set consisting of n observations. Using a random-number generator, I selected n observations from the data with replacement; these n random values constituted a bootstrap sample, X^*_i. That is, each individual observation was independently and randomly drawn and subsequently replaced into the original data before another observation was drawn. A consequence of this sampling scheme was that an observation could be represented more than once or not at all in any bootstrap sample. The data were resampled a large number of times, which resulted in m bootstrap samples. Next, the statistic of interest was computed for each of the m bootstrap samples. In the present study, I calculated bootstrap estimates of the eigenvalues and eigenvectors only from a PCA. Let L^{*i}_j designate the ith bootstrap calculation of the jth eigenvalue or eigenvector. Then the bootstrap estimate of either statistic and the associated standard error is

$$L_B = 1/m \sum L^{*i}_j$$

$$\text{SE}(L_B{}^j) = \sqrt{s^2_L},$$

where s^2 = the variance of the m bootstrap L^{*i}_j samples, i.e., (L^{*1}_j, L^{*2}_j, ..., L^{*m}_j). The estimated mean and standard deviation of the PCA statistics were based on 200 bootstrap replications. This bootstrap sample size was the first from a range of sample sizes (100, 200, 300, 400, and 500) to exhibit a stable convergence with the bootstrap calculations based on larger replicates.

In this study, the correspondence analysis, factor analysis, principal component analysis and bootstrap analysis were performed on an IBM 4381 using the following programs: CA, CORRAN (modified from Lebart et al. 1984), PCA and FA, SAS (SAS 1985). The program to compute the jackknifed statistics was written in QuickBASIC (version 3.0) and was performed using and IBM PC compatible computer.

RESULTS

Percent variance explained

FA and PCA arrived at a similar set of eigenvalues, so results for only the latter analysis are provided. Percent variance explained by the first two axes was generally higher for CA than PCA (Table 2), although PCA explained a higher amount of variation than CA in the first axis for the Saguaro data, and PCA had a higher percent variation value than CA in the second axis for the Mt. Moosilauke data. Along the third, fourth and fifth axes, PCA had higher values of percent variance extracted than CA for most data sets (Table 2). However, several of the comparisons were very similar (e.g., axis IV for the Saguaro data set and axis V for the Mt. Moosilauke data). The tendency for CA to capture more variation in the first few axes was biologically meaningful, for it suggests that CA may be more efficient at describing the underlying continuum that may characterize foraging behavior.

Cumulative variance for the first seven axes ranged from 97% to 99% for the CA results and 90% to 95% for the PCA (Fig. 1). CA would retain the first four or five axes to explain 90% of the variation (one criterion for determining the number of axes to retain and interpret), while PCA would require at least six axes and in one case seven axes. Thus, based on these results, CA preserves most of the original information in a reduced number of axes.

A strong negative correlation existed between species number and percent variance explained by the first axis for the PCA ($r_s = -0.90$, $P < 0.07$; Fig. 2A); the relationship was weaker in the CA ($r_s = -0.40$, $P < 0.42$). The cumulative percent variance associated with the first five axes was also negatively related to the number of species in the sample data matrix ($r_s = -0.90$, $P < 0.07$) for both the PCA and CA (Fig. 2B). A strong positive, but nonsignificant correlation was shown between the number of variables and the percent variance explained by the first PC axis ($r_s = 0.79$; Fig. 3A). A lower positive correlation was exhibited by the first CA axis and the number of variables ($r_s = 0.52$). However, there was a significant positive correlation between the cumulative percent variance explained by the first five PC axes and number of variables ($r_s = 0.95$, $P < 0.05$; Fig. 3B). The correlation shown for the CA was lower and nonsignificant ($r_s = 0.73$, $P < 0.15$). Thus, PCA shows a greater sensitivity to changes in the number of foraging variables included in an analysis than CA.

Presence of the arch effect

In this study, the distortion of the second and higher axes was present in all three multivariate methods (Figs. 4, 5, 6; see also Fig. 1 in Moser et al., this volume). The positions of species along the first two axes from a CA, PCA and FA exhibited a characteristic v-shaped pattern or arch effect. The degree of distortion also was similar for all three analyses. One frequent criticism of CA is the tendency for the distribution of species to be compressed towards the terminal portions of the axes. However, the plot of CA axes 1 and 2 failed to demonstrate any compression of points along the axes.

Differences in interpretation of resource axes

The interpretations derived from one analysis of the foraging data were not necessarily substantiated or similar when applying a second multivariate method. As an example, the second axis from a CA of the Santa Rita data (Table 3) described a gradient with gleaning at leaves and twigs at one end and gleaning and probing of trunks, branches, and ground at the other. However, the interpretation from FA revealed that the axis described a contrast between hovering at leaves, twigs, and branches against gleaning maneuvers. Although not presented, dissimilarities in the biological interpretation among the three multivariate techniques were also evident in the other four data sets.

Greater than 73% (11/15) of the paired comparisons between CA and PCA were statistically significant based on Kendall's rank order correlation coefficient (Table 4). Fewer than 50% of the correlations between CA and FA were significant (7/15). The degree of concordance between PCA and FA was also low; only 53% of the comparisons showing significant correlations.

Jackknife and bootstrap variance estimates

CA and PCA exhibited similar results of jackknife and bootstrap analyses for all three axes (Tables 5 and 6). Because the results from all five data sets were the same, I present the jackknifed coefficients from only the Santa Rita data set. A coefficient was considered to be significantly different from zero if the error ratio exceeded 3.0 ($\alpha < 0.01$). Using this criterion, the first axis of CA and PCA both had 73% of the coefficients significantly different from zero. Inspection of the coefficients revealed that the variables considered significant in the CA and PCA were identical. This supports the conclusion that foraging

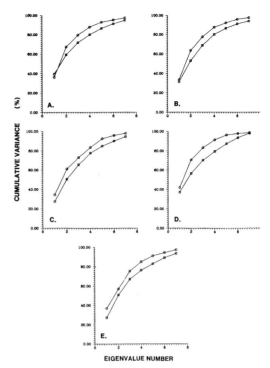

FIGURE 1. Cumulative variance "explained" by the first seven eigenvalues. A comparison of the results from principal component (open boxes) and correspondence analyses (open circles). Note: Factor analysis and principal component analysis gave similar eigenvalues, hence only the latter results were plotted. Results from: A. Saguaro sample; B. Santa Rita sample; C. Hubbard Brook sample; D. Purica sample; and E. Mt. Moosilauke sample.

gradients described by CA 1 and PCA 1 were the same. Nevertheless, PCA and CA differed slightly in the number of coefficients whose error ratios exceeded the critical value of 3.0 for axes 2 and 3. Nearly 50% (7/15) of the coefficients associated with CA 2 were significant, whereas 67% from PCA 2 had error ratios greater than 3.0. Of the variables that were not significant, approximately 63% were common to CA and PCA. Thus, the results for the second axis indicated that PCA and CA described similar trends of variation. While CA 3 had 53% (8/15) of the coefficients exceeding 3.0, PCA 3 had 87% of the coefficients significantly different from zero. Estimates of the eigenvalues corroborated the patterns shown by analysis of the coefficients. The first three eigenvalues of CA and PCA had error ratios that were larger than 3.0.

The jackknifed estimates for the FA statistics revealed a very different pattern (Table 7). Although the percentage of coefficients having an error ratio greater than 3.0 was close to 100%

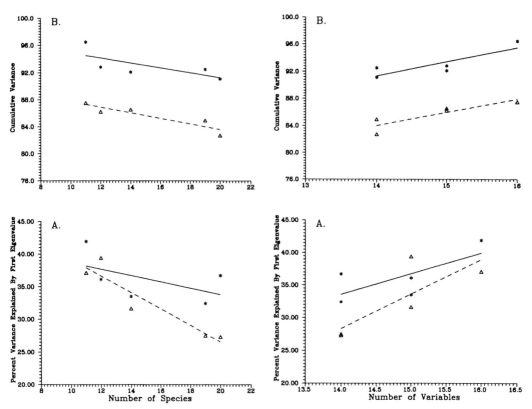

FIGURE 2. Relationship between percent variance explained by the first eigenvalue (A) and cumulative variance explained by the first five eigenvalues (B) with number of species in sample data. Star symbols and solid line present results from the correspondence analysis; open triangles and dashed line present the principal component analysis.

FIGURE 3. Association of number of variables with percent variance explained (A) by the first eigenvalue and (B) the cumulative variance explained by first five eigenvalues. Symbols as in Figure 2.

for each axis (100% for FA 1, 86% for FA 2, and 80% for FA 3), nearly all the estimates were greater than 1.0. For example, 80% of the coefficients characterizing FA 1 were above 1.0. The percentages for FA axes 2 and 3 were 73% and 60%, respectively.

Bootstrap estimates of the coefficients for PCA 1–3 were lower than jackknifed estimates, but were close to the observed values from the original data set (compare Tables 3 and 8). Using the critical value of 3.0 for the error ratio resulted in only approximately 30% of the coefficients from PCA 1 showing a value significantly different from zero. However, nine coefficients (60%) were significant for the second PC axis, but only three coefficients from the third axis were significant. Bootstrap estimates of the eigenvalues corroborate the jackknife analysis. Eigenvalues for all three axes were highly significant, indicating that the axes were associated with significant trends of variation and not simply the random orientation of vectors through a spherical cloud of points.

DISCUSSION

COMPARISONS OF THE MULTIVARIATE STATISTICAL METHODS

In this analysis, the number of axes that extract a "significant" amount of variation differed between CA and PCA or FA. Fewer axes were needed to explain a larger percentage (90%) of variation with CA than with PCA. The primary difference involved the amount of variance associated with the first two axes. Subsequent eigenvalues were either larger for PCA relative to CA or not different. Assuming that the first few, large eigenvalues represented structure (i.e., valid correlations among the variables that correspond with species interactions) and the small eigenvalues depicted noise (i.e., unique species foraging behaviors or repertoires [Gauch, 1982b]),

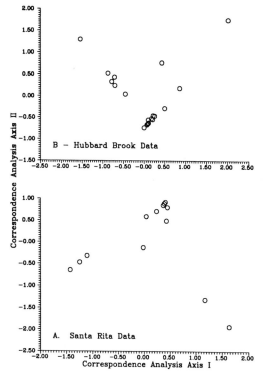

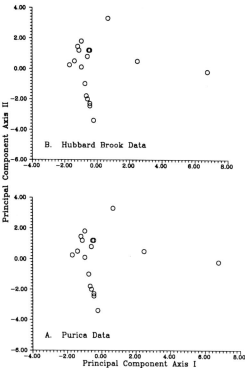

FIGURE 4. Ordination of species' foraging behavior by correspondence analysis. Plot of species' positions along the first two axes for (A) the Santa Rita site and (B) the Hubbard Brook site.

FIGURE 5. Position of species' on the first two axes from a principal component analysis: results from Purica site (A) and Hubbard Brook site (B).

then CA extracted more structure than PCA. Consequently, CA characterized the species' relations with only three or four axes, compared to the five or six necessary for PCA or FA. This held true regardless of whether I retained all axes whose eigenvalues were greater than one or the number of axes needed to account for 90% of the variation.

Miles and Ricklefs (1984) suggested that the analysis of foraging categories by PCA was inappropriate. They argued that the arbitrary subdivision of each foraging technique increased the dimensionality of the data by artificially inflating the number of foraging variables. Because CA maximizes the correlation between the positioning of the variables based on their use by birds and the position of species based on their use of foraging variables to determine the major gradients of variation, they suggested that CA would be more robust to changes in number of variables. It follows that a positive correlation should exist between the number of variables and the percent variance explained by the first axis and cumulative variance in the first few axes. This study supported that conclusion, as the amount of variation packaged in the eigenvalues CA was less sensitive than those of PCA to changes in the number of variables. Therefore, including additional variables in a PCA increased the number of dimensions and diminished the explanatory power of the first few axes. Because these conclusions are based on a small difference in the number of variables, further investigation is necessary. In particular, a sensitivity analysis should be performed where the number of variables within a data set is altered and the resulting change in the magnitude of variation explained by the eigenvalues compared.

Based on the results of the analysis of concordance, similar conclusions about the patterns of foraging among birds would be drawn whether using CA and PCA. However, little concordance was found when comparing the results between CA and FA or PCA and FA. This is a crucial point, for it suggests that the biological interpretation derived for each axis depends on the type of analysis with which the data were summarized. Rotation of the factor axes in FA produced

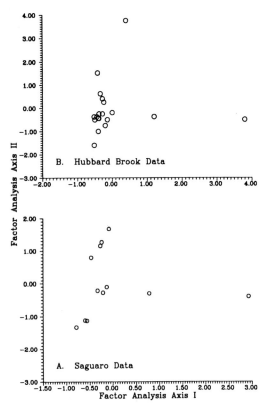

FIGURE 6. Results from a factor analysis with Varimax rotation. The position of species along the first two axes are presented: (A) Hubbard Brook data, (B) Saguaro data.

a unique factor solution that was not comparable to the PCA or CA results.

THE ARCH EFFECT

Previous reviews of multivariate techniques used in analyses of species abundance patterns along an elevational or environmental gradient have recognized the presence of distortion between the first axis and subsequent axes, which is commonly called the "arch effect" or "horseshoe effect" (Gauch et al. 1977, Austin 1985, Pielou 1984, Wartenberg et al. 1987). However, studies that incorporate multivariate analyses of foraging behavior rarely examine the data for the arch effect. Several analyses of guild structure exhibited curvilinearity when species were plotted along the first two axes of a PCA or CA (e.g., Sabo and Holmes 1983; Miles and Ricklefs 1984; and Fig. 1 of Moser et al., this volume).

In this study, all three ordination methods exhibited a similar, consistent positioning of the species within the two-space that may not be associated with biological processes. Some curvature of the data points was evident for the CA and PCA and to a certain extent FA. Previous investigations suggested that this arch effect resulted from sampling species that were distributed along an environmental gradient in a non-monotonic fashion. Because PCA, FA and CA assume that the data are linear, the collapsing of a high-dimensional data matrix to a few axes results in the involution of the second and higher axes relative to the first. For example, Gauch et al. (1977) argued that the arch characterizing CA and PCA was attributable to the sampling of a long gradient in which the distribution of many species was attenuated. Several methods have been proposed to "correct" the arch effect (Pielou 1984). The prevailing technique, Detrended Correspondence Analysis, involves an arbitrary rescaling of the second and higher axes, relative to the first axis (Pielou 1984). However, this procedure has been criticized as being an ad hoc transformation rather than a method for direct analysis of curvilinear data (Wartenberg et al. 1987).

Foraging data are rarely sampled over an environmental gradient. Thus the underlying causes of the curvilinearity may remain obscure, although two possible sources may be considered: (1) The curvature may be a consequence of the constant-sum constraint (Aitchison 1981). Because foraging data are often expressed as frequencies, they must sum to unity for each species. The data therefore are restricted to lie between the values 0 and 1. Consequently, the estimated correlations tend to be negative and the cloud of points in the n-dimensional space is curvilinear. (2) The curvature may represent a nonlinear response of the species to differences in the vegetation structure or prey distribution within the habitat. Regardless of the cause of the curvilinearity, it should be regarded as a structural feature of foraging data. Therefore, special effort should be made to avoid the interpretation of nonlinear relationships within the reduced multivariate space.

BIAS IN THE INTERPRETATION OF THE MULTIVARIATE ANALYSES

At least three axes from CA and PCA should be retained for subsequent interpretation. Thus we can reject the hypothesis that each axis represents an arbitrary and random rotation of orthogonal axes through an n-dimensional spherical cloud of points. This conclusion was corroborated by the highly significant values obtained for the eigenvalues from the bootstrap and jackknife. The jackknifed coefficients for the first axis of CA and PCA showed concordant patterns of organization along that dimension. There was complete overlap of coefficients that differed sig-

TABLE 3. COMPARISON OF THE RESULTS FROM THE THREE MULTIVARIATE ANALYSES. THE COEFFICIENTS PRESENTED BELOW ARE (1) THE SCORES FOR EACH OF THE 15 VARIABLES FROM A CA; (2) THE NORMALIZED LOADINGS FROM A PCA; AND (3) THE ROTATED FACTOR LOADINGS FROM FA. ANALYSES WERE BASED ON THE SANTA RITA DATA

	Coefficients								
	Axis 1			Axis 2			Axis 3		
Variable[a]	CA	PCA	FA	CA	PCA	FA	CA	PCA	FA
GLLF	0.37	−0.210	−0.698	0.85	−0.424	−0.260	0.11	0.119	−0.421
GLTW	0.48	−0.247	−0.705	0.69	−0.331	−0.275	0.07	0.135	−0.253
GLBR	0.47	−0.248	−0.577	0.32	−0.099	−0.176	−0.34	0.086	−0.068
GLTR	0.65	−0.188	−0.108	−0.52	0.297	−0.130	−1.11	−0.084	0.321
GLGR	0.00	−0.072	0.069	−0.33	0.159	−0.130	−2.36	−0.222	−0.080
PRBR	1.56	−0.183	−0.043	−1.78	0.455	−0.093	0.37	0.154	0.984
PRTR	1.53	−0.185	−0.048	−1.74	0.449	−0.098	0.34	0.145	0.960
PRGR	1.43	−0.169	−0.093	−1.59	0.384	−0.055	0.06	0.157	0.834
SATW	−1.09	0.294	0.756	−0.45	0.085	0.125	−0.17	−0.305	−0.241
SABR	−1.41	0.320	0.926	−0.64	0.079	−0.018	0.13	−0.386	−0.187
HAWK	−1.25	0.296	0.798	−0.52	0.064	0.045	−0.02	−0.312	−0.186
HVLF	−0.64	0.316	0.069	0.08	−0.040	0.879	0.62	0.356	−0.179
HVTW	−1.24	0.335	0.131	−0.40	0.051	0.958	1.10	0.388	−0.039
HVBR	−1.04	0.351	0.333	−0.23	0.013	0.722	0.72	0.203	−0.129
HVTR	−0.91	0.289	0.058	−0.50	0.086	0.933	1.00	0.419	0.057

[a] Codes are: GLLF, glean at leaf; GLTW, glean at twig; GLBR, glean at branch; GLTR, glean at trunk; GLGR, glean at ground; PRBR, probe at branch; PRTR, probe at trunk; PRGR, probe at ground; SATW, sally from twig; SABR, sally from branch; HAWK, aerial manuever; HVLF, hover at leaf; HVTW, hover at twig; HVBR, hover at branch; and HVTR, hover at trunk.

nificantly from zero. Thus both analyses arrived at a similar group of variables that structured the foraging behavior of species within the community. However, the second and third axes tended to exhibit unique patterns of variation specific to each analysis, but overlap in the categories that were significant remained relatively high. Most importantly, the results from the jackknife and bootstrap analyses reinforced the interpretations from an analysis of the original Santa Rita data set.

The disparity between PCA and CA in the number of variables that were significantly different from zero in the last two axes may in part be a consequence of the difference in the scaling of the eigenvectors. Because each eigenvector from a PCA is normalized (i.e., the square of the eigenvector equals 1.0), the coefficients are less

TABLE 4. AN ANALYSIS OF THE DEGREE OF CONCORDANCE OF LOADINGS AMONG THE THREE ORDINATION TECHNIQUES. THE ANALYSIS IS BASED ON KENDALL'S RANK ORDER CORRELATION COEFFICIENT

		Comparison		
		CA with		
Sample	Axis	PCA	FA	PCA with FA
Mt. Moosilauke	I	0.26	−0.18	0.76***
	II	0.55**	−0.38	−0.62**
	III	0.24	0.28	−0.29
Hubbard Brook	I	0.74**	0.67**	0.76**
	II	−0.65**	−0.08	0.34
	III	−0.60**	−0.24	0.18
Purica	I	−0.62**	−0.44*	0.67**
	II	0.03	−0.73**	−0.14
	III	0.38	0.17	0.38
Santa Rita	I	−0.75**	−0.55**	0.74**
	II	−0.90***	−0.08	0.03
	III	0.76**	0.12	0.28
Saguaro	I	−0.78**	−0.60**	0.73**
	II	−0.77**	−0.77**	0.83**
	III	0.87***	0.61**	0.81***

* $P < 0.05$, ** $P < 0.01$, *** $P < 0.001$.

TABLE 5. JACKKNIFED CORRESPONDENCE ANALYSIS OF THE SANTA RITA DATA. VALUES ARE COEFFICIENTS OF THE FIRST THREE EIGENVECTORS (COEFF.), THEIR STANDARD ERRORS (SE), AND THE ERROR RATIOS (ER = COEFF./SE). ESTIMATES OF THE FIRST THREE EIGENVALUES ARE GIVEN AT THE BOTTOM OF EACH COLUMN

Variable	Axis I			Axis II			Axis III		
	COEFF.	SE	ER	COEFF.	SE	ER	COEFF.	SE	ER
GLLF	0.778	0.161	4.94	0.943	0.103	9.18	0.267	0.063	4.27
GLTW	0.981	0.160	6.11	0.745	0.149	4.98	0.144	0.084	1.71
GLBR	0.962	0.135	7.09	0.413	0.196	2.09	−1.467	0.095	15.49
GLTR	0.557	0.134	4.14	−0.836	0.175	4.76	−1.299	0.254	5.11
GLGR	0.037	0.109	0.34	0.255	0.203	1.25	−4.597	0.427	10.71
PRBR	0.984	0.462	2.12	−3.117	0.374	8.34	0.267	0.187	1.43
PRTR	1.047	0.432	2.41	−3.161	0.337	9.35	0.405	0.144	3.63
PRGR	0.436	0.515	0.85	−2.119	0.457	4.63	−0.942	0.259	3.62
SATW	−1.721	0.162	10.60	−0.134	0.199	0.67	−0.105	0.063	1.66
SABR	−2.673	0.201	13.32	−0.138	0.287	0.48	0.576	0.162	3.55
HAWK	−1.816	0.167	10.87	−0.032	0.244	1.29	0.296	0.094	3.14
HVLF	−0.946	0.140	6.74	0.331	0.179	1.85	0.097	0.251	0.36
HVTW	−1.639	0.172	9.55	0.538	0.232	2.30	0.153	0.447	0.34
HVBR	−1.727	0.137	12.64	0.048	0.255	0.19	0.172	0.223	0.77
HVTR	−1.021	0.107	9.48	−1.772	0.52	3.43	0.201	0.458	0.44
Eigenvalue	0.753	0.019	38.83	0.829	0.045	18.43	0.329	0.025	10.26

than one by definition. Hence, they tend to have lower standard errors and consequently higher error ratios. However, the magnitude of the coefficients in CA depends on the degree to which the species employs each category; the longer the gradient (i.e., various species specialize on certain foraging categories and therefore recognize each category as distinct), the greater the values for each coefficient. In short, the coefficients are not required to be less than one. This results in higher standard errors and lower error ratios.

Suprisingly, the jackknife estimates of the rotated factor loadings produced rather poor results. While the results based on the eigenvalues suggested that at least three axes should be retained, estimates of the coefficients were highly biased. Because the coefficients from the jackknife analysis exceeded 1.0, it is difficult to evaluate the confidence one should place on an analysis using all data points. The pattern shown in the jackknifed values presented in Table 7 was not unique to the Santa Rita data. Similar trends were evident in all five of the jackknifed factor analyses. Thus, it is possible to discount any artifact due to the data. Most probably, the inflated parameter estimates were a consequence of the

TABLE 6. JACKKNIFED PRINCIPAL COMPONENT ANALYSIS OF THE SANTA RITA DATA. VALUES ARE COEFFICIENTS OF THE FIRST THREE EIGENVECTORS (COEFF.), THEIR ESTIMATED STANDARD ERRORS (SE), AND THE ERROR RATIOS (ER = COEFF./SE). ESTIMATES OF THE FIRST THREE EIGENVALUES ARE GIVEN AT THE BOTTOM OF EACH COLUMN

Variable	Axis I			Axis II			Axis III		
	COEFF.	SE	ER	COEFF.	SE	ER	COEFF.	SE	ER
GLLF	−0.398	0.039	10.14	−0.449	0.032	13.96	0.026	0.108	0.92
GLTW	−0.403	0.027	14.81	−0.347	0.039	9.01	0.130	0.029	4.47
GLBR	−0.291	0.026	11.29	−0.056	0.066	0.84	0.102	0.033	3.06
GLTR	−0.149	0.032	4.65	0.591	0.041	14.35	0.438	0.131	3.35
GLGR	−0.042	0.017	2.37	0.257	0.032	7.95	0.445	0.116	3.82
PRBR	−0.040	0.036	1.12	0.606	0.029	20.62	−0.570	0.171	3.32
PRTR	−0.039	0.033	1.17	0.595	0.031	18.93	0.139	0.036	3.87
PRGR	0.004	0.038	0.13	0.345	0.038	8.89	0.188	0.044	4.33
SATW	0.314	0.017	18.07	−0.007	0.032	0.23	−0.382	0.040	9.45
SABR	0.369	0.025	14.77	−0.070	0.043	1.62	−0.884	0.096	9.17
HAWK	0.324	0.031	10.34	−0.074	0.036	2.01	−0.618	0.059	10.46
HVLF	0.415	0.033	12.45	−0.016	0.012	1.25	0.628	0.054	11.58
HVTW	0.378	0.022	17.10	−0.164	0.035	4.72	0.711	0.048	14.68
HVBR	0.391	0.018	20.74	−0.186	0.032	5.85	0.072	0.053	1.36
HVTR	0.322	0.032	10.21	−0.163	0.052	3.18	0.919	0.065	14.08
Eigenvalue	3.425	0.217	15.71	2.263	0.153	14.81	2.388	0.157	15.19

TABLE 7. JACKKNIFED FACTOR ANALYSIS OF THE SANTA RITA DATA. VALUES ARE COEFFICIENTS OF THE FIRST THREE EIGENVECTORS (COEFF.), THEIR ESTIMATED STANDARD ERRORS (SE), AND THE ERROR RATIOS (ER = COEFF./SE)

Variable	Axis I			Axis II			Axis III		
	COEFF.	SE	ER	COEFF.	SE	ER	COEFF.	SE	ER
GLLF	−2.181	0.178	12.23	1.242	0.148	8.37	−0.299	0.145	2.06
GLTW	−2.841	0.213	13.31	0.301	0.124	2.41	0.186	0.127	1.46
GLBR	−2.926	0.225	12.99	0.163	0.136	1.19	0.163	0.136	1.19
GLTR	0.763	0.191	3.99	0.623	0.082	7.63	0.622	0.081	7.63
GLGR	−0.097	0.028	3.43	−0.146	0.031	4.67	−0.145	0.031	4.67
PRBR	1.875	0.297	6.30	3.167	0.387	8.17	3.167	0.387	8.18
PRTR	1.818	0.298	6.27	3.076	0.376	8.17	3.076	0.377	8.17
PRGR	1.121	0.257	4.34	3.076	0.363	8.48	3.076	0.363	8.48
SATW	2.922	0.268	10.90	−1.326	0.230	5.76	−1.326	0.230	5.76
SABR	4.617	0.372	12.40	2.276	0.296	7.70	2.276	0.296	7.70
HAWK	3.905	0.330	11.82	−1.973	0.231	8.54	−1.974	0.231	8.54
HVLF	−2.391	0.308	7.77	4.881	0.354	13.78	−3.616	0.395	9.16
HVTW	−1.977	0.337	5.85	5.781	0.417	13.83	−3.185	0.377	8.45
HVBR	−0.517	0.218	2.36	3.711	0.263	14.09	−3.722	0.362	10.27
HVTR	−3.001	0.317	8.09	5.811	0.418	13.90	−1.537	0.299	5.12

factor analytic procedure, in particular the Varimax rotation of the factor axes. The factor model emphasizes the importance of partitioning common variance from unique variance among the variables. Each recalculation of the FA based on an iterative deletion of a species from the data matrix may produce a unique representation of the correlation structure, which is specific to the suite of remaining species included in the analysis. Consequently, the factor loadings vary drastically among the pseudovalues. Therefore, each recalculation produces dramatic changes in magnitude and sign of the rotated factor loadings, rather than a small deviation by deleting an observation from the data set. Thus, two conclusions emerge from this analysis: either the jackknife analysis of FA was inappropriate, or the estimates from FA were unique to specific groups of species, or both.

IMPLICATIONS OF THE PRESENT STUDY AND SUGGESTIONS FOR FUTURE STUDIES

CA is the preferred method of analyzing foraging data based on this study. PCA resulted in a similar interpretation of foraging data, but proved less efficient at recovering most of the original variation in the first five axes. These results parallel the study of Gauch et al. (1977),

TABLE 8. BOOTSTRAPPED PRINCIPAL COMPONENT ANALYSIS OF THE SANTA RITA DATA. VALUES ARE COEFFICIENTS OF THE FIRST THREE EIGENVECTORS (COEFF.), THEIR ESTIMATED STANDARD ERRORS (SE), AND THE ERROR RATIOS (ER = COEFF./SE). ESTIMATES OF THE FIRST THREE EIGENVALUES ARE GIVEN AT THE BOTTOM OF EACH COLUMN

Variable	Axis I			Axis II			Axis III		
	COEFF.	SE	ER	COEFF.	SE	ER	COEFF.	SE	ER
GLLF	−0.142	0.033	4.24	−0.296	0.038	7.62	0.026	0.189	0.14
GLTW	−0.119	0.038	3.13	−0.218	0.034	6.23	0.027	0.051	0.61
GLBR	−0.052	0.058	0.89	−0.033	0.049	0.66	0.153	0.058	2.61
GLTR	−0.001	0.045	0.02	0.208	0.031	6.63	0.171	0.133	2.55
GLGR	0.043	0.016	2.60	0.123	0.044	6.04	0.084	0.133	0.63
PRBR	−0.001	0.256	0.00	0.269	0.056	4.78	0.020	0.027	0.74
PRTR	−0.001	0.257	0.00	0.307	0.039	6.94	0.044	0.029	1.54
PRGR	0.005	0.069	0.02	0.275	0.041	6.74	0.029	0.022	1.27
SATW	0.137	0.067	1.98	0.093	0.060	3.92	0.009	0.074	0.12
SABR	0.123	0.082	1.49	0.056	0.192	0.29	−0.156	0.059	2.64
HAWK	0.123	0.071	1.73	0.040	0.178	0.22	−0.173	0.050	3.44
HVLF	0.147	0.036	4.08	0.063	0.186	0.33	0.263	0.049	5.33
HVTW	0.151	0.056	2.66	0.082	0.128	0.64	0.167	0.061	2.73
HVBR	0.140	0.067	2.08	0.038	0.146	0.26	0.116	0.087	1.33
HVTR	0.208	0.031	6.71	0.121	0.020	5.91	0.134	0.059	2.25
Eigenvalue	5.374	0.059	91.34	3.58	0.032	111.03	2.451	0.032	75.16

who found that CA produced ordinations of simulated community patterns superior to those from PCA. Several points make compelling the use of CA in foraging studies. It recovers a high amount of the original variation in the data, despite the curvilinear nature of foraging data. A large proportion of jackknifed coefficients from the first three axes were significantly different from zero. In addition, the estimates of the coefficients exhibited low bias (i.e., the observed coefficients fell within ±2 SE of the jackknifed coefficient). Thus, the interpretations of the patterns of variation in foraging behavior are not based on an arbitrary rotation of axes through a cloud of points. Finally, the absence of the most commonly cited disadvantage of CA, the compression of species at the terminal portions of each axis, provides additional evidence supporting the use of CA in foraging studies.

Factor analysis of foraging data produced relatively unsatisfactory results. While the amount of variance extracted was similar to PCA, FA exhibited a low degree of correspondence with the results from CA and PCA. The presence of the arch effect after rotation of the axes suggests that extreme caution must be exercised in interpretation of the rotated axes. This is especially true because most rotations involve an orthogonal transformation of the axes, and the decisively curvilinear nature of the data may violate the assumptions of the technique. The premise of the FA model—to extract variation from among a group of highly correlated variables after removing the variation attributable to the unique factors—precludes generalizing or comparing results from other studies. This is compounded by conducting the analysis on a correlation matrix. Standardization of the variables by their standard deviation distorts the ecological space, and consequently any patterns that emerge are specific to the particular data set and group of species (Miles and Ricklefs 1984). However, the practice of using a correlation matrix must be balanced by the need to use scale-invariant data with PCA and FA. Yet, this argues more forcefully for using CA, because the standardization of the data is not necessary. A majority of the jackknifed coefficients, although significantly different from zero, exceeded 1.0. Between 80 and 100% of the estimated coefficients were biased. The general conclusion is that FA is inappropriate for the analysis of foraging data.

Further caution must be emphasized in drawing generalizations from multivariate analyses. Most foraging data consist of many observations recorded for a small number of species. Often the number of categories is greater than the number of species. The results from CA, PCA and FA calculated with small sample sizes may be highly sensitive to additions or deletions of foraging categories, random variation in foraging behavior, and the presence of empty cells in the data matrix.

ALTERNATIVE MULTIVARIATE METHODS

The three multivariate methods evaluated in this study all assume that the data were approximately linear. While several studies have demonstrated that CA is less sensitive to curvilinearities within the data (e.g., Gauch et al. 1977, Pielou 1984) than PCA, any interpretations about underlying patterns will be hindered by the presence of the arch. Consequently, nonparametric multivariate methods should prove to be appropriate alternative modes of analysis. Earlier studies that compared nonparametric methods, in particular nonmetric multidimensional scaling (NM-MDS), with PCA or CA found that the former method extracted pattern with lower distortion due to curvilinearities present in the data (Fasham 1977). Techniques such as NM-MDS, psychophysical unfolding theory, and nonparametric mapping have proven to be effective in describing guild structure (e.g., Adams 1985) and resource axes (Gray 1979, Gray and King 1986). Subsequent analyses of avian foraging data should incorporate these underused methods.

ACKNOWLEDGMENTS

I thank R. E. Ricklefs and A. E. Dunham for stimulating discussions about various aspects of the analyses. The critical comments of Barry Noon and Jared Verner greatly clarified many aspects of the manuscript. Peter Landres graciously shared unpublished data. Computer time was provided by Ohio University. David Althoff assisted in various stages of the data analysis. Various stages of the research were supported by the Frank M. Chapman Memorial Fund, Sigma Xi, and the Explorers Club. The writing of this research was sponsored in part by NSF grant BSR-8616788.

AN EXPLORATORY USE OF CORRESPONDENCE ANALYSIS TO STUDY RELATIONSHIPS BETWEEN AVIAN FORAGING BEHAVIOR AND HABITAT

EDGAR BARRY MOSER, WYLIE C. BARROW, JR., AND ROBERT B. HAMILTON

Abstract. Correspondence analysis was used to investigate foraging behaviors of an insectivorous bird community in a bottomland hardwood in Louisiana. The graphical summaries of correspondence analysis depicted the relationships among the species and the habitat variables in an easily interpretable manner. The correspondence analysis ordinated the birds of this community along a foraging-maneuver gradient from hang to perch-glean to flush-chase to sally-glean to aerial-hawk. A foraging-height gradient as well as bird-species relationships with habitat substrates were also identified. The correspondence analysis led to log-linear and logistic regression models that further aided in the exploration of data from this bird community.

Key Words: Bottomland hardwoods; community structure; exploratory analysis; reciprocal averaging; resource partitioning.

Many forest-inhabiting birds are extremely sensitive to habitat change. To understand which habitat variables are most important to a species' distribution, it is necessary to understand how each species uses its habitat and which components influence abundance and survival. In most studies of bird-habitat relationships, many variables are measured, necessitating multivariate approaches to the data analysis (see, e.g., Robinson and Holmes 1982, Airola and Barrett 1985, Lebreton and Yoccoz 1987).

In testing and exploring multivariate hypotheses many researchers found factor analysis, principal component analysis, cluster analysis, or discriminant function analysis to be useful (see James 1971, Morrison 1981, Landres and MacMahon 1983, Holmes and Recher 1986b). Variants of correspondence analysis have also been used including reciprocal averaging (Landres and MacMahon 1983) and detrended correspondence analysis (Sabo and Holmes 1983).

Multivariate techniques often require the distributional assumption of multivariate normality. Further, large sample sizes are often needed to provide sufficient power to detect real relationships (Morrison 1984b, 1988). In many cases, the relationships among the variables are complex and may be nonlinear, resulting in incorrect and inappropriate model specifications (see Noon 1986). Sometimes an analysis consists of so many tests that some of them will appear significant by chance. Thus, we may declare as important factors that are not, or we may overlook important relationships that the methods may be insensitive to.

Tukey (1980) and others (e.g., Hoaglin et al. 1983, James and McCulloch 1985, Cleveland and McGill 1987) have stressed the need for exploratory probing of data sets to aid in the interpretation of results and in generating hypotheses. In this paper we demonstrate how exploratory correspondence analysis can clarify relationships among bird species and their foraging attributes and habitat substrates. In addition we show how log-linear and logistic regression models can be used to supplement the correspondence analysis results.

CORRESPONDENCE ANALYSIS

Variables measured on bird-habitat surveys are often categorical, such as the species of bird or type of substrate, or are easily converted to interpretable categorical values, such as to foraging-height classes, with little loss of information. Techniques that have become popular for exploring the cross-classification of categorical variables or contingency tables are correspondence analysis (Greenacre 1984, Greenacre and Hastie 1987), log-linear models (see Bishop et al. 1975) or a combination of the two (Van der Heijden and Leeuw 1985). Correspondence analysis has been a popular ordination technique for vegetation data (Oksanen 1983, Brown et al. 1984, Fowler and Dunlap 1986), especially detrended correspondence analysis (Hill and Gauch 1980). These techniques ordinate the vegetation along a set of environmental gradients by determining the relative abundances, often presence-absence or a relative frequency score, of plant species occurring on sampled plots. Usually correspondence analysis is performed on two-way tables, although the technique can be used to explore Burt tables (see Greenacre 1984:140–143). A Burt table contains each variable in both the rows and columns of the table and thus contains all of the component two-way tables in a single two-way table. Gauch et al. (1981) discuss the relative merits of correspondence analysis for ordination of ecological data, especially for environmental gradient analysis.

Correspondence analysis identifies a low-dimensional subspace to represent the rows and columns of the two-way table as points in Euclidean space, and therefore is useful for exploring the table graphically while still preserving most of the original information. Row profiles or row points are constructed by dividing each cell frequency of the table by its corresponding row total. Each row profile is assigned a weight called a row mass by dividing the row total by the grand total of the table frequencies. The subspace that has the closest fit, in this case minimizing the weighted chi-square distances of the points to the subspace, is then found. Column profiles (points) and masses can be similarly constructed for the columns and a subspace of closest fit can be found for these as well. Both problems, however, are related by the singular value decomposition of the table that results in the correspondence between the row and column solutions, and therefore either variable may be taken as the row or column variable. This further permits the simultaneous display of the row and column profiles through the biplot (Gabriel 1971, see Greenacre and Hastie 1987). The theoretical development of correspondence analysis along with examples can be found in Greenacre (1984).

The biplot is probably the most useful exploratory result of the correspondence analysis. In a biplot the rows and columns of the table are simultaneously plotted with respect to the principal axes. The amount of variation associated with each axis gives an indication to the dimensionality of the subspace needed to accurately describe the table. Often the first two principal axes are sufficient. For the Burt table analyses, either the row or column solution is displayed, but not both, and the percentage variation explained by each axis computed using the standard formulas needs adjustment based upon the number of variables in the table (Greenacre 1984: 145).

The interpretation of the biplot is based upon the relative association of row and column points on the graph. For a column of the table where a row profile is large, both the column and the row point will be found relatively close together, and vice versa. Distances between row points and the origin and between column points and the origin are interpreted as chi-square contributions to the hypothesis of the independence between the rows and columns. However, the distance between a row and a column point are meaningless, since different scales (metrics) are used for the axes of each point type; rather it is the relative positioning of row with column points and column with row points that is interpreted.

When the table can be sufficiently represented in three or fewer dimensions, the association of row and column points can be found through the directions of the points from the origin or centroid on a plot containing these dimensions. Points lying in the same direction from the centroid are associated by having large profiles in the corresponding rows and columns of the table identified by those points. When the table cannot be sufficiently represented in three or fewer dimensions, then plots consisting of the projections of the points from the higher dimensional space to the lower dimensional subspaces (e.g., two- or three-space) are used. Directions on these plots may not be sufficient to indicate associations, since the correct directions may require use of the other principal axes. Points that appear to be in the same direction from the origin may be far apart when viewed using other important principal axes. However, since the higher dimensional table is projected onto a subspace, points that lie in the same direction in the full space will usually appear spatially close on the plot of the subspace. In these situations, plots of several different subspaces (combinations of axes) should be considered. When three-dimensional representations are used, they should be rotated about the axes or plotted from several different angles so that the relationship among the points is clear.

STUDY AREA AND METHODS

Foraging observations and habitat variables were measured in a bottomland hardwood forest of the Tensas River National Wildlife Refuge in northeastern Louisiana during March through July of 1984–1987. The refuge is described elsewhere (U.S. Fish and Wildlife Service 1980). Three broad habitat types were selected for study. The first consisted of a first terrace flat or backswamp totaling 80 ha. These areas are poorly drained flats of the floodplain with water standing well into the growing season. The dominant forest type is overcup oak-water hickory (*Quercus lyrata-Carya aquatica*) with green ash (*Fraxinus pennsylvanica*), sugarberry (*Celtis laevigata*), American elm (*Ulmus americana*), honey locust (*Gleditsia triacanthos*), Nuttall oak (*Q. nuttallii*), and swamp privet (*Forestiera acuminata*). The understory is restricted to small trees and shrubs. The area will be identified as the flat habitat type.

The forest habitat type is a second terrace flat and is found on slightly higher elevations than the flat habitat type. The area sampled consisted of approximately 160 ha. This habitat type is not seasonally flooded and is dominated by sweetgum (*Liquidambar styraciflua*) and willow oak (*Q. phellos*). Sugarberry, green ash, American elm, and Nuttall oak are also major components while overcup oak, water hickory, cedar elm (*U. crassifolia*), red maple (*Acer rubrum*), and bald cypress (*Taxodium distichum*) occur less frequently. The undergrowth includes greenbrier (*Smilax* sp.), swamp palmetto (*Sabal minor*), switchcane (*Arundinaria gigantea*), and several vines: peppervine (*Am-*

TABLE 1. BIRD SPECIES SURVEYED IN THE BOTTOMLAND HARDWOODS OF THE TENSAS RIVER NATIONAL WILDLIFE REFUGE DURING MARCH–JULY OF 1984–1987

Species	Code	Sample size	Foraging height (m)[a]
Eastern Wood-Pewee (*Contopus virens*)	EP	66	10.8 ± 6.3
Acadian Flycatcher (*Empidonax virescens*)	AF	131	6.6 ± 3.1
Carolina Chickadee (*Parus carolinensis*)	CC	112	7.9 ± 4.0
Tufted Titmouse (*Parus bicolor*)	TT	79	7.1 ± 4.9
Carolina Wren (*Thryothorus ludovicianus*)	CW	54	2.3 ± 2.3
Blue-gray Gnatcatcher (*Polioptila caerulea*)	BG	74	12.1 ± 5.2
White-eyed Vireo (*Vireo griseus*)	WV	98	5.1 ± 3.0
Yellow-throated Vireo (*Vireo flavifrons*)	YV	47	16.3 ± 5.0
Red-eyed Vireo (*Vireo olivaceus*)	RV	85	10.4 ± 4.2
Northern Parula (*Parula americana*)	NP	218	9.7 ± 5.2
Yellow-throated Warbler (*Dendroica dominica*)	YW	100	14.9 ± 5.2
American Redstart (*Setophaga ruticilla*)	AR	52	11.3 ± 3.7
Prothonotary Warbler (*Protonotaria citrea*)	PW	146	3.6 ± 3.5
Swainson's Warbler (*Limnothlypis swainsonii*)	SW	17	0.4 ± 0.5
Kentucky Warbler (*Oporornis formosus*)	KT	50	1.4 ± 1.8
Hooded Warbler (*Wilsonia citrina*)	HD	90	5.4 ± 3.9

[a] Mean ± SD.

pelopsis arborea), rattan (*Berchemia scandens*), poison ivy (*Toxicodendron radicans*), and Virginia creeper (*Parthenocissus quinquefolia*).

The oxbow habitat type occurs along the edges of oxbow lakes. Approximately 8 km of water-forest edge were selected for study. Bald cypress is the dominant species, with associates of water hickory, overcup oak, and cedar elm. Common buttonbush (*Cephalanthus occidentalis*) is the prominent shrub.

FIELD METHODS

We recorded foraging behaviors as we regularly and repeatedly traversed the study areas, moving from one foraging bird to another. For each individual we recorded: the species of bird, sex, time of day, type of foraging maneuver, height at which the maneuver took place, substrate (usually plant species) at which the maneuver was directed, and a general classification of the substrate. The substrate was classified as air, branch, flower, leaf, moss, trunk, or twig, where air indicates aerial foraging and moss indicates foraging in Spanish moss (*Tillandsia usneoides*). The substrate species were classified into habitat management categories of bald cypress and Spanish moss; ground litter, herbs, and fallen logs; overstory including midstory species; snags; understory, particularly shrubs; and vines. The categories were intended to represent habitat characteristics that could be addressed through habitat management. Bird foraging maneuvers were defined as sally-glean, a bird in flight takes a prey item from a substrate; perch-glean, the prey is taken from vegetation while the bird is perched or slowly moving; flush-chase, the prey is flushed from a substrate and is pursued; hang, the bird clutches a leaf or twig and hangs in order to glean prey from the surface; aerial-hawk, a sally into the air in pursuit of a flying prey; and ground-forage, any of the above maneuvers, initiated while the bird is on the ground. A bird was followed until 10 foraging maneuvers were observed or until it was lost from sight. In this analysis, only the first foraging maneuver was used so as to avoid serial correlation problems. Raphael (this volume), however, discusses a Markov chain approach that could be used to model these serially correlated data. Foraging heights were classified as: ground (0–0.5 m); shrub (0.5–2.0 m); midstory (2–10 m); and canopy (>10 m). Foraging and microhabitat data collections were restricted to the bird species listed in Table 1.

The foraging microhabitat was characterized at locations directly under or on the site where a bird's first foraging maneuver was observed. An imaginary cylinder centered at the location with a diameter of 2 m was divided into the four height layers described above. The radius of the cylinder in the canopy layer was extended to 10 m. The percentage of vegetation density was determined for each of the four strata. Additionally, the height of the canopy was also estimated with a range-finder. Availability of habitats was estimated by using the above method at randomly located plots.

DATA ANALYSIS

Correspondence analysis was performed using the CORRESP procedure of Young and Kuhfeld (1986). The principal axes and corresponding coordinates were saved for constructing biplots. In a purely exploratory framework, no assumptions about the data are required. However, since the interpretation of the graphical analysis will refer to the dependencies among the variables, and the results will be used to help specify log-linear models, the log-linear model assumptions discussed below are required. Log-linear models of the contigency tables were fit using the CATMOD procedure of SAS/STAT (SAS 1985). The log-linear models provide methods for examining the dependencies among variables in a contingency table. These models assume independent observations usually from multinomial, product-multinomial, or poisson distributions, and depend upon the large sample, asymptotic properties of maximum likelihood (see Bishop et al. 1975:435–530). Roscoe and Byars (1971) suggested that

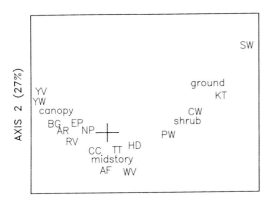

FIGURE 1. Correspondence analysis of bird species with foraging-height class: ground = <0.5 m, shrub = 0.5–2 m, midstory = 2–10 m, and canopy = >10 m. See Table 1 for bird species codes. The origin is located at the crosshairs.

the average expected frequency in the contingency table be at least five for reliable tests of hypotheses, although they found that an average expected frequency of one to two was satisfactory in some instances. Habitat use-availability comparisons were made using logistic regression (see Kleinbaum et al. 1982:419–446) as implemented in the CATMOD procedure of SAS/STAT (SAS 1985). Logistic regression does not require multivariate normality of the explanatory variables. The usual assumptions require that the dichotomous responses be from independent Bernoulli distributions (or binomial counts of "successes" in a known number of trials) and that the probability parameter of these distributions can be modeled as a logistic function of the explanatory variables (see Kleinbaum et al. 1982: 419–446, Weisberg 1985:267–271). Since maximum likelihood was used to estimate the parameters of our logistic models, the large sample, asymptotic properties of maximum likelihood are again assumed to hold. Logistic regression has been found to be more robust than discriminant analysis, probably because its formulation arises from many types of modeling assumptions (Press and Wilson 1978).

The foraging data provided a variety of categorical variables that could have been explored with correspondence analysis, but only the relationships of the bird species with foraging-height classes, substrate types, habitat management categories, habitat types, and foraging maneuvers were explored in this paper. The Tufted Titmouse and Carolina Chickadee were widespread on this study area and so specific hypotheses concerning their respective niches, generated as a result of the correspondence analyses, were examined using log-linear and logistic regression techniques. Log-linear models were fit using the cross-classification of these two bird species with the foraging maneuver, substrate type, habitat management category, and foraging-height class variables to determine factors that might separate their foraging patterns. To simplify the analysis and to insure that the average expected cell frequencies of the table

were at least five, only the predominant factor levels of the variables were included. They were the perch-glean and hang maneuvers; the branch, leaf, and twig substrate types; the overstory, understory, and vines habitat management categories; and the midstory and canopy foraging-height classes. The substrate type and habitat management category variables were not included together in a model because of the resulting small cell frequencies. A logistic regression model was used to discriminate between the microhabitat measurements made at the species' foraging locations. Logistic regression was also used to compare the microhabitat characteristics measured at the bird foraging locations with those measured at random locations within the forest.

RESULTS

CORRESPONDENCE ANALYSIS

The first two principal axes from the correspondence analysis of the cross-classification of the bird species with the foraging-height classes explained 93% of the table chi-square variation. This indicated that the contingency table could be projected from three dimensions to two, with little loss of information. The bird species (rows) and the foraging-height classes (columns) were plotted simultaneously using the first two principal axes to produce a biplot (Fig. 1). Since the row profile (Table 2) for Swainson's Warbler was large in the ground column of the table, Swainson's Warbler was positioned in the direction of the ground value of the foraging-height variable. The Prothonotary Warbler profile was large in both the shrub and midstory columns of the table and so was ordinated between them on the plot. The remaining species were ordinated according to their row profiles indicating their positions along the foraging-height gradient. Since there were no birds with large profiles for both ground and canopy values, the region of the plot opposite midstory is empty.

The sightings of species of birds were then cross-classified with the habitat-management categories. The first three principal axes from the correspondence analysis of this table explained 90% of the total table variation. This analysis indicated that the Yellow-throated Warbler was strongly associated with the bald cypress-Spanish moss category (Fig. 2) and a closer examination of the specific chi-square contributions made by each bird species in the table showed that most of the chi-square variation was due to this particular association. The ground and understory categories were ordinated in a similar direction from the centroid, but the Prothonotary Warbler, for example, was more associated with the understory than with the ground category. Swainson's Warbler used the understory species as well as the ground debris, as expected

TABLE 2. ROW PROFILES AND ROW MASSES FOR THE CROSS-CLASSIFICATION OF BIRD SPECIES WITH FORAGING-HEIGHT CLASS FOR 1419 BIRD FORAGING OBSERVATIONS

Bird species	Foraging-height class				Row mass
	Ground	Shrub	Midstory	Canopy	
Eastern Wood-Pewee	0.00	0.11	0.32	0.58	0.05
Acadian Flycatcher	0.00	0.05	0.77	0.18	0.09
Carolina Chickadee	0.01	0.04	0.64	0.30	0.08
Tufted Titmouse	0.09	0.06	0.58	0.27	0.06
Carolina Wren	0.15	0.50	0.35	0.00	0.04
Blue-gray Gnatcatcher	0.00	0.04	0.28	0.68	0.05
White-eyed Vireo	0.02	0.16	0.73	0.08	0.07
Yellow-throated Vireo	0.00	0.00	0.09	0.91	0.03
Red-eyed Vireo	0.00	0.00	0.51	0.49	0.06
Northern Parula	0.06	0.05	0.43	0.47	0.15
Yellow-throated Warbler	0.00	0.00	0.13	0.87	0.07
American Redstart	0.00	0.00	0.40	0.60	0.04
Prothonotary Warbler	0.16	0.29	0.50	0.05	0.10
Swainson's Warbler	0.71	0.29	0.00	0.00	0.01
Kentucky Warbler	0.30	0.54	0.14	0.02	0.04
Hooded Warbler	0.07	0.21	0.56	0.17	0.06

from the previous correspondence analysis, but the Kentucky Warbler and Carolina Wren were additionally identified as using vines as well. The White-eyed Vireo and Hooded Warbler were also important users of vines. The Eastern Wood-Pewee was associated more often with snags and the overstory, more specifically in bald cypress and water hickory. This explains why the Eastern Wood-Pewee was ordinated between these categories and more in the direction of the bald cypress–Spanish moss category (Fig. 2). The remaining species were generally associated with the overstory category.

The bird species were then ordinated by their sample sizes in the three major habitat types to explore the relative number of encounters in each habitat type (Fig. 3). This table could be exactly represented in two dimensions, as it consisted of only three columns defined by the habitat types. The Yellow-throated Warbler was almost exclusively found in the oxbow habitat type, whereas the Hooded Warbler and the Swainson's Warbler were only observed in the forest habitat type. The Northern Parula, Prothonotary Warbler, and the Eastern Wood-Pewee were also highly associated with the oxbow habitat type. The majority of the other species were sighted most often in the forest and flat habitats.

The midstory and canopy foragers were then subjected to a correspondence analysis with the substrate types. The Eastern Wood-Pewee was found foraging almost entirely on insects in the air; these data contributed to most of the chi-square variation in the table (Fig. 4). The Acadian Flycatcher and American Redstart were often foraging on insects in the air but, just as important, they were identified here as sally-gleaning arthropods from leaves. The Yellow-throated Warbler was again shown strongly associated with Spanish moss. The Yellow-throated Vireo, Tufted Titmouse, and Carolina Chickadee all foraged on foods associated with tree trunks, branches, and twigs of plants, whereas the remaining species appeared associated more with the leaves of the substrate. Although the Yellow-throated Vireo and Yellow-throated Warbler were found high in the canopy, each appeared to differ in their selection of habitat substrates.

The dominant source of variation in the correspondence analysis of the bird species with the first encountered foraging maneuver was produced by the almost exclusive use of aerial-hawking by the Eastern Wood-Pewee. The analysis, however, identified a gradient, primarily along axis 2, from aerial maneuvers to flush-chasing to perching to hanging while foraging (Fig. 5). This three-dimensional ordination explained 97% of the chi-square variation in the table. An analysis, not shown, with the Eastern Wood-Pewee removed from the table, resulted in a single important axis (71%), with aerial-hawking at the extreme end of the axis beyond sally-gleaning.

LOG-LINEAR MODELS

The first log-linear model comparing the Tufted Titmouse and Carolina Chickadee used the variables bird species, foraging maneuver, substrate type, and foraging-height class. The correspondence analysis of the Burt table (Table 3) showed no strong associations of the foraging variables with the species (Fig. 6). The resulting log-linear model, however, identified a depen-

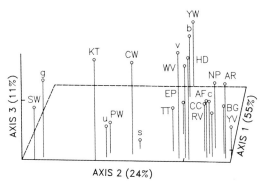

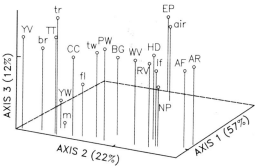

FIGURE 2. Correspondence analysis of bird species with habitat-management category: b = bald cypress and Spanish moss, g = ground litter, c = overstory, s = snags, u = understory, and v = vines. See Table 1 for bird species codes. The origin is located at the intersection of the axis tic marks.

FIGURE 4. Correspondence analysis of bird species with substrate type: air = air, br = branch, fl = flower, lf = leaf, m = Spanish moss, tr = trunk, and tw = twig. See Table 1 for bird species codes. The origin is located at the intersection of the axis tic marks.

dency between the species and the foraging maneuver (Table 4, P = 0.0173), with only a suggestion that the bird species and substrate type were dependent (P = 0.1689). This particular log-linear model corresponded to a logit model for bird species containing only the substrate type and foraging maneuver variables. This logit model was a test of the ability of the variables substrate type and foraging maneuver to discriminate between the two bird species' frequencies of usage. These results suggested that the chickadee did relatively more hanging than the titmouse, but the titmouse did relatively more perch-gleaning. A secondary result was that the hanging maneuver was associated more often with leaf substrates, and perch-gleaning more often with branch substrates.

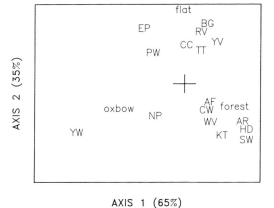

FIGURE 3. Correspondence analysis of bird species with habitat type. See Table 1 for bird species codes. The origin is located at the crosshairs.

When the substrate type was replaced with the habitat-management category in the log-linear models analysis, the bird species were found to be related to the foraging maneuver interacting with the habitat-management category (Table 5, P = 0.0091). This model corresponded to a logit model containing habitat-management category, foraging maneuver, and their interaction. Thus, there appeared to be some differences in the foraging behaviors of these two species, particularly in their maneuvers.

LOGISTIC REGRESSION

None of the variables in the logistic regressions comparing the microhabitat selection of chickadee and titmouse, including substrate height, were good discriminators of foraging microhabitats. Further, there is only an indication that the proportion of canopy vegetation was less at random sites than at sites selected by the titmouse (P = 0.09). The chickadee, however, appeared to select sites with a smaller percentage of ground litter (P = 0.005) and a larger percentage of bare ground (P = 0.05) than random sites. There was also an indication that the proportion of canopy vegetation at chickadee foraging locations was less than at the random plots (P = 0.07). Thus, there appeared to be some differences between foraging sites selected by these two birds with sites selected at random, although no differences were detected between the birds when comparing the two species alone.

DISCUSSION

The foraging-height and foraging-maneuver variables were important in distinguishing the bird species of this community. Examination of the substrate types, habitat-management categories, and habitat types further helped to iden-

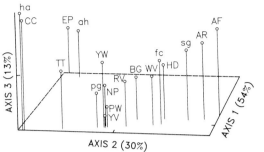

FIGURE 5. Correspondence analysis of bird species with foraging maneuver: ah = aerial-hawk, fc = flush-chase, ha = hang, pg = perch-glean, and sg = sally-glean. See Table 1 for bird species codes. The origin is located at the intersection of the axis tic marks.

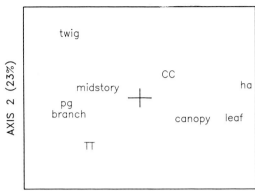

FIGURE 6. Correspondence analysis of the Burt table for bird species, substrate type, foraging maneuver, and foraging-height class. See Table 1 for bird species codes and Figures 1 and 5 for foraging-height class and foraging maneuver codes. The origin is located at the crosshairs.

tify species with specific habitat associations. For example, the Eastern Wood-Pewee foraged on flying insects by aerial-hawking from snags, bald cypress, and water hickory. The Yellow-throated Warbler was associated with bald cypress and Spanish moss.

In building a model (e.g., a log-linear model) of associations between the bird species and the habitat variables, where one species' association with a specific habitat variable or variables dominates the correspondence analysis, one would probably include a separate parameter or set of parameters to account for the association. Our analysis of the bird species with the habitat-management categories suggested that a model for these data should include a parameter to account for the relationship between the Yellow-throated Warbler and the bald cypress-Spanish moss category. In the analysis of the bird species with the foraging maneuvers, we took the graphical analysis one step further through the deletion of the influential Eastern Wood-Pewee. This resulted in a much simplified table, reducing the variation to along one axis, and was probably a more successful summary than that of displaying the complete table analysis. The deletion of the Yellow-throated Warbler in the analysis of the habitat-management categories would have permit-

TABLE 3. BURT TABLE FOR THE CAROLINA CHICKADEE AND TUFTED TITMOUSE CONTAINING THE SELECTED SUBSTRATE TYPES BRANCH, LEAF, AND TWIG; THE FORAGING MANEUVERS HANG AND PERCH-GLEAN; AND THE FORAGING-HEIGHT CLASSES MIDSTORY AND CANOPY

	Bird species		Substrate type			Foraging maneuver		Foraging-height class	
	CC	TT	Branch	Leaf	Twig	HA	PG	Midstory	Canopy
Bird species									
Carolina Chickadee (CC)	73	0	28	30	15	34	39	53	20
Tufted Titmouse (TT)	0	41	18	19	4	11	30	29	12
Substrate type									
Branch	28	18	46	0	0	14	32	36	10
Leaf	30	19	0	49	0	27	22	32	17
Twig	15	4	0	0	19	4	15	14	5
Foraging maneuver									
Hang (HA)	34	11	14	27	4	45	0	32	13
Perch-glean (PG)	39	30	32	22	15	0	69	50	19
Foraging-height class									
Midstory	53	29	36	32	14	32	50	82	0
Canopy	20	12	10	17	5	13	19	0	32

TABLE 4. Maximum Likelihood Log-linear Models Analysis of the Cross-classification of Bird Species (SPECIES) with Substrate Type (SUBTYPE), Foraging Maneuver (MANEUVER), and Foraging-height Class (FORHTCL) for the Carolina Chickadee and Tufted Titmouse

Source	df	Chi-square[a]	P
SUBTYPE	2	15.43	0.0004
MANEUVER	1	9.58	0.0020
FORHTCL	1	12.67	0.0004
SPECIES	1	13.14	0.0003
SPECIES·SUBTYPE	2	3.56	0.1689
SPECIES·MANEUVER	1	5.66	0.0173
SUBTYPE·MANEUVER	2	9.90	0.0071
SUBTYPE·FORHTCL	2	3.16	0.2064
MANEUVER·FORHTCL	1	0.00	0.9959
SUBTYPE·MANEUVER·FORHTCL	2	3.63	0.1627
LIKELIHOOD RATIO[b]	8	2.19	0.9746

[a] Wald Statistics.
[b] Test for lack-of-fit comparing the current model to the saturated or full model.

ted a more detailed exploration of the remaining bird species of that table, which would then be modeled by other parameters in the log-linear model. A modeling approach alone might have required several steps to isolate these individual sources of variation, although they were clear in the biplots. This illustrates how correspondence analysis can provide support to modeling. A further important outcome was that axes two and three often provided considerable detail about many of the bird relationships with the habitat variables, since a single species was often responsible for the variation along the first axis. Thus, axes associated with the smaller singular values can be informative, and approaches using only the first or first two axes may be inadequate.

Our foraging data were not sampled with the goal of estimating species relative abundance. Therefore, caution should be applied when interpreting the association between the bird species and the habitat types in which they were found, because some associations could be an artifact of the sampling process. However, encounters of birds within a habitat are assumed to be random so that stratified modeling approaches are possible. We think that the correspondence analysis of these variables was useful because some species were associated with specific habitat types, such as the Hooded Warbler and Yellow-throated Warbler. Further, the analysis provides a basis for developing hypotheses about distribution that can be examined in subsequent field studies.

In general, correspondence analysis was useful for the examination of two-variable models and for interpreting the log-linear model results from more complex tables, as was done for the Burt tables. Correspondence analysis appeared, however, to provide much less insight into the actual

TABLE 5. Maximum Likelihood Log-linear Models Analysis of the Cross-classification of Bird Species (SPECIES) with Habitat-management Category (HABCAT), Foraging Maneuver (MANEUVER), and Foraging-height Class (FORHTCL) for the Carolina Chickadee and Tufted Titmouse

Source	df	Chi-square[a]	P
HABCAT	2	267.93	0.0001
MANEUVER	1	0.00	0.9874
FORHTCL	1	161.43	0.0001
SPECIES	1	4.03	0.0447
HABCAT·MANEUVER	2	1.33	0.5145
HABCAT·FORHTCL[b]	2		
MANEUVER·FORHTCL	1	0.00	0.9904
SPECIES·HABCAT	2	1.00	0.6060
SPECIES·MANEUVER	1	0.21	0.6487
HABCAT·MANEUVER·FORHTCL	2	1.20	0.5501
SPECIES·HABCAT·MANEUVER	2	9.41	0.0091
LIKELIHOOD RATIO[c]	6	1.69	0.9462

[a] Wald Statistics.
[b] One or more parameter estimates are infinite.
[c] Test for lack-of-fit comparing the current model to the saturated or full model.

specification of a model for the Burt tables. Part of this difficulty may be due to the design of the Burt table itself, since it only contains pairwise relationships among the variables, and therefore, higher-order dependencies in the data are not preserved in the table. More research is needed into graphical ways for exploring complex dependencies in contingency tables.

The logistic regression results must be carefully interpreted. The differences between the microhabitat variables at Carolina Chickadee locations and those from random plots do not necessarily indicate that the habitat use of these birds was based upon the variables declared significant; these variables might have been related to unmeasured qualities that the birds were using. Further, our ability to distinguish between foraging sites and random plots may not have been very powerful, due to the large variation associated with random sites. The power for discriminating between two species' sites should be at least that for discriminating between a particular species' sites and randomly located sites, since the variation in measurements made at sites selected by a species would tend to be no greater than those from randomly selected sites. Determining how sample size affects the power of this analysis, given the amount of random variation, would be a desirable next step (see Morrison 1988). Further exploration of our tables could proceed by using the actual substrate species and by separating the overstory according to the foraging-height classes. The generality of our conclusions requires repetition of the study both temporally and spatially until relationships are clear and stable.

We have demonstrated the power of correspondence analysis for exploring and illustrating graphically the relationships among bird species with habitat variables. Miles (this volume) also found correspondence analysis to be valuable for analyzing foraging behavior. Greenacre and Vrba (1984) demonstrated its usefulness for exploring ecological relationships among African antelopes. Our exploratory analysis suggested that we may associate the bird species with specific habitat conditions as well as with specific foraging behaviors. The graphical displays of these associations suggested hypotheses that we explored using other techniques such as log-linear modeling and logistic regression. This combination of several exploratory (and confirmatory) techniques resulted in a better understanding of the data than the use of a single technique alone, mainly because each technique is only sensitive to particular kinds of dependencies among the variables.

A visual examination of categorical data through correspondence analysis provides valuable insight and confidence in the analysis. Further, the biplot graphics make convenient devices for explaining complex relationships among species to those not trained in avian ecology. With the abundance of categorical data collected during bird foraging studies, the use of exploratory techniques aimed specifically at categorical variables must be encouraged.

ACKNOWLEDGMENTS

We thank L. A. Escobar for many insightful discussions of the topic and for the many helpful comments and suggestions for improving the manuscript. Additional reviews by A. M. Saxton, J. P. Geaghan, W. M. Block, and two anonymous reviewers are also appreciated. The Department of Experimental Statistics, Louisiana State University, and the Louisiana Agricultural Experiment Station of the Louisiana State University Agricultural Center provided support for this project. This paper is approved for publication by the Director of the Louisiana Agricultural Experiment Station as manuscript number 88-19-2239.

ANALYZING FORAGING USE VERSUS AVAILABILITY USING REGRESSION TECHNIQUES

KEVIN M. DODGE, ROBERT C. WHITMORE, AND E. JAMES HARNER

Abstract. Most analyses of bird foraging use versus availability of resources, such as tree species, have employed goodness-of-fit techniques. Rather than pooling observations from all birds and portions of a study area into one goodness-of-fit analysis, we recommend collecting use and availability data on a number of territories, and using the data from each territory as an independent observation in a simple, multiple, or multivariate regression analysis, with availability values as independent variables and use values as dependent variables. In a demonstration, the two methods yield slightly differing results. The advantages of the regression method are discussed with respect to biological scale, sampling considerations, analysis requirements, and interpretation. Regression methods appear superior to goodness-of-fit techniques in each respect, particularly given sufficient sample size, and provide greater promise to researchers examining use versus availability problems.

Key Words: Use versus availability; goodness-of-fit; regression; Solitary Vireo; *Vireo solitarius*; tree species preference.

Patterns of resource use and availability are commonly examined in avian ecology studies. The fit of a bird's apparent preferences to the availability of its potential resources can provide insight into a species' ability to successfully populate an area, adapt to changing conditions, and limit the populations of prey species. In this paper, we propose that regression techniques can replace other methods to analyze use versus availability data. Although we concentrate on a woodland bird's use of different tree species versus their availability, we note that these methods may be extended to other resources as well.

A variety of techniques have been applied to the analysis of use-availability data. These include the forage ratio and modifications (Williams and Marshall 1938, Chesson 1978), the index of electivity (Ivlev 1961), and goodness-of-fit. Johnson (1980) criticized these methods and proposed an alternative (PREFER), involving the use of ranks. Goodness-of-fit, however, continues to be widely used (e.g., Balda 1969; Franzreb 1978, 1983a; Holmes and Robinson 1981; Maurer and Whitmore 1981; Askins 1983; Rogers 1985), usually by the following procedure: the study area (generally a vegetationally homogeneous stand) is traversed in some regular fashion, and observations are taken on all foraging birds (Franzreb 1978, 1983a; Holmes and Robinson 1981; Askins 1983; Rogers 1985). The observations recorded may be prey attacks (e.g., Holmes and Robinson 1981, Maurer and Whitmore 1981), or simply the location of the bird during active foraging behavior (e.g., Balda 1969, Rogers 1985). Some workers have recorded the amount of time spent in each tree species or vegetation category (e.g., Askins 1983). Tree species availability is assessed by sampling throughout the study area (Balda 1969; Franzreb 1978, 1983a; Holmes and Robinson 1981; Maurer and Whitmore 1981). In the analysis, use is quantified by tallying the total number of observations in each tree species, which is multiplied by the relative availability of each tree species to obtain the availability value for that species. An r × 1 (where r denotes the number of tree species being considered) table is then constructed, where the use estimate equals the observed value and the availability estimate equals the expected value in each cell. The null hypothesis of homogeneity of the two populations (use and availability) is tested using the chi-square or G-statistic (Franzreb 1978, 1983a; Holmes and Robinson 1981; Maurer and Whitmore 1981; Askins 1983; Rogers 1985). A significant statistic indicates that use of the suite of tree species differs from availability, but offers no information as to which tree species are responsible for the difference.

We suggest regression procedures might be better suited to the analysis of such data. To our knowledge, regression methods have not been previously applied to this problem in the avian literature, except by Rogers (1985). In a study of Least Flycatcher (*Empidonax minimus*) foraging behavior, Rogers (1985) used the Spearman rank correlation procedure to compare relative tree species use to relative tree species frequency. However, because each observation in the analysis is composed of the use and concomitant availability of a different tree species within a bird's territory, and the analysis is limited to one territory, the design violates the assumption of independence of observations involved in regression-correlation analyses. The proper use of regression techniques appears to be a novel approach to the analysis of avian foraging use versus availability data. The sampling and analysis scenario for a study using regression is different from that used in goodness-of-fit analyses. Rather than lumping together samples from a number of different locations, individual territories are delineated, and separate data are collected on each territory. In each territory, the amount of time spent actively foraging in each specified tree species is recorded (although number of prey attacks may be used if preferred). Territories may be visited more than once through the period of interest, as long as visits are randomized to avoid diurnal and seasonal bias. At the end of the sampling period, all data for a given territory are combined by tree species. Tree species availability is also determined for each territory. Availability may be defined by a variety of measures, including basal area, foliage volume, or percent coverage. In the analysis, all data collected in each territory are combined into one statistical observation, so that the number of statistical observations in the analysis will be equal to the number of territories sampled. Hence, a number of territories must be sampled in order to provide an adequate sample size. Relative values are used in the analysis. Availability values are obtained by dividing the amount (e.g., basal area) of each tree species in each territory by the total amount (e.g., total basal area) of all tree species combined in that territory. Use values are calculated in a similar fashion by dividing the amount of seconds the bird spends in each tree species by the total number of seconds of observation collected in that territory.

Both simple and multiple regression methods may be employed to analyze the data. In all cases, the availability data constitute the independent variable(s), while the use data are used as the dependent variable(s). (Strictly, this is a correlation problem, because the independent variables are not fixed [Dowdy and Wearden 1983].) Use and availability values can be directly compared because they are expressed as relative values, and hence possess the same units of measure. The underlying reasoning for the regression approach is that, if use equals availability, there will be an approximately one-one correspondence between use and availability across all values of availability. To insure that a one-one correspondence can be tested for, the intercept value is always forced to equal 0. Theoretically, the intercept value should always equal 0 since, when availability equals 0, use has to equal 0. However, the actual distribution of data points may produce a computed regression line that deviates from a 0 intercept value. Forcing the regression line to pass through the origin eliminates this problem.

At the most elementary level, simple linear regression can be used to determine the relationship between a single tree species' use and its concomitant availability. The hypothesis that use is proportional to availability is tested by determining, using a t-test, whether or not the slope (b) is significantly different from 1. (Because the intercept is excluded from the equation, the calculation of the standard error of b in the denominator of t is somewhat altered [Afifi and Clark 1984:103]. The particular statistical package being used may or may not incorporate this change, and should be checked.) If b does not deviate significantly from 1, then the tree species is used in proportion to its availability. If b is greater than 1, use is greater than availability. If b is less than 1, availability exceeds use.

Additional information may be obtained by looking at the interaction among several tree species' availabilities in their relation to the use of one tree species (multiple regression). Such relationships can be examined by using raw partial regression coefficients. These values allow the investigator to determine the relative importance of any one availability variable to the use variable while taking the effects of the other availability variables into account by holding their effects constant (Dowdy and Wearden 1983, Afifi and Clark 1984). In this instance, the partial regression coefficient of the tree species availability variable whose use is being examined is tested (using a t-test) to determine whether or not it differs significantly from 1, and if so, in what direction.

At the highest level of complexity is multivariate regression analysis. This technique is characterized by more than one variable on each side of the equation. Multivariate regression may be particularly appropriate for examining the relationship between use and availability of all tree species simultaneously, in that response variables (use of different tree species) may be interrelated (Gnanadesikan 1977, Johnson 1981a). Partial correlation coefficients may be useful in this respect (e.g., Mountainspring and Scott 1985).

METHODS

To demonstrate the use of regression techniques in analyzing foraging use versus availability, and to compare the results of regression analyses with those of the goodness-of-fit technique, Solitary Vireo (*Vireo solitarius*) foraging data were subjected to analysis. Because these data were not collected in independent territories, they are presented solely for illustrative purposes. Unless one is willing to assume that Solitary Vireos do not exhibit individual-specific foraging behavior, the results of these analyses should not be used to develop conclusions regarding this species. The data were collected on a 58 ha study area located in Sleepy Creek Public Hunting and Fishing Area, Berkeley Co., in eastern West Virginia. The study area is situated on the western slope of Third Hill Mountain, in a forest dominated by scarlet oak (*Quercus coccinea*), chestnut

TABLE 1. RESULTS OF SIMPLE REGRESSION ANALYSES OF SOLITARY VIREO FORAGING USE VERSUS AVAILABILITY OF NINE TREE SPECIES OR GROUPS. EACH SPECIES OR GROUP LISTED REPRESENTS A SEPARATE ANALYSIS INVOLVING THE USE OF THAT SPECIES OR GROUP AS THE DEPENDENT VARIABLE, AND THE AVAILABILITY OF THAT SPECIES OR GROUP AS THE INDEPENDENT VARIABLE. LISTED FOR EACH SPECIES OR GROUP IS THE SLOPE (b), STANDARD ERROR OF THE SLOPE, T-VALUE FOR TESTING $H_0: B = 1$ (* = $P < 0.05$; ** = $P < 0.01$), COEFFICIENT OF DETERMINATION (r^2), AND WHETHER USE (U) EQUALS OR IS LESS OR GREATER THAN AVAILABILITY (A)

Tree species or group	b (slope)	Standard error of slope	t ($H_0:b = 1$)	Coefficient of determination	Conclusion
Chestnut oak	0.485	0.111	−4.621**	0.373	U < A
Pines	0.373	0.075	−8.379**	0.437	U < A
Red Maple	0.070	0.146	−6.353**	0.007	U < A
White oak	0.391	0.488	−1.246	0.020	U = A
Scarlet oak	1.577	0.209	2.767**	0.641	U > A
Snags	0.551	0.163	−2.747**	0.262	U < A
Other red oaks	0.408	0.201	−2.952**	0.115	U < A
Hickories	0.882	0.240	−0.494	0.297	U = A
Black gum	0.303	0.140	−4.983**	0.128	U < A

oak (*Q. prinus*), and pitch pine (*Pinus rigida*) in the canopy, and red maple (*Acer rubrum*) and black gum (*Nyssa sylvatica*) in the understory. Lesser amounts of white oak (*Q. alba*), northern red oak (*Q. rubra*), black oak (*Q. velutina*), table mountain pine (*P. pungens*), and several species of hickory (*Carya* spp.) occur. Mountain laurel (*Kalmia latifolia*) and blueberries (*Vaccinium* spp.) compose the majority of the shrub layer. The herbaceous layer is sparse. The area was divided into 50 m × 50 m blocks. From mid-June to late July 1985 the study area was repeatedly traversed from 06:00–13:00 EDT. All vireos encountered were followed for up to approximately 1 hr. The amount of time spent actively foraging in each tree species visited, including snags, was recorded on a cassette recorder and assigned to the block in which the bird was located. Observations were transcribed from cassettes and timed using a stopwatch, and constitute use data. The basal area of each tree species, including snags, was also measured on 0.04 ha circular plots located in the center of each block to characterize tree species availability in that block.

These data were subjected to two different analyses. The nine most abundant tree species (or tree species groups), including snags, were selected for inclusion. All pines were lumped into one variable, as were all hickories, and northern red oak and black oak were combined into "other red oaks." Only those blocks where at least 3 min of foraging observations were recorded were included in the analyses to simulate territories, yielding a sample size of 33 blocks. Although 3 min may seem a minimal amount of time to adequately describe a bird's foraging behavior at a particular location, it is realistic for such a study, considering the difficulty of collecting such data (e.g., Robinson and Holmes 1982). For each block, the number of seconds spent foraging in each tree species or group relative to the total number of seconds of observation in that block was calculated, as was the basal area of each species or group relative to the total basal area of all species in that block combined. Hence, the relative values ranged from 0 to 1. Due to the small sample size, only simple regression analyses were run in the manner described previously. For the comparative goodness-of-fit analysis, foraging observations collected across all the blocks incorporated in the regression analysis were combined for each tree species or group to yield use data (i.e., the total number of seconds spent in each tree species or group across all 33 blocks was calculated). Similarly, tree species availability was determined by combining basal area values for each tree species or group across all 33 blocks, dividing by the total basal area of all tree species, and multiplying by the total number of seconds of foraging observations collected. Any tree species or group not included in the regression analysis was placed in the "other species" category. These data constituted observed and expected values, respectively, as described above. The two values for each tree species or group were then compared, and the chi-square value was calculated across all tree species and groups and checked against the table value for 9 df. To determine the contribution of each tree species or group to the overall results, an analysis of residuals (Everitt 1977:46–48) was performed. Because only one column exists in the table of data, adjusted residuals could not be calculated, so unadjusted values were used for interpretation. Adjusted values are always higher than unadjusted values, so the unadjusted values represent conservative estimates (Everitt 1977:46–48).

RESULTS

Regression

Simple regression analyses indicated that six tree species or groups were underused relative to their availability (chestnut oak, pines, red maple, snags, other red oaks, and black gum), one was used more frequently than expected based on its availability (scarlet oak), and two were used in proportion to their availability (white oak and hickories) (Table 1). Note the fairly inconsistent relationships between use and availability for

TABLE 2. RESULTS OF GOODNESS-OF-FIT ANALYSIS OF SOLITARY VIREO FORAGING USE VERSUS AVAILABILITY OF 10 TREE SPECIES OR GROUPS. EACH SPECIES OR GROUP REPRESENTS ONE ROW OF A 10 × 1 TABLE. LISTED FOR EACH SPECIES OR GROUP ARE THE OBSERVED (USE) AND EXPECTED (AVAILABILITY) VALUES (IN SECONDS), DEVIATION USED IN THE CALCULATION OF THE CHI-SQUARE VALUE, THE PERCENT DEVIATION OF USE FROM AVAILABILITY, THE UNADJUSTED RESIDUAL VALUE (SEE TEXT), AND RESULTS OF Z-TEST OF H_0: RESIDUAL = 0 (* = $P < 0.05$; ** = $P < 0.01$), AND WHETHER USE (U) EQUALS OR IS LESS OR GREATER THAN AVAILABILITY (A). CHI-SQUARE = 8397 ($P < 0.01$).

Tree species or group	Observed (O) (use)	Expected (E) (availability)	$\frac{(O-E)^2}{E}$	% deviation	Unadjusted residual	Conclusion
Chestnut oak	3091	4554	470	−32	−21.68**	U < A
Pines	639	1609	585	−60	−24.18**	U < A
Red maple	254	635	229	−60	−15.12**	U < A
White oak	500	353	61	42	7.82**	U > A
Scarlet oak	8854	3961	6044	124	77.75**	U > A
Snags	1772	3122	584	−43	−24.16**	U < A
Other red oaks	883	1456	226	−39	−15.02**	U < A
Hickories	14	26	6	−46	−2.35*	U < A
Black gum	190	392	104	−52	−10.20**	U < A
Other species	1	90	88	−99	−9.38**	U < A

most species, as exhibited by the relatively low coefficients of determination (r^2 values).

Goodness-of-fit

The goodness-of-fit analysis produced a highly significant chi-square value (Table 2). This indicates that, overall, use of the tree species was different from availability. The analysis of residuals (Everitt 1977) demonstrates that every tree species or group was used disproportionately. Specifically, white oak and scarlet oak were used more than available, while the other species were underused. Percent deviations of use from availability reflect these results. For instance, scarlet oak, according to this analysis, was used 124% more than it was available, while several other species were used at least 50% less than expected.

DISCUSSION

Comparison of methods

Two methods, simple regression and goodness-of-fit analysis, were used to analyze the same data set of foraging use and availability values. The conclusions are the same for seven of nine tree species or groups. Two species, however, yielded different results. Regression showed use to be equal to availability for white oak and hickories, while the goodness-of-fit analysis indicated usage greater and less than availability, respectively. For these two species, variability in use among individual "territories" (blocks) inflated the denominator of the t-test in the regression analysis, making any trend difficult to determine. Because the goodness-of-fit analysis does not incorporate any measure of variability, but rather sums use across individuals, the results appear more definite, but are misleading. If the techniques produce conflicting conclusions, which is better, if not more legitimate? We think the regression method is superior on the basis of four criteria: biological scale, sampling considerations, analysis requirements, and interpretation. We discuss each criterion below (summarized in Table 3).

Biological scale

Wiens (1981) has discussed the importance of selecting the proper scale for examining various ecological questions, and insuring that the same scale is considered for all portions of a particular analysis. Goodness-of-fit studies customarily involve sampling availability randomly throughout the study area. However, birds have already selected certain regions of the area for use. Use and availability are therefore measured at two different scales, in that availability may be determined at points not used by any one individual. Availability must be sampled only at locations possessing potential for use (i.e., within territories). Otherwise, comparisons are invalid. If availability data are collected within areas of known use, the goodness-of-fit method is still invalid if all individuals are lumped together (as is commonly done), because each individual does not possess an equal opportunity to use all areas where availability is measured. The alternative is to run a separate analysis for each individual, but this defeats the purpose of the method, in that general conclusions cannot be readily drawn across all individuals. Regression procedures measure both use and availability at the same

TABLE 3. SUMMARY OF THE ADVANTAGES OF REGRESSION METHODS OVER GOODNESS-OF-FIT TECHNIQUES. THE TWO METHODS ARE COMPARED ON THE BASIS OF BIOLOGICAL SCALE, SAMPLING CONSIDERATIONS, ANALYSIS REQUIREMENTS, AND INTERPRETATION

	Regression	Goodness-of-fit
Biological scale	Use of resources by each individual bird is compared only to the availability of resources within that bird's territory, so that use and availability are measured at the same scale.	Use of resources by each individual bird is compared in part to the availability of some resources to which that bird does not have access (outside its territory), hence use and availability are measured at different scales.
Sampling considerations	For each individual bird, sequential foraging data may be collected, as they will be combined into 1 statistical observation for that individual. This enhances sampling efficiency.	Sequential foraging data may not be collected, as each datum will constitute a statistical observation, and these observations must be independent to satisfy analysis requirements. This reduces sampling efficiency.
Analysis requirements	Though requirements are more rigid (e.g., independent observations, normality, equality of variances, linearity, minimal effect of outliers, and low multicollinearity in multiple regression), they can be met or circumvented with a well-designed study and an adequate sample size.	Few requirements exist, but the effect of nonindependent observations and small expected values can be difficult to overcome.
Interpretation	Use versus availability of different tree species, and deviations of individual birds from the population as a whole, may be readily determined.	Use versus availability of different tree species, and especially the deviation of individual birds from the population as a whole, may be difficult to ascertain.

level. Availability data for a given territory are related only to the individual(s) occurring in that territory. Availability data are collected in some manner throughout each territory because all trees in that territory are potentially accessible.

Sampling considerations

A requirement of goodness-of-fit significance tests is that observations be independent (Everitt 1977, Dowdy and Wearden 1983). Strictly, only one observation can be collected per individual (Peters and Grubb 1983), but workers have employed various strategies in an attempt to sidestep this problem, including separating observations by a specified time interval (the "metronome method") (Wiens et al. 1970, Rusterholz 1981, Morrison 1984a). However, unless the interval between observations in a sequence is sufficiently large, the observations may still be correlated (Hejl et al., this volume). A problem with using only a subset of observations per sampling bout is that a large amount of effort is required to procure a sufficient sample size for analysis (Wagner 1981a). Field time, often limited, is inefficiently used when such a sampling strategy is employed.

The only difficulty with sampling for the regression method is that a relatively large number of territories must be monitored to insure an adequate sample size. The number of samples needed depends on, among other things, the number of variables included in the analysis, and the amount of variability in the data (Johnson 1981b). This hampers the utility of the regression method for less abundant species. Further, sufficient observations must be collected in each territory so that the data are representative of the behavior exhibited by that individual. Otherwise, the regression procedure possesses distinct advantages with respect to sampling. Data are independent, because all foraging observations collected within a territory are incorporated into the same statistical observation. Because territories are separate from one another, these statistical observations are independent. The investigator is therefore free to incorporate all the data recorded for each territory, and time data (time spent in each tree species) may be used. Recording a sequential stream of data, such as time data, lessens the effect of discovery bias, and provides a more complete representation of a species' full range of behaviors (Hertz et al.

1979, Morrison 1984a). By recording all the activity displayed by a bird, rather than single observations, field efficiency is maximized (Wagner 1981a).

Analysis requirements

Goodness-of-fit analysis requirements are less stringent than those for regression methods, but more difficult to satisfy with foraging data. Goodness-of-fit significance tests assume independence of observations, a problem discussed previously. A second problem is small expected values, which will result from including uncommon tree species in the analysis, and may negatively affect the results of both chi-square and G statistic significance tests (Sokal and Rohlf 1969). Some workers have attempted to circumvent this problem by lumping certain categories together following data collection (e.g., Maurer and Whitmore 1981), but this may result in a loss of useful information and affect the randomness of the samples, violating the assumption of random and independent observations (Everitt 1977:40).

Though the assumptions for regression-correlation methods are more rigid, they can be met or circumvented, particularly with larger sample sizes (Green 1979). Normal distributions of variables are required to conduct hypothesis tests and make other inferences (Dowdy and Wearden 1983, Afifi and Clark 1984). Though transformations are available to improve many non-normal distributions prior to analysis, transformations are frequently undesirable, particularly for dependent variables (Johnson 1981a, Afifi and Clark 1984). In general, deviations from normality tend to make statistical tests conservative (Hollander and Wolfe 1973), so that significant results are likely to be truly significant. Variables are also assumed to possess equal variances (homoscedasticity). This assumption is not critical unless differences between variances are large (Afifi and Clark 1984). While the regression procedures discussed here assume a linear relationship between independent and dependent variables, curvilinear relationships are possible. Analysis of such relationships is likely to indicate that use deviates from availability, which should generally be the correct conclusion if substantial curvilinearity exists. Outliers can greatly affect both univariate and multivariate analyses (McDonald 1981, Afifi and Clark 1984); these may be identified before or after the analysis is undertaken. Outlying observations may be removed from the analysis, but such observations often possess important information (Neter et al. 1983, Afifi and Clark 1984). It may be wisest to incorporate all but those observations that are obvious blunders, and investigate the effect, if any, of including any outlying observations after the analysis is conducted.

An additional condition can affect multiple regression analyses. Multicollinearity, or interdependence of variables, can cause regression coefficients to be unstable, hindering interpretation of the results (Afifi and Clark 1984). Techniques are available to locate highly correlated variables, both prior and subsequent to analysis, so that the problem can be minimized (McDonald 1981, Afifi and Clark 1984). Careful variable selection may help this problem, and simple regression analyses may be used to back up multiple regression conclusions.

Interpretation

It is not enough to conclude that a bird does or does not use tree species in relation to their availability. One desires to know just which tree species are preferred or avoided, and which tree species most influence use patterns. In this respect, regression procedures are superior. Although residuals (Everitt 1977:46-48) may be useful in determining which tree species contribute most to a significant chi-square value, goodness-of-fit methods do not allow the investigator to assess the influence of the availability of one tree species on the use of another, or the interaction of different tree species availabilities in determining use trends. Partial regression coefficients may be examined for this purpose in regression analysis.

Another important consideration is whether some individuals provide data divergent from the analysis as a whole, and how such individuals have influenced the analysis. Only by constructing a different table for each territory can individual observations be examined separately using goodness-of-fit techniques. It is possible to examine tables separately, then subsequently combine them for additional analysis, but this approach is relatively awkward (Everitt 1977: 51). Regression methods are more informative and easier to execute and understand in this respect. Various plots and residuals can be examined to identify unusual or outlying observations (Afifi and Clark 1984). A battery of measures (e.g., Cook's distance) are available to ascertain the effect of each observation on the analysis (Afifi and Clark 1984, SAS Institute Inc. 1985). It requires little extra effort to obtain this information using many statistical computer packages (Afifi and Clark 1984).

Extensions

Regression methods may prove useful for other types of food exploitation studies, such as analysis of food use versus availability. The food brought to nestlings might be monitored and related to the prey available on the territory. Nest boxes (e.g., Dahlsten and Copper 1979) would be particularly useful for such a study. Another possible design involves the determination of adult bird diets through emetics or collection and subsequent gut content analysis. Food availability in this case would be ascertained by observing the bird prior to capture or collection and sampling immediately afterward the prey resource in locations in which the bird foraged. The ability of regression techniques to examine territory-territory or bird-bird variation in prey use versus availability makes these methods particularly attractive.

Examination of tree species use versus availability may lead to other investigations. For instance, the prey availability in each tree species might be determined. The amount of prey found in each tree species might be used to weight the abundance of that species, and the analysis rerun. If use approximates availability, it might be concluded that the primary reason some tree species are preferred or avoided is due to the prey they harbor. Tree species use versus availability characteristics may also help explain habitat distribution patterns. The correlation of the presence or abundance of a bird species with the abundance of a particular tree species may be found to be due to the bird's foraging preference for that tree species. The application of information obtained at one scale to phenomena observed at another scale would be valuable.

CONCLUSIONS

Several compelling arguments point to the superiority of the regression technique over the goodness-of-fit method in analyzing use versus availability data. In terms of sampling design, regression procedures are theoretically more appealing. Foraging data are easier to obtain per bird using the regression sampling scheme, and little additional work is demanded to assess availability on each territory. Regression analyses are at least as easy and straightforward to perform as goodness-of-fit methods. Though the statistical assumptions of regression techniques are more rigid, they can be met given careful attention to study design and an adequate sample size. Finally, regression results may be interpreted with greater facility than goodness-of-fit output, and provide more information.

Two additional methods, discussed in this symposium, have recently been applied to the analysis of use versus availability data. Markov chains can be used to produce independent data from sequential observations that are suitable for analysis (Raphael, this volume; Hejl et al., this volume). Although Raphael presents a goodness-of-fit analysis of data pooled across territories, he suggested methods for analyzing data on a territory-by-territory basis. Despite the generation of independent data, the limitations of goodness-of-fit analyses discussed above, particularly in terms of interpretation, still exist. McDonald et al. (this volume) discuss a means of analyzing use versus availability data using selection functions. Success is based in part on the proper choice of a model to fit the distribution of used and available resources, a choice that may not always be easy to make. This method appears to be in the developmental stages, and its applicability and effectiveness remain to be determined.

Of all the techniques, the regression approach appears to us to be the most sound. It makes sense to collect data on a number of territories due to the possibility of individual-specific behavior. Data may be collected in an efficient manner, a variety of statistical packages for analysis are readily available, and the methods and interpretation are straightforward. Conclusions may be drawn across the entire population; yet, the ability to examine the peculiarities of individual territory owners is not lost. We encourage researchers to use the regression method to analyze suitable, presently available data sets, design future studies to accommodate the regression approach, and improve the methodology.

ACKNOWLEDGMENTS

This research was supported by the U.S.D.A. Forest Service, the U.S.D.A. National Agricultural Pesticide Impact Assessment Program, West Virginia University Division of Forestry McIntire-Stennis funds, and a Westvaco Doctoral Fellowship from Westvaco Corporation to Dodge. We thank Larry Hines and Gary Strawn of the West Virginia Department of Natural Resources for use of the study area, Rev. and Mrs. Robert McCarter for accommodations, R. Hawrot, R. Nestor, R. Seiss, D. McConnell, D. Palestra, D. Reckley, J. Herda, and K. Wimer for data collection and entry, S. Smith and C. Lowdermilk for manuscript preparation, R. Cooper for many helpful suggestions, and R. Cooper, D. Fosbroke, G. Hall, R. Hicks, B. Noon, and R. Smith for reviewing the manuscript. This is scientific article number 2133 of the West Virginia University Agricultural Experiment Station.

ANALYZING FORAGING AND HABITAT USE THROUGH SELECTION FUNCTIONS

Lyman L. McDonald, Bryan F. J. Manly, and Catherine M. Raley

Abstract. Methods commonly used for study of natural selection in changing populations are useful in the evaluation of food and habitat selection. In particular, one can derive "selection functions" that allow estimation of the relative probability that a given food item (or habitat class) will be selected next, given equal access to the entire distribution of available food items (or habitat points). Procedures are available for the cases when food items are assigned to qualitative classes and when the items are characterized by measurement of quantitative variables on the item. The primary advantage of these analysis methods is that clear probabilistic statements can be made concerning the likelihood that each of several food or habitat types will be used. Illustrations are given by analysis of prey size selection by Wilson's Warblers (*Wilsonia pusilla*) and by Tree Swallows (*Tachycineta bicolor*).

Key Words: Food exploitation; fitness functions; natural selection; selection functions; selectivity indices; weighted distributions; habitat selection.

Evaluating food and habitat selection by a population of animals is an important aspect of ecological studies. Analysis methods in the literature often assume that food or habitat resources can be classified into one of several categories defined by the researcher (e.g., Carson and Peek 1987). The chi-square goodness-of-fit tests, the chi-square test of homogeneity with Bonferroni z-tests (Marcum and Loftsgaarden 1980), or a rank-order procedure involving the relative ranks of available and used food items (or habitat classes) (Johnson 1980, Alldredge and Ratti 1986), have been used to analyze availability and usage data. Other analysis procedures include common univariate (e.g., Quinney and Ankney 1985) and multivariate statistical methods. In particular, discriminant analysis has been used to make inferences toward food (habitat) selection in the multivariate case (Williams 1981).

When food items are classified into one of several categories, Manly's (1974) selectivity indices can be used to estimate relative probabilities of selection (Heisey 1985). In the study of natural selection on a population consisting of qualitatively distinct morphs, the selectivity indices are the relative "fitness" values of the morphs. In the present application, the indices are relative probabilities of selection of food items (or habitat classes) from categories defined by the researcher.

We consider the case where one or more quantitative variables $\{x_1, x_2, \ldots, x_q\}$ can be measured on each available food item (or habitat point) and adopt parametric methods that have been developed for the study of natural selection in changing populations (Manly 1985). It is assumed that the relative probability of an animal selecting a food item (or a habitat point) given access to the entire population of available items (or habitat points) can be modeled by a function of the variables, $w(x_1, x_2, \ldots, x_q)$. This function is defined to be the selection function (it is called the fitness function in the study of natural selection). It is a function such that if $f(x_1, x_2, \ldots, x_q)$ is the frequency of available items (points) with X-values $X_1 = x_1, X_2 = x_2, \ldots, X_q = x_q$ before selection, then the expected frequency of these food items in the diet (habitat points used) is

$$w(x_1, x_2, \ldots, x_q) f(x_1, x_2, \ldots, x_q).$$

In other words, given access to all food in the population of available items, the probability that an individual item with X-values $X_1 = x_1, X_2 = x_2, \ldots, X_q = x_q$ is selected is proportional to the selection function, $w(x_1, x_2, \ldots, x_q)$. The value of the selection function can be thought of as a weighting factor that represents deviation of resource use from purely random use. The case when the function $w(x_1, x_2, \ldots, x_q)$ is constant over the range of $(x_1, x_2, \ldots, x_q)$ corresponds to the situation in which selection is purely random. Selection functions are particularly applicable to the study of food (habitat) exploitation because of the ease of biological interpretation of relative probabilities.

One further requirement is that the population of resource items must be considered to be infinitely large or that samples of the available and used items are collected instantaneously. In practice, these requirements are never totally satisfied. They must be replaced by the assumption that the proportion of the population withdrawn by the sampling is so small that the basic characteristics of the original population remain unchanged. This requirement is not unique to the present method, and represents a major obstacle to the study of resource selection in natural systems.

The method is illustrated by its application to two data sets: (1) size selection of leafhoppers (family Cicadellidae) by Wilson's Warblers (*Wilsonia pusilla*) (Raley and Anderson, in press), and (2) prey selection by Tree Swallows (*Tachycineta bicolor*). Data for the second illustration are approximated from figures and tables in Quinney and Ankney (1985). Although not illustrated, application to the study of selection of habitat points in a study area is straightforward. For example, Harris (1986) studied nest site selection by Fernbirds (*Bowdleria punctata*) in Otago, New Zealand. He measured nine variables at each nest site and at the closest clump of vegetation to randomly selected points. If we assume the variables he measured influenced the probability of selection of clumps of vegetation for nest sites, and the Fernbirds had access to the entire distribution of vegetation clumps, then our method could be used to estimate the relative probability that a randomly located clump with $X_1 = x_1$, $X_2 = x_2, \ldots, X_9 = x_9$ was selected as a nest site.

DEFINITIONS AND STATISTICAL MODELS

The distribution of $X = \{X_1, X_2, \ldots, X_q\}$ for food items (habitat points) in a study area is defined to be the *distribution of available items* and is denoted $f(x)$. A subset of the items is used by a population of animals under study during a certain period of time. The distribution of X for the subset of items used is defined to be the *distribution of used items* and is denoted by $g(x)$. In the following, reference will be made only to selection of food items with the understanding that results are equally applicable to the study of habitat selection.

We follow the notation and models reviewed in Manly (1985:55–75), where applications to the theory of natural selection of animals in changing populations are considered. McDonald and Manly (1989) also used the mathematical and statistical results (of Manly [1985]) to develop a theory for calibration of biased sampling procedures.

We assume that animals are using food items from the available distribution such that the probability of selection of an item depends only on the variables measured in X and is proportional to the *selection function*,

$$w(x) = w(x_1, x_2, \ldots, x_q). \quad (1)$$

The distribution of used items is proportional to the product of the selection function and the distribution of available items. When the proportionality constant is needed, it is denoted by

$$E_f(w[x]),$$

the expected value of the selection function with respect to the distribution of available items. The distribution of used items can then be written as

$$g(x) = w(x)f(x)/E_f(w[x]). \quad (2)$$

In general, the proportionality constant cannot be estimated unless the sizes of the populations of available items and used items are estimated. The distribution $g(x)$ is known as a "weighted distribution" in the mathematical statistics literature.

Assume that a random sample of n_0 items is selected from the study area and the vector of variables is measured on each to yield the sample of available items $\{x_{01}, x_{02}, \ldots, x_{0n_0}\}$. Similarly, assume that a sample $\{x_{11}, x_{12}, \ldots, x_{1n_1}\}$ of n_1 items is randomly obtained from the population of used food items.

Given the samples of available and used items, the selection function is estimated by appropriate formulae. One can then graph the relative probability of selection in two and three dimensions and test hypotheses concerning the significance of parameters in the selection function. If $w(x)$ does not depend on x (e.g., the constants in the following model are both zero), then the distributions of available and used items are equal. This case is equivalent to the conclusion that the animals are selecting food items from the population of available items at random. In the case that the relative probability of selection of points depends on the variables $X_1, X_2, \ldots, X_q$, one can evaluate whether the selection results mainly in changes in the mean or the variance of a variable.

Manly (1985:55–75) reviewed cases where distributions of available and used items follow univariate normal, multivariate normal, or gamma distributions, and considered robust models for selection functions. He also developed a procedure for estimation of a general multivariate selection function without assuming any particular parametric form for the distributions. Here we give a brief outline of the use of the existing models by illustrating formulae for the case when both distributions are normal with a single variable in the selection function.

Assume that the variable X has a normal distribution with mean μ_0 and variance V_0 for the population of available items, and that the probability of selection of an item with $X = x$ is proportional to the selection function

$$w(x) = \exp\{kx + mx^2\}, \quad (3)$$

where k and m are constants. Under these assumptions, it is known that the distribution of used items will be normally distributed with mean μ_1 and variance V_1, where

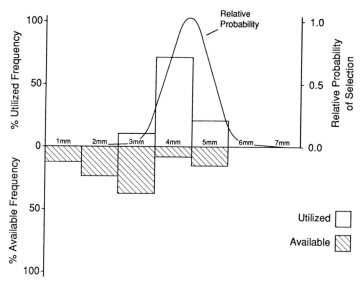

FIGURE 1. Relative probability of selection of leafhoppers by Wilson's Warblers plotted as a function of prey length. The graph is superimposed on the distributions of leafhopper lengths in the samples of available and used prey.

$$\mu_1 = (\mu_0 + kV_0)/(1 - 2mV_0), \text{ and} \quad (4)$$
$$V_1 = V_0/(1 - 2mV_0). \quad (5)$$

These equations can be solved for the constants in the selection function to yield

$$k = (\mu_1/V_1) - (\mu_0/V_0), \text{ and} \quad (6)$$
$$m = ([1/V_0] - [1/V_1])/2. \quad (7)$$

Also, if the distribution of X is normal in both populations, then the selection function must be of the form $w(x) = \exp\{kx + mx^2\}$.

One can denote the usual sample means and variances by $(\hat{\mu}_0, \hat{V}_0)$ and $(\hat{\mu}_1, \hat{V}_1)$ for the samples of available and used items, respectively. The reciprocal of the sample variance should be adjusted slightly when used to estimate the reciprocal of the corresponding parameter. For the two samples, $j = 0, 1$, let

$$\hat{B}_j = (n_j - 3)/(n_j\hat{V}_j), \quad (8)$$

with estimated variance

$$\text{var}(\hat{B}_j) = 2(\hat{B}_j)^2/(n_j - 5), \quad (9)$$

and let

$$\hat{a}_j = \hat{\mu}_j\hat{B}_j, \quad (10)$$

with estimated variance

$$\text{var}(\hat{a}_j) = (\hat{B}_j/n_j) + (\hat{\mu}_j)^2(\text{var}[\hat{B}_j]). \quad (11)$$

The estimators of k and m are

$$\hat{k} = (\hat{a}_1 - \hat{a}_0), \quad (12)$$

with estimated variance

$$\text{var}(\hat{k}) = \text{var}(\hat{a}_1) + \text{var}(\hat{a}_0) \quad (13)$$

and

$$\hat{m} = (\hat{B}_0 - \hat{B}_1)/2, \quad (14)$$

with estimated variance

$$\text{var}(\hat{m}) = (\text{var}[\hat{B}_0] + \text{var}[\hat{V}_1])/4. \quad (15)$$

ILLUSTRATIONS

We first consider a subset of data analyzed by Raley and Anderson (in press), who sought to quantify the relationship between availability and use of invertebrate food resources by riparian birds. They used Johnson's ranking procedure (Johnson 1980) to compare availability and selection of ten orders of invertebrates by the bird community. They evaluated size selection using the Kolmogorov-Smirnov two-sample test to compare the distributions of available and used items. We consider selection of one family of insects (leafhoppers) by one species of bird (Wilson's Warblers) during 15 June to 12 July 1986 (Fig. 1).

The distributions of available and used items are approximately symmetric and the assumption of normality of the lengths of leafhoppers in the populations will be made for this illustration. Under this assumption, and given that the birds have access to the entire distribution of

TABLE 1. FREQUENCY DISTRIBUTIONS OF PREY LENGTHS (MM) WITH SAMPLE MEANS AND STANDARD DEVIATIONS IN THE SAMPLES OF AVAILABLE AND USED PREY (APPROXIMATED FROM QUINNEY AND ANKNEY 1985)

Length (mm)	Available (frequency)	Used (frequency)
1	216	44
2	708	133
3	765	208
4	401	347
5	676	493
6	444	376
7	132	208
8	132	94
9	34	38
10	34	56
$\bar{X}$	4.01 mm	5.08 mm
SD	1.94 mm	1.89 mm

available leafhoppers, the estimated values for the constants in the selection function are

$m = -1.4534$ with $\text{SE}(m) = 0.02005$, and
$k = 12.7481$ with $\text{SE}(k) = 1.6565$.

Both estimates are large with respect to their standard errors, indicating that both the mean and variance of used items are different from the mean and variance of available items. The estimated selection function (scaled by dividing by the largest selection value $\exp[27.9351]$) is

$$w(x) = \exp(12.7481x - 1.4534x^2 - 27.9351).$$

We divide by the constant $\exp(27.9351)$ so that the relative probability of selection of the most "preferred" length 4.5 mm is 1.0. Under the stated assumptions, the relative probability of selection was strongest for leafhoppers of length 4.5 mm by Wilson's Warblers during the period 15 June to 12 July 1986 (Fig. 1). In comparison, a leafhopper of length 3.5 mm was selected approximately one-third as often, whereas a leafhopper of length 5.5 mm was selected approximately one-fifth as often.

Quinney and Ankney (1985) reported size of prey selected (orders Diptera and Homoptera) by Tree Swallows. Their primary objective was to draw conclusions concerning optimal foraging theory. Data from one of their study sites, the sewage lagoon, were approximated from their Figure 1 and Table 3 (Quinney and Ankney 1985) for our second illustration (Table 1).

There was a significant shift in the mean length of used prey compared to available prey, but no significant ($P > 0.05$) difference in the variances. The pooled estimate of the common variance for the two distributions is $\hat{V} = 3.70$. Again, the distributions are approximately normal and estimation of the selection function follows the theory reported in eq. (3) to (15). From eq. (14), the constant m is judged to be zero because the variances are not significantly different. Using the common variance, $\hat{V} = 3.70$, in eq. (12) the constant k is estimated to be

$k = 0.2892$ with $\text{SE}(k) = 0.03469$.

The relative probability of selection of a prey item of length x given access to the entire distribution of available insects is estimated by the selection function

$$w(x) = \exp(0.2892x - 2.892).$$

Again, the original function $w(x) = (0.2892)x$ has been scaled by dividing by $\exp(2.892)$ so that the relative probability of selection of the most "preferred" length 10 mm is the number 1.0 (Fig. 2). An insect of length 9 mm is selected with approximately three-fourths the probability of selection of an insect of length 10 mm, while an insect of length 5 mm has approximately one-third the chance of being eaten.

We consider the classes in Table 1 as qualitative to illustrate the computation of Manly's selectivity indices (Manly 1974) for qualitative variables. The selectivity indices are relative probabilities of selection and are interpreted in exactly the same manner as a particular value of the quantitative selection function considered above. Estimators of the selectivity indices (relative probabilities of selection) for length $x = 1, 2, 3, \ldots, 10$ mm, are

$$w(x) = f_{ux}/f_{ax}, \quad (16)$$

where f_{ux} is the frequency of length x insects in the sample of used prey, and f_{ax} is the frequency of length x insects in the sample of available prey. If the frequency of length x insects in the sample of available prey is $f_{ax} = 0$, then some of the classes must be combined to avoid division by zero. Heisey (1985) recently developed procedures to estimate the selectivity indices under various hypotheses concerning how they depend on other attributes such as age and sex of the birds. Table 2 contains the selectivity indices for the 10 lengths of prey computed by eq. (16) and standardized by multiplying by $0.6071 = 1.0/w(10)$, so that the relative probability of selection of an insect of length 10 mm is 1.0.

The selectivity indices in Table 2 are presented only for the purpose of illustrating the analysis of qualitative classifications for habitat points or food items. The approximations of the frequencies of length 7–10 mm insects from Figure 1 and Table 3 of Quinney and Ankney (1985) are not very precise. Errors in the approximations will influence the selectivity indices in Table 2

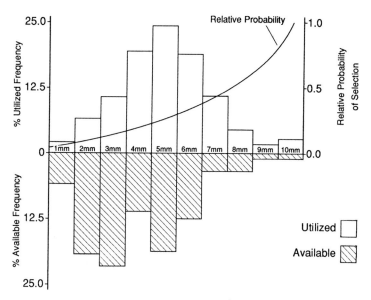

FIGURE 2. Relative probability of selection of insects by Tree Swallows plotted as a function of prey length. The graph is superimposed on the distributions of insect lengths in the samples of available and used prey.

relatively more than the selection function values graphed in Figure 2.

Our analyses agree with the conclusions of Quinney and Ankney (1985): "Swallows were selective in the sizes of insects that they captured.... distribution of sizes of insects captured by the birds from the 10 size classes was significantly different from the distribution of sizes present.... The two smallest classes (1 and 2 mm) were most underrepresented in the diets... in relation to their abundance in the nets."

DISCUSSION

If items are assigned to classes (perhaps purely qualitative), chi-square analysis sometimes leads to interpretation problems (Johnson 1980). If there are relatively large classes of available items that are seldom used, then the relationship of the expected frequency of use to observed frequency of use depends on whether or not the researcher includes those classes in the analysis. One of Johnson's objectives was to develop a procedure to overcome this problem. A major advantage of the selectivity indices is that the estimates are not sensitive to whether or not the large available but seldom used classes are included. For example, the 1 mm insects in Table 1 could be dropped and the selectivity indices (relative to selection of 10 mm insects) in Table 2 do not change. Another common practice is to standardize the selectivity indices so that they sum to 1.0, in which case dropping the large but seldom used class will have little effect.

The power of the selectivity indices and the selection function (univariate and multivariate) is that they provide clear probabilistic statements in the study of resource use. There may be several confounding reasons why an estimated selection function assigns significantly different relative probabilities of selection to items with different characteristics. Discussion of the biological reasons is beyond the scope of this paper; however, violation of the required assumptions may also contribute to a seemingly significant result or hide an important result. The basic assumptions required are: (1) the correct models have been used for the selection function and the distributions of available and utilized samples, (2) sampling

TABLE 2. SELECTIVITY INDICES COMPUTED AS IF THE CLASSES IN TABLE 1 WERE QUALITATIVE. INDICES ARE STANDARDIZED SO THAT THE INDEX FOR LENGTH 10 MM IS 1.000

Length (mm)	Selectivity index
1	0.124
2	0.114
3	0.165
4	0.525
5	0.443
6	0.514
7	0.957
8	0.432
9	0.679
10	1.000

is instantaneous, so that the characteristics of the distributions do not change during the sampling period, (3) the basic sampling unit is the individual food item (habitat point) and must be independently collected, and (4) the researcher has identified those classes or measured those variables that actually influence the probability of selection. An alternative to assumption (3) is that estimates of the selection function are independently replicated over the study area (or time).

The assumption that the correct models have been used can be tested with one of the common statistical tests (e.g., chi-square goodness-of-fit); however, power will be low unless the sample sizes are large. In the illustrations presented the distributions were approximately symmetric except for the available distribution of prey lengths to Tree Swallows (a chi-square goodness-of-fit test comes close to rejecting the hypothesis of normality for this data set). Although the theory is not yet available to defend a formal statement, we think that mild departures from normality will have little effect on the estimated selection function presented in eq. (3).

Estimation procedures are available for one more parametric situation in addition to normality. This case arises when both the distributions are skewed in the same direction and gamma distributions satisfy assumption (1). Estimation formulae appear in Manly (1977).

O'Donald's general quadratic selection function (reported in Manly [1985]) can be fitted between any univariate distributions, normal or otherwise. Use of this selection function diminishes the importance of assumption (1). It is also simple to use, but has disadvantages because the quadratic function may not fit the selection function over its entire range (in fact, it may be negative for extreme values of the variables) and procedures for drawing statistical inferences are not available. A second procedure that does not make assumptions about the parametric form of the distributions is Manly's general multivariate technique (Manly 1985). A robust exponential model is fitted to the selection function. The estimation of fitness functions in the study of natural selection assumes that two or more samples are available over time from the changing population. In the estimation of selection functions for food or habitat exploitation only two samples are available. Further theoretical research is needed to evaluate the statistical properties of Manly's general multivariate technique for the case of two samples and to develop new numerical (nonparametric) estimation procedures.

The second assumption (that sampling is instantaneous) is difficult to satisfy in practice. Whether or not this assumption is reasonable also depends on the "basic sampling unit," discussed below. If the population of available units is very large with respect to the population of used items during the sampling period, then this assumption is not critical. Inferences will be to the "average" distributions during the study. But, if the population of available units (points) is limited and "preferred" units are quickly selected, utilization is changing the available distribution and the present techniques are not applicable. Again, further theoretical research is needed to evaluate selection from a changing population.

In the illustration of selection of leafhoppers by Wilson's Warblers, the population of leafhoppers was judged to be very large with respect to use by warblers. However, if this were not the case and the shape of the available distribution was changing, the low relative frequency of 4 mm leafhoppers (Fig. 1) might be exaggerated by selection for insects of that length. Consequently, this would exaggerate the estimated height of the selection function in this region.

Our analyses were made under the assumption that the samples of food are equivalent to random samples from the populations (i.e., food items in the samples are independently collected). It is rare that studies can be designed so that individual food items are the basic sampling units. It is important to keep track of the different sources of sampling variance and to avoid the infamous pseudo-replication problems (Hurlbert 1984). This is exceptionally difficult in studies of food selection because different collection methods are generally used to obtain the samples of available and used items. One procedure followed by Raley and Anderson (in press) was to also collect invertebrates from the shrub on which the bird was observed to feed. If birds are independently located, then it may be appropriate to consider the bird as the basic sampling unit. The sample of available invertebrates from a unique shrub is paired with the sample of used invertebrates taken from the bird. If there are sufficient numbers of insects in the paired samples, then replicate estimates of the parameters in the selection function could be obtained for each bird. Alternatively, estimates of the parameters in the selection function might be replicated over some larger unit of space or time. For example, data for birds collected in one of several independently sampled quadrats might be pooled. The selection function could be fitted in each quadrat to obtain replicate estimates of the parameters. Statistical inferences toward the mean values for the parameters of the "average" selection function over the study area (or time) would be made by considering the variances of the replicate estimates.

In the study of habitat selection we envision that the basic sampling unit will be a point or a small area associated with a point. A random sample of available points (or a systematic sample with random starts) might be selected with a vector of variables to be measured at each point. Points in this sample that happen to be used are left in the sample of available points. Obtaining a random sample of used points will be more difficult and highly dependent on the particular application.

ACKNOWLEDGMENTS

Work on this project was partially supported by a grant from the Beverly Fund, University of Otago, Dunedin, New Zealand. Field studies were partially supported by the U.S.D.A. Forest Service, Rocky Mountain Forest and Range Experiment Station, Laramie, Wyoming. Significant improvements in the manuscript are due to the review by the referees: Douglas H. Johnson, Martin G. Raphael, Dana L. Thomas, and the editors. Their contributions are gratefully acknowledged.

SECTION III

SPECIALISTS VERSUS GENERALISTS

Overview

SPECIALIST OR GENERALIST: AVIAN RESPONSE TO SPATIAL AND TEMPORAL CHANGES IN RESOURCES

HARRY F. RECHER

Under what conditions should a species specialize on a particular set of resources and when is being a generalist the most successful strategy? These questions have been central to the development of community ecology as a science since MacArthur and Levins developed theories of resource allocation and limiting similarity (Levins 1968; MacArthur 1970; MacArthur and Levins 1964, 1967). These early models assumed that species competed for resources that were presented as a continuum along which species segregated. The degree of specialization or the extent of segregation depended upon the similarity of resources and their abundances. Specialization was favored if resources were abundant or very different. If resources were similar or scarce, being a generalist or a jack-of-all-trades was deemed the best strategy.

The models predicted other responses to a changing resource spectrum. For example, as species specialized on particular resources, more species could co-exist and community diversity would increase. If a generalist dominated the available resources, there would be fewer opportunities for co-existence and diversity would decrease. Models were not mutually exclusive and arguments were raised for a range of alternatives. Thus, in a community of generalists, species diversity could increase if overlap in the use of resources was possible. This might occur if resources were superabundant relative to the demands of the species using them, or if other factors (e.g., predation, chance climatic events) prevented competition from going to completion, with one species excluding another.

These mathematically elegant, albeit simple, models provided a conceptual framework on which a generation of ecologists based their studies of avian foraging ecology and community structure. This was true for those who rejected the assumption that species necessarily competed for resources (e.g., Wiens 1977, Simberloff and Boecklen 1981) and for those who accepted competition as the driving force in the evolution of differences among species (e.g., Cody 1974, Diamond 1978).

There is no doubt that the models of MacArthur and Levins launched an extremely valuable period of scientific enquiry. There is now a considerable literature on the foraging ecology of terrestrial birds and a number of these studies present valuable empirical descriptions of community structure. Nonetheless, gaps remain in our knowledge of terrestrial birds. If I wanted to be glib, I would say that most studies of the foraging ecology of terrestrial birds have focused on the breeding season; most have compared a few species of birds in one or at most a few places; most have treated the individuals of populations as the same; most have combined data collected through the day or over a season or over a year. Few studies have attempted to measure the kinds and abundances of resources available to birds or to directly measure their use by birds. When resources have been measured, emphasis has been on the abundance of prey and has generally failed to distinguish between abundance (the total amount of the resource) and availability (the amount of the resource that birds can use). There have been few attempts to measure either the abundance or availability of different foraging substrates (e.g., amount of different kinds of bark available to bark foraging birds). A consequence of this narrow data base is that questions of temporal and spatial changes in the use of resources by birds have never been satisfactorily answered. Nor, apart from a small number of studies, such as those of Darwin's finches by Peter Grant and his colleagues (see Grant 1986), are there adequate data that describe individual variation in foraging habits in a way that allows the separation of the effects of learned behavior and environmental factors from genetic differences.

Although ornithologists have often described the ways that co-existing species apportion resources, questions of when to be a specialist and when to be a jack-of-all-trades remain a challenge. There are not only interesting ecological questions of when, where, and why species specialize (or generalize) on particular resources, but there are practical considerations. The conservation and management of terrestrial avifaunas requires more detailed information on temporal and spatial differences in the use of resources by species than is available for the majority of species.

I am not the first to recognize these omissions in foraging ecology. Papers in this section of the symposium focused on questions of "Specialist or generalist? Avian response to spatial and temporal changes in resources." Harry Bell and Hugh Ford presented data on the changes in the diets of Australian warblers (Acanthizidae) as the abundance of food decreased during a long drought. Thomas Martin and James Karr studied the foraging behavior of North American wood warblers (Parulinae) during migration and contrasted this with the behavior of the same species during the breeding season and on their wintering grounds. Kenneth Rosenberg described the foraging ecology of specialized dead-leaf foragers in Amazonian forest understory in relation to resource (food and substrate) abundance and the presence-absence of potential competitors. The papers by Stephen et al. and Kellner et al. investigated the response of birds to a superabundant food resource (periodical cicadas) in Ozark forests and the effects that changes in the abundance of a major food item might have on patterns of avian predation on other prey organisms. Thomas Sherry presented an overview of the importance of distinguishing ecological and evolutionary processes in studies of avian foraging ecology. Many of his ideas were derived from studies of the Cocos Finch (*Pinarolaxias inornata*), a species that exists in the absence of competitors (Grant 1986). Each paper in this section represents the kinds of studies that are required for the continued development of our understanding of the foraging ecology of terrestrial birds. By focusing on the behavior of individuals within a population (Sherry), following changes in the behavior of birds over long periods (Bell and Ford) and between seasons (Bell and Ford, Martin and Karr), and studying the behavior of birds in response to known abundances of prey (Kellner et al., Stephen et al., Rosenberg) each of the major areas of avian foraging ecology where more information is required was identified.

SPECIALIST OR GENERALIST?

What is a specialist and what is a generalist? The answer depends on the design of the research, the hypotheses tested, and the system studied. In simple terms a specialist is a species that uses a narrow range of resources and a generalist is one that uses a wide range of resources. Whether any particular species qualifies as a specialist or generalist depends on the species to which it is compared (do they use more or fewer kinds of resources?) and the resources in question (are they diverse or restricted?).

Whether or not particular resources are used depends on morphological and behavioral limitations, learned patterns of behavior, and physiological requirements. For example, the different conclusions reached by Martin and Karr (this volume) and Greenberg (1984a, c) in describing the behavioral plasticity of wood warblers appear to result from the different foraging behaviors investigated. Martin and Karr studied the behaviors used by warblers in taking prey, whereas Greenberg investigated their response to different substrates. Species that had a diverse foraging repertoire were conservative in their use of substrates.

The response of a bird to its environment and the resources it uses depends on the availability of particular resources, the individual's needs (e.g., its physiological requirements), and the presence or absence of other individuals and species. Responses to any variable are graded. Not only do species differ in their use of resources through time and in different places, but the extent to which they specialize or generalize in their use of resources may change. A pattern of change in resource use is as significant a part of the ecology of a species as its use of resources at any particular time or place. Equally important are individual differences within a population in the use of resources.

TEMPORAL PATTERNS

An individual studied intensively for 24 hours may use a narrower range of resources than the same individual studied over a season or from year to year. Similarly, there will be changes in behavior and in the use of resources between seasons and from year to year. These changes will occur in response to weather, to changes in resource abundance and availability, to the differing physiological requirements of birds as they proceed through their molt and reproductive cycles, to the different demands of migration and breeding, and to changes in the species composition of avian communities. In part these changes will be shown by increased or decreased specialization on particular resources.

As resources become more abundant, many species use a broader range of resources and niche overlap increases (e.g., Bell 1985b, Recher 1989b). Often these changes are associated with seasonal patterns of prey abundance: with increased food abundance during spring and summer, niche overlap increases; with decreased food abundance during autumn and winter, niche overlap decreases (e.g., Bell and Ford, this volume; Recher 1989b).

Early work on the ecology of terrestrial birds focused on species relationships during the breeding season with emphasis on the use of food resources. The argument was that, with the need to obtain large amounts of food for egg production and feeding young, breeding placed the greatest demands on birds (e.g., Lack 1968b). It was therefore assumed that competition for resources would be greatest during the breeding season and that species would be most different at this time. Food was assumed to be the critical resource (Martin 1987).

The emphasis on breeding, food resources, and competitive interactions between co-existing species restricted the diversity of studies undertaken. Perhaps because of the practical difficulties in working with mobile populations, little work has been done on terrestrial birds during migration. However, as demonstrated by Martin and Karr (this volume), migration places considerable demands on birds and may be a significant factor in the evolution of specific behavioral and morphological traits. Migration often occurs when food resources are restricted and weather (particularly low temperatures) limits foraging opportunities or requires increased energy for survival. Species interactions at this time may be more significant than those on breeding or wintering areas where food may be abundant, individuals occupy familiar territory, and the energy requirements of individuals are less demanding (e.g., Fretwell 1972, Martin 1987).

Changes in resource abundance not only occur between seasons, but may vary significantly between years. Severe drought conditions in southeastern Australia during 1982–1983 led to almost total reproductive failure of forest and woodland birds and to increased mortality (Ford et al. 1985, Recher and Holmes 1985). Bell and Ford (this volume) showed how birds first specialized on particular resources with decreased niche overlap and then used a wider range of resources with increased overlap as food abundance decreased during prolonged drought.

SPATIAL PATTERNS

The distribution and abundance of resources not only changes with time, but varies significantly between habitats and regions. Kellner et al. (this volume) and Stephen et al. (this volume) used the presence or absence of periodical cicadas to study the response of birds to an abundant food resource. Cicadas were an important food where they occurred. However, it was difficult to demonstrate either a significant shift in avian foraging behavior or to find a response in other prey organisms that may have been released from high levels of avian predation as birds obtained more of their requirements from cicadas.

Demonstrating a response to spatial patterns in resource abundance is difficult. Martin and Karr (this volume) suggested that birds have a characteristic foraging signature, which they define as the ranked abundance of different kinds of foraging maneuvers (e.g., relative proportions of hawks, snatches, and gleans). The signature remains unchanged, although the proportions of particular behaviors may vary, despite changes in resource abundance and physiological requirements. A problem with demonstrating a response to changing patterns of resource abundance is the difficulty in measuring resource availability. Rosenberg (this volume) emphasized the importance of studying resources that can be accurately and precisely measured. He demonstrated that birds selected the most profitable foraging substrates and shifted between substrates as resource abundance changed between habitats. Competition for resources may also have affected the kinds of substrates birds used.

The presence or absence of potential competitors is often assumed to affect spatial variation in the use of resources by terrestrial birds. Keast (1976) drew attention to changes in the foraging behavior of some Australian birds in Tasmania and southwestern Australia and suggested this was in response to the absence of certain competitors. For instance, Brown Thornbills (*Acanthiza pusilla*) appeared to forage higher in the forest canopy in places where the canopy foraging Striated (*A. lineata*) and Little (*A. nana*) thornbills were absent (Keast 1976). Studies that I have recently completed suggest that changes in the foraging behavior of Brown Thornbills in the absence of other acanthizids result from differences in the spatial and temporal patterns of resource abundance, including kinds of prey and foraging substrates, rather than a release from competitive pressures (Recher et al. 1987, Recher unpubl. data).

CONCLUDING REMARKS

Papers in this section addressed questions of spatial and temporal variability. They demonstrated that it was potentially misleading to characterize a species as either a foraging specialist or generalist without defining the resources being used, describing the temporal and spatial scale of the measurements made, and presenting some measure of the degree of individual variation within the population studied. Evolutionary and phylogenetic relationships also need to be considered, along with resource abundances, the physiological requirements of individuals, reproductive condition, and possible interactions with other individuals or species.

Greater understanding of foraging ecology requires a redirection of research. There is a need for long-term studies on temporal changes of foraging patterns in response to changes in resource abundance and the numbers and kinds of birds occurring together. Resources should be defined more broadly than food alone and a distinction made between abundance and availability (see Hutto, this volume), and the availability of various foraging substrates needs to be related to avian foraging patterns and community organization. More work on the ecology of birds during migration or away from their breeding and wintering areas is required. This is particularly important for a balanced approach to the conservation and management of species.

Manipulative studies that change the abundance and distribution of resources will be increasingly important in defining factors affecting avian foraging ecology. However, comparative studies that use natural experiments and contrast behaviors of the same species at different places or times will continue to make a significant contribution in describing patterns of variation in foraging ecology. Regardless of the approach taken, it is necessary to document the existence of individual variation. To what extent do patterns result from genotypic differences, response to differences in resource distribution, learning, or social interactions?

The ways birds respond to changes in the kinds or abundances of prey and their own physiological needs require more attention. Immediate and often short-term adjustments in foraging behavior are probably of greater significance to the survival of individuals and their reproductive success than the possibility of competition for resources between individuals of different species. As such there is a need to re-evaluate the reasons co-existing species differ in their use of resources. Selection is at least as likely to be for efficient foraging with the necessary flexibility to adjust to short-term changes in resources as it is to avoid competition.

WHEN ARE BIRDS DIETARILY SPECIALIZED? DISTINGUISHING ECOLOGICAL FROM EVOLUTIONARY APPROACHES

Thomas W. Sherry

Abstract. Definitions of degree of dietary specialization are motivated by theories of the niche, optimal foraging, predator-prey theory, ecomorphology, comparative morphology, and phylogeny. These methods fall into two fundamentally different, but complementary approaches. The first is ecological (or tactical), emphasizing short-term responses of individual organisms to resource availability and abundance, given phylogenetic constraints. The second approach is evolutionary (or strategic), emphasizing longer-term, genetically based constraints and adjustments of consumers (via adaptive radiation) to patterns in the predictability of resources in both space and time.

Studies of diet specialization have emphasized individuals' tactical approaches to the exclusion of population strategic ones, and have often failed to distinguish between the two approaches. I discuss this distinction in terms of the kinds of information needed to characterize specialists and generalists. I argue that strategic specialists have stereotyped rather than narrow breadth diets, and I discuss the relationships between the two dietary dichotomies of monophagy-polyphagy and stereotypy-opportunism. Three examples illustrate the distinction between strategic and tactical approaches, and problems of failing to separate the two: (1) Cocos Flycatchers (*Nesotriccus ridgwayi,* Tyrannidae) are ecological generalists, but evolutionary specialists; (2) Neotropical flycatchers are specialized dietarily compared with temperate species using a strategic approach (appropriate for this comparison), but the two groups do not differ using the more traditional tactical approach; and (3) particular species of Neotropical frugivores are specialists by strategic definitions, but generalists by tactical ones, a distinction that resolves unnecessary controversy in the literature.

Key Words: Diet breadth, foraging behavior, generalist, niche breadth, opportunism, resource, specialist, stereotypy, Tyrannidae.

Questions concerning ecological specialization continue to fascinate and challenge biologists. Ecologists, for example, ask whether species-rich (especially tropical) communities have relatively specialized species, whether ecological specialists are better competitors than generalists, or whether specialization favors exploitation efficiency. Evolutionary biologists ask questions such as whether evolutionarily derived species are specialized compared with ancestral ones, whether adaptive radiation involves ecological specialization, whether specialization tends to increase over time in fossil lineages, or whether specialists are more extinction-prone than generalists. But what is a specialist? The literature contains a morass of definitions, conceptual approaches, and methods, with no concensus on their applicability.

One prevalent notion is that specialists select a relatively narrow range of foods. By this intuitive notion, some animals are unambiguously specialists: Pandas (*Ailuropoda melanoleuca*) on bamboo, Everglade Kites (*Rostrhamus sociabilis*) on snails. Each has specialized morphology and behavior with which to eat a consistently narrow range of food types throughout the year. However, most species are not so clearly specialized, thus necessitating operational methods to quantify degree of specialization (i.e., position along a hypothetical continuum from specialization to generalization). This need becomes more apparent when we consider the successful evolutionary radiation and abundances of terrestrial birds, particularly Passeriformes, the overwhelming majority of which are relatively generalized insectivores, frugivores, nectarivores, or granivores (Karr 1971, Morton 1973, Rotenberry 1980a).

Categorizing species as specialist or generalist can be ambiguous. The Cocos Flycatcher (*Nesotriccus ridgwayi*), for example, is a specialist or generalist depending on the frame of reference and the methods used to quantify specialization (T. W. Sherry 1985), as elaborated below. Wiens and Rotenberry (1979) and Rotenberry (1980a) equated opportunism with the absence of specialization, and noted that degree of opportunism was ambiguous for some species: The Grasshopper Sparrow (*Ammodramus savannarum*), for example, was opportunistic by the criteria of broad individual diet niches and high overlap with other species, but was relatively specialized based on a relatively narrow population niche breadth and little annual variation. Thus, compared with sympatric species in the shrub-steppe environment, Grasshopper Sparrows were relatively specialized; yet, if one considers that all these shrub-steppe species were at least partially migratory and that their diets varied more seasonally and geographically within than among species (Wiens and Rotenberry 1979), all were opportunists adapted to variable environments. Fox and Morrow (1981) also noted that herbivorous insects feeding on *Eucalyptus* were specialists or generalists depending on scale.

Resolutions to the above ambiguities depend on how we conceptualize and quantify "specialization," and the temporal and spatial scales of concern. Terminology about specialization and related phenomena (stereotypy vs. opportunism, monophagy vs. polyphagy) are used differently by biologists, leading to confusion. In this review I examine how conceptualizations about nature motivate operational methods, and conversely how methods clarify the (often unstated) assumptions of particular investigators.

Ellis et al. (1976) distinguished between "tactical" (ecological) and "strategic" (evolutionary) approaches to diet selectivity. They acknowledged that tactical approaches had received more attention, and then explicitly used a tactical approach. If anything, the emphasis on tactical approaches is greater today than when they wrote. Whatever the actual imbalance, however, the two approaches generate fundamentally different, and sometimes contradictory results, and in failing to distinguish between them one can draw incorrect conclusions. Thus, my second purpose is to distinguish these approaches, and show with explicit examples the dangers of confusing the two.

Although I concentrate on dietary specialization in this review, I ask how dietary specialization is related to an organism's phenotype, especially foraging behavior and morphology, to make the distinction between tactical and strategic approaches. I will focus on methods to quantify specialization, considering examples outside of the bird literature either where they would be useful to avian biologists or where examples are particularly clear.

TACTICAL APPROACHES TO DIETARY SPECIALIZATION

Here I trace some history of theory motivating methods to quantify dietary specialization to understand the necessity of the methods; I then briefly describe and in some cases evaluate these methods with examples wherever possible.

NICHE THEORY

Niche theory has arguably provided the greatest motivation for measures of specialization. Although ecological concepts of the niche were first developed in the early 1900s, G. E. Hutchinson, R. H. MacArthur, and others in the late 1950s and early 1960s formulated an operationally powerful concept (the n-dimensional hypervolume), which had an enormous impact on theoretical and empirical ecological studies (Pianka 1981, Ehrlich and Roughgarden 1987). This variety of studies has coalesced into a school of population ecology, based largely on a concept of the niche centering on competition for resources within a one- (or n-) dimensional space within which each species occupies its own resource space, often represented by a bell-shaped resource-use probability distribution (e.g., McNaughton and Wolf 1970, Vandermeer 1972, MacArthur 1972, Pianka 1981).

Niche breadth in these models is an important parameter used to describe the size of the individual species' niche, or the range of resources or resource states used by that species, and has been related to the number of species within a community. MacArthur (1972), for example, developed a geometric model for species diversity within a community based on the average niche breadth, niche overlap, and total resource spectrum. Many empirical studies have been undertaken to quantify niche parameters, especially niche breadth, based largely on food sizes or types, or on surrogates for food, such as foraging behavior and morphology. All of these quantities are considered substitutes for fitness, the quantity defining the success of a species within Hutchinson's (1957) original niche model.

In this review I divide niche breadth measures into those that are applicable to any consumer population versus those for an entire community. "Single-species measures" may be subdivided depending on whether relative proportions of food categories in the diet or availabilities of different resource categories are used in the calculations. "Multi-species measures," including multivariate statistical procedures, necessitate study of many species simultaneously, and are particularly useful to test hypotheses about multiple communities of organisms.

Single-species measures

Niche theorists view degree of ecological specialization as the inverse of niche breadth (Colwell and Futuyma 1971, Morse 1971b, Hurlbert 1978, E. P. Smith 1982, Holm 1985) and have developed measures to quantify them. The simplest dietary diversity index is the number (or richness) of food taxa in the diet (Herrera 1976, Wheelwright and Orians 1982, Wheelwright 1983, Moermond and Denslow 1985). This measure has the disadvantages of equating frequently with infrequently used foods, and of equating different items, such as adult and larval insects, which may not be functionally equivalent from the perspective of the predator (see below).

A second group of diversity indices incorporates the relative proportion, p_i, of resources or resource categories, where $\Sigma p_i = 1$, and i is one of the r resource states. Among the most widely used are those attributed to Simpson, Levins, Shannon-Wiener, and Brillouin (see Pielou 1975 for a general discussion of their derivation and use). Simpson's index, lambda = Σp_i^2, measures the concentration or dominance of observations (food types, for example) in one or a few categories, and forms the basis of several measures (Pielou 1975). Levins (1968) was the first to use such a measure (B) in a form sometimes (but not always) standardized for comparative purposes by the number (S) of resource states (i.e., $B = 1/[S \cdot \Sigma p_i^2]$). This index has desirable characteristics including simplicity and ease of calculation (e.g., Rotenberry 1980a). Another index is that of Brillouin, given by $H = (1/N) \cdot \log[N!/(n_1! \cdot n_2! \cdot \ldots \cdot n_i!)]$, where there are $n_1, n_2, \ldots, n_i$ prey items in each of i categories, with N total prey items in the collection. It measures the diversity or breadth of a complete collection of items, and the Shannon-Wiener index, $H' = -\Sigma (p_i \cdot \log[p_i])$, measures the diversity of a sample of items, providing an estimate of the unknown actual diversity of an entire population (Pielou 1975). All of these measures equate narrow breadth (specialization) with the use of few resource states, the opposite of information (or entropy) as measured by the information-theory indices. The latter two have the advantage of generalizeability to hierarchical measures of diversity, that is, the diversity weighted by the taxonomic similarity of items within the collection (Pielou 1975).

All these indices based on relative proportions of entities in different categories, whether used to quantify dietary diversity or "species diversity" in a community, confound two quantities, the total number of kinds (richness) of entities and the equitability (= evenness) of their use. The maximum evenness of resource use occurs when all resource categories are used equally, that is, $p_i = 1/r$, and evenness is often measured by the ratio of the actual diversity index to the maximum possible value (see Peet 1974 and Pielou 1975 for further discussion). Herrera (1976) developed a trophic diversity index for use with presence-absence data that reflects the richness rather than evenness component of diet, but is significantly correlated with the Shannon-Wiener diversity index.

None of the above indices provides realistic estimates of diversity when one is sampling a large number of species or resource categories, for which the total richness of entities in the population is unknown (Pielou 1975). For this situation, Pielou developed an asymptotic method in which diversity (calculated with the Brillouin index) is plotted as a function of sample size: With an adequate sample size one obtains a curve increasing from zero but at a decreasing rate towards an asymptote when new samples add little new information about the population. Diversity of the entire collection with unknown S, or number of species, is then estimated from the asymptotic (plateau) part of that curve. Hurtubia (1973) recognized that stomach contents of animals are usually incomplete, and nonrandom, samples of the larger population of potential prey types, and applied Pielou's method to a study of lizard diets. I (T. W. Sherry 1984, 1985) applied it to stomach contents of tropical flycatchers to show that their dietary diversity varied from zero in one flycatcher species, in which every stomach contained essentially one and the same prey type, to a large (unknown) dietary diversity for which available stomachs had too diverse prey types to estimate a population asymptote. This method merits further use with stomach-content data.

Organisms may specialize because few resources are available in a particular environment, or because characteristics of the organism constrain diets. One definition of specialization is thus "a deviation from random feeding by the animal as imposed by its own attributes rather than by the environment" (Holm 1985, after Hengeveld 1980). To measure this one can compare the diet with available resources using simple statistical procedures such as chi-square tests. Such comparisons overwhelmingly show that organisms, including birds, are specialists (e.g., Holmes 1966, Hespenheide 1975b, Morse 1976a, Abbott et al. 1977, Toft 1980, Steenhof and Kochert 1988; see also discussion of electivity and selective predation studies, below). Toft (1980) and T. W. Sherry (1985) documented cases in which an organism consumed prey types in proportions indistinguishable from those deemed available. In the latter case, Cocos Flycatchers ate prey in proportions indistinguishable from those sampled with beating nets in the leafy vegetation where this species feeds. Sampling of available resources in this case did not include tree-trunk insects, fruit, or nectar, foods that Cocos Finches consume in the same habitats (Werner and Sherry 1987), so they are thus available in an evolutionary sense to the flycatcher. Clearly it is operationally difficult to characterize prey effectively "available" from the predator's perspective, leading many (e.g., Ellis et al. 1976, Wiens 1984b, Hutto, this volume) to distinguish between food availability and abundance.

Johnson (1980) used contrived examples to show how inclusion or exclusion of particular resource categories can greatly influence one's conclusions. He developed a method based on differences between ranks of resources used and available, and he developed statistical tests for his method. Craig (1987) presented data on rank this approach, perhaps because of the large effort required to estimate all necessary parameters. Heuristic models, by contrast, attempt to characterize one or a few key processes such as diet selectivity as a function of food abundance (e.g., MacArthur and Pianka 1966).

Experimental approaches

Biologists have used experimental approaches to determine how individual consumers make prey choices over short time periods relative to the animals' generation time. Although such methods may not have been intended to characterize degree of specialization, I mention them here because some have been used to study diet selectivity.

One of the most obvious ways to study selectivity is to present animals with different food types under laboratory or field conditions. This approach has been used to study prey-handling abilities and mimicry (e.g., Smith 1975; Sherry 1982; Chai 1986, 1987), fruit acquisition and handling (review in Moermond and Denslow 1985), and seed-size selection in finches (Hespenheide 1966, Willson 1971, Grant 1986, Benkman 1987a). Chai's (1986, 1987) work, for example, showed that the behavior of an evolutionarily specialized butterfly predator, the Rufous-tailed Jacamar (*Galbula ruficauda*), led to different conclusions about mimicry than studies with nonspecialists that rarely consume butterflies. This experimental approach can elucidate factors involved in the evolution (or coevolution) of both prey and predator characteristics (see also Holmes, this volume), as well as identify tactical responses and capabilities of the predators.

In laboratory experiments behavior must be studied under conditions equivalent to those encountered in the field. In a study of rictal bristle function in flycatchers, for example, Lederer (1972) commendably tested functional morphological hypotheses with an experimental procedure, but performed the experiments under lighting conditions (not adequately specified) bright enough to allow high-speed photography; neither lighting conditions nor prey type (flesh flies, *Sarcophaga*) may have been appropriate to the hypothesis, since those flycatchers with the best-developed rictal bristles are tropical species such as *Terenotriccus, Myiobius,* and *Onychorhynchus,* all of which acrobatically pursue evasive insects (few of which are Diptera) in often poorly and variably lighted tropical rainforest understory (Sherry 1982, 1983).

A widespread approach to diet selectivity and electivity looks at how predators preferentially use or ignore specific food types, usually as a function of food abundance or other characteristics. Tinbergen (1960) showed in a classic paper that titmice (*Parus* spp.) consumption rate varied sigmoidally with caterpillar abundance, and proposed the concept of "specific search images" to explain his results. Ivlev (1961) conducted experimental laboratory studies of fishes, and coined the term electivity for their selecting particular prey in proportions not equal to availability. Subsequent studies have distinguished alternative predatory responses to changing resource abundance, including "switching" (Murdoch 1969, Murdoch and Oaten 1975) and functional responses (Holling 1959b). A popular quantitative approach to questions of electivity is to use indices designed to determine prey preferences when all prey are equally available: Essentially these indices are vectors of m different prey preferences (or aversions) for m prey types under consideration in a particular situation (reviewed by Chesson 1978, 1983). Statistical tests of the null hypothesis that a particular predator's electivities are all zero have also been devised and discussed (Neu et al. 1974, Johnson 1980, Lechowicz 1982). Most electivity studies are done in the lab to control prey types and abundances (e.g., Freed 1980, Chesson 1983, Annett and Pierotti 1984). Steenhof and Kochert (1987) quantified electivity for particular prey types of raptors in the field, and showed that their diets responded most to changes in preferred prey, as predicted by prevailing optimal diet methods, discussed next.

Optimal foraging and optimal diet studies

The voluminous literature on optimal foraging has been extensively reviewed (Krebs et al. 1983, Krebs and McCleery 1984, Stephens and Krebs 1986, Stephens, this volume); here I mention only a few findings relevant to dietary specialization. The first optimal foraging models predicted explicitly that diet specialization should vary positively with food abundance (Emlen 1966, MacArthur and Pianka 1966), and a variety of empirical studies essentially verified this prediction, at least in a qualitative sense (Krebs et al. 1983). More recent models have been developed to address such complicating matters as patch selection, learning and prey-recognition problems, conflicting demands (such as feeding and avoiding predators), and stochastic variation in resources (Krebs and McCleery 1984, Stephens, this volume). These more recent models have also tended to make fewer explicit predictions about diet breadth *per se* than the original models. The main point, however, is that most optimal foraging and optimal diet theories and tests are concerned with short-term (less than generation time) adjustments of behavior of individuals to changing environments. Such approaches explicitly take the phenotype as given, and ask how behavior changes with ecological circumstances given the phenotypic constraints (Krebs et al. 1983, Krebs and McCleery 1984, Stephens and Krebs 1986), rather than asking how the phenotype may have been shaped by ecological circumstances over evolutionary time. Thus optimal foraging or diet approaches have tended to be tactical rather than strategic.

EVOLUTIONARY APPROACHES TO DIETARY SPECIALIZATION

ADAPTATION AS SPECIALIZATION

Evolutionary biologists have often equated specialization with adaptation, often viewed as a "perfecting" force (e.g., Leigh 1971, Holm 1985). At the levels of organization of communities, biomes, or biogeographic realms, adaptive radiation into present-day faunas results from all processes leading to species specialized on nonidentical subsets of the total resources in the environment. Among avian biologists, Leisler (1980), Grant (1986), and Craig (1987) illustrate the use of diverse behavioral, genetic, and ecological methods to examine the evolutionary diversification of related species. The disadvantage of defining specialization simply as adaptation or adaptive radiation is its comprehensiveness: All species are automatically specialized in relation to other species, with no explicit notion about degrees of specialization. Studies of coevolution have added a related concept of specialization, namely the evolutionary interdependence of two species (or more, in the case of diffuse coevolution; Janzen 1980b).

FUNCTIONAL STUDIES

Comparative method

One must study the function of adaptations before asking questions about degrees of evolutionary specialization dependent upon those adaptations. Various methods have been developed to study adaptations, based on comparing different species' phenotypic characteristics (e.g., morphology, anatomy, physiology, behavior) with their ecological ones, such as habitat, feeding behavior, and diet. The "comparative method," perhaps the most flexible and widely used approach to adaptation, compares different adaptations with different ecological circumstances of two or more species to deduce the function of relevant traits, and is most powerful when it deals with instances of convergent evolution (James 1982, Futuyma 1986). Phylogenetic information is required to assess the possibility of convergence, and both experiments and analyses of fitness are necessary to test hypotheses about function (Futuyma 1986). VanderWall and Balda (1981), for example, documented in four corvid species a graded series of behavioral, morphological, and life-history adaptations for exploiting conifer seeds in mountains of the southwestern United States. The four species, ranked in decreasing order of evolutionary specialization on pinyon pine seeds (Clark's Nutcracker [*Nucifraga columbiana*], Pinon Jay [*Gymnorhinus cyanocephalus*], Steller's Jay [*Cyanocitta stelleri*], and Scrub Jay [*Aphelocoma coerulescens*]), showed corresponding reductions in seed selectivity, seed transport volume and distance, flight speed, cache size, bill length, development of seed-carrying structures, and ecological dependence on pine seeds both as adults over winter and as nestlings. The often implicit assumption that all phenotypic characteristics result from natural selection acting directly on particular traits, an operational approach referred to as the "adaptationist programme," has flawed some comparative studies (Gould and Lewontin 1979, Futuyma 1986).

Ecomorphology studies

Associated with niche conceptualizations of communities, ecomorphological studies often use multivariate statistics to explore the meaning of morphological characteristics. A basic premise is that by averaging evolutionary forces over long time periods, morphology provides the best measures of the ecological interactions of species (Karr and James 1975, Ricklefs and Travis 1980). Canonical correlation analysis (e.g., Karr and James 1975, Leisler and Winkler 1985) and correspondence analysis (Miles and Ricklefs 1984) are just two methods used to examine correspondences of morphological and ecological data. Foci of ecomorphological studies have varied (James 1982), but include the correspondence of morphology with behavior and ecology (e.g., Sherry 1982, Leisler and Winkler 1983, Miles and Ricklefs 1984), and "species packing" (Findley 1976, Karr and James 1975, Gatz 1980, Ricklefs and Travis 1980). Species packing should increase with either increased niche overlap or narrower niches (MacArthur 1972), but too few studies have looked at both overlap and packing to get at niche breadth.

Too few studies have paid attention to the function or efficiency of phenotypic characters in comparative studies (for nice recent examples see Greene 1982; Liem and Kaufman 1984; Moermond and Denslow 1985; Benkman 1987a, b; Moermond, this volume). In an elegant experimental study, Laverty and Plowright (1988) showed that naive individuals of a specialized bumblebee species (*Bombus consobrinus*) feed more efficiently on the preferred flower type (Monkshood, *Aconitum* spp.) than do either of two generalist congeners.

Common-garden methods

Any laboratory or field study in which different individuals or populations are exposed, usually experimentally, to the same conditions in one or more environments is a common-garden method, and can potentially provide information on relative performance, ecological efficiency, and fitness. In transplant experiments, for example, James (1983) showed that the environment contributes significantly to size and shape variation of nestling Red-winged Blackbirds (*Agelaius phoeniceus*). Sherry and McDade (1982) showed that a small tyrannid "sit-and-wait" predator (*Attila*) had significantly longer handling times for acridid-tettigoniid Orthoptera than a larger puffbird (*Monasa*) feeding on the same sizes and types of prey. Garbutt and Zangerl (1983) described a general method to analyze results from a common-garden experiment that provides a measure of niche breadth and performance efficiency. Their method can use any measure of performance (such as reproduction, growth, feeding efficiency, or other components of fitness), and could be used with animals.

Comparative psychology

Because species diverge in learning or behavioral traits, comparative psychology provides another class of evolutionary studies with relevance to diet specialization. For example, Greenberg (this volume) has documented differing degrees of "neophobia" among closely related birds. These apparently genetically based differences in fear of approaching novel microhabitats, based on studies of hand-reared individuals, influence the range of microhabitats (and thus diet breadth) of these species in the wild. Neophilia, the complementary behavior, seems particularly well developed in the Cocos Finch (*Pinaroloxias inornata*), living in an almost predator-free environment. Its diet is extremely broad, and individuals appear capable, at least as juveniles, of observing and learning from a diverse array of conspecific and other animals about how and where to feed (Werner and Sherry 1987; see also McKean, this volume). Juvenile Cocos Finches in particular appear to exhibit exploratory behavior towards potential prey objects and substrates, and to observe closely a variety of potential tutors (T. K. Werner and T. W. Sherry, pers. obs.).

A variety of other behavioral attributes can influence diet breadth and stereotypy. The ability of a species to learn from (and to teach) other animals, i.e., culture, is ultimately genetically determined (Bonner 1980) and can influence feeding behavior, as in the case of tool-use (Morse 1980a) and aggregative feeding and nesting behavior (Rubenstein et al. 1977; C. R. Brown 1986, 1988), the efficiency of locating or handling prey types (Waltz 1987), array of foods used (Rubenstein et al. 1977, Giraldeau 1984), acquisition of food aversions (Daly et al. 1982, Mason et al. 1984, Shettleworth 1984), and cooperative hunting (Bednarz 1988). Ability to memorize characteristics of an environment, such as where Clark's Nutcrackers have cached seeds (Kamil and Balda 1985), should facilitate specialization on the seeds. All of these behaviors vary among species, and can influence the range of food types eaten. Biologists have barely begun to explore these influences, let alone genetic constraints involved.

DIETARY INDICES

Dietary homogeneity

A predictable environment is a *sine qua non* of specialized evolutionary relationships such as complex adaptations, obligate mutualism, and other forms of coevolution. In a classic study of ant-plant coevolution, for example, Janzen (1966) stated explicitly the importance of environmental predictability allowing mutualism to evolve and persist in certain environments. Morse (1971b) recognized the importance of stereotypy versus opportunism of resource use patterns in birds. Colwell (1973) specified how certain strategies of species coexistence are favored by the relative predictability of tropical compared with temperate environments. Southwood (1977) noted that individuals, populations, and species should feed more flexibly in disturbed than undisturbed environments. Wiens and Rotenberry (1979) characterized all their shrub-steppe bird species as opportunistic, stressing the unpredictability of these environments from the perspective of birds (see also Futuyma 1986, and literature cited, and Howe and Estabrook 1977). Both empirical and theoretical studies concur that environmental predictability favors the evolution of individual feeding specializations (Werner and Sherry 1987). Glasser (1982, 1984) developed from niche theory a model of trophic specialization based explicitly on resource predictability. It follows that environmental predictability allows some organisms to evolve relatively obligate dependence on resources or on other organisms, and thus to evolve more efficient, elaborate, or complex adaptations appropriate for those specific, predictable environmental circumstances. Resource predictability is probably a function of abundance.

One may test hypotheses about the evolution of dietary specializations by assuming that resource variability over short time periods today is proportional to what the organism has experienced evolutionarily, and then measuring this variability. Direct measures of resource variability have been made in several cases, such as arthropod abundances in rainforest understory versus other tropical habitats (Sherry 1984). Variability in resource types among individuals of a population provides a surrogate measure of resource predictability from the organism's perspective. I (Sherry 1984) thus sampled diets (using stomach contents) of tropical flycatchers across a broad geographic area in Caribbean Costa Rica during the period of year (October–December) inferred to be most food-limiting to these birds, over a 3-year period. I calculated "population dietary heterogeneity" (PDH) from a matrix of prey taxa by stomachs using the G-statistic (Sokal and Rohlf 1981) divided by degrees of freedom. The result was that several of these tropical insectivorous species had extraordinarily homogeneous stomach contents, expected in relatively constant tropical environments (see below). Steenhof and Kochert (1988) used this index to show that diets were most homogeneous within years in the raptor species with the most consistent diets over an 11-year period (encompassing dramatic changes in prey abundance). Werner (1988) also used it to quantify effects on foraging behavior of foraging location, individual bird, time of day, season, and error variation in a Cocos Finch population. Kincaid and Cameron (1982) used a multivariate coefficient of variation in diets, and Roughgarden (1974) partitioned niche width into two components, between-phenotypes (a high value indicating considerable variation among individuals) and within-phenotypes. Other authors have examined dietary correspondence with morphology in species with continuous (Grant 1986) or polymorphic (Smith 1987) morphological variation.

When diets vary among individual animals (e.g., Smith 1987, Werner and Sherry 1987), inferences about resource variation from dietary variation depend on how individual animals feed over long time periods. This is because dietary variability can arise either because environments vary (e.g., Wiens and Rotenberry 1979, Sherry 1984) or because individuals vary independently of each other within constant environments. Foraging behavior of Cocos Finches varied dramatically among individuals within a constant oceanic island environment (Werner and Sherry 1987, Werner 1988), for example, but the foraging consistency of marked individuals year-round indicated that they perceived the environment to be predictable. Conversely, a short-term study might document a misleading degree of dietary homogeneity for the actual variability of the environment, if observations spanned a short time period (e.g., a season) within which all individuals opportunistically fed on the same relatively profitable food (e.g., Fenton and Thomas 1980). Thus studies of dietary homogeneity must span multiple seasons and multiple years to indicate different degrees of variability in long-lived vertebrates such as birds.

Unique food types

Comparatively extreme species along a particular phylogenetic pathway may be identified by relatively unique phenotypic, foraging behavioral, or dietary characteristics. Fitzpatrick (1985) referred to particular tyrannid species that are both highly stereotyped in terms of foraging behavior and represent extreme morphological development in a particular lineage (such as the genus *Todirostrum*). Leisler (1980) spoke of the Lesser Whitethroat (*Sylvia curruca*) as a specialist in this sense (see also Toft 1985). I (Sherry 1982, 1984) showed that a few flycatcher species ate peculiar foods eaten by few other species (e.g., some *Todirostrum* spp. ate relatively alert and agile muscoid Diptera that few other birds appear capable of capturing). Meylan (1988) identified hawksbill turtles (*Eretmochelys imbricata*) as sponge specialists, in part based on how few other vertebrates eat sponges regularly, and in part on the consistency of their diets over much of their geographic range. Multivariate statistical procedures should be appropriate to quantify extreme dietary characteristics (e.g., using the deviation in morphological space from a particular species to the centroid for all species; but see "Multi-species measures" above for dangers inherent in this approach), or distance (in some evolutionary units) from a hypothetical ancestor for the group as a whole. To my knowledge, no one has yet developed quantitative indices for degree of "extremeness," as reflected in dietary or morphological characters.

A special case of specialization on unique prey items, suggested by H. A. Hespenheide (pers. comm.), is specialization on prey types that are distasteful or repugnant to most predators. Some predators have evolved special abilities to overcome this, such as woodpeckers that prey on ants (that contain formic acid), Nunbirds (*Monasa morphoeus*) that prey on aromatic and aposematic stinkbugs (Pentatomidae; Sherry and McDade 1982), orioles (*Icterus* spp.) and the Black-headed Grosbeak (*Pheucticus melanocephalus*) that select palatable parts of unpalatable monarch butterflies (*Danaus plexippus*) (Calvert et al. 1979), and bee-eaters (Meropidae) that devenom bees prior to ingestion (Fry 1969).

INCLUSIVE-NICHE MODEL

A common pattern within guilds of species (Root 1967) is for one species to have its fundamental niche nested within that of another species, and for the socially dominant—usually larger—species to have the smaller niche (Miller 1967, Case and Gilpin 1974, Morse 1974b, Sherry and Holmes 1988, Sherry 1979, Colwell and Fuentes 1974, Rosenzweig 1985). Thus, the dominant species is specialized relative to the other in the range of environmental circumstances tolerated. In the case of diets, we expect dominant species to tolerate a narrower range of food types or show less feeding flexibility and opportunism than subordinates (Morse 1974b, Sherry 1979). Insofar as this nested pattern of niches involves the fundamental, rather than realized, niche, this pattern involves evolutionary responses of one of the organisms to the other (or reciprocal evolutionary responses), but the causes and consequences of such patterns remain unclear.

SUMMARY OF METHODS TO QUANTIFY DIETARY SPECIALIZATION

The foregoing review indicates diverse conceptual approaches to quantifying dietary specialization. Some of this diversity results from the use of different time scales: some indices involve short-term (behavioral, cognitive) responses of organisms; others involve ecological time-periods; yet others involve evolutionary time-scales. These different time scales also involve different levels of organization (e.g., tactical individual vs. strategic population or species approaches) and Sherman (1988) argued that behavioral questions often have different answers at different levels of organization. Approaches to dietary specialization are thus not mutually exclusive, which probably explains why none has emerged as the best under all circumstances.

ECOLOGICAL VERSUS EVOLUTIONARY DIETARY SPECIALIZATION

The preceeding review considered intentions as well as limitations of particular paradigms and studies. In this section I evaluate how these methods quantify either tactical or strategic aspects of dietary specialization, but rarely both.

TACTICAL APPROACHES

Studies of dietary specialization motivated by niche theory have generally characterized foraging behavior or diet by either the range of resources used by a species, or by the degree to which resource use matches availability. Operationally, the procedure is to gather data on some individuals within a population, and pool the data into a population-, or species-specific characterization. These characterizations are then used to study the entire niche space of many species, the packing of species into this space, the overlap of individuals or populations with respect to resource use, and related niche parameters. None of these measures or procedures contains information about the variability of resources experienced by populations at present, let alone over past time periods. Moreover, much of niche theory was developed from the Lotka-Volterra population growth equations (e.g., MacArthur 1972), which describe ecological-scale processes in response to either resource abundance or the competitive influences of other species.

Ignoring differences among individuals provides no perspective about stereotypy and opportunism, information needed in evolutionary approaches to specialization, and pooling data can lead to statistical problems as well (Hurlbert 1984). I emphasize that these sources of individual variation in dietary and other parameters are not only useful statistically, but are critical to strategic questions about populations and communities.

Optimal diet and optimal foraging studies have also tended to examine tactical questions, often by taking the phenotype as a given, thus defining away the question of how the phenotype came to be the way it is. Optimality studies also tend to examine short-term responses, rather than long-term evolutionary responses of organisms to variability and other patterns of resource abundance. Several authors have explicitly recognized this distinction between tactical and strategic approaches to diets (e.g., Ellis et al. 1976, Glasser 1984, Krebs et al. 1983, Holm 1985, Stephens and Krebs 1986), but have usually taken a tactical approach.

Evolutionary questions need not fall outside the domain of optimality studies. The theory of evolutionarily stable strategies models the conditions for evolutionary persistence of alternative strategies (e.g., of resource use or mating tactics). Glasser's (1982, 1984) studies of trophic strategies incorporate resource predictability as well as abundance, thus incorporating an evolutionarily critical parameter. A variety of design and engineering approaches to the analysis of adaptations (e.g., Leigh 1971) are essentially optimality models as well.

STRATEGIC APPROACHES

Strategic approaches to dietary specialization begin with the recognition that some taxa are more specialized than others. Howe and Estabrook (1977), for example, noted that some frugivores are highly specialized in depending on one or a few species of fruiting plants, whereas other frugivores are more opportunistic. Fitzpatrick (1980, 1985) recognized degrees of evolutionary specialization within the adaptive radiation of the Tyrannidae (see also Green 1981). Specialized species often show extreme structures along some evolutionary pathway, for example, and show the greatest degree of foraging stereotypy. In addition, the existence of guilds of organisms with nested ranges of resources, habitats, or other fundamental niche axes—the so-called "inclusive niche" pattern (see above)—means that, even within guilds different species are differentially specialized.

Some of the most important components of evolutionary specialization are illustrated by two hypothetical species, one of which is more specialized than the other (Fig. 1). One intuitive notion is that some combination of phenotypic characteristics confers greater efficiency on specialists than generalists, as illustrated by either (a) the higher maximum benefit:cost ratio of feeding on preferred food items, or (b) the higher benefit:cost ratios on nonpreferred items (Fig. 1b). The Darwin's finches (Geospizinae) illustrate case (a), in which deeper-billed, large-seed specialists are more efficient than shallower-beaked species at handling the most profitable, larger seeds (Grant 1986). Liem (1984) illustrates case (b) with cichlid fishes, in which specialist morphs have greater efficiency than generalists on least preferred foods. Phenotypic characteristics relevant to specialization include anatomical, morphological, behavioral (and psychological; see above), or physiological (for example, Toft 1985) traits. Comparative studies are an important way to compare performance of different phenotypes, populations, or species, and assess the extent of phenotypic diet constraints (Moermond, this volume). The extent of coevolution of the predator and prey also affects specialization, in that a more obligate relationship between the consumer and consumed depends on the predictability of resources and often involves increased efficiency of trophic exploitation by the consumer.

Ultimately, evolutionary notions of specialization must involve genetic and phylogenetic studies, if only to establish the evolutionary units, heritability of feeding behaviors (see Arnold [1981] and Gray [1981]) and the geographic scales on which selection is acting.

Behavioral stereotypy (vs. opportunism) is also relevant. Stereotypy is permitted when critical resources for growth, survival, or reproduction have been predictable in the history of a population. Foraging and dietary stereotypy are thus better measures of evolutionary specialization than tactical measures, most of which are based

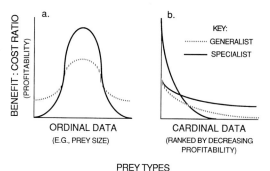

FIGURE 1. Benefit:cost ratios, or prey type profitability, for two hypothetical predators, a generalist and a specialist, when prey are arranged (a) ordinally (e.g., by prey size); or (b) cardinally, by some category of decreasing ranks of benefit:cost ratio. In part (b) the hypothetical specialist could be more efficient than the generalist on higher-ranked food (specialist with highest maximum prey profitability) or on lower-ranked food (specialist with lowest maximum profitability). See text for explanation and examples.

on resource abundance. The dietary heterogeneity index (PDH), discussed previously, is useful for calculating directly ecological opportunism.

Wiens and Rotenberry (1979) defined opportunism as the behaviorally flexible use of abundant and variable resources, and argued that birds breeding in scrub-steppe environments are all relatively opportunistic. Klopfer (1967) conducted laboratory tests of the idea (Klopfer and MacArthur 1960) that tropical birds are more stereotyped in foliage preferences and movement patterns than temperate birds, although his results were inconclusive. As predicted by ecological theory, species living in depauperate island environments have tended to forage in less stereotyped ways than mainland species (e.g., Morse 1980a, Feinsinger and Swarm 1982, Whitaker and Tomich 1983), but exceptions are known. Feinsinger et al. (1988) found mixed support for the relationship between feeding opportunism in hummingbirds and disturbance in Costa Rican forests.

How is feeding opportunism related, if at all, to dietary niche breadth? Morse (1971b, 1980a) proposed that the stereotypy vs. opportunism (= plasticity in his usage) dichotomy is independent of the specialization vs. generalization dichotomy, so that birds can be stereotyped and specialized, stereotyped and generalized, opportunistic and specialized, or opportunistic and generalized (Fig. 2a). Martin and Karr (this volume) found empirical support in migratory warblers for Morse's view. They found that foraging opportunism, determined by seasonal variation

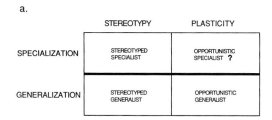

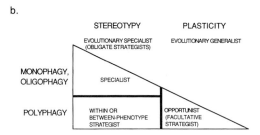

FIGURE 2. Two alternative schemes for the relationships between diet breadth (ecological specialization versus generalization) and diet predictability (stereotypy versus plasticity): (a) Morse's (1971b) scheme, viewing these two dichotomies as independent, and (b) the scheme developed in the present paper, distinguishing ecological from evolutionary specialization. In scheme (b) evolutionary specialization is synonymous with dietary stereotypy (regardless of diet breadth), evolutionary generalization is synonymous with opportunism or dietary plasticity, and ecological specialization versus generalization corresponds with diet breadth (monophagy or oligophagy versus polyphagy). Heavy lines distinguish specialists from generalists, and illustrate the lack of a simple dichotomy in the scheme advocated in the present paper.

in foraging tactics, was not related to foraging generalization, as determined by the range of foraging behaviors. However, the distinction between evolutionary and ecological aspects of specialization made in the present review suggests a different relationship between these two dichotomies (Fig. 2b). Specifically, I argue that an evolutionary generalist is an ecological opportunist (Wiens and Rotenberry 1979, Feinsinger et al. 1988), making it of necessity dietarily broad-niched or polyphagous. However, the converse is not true, because an organism can eat a broad array of foods in a stereotyped way, as illustrated by the Neotropical flycatchers discussed below. Moreover, an organism can have simultaneously broad and stereotyped diets in two fundamentally different ways, namely, by individuals all feeding identically ("within-phenotype strategists" of Roughgarden 1974), or by individuals feeding consistently as specialists relative to one another ("between-phenotype strategists"; e.g., Werner and Sherry 1987). Morse's scheme (Fig. 2a) is also problematic because of his category of "opportunistic specialists," which is an oxymoron by my scheme since evolutionary specialization and opportunism are mutually exclusive. Thus Figure 2b suggests that degree of opportunism and diet breadth may often be correlated, particularly if relatively few species fall into the lower left box (Fig. 2b) of species categorized by both broad and stereotyped diets. Finally, studies of cichlid fish functional anatomy suggest that evolutionary specialists may be more behaviorally versatile and potentially generalized in diet than evolutionary generalists (Liem and Kaufman 1984), contrary to my hypothesized scheme in Figure 2b.

Opportunism must be a widespread phenomenon, judging from its many synonyms. Fenton (1982; Fenton and Fullard 1981) described "short-term specializations" and "mosaic specialization" as widespread, if not predominant, feeding patterns in insectivorous bats. Analogous dietary specializations are termed "local feeding specializations" (Fox and Morrow 1981) or "facultative specializations" (Glasser 1982, 1984). Murdoch (1969) defined "switching generalists" experimentally in a similar way. Greene (1982) used the term "apparent specialists" for species whose specializations are not obviously related to phenotypic characteristics, and he discussed the evolution of behavioral versus phenotypic manifestations of specialization in lizards. Ralph and Noon (1988) used the term "opportunistic specialist" for Hawaiian birds using a narrow range of foraging behaviors, but using different behaviors in different seasons.

Testing ideas on evolutionary versus ecological approaches to diets and the hypothetical scheme on diet breadth in relation to plasticity (Fig. 2) are challenging tasks, and include understanding of patterns in resource variation (Wiens 1984b). Colwell's (1974) suggestions on how to conceptualize and quantify periodic phenomena and time-series analysis are two possible quantitative approaches. A second consideration is how individual animals use resources. A third aspect concerns functional studies, and how different organisms are constrained behaviorally, morphologically, or otherwise to have different capabilities or efficiencies, depending on the particular resources available. A fourth point is that degree of dietary specialization often depends critically on resource abundance, as does niche overlap (e.g., Schoener 1982, Ford, this volume), which has implications for the timing of studies. Finally, the categories specified in Figure 2 are not discrete, but represent endpoints of continuously distributed behavioral patterns. Thus comparative studies (following section) will remain useful to test these ideas.

ECOLOGICAL VERSUS EVOLUTIONARY APPROACHES: THREE EXAMPLES

Three examples below illustrate both different methods to analyze diets and the difference between tactical vs. strategic approaches to diets, by which contradictory conclusions are sometimes reached. The main problem is the use of tactical methods to study strategic questions.

Cocos Flycatcher

The Cocos Flycatcher is one of four year-round resident landbirds, three of which are endemic on the humid (and almost aseasonal), heavily rainforested Cocos Island, isolated approximately 480 km southwest of Costa Rica (5°32'57"N, 86°59'17"W). During a breeding and nonbreeding season visit, I (T. W. Sherry 1982, 1985) quantified diets using stomach samples, available prey with beating nets, foraging behavior, and standard morphological dimensions.

Two widely used tactical approaches to dietary specialization are the diversity of prey types in the diet and the relationship between food consumed and that available. When the diet diversity of Cocos Flycatchers was compared with that of mainland Costa Rican flycatchers occupying species-rich, lowland rainforest, the Cocos Flycatcher had a relatively broad diet based on both prey taxa and especially foraging behaviors (Fig. 3). It also consumed a variety of arthropod taxa in proportions indistinguishable from those available in at least one of the habitats (T. W. Sherry 1984, 1985). The broad array of foraging behaviors and arthropod types in the diet and the close match of diet to available arthropods suggested that the Cocos Flycatcher is a classical ecological generalist, expected on an isolated oceanic island with few competitors.

Strategic approaches provide a different conclusion. The Cocos Flycatcher is closely related to the Yellow Tyrannulet (*Capsiempis flaveola*) and the Mouse-colored Tyrannulet (*Phaeomyias murina*) (Lanyon 1984, Sherry 1986); the latter is an actively foraging perch-gleaner (Traylor and Fitzpatrick 1982), feeding on both insects and fruit in semi-arid scrubland habitats (Fitzpatrick 1980, 1985; pers. comm.). Because of its mostly insectivorous diet, the Cocos Flycatcher is a specialist compared with the tyrannulet. Second, both foraging behavior and prey appear to be constrained by morphology in a variety of Costa Rican flycatchers including the Cocos Flycatcher (Sherry 1982, 1984; Leisler and Winkler 1985; see also Fitzpatrick 1980, 1985). For example,

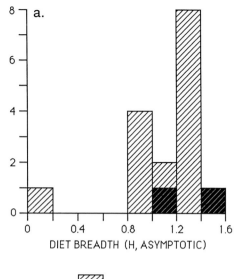

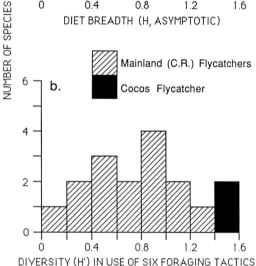

FIGURE 3. Frequency histograms of (a) diet breadth, and (b) diversity or breadth of foraging tactics of the Cocos Flycatcher, based on two different samples—one from a breeding and another from a nonbreeding season—contrasted with mainland (Costa Rican) flycatchers in the Caribbean lowlands of Costa Rica (based on T. W. Sherry 1984, 1985). Mainland flycatcher sample sizes are 16 and 15 for diet breadth and foraging diversity, respectively, because stomach samples were available for a species (*Tolmomyias sulphurescens*) for which foraging behaviors were not observed in this region.

bee and flying ant specialists are relatively large-bodied, narrow-winged, hawking species; Homoptera specialists (including the Cocos Flycatcher) are broad-winged pursuers with long rictal bristles; and generalist flycatchers and specialists on Coleoptera, Hemiptera, and worker ants have intermediate morphological characteristics (Fig. 4). The fact that the Cocos Fly-

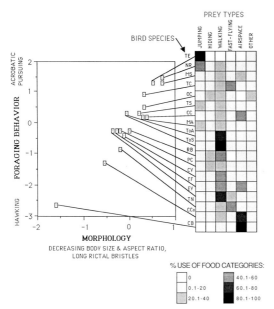

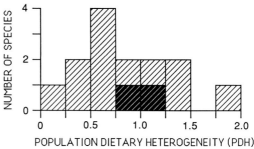

FIGURE 4. Correlations between foraging behavior, morphology, and diet of 18 species of Central American tyrannid flycatchers (based on data and analyses in T. W. Sherry 1982, 1984, 1985; Leisler and Winkler 1985). Both axes are species scores for canonical variate I, based on canonical correlation analysis and principal components analysis of original variables (Leisler and Winkler 1985). Species codes and corresponding names are CB (*Contopus borealis*), CC (*Contopus cinereus*), CCo (*Colonia colonis*), CV (*Contopus virens*), EF (*Empidonax flaviventris*), EV (*Empidonax virescens*), MA (*Myiornis atricapillus*), MS (*Myiobius sulphureipygius*), NR (*Nesotriccus ridgwayi*), OC (*Oncostoma cinereigulare*), PC (*Platyrinchus coronatus*), RB (*Rhynchocyclus brevirostris*), TC (*Todirostrum cinereum*), TE (*Terenotriccus erythrurus*), TN (*Todirostrum nigriceps*), ToA (*Tolmomyias assimilis*), ToS (*Tolmomyias sulphurescens*), and TS (*Todirostrum sylvia*). Prey types and corresponding arthropod taxa are "jumping" = Homoptera; "hiding" = Orthoptera and Lepidoptera larvae; "walking" = Coleoptera, Hemiptera, and non-flying Formicidae; "fast-flying" = Diptera and parasitoid Hymenoptera; "airspace" = Odonata, Apoidea, and flying Formicidae; and "other" = Arachnida, Lepidoptera adults, Dermaptera, and Chilopoda.

catcher has similar foraging behavior and morphology to dietarily similar mainland flycatchers reinforces the conclusion that its specialization on Homoptera is both facilitated and constrained by phenotype. Finally, the population dietary homogeneity of the Cocos Flycatcher is indistinguishable from values for typical mainland flycatchers (Fig. 5), indicating that the Cocos Flycatcher is as stereotyped in diet as flycatchers inhabiting lowland rainforest of Caribbean Costa

FIGURE 5. Comparison of population dietary heterogeneity, *PDH* (a measure of dietary opportunism, based on stomach contents; see text), in 14 species of mainland (Costa Rican) flycatcher and the Cocos Flycatcher. Mainland stomach samples for these calculations come from the breeding season, whereas Cocos Flycatcher samples were taken from both breeding and nonbreeding seasons (data from T. W. Sherry 1984, 1985).

Rica, perhaps because of comparable levels of prey predictability in both environments. These data show that tactical and strategic approaches can lead to contradictory conclusions, and that the Cocos Flycatcher is not the dietary generalist that the tactical approach indicates.

SPECIALIZATION IN TROPICAL VERSUS TEMPERATE INSECTIVOROUS BIRDS

The question of latitudinal gradients is evolutionary, because the comparisons involve species in different biogeographic realms (Nearctic versus Neotropical, in the present case), and because diets are often constrained by phenotypic traits. Most empirical comparisons of tropical and temperate communities have indicated that the majority of avian species added to tropical communities can be accounted for by uniquely tropical resources, and thus by an expanded community niche volume rather than by increased specialization (e.g., Orians 1969b, Terborgh and Weske 1969, Schoener 1971a, Karr 1975, Ricklefs and O'Rourke 1975, Stiles 1978, Askins 1983). Terborgh (1980a) argued instead that increased diversity in a lowland Amazonian bird community in Peru, compared with a south-temperate site in the United States, results from both an expanded tropical resource dimension and greater species packing (implying greater niche specialization). Remsen (1985) reached a similar conclusion.

Stomach-content data for tropical flycatchers (Sherry 1984; unpubl.) show that resident tropical flycatcher species are indeed more specialized than migratory ones, but only if the data are analyzed using a strategic approach (see also Murphy 1987). Niche breadths, calculated using Pielou's asymptotic method, a tactical approach, were not narrower in the thirteen resident than the three migratory species (Mann-Whitney U-test, $P > 0.05$; Fig. 6). Heterogeneity values of stomach contents, by contrast, were lower in the 14 resident species than in four migratory ones (Mann-Whitney U-test, $P < 0.05$, Fig. 6; based on Sherry 1984). The dominance of stomach contents by one or a few arthropod taxa in two flycatcher species—by fulgoroid Homoptera in the Ruddy-tailed Flycatcher (*Terenotriccus erythrurus*) and by *Trigona* bees in the Long-tailed Flycatcher (*Colonia colonis*)—is extraordinary compared with other insectivores (Sherry 1984). Correspondence between morphology, foraging behavior, and diet (Fig. 4) also reinforces the strength of evolutionary contraints to diets in these birds.

Different conclusions resulting from tactical vs. strategic approaches result primarily because some tropical birds have taxonomically broad, but homogeneous diets (Sherry 1984). Cocos Flycatchers, as well as Common Tody Flycatchers (*Todirostrum cinereum*), ate similar prey types in both breeding and nonbreeding periods, based on cluster analysis (T. W. Sherry 1985). Rosenberg (this volume; unpubl.) documented the phenomenon in several antwren species (*Myrmotherula*) inhabiting Peruvian and Bolivian rainforest. Individual antwrens were highly stereotyped in their use of dead leaf foraging microhabitat, from which they took diverse arthropod types. Every individual antwren's stomach contained the same broad array of prey types, indicating a degree of dietary stereotypy only possible in tropical forests where dead-leaf arthropods are relatively predictable (Remsen and Parker 1984).

The homogeneity among tropical insectivorous birds' diets and foraging behavior, both within and between seasons, contrasts sharply with diet data from temperate birds and arctic birds, whose diets are notoriously variable (e.g., Holmes 1966). This is illustrated in community studies in which food abundance or types fluctuate from year to year (Ballinger 1977, Dunham 1980, Kephart and Arnold 1982, Linden and Wikman 1983), and for different species to converge on abundant, preferred food types at a particular time or location (Wiens and Rotenberry 1979, Rotenberry 1980a, Rosenberg et al. 1982, Steenhof and Kochert 1988). Woodpeckers studied by Askins (1983) provide an exception that helps prove the rule about the relationship between specialization and resource predictability. Askins found little difference between tropical and temperate sites in the number of species or degree of dietary specialization, largely because woodpeckers experience similar degrees of seasonal resource stability at different latitudes.

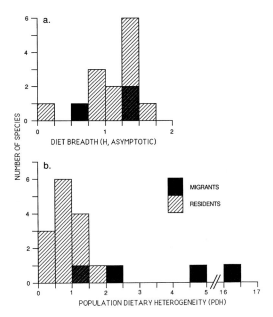

FIGURE 6. Comparison of migrant and resident flycatcher species' diets with respect to (a) diet breadth (using asymptotic Pielou-Hurtubia method, see text; and based on 13 resident and 3 migrant species), and (b) population dietary heterogeneity (based on 14 resident and 4 migrant species—the additional resident and migrant species in part b compared with part a are *Aphanotriccus capitalis* and *Empidonax flaviventris*, respectively). Cocos Flycatcher data are not included in these comparisons. Data from T. W. Sherry (1984, 1985).

Dietary data thus suggest that opportunistic foraging behavior and diets predominate in temperate communities, whereas behavioral stereotypy is more important in the tropics; thus tropical birds appear to be more specialized. However, the question of latitudinal gradients remains unresolved, because of the shortage of evolutionarily meaningful analyses, not to mention the shortage of dietary data from a sufficiently broad spectrum of tropical species.

FRUGIVOROUS BIRDS

McKey (1975) and Howe and Estabrook (1977) proposed that some frugivores are more specialized than others, and that both frugivores and fruits in particular taxa have co-evolved relatively obligate interdependence. Wheelwright and Orians (1982), Wheelwright (1983, 1985), and Moermond and Denslow (1985) questioned aging plasticity among species, (3) to identify possible causes for shifting behaviors, and (4) to determine potential periods of selection on foraging niches of migrants. We emphasize warblers because they are abundant during migration and they include a diversity of relatively closely-related species that employ a diversity of foraging behaviors.

STUDY AREAS AND METHODS

The study site for work during spring and fall migration was Trelease Woods, a 22 ha woodlot northeast of Urbana, Illinois. The forest included mature deciduous tree species and numerous tree-fall gaps that provide patches of understory vegetation that are denser than nongap understory (see Martin and Karr 1986b, Blake and Hoppes 1986 for a more detailed description of the forest). The site for winter work was a young (ca. 25 years old) second-growth forest in Soberania National Park, Panama, where the vegetation was somewhat shorter than in Trelease Woods (see Martin 1985a, Martin and Karr 1986a for more details of study sites). The sites for summer work were early-shrub seral stages (from previous clear-cutting) in northern Ontario (ca. 49°N latitude). Most of the vegetation was deciduous, but some conifers were also present. The foliage was distributed at much shorter heights than on either the migration or winter sites (Martin, unpubl.). Thus, vertical foliage distributions and species of plants varied among the sites.

Foraging maneuvers and other behaviors were observed and recorded on a hand-held tape-recorder for later transcription. Individuals were followed for up to 10 maneuvers, although in practice most individuals could only be followed for one or two observations due to their mobility and obscuring by foliage. Foraging maneuvers we identified included: gleaning (foraging from a substrate from a perched position); hover-gleaning (foraging from a substrate while hovering); sallying (a continuous flight motion while snatching prey from a substrate); and hawking (a flight to snatch an insect in the air). During spring and fall migrations in 1979, only three of these maneuvers (gleaning, hovering, and hawking) were recognized; sallying was categorized as hovering at that time. Consequently, sally maneuvers are absent in figures for 1979 migration seasons.

Foraging behavior was studied in all sites from 1979–1981. Observations in the breeding seasons started in late May and continued into mid-July. Fall migration included late August through early November. Winter foraging behaviors were studied during January and March. Spring migration included mid-April to late May.

Each season was partitioned to allow examination of within-season changes in foraging behaviors. Winter samples were compared between January (middle of the winter season) and March (end of the winter season). Summer was divided into incubation, nestling and fledgling periods. Migration was partitioned into early and late periods; the median date that individuals of each species were observed or captured (see Martin and Karr 1986b for data) during each migration season was used as the cut-off date for grouping observations into early or late categories. A minimum sample size of 25 observations was deemed necessary to provide a representative sample of the foraging behavior. This sample size was derived by using the G-test (see below) to compare the foraging maneuver pattern when sample size was incremented by 5 observations until a sample size was reached where foraging maneuver patterns did not differ statistically. In some cases sample sizes were insufficient ($N < 25$) for one or the other half of a season and such data were not included. In a few cases, such as the Yellow-rumped Warbler (*Dendroica coronata*) during fall migration and Chestnut-sided Warbler (*Dendroica pensylvanica*) during spring migration in 1980, observations were obtained over a brief period that fell in the late or early part of the season, respectively; such data were only displayed for the appropriate seasonal period. The Shannon index of diversity ($H' = -\Sigma p_i \ln[p_i]$) was used to examine the degree to which species were generalized in their foraging. Diversity of foraging maneuvers was not calculated for spring and fall migrations in 1979 because only three of the four maneuvers classified in all other seasons were available for calculations. Differences in foraging within and between seasons were determined based on the log-linear, contingency table, G-test (Sokal and Rohlf 1981).

RESULTS

PLASTICITY DURING MIGRATION

Foraging behavior changed for all of eight warbler species within spring (Fig. 1) and fall (Fig. 2) migrations. Moreover, the changes were relatively consistent among species; most species increased gleaning and decreased hovering and sallying maneuvers from early to late spring (Fig. 1). The exception was the American Redstart (*Setophaga ruticilla*), which increased hawking late in the spring.

Patterns during fall migration were the mirror image of those during spring; species generally decreased gleaning and increased hovering from early to late fall (Fig. 2). Exceptions were the American Redstart, with a mirror image of its foraging maneuver pattern during spring migration, and the Bay-breasted Warbler (*Dendroica castanea*) during fall 1980.

The changes in foraging maneuver patterns also caused consistent shifts in the diversity of foraging maneuvers used by warblers; diversity was greater in early than late spring for all of five species (Fig. 1) and greater in late than early fall for five of six species (Fig. 2).

PLASTICITY DURING WINTER

Three of the four species studied during winter in Panama increased their degree of frugivory from early (January) to late (March–April) in the dry season; the exception was the Magnolia Warbler (*Dendroica magnolia*) which rarely eats fruits (Martin 1985a). The degree to which the other three species shifted to frugivory varied among species in the order: Chestnut-sided Warbler <

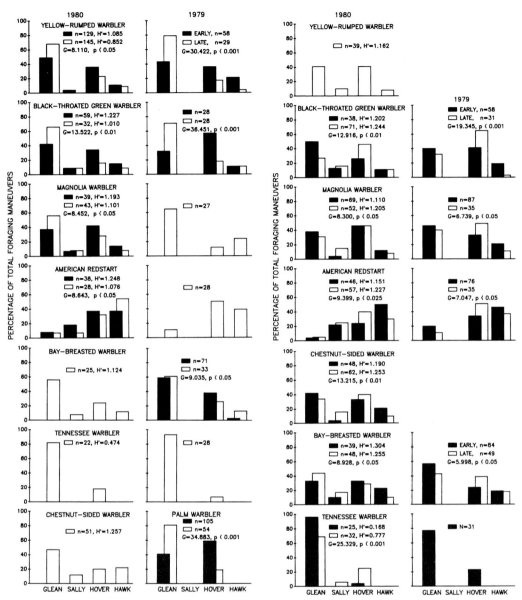

FIGURE 1. The percentage of total foraging maneuvers that was comprised by each type of maneuver during early (solid bars) and late spring (open bars) migrations 1980 and 1979 in Illinois. Only three behaviors were classified in 1979.

FIGURE 2. The percentage of total foraging maneuvers that was comprised by each type of maneuver during early (solid bars) and late fall (open bars) migrations 1980 and 1979 in Illinois. Only three behaviors were classified in 1979.

Bay-breasted Warbler < Tennessee Warbler (*Vermivora peregrina*) (Martin 1985a) as also found by Greenberg (1981b, 1984a). However, if fruits are considered a substrate rather than a maneuver, then our data indicate that foraging maneuvers did not change significantly over the winter in these four species (Fig. 3).

BREEDING SEASON

Of the 1980 data analyzed, Chestnut-sided Warblers exhibited a shift in foraging behavior from the incubation to late nestling-fledgling period ($G = 12.563$, $P < 0.005$), and a marginally significant ($G = 7.616$, $P < 0.06$) shift from incubation to nestling periods (Fig. 4). The net re-

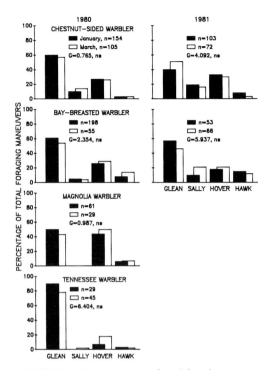

FIGURE 3. The percentage of total foraging maneuvers that was comprised by each type of maneuver during January (solid bars) and March (open bars) for winters in 1980 and 1981 in Panama.

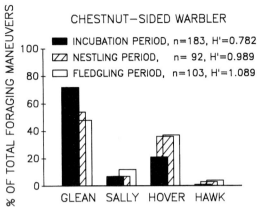

FIGURE 4. The percentage of total foraging maneuvers that was comprised by each type of maneuver during three periods of the breeding cycle of the Chestnut-sided Warbler in Ontario, Canada.

sult was an increase in the diversity of foraging maneuvers due to increased hovering and sallying from incubation through fledgling periods (Fig. 4). Moreover, this pattern was exhibited by each of at least four color-banded individuals included in the 1980 sample (Martin, unpubl.).

BETWEEN-YEAR DIFFERENCES

Frequency and intensity of between-year changes in foraging maneuver patterns varied among species during migration; some species did not change between years (e.g., Yellow-rumped Warbler, Black-throated Green Warbler [*Dendroica virens*]), and others changed frequently (e.g., Bay-breasted Warbler, Magnolia Warbler) (Table 1, Figs. 1, 2). Between-year changes in foraging maneuver patterns also differed among species for breeding and winter seasons; Tennessee and Bay-breasted Warblers changed between years in both of these seasons, whereas Magnolia and Chestnut-sided Warblers did not change between years in either season (Table 1, Fig. 5). Thus, species differed in the relative stability of their foraging behavior during similar periods in different years.

BETWEEN-SEASONS DIFFERENCES

Foraging maneuver patterns were similar between breeding and wintering seasons; 10 of 14 comparisons showed no changes between winter and breeding seasons (Table 2, Fig. 5). Foraging maneuver patterns were generally most different during migration seasons (Table 2). Foraging during migration differed from breeding or winter seasons in 37 cases and did not differ in only 12 cases ($\chi^2 = 12.755$, $P < 0.001$). Moreover, foraging during migration differed from breeding or winter more frequently in early spring or late fall (17 of 19 cases, Table 2) than in late spring or early fall (20 of 30 cases) ($\chi^2 = 4.142$, $P < 0.05$). Similarly, the diversity measures of foraging maneuvers showed that foraging was usually most generalized (greatest H') during early spring and late fall migrations when comparing all seasons (Fig. 5).

DISCUSSION

FORAGING PLASTICITY DURING MIGRATION

All species shifted their foraging maneuver patterns to a variable degree during migration. Most species showed a consistent shift toward increased gleaning and decreased hovering and sallying as spring progressed and the opposite pattern during fall. In part, such shifts can be attributed to shifts in availabilities of insects. Foliage-clinging arthropods and the density of their substrate (leaves) increase through spring and decrease through fall (Kendeigh 1979, Graber and Graber 1983, Martin, pers. obs.). Consequently, more effort is devoted to gleaning foliage-clinging arthropods during late spring and

TABLE 1. SUMMARY OF G-TEST STATISTICS[a] FOR COMPARISONS OF DIFFERENCES IN FORAGING BEHAVIORS BETWEEN YEARS. HOVERING AND SALLYING BEHAVIORS WERE LUMPED FOR MIGRATION SEASONS IN 1980 TO COMPARE WITH MIGRATION SEASONS IN 1979 BECAUSE SALLYING BEHAVIORS WERE NOT SEPARATED FROM HOVERING IN 1979 OBSERVATIONS

	Breeding	Winter	Early spring	Late spring	Early fall	Late fall
Magnolia	4.4	1.6	21.6***	5.7	6.8*	2.9
Tennessee	9.4*	8.7*			29.0***	
Chestnut-sided	3.1	4.4				
Bay-breasted		8.2*		6.4*	12.3**	2.9
Yellow-rumped			3.8	3.7		
Black-throated Green			3.9	1.5	3.3	5.4
Redstart			0.8		14.2***	4.7

[a] * $P < 0.05$, ** $P < 0.01$, *** $P < 0.001$.

early fall (Figs. 1, 2). Moreover, ambient air temperatures are lower during early spring and late fall, which tends to reduce flying insect activity and causes reduced hawking and increased hovering behaviors (Holmes et al. 1978). The increased incidence of hovering and decreased incidence of hawking by American Redstarts during early spring and late fall potentially reflect such effects. Moreover, the increased hovering during early spring and late fall exhibited by several other species may also be partly explained by such effects.

All changes in foraging maneuvers, however, cannot be attributed to changes in food availabilities. Species such as Tennessee and Palm (*Vermivora palmarum*) Warblers used gleaning for 80–90% of their foraging maneuvers during periods of abundant food (Figs. 1, 2). Reductions in flying insect activity during cold periods should thus not greatly influence their foraging behavior. Yet, both of these species increased hovering and sallying in these cold periods during migration and these increases were not acomplished by reduced hawking maneuvers. Similarly, many of the other species, except American Redstart, decreased both hovering and sallying in the warm periods (late spring, early fall, Figs. 1, 2) when flying insects should be most abundant (Kendeigh 1979).

The consistent increase in flying maneuvers during cold periods may reflect thermoregulatory influences. Migratory birds stay in a warm environment most of their life and, as a result, lack the ability to acclimate (Kendeigh et al. 1977). Early spring and late fall represent some of the coldest temperatures and greatest thermoregulatory costs incurred by most migratory species. Temperatures ranged from 0–34°C from early to late spring and the converse in the fall on our Illinois sites (unpubl. data). Since flight metabolism is only about 25% efficient, 75% of the energy is produced as heat which is available for thermoregulation (Calder and King 1974). During cold periods when energy costs of thermoregulation are already high, it may be more efficient to hover and sally because of the increased heat produced by flying movements. Field observations provide some support for this argument. Birds often sat with their feathers fluffed until spotting a prey item that they then flew to eat during early spring and late fall, whereas they were much more active at hopping and walking in late spring and early fall.

Alternatively, the energetically expensive flight maneuvers may simply be used to increase food intake rates (Morse 1973, Bennett 1980) when food is scarce. This possibility is supported by the analyses of the foraging behavior of the Chestnut-sided Warbler during breeding; it increased the incidence of hovering and sallying during later stages of breeding, when food demands of reproduction are apparently greater (see review in Martin 1987). Such shifts cannot be attributed to temperatures, which were also greater (also see below).

In short, foraging maneuvers may vary with changes in available food types, changes in food demands relative to food availability, changes in thermoregulatory costs, or, most likely some combination of these factors.

DIVERSITY OF FORAGING MANEUVERS AND FOOD LIMITATIONS

Increases in the diversity of foraging maneuvers used by a species may reflect decreasing food availability relative to demand (Rabenold 1978), but they could also simply reflect responses to changes in the types of foods that are available. Consequently, comparisons among periods must be interpreted with caution. However, increases in the diversity of foraging maneuvers used by species were generally accomplished by increas-

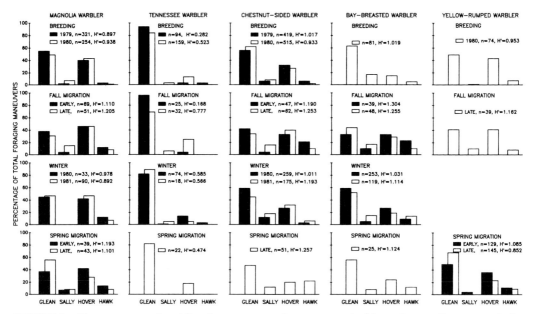

FIGURE 5. The percentage of total foraging maneuvers that was comprised by each type of maneuver during breeding season in Ontario, Canada, fall migration in Illinois, winter in Panama, and spring migration in Illinois. Breeding and winter data are for each of two years and migration data are for early and late in each season for 1980.

ing their use of energetically-expensive maneuvers, such as hovering and sallying. Moreover, increases in both foraging diversity and use of energetically-expensive maneuvers typically occurred when food was reduced relative to demand. For example, species were most generalized in their foraging during early spring and late fall migration periods (Fig. 5), when food abundance was low and energy demands of migration were high (Kendeigh 1979, Graber and Graber 1983). In addition, the increase in diversity of foraging maneuvers of the Chestnut-sided Warbler from incubation to fledgling periods (Fig. 3) also coincides with increasing energy demands of raising young (see Martin 1987 for a review). Thus, foraging diversity seems to provide a crude index to periods of food stress.

The fact that foraging diversity tends to be greatest during early spring and late fall migrations (Fig. 5) suggests that these periods may represent particularly severe periods of food limitation. Moreover, foraging patterns in the food-rich late spring and early fall were more similar to those in winter and breeding, but the patterns differed during the food-poor periods of early spring and late fall when diversities were also greatest. These observations, when taken together, suggest that food is indeed difficult to obtain during these periods.

Migration seasons are not the only periods of food limitation. A variety of correlative and experimental work, as well as the increasing diversity of foraging maneuvers during breeding (Fig. 4), indicates that food is commonly limiting during breeding (reviewed in Martin 1987). Thus, attempts to focus on any single season as the primary determinant of the foraging niche of migratory birds is likely to produce erroneous conclusions.

CONSERVATISM OF FORAGING MANEUVER PATTERNS

Although warblers exhibited statistically significant shifts in their foraging behavior over time, many shifts were basically matters of degree. The general ranking of the four behaviors remained similar among seasons and years, so that the general foraging maneuver pattern of a species was largely conserved (Fig. 5). This conservatism may be expected because foraging maneuvers are so closely tied to morphology (Hutto 1981b).

Studies of communities typically focus on documenting differences among species as a measure of resource partitioning (see Martin 1986, 1988a; Schoener 1986b for reviews). Determination of the conservative nature of traits is important to the way we examine communities. If traits are conservative, many differences among coexisting

TABLE 2. Summary of G-test Statistics[a] for Comparisons of Differences in Foraging Behavior between Seasons. All Migration Data Are from 1980.

	Early spring	Late spring	Early fall	Late fall	Winter 1980	Winter 1981
Magnolia Warbler						
Breeding 1979	8.4*	14.3**	10.0*	20.8***	7.2	3.2
Breeding 1980	9.9*	15.2**	13.3**	13.4**	13.2**	7.7
Winter 1980	11.0*	6.0	2.7	18.3***		
Winter 1981	11.9**	8.9*	4.2	18.1***		
Chestnut-sided Warbler						
Breeding 1979		15.3**	11.3*	12.1**	3.6	7.5
Breeding 1980		23.4***	24.1***	19.4***	1.1	9.2*
Winter 1980		18.7***	22.8***	13.9**		
Winter 1981		13.4**	18.6***	3.8		
Bay-breasted Warbler						
Breeding 1980		8.8*	30.8***	9.6*	11.7**	6.1
Winter 1980		1.4	16.1**	9.2*		
Winter 1981		3.1	11.3*	3.6		
Tennessee Warbler						
Breeding 1979		13.7**	1.2	28.6***	8.9*	4.6
Breeding 1980		1.9	6.9	7.9*	2.1	4.3
Winter 1980		1.9	8.4*	7.9*		
Winter 1981		11.0*	3.3	17.2***		

[a] * $P < 0.05$, ** $P < 0.01$, *** $P < 0.001$.

species may be simply due to differences in their evolutionary histories, rather than the result of interactive processes (Wiens 1983, Martin 1986). Consequently, communities may be noninteractive accumulations of species responding to resources as a function of their individual evolutionary histories (Grinnellian niche approach, *sensu* James et al. 1984). Alternatively, if traits are conservative and communities are structured by interactions, then resource partitioning among coexisting species may be achieved by selection for resource partitioning (permissible combinations, *sensu* Connell 1980; Martin 1988b, c). If traits are more plastic, then individuals of a species may partition resources by modifying their behavior relative to other coexisting species (see Martin 1986). Clearly, we cannot fully understand community dynamics until we understand the dynamics of individuals of the species that make up the community.

ACKNOWLEDGMENTS

R. Greenberg, C. Hunter, M. Morrison, C. J. Ralph, H. Recher, and T. Sherry provided constructive criticisms on an early draft.

DEAD-LEAF FORAGING SPECIALIZATION IN TROPICAL FOREST BIRDS: MEASURING RESOURCE AVAILABILITY AND USE

KENNETH V. ROSENBERG

Abstract. Tropical birds foraging at dead leaves suspended above the ground in forest understory represent a system that potentially overcomes many of the difficulties inherent in measuring resource availability for insectivorous birds. Because the dead leaves are discrete and abundant resource patches, they are easily counted and sampled. I present a scheme for sampling the availability and use of specific substrate types and the abundances of arthropod prey. Availability and use are compared directly for six bird species in three habitats (upland rainforest, low-lying rainforest, and bamboo) at the Tambopata Reserve, southeastern Peru. I conclude that (1) the overall abundance, variety, and high prey productivity of dead leaves helps to maintain extreme specialization in this guild; (2) substrate types are selected nonrandomly by all species, at least partly on the basis of the differential prey availability in each type; (3) individual dead leaves are relatively long-lived and are continually recolonized by arthropods, therefore representing predictable and renewable resource patches to these birds; (4) dead-leaf specialists are exposed to distinctly different prey choices from those of birds that search live foliage. Studies of other insectivorous bird groups should include estimates of substrate availability among habitats, prey availability among substrates, as well as the use of these by the birds.

Key Words: Dead leaves; insectivorous birds; foraging specialization; resource availability.

Understanding of resource availability and distribution, as well as resource-use patterns by birds, is central to the study of foraging specialization and avian community organization. Because of difficulties in measuring arthropod abundance and actual bird diets, these are often inferred for insectivorous birds from general insect sampling, foraging behavior, and morphology. In particular, we know almost nothing of the relative productivities of specific foraging substrates and how these may vary temporally. In tropical communities these problems are often compounded by the increased number of bird species and resource dimensions.

A system that offers great potential for overcoming these difficulties is the foraging by birds among suspended dead foliage in tropical forest understory. Leaves falling from the canopy are often trapped by vines or other vegetation before reaching the ground. They persist either individually or in dense clusters, offering daytime hiding places for nocturnal arthropods. A number of tropical antbirds (Formicariidae), ovenbirds (Furnariidae), and other insectivorous species forage exclusively by extracting arthropods from within these suspended dead leaves (Remsen and Parker 1984). As many as 10–12 species of dead-leaf-searching specialists may occur with other, often congeneric, live-foliage-gleaning species in the same mixed-species foraging flocks (Munn and Terborgh 1979, Munn 1985).

The dead leaves represent abundant, yet discrete, resource patches that are easily counted and sampled for arthropod prey. This contrasts with other substrates, such as live foliage or airspace, that are more generally distributed and that may possess a diverse and highly mobile arthropod fauna. The study of such a well-defined resource system may enable us to discern details of food availability and exploitation that are generalizable to other avian insectivores.

Only one dead-leaf specialist has been studied in detail, the Checker-throated Antwren (*Myrmotherula fulviventris*) in Panama, where it is the only member of this guild (Gradwohl and Greenberg 1980, 1982a, b, 1984). Gradwohl and Greenberg demonstrated the feasibility of measuring resource availability and use for these birds, and they successfully used this foraging system to test ecological as well as behavioral hypotheses. My study of dead-leaf foraging specialization among Amazonian rainforest birds extends these findings to a multi-species assemblage that is part of the world's most diverse avifauna.

My research is aimed at determining how substrate and prey availability promote specialization and how this specialization contributes to the organization of a diverse tropical bird assemblage. In this paper, I describe and evaluate my methods for measuring resource availability and use by these birds. I also assess variability in dead-leaf distribution and prey abundance across habitats and seasons. Then, I provide evidence that individual dead leaves may represent a relatively long-lasting, renewable resource to avian insectivores. Finally, I provide examples of data on several common bird species, comparing available substrates with those actually visited by the birds. My intent is to provide a scheme for quantifying the relevant aspects of a resource system for insectivorous birds, as illustrated with data from one specialized guild.

STUDY AREA AND METHODS

Study site

This study concentrates on the Tambopata Reserve (5500 ha) in the Department of Madre de Dios, southeastern Peru (12°50′S, 60°17′W). The reserve consists of primary lowland rainforest that is typical of a vast portion of southwestern Amazonia. Several forest types are recognized and described by T. L. Erwin (1985). The bird and insect faunas also have been relatively well studied on the reserve (Parker 1982, T. L. Erwin 1985).

I worked at Tambopata from May through July 1987, covering a period from the end of the rainy season to the middle of the dry season. This region is characterized by a 5- to 6-month dry season, punctuated by occasional severe storms from the south that bring strong, cooling winds and sometimes heavy rain. The severe winds are thought to be important in maintaining a broken canopy and a prevalence of gap-inhabiting plants, including bamboo (T. L. Erwin 1985).

My study centered on three habitat types: upland forest, low-lying forest, and bamboo thickets. The upland forest (Upland Type II of T. L. Erwin 1985) is situated on sandy, relatively well-drained soils on ancient alluvial terraces high above the current river levels. This forest has a relatively closed 35- to 40-m canopy and a relatively open understory. Midstory palms and *Cecropia* spp. trees are conspicuously lacking; however, shrub-like understory palms (e.g., *Geonoma* spp.) are common. Low-lying forest (Upland Type I of T. L. Erwin 1985) is the most abundant forest type on the reserve. It occurs on poorly drained clay soils and has an uneven canopy of 30 to 35 m. Subcanopy palms (e.g., *Iriartea* spp., *Socratea* spp.) and *Cecropia* spp. are common, and the understory is often dense with vine tangles and other low vegetation. In places, the understory of this forest consists of nearly pure, dense thickets of bamboo (*Guadua* spp.) that may reach a height of 8–10 m. Primarily because the avifauna associated with this bamboo is often quite distinct from that in the surrounding forest (Parker 1982), I consider the bamboo to be a separate habitat type.

Foraging behavior

The following data were recorded with a microcassette on foraging birds encountered opportunistically on the study site: species, sex and age (if determined), habitat type, height above ground, height of tree, canopy height (all heights estimated to the nearest 1 m), foraging method (e.g., glean, probe), foraging substrate (including specific characteristics, such as leaf size and type), perch type (if different from substrate), and foliage density estimated in a 1-m radius sphere around the bird (scale, 0–5). All dead leaves were further categorized as to type (curled, tattered, or entire), and I noted their position in the vegetation (for example, in vine tangle, suspended from live branch).

Because most species of interest foraged in mixed-species flocks that I could frequently follow for extended periods, I was often able to make repeated but nonconsecutive observations of individuals by rotating my attention among the flock members. In most cases, I recorded 3–5 consecutive foraging attempts before moving on to the next bird, although I did not eliminate longer bouts by species that were difficult to observe.

Dead-leaf abundance

Numbers and distribution of suspended dead leaves were assessed at the end of the rainy season in mid-May and again in July, at the middle of the dry season. I established 10-m line transects perpendicular to existing trails at randomly assigned points, with 10 transects in each habitat type. On each transect, I counted and recorded the size (length and width, estimated to the nearest 1 cm) and type of all dead leaves encountered along a 1-m wide strip, up to 10 m above ground. All palm, *Cecropia*, bamboo, and other "novel" leaf-types were tallied separately. Leaves above 5 m were usually inspected with binoculars. Using these methods, 100 m^3 of the forest understory were sampled, with data recorded separately for each horizontal and vertical 1-m interval. These data yielded the number and surface area (length × width) of dead leaves per cubic meter, with associated variances representing horizontal and vertical patchiness for each plot. Because leaf density was usually high, a large sample of leaf sizes and types was also obtained.

Arthropod abundance

Arthropods were sampled from individual dead leaves collected in areas adjacent to the leaf-sampling transects. For each sample, the first 30–50 leaves encountered within reach, and removable without disturbance, were placed individually into zip-lock plastic bags. Because most arthropods were reluctant to flush from the leaves, escape was minimal. After being killed with insecticide (Raid®), arthropods were separated from the leaves, classified to order, measured to the nearest 1.0 mm, and preserved in 70% ethanol. These voucher specimens will be identified later to lower taxonomic levels, if possible, and deposited at the LSU Entomology Museum. To relate substrate characteristics to arthropod numbers and type, I recorded the size and type of each leaf sampled.

To compare arthropod frequency on live vs. dead leaf substrates, these samples were supplemented with visual searches of an equivalent number of live leaves in the same areas. The type and size of all arthropods encountered on leaf surfaces were recorded during slow passes through understory vegetation, sampling all consecutive leaves clearly visible (upper and lower surfaces) without disturbing the foliage.

Temporal changes in resource availability

As noted above, seasonal change in dead-leaf abundance was assessed on transects censused in May and July 1987. In addition, I individually marked all dead leaves on 2 × 2 × 2-m plots and checked these weekly throughout the season (7–8 weeks) to measure persistence and local accumulation. I established three plots in low-lying forest, two in upland forest, and two in bamboo. These were supplemented by marking additional *Cecropia* leaves and other large leaves that were under-represented on the plots. A total of 1022 leaves was marked, including those recruited into the plots during the study.

TABLE 1. CHARACTERISTICS OF DEAD-LEAF FORAGING BIRDS AT THE TAMBOPATA RESERVE, SOUTHEASTERN PERU. HABITATS ARE UPLAND FOREST (U), LOW-LYING FOREST (L), AND BAMBOO (B)

Species	Body wt. (g)[a]	Habitat	Percent use of dead leaves	Number of foraging observations
Olive-backed Foliage-gleaner	38.8	U	90	124
Brown-rumped Foliage-gleaner	30.7	L, B	97	231
Buff-throated Foliage-gleaner	33.8	L	98	132
Ornate Antwren	9.5	L, B	99	227
White-eyed Antwren	9.3	L, B, U	99	693
Moustached Wren	18.5	B	96	52

[a] Mean of five male and five female specimens.

Finally, to assess turnover and colonization of arthropods at individual leaves, I used a sample of 45 leaves that were easily checked with minimal disturbance. These were monitored every 1–2 days for arthropod inhabitants, for a total of 1305 checks. If the arthropod remained in the leaf (58% of visits), I noted the number of consecutive visits on which it was present. If the arthropod flushed from a leaf during a check, I recorded the time until that leaf was reoccupied. Thus, I simultaneously measured changes in occupancy under conditions of disturbance (perhaps simulating predation) and lack of such disturbance.

RESULTS

AVIAN DEAD-LEAF SPECIALISTS

Data are presented for six bird species that foraged heavily on dead leaves at Tambopata (Table 1). For each species, more than 90% of my observations were at dead-leaf substrates within 10 m of the ground, allowing appropriate comparisons with resource availability measurements. Two additional species of dead-leaf specialists occurred in the understory at Tambopata, but were less common, and up to seven specialists foraged in the subcanopy and canopy.

Antwrens in the genus *Myrmotherula* traveled almost exclusively in mixed-species understory flocks, feeding actively at individually suspended leaves. They often employed acrobatic maneuvers, such as extended reaches or clinging at the tips of leaves, to inspect each leaf carefully for arthropods. The White-eyed Antwren (*M. leucophthalma*) was a habitat generalist at Tambopata, occurring in nearly every foraging flock in all three habitat-types. The Ornate Antwren (*M. ornata*) was restricted to low-lying forest in the vicinity of bamboo (see also Parker 1982) but foraged both inside and away from bamboo thickets.

The larger foliage-gleaners (*Automolus* spp.) also traveled in the same mixed-species flocks, usually moving deliberately along branches or in vine tangles. They probed into individual large leaves or frequently investigated dense clusters of leaves lodged among vines or live foliage. Occasionally, these birds manipulated the substrates with their bills, for example, by picking leaves from a cluster and then dropping them to the ground. Both the Buff-throated (*A. ochrolaemus*) and the Brown-rumped (*A. melanopezus*) foliage-gleaners occurred widely in the low-lying forest, sometimes feeding side by side in the same flocks. All flocks with Brown-rumped Foliage-gleaners were in the vicinity of bamboo thickets and this species is considered a bamboo specialist by Parker (1982) and Terborgh et al. (1984). However, I rarely observed it foraging within bamboo foliage. The Olive-backed Foliage-gleaner (*A. infuscatus*) was largely restricted to the upland forest and more open areas in the low-lying forest far from bamboo.

The sixth species considered here, the Moustached Wren (*Thryothorus genibarbis*), occurred primarily in dense, low, river-edge forest and bamboo thickets. In bamboo, this species foraged in solitary pairs in dense clusters of dead leaves and debris, or at individual large *Cecropia* leaves suspended in dense live foliage. Pairs only temporarily joined mixed-species flocks that passed through their territories.

Species-specific comparisons with respect to foraging height and use of particular dead-leaf types will be presented elsewhere (Rosenberg, unpubl.). In general, species differed most in their use of those leaf types, such as palms, *Cecropia*, and bamboo, that were specific to each habitat. Importantly, no species in any habitat searched leaves classified as entire (<1% of all observations).

DEAD-LEAF ABUNDANCE

The overall height distribution and average density of dead leaves were similar in the three habitats, with most leaves concentrated in the first 3 m above the ground (Fig. 1). Individual transects varied considerably in abundance, however, with density ranging from 2.6/m^3 to 8.7/m^3.

Between May and July, abundance of leaves increased about 50% in two of the three habitats, a difference greater than that between any habitat

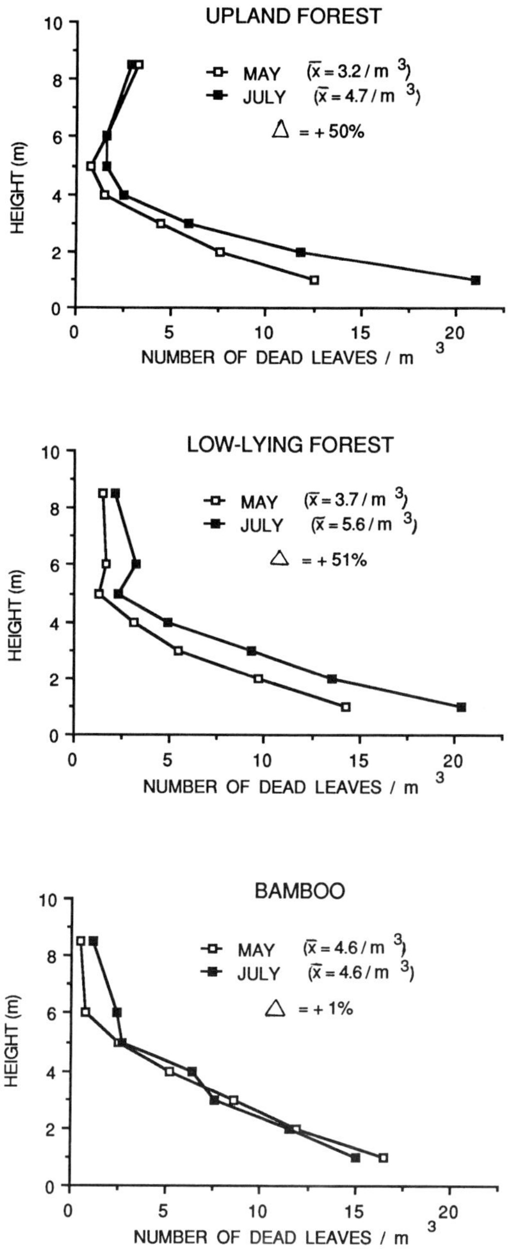

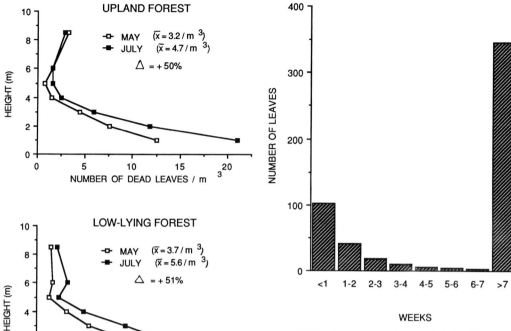

FIGURE 1. Abundance and height distribution of dead leaves in three habitats at Tambopata in May and July 1987 ($\bar{X}$ = average leaf density on 10 transects in each habitat; $\triangle$ = percent change in leaf density between May and July).

FIGURE 2. Persistence of suspended dead leaves at Tambopata (data from 1022 marked leaves on seven plots in three habitats).

types in a single season (14–30%). The steady accumulation of trapped leaves throughout the early dry season was also apparent in the plots with marked leaves. The net number of leaves increased on all plots (36–294%), with the largest increases in upland forest and the smallest in bamboo. The longevity of individual leaves exhibited a bimodal pattern in all three habitats (Fig. 2), with leaves either disappearing shortly after falling or remaining for long periods. Because I could not determine when leaves present at the beginning of the study had first fallen, or when leaves present at the end of the study eventually disappeared, these represent minimum estimates of longevity. However, I can be certain that of all leaves recruited onto the plots during the study period, 20% disappeared in the first week. Similarly, 66% of all leaves marked at the beginning of the study were still present 7 to 8 weeks later.

DISTRIBUTION OF SUBSTRATE TYPES

The distribution of sizes and types of dead-leaf substrates differed greatly among the habitats (Fig. 3). The average leaf size was highest in low-lying forest and lowest in upland forest. In general, leaf sizes exhibited a bimodal pattern with 8- to 10-cm leaves always most abundant, and with the largest leaves in each habitat being "novel" leaves associated with that habitat. For example, understory palm leaflets were numerous in upland forest, larger palm fronds (e.g., *Iriartea* spp.) were common in low-lying forest, and bamboo and *Cecropia* leaves dominated in bamboo thickets. Upland forest also had the

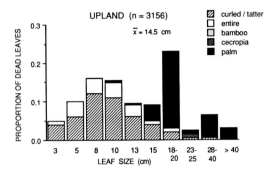

TABLE 2. Prey Densities on Live and Dead Leaf Foliage at Tambopata Reserve

Habitat	Leaf type	Month	Arthropod density (number/leaf)	Number of leaves
Upland forest	dead	May	0.41	380
Bamboo	dead	May	0.53	300
Low-lying forest	dead	May	0.39	320
Low-lying forest	dead	July	0.30	200
Low-lying forest	live	May	0.18	810

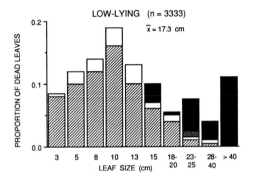

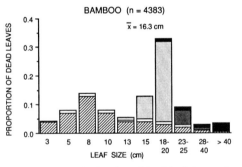

FIGURE 3. Distribution of dead-leaf sizes and types in three habitats at Tambopata in May 1987 ($\bar{X}$ = average leaf size).

greatest proportion of entire leaves (18%). Mean leaf size increased with height above ground in each habitat, as did the proportion of novel and other large leaves in low-lying forest and bamboo.

PREY AVAILABILITY

During May, a total of 1000 dead leaves was sampled for arthropods in the three habitats (Table 2). Prey density ranged from 0.39/leaf in low-lying forest to 0.53/leaf in bamboo. In July, the density of arthropods in 200 dead leaves in low-lying forest was 0.30/leaf. These estimates excluded a large number of 1- 3-mm social ants and their nests concentrated in fewer than 10 leaves (each nest counted as one prey item). In contrast, a search of 810 live-leaf surfaces in low-lying forest in May yielded 0.18 arthropods/leaf. The differences between live- and dead-leaf substrates were even more apparent when the sizes and taxa of the arthropods were considered. Dead-leaf arthropods averaged significantly larger (6.5 mm vs. 3.8 mm, $P < 0.001$, Mann Whitney U-test; Fig. 4). Over 75% of the arthropods on live leaves were 2–4 mm in length and none was >10 mm. In dead leaves, 53% of the arthropods were >5 mm and 16% were >10 mm long. Similarly, nearly two-thirds of the live-leaf arthropods were conspicuously colored ants, flies, and wasps, whereas these made up <10% of the dead-leaf samples. Over two-thirds of the dead-leaf arthropods were cryptically colored beetles, roaches, orthopterans, and spiders (Fig. 5).

The number of arthropods per dead leaf increased sharply with increasing leaf size ($r = 0.944$, $P < 0.01$; Fig. 6). This trend was evident in each of the three habitats. Very small (3–8 cm) leaves and entire leaves had the lowest frequency of arthropods, whereas prey density was extremely high in leaves >40 cm long (regardless of type) and in *Cecropia* leaves (regardless of size). Bamboo and palm leaflets had arthropod densities slightly below the overall average.

Overall, individual dead leaves had a high rate of turnover and renewal of arthropods. Most arthropods that I did not flush remained in a given leaf for only 1–2 days ($\bar{X} = 1.66$, Fig. 7). A few leaf inhabitants stayed longer, however, with the longest being a roach present on nine consecutive visits (12 days) to the same leaf. Given that an arthropod remained in a leaf after a visit, there was a 39% chance of it being there on the next visit, a 44% chance of that leaf being empty, and a 17% chance of a different arthropod being present. In cases in which an arthropod flushed from a leaf, most leaves were reoccupied on the second or third subsequent visit (Fig. 8). In these cases there was a greater chance of the leaf being empty on the next visit (73%); on 16% of my visits, a different arthropod was present.

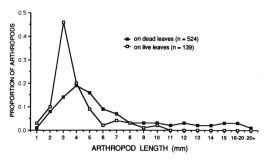

FIGURE 4. Size distribution of arthropods on live and dead leaves at Tambopata.

USE VS. AVAILABILITY

Here, I compare the distributions of dead-leaf sizes and types used by the birds with those available in the appropriate habitats. In this way, I can separate selectivity and avoidance of particular substrate types from simple use. All species selected leaves differently from their availability in their respective habitats (Fig. 9), and all of these differences were highly significant (Kolmogorov-Smirnov and Chi-squared tests; $P < 0.001$). In general, all species selected larger and certain novel types of leaves, and they avoided the smallest leaves in each habitat. Use of *Ce-*

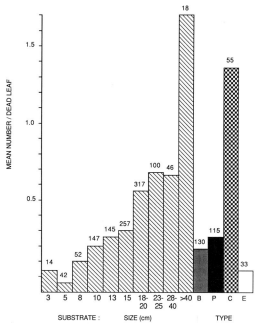

FIGURE 6. Mean number of arthropods in dead leaves of different size and type (B = bamboo, P = palm, C = *Cecropia*, E = entire). Number of leaves sampled, by category, are shown above each bar.

cropia leaves by most species was much greater than their availability, although these leaves were probably under-represented in the transect samples. However, heavy use of some leaf types did not always represent selectivity. For example, use of understory palm leaflets by White-eyed Antwrens in upland forest and of larger palm fronds by Buff-throated Foliage-gleaners in lowlying forest were almost exactly equal to their availability in these two habitats.

To see if selectivity could be explained by the prey productivity of the different sized leaves, I weighted the leaf-availability distribution by the frequency of arthropods in each leaf type (from Fig. 6) and again compared these with substrate use by the birds. Differences were still significant for all species comparisons, except that in most cases use of the very small leaves was now nearly equal to their weighted availability. Thus, low prey density probably explains the avoidance of these small leaves (and of entire leaves), but the larger, and especially *Cecropia*, leaves were still searched more than expected.

DISCUSSION

The empirical data presented here center on one important aspect of the food resource, namely foraging substrate. The exact substrates from which insectivorous birds obtain their prey are

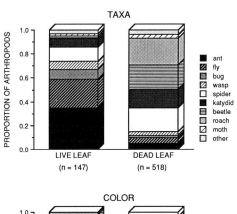

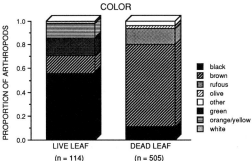

FIGURE 5. Characteristics of arthropods on live and dead leaves at Tambopata.

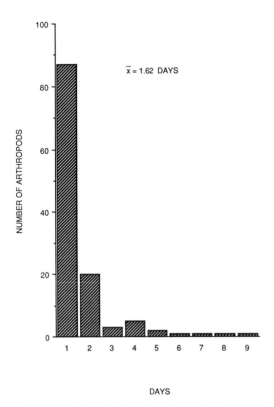

FIGURE 7. Length of stay by arthropods in individual dead leaves at Tambopata (based on sequential checks of leaves from which arthropods did not flush).

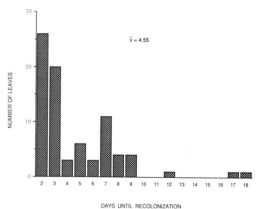

FIGURE 8. Time until recolonization by arthropods at individual dead leaves at Tambopata (based on sequential checks of leaves from which an arthropod had previously flushed).

usually used to define subgroups or guilds within avian communities (e.g., Root 1967, Holmes et al. 1979b). It is largely through substrate choice that prey availability is mediated. It is also likely that overall habitat and foraging-site selection is determined in part by the distribution and productivity of specific foraging substrates. A higher degree of resource specialization and, in particular, substrate subdivision is thought to be one mechanism promoting the higher species diversity in tropical vs. temperate bird communities (Orians 1969b; Karr 1971, 1976; Terborgh 1980a; Remsen 1985). However, critical evaluations of substrate use, even for most temperate communities, are lacking. Substrates are usually measured only in a general way (e.g., bark, foliage, ground), and studies of the arthropod prey available on specific substrates are rarely attempted.

By sampling the availability of particular dead-leaf substrates, I was able to identify finer levels of resource segregation within a guild that was already considered highly specialized with regard to substrate. More importantly, I was able to distinguish between substrate types selected and simple use. Furthermore, by sampling the prey productivity of the individual substrate types, I was able to explain at least part of the observed selectivity. Thus, I can conclude that all species in my study selected foraging sites nonrandomly, avoiding the least productive substrates. Greenberg and Gradwohl (1980) also emphasized the importance of more subtle distinctions in substrate type by demonstrating a large difference in prey availability and avian use between upper and lower leaf surfaces in a Panamanian forest.

In general, this level of understanding has only been possible in studies of guilds such as frugivores or nectarivores in which resources are clearly defined and can be measured precisely. In such studies, the relationship between food availability and community organization has been demonstrated, as has the potential for coadaptation between plants and their specialized avian pollinators (Feinsinger and Colwell 1978, Stiles 1985c) and seed-dispersers (Howe 1977, Moermond and Denslow 1985). Could such strong interactions exist between avian insectivores and their prey? The answer must begin with a detailed knowledge of the distribution and availability of arthropods and their selection by birds exploiting specific foraging sites.

The present study provides clear evidence that birds foraging on dead vs. live foliage are exposed to very different prey choices (cf. Gradwohl and Greenberg 1982a and Greenberg 1987a). The significance of these differences can be assessed, however, only through direct examination of species' diets. Preliminary analysis of stomach contents of several dead-leaf specialist birds from my study areas (Rosenberg, unpubl. data) indicates heavy predation on those taxa (e.g., Orthoptera, spiders) that were most abundant in my dead-leaf samples.

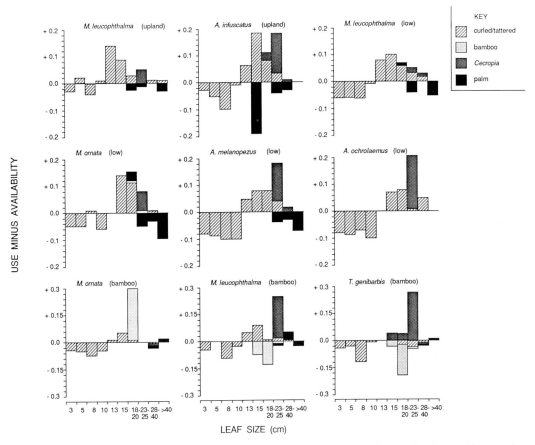

FIGURE 9. Selectivity of dead-leaf substrates by six bird species in three habitats at Tambopata (data are the proportional use of each category by the bird in relation to the availability of that category in that habitat). Bars above the horizontal represent selection and bars below represent avoidance of each category.

For sedentary, permanent-resident birds, foraging specialization may be enhanced where resources exist in predictable patches. The persistence of individual dead leaves and the turnover rates of potential prey in these leaves suggest that antwrens and foliage-gleaners may perceive these leaves as predictable and renewable resources. I suggest that the birds recognize particular leaves within their territories and visit them repeatedly.

Are the patterns discussed here unique to a novel tropical resource or do they have more general applicability for insectivorous birds? To answer this question we require more detailed prey sampling and more detailed observation of substrate and prey choice than has been done to date. For many North American insectivore guilds, for example, we know much about general foraging relationships among species, but we know little about specific diets, how these vary temporally, or how these may be mediated by the differential productivity of specific foraging sites. Certainly, guilds vary in their degree of specialization and the extent to which food availability promotes species interactions. A study design that assesses the relationship between resource availability and use is necessary to address these questions in any system. It should begin with close attention to natural history, so that levels of resource subdivision important to the birds can be determined. The relevant categories of substrate subdivision can then be sampled for potential arthropod prey. In this way, the distribution of specific foraging substrates among the available microhabitats, as well as the relative productivity of each substrate type, can be determined. All these measures may vary geographically and temporally, necessitating replicate samples.

This approach will be easier to apply in cases in which substrates occur in discrete patches, such as the dead leaves. In other systems, innovative methods may be sought to isolate and sample within both sites so that they were selected as sample trees for the canopy arthropod study.

Arthropod sampling

Each site was divided into a grid consisting of 100, 0.16 ha (40 × 40 m) subplots. Data were collected weekly from 1984 through 1986. Sampling began in June 1984 and in April of 1985 and 1986, approximately coincident with appearance of new foliage, and continued through August in all three years of study.

The sampling regime varied slightly each year, as refinements were made to optimize efficiency. In 1984, eight subplots were randomly selected on each site: four along the perimeter of each site and four in the interior. Within those subplots, two crown heights (<10 m and >10 m) were sampled on each of the three tree taxa. Each week, 48 samples (8 subplots × 3 tree taxa × 2 crown heights) were taken, resulting in 1008 samples. In subsequent years, high and low crown heights were merged into one lower mid-crown sample, as we detected no significant differences in arthropod populations as a function of height in 1984. Analyses of 1984 data in this paper used only the 504 lower crown samples, as they were most similar in height to all subsequent samples.

The number of subplots was increased in 1985 to 16 at the control and 24 at the cicada site for a total of 2400 samples. Sampling in 1986 was similar to the previous year, but one week shorter, resulting in 2244 samples.

All sampling was conducted in morning to minimize variation in insect movement within the crown (e.g., Holmes et al. 1978). Each sample unit consisted of three terminal branches from a tree cut with a pole pruner, and dropped into an attached muslin bag. We attempted to standardize each branch cutting to sample similar amounts of foliage for each species through time. Foliage consisted of leaves, petioles, and small twigs (larger stems were clipped and discarded). All foliage was immediately placed into a paper bag (ca. 30 × 17 × 30 cm), which was folded and stapled closed. Bags were returned to the laboratory, weighed to determine wet weight of foliage biomass, then frozen overnight to kill all arthropods. The next day, foliage was shaken, and all stem and leaf surfaces carefully examined. Total arthropod wet weight was measured and specimens were held for identification in 70% ethyl alcohol.

Multiple methods of sampling may be required to properly sample entire canopy arthropod communities (Cooper 1989, Morrison et al. 1989). Methods such as pole pruning miss actively flying insects, while techniques that do catch fliers (e.g., sticky traps) usually do not sample non-flying arthropods, such as spiders and caterpillars. We assumed that potential cicada predators would prey primarily on non-fliers. We also hypothesized that arthropod populations may not be immediately "released" but perhaps release would be best reflected in subsequent immature populations. We assumed that foliage collection maximized catch of immatures, a consistent majority of non-active flying adults, and spiders.

Guild selection

Arthropods were identified to order and, where possible, to family. Size of each specimen was estimated as small (<6 mm), medium (≥6 and <19 mm), or large (≥19 mm). Specimens were categorized into four guilds based loosely on feeding behavior: (1) chewing insect guild, containing all families in orders Lepidoptera and Orthoptera, sawfly families in the order Hymenoptera, and families Chrysomelidae, Scarabaeidae and Curculionidae in the Coleoptera; (2) sucking insect guild, containing all families in orders Homoptera and Thysanoptera, plus families Aradidae, Berytidae, Coreidae, Miridae, Pentatomidae, Scutelleridae, Tingidae, and Thyreocoridae in the order Hemiptera; (3) spider guild, containing all spiders; and (4) medium to large lepidopterous larval guild, a subgroup of the chewing insect guild.

Not all taxa collected are included in our guilds. Choices for guild membership were made to incorporate the two main phytophagous insect feeding types, plus the large, entirely entomophagous, arachnid group. Medium to large lepidopteran larvae (a relatively large-sized and flightless food resource) were examined separately, as they form a common food for insectivorous birds of the forest canopy (Holmes et al. 1979c). Other authors, working with similar canopy arthropod data, have demonstrated that choice of taxa in guild formation can significantly influence results (Stork 1987; Cooper et al., this volume). We are aware that the feeding guilds we specified are broad, but their taxonomic composition remained relatively constant between cicada and non-cicada sites.

Data analyses

We summarized data in two ways: (1) numbers of individuals/sample unit; and (2) numbers of arthropods standardized by kg of tree foliage sampled, calculated by dividing number of individuals by weight of foliage. Exploratory data analysis indicated that most variables were not normally distributed and that mean/variance ratios were not stable. Because most standardized variables approximated a negative binomial or Poisson distribution, for all analyses of variance a square root transformation was used after 3/8 was added to each value, a process that stabilizes the variance of a Poisson distribution regardless of the mean (Anscombe 1948). Standardized data reported in tables are untransformed means, presented with associated sample sizes and standard errors of the mean.

The General Linear Models procedure in SAS (SAS 1982) was used for analysis of variance, with Tukey's mean separation tests ($P \leq 0.05$) where appropriate. Two-way analysis of variance was used to test for differences in mean densities between sites and among years, and, more importantly, to determine if any significant site-year interactions existed. Although arthropod levels could change from year to year and from site to site for many reasons, a significant site-year interaction would suggest that presence of periodical cicadas may have affected population dynamics of other canopy arthropods in 1985. To determine if significant interactions were due to changes between sites in 1985, differences in mean arthropod densities between sites in 1984 and in 1986 were tested against those from 1985 using t-tests (the CONTRAST option in GLM).

We considered two primary factors in categorizing data for further analysis: (1) expression of relative arthropod abundance in a manner that reduces bias associated with variation in size of sample units, and (2) combination of arthropod taxa into appropriate groups for further analysis.

RESULTS

FOLIAGE ANALYSIS AND EXPRESSION OF ARTHROPOD ABUNDANCE

Weight of canopy foliage collected/sample (combined over site, month, and year) varied significantly with tree taxa, and averaged 86.9, 75.1, and 52.6 g for cedar, hickory, and oak, respectively. Amount of foliage collected/sample varied with seasonal phenology: foliage weight of cedar and oak increased as the season progressed, and foliage weight of hickory increased from April to May then remained constant (Table 1). On average, foliage samples collected from the Cassidy site were significantly heavier than those taken at Tillery.

Mean numbers of arthropods collected/sample varied as a function of host tree (Table 2). Mean numbers for chewer and spider guilds were similar in cedar and hickory, but chewers were higher and spiders were lower on oak. Medium to large lepidopterans and sucking insects were least abundant on cedar, with larger numbers on hickory and oak, particularly for members of the sucking insect guild.

Different interpretations of guild abundance among tree species can be made, however, depending on whether or not one uses mean numbers/sample unit or mean numbers/kg of foliage sampled. For example, chewing and sucking insect guilds had significantly different numbers/kg of foliage for each tree taxon, with lowest density on cedar, a greater density on hickory, and the highest on oak (Table 2). On average, more cedar than hickory foliage was collected (Table 1); yet, number of chewers/sample unit was not significantly different between those two tree taxa. However, significant differences are evident when numbers of chewers are expressed per kg of host foliage. We conclude that expression of numbers within guilds on the basis of kg of foliage sampled is more appropriate for valid comparisons among trees, sites, seasons, and years.

GUILD DENSITIES

Approximately 165 taxa were recorded from 5148 samples. In the chewing insect guild, comparison of average numbers/kg of foliage combined over tree taxa and months revealed significant differences among years and between sites (Table 3); however, analysis of variance did not reveal a significant site-year interaction.

Cassidy consistently had a lower average density of chewing insects than Tillery. Densities were similar for 1984 and 1985 on both sites.

TABLE 1. FOLIAGE WEIGHT ANALYSIS BY TREE TAXA AND SITE. DATA ARE NUMBERS OF SAMPLES, MEAN GRAMS OF FOLIAGE, PLUS OR MINUS STANDARD ERROR OF THE MEAN, SEPARATED BY TREE TAXA OR SITE OVER ALL YEARS[a]

	April			May			June			July			August			Overall means		
	N	$\bar{x}$	SE	N	$\bar{x}$	SE	N	$\bar{x}$	SE	N	$\bar{x}$	SE	N	$\bar{x}$	SE	N	$\bar{x}$	SE
Cedar	136	76.7	2.29 A[b]	383	76.1	1.51 A	360	83.3	1.74 B	456	89.1	1.69 C	380	102.1	1.95 D	1715	86.9	0.85
Hickory	136	62.3	2.79 A	384	78.5	1.60 B	360	75.2	1.54 B	455	75.0	1.47 B	380	76.3	1.49 B	1715	75.1	0.74
Oak	136	25.3	1.09 A	384	47.1	1.06 B	360	48.8	1.06 B	456	54.2	1.10 C	380	69.8	1.29 D	1716	52.6	0.60
Cassidy site	216	60.5	2.48 A	719	68.8	1.21 B	624	72.1	1.35 BC	815	76.7	1.27 C	660	85.8	1.40 D	3034	74.7	0.64
Tillery site	192	48.3	2.04 A	432	64.6	1.37 B	456	65.0	1.31 B	552	67.0	1.25 B	480	78.5	1.43 C	2112	67.0	0.66

[a] Comparison of means between each month, as well as overall means between Cassidy and Tillery, are all significantly different ($P < 0.05$).
[b] Means and SE's of the combined tree taxa across each row followed by the same letter are not significantly different ($P < 0.05$).

TABLE 2. Comparison of Mean Numbers/Sample Unit vs. Mean Numbers/kg of Host Foliage for the Four Arthropod Guilds on Each of the Tree Taxa Sampled

	Cedar		Hickory		Oak	
	$\bar{X}$[a]	SE	$\bar{X}$	SE	$\bar{X}$	SE
Chewers						
Numbers	1.12	0.04 A	1.21	0.05 A	1.38	0.05 B
No./kg	13.99	0.49 A	19.14	0.86 B	31.00	1.17 C
Medium & Large Lepidoptera						
Numbers	0.16	0.01 A	0.38	0.02 B	0.34	0.02 B
No./kg	1.47	0.13 A	5.53	0.32 B	7.33	0.45 B
Suckers						
Numbers	0.75	0.04 A	3.16	0.14 B	3.26	0.16 B
No./kg	9.34	0.52 A	45.53	1.88 B	60.49	2.50 C
Spiders						
Numbers	1.62	0.05 A	1.58	0.05 A	1.16	0.06 B
No./kg	20.68	0.74 AB	23.38	0.76 A	22.55	1.03 B

[a] Means across each row followed by the same letter are not significantly different, based on analysis of transformed data ($P < 0.05$). Number of samples for each mean calculated equals 1716.

Chewers increased in 1986 at Cassidy, as would be expected if populations were released in the year following cicada emergence. However, a similar increase occurred at Tillery, negating the idea of a treatment effect at Cassidy.

Lepidopteran larval densities (Table 3) were different from chewers. Lepidopteran larvae were about twice as common in 1984 on Tillery compared to Cassidy. A small, non-significant increase at Cassidy and decrease at Tillery occurred in 1985, and the density at Tillery still was significantly greater than at Cassidy in 1985. A slight increase occurred at Cassidy in 1986, but this site was still less than Tillery. No significant site-year interaction was found, suggesting that periodical cicadas had no impact on population dynamics of medium and large lepidopteran larvae.

Sucking insect densities were substantially higher in 1984 than in either of the two succeeding years (Table 3). Cause of the decline was unknown, but occurred in a similar manner on both sites. Average density of sucking insects was significantly different between the two sites in two of three years, 1985 and 1986. Although means shown in 1985 are close, ANOVA of the transformed data indicated slightly higher densities at Tillery. If cicadas affected bird foraging and alternate prey, one would expect a significant site-year interaction as a result of the cicada emergence, but such an interaction was not found. Comparisons produced highly significant differences among years, but patterns of change (i.e., decrease in density in 1985, which carried forward to 1986) were the same for both sites, which is not compatible with the concept of a treatment effect.

Spider populations were denser at Tillery than Cassidy (Table 3). However, magnitude of density change on each site varied among years, resulting in a significant site-year interaction. Differences in mean population densities between sites were 24 in 1984, about 11 in 1985, and 14.3 in 1986. Using the CONTRAST option in PROC GLM, we found a significant change between mean densities in 1984 and 1985, but not between 1985 and 1986. This difference was not due to more spiders on the Cassidy site during cicada emergence, as spider densities were lowest then. We further analyzed the spider data to discover if a site-month interaction existed during 1985. A changing relationship in mean numbers between sites in months during and after cicada emergence could support the ecological release hypothesis, but none was detected. We suspect that the factors causing changes in spider populations occurred during winter of 1984–1985, thus ruling out the impact of adult cicada emergence.

DISCUSSION

Test of the Hypothesis

If bird predation is a significant mortality factor in dynamics of canopy arthropod populations, results of reduced predation at Cassidy when cicadas were present might be seen in two ways: (1) an immediate increase in arthropods normally preyed upon by birds; and (2) higher populations later in the same year, or during the following year, resulting in increased reproductive output and success. Such changes would not be expected on Tillery where cicadas did not emerge.

TABLE 3. Mean Density/kg Foliage of the Chewing Insect, Medium and Large Lepidopteran, Sucking Insect, and Spider Guilds at Each Study Site, Averaged Over All Tree Taxa and All Months

Guild	Year	Cassidy site				Tillery site				
		N	$\bar{X}$	SE		N	$\bar{X}$	SE		
Chewing	1984	264	11.5	1.01	A[a]	240	24.1	1.94	AB	**[b]
	1985	1440	13.3	0.55	A	960	21.3	1.02	A	**
	1986	1332	23.6	1.10	B	912	33.0	1.86	B	**
Medium	1984	264	3.1	0.54	A	240	6.9	1.14	A	**
& Large	1985	1440	4.0	0.31	A	960	5.6	0.41	A	**
Lepid.	1986	1332	4.5	0.39	A	912	5.5	0.55	A	**
Sucking	1984	264	55.6	4.91	A	240	65.6	6.73	A	**
	1985	1440	34.4	2.20	B	960	36.5	2.14	B	**
	1986	1332	34.0	2.09	B	912	41.4	2.54	B	**
Spiders	1984	264	17.4	1.86	A	240	41.3	4.15	A	**
	1985	1440	12.3	0.48	B	960	23.7	0.96	C	**
	1986	1332	20.7	1.01	A	912	35.0	1.41	B	**

[a] Within each guild, means among years within a site (i.e., within columns) followed by the same letter are not significantly different ($P < 0.05$).
[b] Means between sites during a specific year (i.e., across rows) are all significantly different (**) ($P < 0.05$).

Another possibility considered, which might mask an ecological release resulting from changes in bird foraging patterns, was that arthropod predators could respond functionally or numerically (Holling 1959b), or both, to increases in their canopy arthropod prey. The density-dependent mortality they might cause would conceal the impact of reduced bird predation. We theorized that if that were happening, we should see significant increases in a major predator guild such as spiders (Smith et al. 1987). As evident from Table 3 and the above results, spider populations decreased in 1985 on both sites, and increased in 1986 on both sites, again suggesting that populations of spiders were changing independent of cicada emergence.

Based on the above analyses of site-year interactions, no significant treatment effects from the cicada emergence were evident for chewing, sucking, or lepidopterous larval guilds, and the significant difference found for the spider guild did not appear to be associated with the period when cicadas were present. Thus, we conclude that the hypothesized ecological release did not occur.

Do Birds Affect Prey Population Levels?

The lack of any noticeable effect of periodical cicada emergence on the population dynamics of canopy arthropod prey leads us to consider the general effect of forest birds on canopy arthropod population dynamics. In the entomological literature, the supposition that predators can regulate populations of their arthropod prey formed the basis for the developing concepts of biological control (e.g., Smith 1939, DeBach 1964, Huffaker and Messenger 1976) and the impact of birds on specific insects has been clearly documented in some forest habitats (e.g., Dahlsten et al. 1977, Dahlsten and Copper 1979, Torgersen and Campbell 1982, Torgersen et al. 1983). In general, birds are thought to have greater impacts at endemic rather than epidemic prey densities (reviewed in Buckner 1966), although magnitude of the impact depends on which life stage suffers the greatest predation (e.g., Smith 1985). Many studies have demonstrated that bird predation can be an important source of mortality to overwintering crop pests (e.g., MacLellan 1958, Buckner 1966, Solomon et al. 1976, Stairs 1985).

In the ecological literature, a more general consideration of interactions among trophic levels led some, particularly Hairston et al. (1960), to conclude that animals in higher trophic levels can affect the populations of organisms in lower trophic levels. While that conclusion is not without controversy, it did stimulate interest in the interactions between birds and their arthropod prey. For example, a number of studies have focused on the impact of bird predation on spider populations, concluding that bird predation is important in both tropic and temperate regions (Rypstra 1984), and that winter mortality due to bird predation can be great (Askenmo et al. 1977), birds apparently eating larger individuals (Gunnarsson 1983) of all spider species encountered (Norberg 1978).

It is well documented that some bird species are attracted to arthropod outbreaks (reviewed in Otvos 1979; Kellner et al., this volume). It is less clear, however, that at low densities, territorial forest birds can substantially impact available arthropod resources. The most widely-cited work is that of Holmes et al. (1979c), who found a higher density of lepidopteran larvae inside exclosures designed to eliminate the effect of bird predation in a northern hardwood forest in New Hampshire. However, attempts to replicate that study in the woodlands of eastern Kansas have failed to produce any effect due to bird predation (R. Holt, unpubl.). Our bird census data (K. G. Smith et al., unpubl.) suggest that birds responded to presence of periodical cicadas by congregating in the emergence area. An effect of that might be to maintain high levels of predation on canopy arthropods despite the increased consumption of cicadas by individual birds. Our observational data, however (Steward et al. 1988a, b), indicate this did not happen. Our attempts to study bird predation on lepidopteran larvae by placing caterpillars in the canopy on the Cassidy site in 1984 failed due to heavy predation by vespid wasps (Steward et al. 1988b).

By counting cicada emergence holes in 16 1-m^2 plots in each of the 100 subplots and by using wing traps (see Karban 1982) to collect wings of cicadas that had been eaten by birds, we estimated that over one million adult periodical cicadas emerged on the Cassidy site during 6 May to 3 June 1985 and that birds consumed about 15% of them (Steward 1986; K. S. Williams et al., unpubl.). During that same period, sampling at Tillery yielded no adult cicadas. We suggest those differences in arthropod abundance should have been sufficient to induce a treatment effect if one were to occur.

Initially, we had concerns that high variation in canopy arthropod densities could cause difficulties in determination of treatment-induced differences between sites. However, data presented here indicate that differences associated with such variables as tree species, month, site, and year were detectable in each of the guilds studied. This lends credibility to the suggestion that our extensive sampling effort produced sample sizes sufficient to detect an effect, had treatment-induced differences been present. We conclude that forest birds in the Ozarks may not have a significant impact on the population dynamics of their arthropod prey, and that results from studies conducted in northern forests may not be generalizable to situations in southern forests (see also Rabenold 1978, 1979; Steward et al. 1988b).

ACKNOWLEDGMENTS

We thank M. Cassidy and M. Tillery for allowing this research to be conducted on their lands. We also thank the many dedicated students and staff who helped in field collection and laboratory sorting of foliage samples. We appreciate the assistance of M. Mathis and D. Goldhammer in arthropod identification, R. Sanger for computer manipulation and analysis of data, and J. E. Dunn for statistical advice. We are grateful to R. T. Holmes, C. J. Ralph, and H. Recher for their time and expertise in review of this manuscript. This study was supported, in part, by National Science Foundation grant BSR-84-08090. Published with the approval of the Director, Arkansas Agricultural Experiment Station.

INFLUENCE OF PERIODICAL CICADAS ON FORAGING BEHAVIOR OF INSECTIVOROUS BIRDS IN AN OZARK FOREST

Christopher J. Kellner, Kimberly G. Smith, Noma C. Wilkinson, and Douglas A. James

Abstract. Six aspects of foraging behavior of Tufted Titmouse, Red-eyed Vireo, Acadian Flycatcher, and Blue-gray Gnatcatcher were quantified before, during, and after an emergence of 13-year periodical cicadas in 1985 and during the same three periods in 1986 when no cicadas were present. Comparisons were made among the three periods within years and within periods between the two years to determine the effect of a superabundant food supply on foraging behavior of birds. No obvious effects of cicadas on avian foraging behavior were detected among periods in 1985 and variability in foraging behaviors among periods in 1986 was similar to 1985. In both years, the greatest changes in foraging behavior occurred between the first two periods, suggesting a seasonal component to foraging behavior in the Ozarks. Comparing the period when cicadas were present in 1985 with the same period in 1986 also failed to show an obvious effect of cicadas on foraging behavior. Substantial variability existed between years in all three periods, suggesting annual behavioral flexibility within species for the six variables that we measured. Substantial seasonal and annual variations limited our ability to detect an effect of cicadas on foraging behavior. That suggests that combining data from different seasons and years may bias results and that the traditional approach of defining microhabitat and foraging variables a priori may be inadequate.

Key Words: Arkansas; foraging behavior; predator swamping; periodical cicadas; seasonal variation.

Outbreaks of arthropods provide opportunities for examining the importance of food on many aspects of avian ecology. For instance, irruptions of spruce budworms (*Choristoneura fumiferana*), bark beetles, gypsy moths (*Lymantria dispar*), termites and periodical cicadas (*Magicicada* spp.) provide numerous species of insectivorous birds with a superabundance of food (e.g., Forbush 1924, Morse 1978b, Otvos 1979, Dial and Vaughan 1987, Steward et al. 1988a). It is well known that birds will concentrate in such patches of abundant food.

One of many aspects of avian ecology that could be affected by a superabundance of food is foraging behavior. It is often assumed that insectivorous birds are behaviorally flexible, allowing them to respond opportunistically to changes in arthropod abundance (e.g., Rotenberry 1980a), and that they are able to partition resources by selecting different microhabitats or using different foraging modes (e.g., Hespenheide 1975a). Thus, a common approach to studying foraging behavior of insectivorous birds is to examine foraging mode and microhabitat use (e.g., MacArthur 1958, Morse 1968), often comparing foraging behaviors of sympatric species (e.g., Root 1967, Rice 1978, James 1979).

Two potential problems common to such studies are (1) selection of appropriate variables and (2) combining data collected over more than one season. Avian ecologists often focus on parameters that describe microhabitat, i.e., that subset of available habitat that a species actually uses in searching for and obtaining prey. However, a researcher's definition may not coincide with a bird's perception of a given microhabitat, especially when discrete categories are arbitrarily formed for variables that are actually continuously distributed (e.g., relative height, relative position). Consequently, an investigator's definition of foraging variables may influence both the strength and validity of the conclusions. Secondly, in many studies, foraging modes and microhabitat utilization are examined throughout a season with little attention given to potential, but important, changes in foraging behaviors that occur within and between seasons. Investigators recently have reported substantial variation in behavior both between (e.g., Alatalo 1980, Hutto 1981b, Greenberg 1987b) and within seasons (e.g., Saether 1982; Carrascal 1984; Carrascal and Sanchez-Aguado 1987; Hejl and Verner, this volume). As a result, combining data over several field seasons or even over a single season may lead to biased results and conclusions.

Our objective was to determine the adequacy of "traditional" variables as used in studies of avian foraging mode and microhabitat to describe responses of birds to an outbreak of cicadas. Any conclusions would necessarily take into account variation within seasons and between seasons due to the presence or absence of cicadas.

METHODS

Two study sites in Washington Co., Arkansas, were used: an upland hardwood forest adjacent to hayfields, located northeast of Durham, and an area adjacent to

TABLE 1. Summary of Differences During 1985 and 1986 for the Six Foraging Variables among the Three Periods for the Four Bird Species Studied. All Variables Were Tested Using Likelihood Chi-squared Tests, Except for Absolute Height, for Which a Kruskal-Wallis Test Was Used. Empty Cells Signify $P > 0.05$, * = $0.01 < P < 0.05$, ** = $0.001 < P < 0.01$, *** = $0.0001 < P < 0.001$, **** = $P < 0.0001$

Species	Year	Relative height	Relative position	Foraging mode	Canopy	Absolute height	Substrate
Red-eyed Vireo	1985	***	***	***		****	
	1986		*		***	**	**
Tufted Titmouse	1985	***	*	*	***	****	***
	1986	****			***	****	
Blue-gray Gnatcatcher	1985			****			
	1986	****		****	****		*
Acadian Flycatcher	1985						
	1986	***				****	

the north and west banks of Lake Wilson on the outskirts of Fayetteville. Dominant trees at the Durham site were post oak (*Quercus stellata*) and black oak (*Q. velutina*), shagbark hickory (*Carya ovata*), and eastern redcedar (*Juniperus virginiana*). Dominant trees in upland hardwood forests that surround Lake Wilson were post and black oaks, shagbark and black (*C. texana*) hickories, and winged elm (*Ulmus alata*). Over one million adult cicadas emerged within the forest on the 16 ha study site near Durham during the emergence year (K. G. Smith unpubl.).

Foraging data were collected by one person (Kellner) during spring and summer of 1985 and 1986. Each field season was divided into three periods: 15 April to 10 May (I), 10 May to 10 June (II), and 11 June to 31 July (III). Those periods represented times before, during, and after which adult cicadas were superabundant in 1985.

Data were recorded on only actively hunting birds that attacked prey frequently. Birds that engaged in long, uninterrupted bouts of singing or that were in the company of fledglings were ignored. Once sighted, a foraging bird was followed until lost from sight. Thus, data consist of sequences of observations on individual birds as they searched for and attacked prey. The following variables were noted for each foraging bout: height in meters, relative height (upper, middle, or lower crown, in equal thirds), place in canopy (overstory or understory), relative horizontal position in the crown (inner, middle, or outer, in equal thirds), substrate on which birds were located (branch, twig, trunk, or leaf). Foraging moves, as defined by Robinson and Holmes (1982), were: (1) glean, an attack by a perched bird toward prey that also was perched; (2) hover, an attack in flight toward perched prey; (3) sally, an attack in flight toward flying prey; (4) probe, an attack by a perched bird on prey located beneath the substrate's surface. Although data were collected on a wide variety of species, here we focus on two species that were observed consuming cicadas, Tufted Titmice (*Parus bicolor*) and Red-eyed Vireos (*Vireo olivaceus*), and two species that were not observed consuming cicadas, Acadian Flycatchers (*Empidonax virescens*) and Blue-gray Gnatcatchers (*Polioptila caerulea*).

We compared foraging behavior among the three periods within and between years, allowing detection of foraging differences within each period for each species. We randomly selected approximately one third of the observations for each species and used those subsamples in all statistical analyses. This was done in an attempt to obtain independent samples; however, we realize that this method does not guarantee independence of observations. Likelihood ratio Chi-squared or Fisher's exact tests were used to test for differences for variables with discrete or continuous data, while tests involving height, the only continuous variable, were done using Kruskal-Wallis or Wilcoxon tests. Differences were considered to be statistically significant at $P \leq 0.05$. We also used Schoener's (1970) similarity index to compare foraging behavior of each species between consecutive periods and between the same period in consecutive years.

RESULTS

WITHIN-YEAR COMPARISONS

If cicadas were responsible for variation in foraging behavior among periods, more differences among periods should occur in 1985 than 1986 and those differences would be due to changes in behaviors of vireos and titmice but not flycatchers and gnatcatchers. We found that comparisons between periods within a year were significantly different in 24 of 48 cases (Table 1); however, within-year variation was not restricted to 1985 (the year cicadas were present) for any species. Tufted Titmice showed more significant variations in behavior in 1985 than 1986, but Red-eyed Vireos, which also consumed cicadas, exhibited about the same amount of variation during each year. Acadian Flycatchers and Blue-gray Gnatcatchers both had more within-year variations during 1986, the non-cicada year.

If cicadas were responsible for significant differences in behavior, we would expect shifts in foraging between periods I and II and between periods II and III in 1985, but not in 1986. Titmice exhibited shifts in substrate use that followed that pattern (Table 2). However, shifts most

TABLE 2. Percent of Observations within Each Category and Number of Observations for the Foraging Variables Listed for the Four Species of Birds During the Three Sampling Periods in 1985 and 1986

Species Year	Period	Relative height				Relative position				Foraging mode					Canopy			Substrate		
		Lower	Middle	Upper	N	Inner	Middle	Upper	N	Glean	Hover	Sally	Probe	N	Over	Under	N	Branch	Twig	N
Red-eyed Vireo																				
1985	I	78	9	13	32	38	34	28	32	89	11	0	0	9	100	0	32	42	58	36
	II	22	29	49	41	15	15	70	41	70	30	0	0	10	100	0	41	33	67	36
	III	28	18	54	257	29	19	61	256	34	66	0	0	96	98	2	261	24	76	378
1986	I	46	16	38	132	27	20	53	132	56	44	0	0	41	82	18	133	40	60	111
	II	36	23	41	375	23	19	58	369	56	44	0	0	103	93	7	394	24	76	338
	III	34	22	44	173	23	17	60	172	39	61	0	0	59	91	9	196	24	76	147
Tufted Titmouse																				
1985	I	11	39	50	56	71	11	18	56	92	8	0	0	39	100	0	56	41	59	56
	II	36	28	36	36	56	6	39	36	81	15	2	2	58	100	0	38	63	37	35
	III	39	30	31	74	49	10	41	73	66	19	2	13	62	91	9	76	30	70	73
1986	I	23	15	62	34	26	18	56	34	71	0	0	29	7	100	0	36	56	44	32
	II	62	16	22	131	33	21	46	131	79	10	1	10	61	90	10	142	42	58	107
	III	18	41	41	17	53	13	33	15	50	21	0	29	14	76	24	21	25	75	16
Blue-gray Gnatcatcher																				
1985	I	49	26	25	65	23	28	48	64	38	24	38	0	37	97	3	66	28	72	58
	II	36	26	38	114	32	21	47	114	39	52	9	0	156	97	3	126	18	82	103
	III	35	32	32	68	19	16	65	69	22	68	10	0	135	96	4	71	20	80	66
1986	I	28	19	53	171	19	21	60	172	55	35	10	0	51	93	7	182	9	91	150
	II	44	20	36	313	19	13	68	320	33	50	12	0	127	91	9	356	15	85	251
	III	26	28	46	161	23	14	63	164	20	57	23	0	106	94	6	200	7	93	127
Acadian Flycatcher																				
1985	II	44	34	22	32	42	18	39	33	11	83	6	0	18	89	11	36	42	58	33
	III	54	19	27	154	47	8	45	153	3	90	7	0	245	80	20	167	29	71	133
1986	II	58	18	24	149	42	21	37	150	5	88	7	0	198	73	27	169	27	73	116
	III	68	16	16	68	37	7	56	68	3	85	12	0	61	74	26	90	24	76	66

TABLE 3. MEAN FORAGING HEIGHT, STANDARD DEVIATION, AND SAMPLE SIZE (N) FOR THE FOUR SPECIES OF BIRDS DURING EACH SAMPLING PERIOD IN 1985 AND 1986

Species	Period	1985			1986		
		$\bar{X}$	SD	N	$\bar{X}$	SD	N
Red-eyed Vireo	I	10.1	0.34	31	5.6	0.16	129
	II	7.9	0.72	10	6.2	0.10	361
	III	7.0	0.12	255	6.2	0.13	179
Tufted Titmouse	I	11.6	0.42	52	6.9	0.24	32
	II	5.4	0.45	37	6.1	0.15	131
	III	5.0	0.29	67	3.8	0.49	21
Blue-gray Gnatcatcher	I	5.9	0.39	56	6.0	0.17	167
	II	6.4	0.22	111	5.8	0.13	291
	III	6.5	0.20	71	5.3	0.15	169
Acadian Flycatcher	II	4.5	0.40	32	4.5	0.17	155
	III	3.9	0.17	150	3.2	0.20	83

often occurred between periods I and II, with fewer shifts between periods II and III, in both years suggesting that behaviors tend to change early in the breeding season regardless of cicadas.

The overall pattern of within-year variation is difficult to interpret. Foraging behaviors varied dramatically among periods for each of the six variables we quantified, but no trend was apparent among species. Foraging variables sometimes exhibited significant variation between periods during one year, while exhibiting little variation during another year (Tables 2 and 3). Similarly, within each year, variables often exhibited significant variation between periods for one or more species, but not for all species. Several species did exhibit significant variation between periods for the same variables in 1985 and 1986. However, in all cases save one, significant variation during one year did not follow the same pattern in the following year.

An examination of average similarity indices revealed that, in general, species foraged more similarly across the three periods in 1986 than in 1985. Tufted Titmice were the only exception and foraged more similarly in 1985 than 1986 (Table 4). In 1985, foraging differed more between periods I and II than between periods II and III for all species. No obvious trends in foraging similarities emerged during 1986. This evidence indicates that the three periods differed more in 1985 than 1986. However, this is not evidence that cicadas influenced foraging behavior because trends in foraging similarity were exhibited by the two species that did not consume cicadas, but not by titmice.

BETWEEN-YEAR COMPARISONS

Comparing foraging behavior observed within the same period of both years for each species revealed significant differences in 29 of 66 tests (Table 5). Examining only the number of significant differences, no consistent pattern emerged that would suggest cicadas influenced foraging behavior. No species exhibited more variation between 1985 and 1986 in period II than in I or III. In addition, flycatchers and gnatcatchers exhibited almost as much variation between period II of 1985 and 1986 as did titmice and vireos. Overall, species exhibited substantial variability between each pair of periods indicating that these species are capable of great plasticity for the six variables we quantified, even when comparing similar periods between years.

If cicadas were responsible for significant between-year differences, we would expect significant differences to occur between years for period II and not to occur between years for either periods I or III for those species that ate cicadas. In addition, we would not expect a similar pattern for the two species that did not eat cicadas. Comparing period II for vireos, only one significant difference was found (Table 5). Vireos foraged significantly higher in 1985 than in 1986, perhaps in response to the presence of cicadas in the upper portions of trees. However, vireos also foraged higher during both periods I and III in 1985 (Table 3). Gnatcatchers also foraged significantly higher in 1985 during both periods II and III. Titmice exhibited three significant differences in foraging behaviors that may have resulted from exploitation of cicadas (Table 5), foraging more in upper crowns and inner portions of trees and on branches during the cicada emergence (period II in 1985). Those differences were consistent with our expectations for titmice actively seeking cicadas which are known to concentrate in upper portions of trees and are most abundant along the main trunk and branches to-

TABLE 4. SIMILARITY INDICES FOR EACH SPECIES COMPARING FORAGING BEHAVIORS BETWEEN PERIODS FOR EACH YEAR

Species Period	1985						1986					
	Relative height	Relative position	Foraging mode	Canopy	Substrate	$\bar{X}$	Relative height	Relative position	Foraging mode	Canopy	Substrate	$\bar{X}$
Red-eyed Vireo												
I × II	0.44	0.73	0.81	1.00	0.91	0.78	0.90	0.95	1.00	0.89	0.84	0.92
II × III	0.89	0.95	0.64	0.98	0.91	0.87	0.98	0.98	0.83	0.98	1.00	0.95
Tufted Titmouse												
I × II	0.75	0.80	0.89	1.00	0.78	0.84	0.60	0.90	0.81	0.90	0.86	0.81
II × III	0.95	0.94	0.85	0.91	0.67	0.86	0.56	0.80	0.70	0.86	0.83	0.75
Blue-gray Gnatcatcher												
I × II	0.87	0.92	0.71	1.00	0.90	0.88	0.83	0.92	0.81	0.98	0.94	0.90
II × III	0.94	0.82	0.83	0.99	0.98	0.91	0.82	0.95	0.85	0.98	0.92	0.90
Acadian Flycatcher												
II × III	0.85	0.90	0.66	0.91	0.87	0.84	0.90	0.81	0.95	0.99	0.97	0.92

ward the center of trees. This pattern was not exhibited by any other species during period II, nor did titmice exhibit similar shifts during other periods.

If cicadas influenced foraging behavior, we would expect foraging similarities to be lowest between years for period II for the two cicada consumers, but not for the two non-consumers. However, in both years, a seasonal trend toward increasing foraging stereotypy existed from period I through period III for all species (Table 6).

DISCUSSION

We found substantial within- and between-year variation for all species that we studied, limiting our conclusions regarding the influence of cicadas on foraging titmice and vireos. Similar within- and between-year variation in foraging behavior of birds has been documented by others (e.g., Alatalo 1980, Rabenold 1980, Hutto 1981b, Wagner 1981b, Saether 1982, Carrascal 1984) and may be widespread, making it impossible to pool data over seasons or years. More importantly, unexplained seasonal or yearly variation may make conclusions concerning relationships between species (e.g., Root 1967, Rice 1978, Robinson 1981) or sexes (e.g., Williamson 1971, Bell 1982, Holmes 1986) more tenuous. Consequently, it is important that researchers direct attention toward discovering causes of seasonal and yearly variation in foraging behavior.

TABLE 5. SUMMARY OF DIFFERENCES FOR THE SIX FORAGING VARIABLES WITHIN SAMPLING PERIODS BETWEEN 1985 AND 1986 FOR THE FOUR BIRD SPECIES STUDIED. ALL VARIABLES WERE TESTED USING LIKELIHOOD CHI-SQUARED TESTS, EXCEPT ABSOLUTE HEIGHT, FOR WHICH A KRUSKAL-WALLIS TEST WAS USED, AND PLACE IN CANOPY FOR RED-EYED VIREOS IN THE PRECICADA AND CICADA PERIODS, FOR WHICH FISHER'S EXACT TEST WAS USED. EMPTY CELLS SIGNIFY $P > 0.05$, * = $0.01 < P < 0.05$, ** = $0.001 < P < 0.01$, *** = $0.0001 < P < 0.001$, **** = $P < 0.0001$

Species	Period	Relative height	Relative position	Foraging mode	Canopy	Absolute height	Foraging substrate
Red-eyed Vireo	I	***	*	*	***	****	
	II					****	
	III				***	****	***
Tufted Titmouse	I	*	***	*		****	
	II	*	*		*		*
	III						
Blue-gray Gnatcatcher	I	****		***			***
	II		***		*	*	
	III			*		****	*
Acadian Flycatcher	II			*			
	III					*	*

TABLE 6. SIMILARITY INDICES FOR EACH SPECIES COMPARING FORAGING BEHAVIORS BETWEEN YEARS FOR EACH PERIOD

Species Period	Relative height	Relative position	Foraging mode	Canopy	Substrate	$\bar{X}$
Red-eyed Vireo						
I	0.68	0.76	0.67	0.82	0.98	0.78
II	0.86	0.97	0.86	0.93	0.91	0.91
III	0.90	0.97	0.95	0.93	1.00	0.95
Tufted Titmouse						
I	0.76	0.55	0.71	1.00	0.85	0.77
II	0.74	0.78	0.92	0.90	0.79	0.83
III	0.79	0.93	0.82	0.85	0.95	0.87
Blue-gray Gnatcatcher						
I	0.72	0.89	0.72	0.96	0.81	0.82
II	0.92	0.79	0.95	0.94	0.97	0.91
III	0.87	0.96	0.87	0.98	0.87	0.91
Acadian Flycatcher						
II	0.84	0.98	0.68	0.84	0.85	0.84
III	0.86	0.89	0.95	0.94	0.95	0.92

Several factors may account for seasonal and yearly variation in foraging behavior of birds. First, yearly variation in weather may influence patterns of plant phenology, which, in turn, may influence abundance and availability of arthropods (see Hejl and Verner, this volume). Second, spring migration results in population fluctuations that may influence availability of arthropods, or influence territorial behavior of residents, ultimately resulting in changes in foraging behavior of birds. Third, stage in the breeding cycle will influence foraging behavior of parent birds (e.g., Morse 1968).

Part of our inability to discern the influence of periodical cicadas on foraging behavior may also have resulted from our perspective of microhabitat. Like most researchers, we followed MacArthur (1958) in our definitions and analysis of microhabitat variables. This approach considers a host of mostly discrete variables that are analyzed as separate entities. For example, relative height is analyzed separately from all other variables including relative position. However, there is no reason to assume that a bird's behavior at a particular relative position is not influenced by its relative height. Interactions of this nature between variables could be determined through use of log-linear models (Fienberg 1977). In addition, there is no reason to assume a bird's view of microhabitat consists of discrete compartments (e.g., relative height and relative position as in this paper) and an attempt should be made to redefine microhabitat on a continuous scale. For example, relative height could be defined as a ratio of a bird's absolute height to the total height of the tree in which it is foraging. A similar ratio could be used to describe relative position. Such designations could always be converted back into traditional discrete categories.

We assumed that presence of cicadas would cause a significant change in foraging behaviors of forest birds (see also Hutto, this volume). Perhaps we did not document such a change because cicadas were numerous throughout the study area and birds were able to consume them without shifting microhabitat use. It is also possible that we would have seen a greater effect had we analyzed variables that characterize speed, direction, and distance moved by foraging birds. Morton (1980a) found that such variables were often superior to microhabitat parameters in distinguishing between species. Hutto (this volume) also discussed that notion, contending that changes in food resources (i.e., arthropods) were reflected by changes in foraging movements of insectivorous birds.

ACKNOWLEDGMENTS

R. Hutto, H. Recher, and S. Robinson reviewed the manuscript and provided helpful comments. J. Birdsley provided assistance in the field and K. Golden assisted in data processing. We thank M. Cassidy for kindly allowing us to conduct our research on his property. This research was supported by a grant from the National Science Foundation (BSR 84-08090).

THE INFLUENCE OF FOOD SHORTAGE ON INTERSPECIFIC NICHE OVERLAP AND FORAGING BEHAVIOR OF THREE SPECIES OF AUSTRALIAN WARBLERS (ACANTHIZIDAE)

HARRY L. BELL AND HUGH A. FORD

Abstract. Three species of similar-sized Australian warblers (Acanthizidae) differed markedly in their foraging behavior in eucalypt woodland in northeastern New South Wales. The Brown Thornbill (*Acanthiza pusilla*) is a shrub feeder, the Striated Thornbill (*A. lineata*) is a canopy feeder, whereas the Buff-rumped Thornbill (*A. reguloides*) forages on the ground and on foliage and bark over a range of heights. This study attempted to associate changes in their foraging behavior over three years with changes in the availability of food. Rainfall was well below average in the second and third years of the study. The energy demand of insectivorous birds did not decline during the study, although it was higher in spring and summer than in autumn each year. The abundance of arthropods declined markedly during the drought. Foraging overlaps between the species initially declined as food became scarce. They rose again in winter and spring, 1980, at the height of the drought, when food was particularly scarce. Thornbills appeared to respond to persistent food shortage by expanding their foraging niche and risking greater interspecific competition. Attempts to correlate foraging behavior of insectivorous birds with availability of food are valuable, although measuring arthropod abundance is time-consuming and results are rarely clear cut.

Key Words: Foraging behavior; food shortage; niche overlap; thornbills; *Acanthiza*.

Most studies comparing the foraging behavior of related bird species have made no attempt to measure changes in the abundance of food. Yet, changes in foraging behavior and overlap between species in relation to food abundance provide valuable information on the potential for interspecific competition. For instance if food is superabundant, then two species could overlap completely yet not experience competition. As food becomes scarcer, species should diverge in their foraging behavior and so use different resources to reduce the potential for competition (Lack 1947, Svärdson 1949). However, as intraspecific competition will be stronger than interspecific competition, each species should also broaden its diet (Svärdson 1949, MacArthur and Pianka 1966). As food becomes scarce, an individual should take a wider range of the foods that it encounters, regardless of what other species are doing (Krebs and Davies 1981). In extreme conditions a species may resort to unusual foraging behavior or food.

The foraging behavior of species has been compared during periods of relative abundance and scarcity in many studies (summarized by Smith et al. 1978, Schluter 1981, Schoener 1982). In all but two of these studies overlaps between the species were less when food was judged to be scarce. These studies included animals as diverse as fish (Zaret and Rand 1971), ungulates (Jarman 1971), and doves (Morel and Morel 1974). In none of the studies was foraging behavior and food compared continuously over a period of a year or more. Wiens (pers. comm.) has suggested that as food becomes increasingly scarce one might expect species first to diverge and become more specialized and then to become more opportunistic and exploit whatever food remains. The latter could lead them to overlap more extensively with each other.

The primary objective of this paper was to describe changes in foraging behavior and niche overlap among three small, insectivorous birds through a period of severe food shortage. In particular, the following questions were asked: (1) How do the species differ in their foraging behavior? (2) How does their foraging behavior change seasonally and in successive years? (3) Are seasonal changes related to changes in food abundance? (4) Does food shortage lead to increased or decreased overlap in foraging behavior?

Australian warblers

The Australian warblers (Acanthizidae) are small, insectivorous birds related to the Australian wrens (Maluridae) and honeyeaters (Meliphagidae) (Sibley and Ahlquist 1985). The main genera are *Acanthiza* (thornbills), *Gerygone* (gerygone-warblers) and *Sericornis* (scrub-wrens). Gerygone-warblers forage actively from the foliage of trees, whereas scrub-wrens mostly forage on or near the ground (Recher et al. 1985, Ford et al. 1986). Thornbills, the focus of this study, range from the ground to the canopy. They are found in the temperate parts of Australia or in cool montane forests in the tropics. The three main species groups are the *A. lineata-nana* group

TABLE 1. Rainfall (mm) Each Month of the Study, Compared with 40-Year Mean

	Jan	Feb	Mar	Apr	May	Jun	Jul	Aug	Sep	Oct	Nov	Dec	Total
1978	209	34	137	43	73	42	33	36	30	62	54	176	929
1979	128	14	90	12	63	36	14	9	12	46	81	6	511
1980	63	36	16	0	123	13	9	2	0	56	8	45	371
1981	3	86	16	56	63	8	22	12	53	87	76	92	574
Mean	118	95	68	34	31	45	36	53	38	66	80	95	759

of arboreal feeders, the *pusilla* group of shrub feeders, and the *reguloides* group, which tends toward ground feeding. One member of each group was common in the study area: Striated Thornbill (*A. lineata*), Brown Thornbill (*A. pusilla*) and Buff-rumped Thornbill (*A. reguloides*). General information on habitat, foraging and breeding behavior is summarized in McGill (1970), Frith (1969, 1976) and MacDonald (1973). Recher et al. (1985), Woinarski (1985), Ford et al. (1986) and Recher et al. (1987) have presented data on foraging behavior. Details of the social organization have been published in Bell and Ford (1986). Basically, *lineata* and *reguloides* are cooperative breeders that occur in pairs, trios, or quartets in the breeding season and in clans of up to 20 birds in the nonbreeding season. In contrast, *pusilla* holds territories as pairs throughout the year. All three species are sedentary and of similar size (7 g). Full details of breeding and foraging behavior of the populations studied are presented in Bell (1983) and a summary is provided in Bell (1985a).

STUDY AREA AND METHODS

The work was carried out at Wollomombi Falls Recreation Reserve (30°32'S, 152°02'E, now part of the Oxley Wild Rivers National Park), 40 km east of Armidale in northeastern New South Wales. The site was on the edge of an undulating plateau (920 m) above the gorges of the Macleay River. Steep escarpments provided boundaries to the study area on three sides and cleared land bounded the fourth.

Mean temperatures in Armidale range from 26°C (mean maximum) and 12°C (mean minimum) in January to 12°C (mean maximum) and 1°C (mean minimum) in July. The annual rainfall at Wollomombi averages 759 mm, with a peak in summer. The study coincided with a period of increasing drought (Table 1), which had a severe effect on the vegetation and arthropods.

The vegetation is eucalypt woodland merging in places into open forest. The tree canopy covered 36.7% of the area, with shrubs covering 13.6% (line-transect interception technique, McIntyre 1953). We estimated 84 trees and 316 shrubs per hectare (point-centered quarter sampling method, Cottam et al. 1953). The main trees were stringybarks (*Eucalyptus caliginosa* and *E. laevopinea*—52%), boxes (undescribed species related to *E. cypellocarpa*, *E. conica*, *E. melliodora* and *E. bridgesiana*—28%) and gums (*E. viminalis*, *E. amplifolia* and *E. blakelyi*—19%). The main shrubs were bipinnate *Acacia* (17% of plants, 45% of canopy volume of shrubs), *Cassinia* (18% of plants and volume), *Olearia* (44% and 21%), *Jacksonia* (16% and 9%), and *Exocarpos* (1% of each). Most trees were about 15 m tall, with a few to 30 m. Shrubs were mostly about 2 m high, except for the acacias (typically 5 m). About 3% of the trees and 21% of the shrubs died during the drought. One third of the trees and 40% of surviving shrubs lost most or all of their leaves. Canopy cover of the remainder was thinned.

Arthropods were sampled at monthly intervals from the foliage of *Eucalyptus*, *Acacia*, *Olearia*, *Jacksonia*, *Cassinia* and *Exocarpos*. Samples of insects from ground vegetation were also taken each month. The details of methods and the results are presented in Bell (1985b) and in Ford, Huddy, and Bell (this volume).

Arthropods were also counted monthly from March 1978 to February 1979 on the surface of eucalypt trunks (by Noske 1982). A square 50 cm on each side was checked on eight trees of each of four species and all arthropods seen in a 3-minute period were recorded.

Relative densities of birds were estimated during fine weather each month on transects 600 m long. Distance from the center line of the transect was recorded and the density of insectivorous species estimated by the method of Balph et al. (1977). These data were used mainly to identify changes in relative abundance over the 3 years of the study. Weights for each species were taken from our own banding data. Daily energy requirements were calculated from the formula of King (1974) ($\log_{10}$DER = $\log_{10}$ 317.7 + 0.7052 $\log_{10}$W), in which W = weight in kg, DER multiplied by 4 to convert into kJ). Values were weighted by season, 1.5× in winter and spring, 1.1× in summer and autumn to allow for thermoregulation in cold weather and increased demands due to breeding and molt. Energy demand was calculated from the sum of DER for each species, multiplied by its estimated density.

Foraging data were collected each month from September 1978 to April 1981. For each of the three species, 750 foraging moves were recorded each month, consisting of five successive moves from 150 encounters. A foraging move was one move, in the course of foraging, from one perch to another. In the last year only 450 moves from 90 individuals were recorded monthly. Observations were distributed evenly among morning, the middle of the day and afternoon, and were made in fine weather with no more than light to moderate winds. For each move the following were recorded: substrate, plant species, height and foraging method. For each bird, its identity, location, identity of any

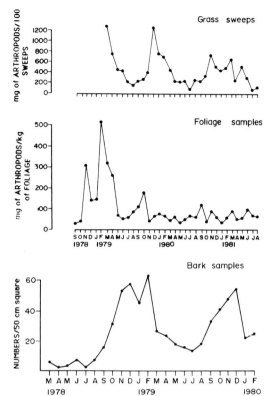

FIGURE 1. Monthly biomass of arthropods from surface vegetation (March 1979 to August 1981), foliage (September 1978 to August 1981), and bark (March 1978 to February 1980).

birds with which it was associated, date, and time of day were recorded. Substrates used were ground (including grass-tussocks, logs, stumps, stones, and cow dung), bark (including trunks, limbs, branches, hollows, lichen, and strips of hanging bark); leaves (including mistletoe clumps, insect nests, and flowers) and air. Plant species included the shrubs mentioned previously; eucalypts were separated into three groups by bark type, though initially the types were not separated (eucalypts grouped). Heights were recorded in the following categories: ground, 0–1 m, 1–2 m, 2–5 m, 5–10 m, 10–15 m, and >15 m. The following foraging methods were used: hawking (bird and prey in air), snatching (bird flying, prey on a substrate), hovering (similar to snatching but bird hovering), gleaning (bird and prey on substrate), and hanging (bird gleaning from substrate upside down).

Contingency tests were used to compare categories among seasons within a year and among years for each species and each foraging dimension. As five sequential moves were recorded the data were not entirely independent. For this reason a level of $P < 0.001$, for the appropriate degrees of freedom, was taken as denoting significance. In fact, all seasonal and yearly comparisons were significant at this level. Cells were examined to determine which contributed most to the large χ^2 value and categories that were greater than expected in one or more seasons or years identified. Overlaps were calculated from Schoener's (1968) equation:

$$O = 1 - \sum |P_{xi} - P_{yi}|$$

in which O = overlap, and P_{xi} and P_{yi} were the frequencies of observations of species x and y in category i. A Spearman's Rank Correlation was used to determine whether overlap was correlated with insect abundance.

FIGURE 2. Mean daily energy demand of insectivorous birds each month (a) and arthropod biomass (b).

RESULTS

Biomass of arthropods on foliage tended to peak in spring and early summer (Fig. 1), as found in other studies in southeastern Australia (e.g., Woinarski and Cullen 1984). Arthropod abundance declined as the drought worsened and in the last year it remained at levels more typical of winter in a normal year. As the amount of foliage declined through the drought, declines of arthropods would have been even greater than indicated in Figure 1. Biomass of arthropods from the ground also peaked in spring and summer and declined during the drought, although not as markedly as those from foliage. Arthropods on the bark surface showed marked peaks in both summers, but counts were terminated early in the drought.

The daily energy demand of insectivorous birds did not appear to decline during the study, despite the drought (Fig. 2). As the density of arthropods on foliage was lower in the last two years than in the first year, the potential for competition would have been higher then. This would have been particularly so in the third year when the amount of foliage had also declined. All three thornbills were sedentary and, as they had high adult survival rates but low breeding success (Bell and Ford 1986), their densities did not change markedly through the study (though Buff-rumped and Striated thornbills declined slightly).

SUBSTRATE

Brown and Striated thornbills mostly foraged from foliage and to some extent on branches (Fig. 3). Buff-rumped Thornbills used a variety of substrates, with about equal amounts of time on branches and the ground and somewhat less time on foliage. Brown Thornbills foraged on the

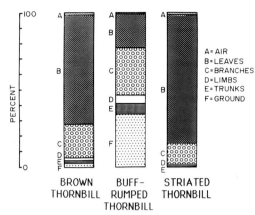

FIGURE 3. Percentage of foraging observations of each thornbill species from each substrate (data for whole study combined, sample sizes 22,222–24,369 moves per species).

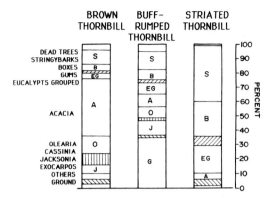

FIGURE 4. Percentage of foraging observations of each thornbill species from each plant species (sample sizes as in Fig. 3). (Eucalypts were not separated by type before April 1979.)

ground far more in autumn and winter than in other seasons in 1980 and more in 1980 than in other years. Buff-rumped Thornbills showed the most regular seasonal change in foraging substrate (Table 2; see Fig. 8 of Ford, Huddy, and Bell, this volume). Leaves were used more in spring and summer than in autumn and winter. Bark was also visited more in the warmer months. The ground was used most in winter, and in autumn in 1980. Ground foraging was not a response to increasing abundance of arthropods on ground vegetation. The amount of ground feeding by Buff-rumped Thornbills was negatively correlated with the abundance of arthropods on the ground ($r_s = -0.707$, $P < 0.05$), and the peak of ground feeding in Brown Thornbills coincided with very low levels of arthropods on the ground. Striated Thornbills did not show consistent seasonal changes in the substrate on which they foraged (Table 2).

PLANT SPECIES

Brown Thornbills fed mostly from shrubs, Buff-rumped Thornbills almost equally on eucalypts, shrubs and the ground, and Striated Thornbills foraged almost exclusively on eucalypts (Fig. 4). Brown Thornbills fed proportionally more on acacias in summer than in other seasons, and

TABLE 2. Substrates Used More Than Expected Each Season within Years and between Years by the Three Thornbills (Contingency Tests, All $P < 0.001$)

Thornbill species	Spring	Summer*	Autumn	Winter
Brown	Ground (1978) Leaves (1979)	Leaves (1979/80)	Ground (1980)	Ground (1980)
Buff-rumped	Leaves (1978) (1980) Bark (1978)	Bark (all years) Leaves (1978/79) (1980/81)	Leaves (1979) (1981) Ground (1980)	Ground (all years)
Striated	Bark (1978)		Bark (1980)	Bark (1980)
	1978/79	1979/80	1980/81	
Brown		Ground		
Buff-rumped	Leaves, bark	Ground	Leaves	
Striated		Bark		

* As summer spans the months December–February, the year is denoted 1979/80, etc.

TABLE 3. Plant Species Used More in a Season within a Year, or in One Year Compared with Other Years by Three Thornbill Species (Contingency Tests, All $P < 0.001$)

Thornbill species	Spring	Summer	Autumn	Winter
Brown	Olearia (1979) Exocarpos (1979) (1980)	Acacia (all years)	Eucalyptus (1979) Ground (1980)	Olearia (1979) Eucalyptus (1980) (1981) Ground (1980) Cassinia (1981)
Buff-rumped	Eucalyptus (1978) Acacia (1979) Jacksonia (1978) (1979)	Acacia (all years) Jacksonia (1979/80)	Ground (1980)	Ground (all years)
Striated	Acacia (1979) (1980) Low shrubs* (all years)	Acacia (1978/79)	Eucalyptus (1979)	Low shrubs (1979)
	1978/79	1979/80	1980/81	
Brown	Olearia	Olearia	Eucalyptus, Cassinia, Exocarpos	
Buff-rumped	Eucalyptus, Acacia	Ground Jacksonia Exocarpos	Eucalyptus	
Striated	Acacia Low shrub	Eucalyptus	Acacia Low shrub	

* All shrubs combined, except *Acacia*.

tended to visit eucalypts more in autumn and winter and other shrubs more in spring (Table 3). Buff-rumped Thornbills consistently visited acacias in summer and *Jacksonia* in spring and summer more than in other seasons, whereas Striated Thornbills visited acacias and other shrubs most in spring. There was no correlation between the proportion of observations of foraging by Brown Thornbills on each plant species and the abundance of arthropods each month (Spearman Rank Correlations, all $P > 0.1$). Buff-rumped Thornbills, however, fed more on acacias and *Olearia* when arthropods were more abundant on these shrubs ($P < 0.05$).

All species showed differences between years in the amount of foraging on each plant species (Table 4), although these were not consistent in the three species.

Height

Brown Thornbills foraged at intermediate levels, as expected from their preference for shrubs (Fig. 5). Buff-rumped Thornbills foraged on the ground and at a wide range of other heights. Striated Thornbills mostly foraged high, on eucalypts. Brown Thornbills fed most often on the ground in the autumn and winter of 1980 and more often above 5 m in autumn and winter of 1981 (based on contingency tests between seasons within years and between years). Buff-rumped Thornbills fed more on the ground in winter each year and in autumn in 1980 and higher in spring and summer in all years. Striated Thornbills tended to forage low during the first year but high in autumn to spring 1980.

Foraging Method

All three species foraged principally by gleaning (Fig. 6). Brown Thornbills snatched and hung more than the Buff-rumped Thornbills. Buff-rumped Thornbills sallied least and gleaned most and Striated Thornbills hung more than the other two species (contingency tests, all $P < 0.01$). There was a tendency for Brown and Buff-rumped

TABLE 4. Seasons within Years and Years in Which the Three Thornbills Used Less Common Foraging Methods More Than Expected (Contingency Tests, All $P < 0.001$)

Thornbill species	Spring	Summer	Autumn	Winter
Brown	Hang (1979, 1980) Snatch (1979)	Hang (1979/80)	Snatch (1979) Hang (1979)	Hover (1979) Hang (1981) Sally (1980) Snatch (1979, 1980)
Buff-rumped	Nonglean* (1978)			Nonglean (1981)
Striated		Snatch (1979/80) Hang (1979/80)	Hang (1979) Snatch (1979)	Hover (1979) (1980) Hang (1981)
	1978/79	1979/80	1980/81	
Brown	Hover Hang		Snatch	
Buff-rumped	Nonglean			
Striated	Sally Hover Hang	Snatch		

* Sally, hover, snatch and hang combined.

thornbills to glean most in summer and use the more active methods more in winter (Table 4). These methods were used by all species most in 1978/79.

Overlaps

All three species showed moderate to substantial differences along each of the foraging dimensions (Table 5). Cody (1974) proposed that overlaps along different niche dimensions could be combined in two ways. Where the dimensions were totally independent then overlaps should be multiplied (product α), but where they were totally dependent they should be averaged (sum α). We found some cases in which the different dimensions were highly interdependent and others in which they were independent to at least some degree. Therefore, a matrix of 2310 categories (6 substrates by 11 plant species by 7 heights by 5 methods) was constructed. In fact many of these categories, (e.g., ground and all but one plant species and height) did not exist and only about 50 ever occurred at frequencies of greater than 1%. Overlaps were then calculated from all categories for which data were available.

The interspecific overlaps in this unidimensional combination are also shown in Table 5. Plant species and either height or substrate were the best separators between species. Combined overall overlaps were not much less than overlaps from plant species alone. Combined overlaps were calculated for each month and compared with the abundance of arthropods from foliage, corrected for relative abundance of different plant species (Fig. 7). Positive correlations existed between some overlaps and arthropod abundance on some substrates (Table 6). When food was more abundant, the thornbills tended to be more similar to each other in their foraging. However only a small proportion of the variance was accounted for by this correlation ($r^2 = 0.08$–0.23). Certainly part of the reason for this was the small sample size and inherent variability in the monthly arthropod samples. When overlaps were compared with the abundance of arthropods on a seasonal basis, the correlation coefficients were higher ($r^2 = 0.50$, 0.53, and 0.42 for Brown/Buff-rumped, Brown/Striated, and Buff-rumped/Striated, respectively, against arthropods on foliage).

This correlation overshadowed a secondary effect, in that overlaps may also have been relatively high when food was still scarce. For instance, in the autumn and winter of 1980, Brown and Buff-rumped thornbills became more similar to each other. Not only were arthropods on

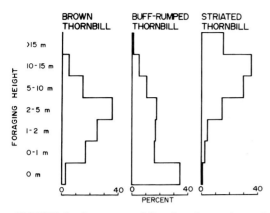

FIGURE 5. Percentage of foraging observations of each thornbill species at each height (sample sizes as in Fig. 3).

foliage and the ground scarce at this time, but the drought was at its height and many trees and shrubs were defoliated. The reason for the increased overlap was that Brown Thornbills started feeding extensively on the ground.

DISCUSSION

This study represents the first attempt to relate the foraging behavior of related and syntopic species of birds to the abundance of their food continuously over such a long period. The worst drought of the century was obviously not anticipated, but it did provide a unique opportunity to relate interspecific overlap to increasing and persistent scarcity of food. Despite the large body of data and the abundance of significant seasonal and year-to-year changes in foraging, we found few relationships between foraging behavior and food abundance. The main reason for this was that arthropod biomass tended to change in parallel on different substrates or plant species. In addition, biomass measures involved large errors and arthropods were highly variable in size and attractiveness to particular bird species. Basically, arthropod biomass was a very crude measure of the availability of food for birds such as thornbills. Except in the case of specialized insectivores where the food is known, attempts to relate behavior to food abundance are unlikely to be highly successful (see Ford, Huddy, and Bell, this volume, for discussion of examples).

The attempt to relate overlap in foraging between species to abundance of food was more successful. The similar trends in biomass of arthropods, on different substrates and plant species, meant that periods of relative overall shortage could be identified. Pooling of data from 3 months each season reduced the error in estimating biomass and allowed reasonable corre-

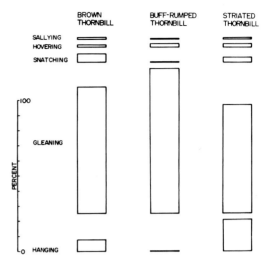

FIGURE 6. Percentage of foraging observations of each thornbill species by each method (sample size as in Fig. 3).

lations between overlap and food abundance to be obtained ($r^2 = 0.42$–0.53). This supported the hypothesis that species diverge in their niches when food becomes scarce (Lack 1947, Svärdson 1949). However, we found some evidence of convergence when food was persistently scarce. This cannot be shown by correlation, but can be revealed by looking at the responses of individual species. In this study the movement of Brown Thornbills onto the ground in autumn and winter 1980 was probably a response to defoliation and scarcity of arthropods on the low shrubs where they usually foraged. Recher et al. (1985) found that ground foraging made up 7% of observations for Brown Thornbills in southern NSW in 1980, where it was also very dry. None of the other studies of foraging in the Brown Thornbill have recorded it foraging on the ground (Woinarski 1985; Ford et al. 1986; Recher et al. 1987). Inev-

TABLE 5. OVERLAPS IN ALL FOUR FORAGING DIMENSIONS AND IN THE UNIDIMENSIONAL COMBINATION BETWEEN EACH PAIR OF THE THREE THORNBILL SPECIES

Thornbill species	Substrate	Plant species	Height	Method	Combined
Brown/Buff-rumped	51	49	66	84	39
Brown/Striated	93	30	38	84	26
Buff-rumped/Striated	36	36	34	74	12

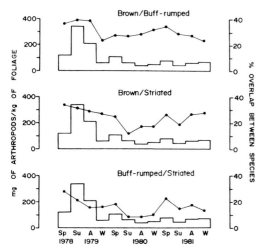

FIGURE 7. Comparison between interspecific overlap in foraging (●) and abundance of foliage arthropods in each season (histograms).

TABLE 6. RELATIONSHIPS BETWEEN INTERSPECIFIC OVERLAPS AND ABUNDANCE OF ARTHROPODS ON FOLIAGE, SURFACE VEGETATION, AND BARK (SPEARMAN RANK CORRELATION)

Thornbill species	Foliage (N = 36)	Ground (N = 30)	Bark (N = 18)
Brown/Buff-rumped	0.284*	0.365*	0.399*
Brown/Striated	0.475**	−0.018	0.123
Buff-rumped/Striated	0.334*	0.128	0.401*

* $P < 0.05$, ** $P < 0.01$.

itably such extreme responses can be recorded only on an opportunistic basis.

Questions concerning interspecific overlap in foraging behavior in relation to food abundance can be answered in several other ways. These should complement rather than replace the type of study described here. Single-species studies can examine foraging method or substrate to food availability in alternative sites (Davies 1977b; Ford, Huddy, and Bell, this volume). Foraging success may be an indirect but more reliable measure of food availability from the bird's viewpoint. However, this can be affected by the level of experience and hunger of the bird. The proportion of time spent foraging from time-budget studies can also indicate the relative abundance of food.

We have attempted to infer interspecific competition from a knowledge of interspecific overlaps and food availability. This is an improvement on interpretations based on overlap alone, or on presumed shortages of food. However, experiments either excluding birds from substrates and measuring arthropod abundance or removing birds from an area and seeking changes in foraging behavior of other species may be more profitable. Preliminary results in Armidale suggest that numbers of insects on foliage enclosed in netting increase in abundance far more than on foliage to which birds have access (Dunkerley and Bridges, unpubl.; also see Torgersen et al., this volume).

In conclusion, no single approach will provide complete answers to the questions relating foraging behavior to food abundance. Long-term monitoring of foraging behavior of insectivorous birds and the abundance of arthropods can provide a valuable overview. Carefully designed experiments may be better at answering specific questions. More studies of both types are needed to indicate the frequency and severity of food shortage experienced by birds, and the extent to which these are due to interspecific competition.

ACKNOWLEDGMENTS

H. Recher, R. T. Holmes, J. Verner and J. Wiens made many helpful comments on an earlier draft of this paper. HB received an Australian Postgraduate Research Scholarship and HF received Internal Research Grants from the University of New England. Sandy Higgins and Vi Watt typed the manuscript.

SECTION IV

ENERGETICS AND FORAGING THEORY

Overview

STUDIES OF FORAGING BEHAVIOR: CENTRAL TO UNDERSTANDING THE ECOLOGICAL CONSEQUENCES OF VARIATION IN FOOD ABUNDANCE

RICHARD L. HUTTO

Patterns at all levels of biological organization can originate as consequences of differences in survival or reproductive success among individuals. Therefore, foraging behavior takes on a special significance in explaining patterns in nature because survival and reproduction depend, ultimately, on an individual's success at acquiring and using energy from food resources. One could choose any of a number of research questions to make the points that I shall raise, but let me focus on the specific problem of understanding whether food abundance is an important determinant of breeding bird community structure.

Are the abundances and kinds of species within a specified area determined primarily by current resource conditions, or by conditions that individuals experienced at some time in the past? Historically, we have viewed communities as being composed of interacting species that somehow adjust themselves in space so that their combined abundances provide the most complete use of current resource production (to paraphrase MacArthur [1969]). The biological reality of such a proposition began to be questioned seriously by Wiens (1977, 1983), who felt that breeding-season food levels are unlikely to play a significant role in determining the local population sizes of most breeding bird species because (1) food is abundant during summer, (2) bird populations are far below food-based carrying capacities, and (3) time lags in the response of populations to changes in the environment are pronounced. Wiens argued that the structure of breeding bird communities may, instead, be determined largely by infrequent events, or ecological "crunches" (as they have come to be known), during which populations *are* limited by food.

Under this view, much of the variablility in community structure from one place to another, or one time to another in a given location, is probably due to stochastic processes acting during the more frequent periods of relaxed selective pressure (Wiens 1983). In fact, Rotenberry and Wiens (1980a) and Wiens (1981) found that the population sizes and territory positions (community composition) of shrubsteppe birds changed independently of annual changes in probable food resource levels. This led them to develop their "checkerboard" model, where changes in the distribution of individual birds on a study plot from one year to the next were suggested to be about as predictable as changes in the distribution of checkers on a checkerboard after it has been given a vigorous shake. They reasoned that in order for food levels to affect the density or distribution of birds, bird populations must be at or near their food-based carrying capacities.

The pendulum has swung back again toward MacArthur's original view with the suggestion that, while food may be abundant overall during the breeding season, there will still be spatial variation in levels of food abundance. Moreover, because a bird's use of time should be strongly influenced by the availability of food (Hutto 1985a, Martin 1986), its breeding success may depend heavily upon whether it has settled in a relatively food-rich or food-poor location. This view has its roots in optimal foraging theory, and emphasizes the fact that food limitation is not an all-or-none phenomenon. Rather, there is a continuum of possible levels of food availability and, therefore, a continuum of amounts of time that must be devoted to feeding activities. Thus, for food abundance to affect bird density or distribution, bird populations do not have to be at or near their food-based carrying capacities (Martin 1986). Even though food may not limit numbers of adults surviving the breeding season, it could still affect the reproductive success of those birds. Moreover, natural selection could lead to a close match between bird population sizes and food abundance if it were to favor those individuals that were flexible enough to settle and forage in a manner that maximized their foraging efficiency and, consequently, their breeding success.

Before we can begin to reconcile these seemingly opposing views and understand the factors that affect community structure, we will need to understand the factors that determine smaller-scale patterns, because the processes ultimately responsible for ecological patterns at the community level may actually go on at a more local level (Rotenberry and Wiens 1980b, Wiens and Rotenberry 1981). Specifically, "future studies of community organization could be strongly benefitted by more detailed studies of foraging behavior and reproductive success of individuals, and less preoccupation with populations" (Martin 1986). Individuals are, after all, the units of natural selection that survive or reproduce differentially.

With this new emphasis on the foraging ecology of individuals, several long-standing questions are being addressed with renewed vigor. Take, for example, the classic question of why no two species occupy the same niche. For 30 years the dogma has been that if species are too similar ecologically, they will compete heavily for food resources and be unable to coexist. In recent years, the importance of such competition has been challenged on the grounds that differences among species could be due to past history or to chance alone, and have nothing at all to do with interactions among species (Connor and Simberloff 1979, 1984, 1986; Strong et al. 1979; Simberloff and Boecklen 1981). While these are viable alternative explanations, the approach that has been used to distinguish between chance and competition has proven unsuccessful because of a failure to focus on the biology of individuals. Rather than look for predicted mechanisms, researchers have tested hypotheses by looking for predicted community-level consequences. Unfortunately, the latter predictions are not inferences that necessarily follow from any of the hypotheses (Diamond and Gilpin 1982; Gilpin and Diamond 1982, 1984; Wright and Biehl 1982; Case and Sidell 1983). A focus on individual organisms may lead us in a more promising direction. Specifically, if competition between two species is powerful enough to cause their divergence along some resource dimension, then individuals that lie inside the zone of ecological overlap along that dimension should do less well than those that lie outside that zone. The "past history" and "chance" models make no such prediction.

Although Wiens' view of competition was perhaps overly skeptical, a valuable consequence of his skepticism has been the present shift toward studies of the behavioral limits on individuals. For example, we are now asking whether individuals have the flexibility to be able to track changes in food resources through space and time. If so, then MacArthur's early view that population sizes of species closely match resource production might be correct after all.

The development of optimal foraging theory has also brought considerable attention to the foraging behavior of individuals. The earliest attempts to model optimal solutions to foraging behavior rarely incorporated realistic physiological, morphological, or behavioral constraints on individuals. Individuals were predicted to use those behaviors that netted the greatest amount of energy per unit time, even if the behaviors were impossible to perform. Nonetheless, these early models led us toward the realization that we need to know more about the range of behaviors that individual organisms can achieve.

The following series of papers provides a splendid example of the new understanding we are gaining through the discovery and incorporation of constraints on the foraging behavior of individuals. For example, Karasov notes that digestive rates may constrain foraging behavior by placing an upper limit on foraging rates. In addition, the presence of significant differences in digestive efficiencies among food types makes it clear that simple tallies of prey density cannot be used as estimates of energy availability. That birds are morphologically and psychologically constrained in their capacity to forage optimally is illustrated exceptionally well in the papers by Moermond and Greenberg. Finally, the papers by Dunning, Maurer, and Stephens give us a preview of the way biologists are beginning to incorporate some of these constraints into a new generation of foraging models.

This is an exciting phase in the study of foraging ecology because foraging constraints may influence everything from habitat use, through mating systems, to community structure. Researchers are beginning to take a more reductionistic approach to the study of ecological patterns by paying close attention to the foraging behavior of individuals. At the same time, they are framing questions in the context of higher levels of biological organization, which gives their studies broader significance relative to earlier studies of foraging behavior.

Energetics of Foraging

DIGESTION IN BIRDS: CHEMICAL AND PHYSIOLOGICAL DETERMINANTS AND ECOLOGICAL IMPLICATIONS

WILLIAM H. KARASOV

Abstract. I review the utilization efficiencies of wild birds on various foods. Average apparent metabolizable energy coefficients (MEC^*; [food energy − excreta energy]/food energy) according to type of food consumed are: nectar, 0.98; arthropods, 0.77; vertebrate prey, 0.75; cultivated seeds, 0.80; wild seeds, 0.62; fruit pulp and skin, 0.64; whole fruits (including seeds), 0.51; herbage, 0.35. The observed differences in MEC^* can be explained largely on the basis of differences in food composition. Fruits and herbage were utilized less efficiently than predicted on the basis of composition alone, possibly because of (1) underestimation of the refractory component of food (i.e., cell wall), (2) the presence of plant secondary chemicals, or (3) features of the digestive system, such as short digesta retention time and/or low enzyme levels.

The digestive system's efficiency in extracting food energy or nutrients is directly related to three variables: (1) digesta retention time; (2) rates of hydrolysis, fermentation, and absorption; and (3) digestive tract surface area and volume. Because these components act in concert, it is best to evaluate digestive system function in an integrated fashion. I present three examples: (1) efficiency is apparently depressed in frugivores because digesta retention time is relatively short and no compensation occurs in rates of hydrolysis and absorption; (2) herbivores must eat large amounts of food, but a compensation appears to be an increase in digestive tract volume; and (3) the presence of caeca in herbivores enhances extraction efficiency by affecting all three variables.

Digestion is important in avian ecology at the level of individuals, populations, and community structure by affecting resource removal rate, and possibly by constraining the rate of production and affecting niche width.

Key Words: Efficiency; food composition; intestine; metabolizable energy; nutrition.

Avian digestion is of interest to biologists because it is one of the factors that mediates birds' interactions with their environment. Foraging time and resource removal rate, for example, are functions of feeding rate. Feeding rate in turn is related to digestion so that for birds in steady state, feeding rate is equal to energy requirement, divided by energy value of food times the efficiency of its utilization. In addition to such ecological relations, avian digestion poses challenging problems for physiologists with the added virtue that certain aspects of the avian digestive system make birds useful models for the study of digestion in general.

I review here four topics related to digestion that have special relevance for avian biologists. First, the utilization efficiencies (a general term I use for various expressions of digestibility and metabolizability; see next section) of wild birds eating wild foods are comprehensively reviewed. The summaries and accompanying analyses should enable biologists to evaluate when they can substitute reasonable estimates for more accurate data obtained at the cost of new feeding trials.

Second, I consider the major chemical features of wild foodstuffs that determine or affect the efficiency with which a bird utilizes foods. I use a simple deterministic model of digestion based on food composition to identify important features of foods that should be measured or studied more intensively in the future.

Utilization efficiency is also affected by properties of the bird. Several recently developed models of digestion identify those particular attributes of the bird that determine digestive efficiency (Sibly 1981, Karasov 1987, Penry and Jumars 1987). Those features, their mode of action, and their interrelations are reviewed in several examples.

While the first three topics deal primarily with utilization efficiency, the fourth topic is broader. I consider how digestion rates might limit energy intake and hence rates of growth and reproduction. Also, the design and degree of adaptability of the gastrointestinal tract may determine diet diversity and hence niche width. These approaches about digestion operating as a possible constraint in ecology may represent an important direction for future research in avian digestion.

METHODS AND TERMS

UTILIZATION EFFICIENCIES

Measuring digestive efficiency in birds is a problem because the feces, which represent primarily undigested residue of food, are mixed in the cloaca with urine. Thus, the difference between the intake and excretory loss rates of dry matter, energy, or nutrients is more

TABLE 1. LIST OF SYMBOLS AND THEIR DEFINITIONS AND UNITS

Symbol	Definition	Units[a]
A_i	Ash concentration food	Proportion of dry mass
AMC^*	Assimilated mass in coefficient (apparent), eq. 2	Proportion of food dry mass
E_e	Endogenous loss of energy	kJ/day
E_m	Endogenous loss of mass	g/day
E_N	Endogenous loss of nitrogen	g/day
F	Fraction absorbed	Proportion
GE_i	Gross energy content of food	kJ/g dry mass
GE_e	Gross energy content of excreta	kJ/g dry mass
GE_R	Gross energy content of refractory material in food	kJ/g dry mass
J	Absorption rate	grams or moles per minute
MEC	Metabolizable energy coefficient (true)	Proportion of food energy
MEC^*	Metabolizable energy coefficient (apparent), eq. 1	Proportion of food energy
MEC^*_p	Predicted metabolizable energy coefficient (apparent), eq. 6	Proportion of food energy
N_i	Nitrogen concentration of food	Proportion of dry mass
Q_e	Rate of excreta production	g/day
Q_i	Rate of food intake	g/day
R_i	Proportion of food refractory to chemical digestion	Proportion of dry mass
T	Mean retention time of digesta	min
V	Amount of nutrient in the gut	grams or moles

[a] Rates can be expressed as g/day or g day^{-1} (kg body mass)$^{-1}$. All masses are dry matter basis.

properly called an apparent assimilable or metabolizable fraction (apparent because it is uncorrected for endogenous losses). In the case of energy, division of this quantity by the gross energy intake yields the apparent metabolizable energy coefficient (MEC^*; Kendeigh et al. 1977):

$$MEC^* = (GE_iQ_i - GE_eQ_e)/GE_iQ_i$$
$$= 1 - (GE_eQ_e/GE_iQ_i) \quad (1)$$

where GE_i and GE_e equal, respectively, the gross energy content (kJ/g dry mass) of the food (intake) and excreta, and Q_i and Q_e equal, respectively, the food intake rate and excreta production rate (g/day) (Table 1). Miller and Reinecke (1984) present a good review of the various expressions of digestibility and metabolizability used in the literature.

In some studies only the flux of dry matter is determined and this yields useful information on the digestive efficiency of the bird; the apparent assimilated mass coefficient (AMC^*):

$$AMC^* = (Q_i - Q_e)/Q_i = 1 - (Q_e/Q_i) \quad (2)$$

One can see that the utilization efficiencies MEC^* and AMC^* differ according to the magnitude of GE_e/GE_i, with MEC^* being the larger value. AMC^*'s are most often reported for digestion trials involving herbage or fruit. From studies where both have been determined (see Appendix 1 and Worthington 1983) I found that, for herbage and fruit, MEC^* could be estimated (on average) from AMC^* by adding 0.03. I used this manipulation in some cases because MEC^* is the more desirable quantity considering our interests in the energetics of feeding.

MEC^* and AMC^* are usually determined in feeding trials with captive birds in which Q_i and Q_e are measured, that is, total collection trials. An alternative method is to use an inert substance as a tracer to relate excreta production to food intake:

$$MEC^* = [GE_i - (\%T_i/\%T_e)GE_e]/GE_i, \text{ or} \quad (3)$$
$$AMC^* = 1 - (\%T_i/\%T_e) \quad (4)$$

where $\%T_i$ and $\%T_e$ equal, respectively, the percent tracer in the food and excreta. In the laboratory one can mix the tracer into the food (e.g., Duke et al. 1968), but it is also possible to use naturally occurring tracers. The virtue of this technique is that it can be applied to a free-living bird if the diet is accurately known and food and excreta can be representatively sampled. Following Marriott and Forbes' (1970) finding that the apparent digestibility of crude fiber in lucerne chaff by Cape Barren Geese (consult the tables in the Appendix for scientific names not presented in the text) was negligible, numerous researchers working with waterfowl have used the inert marker technique and calculated AMC^* using crude fiber (e.g., Halse 1984, Miller 1984), lignin (e.g., Buchsbaum et al. 1986), and cellulose (e.g., Ebbinge et al. 1975) as the inert marker. Because waterfowl do ferment some cell wall (see following sections of this paper), this approach can lead to underestimation of AMC^*. Moss and Parkinson (1972) and Moss (1977) used Mg as an inert marker in a study of captive and free-living Red Grouse eating heather, and concluded that free-living birds digested the heather more efficiently than captives eating the same food. In this case Mg was probably not truly inert, but rather Mg absorption by the intestine was equalled by excretion in urine.

This latter study underscores the difficulty in measuring a utilization efficiency that applies to the ecological situation. Captives fed formulated rations before feeding trials with wild foods need to be conditioned to the new wild foods. For example, when American Robins were first switched from a formulated fruit-mash ration to crickets, their MEC^* was 15% lower than it was after they had fed on crickets for three days (0.59 vs. 0.70, P < 0.001; Levey and Karasov 1989). Such lags in efficiency of digestion following a diet switch might be just a day if adaptations of digestive enzymes or nutrient absorption mechanisms are involved, or many days if changes in gut structure are involved (Miller 1975, Karasov and Diamond 1983). Allowing adequate time for adaptation to a new ration may be especially critical in studies of herbivores, in which changes in gut structure (Savory and Gentle 1976, Hanssen 1979), and hence presumably gut function, may be necessary to utilize a new food efficiently. In the wild, grouse gradually increase their intake of resinous forage well before they must rely upon it during midwinter (Bryant and Kuropat 1980).

The utilization coefficients MEC^* and AMC^* are considered "apparent" because they are not corrected for fecal and urinary endogenous losses of dry matter and energy. The endogenous component of excreta includes endogenous urinary nitrogen (the lowest level of N excretion attained under basal conditions even in the absence of protein intake) and dry matter and energy from bacteria or sloughed-off cells and secretions of the alimentary tract. One can determine the "true" metabolizability of a ration by correcting excretory losses for this endogenous component (Sibbald 1976), and this is often done in poultry science because "true" metabolizability is a more direct measure of energy availability. In chickens the endogenous energy loss (E_e) was about 21 kJ kg$^{-0.75}$ day^{-1}, or expressed as dry mass (E_m) 1.8 g kg$^{-0.75}$ day^{-1} (Guillaume and Summers 1970, Sibbald 1976). In Graylag Geese E_e was 14.4 kJ kg$^{-0.75}$ day^{-1} (Storey and Allen 1982). The correction equation for "true" MEC from MEC^* is $MEC = MEC^* + E_e/([Q_i][GE_i])$ (Guillaume and Summers 1970), while that for "true" AMC from AMC^* would be $AMC = AMC^* + E_m/Q_i$.

Apparent coefficients are generally 0.01–0.03 below "true" coefficients, and if Q_i is well below the level required for maintenance then differences can exceed 0.03 (Miller and Reinecke, 1984). Miller and Reinecke (1984) cautioned investigators to use MEC^*'s only from test birds fed at maintenance levels, though calculations with actual data in Appendix 1 show that this is unnecessarily conservative. They also discussed why the use by ecologists of apparent MEC's in energetics studies is approximately correct.

Retention Time of Digesta in the Gut

There is a certain minimum duration for a digestion trial if utilization efficiency is to be measured accurately. Marked particles of food tend to clear the digestive tract in an exponential fashion in birds eating such diverse foods as nectar (Karasov et al. 1986), fruit and insects (Karasov and Levey 1990), and seeds and herbage (Duke et al. 1968, Herd and Dawson 1984). For exponential clearance, the time to clear 98% of a marked meal is equal to about four times the mean retention time (i.e., the mean residence time of marker particles) (Karasov et al. 1986, Penry and Jumars 1987). More time is required for the metabolic processing of nutrients and excretion of urinary wastes, which are also included in a calculation of MEC^*. As discussed below, the shortest mean retention times found in birds are about 45 min in small frugivores and nectarivores, and these times increase with increasing body mass and for other foods. Thus, digestion trials that begin with fasted birds (even small ones) and last only 4–6 hr have a relatively high likelihood of yielding overestimates of MEC^* with rather high variability (according to differences between birds in the trial in mean retention and metabolic processing time). However, day-long digestion trials with American Robins and European Starlings fed crickets or fruits yield MEC^*'s with the same mean and variance as multi-day trials (Levey and Karasov 1989).

Some researchers record only the first appearance of marked food particles, which may be termed gut-passage time, gut transit time, and gut-passage rate. These and other measures, plus methods for their determination, are discussed in detail in Kotb and Luckey (1972), Warner (1981) and Van Soest et al. (1983).

UTILIZATION EFFICIENCIES OF WILD BIRDS EATING WILD FOODS

Major Patterns According to Food

Appendix 1 shows results from about 250 digestion trials in which either the particular food or the species of bird differs. In some cases a single species or closely related species was fed many different food types (e.g., Northern Bobwhites fed arthropods, seeds, and fruits; grouse species fed seeds, fruit, and herbage; passerine birds fed arthropods, seeds, and fruits). Inspection of those data suggests immediately that a large source of variation in utilization efficiency is the type of food consumed. Indeed, analysis of variance (ANOVA, using the arcsine of the square root of MEC^*) among all trials showed a highly significant effect of food (F = 39.3, P < 0.001). Accordingly, summarized in Figure 1 and Table 2 are estimates of MEC^* organized according to the following major food groups:

Nectar. Studies of nectarivores are in uniform agreement that utilization efficiency is practically 100%. Unfortunately, data are lacking for birds (e.g., passerine frugivores) in which nectar makes up a smaller proportion of the diet.

Arthropods and aquatic invertebrates. About three-fourths of the energy is apparently metabolized (Appendix 1, Bryant and Bryant 1988). Mealworms or domestic crickets have been used in studies with terrestrial arthropods, and the former yield higher utilization efficiencies than the latter, probably due to lower contents of cuticle (see below).

Vertebrates. I could discern no difference in MEC^* among trials where fish, mammal, or bird were offered to carnivorous birds. On average, about three-fourths of the energy in these foods is apparently metabolized.

Seeds. Sixty-two digestion trials were reviewed. Those trials conducted with cultivated seeds yielded significantly higher MEC^*'s (P < 0.001, ANOVA). About four-fifths of their energy was apparently metabolizable. When wild seeds were fed to nonpasserines, less than two-thirds of their energy was apparently metaboliz-

TABLE 2. Utilization Efficiencies and Estimated Metabolizable Energy Contents of Food Types

Food type	N_1, N_2[a]	MEC*			Energy content		
		$\bar{X}$[b]	SD[c]	95% C.I.[d]	GE_c (kJ/g)		(MEC^*) $\times (GE_c)$[f] (kJ/g)
					$\bar{X}$	SD (N)	
Nectar (sucrose)	10, 4	0.98[A]	0.01	0.977–0.983	16.7	(1)	16.4
Cultivated seeds							
Passerines	9, 7	0.80[B]	0.05	0.76–0.83			17.0
Nonpasserines	17, 7	0.80[B]	0.08	0.76–0.83			17.0
All					21.3	4.3 (22)	
Arthropods	7, 6	0.77[B]	0.08	0.72–0.83	25.0	1.9 (4)	19.3
Vertebrates	20, 10	0.75[B]	0.07	0.72–0.79	23.6	2.0 (15)	17.7
Wild seeds							
Passerines	11, 5	0.75[B,C]	0.09	0.70–0.80			15.8
Nonpasserines	25, 7	0.59[D]	0.13	0.54–0.65			12.4
All					21.0	2.8 (27)	
Fruits							
Pulp and skin	31, 5	0.64[C]	0.15	0.59–0.70	19.6	3.4 (28)	12.5
Pulp and skin and seed	22, 9	0.51[D]	0.15	0.44–0.57	21.6	1.6 (10)	11.0
Herbage							
Bulbs and rhizomes	4, 4	0.56[C,D]	0.18	0.38–0.74	17.3	1.6 (2)	9.7
Grouse	19, 10	0.37[E]	0.08	0.33–0.40	21.5	0.8 (8)	8.0
Other	14, 6	0.33[E]	0.12	0.26–0.39	18.5	1.6 (6)[g]	6.1

[a] N_1 = number of feeding studies in which either food or bird species differed; N_2 = number of bird species. In some studies a bird species was fed different foods in separate feeding trials.
[b] Means with the same capitalized letter are not significantly different according to Duncan's Multiple Range Test on arcsine$\sqrt{MEC^*}$.
[c] SD on untransformed values of MEC^* with sample size equal to N_1.
[d] Confidence intervals were established using transformed values of MEC^* (arcsine$\sqrt{MEC^*}$) and the total sample size was taken to equal N_1.
[e] Gross energy content/g dry matter. Mean values from Appendix 1.
[f] Apparent metabolizable energy content/g dry matter.
[g] Excludes aquatic species fed to domestic ducks (Muztar et al. 1977).

able. Passerine species had higher MEC^*'s on wild seeds than nonpasserine species (P < 0.001, ANOVA) (Table 2), whereas there was no significant difference (P > 0.4) between the groups in digestion trials with cultivated seeds. Possible reasons for this might relate to phylogeny or body size.

Fruits. Small frugivores that are seed dispersers either egest or defecate seeds following ingestion of whole fruits. Consequently, most studies with passerine frugivores have determined the utilization efficiency on pulp and skin alone by subtracting the mass and energy value of seeds from that of whole fruit. In some other studies utilization efficiencies were determined on the basis of whole fruits, including seeds. Because seeds can make up a substantial fraction of the mass of the whole fruit (e.g., Sorensen 1984), and because they are relatively indigestible (Servello and Kirkpatrick 1987), one would expect that utilization efficiencies would be lower in the latter kind of digestion trial. This was indeed the case (Table 2). In those trials in which the MEC^* of pulp and skin alone was determined, about two-thirds of the energy was metabolizable; whereas, in those trials where the MEC^* of whole fruit was determined, about half of the energy was metabolizable (P < 0.001). Some larger fruit-eating birds partially digest the seeds, and in those cases MEC^*'s can be quite high (e.g., Willow Grouse eating cowberries apparently digested 81% of the total organic matter; Pullianinen et al. 1968).

Johnson et al.'s (1985) data set on frugivores was omitted from the above analysis because digestion trials were brief; fruit was presented to fasted birds for two hours and excreta were collected during those two hours and for an additional two hours. This might result in overestimation of MEC^*. Indeed, MEC^* for pulp and skin in these trials ($\bar{X} = 0.71$, SD = 0.13, N = 55) was significantly higher (P < 0.01) than for other trials with passerines fed pulp and skin (Table 2).

Herbage. Generally, species of grouse or waterfowl have been used in digestion trials with herbage (Appendix 1). The studies with Ostriches and Emus were excluded from Figure 1 and Table 2 because they were not performed with foods the birds might eat in the wild. There was no significant difference in MEC^*'s in trials with grouse species compared with trials with other species of birds (P > 0.5), except that birds fed bulbs and rhizomes had significantly higher MEC^*'s (P < 0.001) than birds eating other kinds of herbage (leaves, twigs, buds) (Table 2). On average, birds apparently metabolized less than 40% of the gross energy in leaves, twigs, and buds. Sugden (1973) measured much lower MEC^*'s (sometimes negative values) for numerous plants fed to Blue-winged Teal but concluded that his methods yielded questionable values. He fed ducks that were not provided with grit, used test rations mixed with a reference ration, and calculated the MEC^* of the test ration by difference. The technique of mixing test and reference rations, which has been validated with chickens, was also used by Muztar et al. (1977) and they also calculated quite low values of MEC^* for ducks eating wild foods (Appendix 1). This technique may not work effectively for wild foods with low digestibilities.

For all these food groups the metabolizable energy per gram of food is the product of MEC^* and gross energy content (Table 2). On average an herbivore must ingest almost three times as much dry matter as an insectivore or carnivore to obtain the same amount of metabolizable energy.

OTHER FACTORS AFFECTING UTILIZATION EFFICIENCY

There is considerable variation in MEC^*'s within each food type (Fig. 1). Some may be due to differences in composition between particular foods of a given type (e.g., growing vs. senescent vegetation, larval vs. adult arthropods; see Food Chemistry section), and some to individual variability (e.g., in age, reproductive condition; Moss 1983). How great are these effects?

Physiological condition. How might MEC^* vary with age or reproductive condition? When a bird is growing or gaining mass, MEC^* is expected to increase, because much ingested N (protein) is deposited as tissue, rather than being metabolized and excreted. For example, MEC^* was higher (by 0.04) in Long-eared Owls during the period of rapid feather growth (Wijnandts 1984). However, reproductive condition had no significant effect on MEC^* in Willow Ptarmigan (West 1968).

Young, developing birds might be less efficient than adults at extracting energy and nutrients; indeed, several studies have detected lower MEC^*'s in very young birds (e.g., by 0.12 in House Sparrows, Blem 1975; by 0.20 in Black-bellied Whistling-ducks, Cain 1976; see also Myrcha et al. 1973, Penney and Bailey 1970, and Dunn 1975). The MEC^* of Dunlin (*Calidris alpina*) chicks fed mealworms, ground beef, and oats was 0.57 (Norton 1970) which is lower than

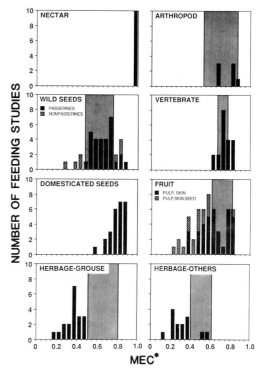

FIGURE 1. Comparison of apparent metabolizable energy coefficients (MEC^*) measured in feeding trials with predicted values based on chemical composition of foods. The frequency histograms (solid black or hatched bars) show the number of feeding trials (y-axis) which yielded the MEC^*'s listed on the x-axis. Data are from Appendix 1. The shaded grey boxes show the range of expected MEC^*'s which were predicted using a simple model which estimates $MEC^*_{predicted}$ based on the chemical composition of the food (see Food Chemistry section). Notice that fruit and herbage appear to be utilized less efficiently than predicted on the basis of food composition alone.

an expected value of 0.75 (Table 2). Thus, there is evidence that very young birds have immature guts and hence lower utilization efficiencies, but see Westerterp (1973) for an apparent exception. It would be interesting to know whether parents assist in digestion by softening food with mucous or predigesting it and then regurgitating it.

None of the digestion trials tabulated in Appendix 1, however, used very young birds and hence this is not an important source of variation in the analysis in Table 2.

Several studies (e.g., Bairlein 1985) have claimed that birds can adaptively modulate the efficiency of food utilization and thus, for example, undergo premigratory fattening without increases in energy intake or decreases in energy expenditure. Data apparently supporting this

TABLE 3. EXAMPLES OF VARIATION IN UTILIZATION EFFICIENCY AMONG DIFFERENT SPECIES EATING THE SAME FOOD[a]

Species	MEC^*
Corn	
Common Pheasant	0.83
Red-winged Blackbird	0.90
Graylag Goose	0.87
Northern Bobwhite	0.86
Sharp-tailed Grouse	0.86
Spur-winged Goose	0.78[b]
Alfalfa meal	
Graylag Goose	0.30
Mallard	0.32
Northern Pintail	0.33
Gadwall/Northern Shoveler	0.34
Spur-winged Goose	0.58[b]
University of Illinois Baby Chick Mash #521	
Hoary Redpoll	0.71
Common Redpoll	0.70
American Tree Sparrow	0.71
Variable Seedeater	0.74
Green-backed Sparrow	0.69
Blue-black Grassquit	0.80
Yellow-bellied Seedeater	0.79
White-throated Sparrow	0.67
Dickcissel	0.68
Sunflower seeds	
Great Tit	0.81
Northern Cardinal	0.74[c]
Evening Grosbeak	0.84
Northern Bobwhite	0.60[d]
Scaled Quail	0.86
House Sparrow	0.76[e]
10 species of passerines	0.82–0.91[e]
Wheat	
House Sparrow	0.72
Graylag Goose	0.78
Northern Bobwhite	0.70
Crickets	
American Robin	0.70
European Starling	0.70
Northern Bobwhite	0.83

[a] Data from Appendices 1 and 3.
[b] MEC^* estimated as $AMC^* + 0.03$ based on results from other species eating corn.
[c] Digestion trial possibly too short.
[d] Unclear whether shells were removed.
[e] S. N. Postnikov and V. R. Dol'nik in Kendeigh et al. (1977).

then increases in food intake will result in increases in apparent utilization efficiency for those components of the food, with no real change in true utilization efficiency. A convincing demonstration of this effect will require measurement of *true* utilization efficiency, or perhaps intestinal extraction efficiencies using isotopes or other methods.

Environmental conditions. There have been numerous studies with wild birds fed both wild foods and assorted poultry "mashes" in which air temperature was changed and sometimes photoperiod (Cox 1961, Brenner 1966, El-Wailly 1966, Brooks 1968, Kontogiannis 1968, West 1968, Owen 1970, Gessaman 1972, Cain 1973, Robel et al. 1979a, Stalmaster and Gessaman 1982, Wijnandts 1984). In about half of the studies, changes in these environmental variables had no significant effect on calculated MEC^*. In those studies in which significant effects were detected, no general patterns emerge except that the changes in MEC^* were generally small (i.e., <0.05). One exception to this generality is the study of Willson and Harmeson (1973) in which they found MEC^* to vary by as much as 0.13 in several digestion trials with seeds fed to passerines. The duration of their digestion trials (5–6 hours), however, was short compared to the probable mean retention time of seeds (>1.5 hours, see below) and this may be the source of the high variation, as discussed above. In studies in which changes in MEC^* have occurred with temperature, the graphical relationships between MEC^* and temperature were sometimes linear, sometimes concave, and sometimes convex. This mixed pattern would seem to rule out any unifying physiological explanation, such as decreased digesta residence time with increasing food intake.

DIFFERENCES IN UTILIZATION EFFICIENCY ASSOCIATED WITH PHYLOGENY

One suggestive piece of evidence that phylogenetic differences exist is that MEC^*'s for passerine species eating wild seeds were significantly higher than for nonpasserine species. But if one compares different species eating the same ration (Table 3), one usually finds that in most cases species have remarkably similar MEC^*'s. There are occasional outliers, some of which may be explained by methodological differences (e.g., Northern Bobwhites on sunflower seeds) or possibly errors (e.g., the substantially higher MEC^* of the Spur-winged Goose eating alfalfa meal is suspect), but others may reflect real physiological differences (e.g., Northern Bobwhites on crickets). Excluding outliers, the standard deviation among species eating the same food is about 0.04.

idea, however, could be artifacts of nutrient retention, as described above, or of the increased energy intake that occurs during premigratory fattening or reproduction. Because *apparent* rather than *true* utilization efficiency is usually measured, if there are endogenous losses of dry mass, energy, fat, carbohydrate, or other items,

This measure of variation may also reflect differences among the studies in environmental conditions or methods.

This analysis corroborates my initial conclusion that probably the largest source of variation in observed utilization efficiencies is due to characteristics of food. Castro et al. (1989) concluded similarly. I do not mean to minimize the importance of the structural and functional characteristics of the birds themselves. Differences in MEC^* as large as 0.15 can occur in different species eating the same food (Table 3), and these may be associated with differences in anatomy and physiology. Also, there are few birds that can eat all types of food. Instead, there are several designs of guts that allow for effective utilization of from one to three of the food types. Presumably, this is the explanation for correspondence between food habits and gut morphology (e.g., Leopold 1953, Kehoe and Ankney 1985, Barnes and Thomas 1987).

In the following two sections of the paper, I elaborate upon the two themes of food chemistry and bird anatomy and physiology as determinants of utilization efficiency. First, I consider the chemical composition of the various food types and the manner in which it can determine utilization efficiency. As one cannot do this without making some assumptions about the physiological characteristics of the birds, I discuss those assumptions, and also attempt to define how particular anatomical and physiological attributes of the gastrointestinal tract affect utilization efficiency and allow its maximation, or optimization within certain constraints.

FOOD CHEMISTRY AS A SOURCE OF VARIATION IN UTILIZATION EFFICIENCY

A Simple Model of Utilization Efficiency Based on Food Chemistry

A simple model of digestion can illustrate the factors contributing to the large variation in utilization efficiencies within and among food types (Fig. 1) and highlight the topics where our knowledge is weakest. Comparison of model estimates with measured utilization efficiencies might reveal digestive adaptations or compromises. The model I present differs from others (e.g., Moss 1983, Servello et al. 1987) in being based upon principles of digestion and metabolism (rather than being empirically derived). Because it is more general (and therefore less accurate), its primary value may be heuristic and not predictive.

The excreta of a bird in steady state consists primarily of material of endogenous origin, undigested components of food (both organic and inorganic), and material of food origin that was absorbed, metabolized, and subsequently excreted by the kidneys (including food protein N as uric acid, urate, or urea where 1 g food N yields from 2.1 to 3 g of N-containing excretory product; see Bell, this volume). Detoxification products of plant secondary chemicals would also be included in this last component and will be considered later. If the food has an ash concentration A_i (proportion of dry mass), a N concentration of N_i, a certain proportion of R_i of its mass that is refractory to chemical digestion and absorption, and if the excretory product is uric acid (3 g excreted/g N consumed), then flux rates for the three components of excreta should be approximately accounted for as follows:

$$Q_e = E_m + Q_i(A_i) + Q_i(R_i) + 3(Q_i[N_i] - E_N) \quad (5)$$

The last component of the equation is the correction for N intake, which is incorporated into the equation primarily for birds eating foods with very high N_i (e.g., predators). It includes a new term E_N, the endogenous N loss. This N-correction is especially necessary for high N_i because the large amounts of N digested and absorbed will yield, after catabolism, appreciable amounts of excretory dry mass. E_N is subtracted from this N-correction because it has already been accounted for in E_m. Multiplying E_N by 3 implies that it is entirely uric acid whereas, in fact, some proportion of E_N might be urea, or endogenous protein N from the alimentary tract (e.g., sloughed cells).

In eq. 5 I have assumed that all of the non-refractory portion of food is digested and absorbed, and this is often the case (some exceptions are discussed below). For example, intestinal extraction efficiencies for amino acids from soybeans averaged 93% in chickens (Achinewhu and Hewitt 1979). For glucose and sucrose, extraction efficiencies are ≥97% in nectarivorous birds (Karasov et al. 1986) and chickens (Sibbald 1976), and apparent extraction efficiencies for fat have been reported to be 93–94% in American Tree Sparrows (Martin 1968), 89–97% in Garden Warblers (Bairlein 1985) and 77–91% in chickens (Mateos and Sell 1981). Several species have been found to assimilate more than 95% of dietary wax (Obst 1986, Place and Roby 1986, Roby et al. 1986).

Substituting eq. 5 into eq. 1 allows one to derive an approximation for MEC^*, but first, energy equivalents must be assigned to R_i, N_i, and E_N (but not A_i because the energy content of ash is zero). To estimate the excretory energy loss per unit N consumed, one can use the energy content of uric acid, 11.5 kJ/g (Bell, this volume). The same energy content will be applied to E_N, though some portion of this is probably protein. The energy content of refractory material in foods becomes a variable, GE_R. Incorporating these into eq. 5 and then substituting into eq. 1 yields:

$$MEC^*_p = 1 - [GE_R]R_i/GE_i - 34.5N_i/GE_i - (E_e - 34.5[E_N])/[GE_i][Q_i] \quad (6)$$

The number 34.5 is the product of 3 (grams uric acid/gram N excreted) and 11.5 (kJ/g uric acid) and thus has units of kJ excreted/g N excreted. If one assumes that only 75% of excreted N is in the form of uric acid (or urate) and 25% in the form of urea (with an energy content of 10.5 kJ/g; Bell, this volume), MEC^*_p is little affected (an increase of ≤0.016, depending upon N_i).

This equation predicts MEC^* based on four characteristics of food (R_i, GE_R, GE_i, N_i) and three characteristics of the bird (E_e, E_N, Q_i). All characteristics of the bird appear in the last term of the equation which tends to have a small effect on the calculation of MEC^*_p. Thus, the model implies that unless the N content of a food is very high, the major determinant of apparent utilization efficiency is the proportion of food that is refractory to chemical digestion. In applying the equation, one can use results from other birds to estimate E_e (e.g., 21 kJ kg$^{-0.75}$ day^{-1} in the chicken, see section Methods and Terms) and E_N (approximately 0.1 g kg$^{-0.75}$ day^{-1} in wild birds, Robbins 1983). The assumption that all birds will share similar E_e's is not unreasonable, considering that other kinds of endogenous losses (e.g., N, creatinine) in birds and mammals are predictable functions of mass$^{0.75}$. Also, even if E_e for a test species did differ substantially from the value for chickens, that usually would not have a large effect on the estimation of MEC^*_p, because the last term of the equation has a small effect on the calculation of MEC^*_p. But use of the chicken data underscores a large gap in our knowledge and emphasizes our current inability to correct accurately MEC^*'s to MEC's for almost any species but the chicken.

FOOD COMPOSITION AS A SOURCE OF VARIATION IN UTILIZATION EFFICIENCY

To understand the role of food composition in determining and affecting utilization efficiency, I will compare predictions of the equation for each food type with measured values of MEC^* (Fig. 1, Table 2). For Q_i in eq. 6, I used average feeding rates from Appendix 1 (in g kg$^{-0.75}$ day^{-1}): leaf and twig eaters, 65 ± 9 (SE); fruit-eaters, 55 ± 6; seed eaters, 52 ± 4; arthropod eaters, 59 ± 7; carnivores, 27 ± 2; and nectar, 74 (Calder and Hiebert 1983). I used average values for GE_i and N_i (Table 4), recognizing that such data may vary according to species of plant or animal sampled, phenological state, time of year, and so on. More difficult is estimating R_i, the proportion of a food that is refractory to chemical digestion. First, no single chemical assay perfectly separates the very digestible components of food from the highly indigestible components. Second, R_i for a food is not solely a function of the food but is also a function of the bird's digestive physiology. As an example, digestion of plant cell wall by geese has been reported to be negligible (Mattocks 1971, but see Buchsbaum et al. 1986), whereas some grouse and emus digest 15–35% of cell wall (Gasaway 1976b, Herd and Dawson 1984, Remington 1990). If we assume that all cell wall is refractory to digestion, we may be able to use eq. 6 to identify those instances when birds appear to digest a substantial fraction of the cell wall, based on comparatively high utilization efficiencies. Thus, developing expectations of extraction efficiency based on food composition is a first step in identifying physiological sources of variation in utilization efficiency.

A discussion of the comparisons of predicted and observed utilization efficiencies for each food type follows.

Nectar. Because nectar has no refractory material, negligible N, and I have assumed that all of the sugar is digested and absorbed, its MEC^*_p (0.986; from eq. 6) is just slightly below 1.0 due to endogenous energy losses. The predicted value is the same as the average observed value measured in 10 feeding trials (Table 2).

Vertebrate prey. I estimated $MEC^*_p = 0.66$–0.76 based on average N contents of vertebrate prey and a range of values for R_i (Table 4). I took R_i to be the proportion of ingested dry matter that was refractory to gastric digestion. This can be estimated based on the pellets egested by carnivores. In strigiforms, which egest pellets following gastric digestion, the ratio of pellet dry mass to ingested dry mass averages 0.13 ± 0.02 (SE) (N = 7 species of owls; Duke et al. 1975, 1976; Kirkwood 1979). In non-strigiform carnivores which pass more bone to the intestine and digest more of it, the ratio is slightly lower, 0.05 ± 0.01 (N = 11 species of hawks, falcons, eagles, and vultures; Duke et al. 1975, 1976; Kirkwood 1979). The energy content of egested material averaged 17.1 kJ/g which was used to estimate GE_R in the model.

All 20 measured values of MEC^* are within 0.09 of the predicted values (Fig. 1), indicating that most of the nonrefractory organic dry matter and hence energy in vertebrate prey can be digested and absorbed by carnivores.

Arthropod prey. I estimated $MEC^*_p = 0.53$–0.86 based on an average N content for arthropods and a range of values for R_i (Table 4). The

TABLE 4. CHEMICAL CHARACTERISTICS OF FOODS THAT AFFECT UTILIZATION EFFICIENCY

Food type	R_i^a	N_i^a	GE_i^b (kJ/g)	GE_R (kJ/g)
Nectar (sucrose)	0	0	16.7	0
Vertebrate prey	0.04–0.17[c]	0.122[d]	23.1	17.1[e]
Arthropods	0.01–0.5[f]	0.086[g]	24.5	18.0
Cultivated seeds		0.02[h]	21.5	16.7[i]
Wild seeds	0.18–0.53[j]	0.01–0.028[j]	21.5	16.7[i]
Fruit, pulp and skin	0.09–0.34[k]	0.01[l]	19.5	16.7[i]
Fruit, pulp and skin and seeds	0.40[m]	0.01[m]	21.6	16.7[i]
Herbage eaten by grouse	0.22–0.6[n]	0.015[o]	21.6	16.7[i]
Herbage eaten by other birds	0.38–0.61[p]	0.015[o]	18.2	16.7[i]

[a] Proportion of dry mass.
[b] From Table 1.
[c] Range for 18 species from Duke et al. (1975, 1976) and Kirkwood (1979).
[d] Average for 12 species of vertebrates from Ricklefs (1974b).
[e] Average from three digestion trials from Duke et al. (1973) and Kirkwood (1979).
[f] Bernays (1986).
[g] Ricklefs (1974b) and Vonk and Western (1984). See also Bell (this volume).
[h] Five species of grains from Ricklefs (1974b).
[i] The average energy content of carbohydrate.
[j] Short and Epps (1976).
[k] Range of NDF's for six species (Levey and Karasov, unpubl., and Servello and Kirkpatrick 1987); average was 0.26.
[l] Average for 18 species from Sorensen (1984), Worthington (1983) and Levey and Karasov (unpubl.) ($\bar{X}$ = 0.013, SD = 0.007).
[m] Average for 50 species from Short and Epps (1976) and Servello et al. (1987).
[n] Gasaway (1976a), Remington (1983, 1990), Servello et al. (1987), Servello and Kirkpatrick (1987).
[o] Most values in the literature for leaves and twigs range 0.01–0.02 (Mattson 1980), though leaves of herbaceous plants sometimes exceed 0.04 (e.g., Servello and Kirkpatrick 1987).
[p] Buchsbaum et al. (1986).

primary material in arthropods refractory to chemical digestion is probably the cuticle, which may comprise 1–50% of dry matter (Bernays 1986). Because cuticle is composed of a mix of chemicals (primarily chitin and protein plus some lipids), I took its energy content to be 18 kJ/g.

Measured values of MEC^* (Fig. 1) for arthropods cluster at the higher end of the range of predicted values. This is not because the arthropods used in the digestion trials had low cuticle contents. Three of the trials used orthopterans, which have cuticle contents of about 50% of dry matter (Bernays 1986). Evidently not all of the cuticle is refractory to digestion, as had been assumed. Some components of cuticle (e.g., lipid waxes and soluble protein) are probably quite digestible while others (e.g., chitin and tanned protein [sclerotin]) are more refractory. The extent to which chitin (up to 60% of the cuticle's dry mass; Fraenkel and Rudall 1947) can be digested by birds has been practically unstudied. One Red-billed Leiothrix (*Leiothrix lutea*) was reported able to digest 56.8% of the chitin in dead mealworm larvae added to its diet (Jeuniaux and Corneluis 1978).

If one assumes that birds digest about 50% of ingested cuticle, then the predicted values of MEC^*_p range from 0.7 to 0.86. This yields very good agreement with measured values of MEC^*. Thus, I conclude that most noncuticular protein and fat in arthropods can be digested and absorbed, as well as a substantial fraction of the cuticle.

Seeds. For wild seeds I estimated MEC^*_p = 0.53–0.83 based on a range of N contents and a range of values for R_i (Table 4). As a reasonable estimate of R_i in vegetation (seeds, fruits, leaves, twigs, buds, storage organs), I used the cell wall content, determined by measuring that proportion of plant dry matter that is insoluble in neutral detergent, and correcting it for its ash content (i.e., neutral detergent fiber [NDF]; Goering and Van Soest 1970, Demment and Van Soest 1985).

Most measured values of MEC^* (Fig. 1) for seeds fall within the predicted range, indicating that little fermentation of cell wall occurs. In those cases where measured MEC^*'s fall below the predicted range, perhaps seeds with even higher cell wall contents were used than I assumed in Table 4.

Because MEC^*'s of wild seeds tend to be lower than for cultivated seeds, we might expect that wild seeds have higher cell wall contents. In fact, the amount of crude fiber (a poorer index to cell wall than NDF) in 20 species of seeds in southern forests (Short and Epps 1976) appears to be about four times greater than in commercially available seeds (Conley and Blem 1978). Possible differences in chemical makeup and hence digestibility between wild and domestic seeds merits further study.

Fruits. For wild fruits I estimated MEC^*_p = 0.67–0.89 based on an average N content and a range of values for R_i (Table 4). Surprisingly, many measured MEC^*'s fall below the predicted range (Fig. 1). That is, the utilization efficiency birds eating insects and seeds, most of mean residence time for the entire gut occurs in the crop.

Measurements of digesta retention in birds eating vertebrates are not directly comparable with the values in Appendix 2. Meal-to-pellet intervals (the time between ingestion of prey and egestion of pellets of undigestible material) are generally 10–20 hours (Duke et al. 1968, Balgooyen 1971, Duke et al. 1976, Rhodes and Duke 1975).

Anatomical measurements of the small intestine. At least three anatomical measurements of the small intestine are useful within the context of equations 7 and 8: intestinal length and surface area, because hydrolytic or absorptive measures are usually expressed per cm length intestine or per cm² nominal area (which excludes the area of villi and microvilli), and gut volume, because of its relation to retention time and intake. In a simple tube the three are related: $(4\pi)(\text{volume})(\text{length}) = (\text{area})^2$. How do these scale with body mass?

In tetraonids, small intestine length scales with mass$^{0.32}$ in species eating the same type of food (calculated from Leopold 1953). Mass of gut contents scales with mass$^{1.0}$ (Moss 1983), and volume probably scales in the same manner. Given these allometries, intestinal surface area might be expected to scale with mass$^{0.66}$. In mammals intestinal nominal surface area has been reported to scale with mass$^{0.63}$ (Karasov 1987) and mass$^{0.75}$ (Chivers and Hladik 1980). Too few data are available for a separate analysis in birds. For purposes of comparing birds of different sizes I shall normalize intestine length to mass$^{0.33}$, intestine surface area to mass$^{0.66}$, and intestine volume to mass$^{1.0}$.

Absorption rate per unit intestine. In mammals, reptiles, and fish rates of absorption of sugar and amino acid/cm² intestine are independent of body size (Karasov 1987). This was also the case in a small sample of birds (7 species) ranging in size from 3.2 to 700 g (Karasov and Levey 1990).

EXAMPLES OF TRADE-OFFS OR ADAPTATIONS IN DIGESTIVE PHYSIOLOGY

Low digesta retention time in frugivores. Retention time is relatively short in frugivorous birds (Herrera 1984b, Karasov and Levey 1990; Appendix 2, Fig. 2). The digestion model predicts that in the absence of a compensatory increase in hydrolysis or absorption rate, a decrease in digesta retention should result in a decrease in digestive efficiency. There is evidence for such a decrease in highly frugivorous Phainopepla (Walsberg 1975), Cedar Waxwings (Martinez del Rio 1989), and manakins (Worthington 1983), as well as in the previous comparison of predicted and observed utilization efficiency (Food Chemistry section).

Table 5 presents a detailed analysis of the effect of short retention on digestive efficiency by comparing the fruit-eating waxwing and nectarivorous hummingbird. These species are compared because they both digest solutions containing monosaccharides and disaccharides (nectar or juice of fruit), and entirely comparable data sets based on identical methodology are available (Karasov et al. 1986, Martinez del Rio et al. 1989). Waxwings, being larger, have longer small intestines with much greater nominal surface area. But when normalized to scaled body mass, intestine lengths are similar, and intestinal surface area is slightly greater in the hummingbird. Waxwings have shorter mean retention times, and, when corrected for body mass, the difference appears to be two-fold. A unit area of hummingbird intestine absorbs glucose seven times faster than that of the waxwing. Given the shorter retention time and lower glucose absorption rate (per cm² or per $g^{0.66}$), one would predict that digestive efficiency in the waxwing may be less than that in the hummingbird when the birds eat meals with very high glucose concentrations.

Digestive efficiencies have been measured in both species using radiolabeled glucose (Karasov et al. 1986, Martinez et al. 1989). When fed high glucose concentrations (585 mM for the hummingbird, 806 mM for the waxwing), the waxwings absorbed significantly less of the glucose than the hummingbirds (0.92 vs. 0.97, $P < 0.001$). The difference is not due to the difference in the glucose concentration; hummingbirds eating even more concentrated sugar solutions still extract more than 97% (Appendix 1). Instead, the difference is due to the relatively lower retention time and absorptive rate in the waxwing. Differences between the two species become even greater for the digestion of sucrose, because it is a two-step process of hydrolysis followed by absorption, and the overall rate is less than that of absorption alone (Martinez del Rio, pers. comm.). Thus, when waxwings and hummingbirds were fed meals containing sucrose (respectively, 439 mM and up to 2000 mM), the former had a much lower digestive efficiency (0.62 vs. 0.98; value for hummingbirds from Hainsworth 1974).

Thus, it appears that frugivores are characterized by relatively short digesta retention times which, in some cases, compromise their ability to extract nonrefractory components of their food. Presumably there is some compensating advantage to short digesta retention. Sibly's (1981) model suggests that the net rate of energy intake (a function of $Q_i \times F$) might be maximized when T is shorter than the time necessary to achieve maximal absorptive efficiency.

Large gut volume in herbivores. It has been argued that, because of the demands of flight, the mass of the digestive tract in birds should be

TABLE 5. Comparison of Digestive System Form and Function in Rufous Hummingbirds and Cedar Waxwings in Relation to Extraction of Glucose from a Meal

	Body mass (g)	Intestine length		Intestine area		Mean retention time		Rate of absorption[a]	Extraction efficiency (%)
		cm	cm/g$^{0.33}$	cm^2	cm^2/g$^{0.66}$	min	min/g$^{0.25}$	nmole min^{-1} cm^{-2}	
Rufous Hummingbird	3.2	5	3.4	1.2	0.6	48	36	942	97
Cedar Waxwing	35	12.4	3.8	17.4	1.7	41	17	127	92

[a] Maximal rate of carrier-mediated glucose uptake across the luminal surface of the gut; average for the proximal, mid, and distal gut (from Karasov et al. 1986, Martinez del Rio et al. 1989, Karasov, unpubl. data).

minimized. But how much gut is enough? It is possible to deduce an answer using models of digestion (Sibly 1981, Penry and Jumars 1987).

Because refractory material lowers the metabolizable energy content per gram food, more must be consumed to obtain the same amount of metabolizable energy. Equations 7 and 8 indicate that if Q_i increases, then to maximize utilization efficiency animals eating food with higher R_i should have larger digestive chambers and a longer digesta retention time. To maximize the rate at which digestive products are formed, digestive chamber size should increase (cf. eq. 5.3, Sibly 1981).

What actually happens when R_i is increased experimentally? Savory and Gentle (1976) added sawdust or cellulose to a conventional ration fed to Japanese Quail (*Coturnix japonica*) and measured feeding rate, utilization efficiency, modal retention time (*sensu* Warner 1981), and digestive tract dimensions after at least 10 weeks. Daily food intake increased and compensated almost exactly for the dilution of the nonrefractory portion of the food, and neither the rate of dry matter digestion (in g/day) nor the utilization efficiency of the nonrefractory portion of the food decreased. These changes were effected without any major change in modal retention time, but the size of the colo-rectum, small intestine, and caeca increased significantly. Other studies with ducks (Miller 1975) and woodpigeons and starlings (reviewed in Sibly 1981) have demonstrated increases in intestinal length of up to 40% when birds were switched to high R_i diets. Thus, as the models predict, a response to increased R_i is larger digestive chambers.

In the wild these changes occur as birds undergo seasonal diet shifts to foods with higher R_i (Davis 1961, Moss 1974, Drobney 1984, Gasaway 1976a). Diet shifts probably account for the differences sometimes seen in intestine lengths between wild and captive birds (reviewed in Sibly 1981) because the captives are usually fed commercial rations with lower R_i.

The proximate mechanism for the intestinal enlargement in most of these cases is probably hyperphagia (Karasov and Diamond 1983), as birds attempt to compensate for caloric dilution (higher R_i) or lower gross energy content in the food. Increased food intake during cold weather or reproduction may have similar effects (Drobney 1984).

The generalization that gut volume should be greater in birds that eat foods with high R_i does not appear to hold for frugivores. For example, four highly frugivorous species in the body mass range 14–35 g (Cedar Waxwings, Phainopepla, and two manakins; Walsberg 1975, Worthington 1983) have small intestine lengths of 13 ± 0.7 (SE) cm, whereas in eight species of less frugivorous or nonfrugivorous birds in that size range they average 19.3 ± 1.2 cm (Herrera 1984b). Herrera (1984b) did not detect a significant difference in gut length between the more frugivorous and less frugivorous species that he studied.

Selective retention of digesta in herbivores. The proportion of refractory material that is microbially fermented is directly related to gut volume and reaction rate and indirectly related to digesta flow rate (Penry and Jumars 1987). The presence of caeca enhances fermentation by affecting all three variables: caeca increase gut volume; decrease digesta flow rate; and increase reaction rate.

Among gallinaceous birds, the proportionally largest variation among species in lower gastrointestinal tract structure is in the caeca, which are generally at least twice as long in browsers as seed-eaters (Leopold 1953). In some species the caeca selectively retain smaller particles and solutes, while larger particles pass down the large intestine. (For a discussion of the evidence for, and mechanism of, this selective retention see Fenna and Boag 1974, Clemens et al. 1975, Gasaway et al. 1975, Bjornhag and Sperber 1977, Hanssen 1979, Sperber 1985.) Thus, in Rock Ptarmigan, which have well-developed caeca, the mean retention time of a liquid marker greatly exceeds that of a solid marker (Appendix 2; see the pheasant also) whereas in the Emu, which lacks enlarged caeca, the markers travel through the digestive tract at approximately the same rate (Appendix 2). Selective retention probably increases the fermentation rate by effectively increasing nutrient concentrations and the surface area available for attack by the microbes.

While a large proportion of the NDF (neutral detergent fiber) that enters the caeca may be fermented there (up to 98% in Blue Grouse; Remington 1990), only a small proportion of the total in the food actually enters (<33% in Blue Grouse). Thus, estimates of the proportion of total dietary NDF actually digested are less than 40% (Food Chemistry section). Because food NDF values are generally less than 50–60% of dry matter, one might expect that NDF digestion provides for less than 25% of the maintenance energy requirements of cecal digesters. Estimates (cf. Gasaway 1976b) have generally been below this.

A caecum is not required for effective digestion of cell walls. Because of the relationship between mean retention time of digesta and body mass (Fig. 2), larger birds will tend to retain digesta long enough for significant fermentation to occur if the symbiotic microbes are present in the small or large intestine. This is the case in Emu in which the major site of fermentation is the distal section of the small intestine (Herd and Dawson 1984), and possibly in geese (Buchsbaum et al. 1986). Additionally, Herd and Dawson (1984) point out that if bonds between hemicellulose and lignin are hydrolyzed by gastric acid or pepsin, the solubilized hemicellulose is fermented more rapidly than other fiber components of the cell wall.

THE INTERPLAY BETWEEN DIGESTION AND ECOLOGY

The discussion so far has emphasized the utilization efficiency of birds consuming their natural foods, and those features of food composition and bird anatomy and physiology that affect that efficiency. The ecological significance of this efficiency is that it influences both the feeding rate and hence foraging time of the bird, as well as the impact of the bird on its environment through its rate of depletion of resources.

My emphasis on efficiency should not be taken to mean that this aspect of digestion is most important with regard to natural selection. Bird guts do not necessarily operate in a manner that maximizes digestive efficiency; the maximization of the rate of energy gain per gram of food and concomitant minimization of digesta volume may sometimes occur at the expense of digestive efficiency (Sibly 1981, Penry and Jumars 1987). Frugivores may provide an example of this. Neither should my emphasis on efficiency be taken to mean that this is the only context in which digestion has implications for ecology. The following are two examples of interplay between digestive physiology and ecology that suggest how digestion can constrain important aspects of an animal's ecology.

DIGESTIVE CONSTRAINTS ON RATES OF PRODUCTION

Because the maximum energy available for growth, storage, and reproduction is the difference between the maximal rate of metabolizable energy intake and the energy expended for maintenance, intake could limit productive processes (Kendeigh 1949, West 1960, Porter and McClure 1984). Unfortunately, data are virtually lacking on the maximal level, what determines it (e.g., food availability, foraging rate, digestion rate), and whether it actually operates as a limit in the wild.

Ruminants are the classic example of animals whose food intake can be limited by digestive anatomy. Similarly, the intake of brassica by Woodpigeons (*Columba palumbus*) (Kenward and Sibly 1977) and nectar by hummingbirds (e.g., Rufous Hummingbird; Karasov et al. 1986) is apparently limited by the rate at which these foods can be processed. Such a digestive bottleneck can explain why hummingbirds spend so much time perching between feeding bouts (>75% of activity time), as they are waiting for their crop to empty. Feeding or foraging rate may also be limited by internal food-processing rate in some frugivores that swallow fruits whole (Levey 1987b).

Drent and Daan (1980) suggested that the evolution of some life history traits reflects in part the maximum energetic or work capacity of parents. For example, if the costs of feeding more nestlings are reflected in higher levels of energy expenditure, then perhaps the maximum intake which must match that expenditure has been an important constraint in evolution of clutch size. To evaluate this idea, one can estimate energy expenditure in the field (using doubly labeled water or time-energy budgets), but there is no established upper bench mark with which to compare field metabolic rates, as there is a lower bench mark (standard or basal metabolic rate). Nor is it clear how one might best measure experimentally the maximal rate.

Kendeigh (1949) and his colleagues used cold stress as an experimental manipulation to measure the maximum rate of metabolizable energy intake. They estimated these maxima in several species of birds maintained at temperatures very near or at the lower limit of their long-term temperature tolerance (Appendix 3). Work with White-throated Sparrows suggested that when the birds were exposed to temperatures below the lower limit of tolerance, they apparently died of starvation (body fat was substantially reduced).

Also, for this species, forced activity increased the lower lethal temperature, but the maximum rate of intake of metabolizable energy did not change (Kontogiannis 1968). While these data suggest that the primary limitation to energy metabolism under these conditions is the rate food can be consumed and digested, at least one other interpretation is possible: that heat generation by muscles is inadequate at the lower limit of tolerance, and the resultant hypothermia causes secondary dysfunction of digestion (Ricklefs 1974b).

Could these measures of maximal intake be used as an upper bench mark against which field metabolic rates could be compared? It may seem incongruous to compare field metabolic rates measured in the breeding season with those measured under conditions of cold and exercise, but cold and exercise should be seen merely as the most practical device for forcing a sustained elevated metabolic and feeding rate. In fact, field metabolic rates of birds in the breeding season tend to fall just below the maximum intake values (Fig. 3). I think that this approach has utility, and that considering the data available, one cannot rule out the possibility that digestive limits on the maximal rate of energy intake were important in the evolution of life history traits.

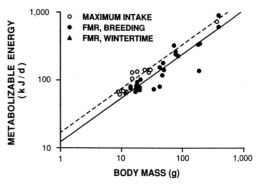

FIGURE 3. Comparison of estimates of maximum rate of energy intake with measures of energy expenditure during two periods of the annual cycle during which expenditure is likely to be particularly high. Maximum intake rates are from Appendix 3. Metabolic rates of free-living birds (field metabolic rates [*FMR*] measured with doubly labeled water) during the breeding season are from Nagy (1987). The single measurement of wintertime FMR is for the Black-capped Chickadee (*Parus atricapillus*) in Wisconsin (Brittingham and Karasov, unpubl. data). The slopes did not differ significantly whereas the intercepts did ($P < 0.005$; analysis of covariance). Data for maximum intake were fit to the equation $Y = 16.4X^{0.65}$, $r = 0.99$; for *FMR*, $Y = 12.1X^{0.65}$, $r = 0.92$.

DIGESTIVE CONSTRAINTS ON NICHE WIDTH

Digestive processes, when rigidly fixed by genotype, can limit a bird's ability to exploit other foraging opportunities via phenotypic adjustment (Karasov and Diamond 1988). Even when adjustment is possible, as in the case of alterations in gut morphology with change in diet, a key question is what are the limits of adjustment, and are they dictated by the foods most frequently eaten (Miller 1975, Barnes and Thomas 1987)? Also, do birds choose foods according to their ability to digest them and, if so, what are the physiological and ecological mechanisms?

Preferences of fruit-eating birds for various sugars may be determined by their abilities to digest them. In behavioral tests, European Starlings and Cedar Waxwings preferred glucose and fructose to sucrose (Schuler 1983, Martinez del Rio et al. 1988, Martinez del Rio et al. 1989). In starlings, the sucrose aversion is associated with an absence of the intestinal enzyme sucrase (Martinez del Rio et al. 1988), which hydrolyses sucrose into fructose and glucose. Too much unabsorbed sucrose in the intestine can cause severe osmotic diarrhea (Sunshine and Kretchmer 1964), and this may provide the sensory cue that leads to aversion. In waxwings the low preference for sucrose is associated with low digestive efficiency due to low levels of sucrase activity relative to digesta retention time (Structure and Function section). Perhaps these low sucrose preferences, which seem to have a physiological basis, affect fruit selection such that the birds favor those containing monosaccharides.

In some mammals the capacity to hydrolyse and absorb sugar and protein is enhanced by greater concentrations of these nutrients in the diet (reviewed in Diamond and Karasov 1987, Karasov and Diamond 1988). If starlings had this regulatory ability, then their sucrase deficiency would not be fixed, they would not necessarily get diarrhea when they eat sucrose, and they would not have an aversion to it that affected their food choice. But because the diet fed the starlings contained some sucrose, they still had negligible sucrase activity; yet, their ability to adaptively increase sucrase activity is apparently limited (Karasov and Diamond 1988). Thus, the starling's inability to digest sucrose, and hence its sucrose aversion, may set a limit to its ecological niche.

Krebs and Harvey (1986) suggested that such digestive constraints in ecology might be more widespread than previously thought. This suggests opportunities for ecologically-oriented research on avian digestion, beyond those studies dealing with the chemical and physiological determinants of digestive efficiency.

ACKNOWLEDGMENTS

It is a pleasure to record my debt to Douglas Levey and Carlos Martinez del Rio, my two primary collaborators in studies on digestive physiology of birds. Kate Meurs and Bruce Darken assisted in library research. Our research was supported by NSF BSR8452089. I had useful discussions with Tom Remington who shared some of his unpublished data. Several people read earlier versions of this manuscript and made valuable comments: Gary Duke, Richard Hutto, Jim King, Jim Luvvorn, Michael Meyer, Charles Robbins, Tom Remington, and the three editors Michael Morrison, C. J. Ralph, and Joseph Jehl, Jr. I thank them all.

APPENDIX I. Utilization Efficiencies of Wild Birds Eating Various Types of Foods

	Body mass		Diet	Q_i^b (g/day)	GE_i^c (kJ/g)	Utilization efficiency[d]		Source
	$\bar{X}$ (g)	%/day[a]				AMC^*	MEC^*	
Nectar								
Black-chinned Hummingbird (*Archilochus alexandri*)	3.2		0.5 M sucrose			0.98		Hainsworth (1974)
			1.0 M sucrose			0.98		
			2.0 M sucrose			0.99		
Rufous Hummingbird (*Selasphorus rufus*)	3.2		0.585 M sucrose			0.97		Karasov et al. (1986)
Blue-throated Hummingbird (*Lampornis clemenciae*)	7.9		0.5 M sucrose			0.98		Hainsworth (1974)
			1.0 M sucrose			0.99		
			2.0 M sucrose			0.98		
Brown Honeyeater (*Lichmera indistincta*)	~9.0		0.8 M sucrose	16.7			0.98	Collins et al. (1980)
			1.2 M sucrose				0.99	
			1.6 M sucrose				0.99	
Arthropods								
Coal Tit (*Parus ater*)	~8.4		Mixed arthropods	2.4	24.4	0.48	0.67	Gibb (1957)
	8.3	0	Mealworms	2.1	27.6	0.71	0.86	
Blue Tit (*P. caeruleus*)	11.3	−0.74	Mealworms	1.9		0.63	0.84	Gibb (1957)
Garden Warbler (*Sylvia borin*)	~20		Mealworms (+elderberries 2×/week)	2.6		0.64		Bairlein (1985)
European Starling (*Sturnus vulgaris*)	71	0	Domestic crickets	5.9	23.2	0.56	0.70	Levey and Karasov (1989)
American Robin (*Turdus migratorius*)	79	0	Domestic crickets	6.5		0.55	0.70	Levey and Karasov (1989)
Northern Bobwhite (*Colinus virginianus*)	178		Domestic crickets		24.7		0.83	Robel et al. (1979a)
Aquatic invertebrates								
African Black Oystercatcher (*Haematopus moquini*)	~50	gaining	Intertidal polychaeta (*Pseudonereis variegata*) and Rock mussels (*Choromytilus meridionalis*)				0.72	Hockey (1984)
	589	~0	Limpet (*Patella granularis*)				0.73	
Lesser Scaup (*Aythya affinis*)	820		Shrimp *Gammarus*		14.7		0.87	Sugden (1973)
Vertebrates								
White Ibis (*Eudocimus albus*)	<100	growing	Sardines plus shrimp				0.85	Kushland (1977)
	700	~0	Anchovies plus shrimp				0.80	
Eurasian Kestrel (*Falco tinnunculus*)	204	0	Day-old cockerel (*Gallus domesticus*)	9.2	24.6	0.51	0.71	Kirkwood (1979)
Common Barn-Owl (*Tyto alba*)	262	0	Day-old cockerel (*Gallus domesticus*)	10.7		0.54	0.73	Kirkwood (1979)
Long-eared Owl (*Asio otus*)	293	0	Lab mice in aviary	~10[e]			0.75[e]	Graber (1962)
			Wood Mouse (*Apodemus sylvaticus*)		21.7		0.79	Wijnandts (1984)

APPENDIX I. Continued

	Body mass					Utilization efficiency[d]		
	$\bar{X}$ (g)	%/day[a]	Diet	Q_i^b (g/day)	GE_i^c (kJ/g)	AMC*	MEC*	Source
			Lab mouse (*Mus musculus*)		25.1		0.79	
			House sparrow (*Passer domesticus*)		21.8		0.68	
			Common vole (*Microtus arvalis*)		23.5		0.68	
			Shrews (*Soricidae*)		22.7		0.62	
			Harvest mouse (*Micromys minutus*)		23.8		0.61	
Broad-winged Hawk (*Buteo platypterus*)	413	0	Lean venison		22.4	0.51	0.74	Mosher and Matray (1974)
Great Horned Owl (*Bubo virginianus*)	1615	0	Mice	26.6	26.3	0.68	0.85	Duke et al. (1973)
			1-day-old turkey poults	26.4	26.8	0.71	0.85	
Snowy Owl indoors (*Nyctea outdoors scandiaca*)	1970 1818	0	Lab rats				0.70 0.76	Gessaman (1972)
Wood Stork (*Mycteria americana*)	2100	0	Frozen whiting fish	64.6	24.6		0.79	Kale (1964)
Cape Gannet (*Morus capensis*)	2755	0	Anchovy	100.0	22.4	0.54	0.74	Cooper (1978)
Bald Eagle (*Haliaeetus leucocephalus*)	3892	−0.2	Chum Salmon (*Oncorhynchus keta*)	63.4	24.4	0.54	0.75	Stalmaster and Gessaman (1982)
	3952	~0	Black-tailed Jackrabbit (*Lepus californicus*)	79.8	19.0	0.54	0.75	
	3924	+0.1	Mallard Duck (*Anas platyrhynchos*)	84.8	24.8	0.67	0.85	
Seeds								
Coal Tit	8.7	−0.3	Scots pine	2.1		0.59	0.81	Gibb (1957)
(*Parus ater*)	9.5	0	Ground nuts (peanuts)	1.8		0.61	0.81	
Blue Tit	10.2	−0.4	Ground nuts (peanuts)	2.2		0.59	0.77	Gibb (1957)
(*Parus caeruleus*)	10.4	−1.25	Scots pine	1.8		0.56	0.75	
Great Tit	17.9	−0.73	Scots pine seeds	3.3	25.7	0.64	0.78	Gibb (1957)
(*Parus major*)	19.2	−1.56	Sunflower	3.1	26.5	0.65	0.81	
	19.6	−0.36	Cob nuts	3.4	27.2	0.65	0.78	
	19.8	−0.71	Ground nuts (peanuts)	3.6	30.1	0.67	0.88	
Song Sparrow (*Melospiza melodia*)	20.8		Foxtail				0.89	Willson and Harmeson (1973)
			Smartweed				0.55	
			Hemp				0.83	
			Pigweed				0.69	
House Sparrow (*Passer domesticus*)	27	~0	Husked wheat	4.7	16.8	0.71	0.72	Weglarczyk (1981)
Eurasian Skylark (*Alauda arvensis*)	40		Barley grain		18.2	0.49	0.81	Green (1978)
Northern Cardinal (*Cardinalis cardinalis*)	44		Foxtails		20.0		0.73	Willson & Harmeson (1973)
			Smartweed		20.1		0.71	
			Hemp		24.7		0.73	
			Ragweed		30.8		0.73	
			Sunflower		22.0		0.74	
Evening Grosbeak (*Coccothraustes vespertinus*)	55.1	~0	Sunflower seeds		30.4		0.84	West and Hart (1966)
Gambel's Quail (*Callipela gambelii*)	144	+0.3	89% commercial grass seed	8.4	19.0		0.60	Goldstein and Nagy (1985)
			9% *Encelia* seed		18.0			
			2% arthropod					

APPENDIX I. Continued

	Body mass		Diet	Q_i^b (g/day)	GE_i^c (kJ/g)	Utilization efficiency[d]		Source
	$\bar{X}$ (g)	%/day[a]				AMC*	MEC*	
Northern Bobwhite	178		Sunflower		25.3		0.60	Robel et al.
(*Colinus virginianus*)			Showy partridgepea		19.4		0.52	(1979a)
			Giant ragweed		23.8		0.76	
			Prostrate lespedeza		20.7		0.69	
			Pin oak acorn meat		21.1		0.55	
			German millet		18.7		0.78	
			Korean lespedeza		20.7		0.63	
			Soybean		23.2		0.68	
			Wheat		18.3		0.70	
			Western ragweed		22.2		0.73	
			Black locust		20.8		0.51	
			Smartweed		18.9		0.51	
			Thistle		23.5		0.48	
	190	−4.1	Partridgepea (*Cassia nictitans*)	7.3	19.4		0.38	Robel and Bisset (1979)
	190	+0.33	Corn	15.0	18.9	0.82	0.86	Robel et al.
		+0.19	Sorghum	16.7	18.0	0.85	0.86	(1979b)
		−0.54	Hemp	14.8	23.3	0.29	0.45	
		−0.91	Shrub lespedeza	14.6	21.0	0.40	0.54	
		−3.53	Acorn	8.6	21.8	0.49	0.57	
		−4.1	Switchgrass	8.2	19.0	0.26	0.41	
Scaled Quail	194	−0.2	Sorghum	13.1	18.0		0.87	Saunders and
(*Callipepla squamata*)		+1.1	Sunflower chips	10.6	25.5		0.86	Parrish (1987)
		+0.4	Hybrid amaranth	14.1	18.8		0.84	
		−0.2	Pearlmillet pennisetum	12.6	18.8		0.84	
		~0	Amaranth	12.8	18.8		0.82	
		−0.6	Dwarf sorghum	12.1	18.4		0.75	
		−1.1	Canary grass	10.5	19.3		0.74	
		−0.4	Sand dropseed	15.2	18.0		0.68	
		−1.5	Blackwell switchgrass	11.2	19.7		0.65	
		−2.1	Bulk switchgrass	10.1	19.7		0.62	
		−0.8	Korean lespedeza	10.7	19.3		0.61	
Northern Shoveler (*Anas clypeata*)	513							
Gadwall (*Anas strepera*)	653	+1.4	Barnyard grass seeds	55	18.7		0.66	Miller (1984)
Northern Pintail (*Anas acuta*)	678							
Black-bellied Whistling-duck (*Dendrocygna autumnalis*)	682		Sorghum	37			0.85	Cain (1973)
Ring-necked Pheasant (*Phasianus colchicus*)								
juvenile hens	753	~0	High lysine corn	33.6	18.6	0.81	0.83	Labisky and
adult hens	900	−0.1	High lysine corn	28.6		0.80		Anderson (1973)
Sharp-tailed Grouse (*Tympanuchus phasianellus*)	950	~0	Corn	31.8	19.1		0.86	Evans and Dietz (1974)
Spur-winged Goose (*Plectropterus gambensis*)	2940	+0.03	Corn	93		0.75		Halse (1984)
Graylag Goose	4600		Ground corn		19.3		0.87	Story and Allen
(*Anser anser*)			Barley		18.3		0.76	(1982)
			Wheat		18.0		0.78	
Eastern Wild Turkey (*Meleagris gallopavo silvestris*) hens	4222	0	Water oak acorns (*Quercus nigra*)			0.57		Billingsley and Arner (1970)
			Wild pecans (*Carya illinoensis*)			0.27		
Fruit								
Red-capped Manakin (*Pipra mentalis*)	14		*Byrsonima crassifolia*		20.3		0.55	Worthington (1983)

APPENDIX I. Continued

	Body mass					Utilization efficiency[d]		
	$\bar{X}$ (g)	%/day[a]	Diet	Q^b (g/day)	GE_f^c (kJ/g)	AMC*	MEC*	Source
			Guatteria amplifolia		18.6		0.40	
			Palicourea elliptica		18.4		0.53	
			Hasseltia floribunda		17.4		0.50	
			Doliocarpus major		16.8		0.78	
			Coccolaba mazanillensis		16.4		0.84	
			Anthurium clavigerum		16.4		0.76	
			Psychotria marginata		16.2		0.65	
			Psychotria horizontalis		15.5		0.61	
			Psychotria deflexa		15.2		0.76	
			Doliocarpus dentata		16.7		0.83	
Golden-collared Manakin (*Manacus vitellinus*)	17		Heliconia latispatha		21.0		0.81	Worthington (1983)
			Byrsonima crassifolia				0.38	
			Guatteria amplifolia				0.58	
			Palicourea elliptica				0.57	
			Hasseltia floribunda				0.51	
			Doliocarpus major				0.79	
			Anthurium brownii				0.49	
			Coccolaba mazanillensis				0.80	
			Anthurium clavigerum				0.37	
			Psychotria marginata				0.70	
			Psychotria horizontalis				0.58	
			Pyschotria deflexa				0.70	
Red-eyed Vireo (*Vireo olivaceus*)	18		Prunus serotina				0.83	Johnson et al. (1985)
			Smilacina racemosa				0.83	
			Sambucus canadensis				0.90	
			Vitis vulpina				0.89	
House Finch (*Carpodacus mexicanus*)	21.4		Mistletoe				0.62[f]	Walsberg (1975)
Phainopepla (*Phainopepla nitens*)	26.7		Mistletoe	4.2[f]	22.1[f]		0.49[f]	Walsberg (1975)
Gray-cheeked Thrush (*Catharus minimus*)	30		Prunus serotina				0.46	Johnson et al. (1985)
			Phytolacca americana				0.78	
Cedar Waxwing (*Bombycilla cedrorum*)	31		Mixed fruits (*Sorbus* sp., *Viburnum* sp., *Ligustrum* sp., *Phellodendron sachalinense*)				0.37[f]	Holthuijzen and Adkisson (1984)
Hermit Thrush (*Catharus guttatus*)	31		Menispermum canadense				0.62	Johnson et al. (1985)
			Smilax lasioneura				0.54	

APPENDIX I. Continued

	Body mass					Utilization efficiency[d]		
	$\bar{X}$ (g)	%/day[a]	Diet	Q^b (g/day)	GE_i^c (kJ/g)	AMC*	MEC*	Source
			Arisaema				0.61	
			Polygonatum commutatum				0.76	
			Prunus serotina				0.74	
			Smilax hispida				0.71	
			Phytolacca americana				0.71	
			Euonymus atropurpurea		23.4		0.75	
			Celtis occidentalis				0.75	
			Smilacina racemosa				0.63	
			Cornus racemosa				0.80	
			Sambucus canadensis		21.1		0.67	
			Vitis vulpina				0.72	
Swainson's Thrush (*Catharus ustulatus*)	32		Polygonatum commutatum				0.65	Johnson et al. (1985)
			Prunus serotina				0.66	
			Lindera benzoin				0.72	
			Phytolacca americana				0.75	
			Smilacina racemosa				0.74	
			Cornus racemosa				0.68	
Veery (*Catharus fuscescens*)	33		Polygonatum commutatum				0.73	Johnson et al. (1985)
			Prunus serotina				0.77	
			Lindera benzoin				0.76	
			Phytolacca americana				0.89	
			Celtis occidentalis				0.74	
			Smilacina racemosa				0.90	
			Sambucus canadensis				0.82	
			Vitis vulpina				0.88	
Gray Catbird (*Dumetella carolinensis*)	39		Parthenocissus quinquefolia				0.42	Johnson et al. (1985)
			Polygonatum commutatum		18.5		0.40	
			Prunus serotina				0.58	
			Lindera benzoin		26.7		0.66	
			Phytolacca americana				0.59	
			Smilacina racemosa		18.6		0.66	
			Cornus racemosa		28.4		0.82	
			Vitis vulpina		18.4		0.83	
Wood Thrush (*Hylocichla mustelina*)	50		Lindera benzoin				0.85	Johnson et al. (1985)
European Starling (*Sturnus vulgaris*)	71	−3.7	Mixed fruits (grape, viburnum, dogwood)	5.2	20.2	0.56	0.55	Levey and Karasov (1989)
Brown Thrasher (*Toxostoma rufum*)	72		Parthenocissus quinquefolia		23.4		0.46	Johnson et al. (1985)
			Prunus serotina		17.0		0.45	
			Lindera benzoin		26.7		0.62	
			Phytolacca americana		18.7		0.81	
American Robin (*Turdus migratorius*)	77		Menispermum canadense		20.8		0.41	Johnson et al. (1985)
			Smilax lasioneura		19.7		0.51	

APPENDIX I. CONTINUED

| | Body mass | | | | | Utilization efficiency[d] | | |
	$\bar{X}$ (g)	%/day[a]	Diet	Q_i[b] (g/day)	GE_i[c] (kJ/g)	AMC*	MEC*	Source
			Polygonatum commutatum				0.69	
			Prunus serotina				0.76	
			Smilax hispida		18.6		0.76	
			Phytolacca americana				0.74	
			Celtis occidentalis		18.6		0.79	
			Smilacina racemosa				0.81	
			Cornus racemosa				0.77	
	79	−2.6	Mixed fruits (grape, viburnum, dogwood)	7.0		0.57	0.57	Levey and Karasov (1989)
Eurasian Blackbird (*Turdus merula*)	91		Elder			0.90	0.82	Sorensen (1984)
			Bramble			0.87	0.80	
			Hawthorn			0.68	0.66	
			Sloe			0.81	0.58	
			Dogrose			0.45	0.47	
			Ivy			0.80	0.83	
Northern Bobwhite (*Colinus virginianus*)	178		Smooth sumac		21.8[f]		0.28[f]	Robel et al. (1979a)
			Rose hips		20.3[f]		0.42[f]	
			Osage orange		23.5[f]		0.63[f]	
			Dogwood		25.1[f]		0.59[f]	
Rock Ptarmigan (*Lagopus mutus*)	420		Berries of *Vaccinium myrtillus*			0.61[f]		Moss (1973)
			Berries of *Empetrum* sp.			0.49[f]		
Willow Ptarmigan (*Lagopus lagopus*)	550		Cowberries (*Vaccinium vitis-idaeu*)	19.2[f]			0.81[f]	Pullianinen et al. (1968)
Ruffed Grouse (*Bonasa umbellus*)	550		Mixed fruits (sumac, grape, autumn eleagnus)				0.48[f]	Servello et al. (1987)
Sharp-tailed Grouse (*Tympanuchus phasianellus*)	950	−3.7	Wood's rose	63.3[f]	20.6[f]		0.72[f]	Evans and Dietz (1974)
		+4.2	Fleshy hawthorn	92.3[f]	19.9[f]		0.39[f]	
		~0	Russian olive	59.6[f]	20.9[f]		0.48[f]	
		~0	Silver buffalo berry	48.9[f]	20.7[f]		0.64[f]	
		+1.8	Western snowberry	39.9[f]	20.6[f]		0.51[f]	
Eastern Wild Turkey (*Meleagris gallopavo silvestris*) hens	4,222	0	Sugarberry (*Celtis laevigata*)			0.23[f]		Billingsley and Arner (1970)
			Chufa (*Cyperus esculeutus*)			0.53[f]		
			Greenbrier (*Smilax rotundifolia*)			0.22[f]		
			Dogwood (*Cornus florida*)			0.30[f]		
			Spicebush (*Lindera benzoin*)			0.56[f]		
			Grape (*Vitis aestrivalis*)			0.41[f]		
Leaves, twigs, buds, bulbs								
Eurasian Skylark (*Alauda arvensis*)	40		Wheat leaf				0.58	Green (1978)
White-tailed Ptarmigan (*Lagopus leucurus*)	360		Willow, birch, alder			0.45		Moss (1973)
Hazel Grouse (*Tetrastes bonasia*)	400		*Betula* sp., *Salix* sp., *Chosenia* sp., *Alnus* sp.			0.38		A. V. Andreev cited in Moss (1983)

APPENDIX I. Continued

	Body mass					Utilization efficiency[d]		
	$\bar{X}$ (g)	%/day[a]	Diet	Q_i^b (g/day)	GE_i^c (kJ/g)	AMC*	MEC*	Source
Rock Ptarmigan (*Lagopus mutus*)	420		Bulbils of *Polygonum*			0.50		Moss (1973)
			Catkins of *Betula pubescens*			0.19		
			Willow and birch			0.37		
	460		*Betula* sp., *Alnus* sp.			0.42		A. V. Andreev cited in Moss (1983)
Willow Ptarmigan/Red Grouse (*Lagopus lagopus*)	500		Willow and birch			0.44		Moss (1973)
hens, wild			Heather (*Calluna vulgaris*)		22.1	0.52	0.50	Moss (1977)
cocks, wild				63		0.46	0.44	
captives			Heather	65		0.26		
captives			Heather	47		0.37		
cocks, captive	600	−0.8	Heather	71	22.2	0.27		Moss and Parkinson (1972)
			Blueberry stems (*Vaccinium myrtillus*)	67.8		0.30	0.31	Pullianinen et al. (1968)
	600		*Chosenia arbutifolia*			0.35		A. V. Andreev cited in Moss (1983)
Northern Shoveler (*Anas clypeata*)	513	+0.9						
Gadwall (*Anas strepera*)	653	+0.9	Alfalfa pellets	43.9	17.6		0.34	Miller (1984)
Northern Pintail (*Anas acuta*)	678	+0.9	Alfalfa pellets		17.6		0.33	Miller (1974)
Ruffed Grouse (*Bonasa umbellus*)	550	−2.8	Aspen male flower buds		20.9		0.18	Hill et al. (1968)
			Grape leaves plus greenbrier leaves				0.43	Servello et al. (1987)
Spruce Grouse (*Dendragapus canadensis*)	575	−0.3[g]	*Pinus contorta* needles	40.4	21.9	0.27	0.30	Pendergast and Boag (1971)
Sharp-tailed Grouse (*Tympanuchus phasianellus*)	950	−5.0	Plains cottonwood buds	21.5	22.5		0.46	Evans and Dietz (1974)
Canvasback (*Aythya valisineria*)	964		American wild celery winter buds (*Vallisneria americana*)	27.6	16.1	0.75	0.79	Takekawa (1987)
Black Grouse (*Tetrao tetrix*)	1000		*Betula* sp.			0.35		A. V. Andreev cited in Moss (1983)
Blue Grouse (*Dendragapus obscurus*)	1040	+0.5	Douglas-fir needles	87	21.0	0.35		Remington (1990)
		−1.2	Lodgepole pine needles	74	21.5	0.34		
		−2.1	Subalpine fir needles	52	21.7	0.30		
		−1.1	Engelmann spruce needles	64	20.1	0.26		
Brant (*Branta bernicla*)	1600		*Spartina patens* (Gramineae)		18.3	0.45	0.51	Buchsbaum et al. (1986)
			S. alterniflora		19.5	0.10	0.34	
Barnacle Goose (*Branta leucopsis*)	1687		*Lolium perenne*		18.7	0.33		Ebbinge et al. (1975)
			Mixed grasses			0.22		
Lesser Snow Goose (*Anser caerulescens*)	2500		Bulrush rhizomes (*Scirpus americanus*)	66–137		0.28	0.36	Burton et al. (1979)
Spur-winged Goose (*Plectropterus gambensis*)	2940	+0.04	Rabbit pellets	164		0.55		Halse (1984)
Mallard (*Anas platyrhynchos*)	3600		Alfalfa		17.4		0.32	Muztar et al. (1977)
			Cladophora		8.3		0.30	
			Duckweed (*Lemna minor*)		17.5		0.15	

APPENDIX I. CONTINUED

	Body mass					Utilization efficiency[d]		
	$\bar{X}$ (g)	%/day[a]	Diet	Q_i^b (g/day)	GE_i^c (kJ/g)	AMC^*	MEC^*	Source
			Watermilfoil		7.6		0.23	
			Pondweed		11.6		0.23	
			Vallisneria americana		12.9		0.22	
Cape Barren Goose (*Cereopsis novaehollandiae*)	3680		Dried lucerne	298		0.26		Marriott and Forbes (1970)
Canada Goose (*Branta canadensis*)	4000	0	*Spartina alterniflora* (Gramineae)		19.0	0.25	0.30	Buchsbaum et al. (1986)
	4000	0	*Juncus gerardi* (Juncaceae)		20.0	0.19	0.40	
Common Capercaillie (*Tetrao urogallus*)	4600		*Pinus sylvestris*			0.33		A. V. Andreev cited in Moss (1983)
Graylag Goose (*Anser anser*)	4600		Dehydrated alfalfa meal		17.5		0.30	Story and Allen (1982)
			Alfalfa haylage		17.6		0.38	
Tundra Swan (*Cygnus columbianus*)	6650	−2.3	Timothy grass		15.5		0.40	McKelvey (1985)
Trumpeter Swan (*C. buccinator*)	10,650	−2.4	Rhizomes of *Carex lyngbei*		18.4		0.56	
Emu (*Dromaius novaehollandiae*)	38,000		Grain and vegetable offal					Dawson and Herd (1983), Herd and Dawson (1984)
			Diet 1	~750		0.60	0.64	
			Diet 2	~459		0.62	0.64	
			Diet 3	~628		0.60	0.68	
Ostrich (*Struthio camelus*)	80,700	−1.4	Lucerne, coarsely milled, H_2O deprived	290	16.6	0.17	0.28	Withers (1983)
	95,400	~0	Lucerne, coarsely milled, ad lib H_2O	1780		0.34	0.43	

[a] Change in body mass during feeding trials.
[b] Feeding rate, g dry matter/day.
[c] Gross energy content per gram dry matter.
[d] Definitions in Table 1.
[e] Recalculated by Wijnandts (1984).
[f] For whole fruit including seeds. All other values in table are for whole fruit minus seeds.
[g] Two other wild-caught birds maintained weight eating *Pinus* needles for 2 mo.

APPENDIX II. Mean Retention Times, or Appearance Times of Digesta Markers in Birds

Species	Body mass (g)	Diet	Appearance time (min) 5%	50%	95%	Source
Leaf and twig eaters						
Common Canary (*Serinus canarius*)	15	Turnip leaves	31[d]	59[f]		Malone (1965)
Rock Ptarmigan (*Lagopus mutus*)	460	Game chow	78[b] 288[c]	114 594	618 1554	Gasaway et al. (1975)
Canvasback (*Aythya valisineria*)	964	Wild celery buds		189[a]		Takekawa (1987)
Red-breasted Goose (*Branta ruficollis*)	1120	Grass	80[d]	91		Owen (1975)
Mallard (*Anas platyrhynchos*)	1150	*Elodea* (algae) cattail	48[d] 84[d]	84[f] 150[f]		Malone (1965)
Ring-necked Pheasant (*Phasianus colchicus*)	1400	Turkey breeder pellets	90[b]	300[f]	510[g] 2100[h]	Duke et al. (1968)
Barnacle Goose (*Branta leucopsis*)	1905	Grass	52[d]	78		Owen (1975)
Lesser Snow Goose (*Chen c. caerulescens*)	2500	Bulrush rhizomes	58[a]	120	192	Burton et al. (1979)
Spur-winged Goose (*Plectropterus gambensis*)	2940	Rabbit pellets	108[a]	138	210	Halse (1984)
Cape Barren Goose (*Cereopsis novaehollandiae*)	3680	Lucerne		78	132	Marriott and Forbes (1970)
Graylag Goose (*Anser anser*)	4600	Grass		120		Mattocks (1971)
Emu (*Dromaius novaehollandiae*)	38,000	Grain plus vegetable offal	132[b] 108[c]	282 234	822 444	Herd and Dawson (1984)
Seed eaters						
Common Canary (*Serinus canarius*)	15	Commercial seeds	58[d]	95[f]		Malone (1965)
Chipping Sparrow (*Spizella passerina*)	11.5	Cracked corn	62[a]			Stevenson (1933)
Field Sparrow (*Spizella pusilla*)	13.7	Cracked corn	101[a]			Stevenson (1933)
Song Sparrow (*Melospiza melodia*)	20.6	Cracked corn	102[a]			Stevenson (1933)
Rufous-sided Towhee (*Pipilo erythrophthalmus*)	41.6	Cracked corn	92[a]			Stevenson (1933)
Mallard (*Anas platyrhynchos*)	1150	Maize Oats Wheat	168[d] 126[d] 90–210[a]	246[f] 192[f] ~210		Malone (1965) Malone (1965) Clark et al. (1986)
Spur-winged Goose (*Plectropterus gambensis*)	2940	Maize	315[a]	384	450	Halse (1984)
Graylag Goose (*Anser anser*)	4600	Corn Wheat Oats Rice hulls		258[a] 168[a] 174[a] 282[a]		Storey and Allen (1982)
Arthropod eaters						
Scarlet Tanager (*Piranga olivacea*)	29	Beetle and moth larvae, mealworms	85[a]			Stevenson (1933)
European Starling (*Sturnus vulgaris*)	71	Crickets		56[c]		Levey and Karasov (unpubl. data)
American Robin (*Turdus migratorius*)	79	Crickets		65[c]		Levey and Karasov (unpubl. data)
American Black Duck (*Anas rubripes*)	904	Blue mussels			50[d]	Grandy (1972)
Mallard (*Anas platyrhynchos*)	1150	Crayfish	66[d]	86[f]		Malone (1965)
Nectar eaters						
Rufous Hummingbird (*Selasphorus rufus*)	3.2	Sugar water	<15[c]	48	180	Karasov et al. (1986)
Fruit eaters						
Red-capped Manakin (*Pipra mentalis*)	14	Tropical fruits	22[d]			Worthington (1983)

APPENDIX II. Mean Retention Times, or Appearance Times of Digesta Markers in Birds

Species	Body mass (g)	Diet	Appearance time (min) 5%	50%	95%	Source
Golden-collared Manakin (*Manacus vitellinus*)	18	Tropical fruits	21[d]			Worthington (1983)
Phainopepla (*Phainopepla nitens*)	26.7	Mistletoe	29[d]			Walsberg (1975)
Cedar Waxwing (*Bombycilla cedrorum*)	31	Dogwood	23[a]			Holthuijzen and Adkisson (1984)
		Red cedar	12[a]			
Cedar Waxwing (*Bombycilla cedrorum*)	35	Fruit mash		41[c]		Martinez del Rio et al. (1989)
European Starling (*Sturnus vulgaris*)	71	Wild grapes	14[c]	53		Karasov and Levey (1990)
American Robin (*Turdus migratorius*)	79	Wild grapes	16[c]	48		Karasov and Levey (1990)
Eurasian Blackbird (*Turdus merula*)	90	Elder	26[d]			Sorensen (1984)
		Bramble	39[d]			
		Hawthorne	32[d]			
		Sloe	19[d]			
		Dogrose	29[d]			
		Ivy	30[d]			

Times shown are times of appearance of 5%, 50%, and 95% of marker fed to animals, or else mean retention time (roughly equivalent to time until appearance of 50% of a marker) determined by another method. The marker or method is indicated by a superscript: [a]dye, [b]particulate marker, [c]liquid marker, [d]fragments of food, [e]meal to pellet interval, [f]midpoint between appearance of first and last marker, [g]portion not digested in caecum, [h]portion digested in caecum.

APPENDIX III. Maximum Rates of Intake of Food and Metabolizable Energy in Birds

Species	Mass	Diet	MEC^*	Q_i Maximum[a] (g/day)	Maximum relative to normal[b]	Metabolizable energy intake Maximum (kJ/day)	Maximum relative to BMR^c	Source
Yellow-bellied Seedeater (*Sporophila nigricollis*)	8.9	Univ. Ill. #521 chick starter feed	0.79	5.1	1.59	68	3.84	Cox (1961)
Blue-black Grassquit (*Volatinia jacarina*)	9.3	Univ. Ill. #521 chick starter feed	0.80	4.4	1.52	62	3.39	Cox (1961)
Variable Seedeater (*Sporophila aurita*)	10.8	Univ. Ill. #521 chick starter feed	0.74	5.0	1.52	69	3.39	Cox (1961)
Zebra Finch (*Poephila guttata*)	12	Laying ration for chickens	0.77	5.4	1.59	67	3.05	El-Wailly (1966)
Hoary Redpoll (*Carduelis hornemanni exilipes*)	15	Univ. Ill. #521 chick starter feed	0.71	10.3	2.94	130	5.03	Brooks (1968)
Common Redpoll (*Carduelis flammea*)	15	Univ. Ill. #521 chick starter feed	0.70	9.1	2.28	105	4.07	Brooks (1968)
American Tree Sparrow (*Spizella arborea*)	18	Univ. Ill. #521 chick starter feed	0.71	10.7	2.05	134	4.55	West (1960)
House Sparrow (*Passer domesticus*)	24	Univ. Ill. #393 chick mash	0.85	8.9	1.85	144	3.97	Kendeigh et al. (1977)
White-throated Sparrow (*Zonotrichia albicollis*)	28	Univ. Ill. #521 chick starter feed	0.67	10.9[d]	1.63	130[d]	3.20	Kontogiannis (1968)
			0.67	11.6	2.14	143	3.52	
Dickcissel (*Spiza americana*)	30	Univ. Ill. #521 chick starter feed	0.68	12.6	1.88	143	3.35	Zimmerman (1965)
Blue-winged Teal (*Anas discors*)	360	Duck Growena	0.75	>53	>2.43	>748	>4.76	Owen (1970)
Black-bellied Whistling-duck (*Dendrocygna autumnalis*)	782	Sorghum	0.85	87	2.29	1282	4.67	Cain (1973)

[a] Highest intake (g dry mass/d) measured at temperature very near or at the lower limits of temperature tolerance.
[b] Normal intake measured at 20–24°C.
[c] Basal metabolic rate (*BMR*) from Lasiewski and Dawson (1967).
[d] Maximum value under experimental condition of low temperature plus forced exercise.

TABLE 1. NUTRITIONAL QUALITY OF INSECTS (FROM REDFORD AND DOREA [1982] UNLESS OTHERWISE INDICATED)

	% H₂O (wet mass)	% ash (dry mass)	% nitrogen (dry mass)	% fat (dry mass)
Orthoptera				
Locust: (*Melanoplus* sp.)		5.6	12.0	7.2
(*Oxya* sp.)		3.8	10.8	7.2
(*Oxya* sp.)		6.5	12.2	5.7
(*Schistocerea paranensis*)		4.2	8.2	18.4
(*S. gregaria*)	70.6	8.7	10.2	13.5
(*Nomadacris septembfasciata*)		8.7	10.2	14.1
(*Locustana* sp.)	57.1		6.8	50.1
Cricket: (*Brachytrypes membranaceus*)	76	8.8	9.1	22.1
(*Gryllus domesticus*)	71	8.28	10.7	16.9
Cockroach: (*Blatta orientalis*)[a]	70.6			
(*Blattella germanica*)[a]				16.3
(*Periplaneta americana*)[a]	61			27.05
Coleoptera				
Tenebrio molitor (mealworm larvae)	66.4	6.9	8.72	32.7
Lachnosterna sp.: (larvae)	79.9	10.0	8.8	15.4
(adults)	69.4	5.2	10.5	16.0
Polycleis equestris (adult weevil)	51.8		10.1	4.6
Sternocera orissa (adults)	60.6		18.6	10.2
Lepidoptera				
Galleria mellonella (larvae)	56.1	1.82	4.92	61.5
Bombyx mori: (adult)	60.7	3.8	9.4	36.1
(adult)	80	5.2	8.7	3.9
(adult)				24
(larvae)	84.5		2.1	
Antherea mylittal (adult)	80	5.3	9.0	7.7
Bombycomorpha pallida (larvae)	82.2		9.4	34.3
Cerina forda (larvae)	79.6		9.3	27.9
Diptera				
Musca domestica: (pupae)		5.3	10.1	15.5
(pupae)		11.9	9.8	9.3
Gasterophilus intestinalis[a]	64.7			19.4
Phaenicia sericata[a]	79		12.5	
Hymenoptera				
Apis mellifera: (larvae)	77	3.0	10.7	16.1
(pupae)	70.2	2.2	14.7	8.0
(adult)[a]			12.2	18.0
Carebara sp.: (alate females)	60.0		1.2	23.8
(alate males)	60.0		4.0	3.3
Carebara vidua: (alate females)	60.0		3.0	59.5
(alate males)	60.0		10.1	8.3
Camponotus rufipes: (alate queen nymph)			22.90	
Iridomyrmex detectus: (alate females)				48.2
(alate males)				9.6
(worker)				18.8
Tetramorium caespitum: (alate females)	44.2			51.3
(alate males)	70.0			8.8
Isoptera				
Harvester termite (sp. unspecified)	77.5	12.09	10.62	7.69
Macrotermes carbonarius: (soldier)	76.5	3.66		
(worker)	72.4	31.67		
(alate)	59.5	1.60		
Dicuspidtermes nemorosus: (soldier)	80.6	3.52		
(worker)	72.2	48.73		
(alate)	56.1	1.76		

TABLE 1. CONTINUED

	% H$_2$O (wet mass)	% ash (dry mass)	% nitrogen (dry mass)	% fat (dry mass)
Homallotermes foraminifer: (soldier)	78.3	21.48		
(worker)	74.0	46.03		
(alate)	52.2			
Orthognathotermes gibberorum: (nymph)				19.72
(worker)	71.35	61.00	2.55	
(soldier)	71.90	26.30	7.54	
Syntermes dirus: (nymph)				22.00
(worker)	79.65	17.05	6.91	3.40
(soldier)	74.0	5.90	11.89	1.75
Grigiotermes metoecus: (nymph)				21.41
(worker)	66.35	59.90	2.99	1.51
Procornitermes araujoi: (nymph)				24.12
(worker)	78.10	16.10	5.42	3.45
(soldier)	78.40	10.05	8.68	4.56
Termes sp. (no caste specified)	44.5	5.3	6.0	50.5
Macrotermes falciger (alates)	34.2	10.9	6.31	41.9
Trinervitermes geminatus: (worker)	80.0	7.2		
(major soldier)	73.0	3.6		
(minor soldier)	80.0	8.9		
(alate)	50.0	2.3		
Basidentitermes potens: (workers)	61.0	71.0		
(soldiers)	69.0	4.5		
Macrotermes bellicosus: (workers)			10.0	<25
(soldiers)			9.4	<35
(alates)			6.5	52.8
Velocitermes paucipilis: (worker)	70.35	12.70	7.32	
(soldier)	69.50	10.85	9.28	
Cortaritermes silvestri: (worker)	77.80	8.50	7.78	6.85
(soldier)	70.95	6.70	6.78	14.40
Nasutermes sp.: (worker)	77.80	11.30	7.14	3.59
(soldier)	72.70	8.75	8.39	2.31
Armitermes euamignathus: (worker)	73.60	46.05	3.34	
(soldier)	73.85	38.00	3.93	
Cornitermes cumulans: (worker)	77.70	36.00	4.27	2.65
(soldier)	78.10	6.45	8.85	4.29

[a] Data from Spector (1958).

metry studies on species with low-protein diets; however, insectivorous species may obtain as much as 60% of their energy from the combustion of protein, and the difference in digestive physiology between birds and mammals becomes more significant.

The assimilation rate (or digestibility) of protein depends upon both the protein and the animal in question. Drodz (1975) gave a range of digestibility of 75–99%. Karasov (1982) indicated that the digestibility of insect (cricket) protein was about 95%.

FATS

The size of the fat body in insects varies by stage of life cycle as well as taxonomic affinity (see Gilbert 1967). Total lipid content may range from 5.3–85.4% total dry mass. In many studies of insectivorous animals the total energy content of an insect meal is derived from analysis of mealworm (*Tenebrio molitor*) samples, in which fat content may be as high as 50–60% dry mass. In social insects, such as wasps and termites, reproductive forms tend to have extremely high fat contents and low water contents compared to larvae or workers. Most adult insects have fat contents in the range of 10–25% dry mass.

Similarly, the fatty acid content of insect fats is variable (summarized in Barlow 1964). For example, most Homoptera appear to have a high proportion of myristic acid (14:0), with very little fats of longer chains and very little unsaturated fatty acid content. In contrast, most coleoptera have a high proportion (50–75%) of unsaturated 18-carbon fatty acids. There are insufficient data on different taxa or on variation in fatty acid content with stage of life cycle to make other generalizations.

TABLE 2. WATER AND ENERGY CONTENT OF VARIOUS INSECTS (FROM CUMMINS AND WUYCHECK [1971] UNLESS OTHERWISE NOTED; ALL ADULT UNLESS OTHERWISE NOTED)

Order / Family	kJ/g (dry mass)	% H$_2$O (wet mass)	Source
Orthoptera	22.18		
Acrididae	21.25		
Gryllidae	25.1		Karasov (1982)
Tettigoniidae	22.80		
Ephemeroptera	22.88		
	22.09		Maxson and Oring (1980)
	23.26		Maxson and Oring (1980)
Heptageniidae	23.37		
Ephemeridae	20.44	85	
Odonata			
Lestidae	20.74	79.6	
Agrionidae	22.59		
Libellulidae	24.52		
Gomphidae	12.69	81.6	
Coleoptera			
Hydrophilidae	22.47		
Tenebrionidae	24.48		
(larvae)	29.71	61	Sall (1979)
(pupae)	28.87	67	Sall (1979)
(adults)	27.61	64	Sall (1979)
(larvae)	27.56	46.6	O'Farrell et al. (1971)
Elateridae	22.76		
Coccinellidae	24.48		
Chrysomelidae	21.85		
Trichoptera			
Limnophilidae	19.30		
Hydropsychidae	22.53	81.2	
Megaloptera			
Corydalidae	21.80		
Diptera	17.89		
Chironomidae	22.69	79.8	
	21.42		Maxson and Oring (1980)
	21.25		Maxson and Oring (1980)
(larvae)	22.25	85.8	
(adults)	23.73	73.8	
Culicidae	20.65		
	23.01		Kunz (1987)
Drosophilidae	24.25		
Calliphoridae	24.13		
Stratiomyidae	12.00		
Tipulidae	25.52		Kunz (1987)
Hemiptera			
Cercopidae	23.59		
Dictyoptera			
Blattellidae	22.58		
Hymenoptera	19.37		
Formicidae	19.03		
Apidae	20.37		
Lepidoptera	21.25		Kunz (1987)
Mixed Insects	22.09		
	23.81	76.3	Nagy et al. (1978)
		63.5	Carpenter (1969)

The caloric value of fatty acids increases with both length of carbon chain and degree of saturation. An average triglyceride contains 75% carbon, 12% hydrogen, and 12% oxygen by weight, and yields 39.54 kJ/g. The end products of the combustion of lipids are carbon dioxide and water, thus

$$\begin{aligned}&\text{100 g average}\\&\text{tricglyceride:} &&6.25\ C + 12\ H + 0.75\ O\\&\text{combustion:} &&+ 8.875\ O_2\\&\text{products:} &&6.2\ CO_2 + 6\ H_2O.\end{aligned} \quad (7)$$

Thus, the digestion of 1 g of average triglyceride yields 39.54 kJ, 1400 ml CO_2, and 1.08 ml metabolic H_2O, and requires 1988 ml O_2. The R.Q. for the combustion of pure fat is about 0.70. The digestibility of fats may also be assumed to be about 95% (Drodz 1975).

CARBOHYDRATES

Carbohydrate is often ignored in insect analysis even though it may occur in large quantities. Kirby (1963) suggested that as much as 33% of the entire dry mass of an insect may be glycogen in some larval forms, that glucose may be found at concentrations of up to 30 mg/100 ml blood, and that trehalose may occur at concentrations of 1500–6000 mg/100 ml blood. The calculations presented above suggest that about 7.2% of dry mass of an average insect is carbohydrate. The combustion of pure carbohydrate results in the production of carbon dioxide and water, and the R.Q. for such reactions is 1.0. The energy yield for the combustion of a variety of carbohydrates shows remarkable uniformity (15.5–17.5 kJ/g) regardless of molecular weight or complexity. Therefore, for our purposes we may assume that all insect carbohydrate is in the form of glycogen $(C_6H_{10}O_5)_n$, thus

$$\begin{aligned}&\text{One mole (162 g):} &&C_6H_{10}O_5\\&\text{combustion:} &&+ 6\ O_2\\&\text{products:} &&6\ CO_2 + 5\ H_2O.\end{aligned} \quad (8)$$

The energy yield for the combustion of glycogen is 17.52 kJ/g. Combustion of 1 g of glycogen yields 0.556 ml of metabolic water and 830 ml of CO_2, and requires 830 ml of O_2. The digestibility of carbohydrate is at least 95% (Drodz 1975).

The results of these calculations for the digestion of protein, fat, and carbohydrate in a typical insect suggest that, while the total energy content of such an insect based on combustion is 24.2 kJ/g dry matter, assimilable energy (EA) is 84.4%, and metabolizable energy (EM) is only 75.9% for mammals and 71.2% for birds (Table 3). Dry matter assimilation is approximately 78%. The combined R.Q. for a bird on a diet of mixed insects is approximately 0.76, and the consumption of 1 g of mixed insects results in the bird's producing 637 ml CO_2. For a mammal on the same diet R.Q. is 0.81, and the CO_2 yield would be 750 ml/g insects. Karasov (this volume) reported a mean EM for birds on an arthropod diet of 77%, but included studies using mealworms. Omitting those studies reduces the mean (N = 4 studies) EM to 73%.

THE EFFECTS OF USING INAPPROPRIATE CONVERSION FACTORS

We can now consider the effects of using inappropriate conversion factors, or ignoring the importance of protein and chitin in a diet of insects. I will use as an example hypothetical data on a 9.0 g little brown bat (*Myotis lucifugus*). If this bat is found, through doubly labeled water studies, to produce 1224 ml CO_2/day we could use the conversions for mammals (Table 3) to estimate its daily energy expenditure (DEE) at

$$(18.35/750) \times 1224 = 29.9 \text{ kJ/day}. \quad (9)$$

We can further estimate that this bat's daily food intake would be

$$\begin{aligned}1224/750 &= 1.63 \text{ g dry insects/day}\\&= 1.63/0.3\\&= 5.4 \text{ g fresh insects/day}.\end{aligned} \quad (10)$$

However, if we were to use the conversion factors for birds by mistake we would obtain values of 33.1 kJ/day (11% too high) for DEE and 6.4 g fresh insects/day (18.5% too high).

Another common mistake is to ignore chitin content. If all of the nitrogen found in insect samples were bound in protein, we would overestimate the proportion of protein in the diet by about 9% (10.4% nitrogen × 6.25 = 65% protein). Using the resulting conversion factors we would estimate that our bat had a DEE of 29.6 kJ/day (only 1% too low), but our estimate of daily food intake would be about 4.8 g fresh insects/day, or about 11% too low.

A third mistake would be to assume, as we often do for plant-eating animals, that no protein is metabolized to produce energy. This would only overestimate DEE by about 9%, but would overestimate daily food intake by about 187%! An appreciation of the nitrogen metabolism of the animal and the digestibility of the diet are just as important as the actual energy content of the diet in obtaining accurate estimates of energy metabolism and food intake.

I present the conversion factors in this paper as a starting point for studies in which the principal prey are adult insects. Similar values may be derived for other diets using the tables or additional data. In some cases more precise con-

TABLE 3. SUMMARY OF CONVERSION FACTORS FOR BIRDS AND MAMMALS ON A DIET OF INSECTS

Component	Mass/g dry mass		kJ/g dry mass			
	Gross	Assimilated	Gross	Assimilated	Metabolized Birds	Metabolized Mammals
Protein	0.595	0.565	14.06	13.36	10.19	11.32
Chitin	0.128	0.000	2.71	0.00	0.00	0.00
Fat	0.155	0.147	6.13	5.82	5.82	5.82
Carbohydrate	0.072	0.069	1.28	1.21	1.21	1.21
Inorganic	0.050	0.000				
Totals	1.000	0.781	24.18	20.40	17.22	18.35

Dry Matter Assimilation (DMA) = 78.1%.
Energy Assimilation (EA) = 20.40/24.18 = 84.4%.
Energy Metabolization (EM) = 17.22/24.18 = 71.2% for birds
 = 18.35/24.18 = 75.9% for mammals.

versions might be obtained through careful feeding studies and diet analysis; however, I suggest that in most cases the errors in such studies are comparable to those in simply deriving conversion factors from published values of insect composition. However, it must be noted that there is a great deal of variation in Tables 1 and 2. Some caution is obviously needed in using the numbers in these tables, and larger sample sizes are needed for many insect groups; what is needed is a broader data base on the general composition of different taxa of insects that can be generally applied to feeding studies. The processes of digestion and nitrogen excretion might be expected to be constant enough that individual studies of digestibility may be unnecessary.

ACKNOWLEDGMENTS

The impetus for this paper came from energetics studies on little brown bats done in collaboration with T. H. Kunz and A. Kurta at Boston University. I am grateful to them for their input. For other ideas and discussion I thank G. A. Bartholomew, K. A. Nagy, A. Collins, and D. W. Thomas.

ENERGETICS OF ACTIVITY AND FREE LIVING IN BIRDS

DAVID L. GOLDSTEIN

Abstract. Knowledge of the energy costs of avian activities, based on studies in both laboratory and field, can be applied to understanding daily energy expenditure (DEE) by free-living birds through the use of time-energy budgets (TEB). In TEB analysis, a compendium of activities is made, and the energy costs are summed. Comparisons of TEB estimates of DEE with those measured directly using doubly-labeled water suggest that the former technique can give accurate results, but can also be misleading. Energy costs of resting and activity should be known for the population under study, and thermoregulatory costs must be properly quantified. Under some conditions energy expenditure by birds reaches a maximum sustainable level. Behavioral flexibility may then be critical to the maintenance of energy balance in the face of changing physical environments and resource availability. Measurements of DEE may provide a quantitative link between foraging ecology (patterns of behavior) and fitness (the ability to survive and reproduce).

Key Words: Foraging energetics; activity costs; time-energy budget; doubly-labeled water; daily energy expenditure.

The costs and benefits of avian foraging can be measured in a variety of currencies. Among these, estimates of energy balance are attractive because of the ability to quantify the energy spent and gained, and because of the fundamental link in which energy gained while foraging can be converted to activity, growth, storage, or reproduction. As such, studies of energy expenditure may provide quantitative tests of a variety of ecological theories regarding such phenomena as foraging strategies, resource competition, or parental investment. Our confidence in these tests rests largely in our ability to assess energy expenditure accurately. In this paper I will address the techniques used to assess rates of energy use, and will discuss some of the implications of the results gained.

THE ENERGY COSTS OF AVIAN ACTIVITIES

The energy costs of avian activities have been estimated in both laboratory and field using a variety of techniques. In the laboratory, analyses of oxygen consumption have been made during resting, alert perching, bipedal locomotion, hovering, gliding, flapping flight, and eating. Such studies provide the bulk of the data available on energy costs of activities.

The cost of a particular activity can be arrived at under certain circumstances (reviewed by Goldstein 1988) using doubly-labeled water (Nagy 1980), which provides a measure of carbon dioxide production (convertible to energy consumption) over an extended (typically several day) period. Such analyses are most applicable to activities with high energy costs; they have been used to calculate the cost of flight in several species (Hails 1979, Turner 1983b, Flint and Nagy 1984, Tatner and Bryant 1986, Masman and Klaasen 1987) and the cost of swimming in Jackass Penguins (*Spheniscus demersus*; Nagy et al. 1984).

Activity costs in unrestrained birds have also been estimated from telemetered heart rates, which may, in well-defined circumstances (see Johnson and Gessaman 1973), provide a reliable index to the rate of oxygen consumption. This approach has been applied rarely to birds, but has yielded estimates of the energy costs of eating and several other activities (Wooley and Owen 1978).

Finally, the energy cost of flight has been estimated for a number of species based on the loss of mass of fat used to fuel the activity (see, e.g., Dolnik and Gavrilov 1973).

Measures of the energy cost of the same activity in the same species using different techniques are few. However, both they and interspecific comparisons reveal consistent ranges of estimates (Table 1), and suggest that each technique is capable of yielding accurate measures of the energy costs of activities.

Our knowledge of the energy costs of some types of activities, such as resting and flight, is now quite good. Flight costs in those aerial species, such as swallows and swifts, that forage in flight are typically 2.5 to 5 times resting energy expenditure (reviewed in Masman and Klaasen 1987). However, sustained flight in other birds is more costly, approximately 11 times resting (Masman and Klaasen 1987), and short flights such as might be used to move between foraging substrates may cost in excess of 20 times resting (Tatner and Bryant 1986).

The energy costs of other activities are less well studied. The energetics of treadmill running are well characterized (Taylor et al. 1982). However, the energy cost of terrestrial foraging and locomotion is complex, and depends on the speed of

TABLE 1. Energy Costs of Avian Activities Measured Using Different Techniques

Activity	Species	Cost	Method	Reference
		Same activity in same species		
Flight	Barn Swallow (*Hirundo rustica*)	1.34[a] 1.30[a]	Mass loss over long flight Doubly-labeled water	Masman and Klaasen (1987)
Flight	House Martin (*Delichon urbica*)	1.01–1.26[a] 0.95–1.08[a]	Doubly-labeled water Mass loss over long flight	Masman and Klaasen (1987)
Flight	Starling (*Sturnus vulgaris*)	9.15[a] 9.0[a]	Oxygen consumption in wind tunnel Doubly-labeled water	Masman and Klaasen (1987)
		Same activity in different species		
Feeding	Loggerhead Shrike (*Lanius ludovicianus*)	2.2[b]	Oxygen consumption	Weathers et al. (1984)
	Black Duck (*Anas rubripes*)	1.7[b]	Heart rate telemetry	Wooley and Owen (1978)
Alert perching	Budgerigar (*Melopsittacus undulatus*)	2.0[b]	Oxygen consumption	Buttemer et al. (1986)
	Black Duck	2.1[b]	Heart rate telemetry	Wooley and Owen (1978)

[a] Cost (watts).
[b] Cost (multiple of basal metabolic rate, the metabolic rate in a resting, post-absorptive animal).

locomotion, slope and evenness of the terrain, and foraging activities that accompany locomotion. The cost of terrestrial foraging has been estimated for just one bird species (Gambel's Quail [*Callipepla gambelii*]: cost was approximately two times predicted resting levels, or 3.5 times actually measured resting energy expenditure [Goldstein and Nagy 1985]).

A number of other activities, including alert perching and food manipulation-eating, have energy costs two to three times resting (Table 2).

THE ENERGY COST OF FREE LIVING

Daily energy expenditure in birds (DEE, the sum of all energy costs incurred in a 24-hour period) has been measured in a number of ways (reviewed by Goldstein 1988), including predominantly the construction of time-energy budgets and the use of doubly-labeled water. Other techniques to measure DEE, such as analysis of sodium turnover or the quantitative collection of excreta, may be applicable to some species or situations (see Nagy 1989), but have been used infrequently or not at all for free-living birds.

In time-energy budget (TEB) analysis, a compendium of activities is made for an animal, and the energy costs of these activities are summed; costs of thermoregulation and production must be added to this. This technique is time consuming, and requires that activities be categorized and accurately timed, that activity costs be estimated, and that thermoregulatory costs be accurately assessed. Yet it requires a minimum of equipment and is inexpensive, and so has been most commonly used to measure DEE in birds.

The doubly-labeled water (DLW) technique provides a more direct and quite accurate ($\pm 8\%$) measure of DEE. However, it provides only a single integrated measure of energy expenditure, and the analyses can be costly. In recent years a number of studies have employed this technique, and its use simultaneously with time-energy budget analysis has provided much insight into the limitations of the TEB technique.

Two particular caveats have emerged from these comparisons. First, resting costs of the study animal must be well known; even subtle seasonal (Goldstein and Nagy 1985) or geographical (Hudson and Kimzey 1966) variation in resting costs can result in significant inaccuracy in calculated energy budgets. Second, a robust analysis of thermoregulatory costs, including accurate assessment of radiative and convective inputs, must be used; again, inattention to these factors can produce significant inaccuracies in the energy budget.

These requirements have been particularly elucidated by a series of comparisons between TEB and DLW analyses of aviary-housed Loggerhead Shrikes (*Lanius ludovicianus*; Weathers et al. 1984) and Budgerigars (*Melopsittacus undulatus*; Buttemer et al. 1986). For the shrikes, substitution into time-energy budgets of metabolic data from a separate population—differing by only 12% in thermal conductance from the study population—produced a 22% increase (inaccuracy) in the calculated rate of energy expenditure. This large effect occurred because the shrikes spent much of their time at temperatures below thermoneutrality. For Budgerigars, ignoring the effects of wind resulted in a similar (15%) underestimate of DEE.

Understanding these details of energetics can significantly affect the interpretation of field observations, as seen in studies of desert phasianids (Goldstein and Nagy 1985, Kam et al. 1987). These birds survive the rigors of the desert summer in part by reducing their activity during the hottest midday hours. This in turn is made possible by the birds' very low resting metabolic rates, which result in low overall energy requirements and hence reduced foraging requirements. Together these factors produce levels of DEE markedly lower than those of other similar-sized birds (Nagy 1987). Time-energy budgets constructed using allometrically predicted, rather than measured, metabolic rates would have significantly over-estimated DEE, and would not have yielded a proper understanding of the forces shaping these birds' activity budgets. Similarly, allometric predictions of DEE have substantial uncertainty (Nagy 1987) and may provide estimates of DEE significantly at variance with actual values.

Accurate continuous assessment of the microclimates occupied by free-living birds is a significant challenge, but has been successfully approached in a number of studies (Mugaas and King 1981, Biedenweg 1983, Goldstein 1984, Stalmaster and Gessaman 1984, Masman 1986). Accurate time-energy budgets also require that time budgets be constructed for individual birds, rather than being compiled from data on many individuals, whose activity patterns may vary considerably (Rijnsdorp et al. 1981). Finally, TEB's require that activities be categorized and recorded; even activities with quite different energy costs, such as restful vs. alert perching, may be difficult to distinguish in the field. Despite these seeming pitfalls, rigorous TEB analyses can yield results very similar to DLW (Goldstein 1988, Nagy 1989). The overall level of accuracy required depends, of course, on the questions being asked by the researcher.

IMPLICATIONS FOR FORAGING ECOLOGY

A foraging bird must choose among behaviors with different energy costs, and must acquire sufficient energy to meet both these costs and the costs of other activities, maintenance, thermoregulation, storage, and production. Studies have demonstrated that a changing physical environment may strongly influence a bird's pattern of time use. Changing weather may alter food availability, thereby necessitating a change in foraging strategies (e.g., Murphy 1987). In more extreme situations, the physical environment imposes such a strenuous thermal load on the bird that foraging is either impossible (excessive heat load; Goldstein 1984) or energetically too expensive (in extreme cold) to be profitable (Evans 1976). In these cases, foraging may cease altogether. An understanding of the energy costs of activities provides a means for evaluating the costs and benefits of these changing behavioral strategies. However, studies of avian daily energy expenditure have demonstrated that more subtle influences, such as convective heat loss and acclimatization of resting metabolic rate, can also have important impacts on overall energy expenditure. It is this overall level of expenditure which must be balanced by the energy gained during foraging.

TABLE 2. ENERGY COSTS OF AVIAN ACTIVITIES

Activity	Energy cost (multiple of BMR)[a]
Flight	
Aerial species	2.7–5.7
Other birds, sustained flight	~11
European robin, short flights	23
Gliding	2
Terrestrial locomotion	Varies with speed and form of locomotion
Perch	
Rest	1.0
Alert	1.9–2.1
Preen	1.6–2.3
Eat	1.7–2.2
Sing-call	2.9
Bathe	2.9

[a] See Hails (1979), Taylor et al. (1982), Masman and Klaasen (1987), and Goldstein (1988) for reviews with complete references.

Under some circumstances, the rate of daily energy expenditure by birds apparently reaches a maximum sustainable level, estimated to be approximately four times the resting (basal) metabolic rate for a variety of species (Drent and Daan 1980, Kirkwood 1983; see also Karasov, this volume). This level may be a consequence of energy processing constraints, such as the ability to transport nutrients through and across the alimentary tract (Karasov et al. 1986). Behavioral responses to changes in weather or resource availability must be critical to balancing the energy budget during such periods of maximal energy expenditure. Studies of these potentially stressful portions of the annual cycle should provide fruitful testing grounds for understanding interactions between behavior and energetics.

Studies of DEE provide a potential link between foraging ecology, energetics, and measures of fitness. Fulfillment of this promise will require accurate assessment of energy expenditure by individual birds. Time-energy budgets can achieve such accuracy, but only if rigorously applied. Energy expenditure depends both on the types of activities employed and on subtle patterns of acclimatization of metabolic rates and thermal balance with the environment. During certain portions of the annual cycle, or in response to changing climatic conditions, behavioral flexibility may be essential to balancing the energy budget. Careful studies of the energetics of individual animals in different circumstances, and of species with similar diets but different behavior patterns (see Goldstein 1988), should help to illuminate the forces governing the patterns of time use by birds.

ACKNOWLEDGMENTS

Thanks to Bill Karasov for reading a draft of this manuscript, and to the editors for several helpful suggestions.

Behavioral and Theoretical Considerations

A FUNCTIONAL APPROACH TO FORAGING: MORPHOLOGY, BEHAVIOR, AND THE CAPACITY TO EXPLOIT

TIMOTHY C. MOERMOND

Abstract. To understand the foraging behavior of birds, one needs to examine the relationships between morphology and foraging behavior, and between foraging behavior and resource use. A basic working principle is that morphological specialization for certain types of foraging maneuvers reduces the ability to perform other maneuvers. A second working principle is that birds select food on a benefit to cost basis. A bird's abilities affect its efficiency in searching for and capturing food items in a given microhabitat. As such, the cost/benefit depends on the context in which food is found. I use three groups of birds as examples of the connection between morphology and foraging behavior and show how this connection can be used to interpret patterns of resource use. Aerial insectivores, such as swallows and swifts, show several dichotomies in morphology that influence their foraging behavior and diet. Foliage insectivores show that resource partitioning is based on subtle differences in wings, legs, and feet that can be correlated with their ability to use particular microhabitats. Studies of fruit-eaters in aviaries have shown that slight differences in ability influence several aspects of food choice. Such results can be used to interpret field observations of food capture behavior to assess resource use by different species. By studying how birds feed in varying contexts, one can infer how morphology restricts their foraging behavior and influences their pattern of resource use.

Key Words: Adaptations; aerial insectivore; ecomorphology; feeding behavior; foliage insectivore; foraging; fruit-eater; jack-of-all-trades; resource partitioning.

Ecomorphology (Leisler and Winkler 1985) is a term for a mechanistic approach to understanding the interface between morphology and ecology (see also Hespenheide 1973b, Karr and James 1975, Ricklefs and Travis 1980, James 1982, Winkler and Leisler 1985, Moermond 1986). Both morphology and habitat structure influence foraging and resource use (e.g., Moermond 1979a, b; Grubb 1979; Robinson and Holmes 1984). With birds, one key to understanding these connections is through functional studies of foraging behavior. The usefulness of foraging behavior for understanding the integration of morphology, behavior, and resource use depends on how well the mechanistic basis for the foraging maneuvers is operationally defined (e.g., Partridge 1976a, b; Norberg 1979, 1981; Robinson and Holmes 1982; Holmes and Recher 1986a, b; Moermond and Howe, in press). In this paper, I discuss the procedures and values of mechanistic approaches, focussing on two important connections between (1) morphology and foraging behavior and (2) foraging behavior and resource use; the latter includes the portion of habitat or microhabitat used. We can ask, then: How tight is the connection between a bird's morphology and its foraging behavior repertoire? How do we recognize limitations or restrictions, and what are the consequences of such constraints?

Studies of resource partitioning in ecologically closely related birds nearly always show differences in the frequency of use of foraging maneuvers (e.g., MacArthur 1958, Root 1967, Lack 1971). Such differences are usually related to differences in the abilities of each species to perform various maneuvers. The basic concept supporting such an assumption is the jack-of-all-trades, master of none principle (MacArthur 1965; Moermond 1979b, 1986). Theory and empirical assessment of adaptations dictate that morphological features designed to perform one type of movement well are unlikely to be well designed for other types of movements. The ecological consequences of this principle can be observed in several different aspects of studies of resource use by birds. For example, one may examine the foraging behavior of a group of species to look for dichotomies in foraging maneuvers. The foraging maneuvers employed by one subgroup may be mutually exclusive of those employed by a different subgroup. Even within a guild, one finds differences in the relative frequencies of foraging maneuvers used. Are such differences important to their relative abilities to exploit the same resources? I shall illustrate at least some possible answers with the series of examples to follow.

To understand the connection between foraging behavior and resource use, one needs to know what foods birds use and what factors influence their selection (Grubb 1979). Optimal foraging theory has shown that birds often select food based on energy, time, and effort (Pyke et al. 1977, Krebs 1978). Optimal foraging predictions can be demonstrated in controlled laboratory situations; however, clear demonstrations in the field are rarely possible and often problematic (Krebs et al. 1983).

AERIAL INSECTIVORES

Aerial insectivores show a dichotomy in morphology and foraging behavior. Swallows (Hirundinidae) and swifts (Apodidae) (which I shall call "screeners" after Emlen 1977) hunt by flying continuously for long periods, often taking multiple prey items per flight. By contrast, sallyers (called "hawkers" by some authors) usually hunt from a perch from which the surrounding air or vegetation can be scanned. Prey are taken by a rapid flight to the item, followed by a return to a perch; usually only one prey item is taken per flight.

Screeners and sallyers have different wing and bill morphologies. The screeners' wings have a higher aspect ratio with a narrow-pointed tip; the sallyer has a broader wing with a rounded, more slotted tip. The screener has a shorter, flatter, wider bill with a large gape; the sallyer has a stronger, longer, narrower bill. Both types of differences are congruent with their different hunting styles: the sallyer's broader, more slotted wings allow rapid acceleration and deceleration; the long, narrow wing of the screener allows more efficient flight at the cruising speed of its extended flights (Burton 1976, Hails 1979).

The dichotomy in foraging behavior between screeners and sallyers is likely to be based on the differences in their wings and the associated differences in costs and effectiveness of different foraging maneuvers. I was unable to find any records of such specialized screeners as swallows and swifts ever sallying. Likewise, most species of sallyers such as tyrannid or muscicapid flycatchers rarely, if ever, hunt like screeners. Although both screeners and sallyers depend on aerial insects for food and both take their prey on the wing, their morphologies and foraging methods are virtually mutually exclusive.

The few exceptions to the screener-sallyer dichotomy provide support for the mechanistic explanation of the dichotomy. I have seen Gray Kingbirds (*Tyrannus dominicensis*) engage in series of long flights in which multiple prey were taken when flying insects were available in unusual aerial swarms (unpublished data). Similar observations have been made for Phainopepla (*P. nitens*) (Walsberg 1977), Swallow-wing (*Chelidoptera tenebrosa*) (Burton 1976), Eastern Kingbird (*Tyrannus tyrannus*), Fork-tailed Flycatcher (*Muscivora tyrannus*) (pers. obs.), and Cedar Waxwings (*Bombycilla cedrorum*) (pers. obs.). Several species of bee-eaters (Meropidae) often employ both screening and sallying (Fry 1984, pers. obs.). All of these exceptions occur in species that sally in open areas with long sallies as compared with forest species, and all have relatively long, narrow, pointed wings for sallyers (e.g., see Fitzpatrick 1985), thereby using intermediate morphologies that incorporate some of the advantages of both screeners and sallyers. The advantages of the typical sallyer's short, broad wing for maneuverability and acceleration may be outweighed by the greater economy and maximum speed of the longer, narrower wing when employed in long sallies. The longer wing of these long-distance sallyers is convergent on that of the typical screener, but not identical. Instances of sallyers using screening or long, multiple-prey sallies should be carefully recorded as indicative of exploitation behavior beyond normal constraints. The descriptions of high aerial food densities that appear to induce screening behavior in the Lewis' Woodpecker (*Melanerpes lewis*) (Bock 1970) suggest the conditions under which the screening may be the more profitable choice.

Within the screeners, swifts have longer, narrower, stiffer wings than swallows that may allow faster, cheaper cruising flight (Hails 1979). Such advantages may allow the long-distance foraging flights observed for some swifts. The advantages of relatively wider, more flexible wings for swallows may be in greater maneuverability, which in turn may mean higher capture rates of certain types of aerial prey or the ability to maneuver closer to obstacles and the ground (Waugh 1978).

Within the swallows, subtle differences in wing and tail shape (Waugh 1978) are apparently related to differences in hunting flight patterns among genera such as the maneuverability of long, fork-tailed *Hirundo* species compared to the straight-line cruising of some square-tailed species (Waugh 1978). Even such subtle differences are associated with differences in resource use such as foraging site and prey type.

Morphological variation in the sallyers also appears to influence resource use (see Fitzpatrick 1980, 1985). The range of foraging behaviors described by Fitzpatrick for tyrannid sallyers appears to be associated with differences in bill, wing, and leg morphology; these differences likely account for observed differences in prey type and diet breadth (Sherry 1984).

FOLIAGE INSECTIVORES

The maneuvers used to take prey from foliage differ substantially among species. For example, many sallyers may snatch or hover-glean prey from foliage (Fitzpatrick 1980), whereas warblers (Sylviinae, Parulinae) primarily glean their prey from a perched position. Birds that habitually glean prey from small twigs and foliage often take only a small percentage of items on the wing, either from the air or from foliage. Species that habitually sally take the great majority of their prey on the wing (Eckhardt 1979, Recher et al. 1985). This dichotomy appears to be quite distinct, even among a single group. For example, Schulenberg (1983) identified a suite of morphology and behavioral characteristics that distinguished the sally-gleaning *Thamnomanes* antshrikes from the more typical perch-gleaning genera. The suite was so evident that he convincingly argued that two atypical *Thamnomanes* were perch-gleaners and not allied to the other sally-gleaning *Thamnomanes*.

Foliage gleaners must move among leaves and twigs arrayed in a variety of patterns that often require special modifications of wing, legs, and feet (e.g., Gaston 1974, Pearson 1977a, Leisler 1980, Winkler and Leisler 1985). Norberg's (1979, 1981) analysis of bark and twig gleaners demonstrated a number of subtle, but important and relevant, differences in morphology that correspond to differences in their use of microhabitats. MacArthur (1958) described differences in movement and microhabitat use for a series of *Dendroica* species (Parulinae) exhibiting only minor differences in morphology. Morrison (1982) attributed differences in wing shape between Black-throated Gray (*D. nigrescens*) and Hermit (*D. occidentalis*) Warblers to differences in habitat use. In my lab, we demonstrated distinct differences in the reaching ability of Yellow-rumped (*D. coronata*) and Palm (*D. palmarum*) warblers that were correlated with minor differences in leg morphology and that corresponded to differences in each species' use of foraging maneuvers in the field (Moermond and Howe, in press). The minor differences in leg morphology were of the same magnitude as those reported by Pearson (1977a) for antwrens (*Myrmetherula* spp.: Formicariidae) occupying different microhabitats.

Winkler and Leisler (1985) demonstrated that for European sylviine warblers that foraged high in trees or shrub vegetation (e.g., *Sylvia* spp.), wing morphology varied considerably with associated habitat differences. For species that used low, dense vegetation (e.g., *Acrocephalus* spp.), differences in leg morphology appeared more critical. This work suggests that the foraging maneuvers that can be successfully applied to these two categories of vegetation are quite different, requiring a different suite of morphological adaptations. The rules that may govern such adaptations appear to include those that influence the performance of birds negotiating different microhabitats. Such differences are likely involved in determining the habitat type and range of birds (Winkler and Leisler 1985).

FRUIT-EATING BIRDS

The taking of fruits provides a clear example of morphological constraints. In the Neotropics,

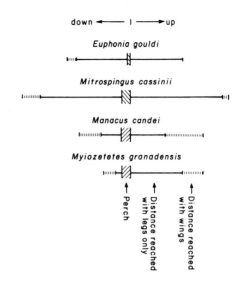

FIGURE 1. The ability of captive individuals of four species of birds to reach for fruits above and below a perch. The solid line shows the distance reached without using the wings. The broken line shows the additional distance reached with the use of the wings. The perch diameter was 3 mm for *Euphonia gouldi* and 12 mm for the others. Two individuals were tested for each species with the maximum reaches shown. Adapted from Moermond et al. (1986).

the dichotomy between birds taking fruits from a perch versus those taking fruits on the wing is sharp (Moermond and Denslow 1985, Moermond, in press). Species taking fruits from a perch apparently use only simple actions. Fruits at the level of a sturdy perch can be taken with only a slight downward lean, but fruits below the perch require a more extreme extension of the body. The ability to reach varies considerably. Among small Neotropical fruit-eating birds we tested, some tanager species (e.g., *Euphonia gouldi*, *Thraupis palmarum*) were able to extend their entire bodies below the perch; whereas a tyrannid flycatcher (*Myiozetetes granadensis*) and a manakin (*Manacus candei*) were unable to reach more than a small distance down from a perch (Fig. 1) (Moermond and Denslow 1985, Moermond et al. 1986). The added cost to a bird of obtaining a particular fruit placed below a perch may cause a switch in preference from that fruit to a fruit of lesser quality that is easier to obtain (Moermond and Denslow 1983). The decision as to which fruits to take or not appears to be a cost/benefit choice that is influenced by the morphological abilities of each bird species (Moermond et al. 1986, 1987).

When a fruit cannot be taken from a perch, then it must be taken on the wing. How restrictive the choice is for such fruits depends on the ability of the bird. In the spring of 1986 in central

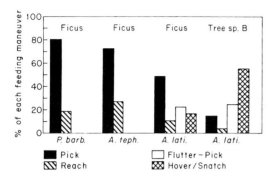

FIGURE 2. Feeding maneuvers used by three species of bulbuls (*Pycnonotus barbatus, Andropadus tephrolaemus,* and *A. latirostris*) feeding on cauliflorous fruits of *Ficus* and pendant fruits of *Prunus africanus* (Tree sp. B). Picks, reaches, and flutter-picks were performed while clinging to a perch. Hover-gleans and snatches of fruits were all done in flight. The total sample sizes for each case shown from left to right are 53, 75, 90, and 48, respectively.

Africa (in the Nyungwe forest of Rwanda), I observed three species of bulbuls (*Pycnonotus barbatus, Andropadus tephrolaemus, A. latirostris*) feeding actively on a *Ficus* with large, cauliflorous fruits (Moermond, in press). *P. barbatus* and *A. tephrolaemus* took all their fruits from a perch while *A. latirostris* relatively frequently used its wings to pick a fruit with a flutter (23%) or while flying by (16%, Fig. 2). The use of wings by *A. latirostris* corresponded to its frequent use of its wings while hunting insects among foliage. The other bulbul species rarely use aerial maneuvers while foraging for insects or fruit.

During the same period, fruits were simultaneously available on a nearby tree; however, these fruits were pendant below thin, flexible perches. *A. latirostris* fed extensively on these fruits also, but it frequently used aerial maneuvers (snatches and hover-gleans) to obtain the fruits (56%, Fig. 2, tree B). During the time I observed over 100 feedings of *A. latirostris* in tree B (*Prunus africanus*), I never observed any feeding by *A. tephrolaemus* and only six feedings by *P. barbatus*. Four of these six feeding maneuvers were aerial snatches. These data suggest that feeding on fruits that require more aerial maneuvers is restricted to the species with more ability to use its wings. Although all three bulbuls are very similar in size and morphology, what appear as minor or subtle differences in frequency of feeding maneuvers of fruit-eaters can be shown to influence food type and diet breadth (Moermond et al. 1986).

CONCLUSIONS

The combination of constraints of morphology on foraging behavior and the influence of foraging on food exploitation has many implications for our understanding of bird communities such as the divisions between species guilds (Ford 1985, Holmes and Recher 1986b, Terborgh and Robinson 1986) or the barriers to niche shifts (e.g., Diamond 1970). On a finer scale, the connections between morphology and foraging ability probably rarely determine diet and habitat use, but act as a directing influence often enabling appropriate responses to environmental context (e.g., Moermond et al. 1987). The conclusion of this approach is that foraging behavior offers important clues to assessing and interpreting the food exploitation patterns and capabilities of birds. These clues will be most insightful when the observations of foraging are keyed to how the bird is responding to resources in a given context.

ACKNOWLEDGMENTS

This paper has benefited from numerous discussions and comments over the course of several years from several of my colleagues and students: Julie S. Denslow, Doug Levey, Cynthia Paszkowski, Bette Loiselle, John Blake, Eduardo Santana C., Michael DeJong, Robert Howe, and Amy Vedder. The manuscript was improved by the editors and reviewers, especially H. Recher, C. J. Ralph, and M. L. Morrison. The work was supported by the Department of Zoology of the University of Wisconsin–Madison. I thank Michael Morrison especially for his encouragement.

ECOLOGICAL PLASTICITY, NEOPHOBIA, AND RESOURCE USE IN BIRDS

RUSSELL GREENBERG

Abstract. Determining the mechanisms that underlie ecological plasticity should be an important focus of avian behavioral ecology. Most attempts to model the responses of birds to changes in food distribution and abundance of potential competitors are based on the assumption that different species sample and track resources in an equivalent manner. However, differences in how readily birds respond to novel resources are difficult to model and may be impossible to predict based on strictly economic approaches. Most observers of wild birds have noted intrinsic differences within and between species in "ecological plasticity," or the tendency to exploit new resources. Furthermore, it has been proposed that the degree of plasticity influences a species' colonizing ability and, ultimately, the probability that it can give rise to other species occupying new adaptive zones. I propose that variation in plasticity is a direct result of variation in neophobia: the fear of feeding on new foods or approaching new situations. This provides natural selection with the raw material for adjusting adaptive levels of neophobia. Where ecological plasticity is favored, selection could act to reduce neophobia.

Key Words: Habitat selection; foraging; warbler; *Melospiza*; *Dendroica*.

Most field ornithologists possess an intuitive feel for the variation in ecological plasticity in species. Species, even closely related ones, often differ strikingly in the range of habitats occupied or foods taken. On the surface, it seems that this variation does not result entirely from differences in morphological adaptations, but also stems from differences in the psychological basis of decision-making.

Although there has been a long history of interest in ecological plasticity, its definition has been vague and has involved a blending of two rather distinct attributes: lack of specialization and flexibility in the face of change. Plasticity has most often been related to the lack of specialization, the observed ecological amplitude of a species (specialist versus generalist). Miller (1942), Klopfer and MacArthur (1960) and others have associated ecological plasticity with the breadth of resources and habitats used by a species. Klopfer (1967), for example, defined stereotypy (the opposite of plasticity) at the level of perception as "a sensitivity to, or an awareness of, or preference for, a limited range of a larger complex of stimuli." He distinguishes this plasticity in preference from locomotory plasticity, which involves the lability of motor patterns used in searching and attacking prey. Since the ability to perform a variety of locomotory skills results from morphological specialization, it is probably best to consider preference, the focus of this paper, and locomotion separately (see Martin and Karr, this volume).

Plasticity is not simply an alternative term for the concepts of generalist versus specialist (Morse 1980a). What separates it is the second attribute, flexibility in the face of change. In general, plasticity can be defined as "the capacity of organisms of the same (= similar) genotype to vary in developmental pattern, in phenotype, or in behavior according to varying environmental conditions" (Merriam-Webster 1986). This attribute of ecological plasticity, then, reflects a bird's ability to respond to changes in food, competition environment, and the presence of novel resources. To separate the static concept of specialization from the dynamic concept of plasticity, Morse (1980a:12) constructed a two by two classification, with examples, based on degrees of ecological amplitude (specialist versus generalist) and ability to respond to changes in resources (stereotyped versus plastic). Although it is useful to divorce these two concepts, it is likely that there is a strong correlation between observed generalization and plasticity in birds. A thorough discussion of the relationship between ecological specialization and predictability through time can be found in Sherry (this volume).

At what level of biological organization should ecological plasticity be analyzed? With the exception of the experiments of Klopfer (1963, 1965, 1967), assessment of plasticity has been based generally on the performance of populations or species. However, if it is to be argued that variation in ecological plasticity is adaptive, then the ways by which plasticity is regulated in individuals need to be established. The purpose of this paper is three-fold: (1) to briefly establish the importance of the study of ecological plasticity of individuals to the understanding of the evolution and ecology of foraging behavior and habitat selection; (2) to propose the neophobia hypothesis as a mechanism for regulating the degree of ecological plasticity that characterizes a particular species of birds; and (3) to summarize

TABLE 1. MEAN NUMBER OF APPROACHES AND FEEDING ATTEMPTS AT TEN OBJECTS WITH HIDDEN MEALWORMS (GREENBERG 1983)

	No. of foraging attempts	No. of approaches	No. of weak approaches[a]
Chestnut-sided Warbler	1.0[b]	5.6	1.7[b]
Bay-breasted Warbler	4.0	5.5	0.3

[a] Weak approaches involved birds that came no closer than 7.5 cm to the object.
[b] Interspecific difference significant P < 0.05 based on Mann-Whitney test.

associative learning (for a discussion of the role of neophobia in limiting learning ability in rats see Holson 1987).

That a neophobic response can affect the evolution of ecological plasticity is clear. Differences in neophobia between laboratory and wild strains of rats and among breeds of dogs, for example, suggest that enough heritable variation in the novelty response exists for artificial selection to shape major differences (Barnett 1958, Barnett and Cowan 1976, Mitchell 1976).

Although a neophobic response potentially can play an important role in determining differences in plasticity, how can it be distinguished from the overall response that animals could have to any feeding situation? Many foraging decisions are probably marked by some ambivalence. Birds are attracted to a particular location based on direct observation of food, expectations derived from past experience, or the presence of other birds. But the presence of predators or competitors adds risk, and may discourage birds from visiting an otherwise attractive site. This continual ambivalence has been the subject of intense study by workers interested in the trade-offs between risk and energy reward in sparrows feeding away from shrubbery, for example (Grubb and Greenwaldt 1982, Schneider 1984). By keeping the expected energy gain constant, but moving the food with respect to cover, one can infer the relative role of fear of predation in shaping the decisions of sparrows. In a similar manner, the role of fear of novelty can be explored by manipulating novelty while keeping expected gain constant.

In the experiments described below, I assumed that by presenting food to hungry birds in a conspicuous and familiar manner the attraction of a feeding site could be adjusted sufficiently high that its contribution to variation in feeding rate is insignificant. Differential latency to feed when novelty is imposed can be safely attributed to an aversion, and the experiment need only distinguish exactly what causes the aversion.

The problem of inferring the experience of wild-caught birds can be obviated by rearing birds under controlled conditions. However, it is more practical to assay the response of wild-caught birds and I have inferred differences in neophobia by presenting them with a wide range of objects unlikely to have been seen previously. These objects are characterized by many types of stimuli. If the birds respond with consistent aversion to all of the various objects, then it is unlikely that the experiments are distinguishing an innate response to a particular stimulus. A consistent response can most parsimoniously be ascribed to a generalized novelty response.

Because the experiments encourage or force the subjects to confront the potentially aversive objects, differences in latency or any other measure can only be compared qualitatively within the experimental paradigm. The fact that it may take one individual only 20 min and another only a minute does not mean that the former would visit the aversive object rapidly in the wild. Further, the objects are selected to be highly divergent from what is normally encountered and may produce aversions greater than one would see from natural habitat features.

EXPERIMENTS WITH WARBLERS

I studied two species of *Dendroica* warblers that winter in Central Panama (Greenberg 1984a). The fact that I could study them in sympatry is important, since the attributes of specialization and plasticity are relative; only by comparing the response of species to the same resources can comparisons be made. Based on my observations during the three winters, and others made by Morton (1980a), I concluded that Bay-breasted Warblers (*D. castanea*) were more flexible and generalized in their foraging behavior than Chestnut-sided Warblers (*D. pensylvanica*). Often Bay-breasted Warblers displayed a high degree of opportunism, feeding on insects attracted to lights, garbage cans, sewage outfalls and dog food dishes. In forests Bay-breasted Warblers were the most variable and generalized of the small foliage-gleaning birds with respect to foraging height, substrate, and gross diet composition (fruit versus insects). Chestnut-sided Warblers consistently ranked as the most specialized.

To study the mechanisms that regulate the degree of apparent plasticity in the two species, I observed the responses of immatures, captured in autumn migration, to novel feeding situations presented in captivity (Greenberg 1983, 1984b).

The first experiment explored how individuals of the two species responded to presumably novel microhabitats that contained hidden prey. Under these circumstances both the intrinsic attractiveness and aversiveness were operating to determine the ultimate success of the bird approaching and capturing the prey. A series of 10

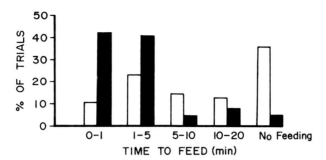

FIGURE 1. Percentages of trials in which Bay-breasted (black bars) and (white bars) Chestnut-sided Warblers took various time intervals to feed. Experiment is the pooled result of eight Bay-breasted and seven Chestnut-sided Warblers offered food at eight different microhabitats. The Chestnut-sided Warblers took longer to feed at all microhabitats. Control trials with no objects averaged less than 30 s for both species.

objects was presented in the home cage of the warblers for 10 minutes, and the numbers of approaches, close approaches, and prey captures were recorded. Although the two species approached a similar number of objects, the Bay-breasted Warblers approached more closely and captured the prey four times more often than did the Chestnut-sided Warblers (Table 1). The actual behavior of the two species seemed even more revealing: Chestnut-sided Warblers approached with a great amount of ambivalence— continually approaching and withdrawing— which was not observed in the Bay-breasted Warblers.

The second experiment introduced the strategy of reducing the uncertainty regarding the intrinsic attractiveness of an object by first depriving the birds of food and then presenting them with a conspicuous food reward (a cup of mealworms). Control trials consisted of presenting the food without the cup and the warblers fed rapidly during these trials (<30 s). A long latency was attributed to the response of the birds to the objects placed next to the familiar food cup. The data showed that regardless of what the object was, Chestnut-sided Warblers took much longer to feed at novel objects than did Bay-breasted Warblers (Fig. 1).

Additional experiments demonstrated that:

(1) The increased latency of the Chestnut-sided Warblers was not due to an increase in their exploratory behavior of the novel objects (i.e., greater "curiosity" in the Chestnut-sided Warbler, Greenberg 1984b).

(2) Naive Chestnut-sided Warblers distinguished objects that they were reared with versus novel objects up to four months after the rearing period (Greenberg 1984c).

(3) Increased hunger and interspecific social stimulation did not decrease the degree of neophobia shown by Chestnut-sided Warblers (Greenberg 1987). Repeated short-term exposure did decrease the latency to feed at novel objects.

EXPERIMENTS WITH SONG AND SWAMP SPARROWS

The experiments with warblers provided the basis for the Neophobia Hypothesis. The experiments I have performed on *Melospiza* sparrows were the first prospective test of one of the major predictions of the hypothesis: a more generalized species should show consistently lower aversion to feeding in the presence of novel objects than a more stereotyped congener.

A prediction was made that the Swamp Sparrow (*M. georgiana*) should be more neophobic than the apparently more generalized and adaptable Song Sparrow (Miller 1956, Wetherbee 1968, Peters et al. 1980), despite their close phylogenetic affinity (Zink 1982). Song Sparrows occur in a wider range of scrub and marsh habitats (Morse 1977, Yeaton and Cody 1977), they are common colonists of small oceanic islands with a variety of habitats, and they occur commonly as a commensal with human. Swamp Sparrows are more restricted to shrub-marsh habitats.

One of the advantages of working with sparrows over most warblers is that they can be baited into feeders in the field. I exploited this to conduct experiments on novel object reactions both in the field and in the lab. The two approaches are complementary: field experiments remove the possibility that the responses are a result of stimulus deprivation and do not reflect responses of birds in the "real world." Caged experiments allow for individual testing of subjects under more controlled conditions.

Field experiments (Greenberg, 1989) were conducted by color-banding sparrows at a marsh along the Potomac in Alexandria, Virginia. After a regular group of Song and Swamp sparrows

TABLE 2. THE NUMBER OF VISITS/30 MIN TO A FEEDING STATION BY SONG AND SWAMP SPARROWS WHEN SURROUNDED BY NOVEL OBJECTS COMPARED WITH PAIRED CONTROL PERIODS

	Song		Swamp	
Object	Control	Experimental	Control	Experimental[a]
Black box	30	20	21	1
Easter grass	22	17	7	0
Tropical leaves	24	29	49	17
Tube	30	25	45	21
Green spikes	42	35	28	7
Orange leaves	29	30	18	9
Totals	177	156	168	55

[a] Difference between experimental and control in Swamp Sparrow is significant based on Wilcoxon paired-rank test ($t = 0$, $P < 0.025$).

was established at the feeders, I placed replicates of novel objects in a circle 0.5 m from the feeder. The feeder was watched for 30 min with the objects and 30 min without the objects, with the control and experimental periods alternated. Bird seed was added prior to each observation period. The number of individual visits was recorded for each species for each period (Table 2). Both predictions of the neophobia hypothesis were confirmed: (1) Swamp Sparrows visited the novel objects less often than the unadorned feeder, whereas there was no significant difference in Song Sparrows; and (2) the difference between the two species was consistent over all of the objects.

Individually housed immatures of both species were tested the next winter in a manner similar to the warbler experiments. Although the Song Sparrows averaged slightly slower in its foraging latency at plain cups (controls), they were consistently and significantly faster than Swamp Sparrows during the experimental trials with novel objects (Table 3). Swamp Sparrows also approached the cup more often prior to feeding than did the Song Sparrows. The hesitancy disappeared in the Swamp Sparrows when they were repeatedly exposed to the objects.

DIRECTIONS FOR FURTHER WORK ON THE NEOPHOBIA HYPOTHESIS

Experiments to this point have established that novelty is an important factor underlying differences in plasticity between some species. Since large variation is found between closely-related species, and within species, in the case of rats and dogs, differences in ecological plasticity caused by changes in novelty responses may have the capability of rapid evolution. If so, closely-related species that rely upon more stable resources should forage more conservatively and hence be more neophobic.

However, these experiments do not yet establish neophobia as a general mechanism for regulating ecological plasticity. The following points might be addressed in future studies:

(1) The physiological correlates of feeding aversion need further work to see if variation in novelty responses are associated with elevated heart rate and circulating steroid levels, which would suggest that the acute stress responses are operating. Experiments could then examine whether interspecific variation in neophobia is correlated with the degree of change in these factors.

(2) Captive experiments with naive birds (Greenberg 1984c) should be pursued to determine if there is a genetic basis to the interspecific differences in neophobia.

(3) Further work should bridge the gap between the qualitative results obtained from the experiments employed so far and the magnitude of the effect of novelty under more natural conditions. This is important for applying this concept to field studies of foraging, since the results of laboratory and feeder studies can only be used

TABLE 3. MEAN LATENCY (SEC) AND NUMBER OF APPROACHES PRIOR TO FEEDING FOR 11 SWAMP AND NINE SONG SPARROWS FEEDING WHEN NOVEL OBJECTS WERE PLACED NEXT TO THE FOOD CUP

	Object						
	Easter grass	Tube	Green spikes	Tropical leaves	Black box	Orange leaves	No objects
Song Sparrow							
Latency[a]	304	95	165	90	90	38	80
Approaches[b]	1.4	0.9	1.0	0.3	0.5	0.3	0
Swamp Sparrow							
Latency	556	508	500	330	417	209	43
Approaches	2.8	5.0	4.4	2.5	2.6	3.1	0

[a] Two-way ANOVA produced a significant species effect $F_{1,98} = 26.7$, $P < 0.0001$.
[b] Two-way ANOVA produced a significant species effect $F = 20.1$, $P < 0.001$.

to compare species in a qualitative sense. Because novelty responses are easily measured, they can be used in a wide variety of experimental studies to test hypotheses concerning the adaptive significance and evolutionary and ecological consequences of differences in plasticity.

ACKNOWLEDGMENTS

I thank Verner Bingman, Richard Hutto, Michael Morrison, Eugene Morton, and C. J. Ralph for reviewing a draft of this manuscript. Support for the research has been provided by a Smithsonian Institution Post Doctoral Fellowship and Friends of the National Zoo. Support for ms preparation has been provided by NSF grant BSR 8705003.

FOOD AVAILABILITY, MIGRATORY BEHAVIOR, AND POPULATION DYNAMICS OF TERRESTRIAL BIRDS DURING THE NONREPRODUCTIVE SEASON

SCOTT B. TERRILL

Abstract. Migration is a phenomenon that has major implications for the spatial dynamics and organization of migrant populations and communities during the nonreproductive season, and food availability appears to be the major factor responsible for migratory behavior. The evolutionary relationship between spatial and temporal characteristics of resource availability and migratory behavior is briefly overviewed. Results of field work and laboratory experiments concerning the proximate relationship between food availability and migratory behavior indicate that some migrants extend autumn migration past the normal migratory period if food becomes scarce, and that some migrant populations can exhibit large-scale distributional shifts in winter in response to food availability. This is an important consideration when attempting to census and monitor wintering populations of migrants. More work is needed to clarify the role of food availability in regulating population size of migrants during the nonreproductive season and to assess the effects of differential migratory distance on individual fitness.

Key Words: Food availability; migration; population dynamics; warbler; *Dendroica*; *Sylvia*; *Junco*; nonreproductive season.

Bird migration comprises a movement from the breeding (natal) grounds followed by a subsequent return for the next reproductive effort. It has long been recognized that food availability probably plays a significant, if not dominant, role in the evolution of migratory behavior (see Gauthreaux 1982 for a recent review) and in the regulation of the distribution and dynamics of migrant populations. Few data, however, actually address this relationship empirically (e.g., Hutto 1980, Rappole and Warner 1980, Greenberg 1986).

I will briefly review and discuss food availability, migratory behavior, and migrant population-level phenomena during the nonreproductive period by: (1) reviewing the diversity of migratory behavior found in birds as a function of large-scale characteristics of resource availability; (2) considering the role of food availability in regulating dispersion, social behavior, and movements of nonbreeding migrants; and (3) outlining laboratory studies of the effect of food availability (including differential access to food as mediated through competition) on migratory behavior. Finally, I suggest how these results may be relevant to censusing and monitoring migrant populations.

RESOURCES AND THE REGULATION OF MIGRATORY BEHAVIOR

Bird migration is generally considered an adaptation that allows birds to exploit abundant food for reproduction in a region subject to harsh conditions between breeding seasons (e.g., Gauthreaux 1982, Cox 1985). Migratory behavior varies from strongly endogenously controlled, with high heritability values (e.g., Berthold 1988b), to environmentally stimulated (Berthold 1975; Gwinner and Czeschlik 1978; Gauthreaux 1982; Myers 1984; Terrill and Ohmart 1984; Terrill 1987, 1988, in press a, b; Gwinner et al. 1988). This variation parallels, and is probably an evolutionary response to, large-scale spatial and temporal characteristics of resource availability during the nonreproductive period.

Food availability is generally considered the fundamentally important determinant of migratory distance (e.g., Gauthreaux 1982). Some population-level trends indicate that birds migrate only as far as necessary to maximize the probability of obtaining adequate resources for survival between reproductive periods while minimizing the distance travelled to do so (e.g., Terborgh and Faaborg 1980, Terrill and Ohmart 1984, Terrill in press c). However, hypotheses concerning relationships between migratory distance, annual survivorship, and reproductive success remain largely untested.

When the probability of overwinter survival on the breeding grounds frequently approaches zero, natural selection has favored individuals (annual migrants) that leave the area *before* food becomes scarce (e.g., Farner 1955; Lack 1968a; Terrill 1987, 1988, in press a, b), an adaptation that enables them to accumulate and maintain substantial fat reserves for migration. Thus, this type of migratory behavior ("obligate" migration—see Terrill and Able 1988) is anticipatory in the sense that a decrease in food availability is ultimately responsible for its occurrence, but is not the proximate factor releasing the behavior (Lack 1968a). Obligate migratory behavior is apparently induced primarily by endogenous mechanisms (see Berthold 1975, 1988a, b, c; Gwinner 1986 for reviews). The duration and distance of obligate migration is theoretically related to the probability of overwinter survival along the migratory route (Terrill and Ohmart 1984; Terrill 1987, 1988, in press a, b). Presumably, individuals that spontaneously migrate across regions with very low probabilities of overwinter survival before resources become scarce for the winter are at an advantage relative to individuals that terminate their migration and attempt to overwinter in the region. Where food availability is more variable, selection has favored a more environmentally sensitive migratory system, "facultative migration," which appears to be a direct response to changes in environmental conditions and may, or may not, occur in any given year.

Obligate and facultative migration appear to represent two ends of a behavioral continuum (Gwinner and Czsechlik 1978). Not only are different species and populations represented along this continuum, but the behavior of even individual migrants can vary (e.g., Perdeck 1964; Terrill 1987, 1988, in press a, b, c; Gwinner et al. 1988). These studies indicate that as the endogenous drive to migrate wanes with time and distance, the stimulus to continue migrating becomes more directly dependent upon environmental conditions such as food availability and social environment (Terrill 1987, 1988; Gwinner et al. 1988; Terrill and Berthold in prep.). More specifically, at least some annual migrants are apparently capable of changing from an "obligate phase" (during which the fundamental stimulus for migration is endogenous) to a "facultative phase" (the stimulus to migrate is directly dependent upon immediate resource availability) with time and distance of autumnal migration. Theoretically, the obligate phase takes migrants across regions where the probability of overwinter survival is consistently very low. As the probability of survival increases, the birds switch to a facultative mode that enables them to track variations in resource distribution and minimize the total distance of migration during any given year. In a sense then, obligate migratory behavior might be considered as the coarse-grained determinant of migratory distance (an evolutionary result of long-term patterns of resource availability), while the facultative phase fine tunes migration during a given year (in response to short-term fluctuations in resources).

FOOD AVAILABILITY AND MIGRANT POPULATIONS DURING THE NONREPRODUCTIVE SEASON: A SURVEY

Nonbreeding migrant spacing behavior ranges from highly territorial to very social and apparently is correlated with a number of factors including habitat, distribution of resources, and predation (Gauthreaux 1982, Pulliam and Millikan 1982, Pulliam and Caraco 1984, Myers 1984). The dispersion and distribution of wintering migrants range from remaining essentially stable (between autumnal and vernal migration) to very dynamic, with movements continuing throughout much of the "wintering period" (e.g., Moreau 1972, Curry-Lindahl 1981, Lack 1983, Terrill 1988). It is often assumed that this behavioral continuum reflects a range in the distribution of resources on the wintering grounds from relatively stable to dynamic within and between winters (e.g., Gauthreaux 1982).

The potential importance of food availability during the nonreproductive period in regulating the overall size of migrant populations has not been ignored (e.g., Lack 1954, 1968a; Fretwell 1972; Schwartz 1980; Ketterson and Nolan 1982; Myers 1984; Berthold 1988b), and the degree to which migrant population size might be regulated during the nonbreeding periods is an important and generally open question.

Apparently, food can be limited during the nonreproductive period of migrants; evidence includes: defense of territories that play no role in reproduction (e.g., Rappole and Warner 1980); large-scale movements within and between winters (e.g., Terrill 1988, in press c); major population declines that appear to be occurring on the wintering grounds (e.g., Berthold 1988a); and differential movements by certain age or sex classes of the same populations (e.g., Kalela 1954; Lack 1954; Gauthreaux 1978, 1982). This last point has potential implications for differential access to food as mediated through competition, which, in turn, has relevance to the structure and dynamics of migrant populations. Theoretically, dominant individuals restrict access to food by subordinates, which forces subordinates to migrate farther to obtain resources. Distributional patterns often (but not always) support this concept; however, rigorous evaluations of the availability of resources per individual are generally lacking. Although it has been demonstrated that individuals migrating farther from the breeding grounds in winter have lower average reproductive success the following breeding season (e.g., Schwabl 1983), I know of no empirical information on the relationship between differential migration, survival rates and *lifetime* reproductive success in any migrant.

During the nonreproductive period, migrants show the full spectrum of social behavior described for birds in the breeding season (see Pulliam and Millikan 1982). Although many migrants that are territorial during the reproductive season become gregarious during the nonreproductive season, others remain territorial throughout the year, establishing territories during migration and on the wintering grounds (e.g., Gauthreaux 1982). [The relationship between resource distribution and spacing and social behavior has been discussed in detail elsewhere, e.g., Brown 1969, Pulliam and Millikan 1982.]

Extended use of resources in the same locality throughout the winter (site tenacity) should occur when food is available for an extended period (throughout a particular winter), and between-year faithfulness to the same wintering site (site fidelity) should occur when food availability is relatively constant between years. Examples of migrants that are often territorial during the winter and frequently exhibit both nonbreeding site tenacity and fidelity include a group of parulid warblers that breed in temperate North America and winter in dense understory in tropical and subtropical regions (Schwartz 1964, Rappole and Warner 1980). This habitat is apparently buffered from the extreme fluctuations in food availability found in higher vegetational strata and in other regional habitats between the wet and dry seasons. Nonbreeding site tenacity and fidelity are not restricted to territorial migrants, but appear in gregarious species as well. For example, there are numerous reports of banded sparrows (that readily associate in flocks during winter) returning to the same wintering sites (e.g., Ketterson and Nolan 1985), especially at feeders. Cases of winter site tenacity and fidelity by migrants are numerous (for example, see Curry-Lindahl 1981 and Gauthreaux 1982) and they indicate (as do some studies on wintering migrant communities, e.g., see Keast and Morton 1980) that migrant assemblages are often stable throughout the wintering period. Alternatively, some studies indicate that movements by migrants can occur throughout the nonbreeding season and that individuals (even large numbers) may occupy different regions within and between winters. These studies indicate a much higher potential for extensive winter movement by migrants than has generally been considered to be the case (cf. Curry-Lindahl 1981, Terrill in press a).

The presence of both winter-site faithfulness and tenacity and winter-site plasticity within the same species provides opportunities to test assumptions about both proximate and ultimate factors responsible for this variation (Ketterson and Nolan 1985). There exists a growing list of species that appear to exhibit the full spectrum of behavior (Curry-Lindahl 1981). One such species is the Yellow-rumped Warbler (*Dendroica coronata*). Some annual migrant populations of this species winter in the Sonoran desert of the southwestern United States and northwestern Mexico. These populations comprise good systems for testing hypotheses concerning the relationship between food availability and winter population dynamics for several reasons. First, these birds are highly restricted to insular patches of lush habitat surrounded by desert, which facilitates accurate censusing of local populations (and greatly decreases the possibility that birds disappearing from a site are dispersing locally rather than actually migrating). Second, numbers of wintering warblers at particular sites (and even within regions) vary greatly within and between winters. Third, there is a general correlation between severity of weather conditions and numbers of overwintering birds in a particular area (the colder the winter the fewer the overwintering warblers), implying either large-scale mortality or changes in winter distribution.

Terrill and Ohmart (1984) found that numbers of Yellow-rumped Warblers were positively correlated with food availability at a series of sites from the northern edge of the winter distribution in Arizona south into northern Mexico (Fig. 1). Transects were established at each site, birds were censused, and insect sweep samples were used to measure food availability throughout two winters. There was a strong positive association between the dominance of certain insect groups in the sweep samples and in the stomach contents of warblers collected at the same sites, indicating that sweep samples adequately reflected warbler diets (Terrill and Ohmart 1984). Changes in insect populations appeared to be strongly influenced by climatic conditions with numbers crashing with the occurrence of relatively severe cold fronts.

The dynamic state of the wintering warbler population was reflected in numerical changes along the north–south transect during two different winters (Fig. 2). Decreases at northerly sites corresponded to increases at southerly sites, suggesting movement, and the magnitude of change was correlated with the availability of insects (Terrill and Ohmart 1984). These population shifts occurred in January, even though these birds were considered winter residents. A subsequent analysis of tower kills of nocturnally migrating Yellow-rumped Warblers in Florida demonstrated that they are capable of migrating throughout the entire winter, although numbers are highly variable between years and large num-

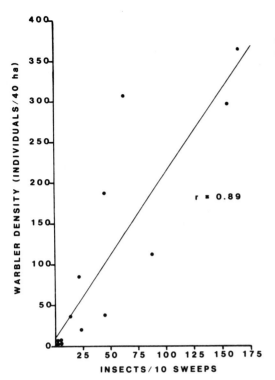

FIGURE 1. Significant (P < 0.001) correlation between numbers of insects and numbers of Yellow-rumped Warblers at Arizona riparian sites from October through early February for two years (from Terrill and Ohmart 1984).

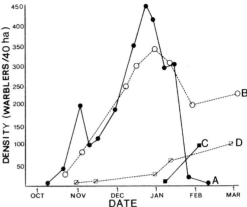

FIGURE 2. Mean densities of Yellow-rumped Warblers in highly isolated riparian habitats near Phoenix, Arizona (33°N) and to the south near Magdelena, Mexico (31°N). Note the dramatic decrease in numbers of birds in midwinter of 1979–80 in the Phoenix area (A) in the wake of a major cold front, relative to a much lesser decrease in 1980–81 (at the same sites) when no major fronts occurred (B). Numerical decreases in the north corresponded to increases to the south, at Magdelena, during 1979–80 (C) and 1980–81 (D).

bers migrate in winter only during, or after, unusually severe cold periods (Terrill and Crawford 1988).

Overall then, these results indicate that: (1) the number of individuals wintering at particular sites (and regions) is a function of food availability; (2) winter migrant communities can change substantially within and between winters; and (3) these dynamics are due, at least in part, to the presence of migratory behavior in response to changing resource availability in winter after the "normal" migration period has ended. Similar results have been found in other species including temperate migrants (e.g., Pulliam and Parker 1979, Niles et al. 1969) and tropical wintering migrants (e.g., Wood 1979). The Yellow Wagtail (*Motacilla flava*) provides a rather spectacular example of a situation similar to that found in Yellow-rumped Warblers. Wood (1979) found a progressive decline in numbers of wintering wagtails at an African study site from about 16,000 in November to 2000–3000 in March, and that food availability and numbers of wagtails in the study area declined concurrently. A southward shift (i.e., extended migration) was supported by ringing recoveries.

FOOD AVAILABILITY AND INDIVIDUAL MIGRANT BEHAVIOR: AN EXPERIMENTAL APPROACH

The relationship between food availability and migratory behavior has been tested in the laboratory in several species, primarily during the autumn migration period. Most studies indicate that food deprivation heightens migratory activity at this time (e.g., Biebach 1985; Gwinner et al. 1985; Terrill in press b, c; Gwinner et al. 1988), especially if a migrant is unable to refuel during its diurnal rest (Gwinner et al. 1988). Several recent studies indicate that food deprivation may also inhibit fall migratory activity in some species, depending upon time of day food is restricted, severity of deprivation, and other factors (Terrill and Berthold in prep., Holberton pers. comm.).

Laboratory evidence for facultative migratory behavior in annual migrants during winter has recently been found in several species (Terrill 1987, Gwinner et al. 1988, Terrill and Berthold in prep.). These experiments have been carried out after the ending of spontaneous, autumnal, migratory activity associated with unlimited food (indicative of the obligate phase; see above).

One such species is the Dark-eyed Junco (*Junco hyemalis*). Juncos were used to test for pos-

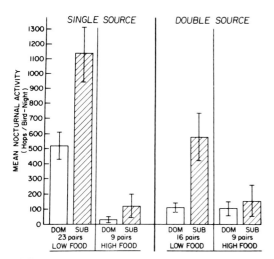

FIGURE 3. The effects of manipulating social environment, food abundance, and number of food sources on migratory activity of Dark-eyed Juncos (from Terrill 1987; see text for details).

sible effects of restricted access to food, as mediated through social dominance, upon migratory behavior in winter. The potential role of social dominance in regulating demography and population dynamics in migrants is of considerable interest. Rigorously testing the effects of social dominance on migratory behavior in the field is difficult. Individual birds must be marked and tracked, dominance hierarchies determined, and access to food on an individual basis measured (Ketterson and Nolan 1985). Perhaps the most formidable aspect is determining whether the birds that disappear from a study site are actually migrating, dispersing locally, or dying.

Juncos were selected for this study for several reasons: (1) they exhibit both winter site fidelity and winter site plasticity (Ketterson and Nolan 1982, 1985); (2) they are highly social during the nonbreeding season and their social interactions have been extensively studied; (3) they are abundant and easily captured and maintained in captivity; (4) migratory juncos show nocturnal migratory activity (*Zugunruhe*) in the laboratory; and (5) females are generally subordinate to males during the nonbreeding season and on average migrate farther (Balph 1975; Ketterson and Nolan 1976, 1982, 1983).

Paired juncos (all but one bird were females) were kept indoors and their nocturnal activity was monitored from November through January 1983–84 and from December through May 1984–85. The dominant member of each of 23 pairs was determined. During the day members of each pair were allowed to interact. At night, a partition was used to divide each cage into two single-bird activity cages, allowing the nocturnal activity of each individual to be measured. I compared migratory activity of dominants and subordinates subjected to several different treatments: (1) "low food" comprised approximately eight g of food per day per pair; (2) "high food" was 14 g; (3) "single source" indicates that the food (either high or low amounts) was placed into a single, centrally located source; and (4) "double-source" indicates that the food was evenly divided between two sources placed at opposite ends of each cage. Combining data from identical treatments over the two experimental periods, the following comparisons yielded significant (paired sample t-tests) differences in migratory activity (Fig. 3): (1) low single-source subordinates showed higher activity than dominants ($t = 3.67$; $P < 0.01$); (2) low double-source subordinates higher than dominants ($t = 2.75$; $P < 0.05$); (3) low single-source subordinates greater than low double-source subordinates ($t = 2.17$; $P < 0.025$); and (4) low single-source dominants greater than low double-source dominants ($t = 3.34$; $P < 0.002$). In general, birds (whether paired or solitary) showed little or no nocturnal activity when they had abundant food during the winter months. This lack of activity contrasts with the high activity in fall when birds have access to unlimited food during the fall migratory period. Although, on average, female juncos migrate farther than males, population-level studies (Ketterson and Nolan 1982, 1985) indicate that social dominance does not explain the differential migration of juncos in that immatures, which are normally thought to be subordinate, do not migrate as far as adults. Further, they consistently find no evidence of differential disappearance during the winter period of any age or sex class at their study sites (e.g., Rogers et al. 1988). Thus, they find conflicting patterns concerning the hypothesis that social dominance might be involved in differential migration in this species.

Experimental results very similar to those involving the juncos (Terrill 1987) have been found in a long-distance migrant, the Garden Warbler (*Sylvia borin*). This species shows spontaneous *Zugunruhe* during the autumnal migratory period (approximately September–December) with access to unlimited food and also shows enhanced migratory activity in response to food deprivation (Gwinner et al. 1985, 1988). In winter, these warblers are generally not active at night; however, nocturnal activity can be stimulated by food deprivation, indicating that migratory activity may be reactivated in birds that have settled for the winter in a certain area but are then confronted with a deteriorating food supply (Gwinner et al. 1988). In such situations further movement in the migratory direction may increase the birds' probability of finding adequate food relative to local or random movements. Similar results have been reported in at least two other members of the Muscicapidae, the Blackcap, *Sylvia atricapilla* (Terrill and Berthold in prep.), and the Pied Flycatcher, *Ficedula hypoleuca* (Thalau and Wiltschko in prep.).

IMPLICATIONS FOR CENSUSING AND INTERPRETING DATA

Although more information is needed concerning resource availability, I conclude that food availability is potentially important in regulating the distribution and dynamics of wintering migrant populations. This general result has relevance to monitoring studies. First, a single winter census may not reflect the population size, or habitat utilization at a given site throughout the winter. Secondly, the possibility of large-scale geographic shifts of annual migrant populations within and between winters should be considered (this is especially important in terms of monitoring and interpreting data on absolute population size). Third, intraspecific competition may limit access to food during the nonreproductive period and be important in determining differential migration and population dynamics in some species of annual migrants.

The spatially complex life histories of migratory birds pose tremendous challenges to the analysis and understanding of avian populations and community dynamics (Bennett 1980). Although the challenge is substantial, the task is important in terms of accurately understanding migrant behavior and ecology, and is vital to proper conservation of migrant species.

ACKNOWLEDGMENTS

I thank C. J. Ralph, K. P. Able, P. Berthold, E. D. Ketterson and an anonymous reviewer for very constructive comments on an earlier draft, and K. V. Rosenberg for extensive help in preparing for this symposium.

FORAGING THEORY: UP, DOWN, AND SIDEWAYS

DAVID W. STEPHENS

Abstract. A large body of evidence is consistent with the idea that foragers tend to choose alternatives that yield more food in less time. But how do animals evaluate alternatives that vary both in time commitment and food gain? Two empirical trends, risk-sensitive foraging preferences and preference for immediacy, suggest that traditional models are incomplete because they ignore the temporal pattern of food acquisition.

Students of foraging theory are stepping *down* one level of organization by asking about the mechanisms of foraging behavior. I give two examples: an argument from foraging theory is used to evaluate the functional or adaptive significance of animal learning; and techniques from animal psychology are used to examine an issue—rules for patch leaving—that arose from arguments about foraging models.

Students of foraging theory are also stepping *up* one level of organization by addressing issues in population and community ecology. Although examples from functional response and resource partitioning show how this might proceed, advocates have yet to explicitly address its most fundamental issue: how accurate must a theory of feeding behavior be for it to be a useful building block of population models. If a precise theory is required, then work towards an accurate theory of behavior will contribute more to an understanding of population and community ecology than immediately applying current models to population processes.

WHAT FORAGING THEORY IS AND IS NOT

In this paper I discuss current and future directions for foraging theory. I pay special attention to directions that empiricists might follow most profitably. By "foraging theory" I mean those models that are sometimes called by the unfortunate name "optimal foraging," and their extensions and elaborations. The phrase, optimal foraging, is unfortunate for two reasons. First, although it is perfectly reasonable to try to distinguish foraging models that use maximization, minimization, or stability arguments from those that do not, "optimal foraging" is easily read to mean some claim about the single best way to forage. Foraging theory makes no such claim.

Second, even if you are an enthusiastic proponent of optimization models, many ideas—e.g., about perception, the development of behavior—must play important roles in any body of theory about foraging behavior and have little or nothing to do with optimization. Indeed, one source of the controversy surrounding optimization models of feeding behavior has been the absurd idea that "optimization" somehow summarizes everything anyone needs to know about foraging behavior.

This paper, reflecting my own biases and interests, is about how empiricists can most effectively influence foraging theory and foraging theorists. I would like to encourage others to do the kind of work that would tell foraging theorists what kinds of new models and new ideas are necessary to build a more accurate and general body of theory. My second motive is answering a question I am often asked: "Where is foraging theory going?" The title reflects the whimsical answer that I usually give to this question. I think foraging theory is going in three directions—up, down, and sideways. I think students of foraging must go *sideways* by pursuing those questions they have traditionally asked: They must continue to ask evolutionarily motivated questions about the costs and benefits of, and constraints on, the foraging behavior of individuals. Students of foraging theory also find themselves stepping *down* one level of organization to ask questions about the mechanisms of foraging behavior. Moreover, many students of foraging theory have as their eventual goal stepping *up* one level of organization by using an understanding of foraging behavior to deduce things about the interactions of predators and their prey, or about population and community dynamics.

Following this logic, I have organized this paper into four sections. The first three sections correspond to my three directions: sideways, down, and up; while in the last section I discuss two components that make empirical studies influential.

SIDEWAYS

In this section I outline the lessons that 20 years of foraging theory have taught, including lessons that encourage my own further interest in foraging models, and lessons that highlight the shortcomings of foraging theory. I am in a curious rhetorical dilemma: I want to convince the reader that foraging models have worked well enough to be worth further study, but no modeler wants to work in a field where all the problems are solved. As a consequence, I divide my review into two parts. First, I review the interaction between theory and data that encourages my further interest. Second, I discuss some more prob-

TABLE 1. ASSUMPTIONS AND PREDICTIONS OF BASIC FORAGING MODELS

Prey Model: Assumptions
Decision
- The set of probabilities of attack upon encounter for each prey type, p_i for prey type i.

Currency
- Maximization of the long-term average rate of net energy intake.

Constraints
- Searching for prey and handling prey are mutually exclusive activities.
- The forager encounters prey one after the other, and prey are encountered according to a Poisson process (a fine-grained environment).
- Three parameters—a net energy value e_i, a handling or involvement time h_i and an encounter rate λ_i—can be associated with each prey type (e.g., with the ith prey type).
- Encounter without attack takes no time and causes no energy gains or losses.
- The forager is completely informed. It "knows" the model's parameters, recognizes prey types upon encounter, and it does not use information it may obtain while foraging.

Prey Model: Predictions
- *Absolute preferences.* Prey types are either always taken upon encounter or never taken upon encounter (this is called the zero-one rule, because it is equivalent to saying that the optimal p_i must be either zero or one).
- Prey types are ranked by their profitabilities (e_i/h_i), and types are added to the "diet" in rank-order.
- The "decision" to include a given prey type depends only on its own profitability and the profitabilities and encounter rates of higher ranked types. Specifically, inclusion should not be affected by a type's own encounter rate.

Patch Model: Assumptions
Decision
- The set of patch residence times for each patch type, t_i for patch type i.

Currency
- Maximization of the long-term average rate of net energy intake.

Constraints
- Searching for patches and hunting within patches are mutually exclusive activities.
- The forager encounters patches one after the other, and patches are encountered according to a Poisson process (a fine-grained environment).
- Two things—a gain function $g_i(t_i)$ that relates the time spent in a patch to the energy acquired there, and an encounter rate λ_i—can be associated with each patch type (e.g., with the ith patch type).
- The gain function has two important characteristics.

TABLE 1. CONTINUED

— It starts at zero ($g(0) = 0$), spending zero time yields zero energy.
— It is initially increasing ($g'(0) > 0$) and it eventually bends down ($g''(t) < 0$ for all t values greater than some fixed t value).

- The forager is completely informed. It "knows" the model's parameters, recognizes patch types upon encounter, and it does not use information it may obtain while foraging.

Patch Model: Predictions
- The patch-residence time should be chosen so that the instantaneous rate ($g'(t)$) of gain at leaving equals the average rate of gain in the whole habitat. (Notice that this is an abstract mathematical condition, it is *not* the same as the leaving rule: "measure the instantaneous rate of gain and leave when it equals the habitat rate of gain.) This condition has a number of interesting implications:
— Whatever leaving rule the forager adopts, it should be one such that the instantaneous rates of gain at leaving are the same in all patch types.
— If the time required to travel between patches increases, the rate-maximizing patch residence will increase. (For some degenerate gain functions it can stay the same.)

---

lematic (and hence more exciting) issues that have arisen during foraging theory's development.

THE ENCOURAGEMENT

Broadly speaking, two models of foraging have been studied widely enough to allow discussion at a general level. These are the "prey model" (sometimes called the diet model) and the "patch model" (sometimes called the marginal-value theorem). Both models take the familiar form of optimality models; i.e., they make some assumption about what is maximized, a currency assumption; they make another assumption about what is controlled, a decision assumption; and they make assumptions about the things that place limits on the decision and currency, the constraint assumptions (see Stephens and Krebs 1986 for detailed discussion). The two models make identical currency assumptions and similar constraint assumptions; but they make very different decision assumptions (Table 1). These models are extremely simple, but they can in principle make detailed quantitative predictions and somewhat weaker qualitative predictions.

Kamil et al.'s (1987) recently published collection of papers on foraging behavior begins with two papers that evaluate the success of these models. The first of these (Schoener 1987:48) concludes that foraging theory "... has often been verified with tests and therefore it *should* be pursued further" (emphasis Schoener's). Although the second (Gray 1987:95) concludes that foraging models (together with all optimality models) are such dismal failures that they "... could be said to weigh like a nightmare on the brain of the living." Krebs and I have addressed the difference between Gray's and Schoener's conclusions (Stephens and Krebs 1986), and while I agree more closely with Schoener than with Gray, I recommend that readers compare both papers and form their own opinions.

Empirical lessons

By and large, the *quantitative* predictions of the patch and prey models have not fared well, with some exceptions. Stephens and Krebs (1986) found only 11 unambiguous, quantitative fits in our tabulation of 125 studies. (We took the authors' interpretations at face value, so even some of these quantitative fits have been criticized; but, on the plus side, many [about 64%] of these studies were not designed to test quantitative predictions.)

However, two astonishingly consistent *qualitative* trends are evident. The first is predicted by the patch model: the time spent exploiting a depleting patch should increase as the time required to travel between patches increases. This prediction has been found to hold practically everywhere it has been studied (the only disagreements I know of are a case in which exploitation time was unaffected by travel time [Waage 1979] and another in which the effect persisted when it should not have [Kacelnik and Cuthill 1987]). Indeed, I think this may be the most general empirical trend to emerge, not just from foraging theory but from the spate of modeling in behavioral and evolutionary ecology that began in the late 1960s.

The second qualitative success of these models is almost as universal: as predicted by the prey model, foragers selectively attack prey items that are most profitable (they have the highest ratio of "energy available," e, to "time required for handling and consumption," h, in symbols e/h). Even Gray (1987) acknowledges the pervasiveness of this trend, and his tabulation shows that this prediction was supported in over 75% of relevant studies. Gray dismisses this by arguing that this prediction is trivially obvious, but Schoener (1987) points out that this obviousness is not reflected in the pre-foraging-theory literature. In fact, ecologists before the advent of foraging theory mainly argued about whether animals were selective at all (references in Schoener 1987). Gray did not review tests of the patch model.

To be sure, these models also have their qualitative failures. The prey model's prediction of absolute preferences (the idea that a given prey type should always be ignored or always be accepted) has been consistently rejected. The prey model's other main prediction (that a type's inclusion "in the diet" does not depend on its own abundance) has sometimes been supported and sometimes rejected. I think this prediction does pretty well, if one considers the relative quality of studies supporting and rejecting it, but at face value the results are clearly mixed.

How, then, can I be encouraged? The answer comes from knowing something about the models behind the predictions, and especially behind modifications of those models. While empiricists have been comparing the models to reality, modelers have been trying to improve them logically, either by making them more general or by making them more appropriate for particular empirical situations. These modeling efforts show us that the two empirically confirmed trends (the patch model's travel time-exploitation time correlation, and the prey model's preference for more profitable prey) are also the two trends that, on *a priori* grounds, we would expect to be the most robust. For example, McNamara (1982) has persuasively argued that the "travel time-patch exploitation time" correlation would be predicted by *any* rate-maximizing model; although the details of rate-maximizing patch-leaving behavior can vary widely, this simple trend should remain. To take an example from the other side of the coin, modelers have also shown that the absolute-preferences prediction is very fragile; indeed a modification as mild as allowing choice behavior to have a variance greater than zero makes the prediction of absolute preferences evaporate (Stephens 1985).

My conclusion is that a surprising amount of data from a wide range of taxa are consistent with the simple notion of rate maximizing. Long-term, average rate maximizing is, of course, a specific way to combine less time and more energy (or simply more food if nutrients other than energy are important). A critic might argue, and I would agree, that many models that somehow place value on options that provide more food in less time would be consistent with these qualitative trends. (Below I will explain why one cannot dogmatically assert rate maximizing.) So, we have as a minimal and conservative conclusion that foraging animals act economically, in the sense that they tend to choose alternatives that yield more food in less time.

Some conceptual lessons

Foraging theory has not only had some empirical successes, but it has also had some important conceptual successes, because it has changed the way students think about this subject. For example, the prey model shows that the choice of a diet (a list of the things an animal eats, and sometimes the proportions of these things) is a consequence of two types of behavioral choices that are logically different. The prey model predicts which items should be attacked, given a fixed and well-defined process of encounters with prey; this *encounter process* characterizes which prey are encountered during search and how frequently. But nothing in theory or logic says that a forager cannot also make decisions that change the encounter process. Foragers might change it by doing obvious things, such as moving from one part of their habitat to another, or by doing subtle things, such as looking up instead of down. Hence, the diet is determined at a minimum by (1) choices that determine the parameters of the encounter process, and (2) the choice of which items to attack and ignore. This simple separation of choices casts doubt on stomach contents studies of selectivity and choice; rather, it suggests that watching foraging behavior may be the most informative way to study diet choice.

A related idea is the concept that a forager's perceptive abilities define what a "prey type" is; to paraphrase a clever phrase maker: animals do not eat Latin binomials. Two species may form a single type, as they do in model-mimic systems, or (more commonly) a single species may form many types; small *Genera generalis* caterpillars and large *Genera generalis* caterpillars may well be different types from a forager's perspective (see Getty 1985 for a sophisticated discussion).

The patch model also has its conceptual successes, but because the phenomenon of patch tenacity is really new, these successes cannot be contrasted with older approaches. Two generations of modeling this problem have suggested two quite different economic reasons to move on to a new patch. The first (originally proposed by Charnov 1976) is that patches usually decline in quality as the forager exploits them; one reason to leave is simply because things are getting worse. The second reason (Oaten 1977) is that experience gained while exploiting a patch may tell the forager that this patch is an inferior one and hence not worth further effort. Both of these reasons sometimes apply (e.g., Lima 1983 for the patch assessment case, and Cowie 1977 for patch depression), but we do not know much about their relative importance in nature.

THE DISCOURAGEMENT: WHAT NEXT?

Many things are wrong with foraging theory as presently constituted; most are aspects of foraging behavior that have been left out of traditional foraging models. A list of aspects that need to be included in a more general theory of feeding can be found in the chapter headings of Stephens and Krebs (1986): *Incomplete information,* including problems of resource assessment and the abilities of foragers to recognize and discriminate prey and other resources; *Tradeoffs,* including tradeoffs between energy and other "nutrients" (including toxins) between foraging and predation, or foraging and reproduction; *Risk-sensitivity,* including general questions about the pattern of resource acquisition in time and how foragers value different "patterns"; *Dynamic tactics,* the problem of allowing decision variables to be functions of other "state variables," so that one can solve for the best trajectory of decisions instead of the single best decision. Rather than discuss each of these here, I will discuss a particular empirical issue that I think addresses some fundamental flaws in the traditional assumptions. I would like to explain why I retreated from stridently advocating rate maximizing to the milder position that foragers value less time and more food (or energy) in some vague way. One might deny strict rate maximizing, because it ignores complementary nutrients and the threat of predation. Indeed these are limitations, but rate maximizing cannot be generally correct even in conditions in which time and energy alone are important because (1) it ignores the variability in food gain (or *risk*), and (2) because it ignores the importance of immediacy in food gain.

Risk

Conventional foraging models were built on the premise of maximizing the "long-term average rate of energy intake," which is a very specific and potentially restrictive assumption. Consider the difficulties inherent in limiting our attention to *averages.* Suppose that a forager can choose between two alternatives. Alternative A provides a mean food gain of 10 joules and standard deviation of 10 joules in a period of 1 min, while alternative B provides a mean food intake of 10 joules and standard deviation of 1000 joules also in a period of 1 min. Because both alternatives take the same time, they obviously provide the same average rate of energy intake. A model based on long-term, average rate maximization would provide no basis for preferring one of these alternatives; instead, any such model predicts that foragers should be indifferent between the high and low variance choices. But a real forager would hardly be indifferent between these two choices that vary so much in variance or risk? Animals have consistent preferences when presented with alternatives that vary only in their degree of riskiness, even when means do not vary (Caraco et al. 1980, Real 1981, Real et al. 1982,

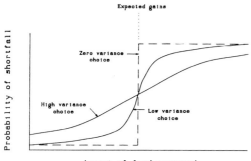

FIGURE 1. How minimizing the probability of an energetic shortfall predicts changing risk-sensitivity. The two curves represent the probability that a forager will fall short as a function of the amount of food required. The solid curve plots the probability of a shortfall for a normal distribution with low variance, and the dotted curve plots the probability of a shortfall for a normal distribution with high variance. The low and high variance distributions have the same mean (= expected gains). If the food requirement is greater than expected gains, then the high variance distribution yields a lower probability of a shortfall, but if food requirements are less than expected gains, the low variance distribution yields the lowest probability of a shortfall (after Pulliam and Millikan 1982).

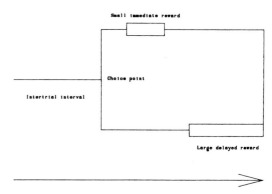

FIGURE 2. The contingencies for a typical experiment (in this case Green et al. 1981) showing preference for immediacy. After a fixed intertrial interval, the forager is presented with two mutually exclusive alternatives. One leads to a small, relatively immediate reward; the other leads to a larger delayed reward. In the experiment shown, reward size was controlled by controlling duration of food access, so a large reward takes more time. Notice that the total "cycle time" is the same regardless of the subject's choice.

Caraco 1983, Barnard and Brown 1985, Stephens and Paton 1986, Wunderle and O'Brien 1986). In my view, conventional foraging models fail because they require that foragers be indifferent over risk (or variance). This requirement is not only counter-intuitive, but it is also an empirical failure.

The work of Caraco and his colleagues provides several important examples of risk-sensitive preferences. Caraco et al. (1980) showed that Yellow-eyed Juncos (*Junco phaeonotus*) kept on positive energy budgets (i.e., fed at a rate that allowed them to maintain their *ad libitum* feeding weight) preferred certain alternatives, whereas those maintained on negative energy budgets preferred variable alternatives. This may mean that juncos are not maximizing the long-term, average rate of gain, but instead are minimizing the chance of falling short of some critical amount of food (see Fig. 1) at some critical time. The presence of risk sensitivity hints that part of the problem with long-term rate maximizing is its failure to consider details of the temporal pattern of food acquisition.

Immediacy

Consider another set of hypothetical alternatives. Suppose that every 2 min a forager is offered two alternatives. Alternative α leads to 1 joule of food delayed by 30 s and alternative β leads to 10 joules of food delayed by 1 min. Because the time between offers is fixed, the forager must wait 1.5 min from being fed until the next offer if it chooses alternative α, but it must wait only 1 min if it chooses alternative β (Fig. 2). If the long term is all that is important, then these alternatives amount to nothing more than 1 joule in 2 min, and 10 joules in 2 min, and β must be a much better choice. Annoyingly, real animals do not always agree. It is easy to find instances in which foragers prefer smaller but more immediate gains, even when they could do better in the long term by waiting for larger gains. This phenomenon is well known among animal psychologists (e.g., Green et al. 1981). Behavioral ecologists are just beginning to investigate this phenomenon in animals other than rats and pigeons, and at least one such study agrees with the psychological results (e.g., Barkan and Withiam [in press]).

Preference for immediacy is a vexing problem and few attempts have been made to explain it, compared to the number of attempts to explain risk sensitivity. The most convincing explanation is that foragers expect to be interrupted (by conspecifics or predators), so that the immediate small thing may actually be better than the delayed large thing (Kagel et al. 1986, McNamara and Houston 1987a). While this is the most reasonable explanation available, I think the effect is too strong and persistent to be explained completely thusly because none of the experiments have included any interruptions. So proponents must argue that animals are "hard-wired" to expect interruptions (or that something external to the experiment itself has created such an expectation). This is possible, but animals react in other ways to the presence and absence of potential interrupters.

Like risk sensitivity, preference for immediacy points out that long-term, average rate maximization ignores some important features of the flow or pattern of food acquisition. Taken together, preference for immediacy and risk sensitivity punch a sizable hole in long-term, average rate maximizing.

Moreover, there are reasons to think that these two phenomena are related. To represent the decrease of food value with delay, suppose that a delay of δ seconds means that an amount of food A is really worth only $Af(\delta)$ (Fig. 3A), where $f(\delta)$ is a *discounting function* that represents the fraction of A's value that remains after a delay of δ. Figure 3B shows that this positively accelerated discounting function also predicts that a forager should prefer a probability distribution of delays to a certain delay with the same mean (for example, the risky choice $\Pr(\delta_1) = \Pr(\delta_2) = 1/2$ should be preferred to the certain choice $\Pr(\frac{\delta_1+\delta_2}{2}) = 1$). Indeed, this trend has been widely observed; rats and pigeons prefer alternatives with variable delays before reward to alternatives with fixed delays (see Hamm and Shettleworth 1987).

A biological time-energy problem. Both risk sensitivity and preference for immediacy suggest that something about the pattern of food acquisition is important and neglected. An enormous body of evidence is consistent with the "more food in less time" postulate, but both risk sensitivity and preference for immediacy show that, when it comes to details, we do not know how or why animals evaluate combinations of time commitment and food gain as they do. The possibility of a link between these two phenomena makes me hopeful that some crucial piece of the puzzle may click into place at any moment. I think this is the most fundamental "sideways" problem in foraging theory, because a solution would change our view of every aspect of feeding behavior.

I have talked about *the* solution of the time-energy problem, but many solutions might exist. Different individuals or species may value time-amount combinations differently at different seasons or phases of their life history. I might hope for a general solution, but I certainly do not insist on one. In fact, there may well be ecological correlations that would be exciting discoveries if they hold up. For example, while birds (pigeons and chickadees, for example) seem to have strong preferences for immediacy, my own work with

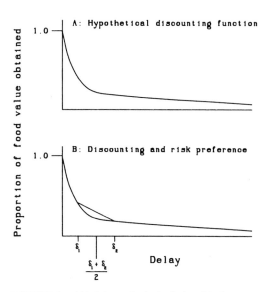

FIGURE 3. (A) A hypothetical relationship between delay and food value: food item that would have value A if obtained immediately will only have value $Af(\delta)$ if it is delayed by δ s. The "discounting" function $f(\delta)$ is shown in this panel. (B) If the "discounting" function is curved as shown here (positively accelerated), then this predicts risk preference over delays. Suppose that the forager can choose between (1) an alternative that yields δ_1 half the time and δ_2 half the time, and (2) an alternative that always yields $\frac{\delta_1+\delta_2}{2}$. The expected food value from the risky alternative will lie halfway along the line segment that connects the points $(\delta_1, f(\delta_1))$ and $(\delta_2, f(\delta_2))$. The expected value of the risky alternative is higher than the expected value of the certain alternative, because this line is always above the curve.

honeybees suggests that the preferences of these social, dawn-to-dusk foraging machines are consistent with long-term rate maximizing, because honeybees will pass up immediate gains to increase longer term gains (Stephens et al. 1986). A similar point can be made about risk sensitivity, and how general the switch from risk preferring to risk avoiding shown by Caraco's juncos may be. Early indications suggest that it may not work for Bananaquits, *Coereba flaveola* (Wunderle et al. 1987).

DOWN

In the last few years, many behavioral ecologists have begun an exciting collaboration with animal psychologists. Why should psychologists care about the evolutionary function of behavior? Stephens and Krebs (1986) answer this question using a slide rule as an example. *The mechanism* of a slide rule is defined in terms of how it accomplishes the function of multiplication. You would interpret this differently if you thought it was a device for digging holes. In more general terms, function is implicit in most arguments about mechanism. It is always a good idea to bring hidden assumptions into the light (see also Kamil and Yoerg 1982, Shettleworth 1983).

A traditional psychologist might agree, but argue that general models of function are sufficient to interpret mechanisms, such as "the function of feeding behavior is to acquire food." Yet, foraging theory has discovered new phenomena such as the "travel time-patch exploitation time" correlation that were not evident from more general arguments. Indeed, the concept of patch-exploitation tactics has fostered new research on psychological questions about rules for patch leaving (see below).

What does animal psychology offer, and why should foraging ecologists care about mechanisms? Simply, more details lead to better models. I can make this point somewhat more formally by returning to the elements of optimality models: currency, constraint, and decision. Early foraging models make unrestrictive and general assumptions about constraints (e.g., foragers cannot search and eat at the same time), which results from psychology can make more sophisticated. For example, Getty and his colleagues (Getty 1985, Getty and Krebs 1985, Getty et al. 1987) have taken some models derived from sensory psychology (signal-detection theory, Egan 1975) to derive predictions about the detection of cryptic prey. The results are impressive; they have refined foraging theory's view of what a prey type is (see Stephens and Krebs 1986, Chapter 3).

A subtle variant on the same theme is what I call the feasibility-of-mechanisms problem. A traditional foraging model can work only if there is some decision mechanism, which in the patch model must allow a link between travel time and patch-leaving decisions. Traditional models ignore these mechanisms, by assuming that one mechanism can be implemented as easily (and as cheaply) as any other. This cannot be generally correct, and may be an issue that psychologists might help resolve.

I think most students of feeding behavior would agree that both laboratory and field studies have something to offer, but persons tend to specialize. Kamil (1988) has addressed this dichotomy by defining the different goals of laboratory and field work. He argues that all studies should have two goals: high internal validity (such issues as the repeatability of results, and avoiding confounding variables) *and* high external validity (how readily one can generalize from the situation studied to others). These goals are usually in conflict: an operational decision that increases internal validity will often decrease external validity and *vice versa*. Laboratory work tends to have high internal validity but compromised external validity, while the reverse is true of field work. Behavioral ecology is a complex subject; it is probably expecting too much for a single study to establish the *general* validity of a result.

TWO EXAMPLES ON THE WAY DOWN

Below I give two examples of important "down" questions. One is a case in which foraging theory seems to say something new about a field in the traditional domain of animal learning. The second is an instance in which a mechanistic perspective is making inroads into a traditional question in behavioral ecology.

Information acquisition and animal learning

An animal's experience often changes its behavior. Animal learning is a central topic in animal psychology, in which psychologists have focused on mechanisms. Recently, behavioral ecologists and students of foraging behavior have begun to look at functional aspects.

Older models supposed that foragers were completely informed and did not need to use experience to improve their foraging decisions. Modelers were initially attracted to learning simply because they wanted to improve their models by allowing foragers to use new information. Because the approach of these "learning" models has been functional, conditions seem ripe for the kind of "function-mechanism" interaction that I advocated above.

With a few exceptions (Hollis 1982, Johnston 1982, Kamil and Yoerg 1982, Staddon 1983), psychologists seldom discuss functional significance of learning. Johnston (1982:74) concludes that "the ability to learn . . . has as its primary selective benefit that it permits adaptation to ecological factors that vary over periods that are short in comparison with the lifetime of an individual." This is a sensible idea; however, recent analyses of "incomplete information" problems in foraging theory show that it is only a part of the functional story.

When I began to model how to track a changing environment, I thought that the inclusion of a term that represented how frequently the environment changed would allow me to make a more quantitative statement than Johnston's. In rough outline, I made the following assumptions (see Stephens 1987 for details). Some varying resource always looks the same even though it can actually be in one of two states, good or bad. Although these states look the same, the forager can easily tell the difference when it exploits the resource. There is an alternative, stable resource whose quality is mediocre. I represented the *persistence* of the varying resource by a conditional probability; hence, persistence is the probability that a good state will be immediately followed by another good state, or that a bad state will be immediately followed by another bad. When my hypothetical forager experiences a bad state, it can switch to exploiting the stable-but-mediocre alternative, but should occasionally return to check out the varying resource.

My hypothetical forager was free to use its experience or to ignore it by adopting an "averaging" tactic—attacking only the varying resource or only the mediocre-stable resource, whichever had the highest average quality. Hence, I was able to look for conditions in which learning was worthwhile.

The first thing I discovered surprised me: when the varying resource has no persistence, an averaging tactic is best. Superficially this seems like a counter-example to Johnston's assertion, a case in which an environmental feature changes over a period that is much shorter than an individual's life, but learning is not an economically sound policy. At first I dismissed this as a special case, believing that if I looked at the whole range of persistences I would find that learning paid off most at some intermediate level of variability. Instead I found that the payoff increased continually, and that the longer states persist the more worthwhile it is to learn about them.

This suggests that there is more to the relationship between environmental variability and the value of learning than Johnston's statement implies. This apparent paradox can be resolved by thinking of two kinds of predictability. Johnston's statement deals with the ability of the previous generation to predict the environment of its offspring (when this predictive link is weak, learning is favored), while my argument has to do with the ability of an individual's experience to predict the future states of its own environment (when this predictive link is strong, learning is favored). This opens up an enormous number of new and fascinating questions about how these two kinds of "predictability" may or may not be related, and how these relationships may affect the value of learning.

Finally, these arguments about the functional significance of learning provide a more serious example of my "slide rule" point. Behaviorists often want to make statements about the presumably mechanistic limitations of what can be learned: A stimulus of type A can be associated with food, but a stimulus of type B cannot. I have concluded that learning may not be worthwhile in some situations. There is a big difference between something that is not learned because of a mechanistic limitation and something that is not learned because it is not worthwhile.

Rules for patch departure

Can animals count, keep track of the time between two events, or integrate information about time and number? These are the kinds of questions that psychologists study.

Recently such questions have become important, for purely theoretical reasons, in foraging theory. Early workers (see Charnov 1976) on patch-leaving models seemed to suggest that a general rule for patch departure has the form: leave the patch when the instantaneous rate of gain drops to some critical value. While this rule may work sometimes, its generality has been widely criticized (Oaten 1977; Green 1980, 1984; Iwasa et al. 1981; McNair 1982). Four types of patch-leaving rules have been presented: (1) a fixed-number rule: leave after finding n prey; (2) a fixed-time rule: leave after spending t seconds in the patch; (3) a giving-up time or run-of-bad-luck rule: leave τ seconds after the last prey capture; (4) a rate rule: leave when the "instantaneous" rate of prey capture drops to some critical rate. Iwasa et al. (1981) have shown that different rules work in different situations. For example, if prey are captured at random intervals and all patches have the same number of prey, then a fixed-number rule makes sense. However, if the number of prey per patch is highly variable, a run-of-bad-luck rule makes more sense. If the number of prey per patch follows a Poisson distribution, then a fixed-time rule works well (Iwasa et al. 1981). Hence, an esoteric argument about models of patch leaving has helped to place some issues from animal psychology (such as counting and timing) in ecological perspective.

More importantly, some of my colleagues at the University of Massachusetts have performed an experiment designed to deduce what kinds of patch-leaving rules animals actually use. Kamil, Yoerg, and Clements (in press) presented feeding Blue Jays (*Cyanocitta cristata*) with a simple patch-leaving problem. Two patchy resources were available to an individual jay. One resource was initially of high quality but eventually depleting, while the other was of low but constant quality. To simplify matters, Kamil et al.'s depleting patch depleted suddenly; hence up to n prey were delivered at a high fixed rate in the depleting patch, but no prey were delivered after the nth. Each bird was exposed to three treatments $n =$ three, six, and nine prey; and each jay experienced a single treatment (two patches per day) for a very long time (often up to two months).

Since in any given treatment there is a fixed number of prey in the good patch, the best rule is obviously a fixed-number rule: exploit the high-quality depleting patch until it has provided all n prey, then switch to the constant, mediocre patch. Kamil et al. examined patch-leaving rules by calculating the relative frequency of patch-leaving events that were preceded by all possible numbers of prey captures. They found that a fixed-number rule did not completely explain the cisely, but can get by with a vague understanding of others, then the future of this interaction will depend on which aspects of feeding behavior are deemed to be critical.

EMPIRICAL DIRECTIONS

In my opinion, the most influential studies are those that force modelers to change their assumptions at the most fundamental level, such as Caraco et al.'s (1980) demonstration that the preferences of juncos are sensitive to variance, and that this preference can change direction with the junco's state of hunger. This study challenges fundamental assumptions, because it casts doubt on the traditional assumption of *average* rate maximizing. Moreover, it provides an interesting new observation, the switch in risk sensitivity, for which alternative models would have to account. Caraco et al. were fortunate that juncos are sensitive to variance, since they might have confirmed the traditional assumption instead of overthrowing it, and that people were beginning to think about variance sensitivity, so that their work got immediate attention. But, two features of this study cannot be ascribed to luck. Caraco et al. tested explicit and meaningful alternatives, and they chose these alternatives and the conditions of the experiment with an understanding of their larger theoretical significance. Unfortunately, many tests of foraging models do not meet these two criteria.

Understanding the theory

Many would-be testers do not understand the theory well enough to perform meaningful tests. For example, the patch model has been tested in cases in which there is no patch depletion (see Dunning, this volume), and the prey-choice model has actually been applied to situations in which the two prey types of interest were found in different parts of the forager's habitat. No one is interested in the "discovery" that a model does not work when it would not be expected to. Understanding the theory pays empirical dividends in another more important way, because it focuses attention on important problems. For example, an attack on an assumption made in 50 foraging models will be more influential than an attack on an assumption made in one model. Caraco et al.'s (1980) demonstration of risk sensitivity challenged an assumption made by practically all models before 1980.

Testing meaningful alternatives

Many earlier tests of foraging models were essentially confirmatory tests. Because they tested only one model, if they compared it to any alternative it was the alternative of some random model. While this approach may once have been justified, that time is past. Alternative models are available, and the failure to use them makes tests of foraging theory difficult to evaluate. The ideal alternatives are contrasting views already present in the literature. See Kamil (1988) for a full discussion of this and other empirical issues in ornithology.

CONCLUSIONS

I think that foraging theory has helped to refine our ideas about feeding behavior, and that its future will be more interesting than its past. It faces some fundamental challenges about how animals evaluate alternatives that vary in both time and amount, about the mechanisms that govern feeding behavior, and about what the answers to such questions will tell us about ecological processes. Each of these directions represents exciting empirical and theoretical opportunities.

Foraging theory is nothing if not controversial. The critics have helped theorists to see the limitations of their approaches, producing a more productive and cautious discipline. But they have failed to convince some that foraging theory is "a complete waste of time" (Pierce and Ollason 1987). In my view foraging theory may be limping along, but it is moving, and its critics have failed to make clear what alternative research programs would provide a more productive approach. Until critics meet this challenge, foraging theorists can take comfort from a north African proverb: "The dogs may bark but the camel train goes on" (quoted by Murray 1981).

ACKNOWLEDGMENTS

I thank Al Kamil, Ted Sargent, and Sonja Yoerg for their comments. A number of the ideas expressed here were formulated and refined in conversations with Tom Getty, whom I thank for patience and sound advice.

EXTENSIONS OF OPTIMAL FORAGING THEORY FOR INSECTIVOROUS BIRDS: IMPLICATIONS FOR COMMUNITY STRUCTURE

BRIAN A. MAURER

Abstract. Optimal foraging theory has been successful in developing specific, testable predictions regarding the behavior of a number of organisms. Useful models must include as much relevant biological detail as possible. Two such models are presented here. The multitactical model predicts that organisms will pursue a given prey with a given tactic if the gain exceeds the cost. By assuming that the probability of capture increases as prey encounter rates increase, predicting the prey densities at which switching tactics is profitable is possible. The interference model predicts that if a bird is choosing an optimal diet it will inevitably face increasing interference costs as prey densities increase. A bird should avoid interference whenever possible. If it is assumed that the encounter rate with other birds increases as prey encounter rates increase, it is possible to predict prey encounter rates at which birds will switch foraging tactics to avoid interference. For birds to make foraging decisions, they must be capable of evaluating profitability of prey items, the probability of capture using a tactic, and the amount of time required to capture them. I present suggestions for testing of the models and consider implications of these models for the generation and maintenance of community structure.

Key Words: Community structure; insectivorous birds; interference competition; multitactical foraging; optimal foraging.

For several reasons, ecologists studying insectivorous birds have been slow to use optimal foraging theory as a predictive tool. First, the assumptions underlying optimal foraging theory have been questioned (e.g., see reviews by Maynard Smith 1978, Krebs et al. 1983, Myers 1983, Stephens and Krebs 1986). Second, many optimal foraging models are not constructed of variables that are easily measured in the field. Finally, field biologists often seem resistant to theorizing, perhaps because theoretical formulations often ignore biological properties of the organisms. The debate regarding assumptions will continue, until someone can explain why the models produce successful predictions. Even if the models fail in some respects, they provide powerful tools for developing specific, testable hypotheses regarding foraging behavior (Stephens and Krebs 1986; Stephens, this volume).

In this paper, I attempt to develop simple extensions of a model of optimal foraging that can be applied to insectivorous birds in forest ecosystems. No attempt will be made to deal with all of the complexities of their foraging behavior, but two important observations will be used to extend optimal foraging theory for insectivorous birds. The first extension is based on the observation that insectivorous birds often use several different tactics to secure the same type prey (Robinson and Holmes 1982). For example, lepidopteran larvae can be caught by either gleaning from the surface of a leaf or snatching from leaves while hovering. The second extension recognizes that interference competition among birds regularly occurs (Morse 1976b, Maurer 1984) and may influence foraging behavior. The models I consider below incorporate both multitactical foraging and the costs of interference.

In this paper I: (1) present several models beginning with the basic optimal foraging model and then add successive considerations for multitactical foraging and interference competition; (2) examine the basic assumptions of the models and describe how model parameters might be estimated in the field; (3) suggest specific experiments to test the models; and (4) consider implications for the maintenance of community structure for forest birds.

THE MODELS

OPTIMAL FORAGING IN A FINE-GRAINED HABITAT

Charnov (1976a) developed a model of optimal foraging that built on work by MacArthur and Pianka (1966), but parameterized the arguments in a way similar to Holling (1959a, b). Stephens and Krebs (1986) demonstrated the generality of the Charnov model and discussed many refinements. The model predicts the choice that a predator will make when it encounters a prey item of given quality. That quality has two components, energy value and handling time. The predator was assumed to maximize the rate of energy intake by behavioral adjustments.

Let E be the rate of energy intake; then, according to Charnov (1976a)

$$R = \frac{E}{T_h + T_s}, \qquad (1)$$

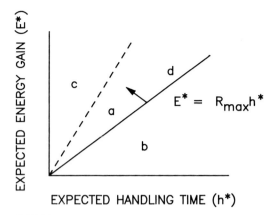

FIGURE 1. In Charnov's (1976a) model of optimal foraging, the solid line indicates the lower limit of the prey acceptability region. A prey type at b would never be selected. If prey become superabundant at c, then the lower limit of the prey acceptability region moves towards the dashed line, so that prey types a and d, which were originally taken, become suboptimal. Under these conditions prey type d is not included because of a greater handling time, even though it has higher energy content.

where R is the long-term rate of energy intake (Stephens and Krebs 1986), T_h is the total amount of time spent in handling the prey, and T_s is the total amount of time spent in searching for prey. Handling time includes the time between when a foraging bird first identifies a prey item until it begins foraging again. It could be further broken down into pursuit time and actual handling time (e.g., Eckhardt 1979).

Charnov (1976a) suggested that if one defined the following quantities: E_i^* = expected energy gain of prey i, h_i^* = expected handling time of prey i, P_i = probability that prey type i is attacked, and λ_i = encounter rate of predator with prey type i, then the individual terms in eq. (1) could be defined as $E = \Sigma \lambda_i E_i^* T_s P_i$ and $T_h = \Sigma \lambda_i h_i^* T_s P_i$. These values, upon substitution into eq. (1) give

$$R = \frac{\Sigma \lambda_i E_i^* P_i}{1 + \Sigma \lambda_i h_i^* P_i}.$$

Charnov (1976a) showed that R is maximized if the following three conditions hold: (1) $P_i = 0$ or $P_i = 1$; that is, the predator always attacks some prey types and never attacks others. (2) If prey types are ranked according to the ratio of expected energy gain to expected handling time (E_i^*/h_i^*), then the inclusion of a prey type in the optimal diet depends only on the density of items of higher ranking. The term E_i^*/h_i^* represents a measure of prey quality, and can be thought of as the expected energy gain per unit time of effort (exclusive of search time). (3) Those prey items that are eaten are those for which the following inequality holds:

$$\frac{E_i^*}{h_i^*} > R_{\max},$$

where $R_{\max}$ is the maximized rate of energy intake. These results are presented graphically in Figure 1. Condition (3) can be interpreted by noticing that the long-term rate of energy intake (R) also includes search time (see eq. 1). Thus, for a prey type to be of sufficient quality to be included in the diet, the energy derived from its consumption must allow the predator to compensate for time that must be spent searching.

MULTITACTICAL FORAGING IN A FINE-GRAINED ENVIRONMENT

To extend Charnov's model to the multitactical situation, first note that the rate of energy intake is assumed to be given as in eq. (1). The quantities E_i^* and λ_i remain as before, but the following new quantities are defined: P_{ik} = probability that prey i is pursued with tactic k, C_{ik} = probability that prey i is captured using tactic k, $h_{i\,k}^*$ = expected handling time of prey i using tactic k. If E_k is the total energy collected using tactic k, then the total energy obtained will be the sum of the energy obtained from each tactic, and the handling time will be the sum of the handling times of each tactic, so:

$$E = \Sigma \Sigma \lambda_i E_i^* T_s P_{ik} C_{ik} \quad (2a)$$

$$T_h = \Sigma \Sigma \lambda_i h_{i\,k}^* T_s P_{ik}. \quad (2b)$$

This formulation assumes that there are some prey items that are pursued but not captured because the probability of capture, C_{ik}, can be <1, but does not appear in the relationship defining total handling time. When these relationships are substituted into eq. (1), then the following relationship is obtained:

$$R = \frac{\Sigma \Sigma \lambda_i E_i^* P_{ik} C_{ik}}{1 + \Sigma \Sigma \lambda_i h_{i\,k}^* P_{ik}}. \quad (3)$$

A similar set of conditions to Charnov's (1976a) holds when R is maximized in this model. These are: (1) $P_{ik} = 0$ or $P_{ik} = 1$, that is, the predator either always or never attacks some prey items using a given tactic. (2) When prey are ranked according to the ratio of expected energy gain to a tactic's expected handling time multiplied by the probability of capture using that tactic ($[E_i^*/h_{i\,k}^*] \times C_{ik}$), then the inclusion of a prey type in the optimal diet depends only on the density of items of higher ranking. The value of a prey item is weighted by its expected probability of capture.

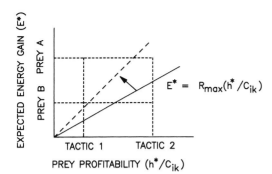

FIGURE 2. In the simplest multitactical model, there are two prey types of differing energy content and two tactics that differ in their profitability. The solid line represents the lower limit of the prey acceptability region. Prey type B is never taken using tactic 2, but is using tactic 1. If prey type B becomes superabundant, then the lower limit of the prey acceptability region moves towards the dashed line. When this occurs, prey type A will not be taken using tactic 2.

(3) Those prey items eaten are those satisfying the following inequality:

$$\frac{E_i^*}{h_{ik}^*/C_{ik}} > R_{max}.$$

The value of a prey type weighted by its probability of capture must exceed the maximal rate of intake. Thus, some items that might otherwise be taken in the diet may not be included if the probability of their capture is too low. These conditions are summarized graphically in Figure 2.

The above argument holds if the probability of capture using a tactic is independent of the density of prey. However, probability of capture may increase as the rate of encounter of prey items increases (Figs. 3A,B). The simplest assumption to make is that there is a linear increase in C_{ik} with λ_i over a certain range of prey encounter rates. If this assumption is made, then prey profitability will be proportional to h_{ik}^*/λ_i (Figs. 3C,D).

Consider two different tactics, each with a different functional relationship of C_{ik} with prey encounter rate. There are two ways for the tactics to be related to encounter rate (Figs. 3A,B). First, one tactic might always be superior to the other, so that the probability of capture using it will always be greater (Fig. 3A). If this condition holds, the less successful tactic will only be used when the prey item has an exceptionally high energy value. Prey items with low energy content will always be taken only with the first tactic (Fig. 3C). However, if one tactic has a lower capture probability at low prey encounter rates, but is

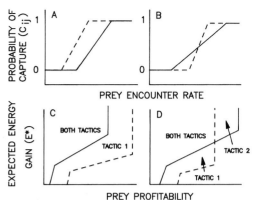

FIGURE 3. Results of the multitactical foraging model when probability of capture is assumed to increase with prey encounter rates. A. An assumed relationship between the probability of capture and prey encounter rate for two tactics. Tactic 1 (dashed line) is always better than tactic 2 (solid line). B. A different relationship where tactic 1 (dashed line) is better than tactic 2 only at higher prey densities. C. The resulting limits for prey acceptability from part A. Tactic 2 is only used for relatively high quality prey, while tactic 1 is the sole tactic used to take prey of low energy content. Note that prey profitability increases to the left on the axis. D. The resulting limits for prey acceptability from part B. Tactic 1 is used exclusively for low quality prey at high prey profitabilities, while tactic 2 is used exclusively for prey of low profitability. Note that prey profitability increases to the left on the axis.

superior at high rates (Fig. 3B), at low encounter rates it will be used exclusively to obtain energy rich prey; energy poor prey will be taken only at high densities by using the first tactic (Fig. 3D).

FORAGING WITH INTERFERENCE

The inclusion of interference interactions into the optimal foraging model described above is accomplished by a redefinition of eq. (1) to include energetic and time costs for interactions. Thus we have

$$R = \frac{E_g - E_c}{T_s + T_h + T_c}, \qquad (4)$$

where E_g is the gross energy intake, E_c is the energy lost in interference interactions, and T_c is the time spent in interactions. For simplicity, assume that the bird is foraging on a single homogeneous prey resource with expected energy content E^*, expected handling time h^*, encounter rate λ, and probability of attack P. Then the quantities in eq. (4) can be defined as: $E_g = \lambda E^* P T_s$; $E_c = \eta E_c^* P_c T_s$; $T_h = \lambda h^* P T_s$; and $T_c = \eta t^* P_c T_s$, where η is the encounter rate with other foragers, E_c^* is the expected energy spent in a single interference interaction, P_c is the probability that an encounter will result in an interference interaction, and t^* is the expected time spent in the interaction. Substituting into eq. (4) gives:

$$R = \frac{\lambda E^*P - \eta E_c^*P_c}{1 + \lambda h^*P + \eta t^*P_c}. \quad (5)$$

Eq. (5) indicates that R will be maximized if either η or P_c are equal to 0. If an individual can predict when another individual will be encountered, then the first bird should act to prevent interference. Territoriality can reduce the amount of interference and may be reinforced in species with high encounter rates. Although some birds defend territories against both conspecifics and individuals of different species, territoriality is usually directed at conspecifics. If this is true, then much of the interference birds experience will be due to encounters with other species. If such encounters are random or show no consistent pattern, adaptations to prevent them may not evolve and, consequently, it may not be possible for natural selection to minimize η or P_c.

If avoidance of interference from other species cannot readily evolve, then what is the ecological cost of interference? In eq. 5, the effects may be examined by making some simplifying assumptions. First, assume that the energetic cost for an interference interaction is negligible, then eq. 5 can be rearranged to give:

$$\frac{T_c}{T_s} = \frac{\lambda P(E^* - Rh^*)}{R} - 1. \quad (6)$$

The ratio of time spent in interference to time spent searching is thus a linear function of prey encounter rate. Notice that for the slope to be positive, $E^*/h^* > R$. Thus, if a bird is foraging optimally, that is it meets the condition $E^*/h^* > R$, it must spend more time in interference interactions as prey density increases (Maurer 1984). The threshold prey encounter rate above which interference will be experienced is given by

$$\lambda_0 = \frac{R}{E^* - Rh^*}. \quad (7)$$

Second, assume that the time spent in interference interactions is negligible, but each encounter is energetically expensive. In this instance, eq. (5) can be rearranged to give

$$E_c = \lambda P(E^* - Rh^*) - R. \quad (8)$$

Again, the cost for interference—this time in energy lost—is a linear function of prey encounter rate, and if the organism is foraging optimally energy lost to interference will increase with prey encounter rates. The threshold prey encounter rate for this cost to be positive is also given by eq. (7).

This simple model suggests that both the time spent and the total amount of energy expended in interference interactions will increase as prey encounters increase. There are at least two ways for the cost of interference to increase (Maurer 1984). First, as the density of prey increases, birds will encounter other birds attracted to the abundant resource more often. Second, prey encounter rates may also be high if insect prey are clumped, and if birds are attracted to such clumps, the amount of time and energy spent in resolving interference will increase.

COMBINING THE MULTITACTICAL AND INTERFERENCE MODELS

In this section a different approach to the cost of interference is taken by asking, "How does a bird make decisions in foraging if using different tactics exposes it to different intensities of interference competition?" In eq. (4) we can take E_g and T_h as in eqs. (2a) and (2b) and make the simplifying assumption that every predator encounter will result in an interference interaction, so $P_c = 1$. The interference terms become: $E_c = \Sigma \Sigma \eta_{ik}E_c^*T_k$, and $T_c = \Sigma \Sigma \eta_{ik}t^*T_k$, where $T_k = \Sigma \lambda_i h_i^* P_{ik}T_s$. The modified conditions that predict the decision of the predator to pursue a particular item are similar to the conditions for previous models: (1) $P_{ik} = 0$ or $P_{ik} = 1$. (2) Prey items and tactics can be ranked according to their profitabilities, which now are given by the expression:

$$\frac{E_i^*C_{ik} - \eta_{ik}E_c^*}{h_k^*(1 + \eta_{ik}t^*)};$$

(3) The profitability of a prey item as given by condition (2) must exceed R_{max} in order to be attacked using a given tactic.

In the expression for profitability, there are several factors that affect the profitability of a prey item pursued with a given tactic. Increasing the probability of capture using a given tactic (C_{ik}) increases the value of a prey type and increasing the expected handling time using that tactic decreases the value of a prey item. Increasing the amount of interference decreases the value of a prey item, making it less likely to be included in the optimal diet. This can be seen by rearranging the inequality implied by conditions (2) and (3):

$$\frac{E_i^*C_{ik} - R_{max}h_{ik}}{R_{max}h_{ik}t^* + E_c^*} > \eta_{ik}. \quad (9)$$

If $\eta_{ik} = 0$, then inequality (9) reduces to the condition that must be met in the multitactical model for a prey item to be included in the diet.

However, if $\eta_{ik} > 0$, then the value of the item has to be greater (i.e., $E_i^*C_{ik} - R_{max}h_{ik}$ has to be larger) for the item to be included in the diet. Therefore, interference forces the optimally foraging organism to pursue items of greater energetic quality than it would have to if there was no interference.

A final variation of the combined model is obtained by assuming that the predator encounter rate is a positive linear function of prey encounter rate (Figs. 4A,B). Assuming that above a threshold prey encounter rate interference with a predator when using a given tactic for a specific prey type increases, then the predator encounter rate is

$$\eta_{ij} = \frac{\lambda_i - \lambda^0_{ij}}{a_{ij}}, \text{ if } \lambda_i > \lambda^0_{ij},$$

otherwise

$$\eta_{ij} = 0,$$

where $(1/a_{ij})$ is the slope of the relationship between η_{ij} and λ_i, and λ^0_{ij} is the threshold prey encounter rate for a tactic j. This model shows that if one tactic is always better than another tactic (Fig. 4A), there will be one set of prey items that will always be taken by the better tactic, and another region where both tactics will be used, but the inferior tactic will never be used exclusively (Fig. 4C). However, if one tactic is inferior at low prey encounter rates but better at high encounter rates (Fig. 4B), then that tactic will be used exclusively at high encounter rates. The other tactic will then be used exclusively to take relatively low quality prey at relatively low encounter rates (Fig. 4D). Further, as prey encounters increase the increased cost of interference induced by high predator encounter rates can allow prey that would be taken at low encounter rates to become suboptimal. Thus, if a prey species can increase its density sufficiently to attract high numbers of predators, it can reduce predation by becoming more costly for individual predators to harvest due to high numbers of interference interactions.

DISCUSSION

VALIDITY OF MODEL ASSUMPTIONS

Optimal foraging models make many implicit and explicit assumptions. Perhaps the most important is that the predator has complete information when deciding whether or not to attack a prey item (Stephens and Krebs 1986). The information needed by a foraging bird in the models above is the energetic content of the prey item (or its average value), the time spent handling the item using different tactics, and the rate of encounter with other birds. It is unlikely that a

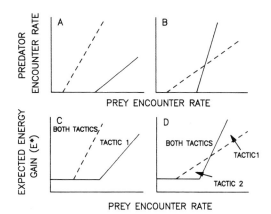

FIGURE 4. Results of the combined model when predator encounter rate is assumed to increase with increasing prey encounter rates. A. Tactic 1 (solid line) is always better than tactic 2 (dashed line) in allowing the predator to avoid other predators. B. Tactic 1 (solid line) is better than tactic 2 (dashed line) only at lower prey encounter rates. C. Prey acceptability lines resulting from conditions specified in A. D. Prey acceptability lines resulting from conditions specified in B.

bird will be able to make fine distinctions between individual prey species. In fact, it will only be profitable to make such distinctions if the energy or time loss for not doing so is larger than the actual value of the prey (Stephens and Krebs 1986:79–80). In making distinctions among edible prey items, birds may estimate energy content by evaluating prey size. To estimate handling time, birds may use the location of a prey item. For example, a lepidopteran larvae under a leaf may be difficult to remove by a perched bird, but more efficiently removed by hovering. Assessment of predator encounter rates can be made visually as the bird forages.

Empirical evidence suggests that birds do discriminate among locations of potential prey items (e.g., Holmes et al. 1979b, Maurer and Whitmore 1981, Robinson and Holmes 1982). Several studies have shown that differences in foraging reflect different prey distribution (Maurer and Whitmore 1981, Franzreb 1983a, Mannan and Meslow 1984). Although insectivorous birds probably make the kinds of distinctions among prey items that are necessary to apply foraging models, it remains to be seen whether field methods of sufficient accuracy can be developed to test model predictions.

TESTING THE MODELS

The multitactical model

This model predicts which of several foraging tactics will be used on different prey. In order to do so, several quantities must be measured. First, the expected energetic content of prey must be estimated. Usually, only relative content, such as prey size, need be measured, so prey items can be ranked. Next, it is necessary to estimate expected handling times of items for each foraging tactic. For example, capture of a lepidopteran larvae under a leaf may require less than a second while hovering, but take longer if picked from the substrate while perched. Finally, it is necessary to estimate the probability of capture of each prey type using each tactic. Based on these quantities, it should be possible to predict whether a prey will be taken, or equivalently, whether a tactic will be used to obtain prey in a specified location.

Handling times and capture probabilities are likely to vary among bird species. For example, American Redstarts (*Steophaga ruticilla*) may be more proficient at hovering beneath leaves than Scarlet Tanagers (*Piranga olivacea*). Hence, the optimal prey set should reflect species-specific behaviors, resulting in a correlation between foraging maneuvers and the types of prey taken (Robinson and Holmes 1982).

If it is possible to measure prey densities, and thus estimate encounter rates, then the multitactical model in Figure 3 can predict when a bird will switch foraging tactics. To do so, the relationship between prey density and capture probability using different techniques would need to be estimated. For example, suppose over the course of a breeding season, lepidopteran larvae under leaves increased in density. If the probabilities of capture for hovering and gleaning increased at different rates as the larvae became more common, then the multitactical model could be used to predict when birds should switch tactics. It should also be possible to design more rigorous tests of the model using laboratory experiments in which prey encounter rates are manipulated and prey are presented in ways that require different tactics.

The interference model

This model predicts quantitatively a threshold prey encounter rate, above which the costs of interference are greater than 0 (eq. [7]). The critical quantities to be measured are the time and energetic cost of interference interactions, the expected energy content and handling time of prey, the prey encounter rate, and the long term energy intake rate.

Observations of interactions among individuals in foraging flocks might be used to test the model. For such studies, the optimal solution for the model is to avoid interference interactions (see eq. [5]). Flocking species must often balance the disadvantages of flocking (which could be parameterized in terms of eqs. [6] and [8]) with advantages in minimizing risks of predation or locating rare food items.

The combined model

The most interesting aspect of this model is its prediction of the prey encounter rates at which a foraging bird should switch tactics to avoid interference from other birds. To test this model, it is necessary to estimate the quantities for both the interference and multitactical models, establish a relationship between prey density and the encounter rate with other birds and show how that relationship varies depending on the type of tactic used. The model does not require that a tactic causes a bird to encounter other birds; it simply assumes some correlation exists between the tactic used and the likelihood of encountering another bird. Thus, the foraging bird can expect to alter the amount of interference it experiences by using different prey capture techniques.

IMPLICATIONS FOR COMMUNITY STRUCTURE

How do patterns of foraging behavior influence community structure? In his classic study, MacArthur (1958) showed that five species of paruline warblers foraged in different locations in coniferous trees. This was used as evidence that the species could not outcompete one another and hence could all persist in the same habitat. These ideas led to the widespread acceptance of the idea of niche partitioning: species had to be sufficiently different in their resource use to allow them to coexist. Since that study, many workers have assumed that differences in foraging behavior are adaptations to permit coexistence (Schoener 1974).

The view of foraging behavior in this paper suggests a different emphasis. If insectivorous birds encountered prey of uniform energetic content and ease of capture, there would be no need to make foraging decisions. However, insects have a wide variety of predator avoidance tactics (e.g., Heinrich 1979c) that in effect create a great deal of spatial variation in insect abundance. Presumably, if a bird used only a single prey type, species representing that type would evolve to reduce predation, so that either the predator would have to evolve to use a different prey type, become more efficient, or go extinct. Hence, a predator should diversify its methods of taking prey so that any one set of prey types will not have too great a selection pressure. Spatial variation in insect populations can also result from predation by other species of birds, mammals, and parasitoids, and by variations in the defensive chemistry of host plants (Cates and Rhoades 1977). In the face of such spatial variation in insect prey, a multitactical strategy would allow a bird to vary its foraging behavior in response to prey dispersion.

If bird species found together have sufficiently variable foraging behaviors, then applying traditional models of community structure to bird communities may be inappropriate (Wiens 1976, 1977; Maurer 1984). In a spatially variable environment, it may be impossible for one species to exclude another. Furthermore, if interference increases with prey density, species may be subjected to many different forms of competition in different ecological settings, each with its own consequences (Maurer 1984, 1985a). Therefore, selection affecting divergence might be variable in intensity and in the phenotypic characters favored (Wiens 1976, 1977). Hence, species may not individually evolve pairwise adaptations, but rather evolve generalized adaptations allowing them to compete effectively with many species. Consequently, communities of insectivorous birds probably are not assemblages of coevolved species, but collections of species that have the right sets of adaptations that allow them to live together. In this approach to community structure, competition is a transient factor in the habitat that varies spatially and temporally in its effects on individual organisms. Community structure is determined by a hierarchical set of factors operating at different spatial and temporal scales (Maurer 1985b, 1987). At the organismic level, decisions made by individual organisms attempting to maximize their long term net energy intake determine how much energy enters the community and thus determine, in part, how the community responds as a unit to changes in its environment.

ACKNOWLEDGMENTS

I thank R. Hutto, C. J. Ralph, J. Rotenberry, D. Stephens, and R. Szaro for reading the manuscript and making suggestions for its improvement.

MEETING THE ASSUMPTIONS OF FORAGING MODELS: AN EXAMPLE USING TESTS OF AVIAN PATCH CHOICE

JOHN B. DUNNING, JR.

Abstract. Birds have been widely used to test predictions of foraging theory. The accuracy of such tests depends on whether birds feed in a manner consistent with the assumptions of the foraging model being tested. If the feeding strategy of a species does not conform to the model's assumptions, the conclusions reached from the test are weakened. I discuss whether birds in general conform to the assumptions of various models of patch choice. These models examine an organism's decision to leave a patch in which it is foraging. Many birds appear to forage in ways consistent with three of five assumptions of the marginal value theorem (MVT). The MVT assumptions most often violated by birds are those of random search and the decline of foraging success with time in patch. Several alternatives to the MVT assume that decisions to leave patches are based on simple measures of the forager's expectation of success in a patch. These "expectation models" contain less explicit assumptions than the MVT, and most assumptions are consistent with the foraging of birds; however, some assumptions are not. For example, an expectation model based on the number of prey in each patch (the "fixed-number hypothesis") assumes that all patches have the same number of prey, which is rarely true. Virtually all assumptions examined have been violated in at least some of the studies discussed; to avoid this in future studies, tests of predictions generated from patch-choice models must be designed with care.

Key Words: Assumptions; foraging models; hunting by expectation; marginal value theorem; patch choice.

Theoretical models have been widely used in conjunction with studies of the foraging behavior of birds (Krebs et al. 1983; Stephens and Krebs 1986; Stephens, this volume). Models are used to generate predictions of how organisms will respond to a given situation. The strongest test occurs when alternative models give contrasting predictions. The behavior of test organisms can then be examined to determine which model most closely predicts the observed behavior. The strength of any conclusions depends on whether the assumptions of each model are met by the test situation. If the test violates important assumptions of one of the alternative models, then the ability of the test to discriminate between models is severely weakened (Stephens, this volume).

In this paper I illustrate this concern by examining whether birds meet the assumptions of one class of foraging models, those dealing with patch choice. I demonstrate that the most prominent patch-choice model, the marginal value theorem, contains restrictive assumptions that many bird species may not meet. Several alternative models seem more appropriate because their assumptions are more consistent with avian foraging behavior.

PATCH CHOICE

A forager faces a hierarchical series of decisions (Gray 1987). It must first select a habitat for foraging, and then select among the patches within that habitat. The organism must then decide which prey items to eat in each patch and, finally, when to move to another patch. This last decision is referred to as patch choice, and models have been developed to predict the rules that foragers should use to leave patches. Because birds are easily observed, and readily maintained in captivity, they are often used to test the predictions of patch-choice models (Krebs et al. 1974; Cowie 1977; Pyke 1978a; Lima 1983, 1984; Ydenberg 1984; Ydenberg and Houston 1986).

What is a patch? A general definition that is applicable to all studies is difficult to describe because patches occur at several different scales. For instance, studies of bumblebees (*Bombus* species) have defined patches at three scales: individual flowers (Whittam 1977, Zimmerman 1983), single inflorescences within plants (Pyke 1978b, Heinrich 1979a, Haynes and Mesler 1984), and individual plants with multiple inflorescences (Zimmerman 1981, Best and Bierzychudek 1982, Hodges 1985). Whether a particular model will predict foraging behavior may depend upon the scale employed. For example, bumblebee studies that defined individual flowers as patches often support predictions of optimal foraging models, while studies at the next higher scale (within inflorescences) rarely do (pers. obs.). A general definition of patch, therefore, must be applicable at a variety of spatial scales. Stephens and Krebs (1986) define a patch as a localized search area in which there is a specified relationship between time spent and energy gained. A predator can control its intake from a patch by controlling its time spent there. Defining a patch this way distinguishes a patch from a prey item, which is assumed to be eaten entirely. Prey items therefore are assumed to yield a fixed energy gain and require a fixed handling time (Stephens and Krebs 1986). Since patches are distinguished by a specific energy-time relationship, leaving a patch is equivalent to accepting a different energy intake rate. Patches are often thought of as discrete areas of foraging substrate separated from other patches by areas which yield less energy per time. Time spent crossing these less suitable areas is defined as between-patch travel time. A forager may visit many patches during a single foraging bout; in fact, multiple-patch visitation within a bout is an important assumption of some patch-choice models.

THE MARGINAL VALUE THEOREM

The most widely studied patch-choice model is the marginal value theorem (MVT; Charnov 1976b). The derivation of the MVT has been described in detail (Pyke 1984, Stephens and Krebs 1986, Schoener 1987). The MVT proposes that foragers should leave a patch when intake rates decline to the "average capture rate for the habitat" (Charnov 1976b:132). Thus, a forager should stay in a patch as long as its foraging rate in that patch is greater than that attainable, on average, elsewhere. This prediction is derived from a model based on a series of explicit assumptions (Stephens and Krebs 1986). For the purpose of this review, five assumptions are relevant: (1) the environment is repeating; (2) organisms forage randomly; (3) organisms exhibit behavioral choice among foraging options; (4) foraging success declines with time in patch; and (5) foragers maximize net rates of energy intake. I will discuss whether birds in general, and patch-choice experiments using birds, meet these assumptions.

REPEATING ENVIRONMENT

Charnov (1976b) assumed that patch types within a habitat are distributed at random, and that foragers visit many patches of different types during a single foraging bout. This type of habitat (called a "repeating" environment since the forager repeatedly experiences the various patch types as it forages, MacArthur 1972) is a necessary assumption of the MVT because this type of environment is an assumption of renewal theory, from which the MVT is derived.

Many species probably visit a variety of patches during each foraging bout, and therefore meet the assumption of a repeating environment. Towhees, for instance, often move slowly along the ground during a foraging bout, feeding from numerous spots in the leaf litter. If the litter contains areas that vary in prey richness (Greenlaw 1969), and if the birds' behavior at each spot is considered an independent sample within the overall foraging bout (e.g., Hailman 1974, Burtt and Hailman 1979), then the towhees are sampling many patches in a repeating environment.

The foraging of other groups of birds may also conform to this assumption. Hummingbirds feed from many flowers, often on different plants, during a single bout. Some, however, concentrate on a particular plant species, reducing the types of patches visited within a bout. A third example of a repeating environment is Gibb's (1958, 1962) description of the foraging of tits on pine cones. The tits concentrated their search on a single cone at a time, and visited many cones in a foraging bout. Cones varied in prey richness both within and between trees. Thus if the cones are considered separate patches, the tits experienced a repeating environment. Other groups of birds for which the foraging habitat may be repeating include shorebirds (Goss-Custard 1970), pelicans (Brandt 1984), and egrets (R. M. Erwin 1985), all of which have been shown to move through many patches in a single feeding bout.

Sit-and-wait predators tend to survey their hunting grounds from a single spot. Shrikes and buteonine hawks often hunt from a single perch, returning repeatedly to that perch after attacking prey. These birds are not moving through patches in a conventional sense, although it can be argued that "patches" of mobile prey are moving past the predator. More active sit-and-wait predators, such as flycatchers and motmots, change perches relatively frequently. In order to meet the MVT assumption of repeating environments, bout length for sit-and-wait predators should be defined so as to ensure that patches of different types have been sampled during each bout. This could be accomplished by including many perch shifts within each bout, or by extending bout length at a single perch to include the passage of many "patches" of prey.

Experimental designs often do not incorporate a repeating environment. At artificial feeders, hummingbirds (Montgomerie et al. 1984, Pimm et al. 1985) and sparrows (Schneider 1984, Dunning 1986) usually stay at one "patch" (feeder) throughout an entire bout, even if alternative patches are provided. Early laboratory experiments with tits offered the birds artificial "trees" with patches of different prey density on the branches (Krebs et al. 1974, Cowie 1977). These designs constitute a repeating environment, since both rich and poor patches were encountered during bouts. More recent laboratory experiments have presented tits with food appearing along a conveyor belt (Ydenberg 1984, Ydenberg and Houston 1986). In these experiments the birds do not move through more than one patch per bout, unless the birds are considered sit-and-wait predators with patches of prey moving in front of them on the conveyor belt.

Because birds often forage in a manner consistent with the assumption of a repeating environment, they can be used to test predictions of the MVT. However, care must be taken that the experimental design of such tests, especially laboratory studies, incorporate a repeating environment.

RANDOM SEARCH

The MVT assumes that foragers are as likely to return to a previously examined patch as they are to move to a new patch; that is, they search randomly (Green 1987). An alternative is systematic search, in which the probability of visiting a previously-searched patch is reduced (Kamil 1978, Baum 1987). Most birds that feed on dispersed prey are systematic searchers and tend not to recross their foraging path (Cody 1971, Lima 1983, Eichinger and Moriarty 1985). Similarly, if prey are strongly clumped, birds often return repeatedly to patches that were particularly profitable (Smith and Dawkins 1971, Krebs 1974, Zach and Falls 1976b). Thus, birds feeding on either dispersed or strongly clumped prey are usually not random searchers. Hummingbirds, the subject of many early patch-choice experiments, also show nonrandom search, since many species often move to patches of flowers in a regular, repeated order, skipping many available flowers (traplining, *sensu* Feinsinger 1976).

I know of no studies that have demonstrated (or imposed) random search with their avian subjects. Birds are not the only group that violate this assumption. Other common subjects for patch-choice experiments also forage nonrandomly, including bees (Thomson et al. 1982, Heinrich 1979b, Marden 1984, Wetherwax 1986) and invertebrate stream predators (Waage 1979).

This is an important deviation from the assumptions of the MVT for two reasons. First, systematic searchers experience a constant rate of finding prey within a patch, since they do not search areas already depleted (Green 1987). The MVT assumes that rates of finding prey within patches decrease exponentially, and that this decreased rate of success triggers the decision to leave a patch (see below). Second, foragers using nonrandom search often should leave patches using different rules than do foragers searching randomly (Green 1987). This difference between the rules may be quantitative in some cases, but the predicted rules can be qualitatively very different. Since the way foragers search within patches can have a major effect on the predictions being tested (Green 1987), species that search systematically seem inappropriate for testing MVT predictions.

BEHAVIORAL CONTROL

An implicit assumption of the MVT is that the forager is able to forage in more than one patch type, and has the ability to compare intake rates in order to decide when to abandon a patch. In other words, the predicted response is assumed to be within the behavioral repertoire of the forager. That this assumption is true for most birds seems trivial; however, an instructive example exists within the patch-choice literature.

Bumblebees collect nectar from inflorescences in a stereotyped manner: they start at the lowest flower, move straight up the inflorescence, and quit before reaching the top. Since the lowest flowers usually hold the most nectar, this strategy seems consistent with energy maximization, and was cited as confirmation of one optimal foraging model (Pyke 1978b). However, the same behavior is used by bees on inflorescences in which the bottom flowers do not have the most nectar, and in flowers in which the nectar gradient was experimentally reversed (Waddington and Heinrich 1979, Best and Bierzychudek 1982, Corbet et al. 1981). In fact, pollen-collecting bumblebees also move from the bottom up, even though pollen levels are highest in the topmost flowers (Haynes and Mesler 1984). The stereotyped path taken by bees appears to be an invariant response, shaped not just by distributions of nectar, but also by the position assumed by bees while foraging and the need to reduce revisits to the same flower. Since the bees are apparently not responding to differences in intake rates, this system is not really appropriate for testing MVT models.

Many birds show behavioral flexibility in their foraging repertoire and quickly adapt their strategy to take advantage of temporary or novel sources of food. Some species are quite stereotyped, however. The avian equivalent of the bumblebee might be the Brown Creeper (*Certhia americana*). This surface-gleaner flies to the bottom of tree trunks, moves upward as it searches for insects among the cracks in bark, and leaves the trunk before reaching the top. Although this search pattern might be an optimal response to some particular distribution of insects on the surface of the tree, it is more probably a consequence of the posture adopted by the bird while feeding, and the potential interference from branches at the top of the tree (Franzreb 1985). The behavioral repertoire of the bird must be considered carefully when designing patch-choice experiments using stereotyped foragers such as Brown Creepers, crossbills (Benkman 1987a), or wintering Worm-eating Warblers (*Helmitheros vermivorus,* a dead-leaf specialist, Greenberg 1987b).

Interspecific comparisons of patch-choice behavior can be affected by the degree of foraging specialization shown by the species being compared. Species whose foraging is relatively specialized may not respond to changes in foraging success in the same manner as generalist foragers. Thus, specialist and generalist species may show differences in foraging not predicted by a model which does not incorporate such variation. I found an example of such differences in my study of patch choice in towhees (Dunning 1986). I placed individuals of three towhee species in an artificial foraging arena that contained patches of different litter types. The towhees preferred to feed in one litter type, initially ignoring other available litters. I used this preference to examine foraging success rates before and after birds changed patches.

In each of a series of trials spread over consecutive mornings, a bird had a choice of feeding on a variable amount of seed under a preferred litter type, or a constant, abundant amount of seed under a non-preferred litter. Preferred and non-preferred litters were selected for each individual bird during preliminary trials. Initially 35 g of seed were available under the litter in both patches at the start of each trial. Each morning I reduced the amount of seed under the preferred litter by 5 g per trial, while maintaining the abundant levels under the non-preferred litter. Thus each bird experienced increasingly lower seed densities if it remained in the preferred litter. Eventually all birds switched to using the non-preferred litter.

The three species differed in the timing of the switch from preferred to non-preferred litter. Canyon Towhees (*Pipilo fuscus*), a foraging generalist found in Arizona in relatively food-poor desert washes and canyons, started using the non-preferred patches relatively quickly in the series of trials (Fig. 1). I also studied Rufous-sided Towhees (*P. erythrophthalmus*), montane sparrows of oak and pine-oak woodlands, and Abert's Towhees (*P. aberti*), which are restricted to desert riparian systems in southern Arizona. Each of these two species depends more on a specialized foraging technique, double-scratching in leaf litter (C. J. Harrison 1967), than does the Canyon Towhee (Davis 1957; Marshall 1960, pers. obs.). All individuals of the two relatively specialized species began using the non-preferred patches on trials later than the slowest-switching Canyon Towhee.

The differences in timing of patch switching among the species were not due to interspecific differences in foraging success rates experienced in the trials (Dunning 1986). Instead, I believe that the two more specialized species reacted in a qualitatively different manner to the changes in foraging success within the preferred patch, perhaps by using different decision rules (Dun-

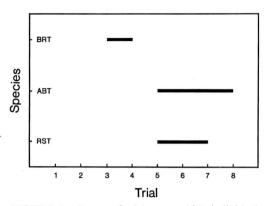

FIGURE 1. Range of trials over which individual towhees switched from use of preferred litter to use of non-preferred litter. BRT = Canyon Towhee, ABT = Abert's Towhee, RST = Rufous-sided Towhee. Notice that the range for Canyon Towhee did not overlap that of the other two species.

ning, in prep.). Since different decision rules can lead to different predictions of patch-choice behavior (Green 1987), my ability to test specific patch-choice predictions through comparison between towhee species was reduced. Studies comparing foraging behavior between species should therefore consider the effect of interspecific differences in foraging specialization on the predictions being tested.

DECREASING REWARD WITH TIME IN PATCH

The MVT assumes that as a predator stays in a patch, it depletes the patch of food items, and correspondingly, its intake rate declines. This decline is crucial to the model, since it is the decline in intake rate that motivates the forager to move to another patch. Birds may not conform to this assumption in several ways. As noted before, systematic searchers (such as many birds) may not experience a decrease in intake rate during search within a patch. In addition, not all patch types would be expected to be depleted by predation, regardless of the type of searching by the predator. For example, prey density in the area scanned by a sit-and-wait predator is variable with time since mobile prey move in and out of the predator's range. Also, a predator in a patch with superabundant resources might not have its intake rate decrease as it feeds. Hummingbirds, for example, feeding at large artificial feeders probably experience no decrease in intake rate until the feeder is drained. A forager feeding on superabundant resources may leave a patch only when satiated.

Data demonstrating that capture success decreases with time in a patch for wild birds are limited. Capture success may have declined with time for wild Snowy Egrets (*Egretta thula*), but did not for Great Egrets (*Casmerodius alba*) (R. M. Erwin 1985). Intercapture intervals at the end of foraging bouts by American Kestrels (*Falco sparverius*) are longer than earlier intervals, implying that foraging success declined during the bout (Rudolph 1982). However, inspection of the data presented by Rudolph shows that intercapture intervals actually were constant throughout most of the bout, dropping only at the end. This same pattern has been shown in laboratory studies of White-throated Sparrows (*Zonotrichia albicollis*; K. Johnson, pers. comm.).

Sparrows feeding on patches of seed deplete the seed levels with time, eventually prompting patch switching (Schneider 1984, Dunning 1986). However, this depletion may occur over many foraging bouts, and so is qualitatively different than the depletion assumed by the MVT. (Note that with the assumption of randon search, a forager is unlikely to return to one patch repeatedly until depletion.) Foragers showing this type of systematic search may be using an expectation patch-choice rule to determine when to stop returning to previously-used patches (Dunning 1986).

The use of microcomputers in laboratory studies to control delivery of food items allows the researcher to control the depletion of patches (Ydenberg 1984, Kamil et al. 1985, Ydenberg and Houston 1986, Hanson 1987). In these studies, a bird initiates the beginning of a foraging bout by landing on a feeding perch, or striking a control key. With the initiation of the bout, food is delivered at a decreasing rate until the bird ends the bout. Delivery rates are reset to the original starting rate for the initiation of the next bout. Thus, in certain controlled situations, use of this kind of apparatus can ensure that intake rates decline when a forager stays in a patch.

The importance of meeting this assumption may vary with the specific test of the MVT. Species which feed on superabundant resources (e.g., hummingbirds at feeders) are clearly not suitable for MVT tests, since their feeding rates are constant over time. Species which experience a constant intake rate initially upon entering a patch may conform to a modified version of the MVT (the "combined patch and prey model," Stephens and Krebs 1986). Predictions of Charnov's (1976b) version of the MVT require a forager that experiences a decreasing intake rate as it feeds in a patch; testing the model with foragers that do not meet this requirement weakens the conclusions reached from the test.

MAXIMIZATION OF ENERGY INTAKE

The most prominent assumption incorporated into the MVT is that organisms seek to maximize their net rate of energy intake (E_n). The role that this assumption has played in the development of foraging theory is considered at length by Stephens (this volume). I would like to add one point to his discussion. Strictly speaking, organisms should seek to maximize fitness, and the maximization assumption essentially assumes that maximizing E_n is the short-term equivalent of maximizing fitness (Sih 1982). This should not be accepted automatically in all cases, as the following example demonstrates.

A field test of the relationship between fitness and intake in birds examined diet and reproductive success in breeding Herring Gulls (*Larus argentatus*; Pierotti and Annett 1987). This study addressed which habitat a gull should forage in, and which prey items to eat. These are different hierarchical foraging decisions than patch choice; however, I discuss the study here because it is a particularly elegant example of how energy maximization may not maximize fitness. Individual gulls feeding in different areas specialized on mussels, garbage, or storm-petrels. Garbage provided the greatest E_n, while mussels provided the least. In spite of this, mussel specialists fledged more than double the number of young fledged by garbage specialists. Pierotti and Annett suggest that mussels provided limiting nutrients to the egg-laying females, allowing them to lay more clutches and hatch more young. Thus, in this system, gulls with the highest intake rates did not have the highest fitness.

Nevertheless, some studies examining use of patches have demonstrated that some birds adopt a strategy that seems to maximize intake rates. Ydenberg and Houston (1986) compared intake rates of captive tits at the start of foraging periods with the rates of the same birds later in the period. The birds' combination of handling, travel and patch residence times at the start of each period maximized intake rates relative to other combinations of these variables. Intake rates declined later in the period due to conflicting demands for territorial defense. Studies of diet composition of birds are sometimes able to demonstrate that observed diets conform to the energy maximization assumption of optimal diet theory (e.g., Pulliam 1980, Benkman 1987b). Fewer studies of patch choice examine the assumption of E_n maximization, perhaps because of the difficulty with which intake rates in a variety of patches can be estimated.

Montgomerie et al. (1984) investigated which of two functions were maximized by hummingbirds. They suggested that in most situations, maximizing net energy per volume consumed (NEVC) would yield more energy than maximizing E_n. They devised a test which showed that hummingbirds preferred patches that yielded high NEVC over those yielding high E_n. However, Montgomerie et al. concluded that both functions would be maximized simultaneously in most situations.

Some studies examining energy maximization as a foraging goal have been unable to demonstrate that this goal is attained (see Stephens, this volume). However, it is now recognized that most organisms face multiple demands, and it may be rare that a forager can adopt a strategy of unconstrained energy maximization. Recent developments in foraging theory have added realistic constraints to foragers' ability to maximize intake rates, and examine how these constraints change patch-choice decisions (Caraco 1982; Getty and Krebs 1985; McNamara and Houston 1985; Lima and Valone 1986; Stephens, this volume). These extensions of the original patch-choice models may be more useful in examining patch choice in species for which the maximization assumption is not valid. Testing of models assuming energy maximization is still important, however, because deviations from such models' predictions can identify important constraints.

ALTERNATIVE MODELS OF PATCH CHOICE

Fixed-number and Fixed-time Hypotheses

The main alternatives to the MVT have been models in which the decision to leave a patch is based on the forager's expectation of success. These expectations of success are based on simple measures of the environment, such as the number of prey in a patch or the amount of time normally needed to deplete a patch. The first such "expectation models" were the fixed-number and fixed-time hypotheses (Gibb 1962, Krebs 1973, Krebs et al. 1974), which proposed that foragers stay in a patch until they found a certain number of prey (fixed-number) or until a certain time had elapsed (fixed-time).

The fixed-number hypothesis was suggested by Gibb (1962) to explain certain unusual foraging characteristics shown by wintering tits. The only explicit assumption of this hypothesis is that patches contain a specific number of prey, such that the forager can develop an accurate expectation of the number of prey within a patch. This expectation could be learned from past experience, or be genetically programmed by natural selection. It is unlikely that wild birds routinely forage on prey distributed in such a regular manner. In fact, Gibb's earlier studies (1958, 1960) demonstrated that the resource base of the tits, for which Gibb originally suggested the fixed-number hypothesis, did not meet this assumption (Krebs 1973, Krebs et al. 1974). Green (1987) proposed that a fixed-number strategy is optimal for foragers that search randomly for regularly-distributed prey, but systematic searchers should vary their patch-leaving rules as time in a patch increases. Since most birds probably violate both the fixed-number hypothesis' assumption of regular prey distribution, and the related implicit assumption of random search, birds are not appropriate for testing predictions of this model. I know of no experimental study with birds that supports a fixed-number hypothesis.

The fixed-time hypothesis was proposed by Krebs (1973) to explain the foraging patterns observed by Gibb. Krebs suggested that Gibb's tits stayed in each patch for a set period of time, rather than until a set number of prey were captured, basing this suggestion on the logic that animals are better at measuring time than at counting (Krebs et al. 1974). Recent theoretical models have demonstrated that leaving a patch after a fixed time can be a profitable strategy when prey are distributed randomly among patches (i.e., when prey have a Poisson distribution; Iwasa et al. 1981, Green 1987). This is true regardless of whether the forager searches randomly or systematically; however, random searchers should quit patches before systematic searchers, all else being equal (Green 1987).

Although many prey types for avian foragers may be strongly clumped (i.e., some seed types, fish schools, flowers), some prey types may approximate a Poisson distribution. Prey of foliage-gleaners, for instance, may be randomly distributed among leaves of a tree. Several studies of birds have at least partially supported a fixed-time hypothesis. Krebs et al. (1974) were unable to conclusively reject the hypothesis that tits in an artificial arena were leaving patches at a fixed time. Although the tits tended to spend more time in richer patches (a result allowing rejection of the fixed-time hypothesis if significant) in four different experiments, the differences between patches were significant in only one experiment. Since the birds tended to spend more time in richer patches, Krebs et al. concluded that an optimal foraging model based on giving-up times was better supported by the data (but see McNair 1982 for a reinterpretation of the data).

Zach and Falls (1976d) examined the movements of Ovenbirds (*Seiurus aurocapillus*) hunting for dead flies on an artificial feeding board. They compared predictions of a fixed-number hypothesis, a fixed-time hypothesis, and a giving-up time hypothesis based on the MVT. The fixed-number and giving-up time predictions were clearly rejected. Search by the Ovenbirds violated both the random search and probably the patch depletion assumptions of the MVT. One of three predictions from a fixed-time hypothesis was supported. Zach and Falls conclud-

TABLE 1. EVIDENCE THAT BIRDS USE PRIOR KNOWLEDGE FROM PREVIOUS WITHIN-PATCH FORAGING TO DETERMINE CURRENT FORAGING STRATEGY. PRIOR KNOWLEDGE IS INFORMATION POTENTIALLY AVAILABLE TO BIRDS FROM PREVIOUS EXPERIENCE. EVIDENCE OF EXPECTATION ARE OBSERVED BEHAVIORS CONSISTENT WITH USE OF WITHIN-PATCH EXPERIENCE

Source	Organism	Prior knowledge	Evidence of expectation
Smith and Dawkins (1971)	Great Tit	Estimate of variability in patch quality	Tits showed time lag in response to changes in patch richness.
Smith and Sweatman (1974)	Great Tit	Location of four patches of variable quality	When locations of richest and poorest patches were switched, tits moved to second richest patch.
Zach and Falls (1976b)	Ovenbird	Location and quality of patches	Search patterns based on previous, not current, prey distributions.
Lima (1983)	Downy Woodpecker	Distribution of seeds in feeders	Search pattern based on previous seed distributions; time lags before changes in searching to match current seed distributions.
Dunning (1986)	Towhees	Quality of two types of leaf litter	Birds left preferred patches after large decrease from previous within-patch success.

ed the data weakly supported a time expectation model.

Valone and Brown (unpubl.) used the amount of seed left in a patch after visits by a forager (the "giving-up density," J. S. Brown 1986) to compare predictions of four foraging strategies potentially used by a variety of desert granivores. The giving-up densities of Gambel's Quail (*Callipepla gambelii*) and possibly of Mourning Doves (*Zenaida macroura*) were most consistent with predictions of a fixed-time hypothesis.

WITHIN-PATCH HYPOTHESIS

Another expectation hypothesis proposes that a forager changes patches when success in the current patch has declined to a threshold (as envisioned by the MVT), but that the thresholds are based only on the forager's past and present experience within its current patch type. I call this the within-patch hypothesis, since only information gained by the forager in a single patch type is used to determine when to leave. Previous foraging within the patch type establishes an expected rate of intake from patches of that type. The forager leaves its current patch when success there drops below expectations. A forager's expectation (and therefore its threshold) is altered by large scale changes in patch characteristics, since expectations are updated with new information during each foraging bout. When success drops below the threshold, the forager leaves the patch and samples other available patches to determine if a higher rate is available elsewhere.

This hypothesis is very similar to the MVT. In both models, foragers are assumed to be able to monitor current success, and to compare this success with a threshold rate based on past experience. The main difference between the two models is that the MVT predicts thresholds based on experience in all available patch types, usually in the form of an average habitat success rate. Thus, the MVT predicts that foragers compare success rates between patch types, while the within-patch hypothesis predicts that patch switching is based on changes in foraging success within the same patch type. The within-patch hypothesis may be more realistic for organisms that forage in complex or variable habitats, in which the information needed to make comparisons of foraging success between all potential patches may require an omniscient forager.

The major explicit assumption of the within-patch hypothesis is that foragers respond primarily to changes in prey distribution or foraging rates within patches. There is extensive evidence supporting this assumption (Table 1). I will discuss two studies as examples. Lima (1983, 1984) examined the use of artificial feeding logs by wild Downy Woodpeckers (*Picoides pubescens*). He found that, with experience, the woodpeckers matched their searching pattern to the distribution of prey in the logs. The birds generally started at the bottom of each log, and moved up, leaving when their expectation of prey distribution indicated that no more prey were available. When Lima abruptly changed the prey distribution, the woodpeckers continued to search the logs in a manner consistent with the previous prey distribution, clearly showing the birds were reacting to their expectations of where the prey were. Within several days (the time frame of the trials), the woodpeckers altered their foraging to match the new prey distributions.

In my study of patch use by captive towhees, I looked specifically for evidence of thresholds based on between-patch or within-patch com-

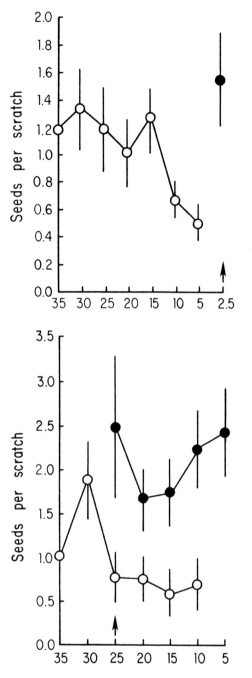

FIGURE 2. Foraging efficiency per trial for two representative towhee individuals: Upper graph is data for an Abert's Towhee; lower graph is data for a Canyon Towhee. Open circles = seeds per scratch in preferred litter; closed circles = seeds per scratch in non-preferred litter. Vertical lines = 95% confidence limits. Trials 1–8 are listed on the x-axis by the amount of seed present in the preferred litter patch during that trial; arrow indicates trial during which bird switched patches. Notice that switching followed large drops in foraging efficiency in preferred patch.

parisons of foraging success (Dunning 1986, in prep.). As described before, individual towhees of three species were placed in a foraging arena where they could feed on a variable amount of seed under a preferred litter type, or on a constant, abundant amount of seed under a non-preferred litter. All birds initially concentrated their foraging in the preferred patch, then eventually shifted their foraging to the non-preferred patch. If the towhees were comparing success rates in both patches, then the birds should have switched patches when success in the preferred litter dropped below that available in the non-preferred patch. None of the 12 birds that I observed fit this pattern (see examples in Fig. 2). Instead, changes in patch choice appeared to be associated with large drops in success within the preferred patch itself. This was especially true for Canyon Towhees (Fig. 2b), since three of four individuals of this species switched at the first significant drop in within-patch success (Dunning 1986). Since the patches were adjacent to each other within the foraging arena, and did not appear to differ in exposure or any other factor, I concluded that the birds were not using information from previous trials on success rates in non-preferred patch to determine when to leave the preferred litter. Patch choice by the towhees was most consistent with a decision rule based on changes in within-patch foraging success.

SYNOPSIS

Models of patch choice predict that foragers leave patches based on particular decision rules. These models incorporate assumptions about how foragers behave. The suitability of a particular species for testing predictions of these models depends on how well the foraging of the organism meets the assumptions (Maurer, this volume). Care must be taken in designing experiments that test the predictions of theoretical models, because the conclusions are weakened if underlying assumptions are violated.

Of the assumptions considered in this paper, the one that is probably met least often by birds is the random search assumption of the MVT and (implicitly) the fixed-number hypothesis. Since many bird species studied do not search patches randomly, birds may not be appropriate test organisms for MVT predictions. Green (1987) suggests the theoretical emphasis on random foraging is misplaced, and develops several patch decision rules based on systematic search. These rules deserve empirical testing.

A variety of birds also do not appear to forage in a manner consistent with the MVT assumption that foraging success declines with time in patch. Patch-choice in these species may be better studied by considering modified versions of the MVT that reflect different patterns of how intake rate changes with time in a patch (Stephens and Krebs 1986).

Some expectation models of patch choice incorporate assumptions which may be realistic for many bird species. The fixed-time hypothesis is most appropriate for birds that feed on randomly-dispersed prey, while the within-patch hypothesis was proposed for foragers that use decision rules based on within-patch changes in success. The latter model may be especially appropriate for birds in habitats that change rapidly, where information from all patches in the habitat may be difficult to gather. The uniform environment assumption of the fixed-number hypothesis makes this expectation model less applicable to birds.

One area currently being explored is the incorporation of realistic constraints into patch-choice models (Stephens, this volume). Constraints have been added to optimal diet theory, leading to a better understanding of diet selection under realistic conditions. Constraints on foragers' ability to collect and use information, for instance, lead to predictions of partial preferences in diet selection, a commonly observed phenomenon (Getty and Krebs 1985, McNamara and Houston 1987b). Incorporation of realistic constraints into optimal patch theory is leading to better understanding of time lags in patch switching (Lima 1984) and sampling strategies (Lima 1985).

Birds have proven useful for testing a variety of foraging models. As illustrated by these models of patch choice, careful consideration of the assumptions underlying theoretical models will improve a researcher's ability to use birds, or any other organism, to understand foraging better through the use of theory.

ACKNOWLEDGMENTS

I thank Brent Danielson, Michael Kaspari, Ron Pulliam, Richard Hutto, C. J. Ralph, Joseph Jehl, and Michael Morrison for constructive comments on this paper. The Institute of Ecology, University of Georgia, provided logistic and financial support during the preparation of the manuscript.

LITERATURE CITED

ABBOTT, I. 1980. Theories dealing with the ecology of landbirds on islands. In A. MacFadyen [ed.], Advances in ecological research. Vol. 2. Academic Press, New York.

ABBOTT, I., AND P. VAN HEURCK. 1985. Tree species preference of foraging birds in jarrah forest of Western Australia. Aust. Wildl. Res. 12:461-466.

ABBOTT, I., L. K. ABBOTT, AND P. R. GRANT. 1977. Comparative ecology of Galapagos ground finches (Geospiza Gould): evaluation of the importance of floristic diversity and interspecific competition. Ecol. Monogr. 47:151-184.

ACHINEWHU, S. C., AND D. HEWITT. 1979. Assessment of the nutritional quality of proteins: the use of "ileal" digestibilities of amino acids as measures of their availabilities. Br. J. Nutr. 41:559-571.

ADAMS, J. 1985. The definition and interpretation of guild structure in ecological communities. J. Anim. Ecol. 54:43-59.

AFIFI, A. A., AND V. CLARK. 1984. Computer-aided multivariate analysis. Wadsworth, Inc., Belmont, CA.

AIROLA, D. A., AND R. H. BARRETT. 1985. Foraging and habitat relationships of insect-gleaning birds in a Sierra Nevada mixed-conifer forest. Condor 87:205-216.

AITCHISON, J. 1981. A new approach to null correlations of proportions. Math. Geol. 13:175-189.

AITCHISON, J. 1983. Principal component analysis of compositional data. Biometrika 70:57-65.

ALATALO, R. V. 1978. Resource partitioning in Finnish woodpeckers. Ornis Fenn. 55:49-59.

ALATALO, R. V. 1980. Seasonal dynamics of resource partitioning among foliage-gleaning passerines in northern Finland. Oecologia 45:190-196.

ALATALO, R. V. 1981. Interspecific competition in tits Parus spp. and the Goldcrest Regulus regulus: foraging shifts in multispecific flocks. Oikos 37:335-344.

ALATALO, R. V. 1982. Multidimensional foraging niche organization of foliage gleaning birds in northern Finland. Ornis Scand. 13:56-71.

ALATALO, R. V., AND R. H. ALATALO. 1979. Resource partitioning among a flycatcher guild in Finland. Oikos 33:46-54.

ALDRICH, S. A. 1984. Ecogeographic variation in size and proportions of the Song Sparrow. Ornithol. Monogr. 35:1-134.

ALERSTAM, T., AND P. H. ENCKELL. 1979. Unpredictable habitats and evolution of bird migration. Oikos 33:228-232.

ALI, A., B. H. STANLEY, AND P. K. CHAUDHURI. 1986. Attraction of some adult midges (Diptera: Chironomidae) of Florida to artificial light in the field. Florida Entomol. 69:644-650.

ALLAIRE, P., AND C. D. FISHER. 1975. Feeding ecology of three resident sympatric sparrows in eastern Texas. Auk 92:260-269.

ALLARD, H. A. 1937. Some observations in the behavior of the periodical cicada Magicicada septendecim L. Am. Nat. 71:588-609.

ALLDREDGE, J. R., AND J. T. RATTI. 1986. Comparison of some statistical techniques for analysis of resource selection. J. Wildl. Manage. 50:157-165.

ALLEN, J. A. 1974. Further evidence for apostatic selection by wild passerine birds: training experiments. Heredity 33:361-372.

ALLEN, T. F. H., AND T. B. STARR. 1982. Hierarchy: perspectives for ecological complexity. Univ. Chicago Press, Chicago, IL.

ALTMANN, J. 1974. Observational study of behavior: sampling methods. Behaviour 49:227-267.

AMERICAN ORNITHOLOGISTS' UNION. 1983. Check-list of North American birds. 6th ed. Amer. Ornithol. Union, Washington, DC.

AMLANER, C. J., JR., AND D. W. MACDONALD. 1980. A handbook on biotelemetry and radio tracking. Pergamon Press, Oxford.

ANDERSON, J. R., AND R. W. MERRITT. 1977. The impact of foraging meadowlarks, Sturnella neglecta, on the degradation of cattle dung pads. J. Applied Ecol. 14:355-362.

ANDERSON, J. W., AND S. R. BRIDGES. 1988. Dietary fiber content of selected foods. Am. J. Clin. Nutr. 47:440-447.

ANDERSON, T. R. 1977. Reproductive responses of sparrows to a superabundant food supply. Condor 79:205-208.

ANDERSSON, M. 1980. Nomadism and site tenacity as alternative reproductive tactics in birds. J. Anim. Ecol. 49:175-184.

ANDERSSON, M., AND J. R. KREBS. 1978. On the evolution of hoarding behaviour. Anim. Behav. 26:707-711.

ANDRLE, R. F., AND P. R. ANDRLE. 1976. The Whistling Warbler of St. Vincent, West Indies. Condor 78:236-243.

ANGERS, C. 1979. Sample size estimation for multinomial populations. Am. Stat. 33:163-164.

ANGERS, C. 1984. Large sample sizes for the estimation of multinomial frequencies from simulation studies. Simulation 43:175-178.

ANNETT, C., AND R. PIEROTTI. 1984. Foraging behavior and prey selection of the leather seastar Dermasterias imbricata. Marine Ecol. Prog. Ser. 14:197-206.

ANSCOMBE, F. J. 1948. The transformation of poisson, binomial, and negative binomial data. Biometrika 35:246-254.

ARNOLD, S. J. 1981. The microevolution of feeding behavior, p. 409-453. In A. C. Kamil and T. D. Sargent [eds.], Foraging behavior: ecological, ethological, and psychological approaches. Garland STPM Press, New York.

ARNOLD, S. J., AND M. J. WADE. 1984. On the measurement of natural and sexual selection: applications. Evolution 38:720-734.

ASHMOLE, N. P., AND S. H. TOVAR. 1968. Prolonged parental care in Royal Terns and other birds. Auk 85:90-100.

ASKENMO, C., A. VON BROMSSON, J. EKMAN, AND C. JANSSON. 1977. Impact of some wintering birds on spider abundance in spruce. Oikos 28:90-94.

ASKINS, R. A. 1981. Survival in winter: the importance of roost holes to resident birds. Loon 53:179-184.

ASKINS, R. A. 1983. Foraging ecology of temperate and tropical woodpeckers. Ecology 64:945-956.

AULT, S. C., AND F. A. STORMER. 1983. A suction device for sampling seed availability. Wildl. Soc. Bull. 11:63-65.

AUSTIN, G. T. 1976. Sexual and seasonal differences in foraging of Ladder-backed Woodpeckers. Condor 78:317-323.

AUSTIN, G. T., AND E. L. SMITH. 1972. Winter foraging ecology of mixed insectivorous bird flocks in oak woodland in southern Arizona. Condor 74:17-24.

AUSTIN, M. P. 1985. Continuum concept, ordination methods and niche theory. Ann. Rev. Ecol. Syst. 16:39-61.

AVERY, M. I., AND J. R. KREBS. 1984. Temperature and foraging success of Great Tits Parus major hunting for spiders. Ibis 126:33-38.

BAIRD, J. W. 1980. The selection and use of fruit by birds in an eastern forest. Wilson Bull. 92:63-73.

BAIRLEIN, F. 1985. Efficiency of food utilization during fat deposition in the long distance migratory Garden Warbler, Sylvia borin. Oecologia 68:118-125.

BAKER, H. G., AND I. BAKER. 1983. Floral nectar sugar constituents in relation to pollinator types, p. 117-141. In C. E. Jones and R. J. Little [eds.], Handbook of experimental pollination biology. Van Nostrand Reinhold, New York.

BAKER, J. A. AND R. J. BROOKS. 1981. Distribution patterns of raptors in relation to density of meadow voles. Condor 83:42-47.

BAKER, M. C. 1974. Foraging behavior of Black-bellied Plovers (Pluvialis squatarola). Ecology 55:162-167.

BAKER, M. C., AND A. E. M. BAKER. 1973. Niche

relationships among six species of shorebirds on their wintering and breeding ranges. Ecol. Monogr. 43:193-212.

BAKER, M. C., C. S. BELCHER, L. C. DEUTSCH, G. L. SHERMAN, AND D. B. THOMPSON. 1981. Foraging success in junco flocks and the effects of social hierarchy. Anim. Behav. 29:137-142.

BAKER, R. R. 1970. Bird predation as a selective pressure on the immature stages of cabbage butterflies, Pieris rapae and P. brassicae. J. Zool. (Lond.) 162:43-59.

BALDA, R. P. 1969. Foliage use by birds of the oak-juniper woodland and ponderosa pine forest in southeastern Arizona. Condor 71:399-412.

BALDWIN, I. T., AND J. C. SCHULTZ. 1983. Rapid changes in tree leaf chemistry induced by damage: evidence for communication between plants. Science 221:277-278.

BALGOOYEN, T. G. 1971. Pellet regurgitation by captive Sparrow Hawks (Falco sparverius). Condor 73:382-385.

BALLING, S. S., AND V. H. RESH. 1982. Arthropod community response to mosquito control ditches in San Francisco Bay salt marshes. Environ. Entomol. 11:801-808.

BALLINGER, R. E. 1977. Reproductive strategies: food availability as a source of proximal variation in a lizard. Ecology 58:628-635.

BALPH, D. F., AND M. H. BALPH. 1983. On the psychology of watching birds: the problem of observer-expectancy bias. Auk 100:755-757.

BALPH, D. F., AND H. C. ROMESBURG. 1986. The possible impact of observer bias on some avian research. Auk 103:831-832.

BALPH, M. H. 1975. Wing length, hood coloration and sex ratio in Dark-eyed Juncos wintering in northern Utah. Bird-Banding 46:126-130.

BALPH, M. H. 1979. Flock stability in relation to dominance and agonistic behavior in wintering Dark-eyed Juncos. Auk 96:714-722.

BALPH, M. H., L. C. STODDARD, AND D. F. BALPH. 1977. A simple technique for analyzing bird transect counts. Auk 94:606-607.

BARKAN, C. P. L., AND M. L. WITHIAM. In press. A test of the simultaneous encounter model of prey choice in Black-capped Chickadees. Am. Nat.

BARKAN, C. P. L., J. L. CRAIG, S. D. STRAHL, A. M. STEWART, AND J. L. BROWN. 1986. Social dominance in communal Mexican Jays, Aphelocoma ultramarina. Anim. Behav. 34:175-187.

BARKER, R. D. 1986a. An investigation into the accuracy of herbivore diet analysis. Aust. Wildl. Res. 13:559-568.

BARKER, R. D. 1986b. A technique to simplify herbivore diet analysis. Aust. Wildl. Res. 13:569-573.

BARLOW, J. S. 1964. Fatty acids in some insect and spider fats. Can. J. Biochem. 42:1365-1374.

BARNARD, C. J., AND C. A. J. BROWN. 1985. Risk-sensitive foraging in common shrews (Sorex araneus L.). Behav. Ecol. Sociobiol. 16:161-164.

BARNARD, D. R. 1980. Effectiveness of light-traps for assessing airborne Culicoides variipennis populations. J. Econ. Entomol. 73:844-846.

BARNES, G. G., AND V. G. THOMAS. 1987. Digestive organ morphology, diet, and guild structure of North American Anatidae. Can. J. Zool. 65:1812-1817.

BARNETT, S. A. 1958. Experiments on "neophobia" in wild and laboratory rats. Br. J. Psych. 49:195-201.

BARNETT, S. A., AND P. E. COWAN. 1976. Activity, exploration, curiosity, and fear: an ethological study. Interdisc. Sci. Rev. 1:43-62.

BASAWA, I. V., AND B. L. S. PRAKASA RAO. 1980. Statistical inference for stochastic processes. Academic Press, New York.

BAUM, W. M. 1987. Random and systematic foraging, experimental studies of depletion, and schedules of reinforcement, p. 587-607. In A. C. Kamil, J. R. Krebs, and H. R. Pulliam [eds.], Foraging behavior. Plenum Press, New York.

BAYER, R. D. 1985. Bill length of herons and egrets as an estimator of prey size. Colonial Waterbirds 8:104-109.

BEALS, E. W. 1960. Forest bird communities in the Apostle Islands of Wisconsin. Wilson Bull. 72:156-181.

BEALS, E. W. 1973. Mathematical elegance and ecological naivete. J. Ecol. 61:23-35.

BEALS, E. W. 1984. Bray-Curtis Ordination: an effective strategy for analysis of multivariate ecological data. Adv. Ecol. Res. 14:1-55.

BEAMER, R. H. 1931. Notes on the 17-year cicada in Kansas. J. Kansas Entomol. Soc. 4:53-58.

BEAVER, D. L., AND P. H. BALDWIN. 1975. Ecological overlap and the problem of competition and sympatry in Western and Hammond's flycatchers. Condor 77:1-13.

BECHARD, M. J. 1982. Effect of vegetative cover on foraging site selection by Swainson's Hawk. Condor 84:153-159.

BEDNARZ, J. C. 1988. Cooperative hunting in Harris' hawks (Parabuteo unicinctus). Science 239:1525-1527.

BEECHER, W. J. 1951. Adaptations for food getting in the American blackbirds. Auk 68:411-440.

BEECHER, W. J. 1978. Feeding adaptations and evolution in the starlings. Bull. Chicago Acad. Sci. 11:269-298.

BECHINSKI, E. J., AND L. P. PEDIGO. 1982. Evaluation of methods for sampling predatory arthropods in soybeans. Environ. Entomol. 11:756-761.

BEEHLER, B. 1983. Frugivory and polygamy in birds of paradise. Auk 100:1-12.

BEGON, M., J. L. HARPER, AND C. R. TOWNSEND. 1986. Ecology: individuals, populations, and communities. Sinauer Assoc., Sunderland, MA.

BELL, G. P., G. A. BARTHOLOMEW, AND K. A. NAGY. 1986. The roles of energetics, water economy, foraging behavior, and geothermal refugia in the distribution of the bat, Macrotus californicus. J. Comp. Physiol. B. 156:441-450.

BELL, H. L. 1980. Composition and seasonality of mixed-species feeding flocks of insectivorous birds in the Australian Capital Territory. Emu 80:227-232.

BELL, H. L. 1982. Sexual differences in the foraging behavior of the Frill-necked Flycatcher Arses telescopthalmus in New Guinea. Aust. J. Ecol. 7:137-147.

BELL, H. L. 1983. Resource partitioning between three syntopic thornbills Acanthizidae: Acanthizae (Vigors and Horsefield). Ph.D. thesis, Univ. of New England, Armidale, N.S.W., Australia.

BELL, H. L. 1984. A bird community of lowland rainforest in New Guinea. 6. Foraging ecology and community structure of the avifauna. Emu 84:142-158.

BELL, H. L. 1985a. Seasonal variation and the effects of drought on the northern Tablelands of New South Wales. Aust. J. Ecol. 10:207-221.

BELL, H. L. 1985b. The social organization and foraging behavior of three syntopic thornbills Acanthiza spp., p. 151-163. In J. A. Keast, H. F. Recher, H. A. Ford, and D. Saunders [eds.], Birds of eucalypt forests and woodlands: ecology, conservation, management. Royal Aust. Ornithol. Union, Surrey Beatty, Sydney.

BELL, H. L., AND H. A. FORD. 1986. A comparison of the social organization of three syntopic species of Australian thornbill, Acanthiza. Behav. Ecol. Sociobiol. 19:381-392.

BELWOOD, J. J., AND M. B. FENTON. 1976. Variation in the diet of Myotis lucifugus (Chiroptera: Vespertilionidae). Can. J. Zool. 54:1674-1678.

BELWOOD, J. J., AND J. H. FULLARD. 1984. Echolocation and foraging behavior in the Hawaiian hoary bat, Lasiurus cinereus semotus. Can. J. Zool. 62:2113-2120.

BENDELL, B. E., P. J. WEATHERHEAD, AND R. K. STEWART. 1981. The impact of predation by Red-winged Blackbirds on European corn borer populations. Can. J. Zool. 59:1535-1538.

BENKMAN, C. W. 1987a. Crossbill foraging behavior, bill structure, and patterns of food productivity. Wilson Bull. 99:351-368.

BENKMAN, C. W. 1987b. Food profitability and the foraging

ecology of crossbills. Ecol. Monogr. 57:251-267.
BENNETT, S. E. 1980. Interspecific competition and the niche of the American Redstart (Setophaga ruticilla) in wintering and breeding communities, p. 319-335. In A. Keast and E. S. Morton [eds.], Migrant birds in the Neotropics: ecology, behavior, distribution and conservation. Smithsonian Institution Press, Washington, D.C.
BENT, A. C. 1958. Life histories of North American blackbirds, orioles, tanagers, and allies. U.S. Natl. Mus. Bull. 211.
BERGELSON, J. M., AND J. H. LAWTON. 1988. Does foliage damage influence predation on the insect herbivores of birch? Ecology 69:434-445.
BERGELSON, J. M., S. FOWLER, AND S. HARTLEY. 1986. The effects of foliage damage on casebearing moth larvae, Coleophora serratella, feeding on birch. Ecol. Entomol. 11:241-250.
BERLYNE, P. E. 1950. Novelty and curiosity as determinants of exploratory behavior. Br. J. Psych. 41:68-80.
BERNAYS, E. A. 1986. Evolutionary contrasts in insects: nutritional advantages of holometabolous development. Physiol. Entomol. 11:377-382.
BERRYMAN, A. A. 1964. Identification of insect inclusions in x-rays of ponderosa pine bark infested by the western pine beetle, Dendroctonus brevicomis LeConte. Can. Entomol. 96:883-888.
BERRYMAN, A. A. 1970. Evaluation of insect predators of the western pine beetle, p. 102-112. In R. W. Stark and D. L. Dahlsten, [eds.], Studies on the population dynamics of the western pine beetle, Dendroctonus brevicomis LeConte (Coleoptera: Scolytidae). Univ. California Div. Agric. Sci., Berkeley.
BERRYMAN, A. A. 1986. Forest insects: principle and practice of population management. Plenum Press, New York.
BERRYMAN, A. A. 1987. The theory and classification of outbreaks, pp. 3-30. In P. Barbosa and J. C. Schultz [eds.], Insect outbreaks. Academic Press, New York.
BERRYMAN, A. A., AND R. W. STARK. 1962. Radiography in forest entomology. Ann. Entomol. Soc. Am. 55:456-466.
BERTHOLD, P. 1975. Migration: control and metabolic physiology, p. 77-128, In D. S. Farner and J. R. King [eds.], Avian biology. Vol. 5. Academic Press, New York.
BERTHOLD, P. 1988a. The biology of the genus Sylvia: a model and a challenge for Afro-European cooperation. Tauraco 1:3-28.
BERTHOLD, P. 1988b. Evolutionary aspects of migratory behavior in European warblers. J. Evol. Biol. 1:195-209.
BERTHOLD, P. 1988c. The control of migration in European warblers, p. 216-249. in H. Ouellet [ed.], Acta XIX Congressus Internationalis Ornithologici. Vol. 1, University of Ottawa Press, Canada.
BEST, L. B. 1977. Nesting biology of the Field Sparrow. Auk 94:308-319.
BEST, L. S., AND P. BIERZYCHUDEK. 1982. Pollinator foraging on foxglove (Digitalis purpurea): a test of a new model. Evolution 36:70-79.
BETTS, M. M. 1955. The food of titmice in oak woodland. J. Anim. Ecol. 24:282-323.
BIEBACH, H. 1985. Sahara stopover in migratory flycatchers: fat and food affect the time program. Experientia 41:695-697.
BIEDENWEG, D. W. 1983. Time and energy budgets of the Mockingbird (Mimus polyglottos) during the breeding season. Auk 100:149-160.
BIERMANN, G. C., AND S. G. SEALY. 1982. Parental feeding of nestling Yellow Warblers in relation to brood size and prey availability. Auk 99:332-341.
BILCKE, G., R. MERTENS, M. JEURISSEN, AND D. A. DHONDT. 1986. Influences of habitat structure and temperature on the foraging niches of the pariform guild in Belgium during winter. Gerfaut 76:109-130.

BILLINGSLEY, B. B., AND D. H. ARNER. 1970. The nutritive value and digestibility of some winter foods of the eastern Wild Turkey. J. Wildl. Manage. 34:176-182.
BISHOP, Y. M. M., S. E. FIENBERG, AND P. W. HOLLAND. 1975. Discrete multivariate analysis: theory and practice. MIT Press, Cambridge, MA.
BISHTON, G. 1986. The diet and foraging behaviour of the Dunnock Prunella modularis in a hedgerow habitat. Ibis 128:526-539.
BJORNHAG, G., AND I. SPERBER. 1977. Transport of various food components through the digestive tract of turkeys, geese and guinea fowl. Swed. J. Agric. Res. 7:57-66.
BLAKE, J. G. 1983. Trophic structure of bird communities in forest patches in east-central Illinois. Wilson Bull. 95:416-430.
BLAKE, J. G., AND W. G. HOPPES. 1986. Influence of resource abundance on use of tree-fall gaps by birds in an isolated woodlot. Auk 103:328-340.
BLAKE, J. G., F. G. STILES, AND B. A. LOISELLE. In press. Birds of La Selva: habitat use, trophic composition, and migrants. In A. Gentry, [ed.], Four neotropical forests: a comparison of La Selva, Costa Rica; Barro Colorado Island, Panama; the minimum critical size of ecosystems area, Brazil; and Manu National Park, Peru. Yale Univ. Press, New Haven, CT.
BLANCHER, P. J., AND R. J. ROBERTSON. 1987. Effect of food supply on the breeding biology of Western Kingbirds. Ecology 68:723-732.
BLAXTER, K. L., AND J. A. F. ROOK. 1953. The heat of combustion of the tissues of cattle in relation to their chemical composition. Br. J. Nutr. 7:83.
BLEM, C. R. 1975. Energetics of nestling house sparrows Passer domesticus. Comp. Biochem. Physiol. 52A:305-312.
BLEM, C. R. 1976. Patterns of lipid storage and utilization in birds. Am. Zool. 16:671-684.
BLENDEN, M. D., M. J. ARMBRUSTER, T. S. BASKETT, AND A. H. FARMER. 1986. Evaluation of model assumptions: the relationship between plant biomass and arthropod abundance, p. 11-14. In J. Verner, M. L. Morrison, and C. J. Ralph [eds.], Wildlife 2000: modeling habitat relationships of terrestrial vertebrates. Univ. Wisconsin Press, Madison.
BLEST, A. D. 1956. Protective coloration and animal behaviour. Nature 178:1190-1191.
BLOCK, W. M., AND M. L. MORRISON. 1987. Conceptual framework and ecological considerations for the study of birds in oak woodlands, p. 163-173. In T. R. Plumb and N. H. Pillsbury [tech. coords.]. Proceedings of the symposium on multiple-use management of California's hardwoods resources. U.S.D.A. For. Ser. Gen. Tech. Rep. PSW-100, Berkeley, CA.
BLOCK, W. M., K. A. WITH, AND M. L. MORRISON. 1987. On measuring bird habitat: influence of observer variability and sample size. Condor 89:241-251.
BLOM, E. 1987. The changing seasons. Am. Birds 41:248-252.
BLUS, L. J., AND J. A. KEAHEY. 1978. Variation in reproductivity with age in the Brown Pelican. Auk 90:128-134.
BOAG, P. T., AND P. R. GRANT. 1984. Darwin's finches (Geospiza) on Isla Daphne Major, Galapagos: breeding and feeding ecology in a climatically variable environment. Ecol. Monogr. 54:463-489.
BOCK, C. E. 1970. The ecology and behavior of the Lewis' Woodpecker (Asyndesmus lewis). Univ. Calif. Publ. Zool. 92:1-91.
BOCK, W. J., AND W. D. MILLER. 1959. The scansorial foot of the woodpeckers, with comments on the evolution of perching and climbing feet in birds. Am. Mus. Novitates No. 1931.
BOIVIN, G., AND R. K. STEWART. 1983. Sampling techique and seasonal development of phytophagous mirids (Hemiptera: Miridae) on apples in southwestern Quebec. Ann. Entomol. Soc. Am. 76:359-364.

BONNER, J. T. 1980. The evolution of culture in animals. Princeton Univ. Press, Princeton, NJ.
BORROR, D. J., D. M. DELONG, AND C. A. TRIPPLEHORN. 1976. An introduction to the study of insects. 4th ed. Holt, Rinehart and Winston, New York.
BORROR, D. J., D. M. DELONG, AND C. A. TRIPPLEHORN. 1981. An introduction to the study of insects. 5th ed. Saunders College Publ., New York.
BOSWALL, J. 1977. Tool-using by birds and related behaviour. Avic. Mag. 83:88-97, 146-159, 220-228.
BOUCHER, C. 1978. Cape Hangklip area. II. The vegetation. Bothalia 12:455-497.
BOWERS, M. D., I. L. BROWN, AND D. WHEYE. 1985. Bird predation as a selective agent in a butterfly population. Evolution 39:93-103.
BRADBURY, J. W. 1977. Lek mating behavior in the hammer-headed bat. Z. Tierpsychol. 45:225-255.
BRADBURY, J., D. MORRISON, E. STASHKO, AND R. HEITHAUS. 1979. Radio-tracking methods for bats. Bat Res. News 20:9-17.
BRADLEY, D. W. 1985. The effects of visibility bias on time-budget estimates of niche breadth and overlap. Auk 102:493-499.
BRAFIELD, A. E., AND M. J. LLEWELLYN. 1982. Animal energetics. Blackie, London.
BRANDT, C. A. 1984. Age and hunting success in the Brown Pelican: influence of skill and patch choice on foraging efficiency. Oecologia 62:132-137.
BRAWN, J. D., AND F. B. SAMSON. 1983. Winter behavior of Tufted Titmice. Wilson Bull. 95:222-272.
BRAWN, J. D., W. J. BOECKLEN, AND R. P. BALDA. 1987. Investigations of density interactions among breeding birds in ponderosa pine forests: correlative and experimental evidence. Oecologia 72:348-357.
BRAY, O. E., AND G. W. CORNER. 1972. A tail clip for attaching transmitters to birds. J. Wildl. Manage. 36:640-642.
BRAY, O. E., K. H. LARSEN, AND D. F. MOTT. 1975. Winter movements and activities of radio-equipped starlings. J. Wildl. Manage. 39:795-801.
BRAY, O. E., W. C. ROYALL, JR., J. L. GUARINO, AND R. E. JOHNSON. 1979. Activities of radio-equipped Common Grackles during fall migration. Wilson Bull. 91:78-87.
BREELAND, S. G., AND E. PICKARD. 1965. The Malaise trap-an efficient and unbiased mosquito collecting device. Mosquito News 25:19-21.
BREITWISCH, R., M. DIAZ, AND R. LEE. 1987. Foraging efficiencies and techniques of juvenile and adult Northern Mockingbirds (Mimus polyglottos). Behaviour 101:225-235.
BRENNER, J. F. 1966. Energy and nutrient requirements of Red-winged Blackbirds. Wilson Bull. 78:111-120.
BRENSING, D. 1977. Nahrungsokologische Untersuchungen an Zugvogeln in einem Sudwestdeutschen Durchzugsgebiet wahrend des Wegzuges. Die Vogelwarte 29:44-56.
BROADBENT, L. 1948. Aphis migration and the efficiency of the trapping method. Ann. Appl. Biol. 35:379-394.
BROCKE, M. DE L., D. A. SCOTT, AND D. M. TEIXEIRA. 1983. Some observations made at the first recorded nest of the Sharpbill Oxyruncus cristatus. Ibis 125:259-261.
BROCKMAN, H. J., AND C. J. BARNARD. 1979. Kleptoparasitism in birds. Anim. Behav. 27:487-514.
BRODY, S. 1945. Bioenergetics and growth. Hafner, New York.
BRONSTEIN, J. L., AND K. HOFFMANN. 1987. Spatial and temporal variation in frugivory at a Neotropical fig, Ficus pertusa. Oikos 49:261-268.
BROOKE, M. DE L. 1979. Differences in the quality of territories held by Wheateaters (Oenanthe oenanthe). J. Anim. Ecol. 48:21-32.
BROOKE, M. DE L., D. A. SCOTT, AND D. M. TEIXEIRA. 1983. Some observations made at the first recorded nest of the Sharpbill Oxyruncus cristatus. Ibis 125:259-261.
BROOKES, M. H., R. W. STARK, AND R. W. CAMPBELL, eds. 1978. The Douglas-fir Tussock Moth: A synthesis. USDA For. Serv. Science and Education Agency, Tech. Bull. 1585, 331 pp.
BROOKS, J. M. 1986. Observations of the feeding of Black-throated Blue Warblers in Florida during migration. Florida Field Nat. 14:96-97.
BROOKS, W. S. 1968. Comparative adaptations of the Alaskan redpolls to the arctic environment. Wilson Bull. 80:253-280.
BROWER, L. P. 1963. Mimicry, a symposium. Proc. XVI Intern. Zool. Congr.:145-186.
BROWER, L. P., J. V. Z. BROWER, F. G. STILES, H. J. CROZE, AND A. S. HOWER. 1964. Mimicry: differential advantage of color patterns in the natural environment. Science 144:183-185.
BROWN, C. R. 1986. Cliff Swallow colonies as information centers? Science 234:518-519.
BROWN, C. R. 1988. Enhanced foraging efficiency through information centers: a benefit of coloniality in cliff swallows. Ecology 69:602-613.
BROWN, H. E., M. M. BAKER, JR., AND J. J. ROGERS. 1974. Opportunities for increasing water yields and other multiple use values on ponderosa pine forest lands. USDA For. Serv. Res. Pap. RM-129, Fort Collins, CO.
BROWN, J. H., AND M. A. BOWERS. 1984. Patterns and processes in three guilds of terrestrial vertebrates, p. 282-296. In D. R. Strong Jr., D. Simberloff, L. G. Abele, and A. B. Thistle, [eds.], Ecological communities. Princeton Univ. Press, Princeton, NJ.
BROWN, J. J., AND G. M. CHIPPENDALE. 1973. Nature and fate of the nutrient reserves of the periodical (17 year) cicada. J. Insect Physiol. 19:607-614.
BROWN, J. L. 1963. Social organization and behavior of the Mexican Jay. Condor 65:126-153.
BROWN, J. L. 1964. The evolution of diversity in avian territorial systems. Wilson Bull. 76:160-169.
BROWN, J. L. 1969. Territorial behavior and population regulation in birds. A review and re-evaluation. Wilson Bull. 81:293-329.
BROWN, J. L. 1975. The evolution of behavior. Norton, New York.
BROWN, J. L. 1987. Helping and communal breeding in birds: ecology and evolution. Princeton Univ. Press, Princeton, NJ.
BROWN, J. S. 1986. Coexistence on a resource whose abundance varies: a test with desert rodents. Ph.D. Diss., Univ. of Arizona, Tucson.
BROWN, L., AND J. BROWN. 1984. Dispersal and dispersion in a population of soldier beetles (Coleoptera: Cantharidae). Environ. Entomol. 13:175-178.
BROWN, M. J., D. A. RATKOWSKY, AND P. R. MINCHIN. 1984. A comparison of detrended correspondence analysis and principal co-ordinates analysis using four sets of Tasmanian vegetation data. Aust. J. Ecol. 9:273-279.
BROWN, R. G. B. 1969. Seed selection by pigeons. Behaviour 34:115-131.
BRUNS, M. 1960. The economic importance of birds in forests. Bird Study 7:193-208.
BRUSH, T., AND E. W. STILES. 1986. Using food abundance to predict habitat use by birds, p. 57-63. In J. Verner, M. L. Morrison, and C. J. Ralph [eds.], Wildlife 2000: modeling habitat relationships of terrestrial vertebrates. Univ. Wisconsin Press, Madison.
BRYANT, D. M. 1973. The factors influencing the selection of food by the House Martin (Delichon urbica (L.)). J. Anim. Ecol. 42:539-564.
BRYANT, D. M. 1975a. Breeding biology of House Martins Delichon urbica in relation to aerial insect abundance. Ibis 117:180-216.
BRYANT, D. M. 1975b. Reproductive costs in the House Martin (Delichon urbica). J. Anim. Ecol. 48:655-675.
BRYANT, D. M., AND V. M. T. BRYANT. 1988. Assimilation efficiency and growth of nestling insectivores. Ibis 130:268-274.

BRYANT, D. M., AND K. R. WESTERTERP. 1981. The energy budget of the House Martin (Delichon urbica), p. 91-102. In H. Klomp and J. W. Woldendorp [eds.], The integrated study of bird populations. North Holland Publishing Co., Amsterdam.

BRYANT, D. M., AND K. R. WESTERTERP. 1983. Time and energy limits to brood size in House Martins (Delichon urbica). J. Anim. Ecol. 52:906-925.

BRYANT, J. P., AND P. J. KUROPAT. 1980. Selection of winter forage by subarctic browsing vertebrates: the role of plant chemistry. Ann. Rev. Ecol. Syst. 11:261-285.

BUCHLER, E. R. 1976. Prey selection by Myotis lucifugus (Chiroptera: Vespertilionidae). Am. Nat. 110:619-628.

BUCHSBAUM, R., J. WILSON, AND I. VALIELA. 1986. Digestibility of plant constituents by Canada Geese and Atlantic Brant. Ecology 67:386-393.

BUCKNER, C. H. 1966. The role of vertebrate predators in the biological control of forest insects. Ann. Rev. Entomol. 11:449-470.

BUCKNER, C. H. 1967. Avian and mammal predators of forest insects. Entomophaga 12:491-501.

BUCKNER, C. H., AND W. J. TURNOCK. 1965. Avian predation on the larch sawfly, Pristiphora erichsonii (HTG), (Hymenoptera: Tenthredinidae). Ecology 46:223-236.

BURNHAM, K. P., D. R. ANDERSON, AND J. L. LAAKE. 1980. Estimation of density from line transect sampling of biological populations. Wildl. Monogr. 72.

BURTON, B. A., R. J. HUDSON, AND D. D. BRAGG. 1979. Efficiency of utilization of bulbrush rhizomes by Lesser Snow Geese. J. Wildl. Manage. 43:728-735.

BURTON, P. J. K. 1976. Feeding behavior in the Paradise Jacamar and the Swallow-wing. Living Bird 15:223-238.

BURTT, E. H., JR., AND J. P. HAILMAN. 1979. Effect of food availability on leaf-scratching by the Rufous-sided Towhee: test of a model. Wilson Bull. 91:123-126.

BURTT, E. H., JR., B. D. SUSTARE, AND J. P. HAILMAN. 1976. Cedar Waxwing feeding from spider web. Wilson Bull. 88:157-158.

BUSBY, D. G., AND S. G. SEALY. 1979. Feeding ecology of a population of nesting Yellow Warblers. Can. J. Zool. 57:1670-1681.

BUTTEMER, W. A., A. M. HAYWORTH, W. W. WEATHERS, AND K. A. NAGY. 1986. Time-budget estimates of avian energy expenditure: physiological and meteorological considerations. Physiol. Zool. 59:131-149.

BYERLY, K. F., A. P. GUTIERREZ, R. E. JONES, AND R. F. LUCK. 1978. A comparison of sampling methods for some arthropod populations in cotton. Hilgardia 46:257-281.

CABLE, D. R. 1975. Influence of precipitation on perennial grass production in the semi-desert southwest. Ecology 56:91-96.

CACCAMISE, D. F., AND R. S. HEDIN. 1985. An aerodynamic basis for selecting transmitter loads in birds. Wilson Bull. 97:306-318.

CAIN, A. J., AND P. M. SHEPPARD. 1954. Natural selection in Cepaea. Genetics 45:89-116.

CAIN, B. W. 1973. Effect of temperature on energy requirements and northward distribution of the Black-bellied Tree Duck. Wilson Bull. 85:309-312.

CAIN, B. W. 1976. Energetic of growth for Black-bellied Tree Ducks. Condor 78:124-128.

CALDER, W. A. III. 1984. Size, function, and life history. Harvard Univ. Press, Cambridge, MA.

CALDER, W. A. III, AND S. M. HIEBERT. 1983. Nectar feeding, diuresis, and electrolyte replacement of hummingbirds. Physiol. Zool. 56:325-334.

CALDER, W. A., AND J. R. KING. 1974. Thermal and caloric relations of birds, p. 259-413. In D. S. Farner, and J. R. King [eds.], Avian biology. Vol. 4. Academic Press, New York.

CALVER, M. C., AND R. D. WOOLLER. 1982. A technique for assessing the taxa, length, dry weight, and energy content of the arthropod prey of birds. Aust. Wildl. Res. 9:293-301.

CALVERT, W. H., L. E. HEDRICK, AND L. P. BROWER. 1979. Mortality of the monarch butterfly (Danaus plexippus L.): avian predation at five overwintering sites in Mexico. Science 204:847-851.

CAMPBELL, R. W. 1973. Numerical behavior of a gypsy moth population system. For. Sci. 19:162-167.

CAMPBELL, R. W. 1987. Population dynamics, p. 80-81. In Brookes, M. H, R. W. Campbell, J. J. Colbert, R. G. Mitchell, and R. W. Stark [tech. coords.], Western spruce budworm. USDA. For. Serv. Tech. Bull. 1694, Washington, DC.

CAMPBELL, R. W., AND R. J. SLOAN. 1977. Natural regulation of innocuous gypsy moth populations. Environ. Entomol. 6:315-322.

CAMPBELL, R. W., AND T. R. TORGERSEN. 1982. Some effects of predaceous ants on western spruce budworm pupae in north central Washington. Environ. Entomol. 11:111-114.

CAMPBELL, R. W., AND T. R. TORGERSEN. 1983a. Compensatory mortality in defoliator population dynamics. Environ. Entomol. 12:630-632.

CAMPBELL, R. W., AND T. R. TORGERSEN. 1983b. Effect of branch height on predation of western spruce budworm (Lepidoptera: Tortricidae) pupae by birds and ants. Environ. Entomol. 12:697-699.

CAMPBELL, R. W., T. R. TORGERSEN, S. C. FORREST, AND L. C. YOUNGS. 1981. Bird exclosures for branches or whole trees. USDA For. Serv. Gen. Tech. Rep. PNW-125, Portland, OR.

CAMPBELL, R. W., T. R. TORGERSEN, AND N. SRIVASTAVA. 1983. A suggested role for predaceous birds and ants in the population dynamics of the western spruce budworm. For. Sci. 29:779-790.

CANE, V. R. 1959. Behavior sequences as semi-Markov chains. J. Royal Stat. Soc. Ser. 21:36-58.

CANE, V. R. 1978. On fitting low-order Markov chains to behavior sequences. Anim. Behav. 26:332-338.

CAPEN, D. E., J. W. FENWICK, D. B. INKLEY, AND A. C. BOYNTON. 1986. Multivariate models of songbird habitat in New England forests, p. 171-175. In J. Verner, M. L. Morrison, and C. J. Ralph [eds.], Wildlife 2000: modeling habitat relationships of terrestrial vertebrates. Univ. Wisconsin Press, Madison.

CARACO, T. 1980. On foraging time allocation in a stochastic environment. Ecology 61:119-128.

CARACO, T. 1981a. Energy budget, risk and foraging preferences in Dark-eyed Juncos (Junco hyemalis). Behav. Ecol. Sociobiol. 8:820-830.

CARACO, T. 1981b. Risk-sensitivity and foraging groups. Ecology 62:527-531.

CARACO, T. 1982. Aspects of risk-aversion in foraging White-crowned Sparrows. Anim. Behav. 30:719-727.

CARACO, T. 1983. White-crowned Sparrows (Zonotrichia leucophrys): foraging preferences in a risky environment. Behav. Ecol. Sociobiol. 12:63-69.

CARACO, T., S. MARTINDALE, AND T. S. WHITHAM. 1980. An empirical demonstration of risk-sensitive foraging preferences. Anim. Behav. 28:820-830.

CARBON, B. A., G. A. BARTLE, AND A. M. MURRAY. 1979. A method for visual estimation of leaf area. For. Sci. 25:53-58.

CARNES, B. A., AND N. A SLADE. 1982. Some comments on niche analysis in canonical space. Ecology 63:888-893.

CAROLIN, V. M., AND W. K. COULTER. 1971. Trends of spruce budworm and associated insects in Pacific northwest forests sprayed with DDT. J. Econ. Entomol. 64:291-297.

CAROLIN, V. M., AND F. W. HONING. 1972. Western spruce budworm. USDA For. Serv. Pest Leaflet No. 53, Washington, DC.

CARPENTER, F. L. 1976. Ecology and evolution of an Andean hummingbird. Univ. Calif. Publ. Zool. 106:1-74.

CARPENTER, F. L. 1983. Pollination energetics in avian communties: simple concepts and complex realities, p.

215-234. In C. E. Jones and R. J. Little [eds.], Handbook of experimental pollination biology. Van Nostrand Reinhold, New York.

CARPENTER, F. L. 1987. Food abundance and territoriality: to defend or not to defend? Am. Zool. 27:387-399.

CARPENTER, F. L., AND J. L. CASTRONOVA. 1980. Maternal diet selectivity in Calypta anna. Am. Midl. Nat. 103:175-179.

CARPENTER, F. L., AND R. E. MACMILLEN. 1976. Threshold model of feeding territoriality and test with a Hawaiian honeycreeper. Science 194:639-642.

CARPENTER, F. L., D. C. PATON, AND M. A. HIXON. 1983. Weight gain and adjustment of feeding territory size in migrant hummingbirds. Ecology 80:7259-7263.

CARPENTER, R. E. 1969. Structure and function of the kidney and the water balance of desert bats. Physiol. Zool. 42:288-302.

CARRASCAL, J. P. 1984. Cambios en el uso del espacio en un gremio de aves durante el periodo primavera-verano. Ardeola 31:47-60.

CARRASCAL, J. P., AND F. J. SANCHEZ-AGUADO. 1987. Spatio-temporal organization of the bird communities in two Mediterranean Montane forests. Holarctic Ecol. 10:185-192.

CARSON, R. G., AND J. M. PEEK. 1987. Mule deer habitat selection patterns in northcentral Washington. J. Wildl. Manage. 51:46-51.

CASE, T. J., AND M. E. GILPIN. 1974. Interference competition and niche theory. Proc. Natl. Acad. Sci. 71:3073-3077.

CASE, T. J., AND R. SIDELL. 1983. Pattern and chance in the structure of model and natural communities. Evolution 37:832-849.

CASTILLO, J. A., AND W. G. EBERHARD. 1983. Use of artificial webs to determine prey available to orb weaving spiders. Ecology 64:1655-1658.

CASTRO, G., N. STOYAN, AND J. P. MYERS. 1989. Assimilation efficiency in birds: a function of taxon or food type? Comp. Biochem. Physiol. 92A:271-278.

CATES, R. G., AND D. F. RHOADES. 1977. Patterns in the production of anti-herbivore chemical defenses in plant communities. Biochem. System. Ecol. 5:185-193.

CATTERALL, C. P. 1985. Winter energy deficits and the importance of fruit versus insects in a tropical island bird population. Aust. J. Ecol. 10:265-279.

CAVALLI-SFORZA, L. L., AND M. W. FELDMAN. 1981. Cultural transmission and evolution: a quantitative approach. Monogr. Popul. Biol. No. 16. Princeton Univ. Press, Princeton, NJ.

CAVÉ, A. J. 1968. The breeding of the Kestrel, Falco tinnunculus L. in the reclaimed area Ostelijk Flevoland. Arch. Neerl. Zool. 18d:313-407.

CAYONETTE, R. 1947. Starlings catching insects on the wing. Auk 64:458.

CHAI, P. 1986. Field observations and feeding experiments on the responses of Rufous-tailed Jacamars (Galbula ruficauda) to free-flying butterflies in a tropical rain forest. Biol. J. Linn. Soc. 29:161-189.

CHAI, P. 1987. Wing coloration of free-flying Neotropical butterflies as a signal learned by a specialized avian predator. Biotropica 20:20-30.

CHAPMAN, J. A., AND J. M. KINGHORN. 1955. Window-trap for flying insects. Can. Entomol. 82:46-47.

CHARLES-DOMINIQUE, P., M. ATRAMENTOWICZ, M. CHARLES-DOMINIQUE, H. GERARD, A. HLADIK, C. M. HLADIK, AND M. F. PREVOST. 1981. Les mammifères frugivores arboricoles nocturnes d'une foret Guyanaise: inter-relations plantes-animaux. Revue d'Ecologie 35:341-435.

CHARNOV, E. L. 1976a. Optimal foraging: attack strategy of a mantid. Am. Nat. 110:141-151.

CHARNOV, E. L. 1976b. Optimal foraging: the marginal value theorem. Theor. Popul. Biol. 9:129-136.

CHATFIELD, C. 1973. Statistical inference regarding Markov chain models. Appl. Stat. 22:7-20.

CHATFIELD, C., AND R. E. LEMON. 1970. Analyzing sequences of behavioral events. J. Theor. Biol. 29:427-445.

CHERRETT, J. M. 1964. The distribution of spiders on the Moor House National Nature Reserve, Westmorland. J. Anim. Ecol. 33:27-48.

CHESSON, J. 1978. Measuring preference in selective predation. Ecology 59:211-215.

CHESSON, J. 1983. The estimation and analysis of preference and its relationships to foraging models. Ecology 64:1297-1303.

CHIANG, C. L. 1980. An introduction to stochastic processes and their applications. Robert E. Krieger Publ. Co., Huntington, New York.

CHIVERS, D. J., AND C. M. HLADIK. 1980. Morphology of the gastrointestinal tract in primates: comparisons with other mammals in relation to diet. J. Morph. 166:337-388.

CLARK, C. W., AND M. MANGEL. 1984. Foraging and flocking strategies: information in an uncertain environment. Am. Nat. 123:626-641.

CLARK, G. A., JR. 1971. The occurrence of bill-sweeping in the terrestrial foraging of birds. Wilson Bull. 83:66-73.

CLARK, G. A., JR. 1983. An additional method of foraging in litter by species of Turdus thrushes. Wilson Bull. 95:155-157.

CLARK, L. 1987. Thermal constraints on foraging in adult European Starlings. Oecologia 71:233-238.

CLARK, R. G., H. GREENWOOD, AND L. G. SUGDEN. 1986. Preliminary estimates of rate of grain passage through the digestive tract of Mallards. Can. Wildl. Serv. Publ. No. 160. March.

CLARKSON, K., S. F. EDEN, W. J. SUTHERLAND, AND A. I. HOUSTON. 1986. Density dependence and magpie food hoarding. J. Anim. Ecol. 55:111-121.

CLEMENS, E. T., C. E. STEVENS, AND M. SOUTHWORTH. 1975. Sites of organic acid production and pattern of digesta movement in the gastrointestinal tract of geese. J. Nutr. 105:1341-1350.

CLEVELAND, W. S., AND M. E. MCGILL. 1987. Dynamic graphics for statistics. Wadsworth, Monterey.

CLIFFORD, H. T., AND W. STEPHENSON. 1975. An introduction to numerical classification. Academic Press, New York.

CLUTTON-BROCK, T. H., AND P. H. HARVEY. 1984. Comparative approaches to investigating adaptation, p. 7-30. In J. R. Krebs and N. B. Davies [eds.]. Behavioural ecology: an evolutionary approach. Blackwell, London, England.

CLUTTON-BROCK, T. H., F. E. GUINNESS, AND S. D. ALBON. 1981. Red Deer: behavior and ecology of two sexes. Univ. of Chicago Press, Chicago, IL.

COATES-ESTRADA, R., AND A. ESTRADA. 1986. Fruiting and frugivores at a strangler fig in the tropical rain forest of Los Tuxtlas, Mexico. J. Trop. Ecol. 2:349-357.

COCHRAN, W. G. 1954. Some methods for strengthening the common X^2 tests. Biometrics 10:417-441.

COCHRAN, W. G. 1977. Sampling techniques. 3rd ed. John Wiley and Sons, New York.

COCHRAN, W. W. 1980. Wildlife telemetry, p. 507-520. In S. D. Schemnitz [ed.], Wildlife management techniques. The Wildlife Society, Washington, D.C.

COCHRAN, W. W., AND C. G. KJOS. 1985. Wind drift and migration of thrushes: a telemetry study. Ill. Nat. Hist. Surv. Bull. 33:297-330.

COCHRAN, W. W., G. G. MONTGOMERY, AND R. R. GRABER. 1967. Migratory flights of Hylocichla thrushes in spring: a radiotelemetry study. Living Bird 6:213-225.

CODY, M. L. 1968. On the methods of resource division in grassland bird communities. Am. Nat. 102:107-146.

CODY, M. L. 1971. Finch flocks in the Mohave Desert. Theor. Popul. Biol. 2:142-158.

CODY, M. L. 1974. Competition and the structure of bird communities. Princeton Monogr. in Population Biology 7, Princeton Univ. Press, Princeton, NJ.

CODY, M. L. 1981. Habitat selection in birds: the roles of vegetation structure, competitors, and productivity. Bioscience 31:107-113.

COLEMAN, J. D. 1974. Breakdown rates of foods ingested by starlings. J. Wildl. Manage. 38:910-912.

COLGAN, P. W. 1978. Quantitative ethology. Wiley Interscience. John Wiley and Sons, New York.

COLLINS, B. G. 1983a. Pollination of Mimetes hirtus (Proteaceae) by Cape Sugarbirds and Orange-breasted Sunbirds. J. S. Afr. Bot. 49:125-142.

COLLINS, B. G. 1983b. A first approximation of the energetics of Cape Sugarbirds (Promerops cafer) and Orange-breasted Sunbirds (Nectarinia violacea). S. Afr. J. Zool. 18:363-369.

COLLINS, B. G. 1985. Energetics of foraging and resource selection by honeyeaters in forest and woodland habitats of Western Australia. N.Z. J. Zool. 12:577-587.

COLLINS, B. G., AND P. BRIFFA. 1982. Seasonal variation of abundance and foraging of three species of Australian honeyeaters. Aust. Wildl. Res. 9:557-569.

COLLINS, B. G., AND P. BRIFFA. 1983. Seasonal and diurnal variations in the energetics and foraging activities of the Brown Honeyeater, Lichmera indistincta. Aust. J. Ecol. 8:103-111.

COLLINS, B. G., AND C. NEWLAND. 1986. Honeyeater population changes in relation to food availability in the jarrah forest of Western Australia. Aust. J. Ecol. 11:63-76.

COLLINS, B. G., G. CARY, AND G. PACKARD. 1980. Energy assimilation, expenditure, and storage by the Brown Honeyeater, Lichmera indistincta. J. Comp. Physiol. 137:157-163.

COLLINS, B. G., AND A. G. REBELO. 1987. Pollination biology and breeding systems in the Proteaceae of Australia and southern Africa. Aust. J. Ecol. 12:387-421.

COLLINS, B. G., C. NEWLAND, AND P. BRIFFA. 1984. Nectar utilization and pollination by Australian honeyeaters and insects visiting Calothamnus quadrifidus (Myrtaceae). Aust. J. Ecol. 9:353-365.

COLLINS, S. L. 1983. Geographic variation in habitat structure of the Black-throated Green Warbler (Dendroica virens). Auk 100:382-389.

COLLOPY, M. W. 1983. A comparison of direct observations and collections of prey remains in determining the diet of Golden Eagles. J. Wildl. Manage. 47:360-368.

COLWELL, R. K. 1973. Competition and coexistence in a simple tropical community. Am. Nat. 107:737-760.

COLWELL, R. K. 1974. Predictability, constancy, and contingency of periodic phenomena. Ecology 55:1148-1153.

COLWELL, R. K., AND D. J. FUTUYMA. 1971. On the measurement of niche breadth and overlap. Ecology 52:567-576.

COLWELL, R. K., AND E. R. FUENTES. 1975. Experimental studies of the niche. Ann. Rev. Ecol. Syst. 6:281-310.

COLWELL, R. K., B. J. BETTS, P. BUNNELL, F. L. CARPENTER, AND P. FEINSINGER. 1974. Competition for the nectar of Centropogon valerii by the hummingbird Colibri thalassinus and the flower-piercer Diglossa plumbea and its evolutionary implications. Condor 76:447-452.

COMINS, H. N., AND M. P. HASSELL. 1979. The dynamics of optimally foraging predators and parasitoids. J. Anim. Ecol. 48:335-351.

CONFER, J. L., C. J. HIBBARD, AND D. EBBETS. 1986. Downy Woodpecker reward rates from goldenrod gall insects. Kingbird 36:188-192.

CONFER, J. L., AND P. PAICOS. 1985. Downy Woodpecker predation at goldenrod galls. J. Field Ornithol. 56:56-64.

CONLEY, J. B., AND C. R. BLEM. 1978. Seed selection by Japanese Quail, Coturnix coturnix japonica. Am. Midl. Nat. 100:135-140.

CONNELL, J. H. 1975. Some mechanisms producing structure in natural communities: a model and evidence from field experiments, p. 460-490. In M. L. Cody and J. M. Diamond [eds.], Ecology and evolution of communities. Belknap Press, Cambridge, MA.

CONNELL, J. H. 1980. Diversity and the coevolution of competitors, or the ghost of competition past. Oikos 35:131-138.

CONNER, R. N. 1977. Seasonal changes in the foraging methods and habits of six sympatric woodpecker species in southwestern Virginia. Ph.D. diss., Virginia Polytechnic Institute and State Univ., Blacksburg.

CONNER, R. N. 1980. Foraging habitats of woodpeckers in southwestern Virginia. J. Field Ornithol. 51:119-127.

CONNER, R. N. 1981. Seasonal changes in woodpecker foraging patterns. Auk 98:562-570.

CONNER, R. N., M. E. ANDERSON, AND J. G. DICKSON. 1986. Relationships among territory size, habitat, song, and nesting success of Northern Cardinals. Auk 103:23-31.

CONNOR, E. F., AND D. SIMBERLOFF. 1979. The assembly of species communities: chance or competition? Ecology 60:1132-1140.

CONNOR, E. F., AND D. SIMBERLOFF. 1984. Neutral models of species' co-occurrence patterns, p. 316-331. In D. R. Strong, Jr., D. Simberloff, L. G. Abele, and A. B. Thistle [eds.], Ecological communities: conceptual issues and the evidence. Princeton Univ. Press, Princeton, NJ.

CONNOR, E. F., AND D. SIMBERLOFF. 1986. Competition, scientific method, and null models in ecology. Am. Sci. 74:155-162.

CONOVER, W. J. 1971. Practical nonparametric statistics. John Wiley and Sons, New York.

CONOVER, W. J. 1980. Practical nonparametric statistics. 2nd ed. John Wiley and Sons, New York.

COOK, L. M., G. S. MANI, AND M. E. VARLEY. 1986. Postindustrial melanism in the peppered moth. Science 231:611-613.

COOPER, J. 1978. Energetic requirements for growth and maintenance of the Cape Gannet (Aves: Sulidae). Zool. Africana 13:305-317.

COOPER, R. J. 1988. Dietary relationships among insectivorous birds of an eastern deciduous forest. Ph.D. diss., West Virginia Univ., Morgantown.

COOPER, R. J. 1989. Sampling forest canopy arthropods available to birds as prey, p. 436-444. In L. L. McDonald, B. F. J. Manly, J. A. Lockwood, and J. A. Logan [eds.], Estimation and analysis of insect populations. Springer-Verlag, New York.

COOPER, R. J., K. M. DODGE, S. B. DONAHOE, P. J. MARTINAT, AND R. C. WHITMORE. In press. Effect of Dimilin application on eastern deciduous forest birds. J. Wildl. Manage.

COPPINGER, R. 1970. The effect of experience and novelty on avian feeding behavior with reference to the evolution of warning coloration in butterflies. II. Reactions of naive birds to novel insects. Am. Nat. 104:323-333.

CORBET, S. A., I. CUTHILL, M. FALLOWS, T. HARRISON, AND G. HARTLEY. 1981. Why do nectar-foraging bees and wasps work upwards on influorescences? Oecologia 51:79-83.

COREY, D. T. 1978. The determinants of exploration and neophobia. Neur. Biobehav. Rev. 2:235-253.

CORY, E. N., AND P. KNIGHT. 1937. Observations on Brood X of the periodical cicada in Maryland. J. Econ. Entomol. 30:287-289.

COTT, H. B. 1940. Adaptive colouration in animals. Methuen and Co., London.

COTTAM, G., AND J. CURTIS. 1956. The use of distance measures in phytosociological sampling. Ecology 37. 451-460.

COTTAM, G., J. T. CURTIS, AND B. W. HALE. 1953. Some sampling characteristics of randomly dispersed individuals. Ecology 34:741-757.

COWAN, P. E. 1977. Neophobia and neophilia: new-object and new-place reactions in three Rattus species. Journ. Comp. Phys. Psych. 91:63-71.

COWIE, R. J. 1977. Optimal foraging in Great Tits (Parus

major). Nature 268:137-139.
Cowie, R. J., AND J. R. Krebs. 1979. Optimal foraging in patchy environments, p. 183-205. In R. M. Anderson, B. D. Turner, and L. R. Taylor [eds.], The British Ecological Society Symposium, Vol. 20, Population dynamics. Blackwell Scientific Publ., Oxford, England.
Cowie, R. J., J. R. Krebs, AND D. F. Sherry. 1981. Food storing by Marsh Tits. Anim. Behav. 29:1252-1259.
Cox, G. C. 1961. The relation of energy requirements of tropical finches to distribution and migration. Ecology 42:253-266.
Cox, G. W. 1983. Foraging behavior of the Dune Lark. Ostrich 54:113-120.
Cox, G. W. 1985. The evolution of avian migration systems between temperate and tropical regions of the New World. Am. Nat. 126:451-474.
Craig, J. L. 1984. Wing noises, wing slots and aggression in New Zealand honeyeaters (Aves: Meliphagidae). N. Z. J. Zool. 11:195-200.
Craig, J. L. 1985. Status and foraging in New Zealand honeyeaters. N.Z. J. Zool. 12:589-597.
Craig, R. B. 1978. An analysis of the predatory behavior of the Loggerhead Shrike. Auk 95:221-234.
Craig, R. J. 1984. Comparative foraging ecology of Louisiana and Northern waterthrushes. Wilson Bull. 96:173-183.
Craig, R. J. 1987. Divergent prey selection in two species of waterthrushes (Seiurus). Auk 104:180-187.
Craighead, J. J., AND F. C. Craighead, Jr. 1956. Hawks, owls, and wildlife. Stackpole Co., Harrisburg, PA.
Cramp, S., AND K. E. L. Simmons. 1982. The birds of the Western Palearctic, Vol. VIII. Oxford Univ. Press, Oxford.
Crawford, H. S., R. W. Titterington, AND D. T. Jennings. 1983. Bird predation and spruce budworms populations. J. For. 81:433-435, 478.
CRC. 1986. A handbook of chemistry and physics. CRC Press, Cleveland, OH.
Croat, T. B. 1969. Seasonal flowering behavior in Central Panama. Ann. Miss. Bot. Gard. 56:295-307.
Croat, T. B. 1975. Phenological behavior of habit and habitat classes on Barro Colorado Island (Panama Canal Zone). Biotropica 7:270-277.
Crome, F. H. J. 1975. The ecology of fruit pigeons in tropical northern Queensland. Aust. Wildl. Res. 2:155-185.
Crook, J. H. 1964. The evolution of social organization and visual communication in the weaver birds (Ploceinae). Behav. Suppl. 10:1-178.
Crowell, K. L. 1962. Reduced interspecific competition among the birds of Bermuda. Ecology 43:75-88.
Croxall, J. P. 1977. Feeding behaviour and ecology of New Guinea rainforest insectivorous passerines. Ibis 119:113-146.
Croze, H. 1970. Searching image in Carrion Crows. Hunting strategy in a predator and some anti-predator devices in camouflaged prey. Z. Tierpsychol. 5:1-86.
Cruden, R. W., S. M. Hermann, AND S. Peterson. 1983. Patterns of nectar production and plant-pollinator coevolution, p. 80-125. In B. Bentley and T. Elias [eds.], The biology of nectaries. Columbia Univ. Press, New York.
Cruz, A. 1977. Ecology and behavior of the Jamaican Woodpecker. Bull. Florida St. Mus. 22:149-204.
Cruz, A. 1978. Adaptive evolution in the Jamaican Blackbird Nesopsar nigerrimus. Ornis Scand. 9:130-137.
Cruz, A., AND D. W. Johnston. 1979. Occurrence and feeding ecology of the Common Flicker on Grand Cayman Island. Condor 81:370-375.
Cruz, A., AND D. W. Johnston. 1984. Ecology of the West Indian Red-bellied Woodpecker on Grand Cayman: distribution and foraging. Wilson Bull. 96:366-379.
Culin, J. D., AND K. V. Yeargan. 1983. Comparative study of spider communities in alfalfa and soybean ecosystems: foliage dwelling spiders. Ann. Entomol. Soc. Am. 76:825-831.
Cummins, K. W., AND J. C. Wuycheck. 1971. Caloric equivalents for investigation in ecological energetics. Mitt. Interat. Verein. Limnol. 18:1-158.
Curio, E. 1976a. The ethology of predation. Springer-Verlag, New York.
Curio, E. 1976b. The ethology of predation. Zoophysiol. Ecol. 7:1-25.
Curry-Lindahl, K. 1981. Bird migration in Africa. Academic Press, London.
Custer, T. W., AND F. A. Pitelka. 1975. Correction factors for digestion rates for prey taken by Snow Buntings (Plectrophenax nivalis). Condor 77:210-212.
D'Arcy-Burt, S., AND R. P. Blackshaw. 1987. Effects of trap design on catches of grassland Bibionidae (Diptera: Nematocera). Bull. Entomol. Res. 77:309-315.
Dahlsten, D. L. 1967. Preliminary life tables for pine sawflies in the Neodiprion fulviceps complex (Hymenoptera: Diprionidae). Ecology 48:275-289.
Dahlsten, D. L. 1976. The third forest. Environment 18:35-42.
Dahlsten, D. L., AND W. A. Copper. 1979. The use of nesting boxes to study the biology of the Mountain Chickadee (Parus gambeli) and its impact on selected forest insects, p. 217-260. In J. G. Dickson, R. N. Conner, R. R. Fleet, J. A. Jackson, and J. C. Kroll [eds.], The role of insectivorous birds in forest ecosystems. Academic Press, New York.
Dahlsten, D. L., AND S. G. Herman. 1965. Birds as predators of destructive forest insects. Calif. Agric. 19:8-10.
Dahlsten, D. L., AND F. M. Stephen. 1974. Natural enemies and insect associates of the mountain pine beetle, Dendroctonus ponderosae (Coleoptera: Scolytidae), in sugar pine. Can. Entomol. 106:1211-1217.
Dahlsten, D. L., AND N. J. Mills. In press. Biological control of forest insects, in T. W. Fisher, et al. [eds.], Biological Control of Insect Pests and Weeds. Univ. of Calif. Press, Berkeley.
Dahlsten, D. L., R. F. Luck, E. I. Schlinger, J. M. Wenz, AND W. A. Copper. 1977. Parasitoids and predators of the Douglas-fir tussock moth, Orygia pseudotsugata (Lepidoptera: Lymantriidae), in low to moderate populations in central California. Can. Entomol. 107:727-746.
Daly, M., J. Rauschenberger, AND P. Behrends. 1982. Food aversion learning in kangaroo rats: a specialist-generalist comparison. Anim. Learn. Behav. 10:314-320.
Dammon, H. 1987. Leaf quality and enemy avoidance by larvae of a pyralid moth. Ecology 68:88-97.
Daudin, J. J. 1986. Selection of variables in mixed-variable discriminant analysis. Biometrics 42:473-481.
Davies, N. B. 1976. Food, flocking and territorial behavior of the Pied Wagtail (Motacilla yarrellii yarellii Gould) in winter. J. Anim. Ecol. 45:235-253.
Davies, N. B. 1977a. Prey selection and social behavior in wagtails (Aves: Motacillidae). J. Anim. Ecol. 46:37-57.
Davies, N. B. 1977b. Prey selection and the search strategy of the Spotted Flycatcher (Muscicapa striata): a field study on optimal foraging. Anim. Behav. 25:1016-1033.
Davies, N. B. 1980. The economics of territorial behavior in birds. Ardea 68:63-74.
Davies, N. B., AND R. E. Green. 1976. The development and ecological significance of feeding techniques in the Reed Warbler Acrocephalus scirpaceus. Anim. Behav. 24:213-229.
Davies, N. B., AND A. I. Houston. 1981. Owners and satellites: the economics of territory defence in the Pied Wagtail, Motacilla alba. J. Anim. Ecol. 50:157-180.
Davies, N. B., AND A. I. Houston. 1983. Time allocation between territories and flocks and owner-satellite conflict in foraging Pied Wagtails, Motacilla alba. J. Anim. Ecol. 52:621-634.

LITERATURE CITED

Davies, N. B., and A. Lundberg. 1985. The influence of food on time budgets and timing of breeding of the Dunnock, Prunella modularis. Ibis 127:100-110.

Davis, C. M. 1979. A nesting study of the Brown Creeper (Certhia familiaris). Living Bird 17:237-264.

Davis, D. E. 1945. The annual cycle of plants, mosquitos, birds, and mammals in two Brazilian forests. Ecol. Monogr. 15:243-295.

Davis, J. 1957. Comparative foraging behavior of the Spotted and Brown towhees. Auk 74:129-166.

Davis, J. 1961. Some seasonal changes in morphology of the Rufous-sided Towhee. Condor 63:313-321.

Davis, J. 1965. Natural history, variation, and distribution of the Strickland's Woodpecker. Auk 82:537-590.

Davison, G. W. H. 1981. Diet and dispersion of the great argus Argusianus argus. Ibis 123:485-494.

Dawson, T. J., and R. M. Herd. 1983. Digestion in the Emu: low energy and nitrogen requirements of this large ratite bird. Comp. Biochem. Physiol. 75A:41-45.

DeBach, P. [ed.]. 1964. Biological control of insect pests and weeds. Reinhold, New York.

DeBenedictus, P. A. 1966. The bill-brace feeding behavior of the Galapagos finch Geospiza conirostris. Condor 68:206-208.

DeLaet, J. 1984. Site-related dominance in the Great Tit Parus major major. Ornis Scand. 15:73-78.

DeLotelle, R. S., R. J. Epting, and J. R. Newman. 1987. Habitat use and territory characteristics of Red-cockaded Woodpeckers in central Florida. Wilson Bull. 99:202-217.

DeMars, C. J. 1970. Frequency distributions, data transformation and analysis of variation used in determination of optimum sample size and effort for broods of the western pine beetle, p. 42-65. In R. W. Stark and D. L. Dahlsten, [eds.], Studies on the population dynamics of the western pine beetle, Dendroctonus brevicomis LeConte (Coleoptera: Scolytidae). Univ. California Div. Agric. Sci., Berkeley.

Demment, M. W., and P. J. van Soest. 1985. A nutritional explanation for body-size patterns of ruminant and nonruminant herbivores. Am. Nat. 125:641-672.

Dempster, J. P. 1983. The natural control of populations of butterflies and moths. Biol. Rev. 58:461-481.

Denslow, J. S., and T. C. Moermond. 1982. The effect of accessibility on rates of fruit removal from neotropical shrubs: an experimental study. Oecologia 54:170-176.

Denslow, J. S., and T. C. Moermond. 1985. The interaction of fruit display and the forging strategies of small frugivorous birds, p. 245-253. In W. D'Arcy and M. D. Correa A. [eds.], The botany and natural history of Panama: la botanica y historia natural de Panama. Monogr. Syst. Botany, Vol. 10, Missouri Botanical Garden, St. Louis, MO.

Denslow, J. S., T. C. Moermond, and D. J. Levey. 1986. Spatial components of fruit display in understory trees and shrubs, p. 37-44. In A. Estrada and T. H. Fleming [eds.], Frugivores and seed dispersal. Dr. W. Junk, Publishers, Dordrecht, The Netherlands.

De Ruiter, L. 1952. Some experiments on the camouflage of stick caterpillars. Behaviour 4:222-232.

Dethier, V. G. 1970. Chemical interactions between plants and insects, p. 83-102. In E. Sondheimer and J. B. Simeone [eds.], Chemical ecology. Academic Press, New York.

De Weese, L. R., C. J. Henny, R. L. Floyd, K. A. Bobal, and A. W. Schultz. 1979. Response of breeding birds to aerial sprays of Trichlorofon (Dylox) and Carbaryl (Sevin-4-oil) in Montana forests. USDI Fish and Wildl. Ser. Spec. Sci. Rep. Wildl. No. 224, Washington, D.C.

Dial, K. P., and T. A. Vaughan. 1987. Opportunistic predation on alate termites in Kenya. Biotropica 19:185-187.

Diamond, J. M. 1970. Ecological consequences of island colonization by southwest Pacific birds. I. Types of niche shifts. Proc. Natl. Acad. Sci. 67:529-536.

Diamond, J. M. 1975. Assembly of species communities, p. 342-444. In M. Cody and J. M. Diamond [eds.], Ecology and evolution of communities. Harvard Press, Cambridge, MA.

Diamond, J. M. 1978. Niche shifts and the rediscovery of interspecific competition. Am. Sci. 66:322-331.

Diamond, J. M., and M. E. Gilpin. 1982. Examination of the "null" model of Connor and Simberloff for species co-occurrences on islands. Oecologia 52:64-74.

Diamond, J. M., and W. H. Karasov. 1987. Adaptive regulation of intestinal nutrient transporters. Proc. Natl. Acad. Sci. 84:2242-2245.

Dietrick, E. J. 1961. An improved back pack motor fan for suction sampling of insect populations. J. Econ. Entomol. 54:394-395.

Dietrick, E. J., E. I. Schlinger, and R. van der Bosch. 1959. A new method for suction sampling arthropods using a suction collecting machine and modified Berlese funnel separator. J. Econ. Entomol. 52:1085-1091.

Dilger, W. C. 1956. Adaptive modifications and evological isolating mechanisms in the thrush genera Catharus and Hylocichla. Wilson Bull. 68:171-199.

Dillery, D. G. 1965. Post-mortem digestion of stomach contents in the Savannah Sparrow. Auk 82:281.

Dillon, W. R., and M. Goldstein. 1984. Multivariate analysis: methods and applications. John Wiley and Sons, New York.

Dinerstein, E. 1986. Reproductive ecology of fruit bats and the seasonality of fruit production in a Costa Rican cloud forest. Biotropica 18:307-318.

Dixon, K. L. 1963. Some aspects of social organization in the Carolina Chickadee. Proc. XIII Int. Ornithol. Congr. 240-258.

Dixon, K. L. 1965. Dominance-subordinance relationships in Mountain Chickadees. Condor 67:291-299.

Dixon, W. J. 1983. BMDP Statistical Software. Univ. of California Press, Berkeley.

Dixon, W. J., M. B. Brown, C. Engelman, J. W. Frane, M. A. Hill, R. I. Jennrich, and J. D. Toporek. 1985. BMPD statistical software. Univ. of California Press, Berkeley.

Doane, J. F., and C. D. Dondale. 1979. Seasonal captures of spiders (Araneae) in a wheat field and its grassy borders in central Saskatchewan. Can. Entomol. 111:439-445.

Dobson, A. J. 1983. An introduction to statistical modelling. Chapman and Hall, New York.

Dolnik, V. R., and V. M. Gavrilov. 1973. Energy metabolism during flight of some passerines, p. 288-296. In B. E. Bykovskii [ed.], Bird migrations, ecological and physiological features. Halstead Press, New York.

Douglass, J. F. 1977. Prairie Warbler feeds from spider web. Wilson Bull. 89:158-159.

Douthwaite, R. J. 1976. Fishing techniques and food of the Pied Kingfisher on Lake Victoria in Uganda. Ostrich 47:153-160.

Dowden, P. B., H. A. Jaynes, and V. M. Carolin. 1953. The role of birds in a spruce budworm outbreak in Maine. J. Econ. Ent. 46:307-312.

Dowdy, S., and S. Wearden. 1983. Statistics for research. John Wiley and Sons, New York.

Dowell, R. V., and R. H. Cherry. 1981. Survey traps for parasitoids, and coccinellid predators of the citrus blackfly, Aleurocanthus woglum. Ent. Exp. and Appl. 29:356-362.

Drent, R. H., and S. Daan. 1980. The prudent parent: energetic adjustments in avian breeding. Ardea 68:225-252.

Driver, E. 1949. Mammal remains in owl pellets. Am. Midl. Nat. 41:139-142.

Drobney, R. D. 1984. Effect of diet on visceral morphology of breeding Wood Ducks. Auk 101:93-98.

Drodz, A. 1975. Food habits and food assimilation in mammals, p. 325-332. In W. Grodzinski, R. Z. Klekowski, and A. Duncan [eds.], Methods for ecological bioenergetics. IBP Handbook No. 24, Blackwell Sci. Publ.,

Oxford.
DRYDEN, P. A., G. P. JONES, E. BURCHER, AND R. S. D. READ. 1985. Effect of chronic ingestion of dietary fibre on the rate of glucose absorption in rats. Nutr. Rep. Int. 31:609-614.
DUESER, R. D., AND H. H. SHUGART. 1978. Microhabitats in a forest-floor small mammal fauna. Ecology 59:89-98.
DUESER, R. D., AND H. H. SHUGART. 1979. Niche pattern in a forest-floor small-mammal fauna. Ecology 60:108-118.
DUDLEY, C. O. 1971. A sampling design for eggs and first instar larval populations of the western pine beetle Dendroctonus brevicomis LeConte (Coleoptera: Scolytidae). Can. Entomol. 103:1291-1313.
DUFFY, D. C., AND S. JACKSON. 1986. Dietary studies of seabirds: a review of methods. Colonial Waterbirds 9:1-17.
DUFTY, A. M., JR. 1982. Movements and activities of radio-tracked Brown-headed Cowbirds. Auk 99:316-327.
DUKE, G. E., J. G. CIGANEK, AND O. A. EVANSON. 1973. Food consumption and energy, water, and nitrogen budgets in captive Great Horned Owls (Bubo virginianus). Comp. Biochem. Physiol. 44A:283-292.
DUKE, G. E., O. A. EVANSON, AND A. A. JEGERS. 1976. Meal to pellet intervals in 14 species of captive raptors. Comp. Biochem. Physiol. 53A:1-6.
DUKE, G. E., A. A. JEGERS, G. LOFF, AND O. EVANSON. 1975. Gastric digestion in some raptors. Comp. Biochem. Physiol. 50A:649-656.
DUKE, G. E., G. A. PETRIDES, AND R. K. RINGER. 1968. Chromium 51 in food metabolizability and passage rate studies with the ring-necked pheasant. Poult. Sci. 47:1356-1364.
DUNHAM, A. E. 1980. An experimental study of interspecific competition between the iguanid lizards Sceloporus merriami and Urosaurus ornatus. Ecol. Monogr. 50:309-330.
DUNN, E. H. 1975. Caloric intake of nestling Double-Crested Cormorants. Auk 92:553-565.
DUNNING, J. B. 1984. Body weights of 686 species of North American birds. Western Bird Banding Association, Monograph. No. 1.
DUNNING, J. B. 1986. Foraging choice in three species of Pipilo (Aves: Passeriformes): a test of the threshold concept. Ph.D. diss., Univ. of Arizona, Tucson.
DUNNING, J. B., JR., AND J. H. BROWN. 1982. Summer rainfall and winter sparrow densities: a test of the food limitation hypothesis. Auk 99:123-129.
DURBIN, J., AND G. S. WATSON. 1951. Testing for serial correlation in least square regression, II. Biometrika 38:159-178.
DURKIS, Y. J., AND R. M. REEVES. 1982. Barriers increase efficiency of pitfall traps. Entomol. News. 93:8-12.
DYBAS, H. S., AND D. D. DAVIES. 1962. A population census of seventeen-year periodical cicadas (Homoptera: Cicadidae: Magicicada). Ecology 43:444-459.
EARLÉ, R. A. 1983. Foraging overlap and morphological similarity among some insectivorous arboreal birds in an eastern Transvaal forest. Ostrich 54:36-42.
EARLÉ, R. A. 1985. Foraging behavior and diet of the South African Cliff Swallow Hirundo spilodera (Aves, Hirundinidae). Navors. Nas. Mus. (Bloemfontein) 5:53-56.
EAST, M. 1980. Sex differences and the effect of temperature on the foraging behavior of Robins Erithacus rubecula. Ibis 122:517-520.
EAST, M. L, AND H. HOFER. 1986. The use of radio-tracking for monitoring Great Tit Parus major behaviour: a pilot study. Ibis 128:103-114.
EATON, S. W. 1953. Wood warblers wintering in Cuba. Wilson Bull. 65:169-175.
EBBINGE, B., K. CANTERS, AND R. DENT. 1975. Foraging routines and estimated daily food intake in Barnacle Geese wintering in the northern Netherlands. Wildfowl 26:5-19.

ECKHARDT, R. C. 1979. The adaptive syndromes of two guilds of insectivorous birds in the Colorado Rocky Mountains. Ecol. Monogr. 49:129-149.
EDINGTON, J. S. L. 1983. White's Thrush: some aspects of its ecology and feeding behavior. S. Austr. Ornithol. 29:57-59.
EDMUNDS, M. 1974. Defence in animals. Longman, London.
EFRON, B. 1982. The jackknife, the bootstrap and other resampling plans. Society for Industrial and Applied Mathematics, Philadelphia, PA.
EFRON, B., AND G. GONG. 1983. A leisurely look at the bootstrap, the jackknife, and cross-validation. Am. Stat. 37:36-48.
EGAN, J. P. 1975. Signal detection theory and ROC analysis. Academic Press, New York.
EHRLICH, P. R., AND J. ROUGHGARDEN. 1987. The science of ecology. MacMillan Publ. Co., New York.
EICHINGER, J., AND D. J. MORIARTY. 1985. Movement of Mojave Desert sparrow flocks. Wilson Bull. 97:511-516.
EISENMANN, E. 1961. Favorite foods of neotropical birds: flying termites and Cecropia catkins. Auk 78:636-637.
EISNER, T. 1970. Chemical defense against predation in arthropods, p. 157-217. In E. Sondheimer and J. B. Simeone [eds.], Chemical ecology. Academic Press, New York.
EISNER, T., J. S. JOHNESSEE, J. CARREL, L. B. HENDRY AND J. MEINWALD. 1974. Defense use by an insect of a plant resin. Science 184:996-999.
EKMAN, J. 1979. Coherence, composition and territories of winter social groups of the Willow Tit Parus montanus and the Crested Tit P. cristatus. Ornis Scand. 10:56-68.
EKMAN, J. 1986. Tree use and predator vulnerability of wintering passerines. Ornis Scand. 17:261-267.
EKMAN, J. 1987. Exposure and time use in Willow Tit flocks: the cost of subordination. Anim. Behav. 35:445-452.
EKMAN, J., AND C. ASKENMO. 1984. Social rank and habitat use in Willow Tit groups. Anim. Behav. 32:508-514.
EKMAN, J., G. CEDERHOLM, AND C. ASKENMO. 1981. Spacing and survival in winter groups of Willow Tit Parus montanus and Crested Tit P. cristatus. A removal study. J. Anim. Ecol. 50:1-9.
ELEY, J. W., G. R. GRAVES, T. A. PARKER, III, AND D. R. HUNTER. 1979. Notes on Siptornis striaticollis (Furnariidae) in Peru. Condor 81:319.
ELLINGTON, J., K. KISER, G. FERGUSON, AND M. CARDENAS. 1984. A comparison of sweep-net, absolute, and Insectivac sampling methods in cotton ecosystems. J. Econ. Entomol. 77:599-605.
ELLIOT, B. G. 1969. Life history of the Red Warbler. Wilson Bull. 81:184-195.
ELLIOTT, W. M., AND W. G. KEMP. 1979. Flight activity of the green peach aphid (Homoptera: Aphididae) during the vegetable growing season at Harrow and Jordan, Ontario. Proc. Entomol. Soc. Ont. 110:19-28.
ELLIS, J. E., J. A. WIENS, C. F. RODELL, AND J. C. ANWAY. 1976. A conceptual model of diet selection as an ecosystem process. J. Theor. Biol. 60:93-108.
EL-WAILLY, A. J. 1966. Energy requirements for egg laying and incubation in the Zebra Finch (Taeniopygia castanotis). Condor 68:582-594.
EMLEN, J. M. 1966. The role of time and energy in food preference. Am. Nat. 100:611-617.
EMLEN, J. T. 1971. Population densities of birds derived from transect counts. Auk 88:323-342.
EMLEN, J. T. 1977. Land bird communities of Grand Bahama Island: the structure and dynamics of an avifauna. Ornithol. Monogr. 24:1-129.
EMLEN, J. T. 1981. Divergence in the foraging responses of birds on two Bahama Islands. Ecology 62:289-295.
EMLEN, J. T., M. J. DEJONG, M. J. JAEGER, T. C. MOERMOND, K. A. RUSTERHOLZ, AND R. P. WHITE. 1986. Density trends and range boundary constraints of forest birds along a

latitudinal gradient. Auk 103:791-803.
ENDLER, J. A. 1986. Natural selection in the wild. Princeton Univ. Press, Princeton, NJ.
ENGELS, W. L. 1940. Structural adaptations in thrashers (Mimidae; genus Toxostoma) with comments on interspecific relationships. Univ. Calif. Publ. Zool. 42:341-400.
ERRINGTON, P. L. 1930. The pellet analysis method of raptor food habits study. Condor 32:292-296.
ERRINGTON, P. L. 1932. Techniques of raptor food habits study. Condor 34:75-86.
ERWIN, R. M. 1985. Foraging decisions, patch use, and seasonality in egrets (Aves: Ciconiiformes). Ecology 66:837-844.
ERWIN, T. L. 1985. Tambopata Reserve Zone, Madre de Dios, Peru: history and description of the reserve. Rev. Peruana Entomol. 27:1-8.
ESJBERG, P. 1987. The influence of diurnal time and weather on sex trap catches of the turnip moth (Agrotis segetum Schiff.) (Lep. Noctuidae). J. Appl. Entomol. 103:177-184.
ESTRADA, A., AND R. COATES-ESTRADA. 1986. Frugivory in howling monkeys (Alouatta palliata) at Los Tuxtlas, Mexico: dispersal and fate of seeds, p. 93-104. In A. Estrada and T. H. Fleming [eds.], Frugivores and seed dispersal. Dr. W. Junk Publishers, Dordrecht, The Netherlands.
EVANS, F. C., AND W. W. MURDOCH. 1968. Taxonomic composition, trophic structure, and seasonal occurrence in a grassland insect community. J. Anim. Ecol. 37:259-273.
EVANS, K. E., AND D. R. DIETZ. 1974. Nutritional energetics of Sharp-tailed Grouse during winter. J. Wildl. Manage. 38:622-629.
EVANS, P. R. 1976. Energy balance and optimal foraging strategies in shorebirds: some implications for their distributions and movements in the non-breeding season. Ardea 64:117-139.
EVERITT, B. S. 1977. The analysis of contingency tables. Chapman and Hall, London.
FARNER, D. S. 1955. The annual stimulus for migration: experimental and physiologic aspects, p. 198-237. In A. Wolfson [ed.], Recent studies in avian biology. Univ. Illinois Press, Urbana.
FASHAM, M. J. R. 1977. A comparison of nonmetric multidimensional scaling, principal components and reciprocal averaging for the ordination of simulated coenoclines and coenoplanes. Ecology 58:551-561.
FAY, H. A. C., AND B. M. DOUBE. 1987. Aspects of the population dynamics of adults of Haematobia thirouxi potans (Bezzi) (Diptera: Muscidae) in southern Africa. Bull. Entomol. Res. 77:135-144.
FEDDUCIA, A. 1973. Evolutionary trends in the neotropical ovenbirds and woodhewers. Ornithol. Monogr. 13:1-69.
FEENY, P. P. 1970. Seasonal changes in oak leaf tannins and nutrients as a cause of spring feeding by winter moth caterpillars. Ecology 51:565-581.
FEINSINGER, P. 1976. Organization of a tropical guild of nectarivorous birds. Ecol. Monogr. 46:257-291.
FEINSINGER, P. 1978. Ecological interactions between plants and hummingbirds in a successional tropical community. Ecol. Monogr. 48:269-287.
FEINSINGER, P. 1983. Variable nectar secretion in a Heliconia species pollinated by Hermit Hummingbirds. Biotropica 15:48-52.
FEINSINGER, P., AND R. K. COLWELL. 1978. Community organization among neotropical nectar-feeding birds. Am. Zool. 18:779-795.
FEINSINGER, P., AND L. A. SWARM. 1982. "Ecological release," seasonal variation in food supply, and the hummingbird Amazilia tobaci on Trinidad and Tobago. Ecology 63:1574-1587.
FEINSINGER, P., L. A. SWARM, AND J. A. WOLFE. 1985. Nectar-feeding birds on Trinidad and Tobago: comparison of diverse and depauperate guilds. Ecol. Monogr. 55:1-28.
FEINSINGER, P., W. H. BUSBY, K. G. MURRAY, J. H. BEACH, W. Z. POUNDS, AND Y. B. LINHART. 1988. Mixed support for spatial heterogeneity in species interactions: hummingbirds in a tropical disturbance mosaic. Am. Nat. 131:33-57.
FELLIN, D. G., AND J. E. DEWEY. 1982. Western spruce budworm. USDA For. Serv. Forest Insect and Disease Leaflet No. 53, Washington, D.C.
FENNA, L., AND D. A. BOAG. 1974. Filling and emptying of the galliform caecum. Can. J. Zool. 52:537-540.
FENTON, F. A., AND D. E. HOWELL. 1957. A comparison of five methods of sampling alfalfa fields for arthropod populations. Ann. Entomol. Soc. Am. 50:606-611.
FENTON, M. B. 1982. Echolocation, insect hearing, and feeding ecology of insectivorous bats, p. 261-285. In T. H. Kunz [ed.], Ecology of bats. Plenum Press, New York.
FENTON, M. B., AND J. H. FULLARD. 1981. Moth hearing and the feeding strategies of bats. Am. Sci. 69:265-274.
FENTON, M. B., AND D. W. THOMAS. 1980. Dry-season overlap in activity patterns, habitat use, and prey selection by sympatric African insectivorous bats. Biotropica 12:81-90.
FERGUSON, T. S. 1967. Mathematical Statistics. Academic Press, New York.
FICKEN, M. S., AND R. W. FICKEN. 1968. Ecology of Blue-winged Warblers, Golden-winged Warblers and some other Vermivora. Am. Midl. Nat. 79:311-319.
FICKEN, M. S., S. R. WITKIN, AND C. M. WEISE. 1981. Associations among members of a Black-capped Chickadee flock. Behav. Ecol. Sociobiol. 8:245-249.
FICKEN, R. W., M. S. FICKEN, AND D. H. MORSE. 1968. Competition and character displacement in two sympatric pine-dwelling warblers (Dendroica, Parulidae). Evolution 22:307-314.
FIENBERG, S. E. 1970. The analysis of multidimensional contingency tables. Ecology 51:419-433.
FIENBERG, S. E. 1977. The analysis of cross-classified categorical data. MIT Press, Cambridge, MA.
FIENBERG, S. E. 1980. The analysis of cross-classified categorical data. 2nd ed. MIT Press, Cambridge, MA.
FILLMAN, D. A., W. L. STERLING, AND D. A. DEAN. 1983. Precision of several sampling techniques for foraging red imported fire ant (Hymenoptera: Formicidae) workers in cotton fields. J. Econ. Entomol. 76:748-751.
FINDLEY, J. S. 1976. The structure of bat communities. Am. Nat. 110:129-139.
FISCHER, D. H. 1981. Wintering ecology of thrashers in southern Texas. Condor 83:340-346.
FISCHER, D. H. 1983. Growth, development, and food habits of nestling mimids in south Texas. Wilson Bull. 95:97-105.
FISHER, J., AND R. A. HINDE. 1949. The opening of milk bottle by birds. Br. Birds 42:347-357.
FITZPATRICK, J. W. 1980. Foraging behavior of Neotropical tyrant flycatchers. Condor 82:43-57.
FITZPATRICK, J. W. 1981. Search strategies of tyrant flycatchers. Anim. Behav. 29:810-821.
FITZPATRICK, J. W. 1985. Form, foraging behavior, and adaptive radiation in the Tyrannidae, p. 447-470. In P. A. Buckley, M. S. Foster, E. S. Morton, R. S. Ridgely, and F. G. Buckley [eds.], Neotropical ornithology. Ornithol. Monogr. No. 36. American Ornithologists' Union, Washington, D.C.
FJELDSA, J., N. KRABBE, AND T. A. PARKER, III. 1987. Rediscovery of Cinclodes excelsior aricomae and notes on the nominate race. Bull. Br. Ornithol. Club 107:112-114.
FLEISCHER, S. J., W. A. ALLEN, J. M. LUNA, AND R. L. PEINKOWSKI. 1982. Absolute-density estimation from sweep sampling, with a comparison of absolute-density sampling techniques for adult potato leafhopper in alfalfa. J. Econ. Entomol. 75:425-430.
FLEISS, J. L. 1981. Statistical methods for rates and proportions. 2nd ed. John Wiley and Sons, New York.

FLEMING, T. H. 1979. Do tropical frugivores compete for food? Am. Zool. 19:1157-1172.

FLEMING, T. H. 1981. Fecundity, fruiting pattern, and seed dispersal in Piper amalgo (Piperaceae), a bat-dispersed tropical shrub. Oecologia 51:42-46.

FLINT, E. N., AND K. A. NAGY. 1984. Flight energetics of free-living sooty terns. Auk 101:288-294.

FOLSE, L. J., JR. 1982. An analysis of avifauna-resource relationships on the Serengeti Plains. Ecol. Monogr. 52:111-127.

FORBUSH, E. H. 1924. Gulls and terns feeding on the seventeen-year cicada. Auk 41:468-470.

FORD, H. A. 1979. Interspecific competition in Australian honeyeaters--a depletion of common resources. Aust. J. Ecol. 4:145-164.

FORD, H. A. 1985. A synthesis of the foraging ecology and behavior of birds of eucalypt forests and woodlands, p. 249-254. In A. Keast, H. F. Recher, H. Ford, and D. Saunders [eds.], Birds of eucalypt forests and woodlands: ecology, conservation, management. Surrey Beatty & Sons Pty. Ltd., Chipping Norton, N.S.W., Australia.

FORD, H. A., AND D. C. PATON. 1982. Partitioning of nectar sources in an Australian honeyeater community. Aust. J. Ecol. 7:149-159.

FORD, H. A., L. BRIDGES, AND S. NOSKE. 1985. Density of birds in eucalypt woodland near Armidale, north-eastern New South Wales. Corella 9:97-107.

FORD, H. A., N. FORDE, AND S. HARRINGTON. 1982. Non-destructive methods to determine the diets of birds. Corella 6:6-10.

FORD, H. A., S. NOSKE, AND L. BRIDGES. 1986. Foraging behaviour of birds in eucalypt woodland in north-eastern New South Wales. Emu 86:168-179.

FORSHAW, J. M. 1973. Parrots of the world. Doubleday, New York.

FORSHAW, J. M., AND W. T. COOPER. 1979. The birds of paradise. Godine, Boston.

FOSTER, M. S. 1977. Ecological and nutritional effects of food scarcity on a tropical frugivorous bird and its fruit source. Ecology 58:73-85.

FOSTER, M. S. 1987. Feeding methods and efficiencies of selected frugivorous birds. Condor 89:566-580.

FOSTER, R. B. 1982a. The seasonal rhythm of fruit fall on Barro Colorado Island, p. 151-172. In E. Leigh, Jr., A. S. Rand, and D. Windsor [eds.], The ecology of a tropical forest. Smithsonian Instit. Press, Washington, D.C.

FOSTER, R. B. 1982b. Famine on Barro Colorado Island, p. 201-212. In E. Leigh, Jr., A. S. Rand, and D. Windsor [eds.], The ecology of a tropical forest. Smithsonian Instit. Press, Washington, D.C.

FOWLER, N. L., AND D. W. DUNLAP. 1986. Grassland vegetation of the eastern Edwards plateau. Am. Midl. Nat. 115:146-155.

FOX, L. R., AND P. A. MORROW. 1981. Specialization: species property or local phenomenon? Science 211:887-893.

FRAENKEL, G., AND K. M. RUDELL. 1947. The structure of insect cuticles. Proc. Royal Soc. Lond. B 134:111-144.

FRAKES, R. A., AND R. E. JOHNSON. 1982. Niche convergence in Empidonax flycatchers. Condor 84:286-291.

FRANKIE, G. W., H. G. BAKER, AND P. A. OPLER. 1974. Comparative phenological studies of trees in tropical wet and dry forests in the lowlands of Costa Rica. J. Ecology 62:881-919.

FRANKLIN, J. F., AND C. T. DYRNESS. 1973. The natural vegetation of Oregon and Washington. USDA For. Serv. Gen. Tech. Rep. PNW-8, Portland, OR.

FRANZBLAU, M. A., AND J. P. COLLINS. 1980. Test of a hypothesis of territory regulation by an insectivorous bird by experimentally increasing prey abundance. Oecologia 46:164-170.

FRANZREB, K. E. 1978. Tree species used by birds in logged and unlogged mixed-coniferous forest. Wilson Bull. 90:221-238.

FRANZREB, K. E. 1983a. A comparison of avian foraging behavior in unlogged and logged mixed-coniferous forest. Wilson Bull. 95:60-76.

FRANZREB, K. E. 1983b. Intersexual habitat partitioning in Yellow-rumped Warblers during the breeding season. Wilson Bull. 86:146-150.

FRANZREB, K. E. 1984. Foraging habits of Ruby-crowned and Golden-crowned kinglets in an Arizona montane forest. Condor 86:139-145.

FRANZREB, K. E. 1985. Foraging ecology of Brown Creepers in a mixed-coniferous forest. J. Field Ornithol. 56:9-16.

FRASER, D. F. 1976. Coexistence of salamanders in the genus Plethodon: a variation of the Santa Rosalia theme. Ecology 57:238-251.

FRASER, W. 1983. Foraging patterns of some South African flycatchers. Ostrich 54:150-155.

FREED, A. N. 1980. Prey selection and feeding by the green tree frog (Hyla cinerea). Ecology 61:461-465.

FREEMAN, H. J. 1984. Amino acid and dipeptide absorption in rats fed chemically defined diets of differing fiber composition. Can. J. Physiol. Pharmacol. 62:1097-1101.

FRETWELL, S. D. 1968. Habitat distribution and survival in the Field Sparrow (Spizella pusilla). Bird Banding 40:1-25.

FRETWELL, S. D. 1969. Dominance behavior and winter habitat distribution in juncos (Junco hyemalis). Bird Banding 40:1-25.

FRETWELL, S. D. 1972. Populations in a seasonal environment. Monogr. Popul. Biol. No. 5. Princeton Univ. Press, Princeton, N.J.

FRETWELL, S. D., AND J. L. LUCAS, JR. 1970. On territorial behavior and other factors influencing habitat distribution in birds. I. Theoretical development. Acta Biotheor. 19:16-36.

FRITH, D. W. 1984. Foraging ecology of birds in an upland tropical rainforest in north Queensland. Aust. Wildl. Res. 11:325-347.

FRITH, H. J. 1969. Birds in the Australian high country. Reed, Sydney, Australia.

FRITH, H. J. 1976. Reader's Digest complete book of Australian birds. Reader's Digest, Sydney.

FRY, C. H. 1969. The recognition and treatment of venomous and non-venomous insects by small bee-eaters. Ibis 111:23-29.

FRY, C. H. 1980. The evolutionary biology of kingfishers (Alcedinidae). Living Bird 18:113-160.

FRY, C. H. 1984. The bee-eaters. Calton, London.

FUTUYMA, D. H. 1986. Evolutionary biology. 2nd ed. Sinauer Assoc., Sunderland, MA.

GABRIEL, K. R. 1971. The biplot graphic display of matrices with application to principal component analysis. Biometrika 58:453-467.

GAGE, S. H., C. A. MILLER, AND L. J. MOOK. 1970. The feeding response of some forest birds to the black-headed budworm. Can. J. Zool. 48:359-366.

GARBUTT, K., AND A. R. ZANGERL. 1983. Application of genotype-environment interaction analysis to niche quantification. Ecology 64:1292-1296.

GARRETT, W. N., AND N. HINMAN. 1969. Re-evaluation of the relationship between carcass density and body composition of beef steers. J. Anim. Sci. 28:1-5.

GARTON, E. O. 1987. Habitat requirements of avian predators, p. 82-85. In M. H. Brookes, R. W. Campbell, J. J. Colbert, R. G. Mitchell, and R. W. Stark [tech. coords.], Western spruce budworm. USDA For. Serv. Tech. Bull. 1694, Washington, D.C.

GARTSHORE, R. G., R. J. BROOKS, J. D. SOMERS, AND F. F. GILBERT. 1979. Temporal change on gullet food passage in penned Red-winged Blackbirds (Agelaius phoeniceus): significance for research in feeding ecology. Can. J. Zool. 57:1592-1596.

GASAWAY, W. C. 1976a. Seasonal variation in diet, volatile

fatty acid production, and size of the cecum of Rock Ptarmigan. Comp. Biochem. Physiol. 53A:109-114.
GASAWAY, W. C. 1976b. Volatile fatty acids and metabolizable energy derived from cecal fermentation in the Willow Ptarmigan. Comp. Biochem. Physiol. 53A:115-121.
GASAWAY, W. C., D. F. HOLLEMAN, AND R. G. WHITE. 1975. Flow of digesta in the intestine and cecum of the Rock Ptarmigan. Condor 77:467-474.
GASS, C. L. 1978. Rufous Hummingbird feeding territoriality in a suboptimal habitat. Can. J. Zool. 56:1535-1539.
GASS, C. L. 1979. Territory regulation, tenure and migration in Rufous Hummingbirds. Can. J. Zool. 57:914-923.
GASS, C. L., AND K. P. LERTZMAN. 1980. Capricious mountain weather: a driving variable in hummingbird territorial dynamics. Can. J. Zool. 58:1964-1968.
GASTON, A. J. 1974. Adaptation in the genus Phylloscopus. Ibis 116:432-450.
GATZ, A. J., JR. 1980. Phenetic packing and community structure: a methodological comment. Am. Nat. 116:147-149.
GAUCH, H. G., JR. 1982a. Multivariate analysis in community ecology. Cambridge Univ. Press, Cambridge, England.
GAUCH, H. G., JR. 1982b. Noise reduction by eigenvector ordination. Ecology 63:1643-1649.
GAUCH, H. G., JR., R. H. WHITTAKER, AND S. B. SINGER. 1981. A comparative study of nonmetric ordinations. J. Ecol. 69:135-152.
GAUCH, H. G., JR., R. H. WHITTAKER, AND T. R. WENTWORTH. 1977. A comparative study of reciprocal averaging and other ordination techniques. J. Ecol. 65:157-174.
GAUTHREAUX, S. A., JR. 1978. The ecological significance of behavioral dominance, p. 17-54. In P. P. G. Bateson and P. H. Klopfer [eds.], Perspectives in ethology. Vol. 3. Plenum Press, New York.
GAUTHREAUX, S. A., JR. 1982. The ecology and evolution of avian migration systems, p. 93-168. In D. S. Farner, J. R. King, and K. C. Parkes [eds.], Avian biology. Vol. 6. Academic Press, New York.
GAUTHREAUX, S. A., JR. 1985. The temporal and spatial scales of migration in relation to environmental changes in time and space, p. 503-515. In M. A. Rankin [ed.], Migration: mechanisms and adaptive significance. Contributions in Marine Science Suppl. vol. 27. Univ. of Texas Marine Sci. Instit. Port Aransas, Texas.
GAUTIER-HION, A., J. P. GAUTIER, AND R. QURIS. 1981. Forest structure and fruit availability as complementary factors influencing habitat use by a troop of monkeys (Cercopithecus cephus). Revue d'Ecologie 35:511-536.
GAUTIER-HION, A., J.-M. DUPLANTIER, R. QURIS, F. FEER, C. SOURD, J.-P. DECOUX, G. DUBOST, L. EMMONS, C. ERARD, P. HECKETSWEILER, A. MOUNGAZI, C. FOUSSILHON, AND J. M. THIOLLAY. 1985. Fruit characters as a basis of fruit choice and seed dispersal in a tropical forest vertebrate community. Oecologia 65:324-337.
GAVETT, A. P., AND J. S. WAKELEY. 1986. Diets of House Sparrows in urban and rural habitats. Wilson Bull. 98:137-144.
GENDRON, R. F., AND J. E. R. STADDON. 1983. Searching for cryptic prey: the effect of search rate. Am. Nat. 121:172-186.
GENTRY, A. H. 1982. Patterns of neotropical plant species diversity. Evol. Biol. 15:1-84.
GESSAMAN, J. A. 1972. Bioenergetics of the Snow Owl (Nyctea scandiaca). Arctic and Alpine Res. 4:223-238.
GETTY, T. 1985. Discriminability and the sigmoid functional response: how optimal foragers could stabilize model-mimic complexes. Am. Nat. 125:239-256.
GETTY, T., AND J. R. KREBS. 1985. Lagging partial preferences for cryptic prey: a signal detection analysis of Great Tit foraging. Am. Nat. 125:39-60.
GETTY, T., A. C. KAMIL, AND P. G. REAL. 1987. Signal detection theory and foraging for cryptic and mimetic prey, p. 525-548. In A. C. Kamil, J. R. Krebs, and H. R. Pulliam [eds.], Foraging behavior. Plenum Press, New York.
GIBB, J. 1950. The breeding biology of Great and Blue titmice. Ibis 92:507-539.
GIBB, J. 1954. Feeding ecology of tits, with notes on Treecreeper and Goldcrest. Ibis 96:513-543.
GIBB, J. A. 1957. Food requirements and other observations on captive tits. Bird Study 4:207-215.
GIBB, J. A. 1958. Predation by tits and squirrels on the eucosmid Ernarmonia conicolana (Heyl.). J. Anim. Ecol. 27:375-396.
GIBB, J. A. 1960. Populations of tits and Goldcrests and their food supply in pine plantations. Ibis 102:163-208.
GIBB, J. A. 1962. L. Tinbergen's hypothesis of the role of specific search images. Ibis 104:106-111.
GIBB, J. A., AND M. M. BETTS. 1963. Food and food supply of nestling tits (Paridae) in Breckland Pine. J. Anim. Ecol. 32:489-533.
GIBSON, A. R., A. J. BAKER, AND A. MOEED. 1984. Morphometric variation in introduced populations of the common myna (Acridotheres tristis): an application of the jackknife to principal component analysis. Syst. Zool. 33:408-421.
GILBERT, L. I. 1967. Lipid metabolism and function in insects. Adv. Insect Physiol. 4:70-211.
GILL, F. B. 1971. Ecology and evolution of the sympatric Mascarene White-eyes, Zosterops borbonica and Zosterops olivacea. Auk 88:35-60.
GILL, F. B., AND L. L. WOLF. 1975a. Economics of territoriality in the Golden-winged Sunbird. Ecology 56:333-345.
GILL, F. B., AND L. L. WOLF. 1975b. Foraging strategies and energetics of East African sunbirds at mistletoe flowers. Am. Nat. 109:491-510.
GILL, F. B., AND L. L. WOLF. 1977. Nonrandom foraging by sunbirds in a patchy environment. Ecology 58:1284-1296.
GILL, F. B., AND L. L. WOLF. 1978. Comparative foraging efficiencies of some montane sunbirds in Kenya. Condor 80:391-400.
GILL, F. B., AND L. L. WOLF. 1979. Nectar loss by Golden-winged Sunbirds to competitors. Auk 96:448-461.
GILLIAM, J. F. 1982. Foraging under mortality risk in size-structured populations. Ph.D. diss., Michigan State Univ., East Lansing, MI.
GILPIN, M. E., AND J. M. DIAMOND. 1982. Factors contributing to non-randomness in species co-occurrences on islands. Oecologia 52:75-84.
GILPIN, M. E., AND J. M. DIAMOND. 1984. Are species co-occurrences on islands non-random, and are null hypotheses useful in community ecology? p. 297-315. In D. R. Strong, Jr., D. Simberloff, L. G. Abele, and A. B. Thistle [eds.], Ecological communities: conceptual issues and the evidence. Princeton Univ. Press, Princeton, NJ.
GIRALDEAU, L.-A. 1984. Group foraging: the skill-pool effect and frequency-dependent learning. Am. Nat. 124:72-79.
GIRALDEAU, L.-A., AND L. LEFEBVRE. 1987. Scrounging prevents cultural transmission of food-finding behaviour in pigeons. Anim. Behav. 35:387-394.
GIUNTOLI, M., AND L. R. MEWALDT. 1978. Stomach contents of Clark's Nutcrackers collected in western Montana. Auk 95:595-598.
GLASE, J. C. 1969. Ecology of social organization in the Black-capped Chickadee. Living Bird 12:235-267.
GLASSER, J. W. 1982. A theory of trophic strategies: the evolution of facultative specialists. Am. Nat. 119:250-262.
GLASSER, J. W. 1984. Evolution of efficiencies and strategies of resource exploitation. Ecology 65:1570-1578.
GLEN, D. M., N. F. MILSON, AND C. W. WILTSHIRE. 1981.

The effect of predation by Blue Tits (Parus caerulus) on the sex ratio of codling moths (Cydia pomonella). J. Appl. Ecol. 18:133-140.

GLICKMAN, S., AND R. SROGES. 1966. Curiosity in zoo animals. Behaviour 22:151-188.

GNANADESIKAN, R. 1977. Methods for statistical data analysis of multivariate observations. John Wiley and Sons, New York.

GOCHFELD, M., AND J. BURGER. 1984. Age differences in foraging behavior of the American Robin (Turdus migratorius). Behaviour 88:227-239.

GOERING, H. K., AND P. J. VAN SOEST. 1970. Forage fiber analyses: apparatus, reagents, procedures, and some applications. USDA Agric. Handb. 379, Washington, D.C.

GOLDSTEIN, D. L. 1984. The thermal environment and its constraint on activity of desert quail in summer. Auk 101:542-550.

GOLDSTEIN, D. L. 1980. Estimates of daily energy expenditure in birds: the time-energy budget as an indicator of laboratory and field studies. Am. Zool. 28:829-844.

GOLDSTEIN, D. L., AND K. A. NAGY. 1985. Resource utilization by desert quail: time and energy, food and water. Ecology 66:378-387.

GOLLEY, F. B. 1961. Energy values of ecological materials. Ecology 42:581-584.

GOODMAN, L. A. 1965. On simultaneous confidence intervals for multinomial proportions. Technometrics 7:247-254.

GORDON, M. S., G. A. BARTHOLOMEW, A. D. GRINNELL, C. B. JORGENSON, AND F. N. WHITE. 1968. Animal function: principles and adaptations. MacMillan Publ. Co., Inc., New York.

GOSS-CUSTARD, J. D. 1969. The winter feeding ecology of the Redshank Tringa totanus. Ibis 111:388-356.

GOSS-CUSTARD, J. D. 1970. The responses of Redshanks (Tringa totanus (L.)) to spatial variations in the density of their prey. J. Anim. Ecol. 39:91-113.

GOSS-CUSTARD, J. D. 1977a. The energetics of prey selection by Redshank Tringa totanus (L.) in relation to prey density. J. Anim. Ecol. 46:1-19.

GOSS-CUSTARD, J. D. 1977b. Responses of Redshank, Tringa totanus, to the absolute and relative densities of two prey species. J. Anim. Ecol. 46:867-874.

GOSS-CUSTARD, J. D. 1977c. Optimal foraging and the size selection of worms by Redshank Tringa totanus. Anim. Behav. 25:10-29.

GOSS-CUSTARD, J. D., AND S. E. A. L. DURRELL. 1987. Age-related effects in Oystercatchers, Haematopus ostralegus, feeding on mussels, Mytilus edulis. 1. Foraging efficiency and interference. J. Anim. Ecol. 56:537-549.

GOSS-CUSTARD, J. D., J. T. CRAYFORD, J. T. BOATES, AND S. E. A. LE V. DIT DURELL. 1987. Field tests of the accuracy of estimating prey size from bill length in oystercatchers, Haematopus ostralegus, eating mussels, Mytilus edulis. Anim. Behav. 35:1078-1083.

GOULD, J. L. 1974. Genetics and molecular ethology. Z. Tierpsychol. 36:267-292.

GOULD, S. J., AND LEWONTIN, R. C. 1979. The spandrels of San Marco and the panglossian paradigm: a critique of the adaptationist programme. Proc. Roy. Soc. Lond. B. Biol. 205:581-598.

GOWER, J. C., AND P. G. N. DIGBY. 1981. Expressing complex relationships in two dimensions, p. 83-118. In V. Barnett [ed.], Looking at multivariate data. John Wiley and Sons, New York.

GRABER, J. W., AND R. R. GRABER. 1983. Feeding rates of warblers in spring. Condor 85:139-150.

GRABER, R. R. 1962. Food and oxygen consumption in three species of owls (Strigidae). Condor 64:473-487.

GRABER, R. R., AND S. L. WUNDERLE. 1966. Telemetric observations of a robin (Turdus migratorius). Auk 83:674-677.

GRACE, J. B., AND R. G. WETZEL. 1981. Habitat partitioning and competitive displacement in cattails (Typha): experimental field studies. Am. Nat. 118:463-474.

GRADWOHL, J., AND R. GREENBERG. 1980. The formation of antwren flocks on Barro Colorado Island, Panama. Auk 97:385-395.

GRADWOHL, J., AND R. GREENBERG. 1982a. The breeding season of antwrens on Barro Colorado Island, p. 345-351. In E. G. Leigh, A. S. Rand, and D. Windsor [eds.], The ecology of a tropical forest: seasonal rhythms and long-term changes. Smithsonian Instit. Press, Washington, D.C.

GRADWOHL, J., AND R. GREENBERG. 1982b. The effect of a single species of avian predator on the arthropods of aerial leaf litter. Ecology 63:581-583.

GRADWOHL, J., AND R. GREENBERG. 1984. Search behavior of the checker-throated antwren foraging in aerial leaf litter. Behav. Ecol. Sociobiol. 15:281-185.

GRAJAL, A., S. D. STRAHL, R. PARRA, M. G. DOMINGUEZ, AND A. NEHER. 1989. Foregut fermentation in the Hoatzin, a neotropical leaf-eating bird. Science 245:1236-1238.

GRANDY, J. W., IV. 1972. Digestion and passage of blue mussels eaten by Black Ducks. Auk 89:189-190.

GRANT, P. R. 1966. Further information on the relative length of the tarsus in land birds. Postilla 98:1-13.

GRANT, P. R. 1985. Climatic fluctuations on the Galapagos Islands and their influence on Darwin's finches, p. 471-483. In P. A. Buckley, M. S. Foster, E. S. Morton, R. S. Ridgely, and F. G. Buckley [eds.], Neotropical ornithology. Ornithol. Monogr. No. 36.

GRANT, P. R. 1986. Ecology and evolution of Darwin's finches. Princeton Univ. Press, Princeton, NJ.

GRANT, P. R., AND B. R. GRANT. 1980. Annual variation in finch numbers, foraging and food supply on Isla Daphne Major, Galapagos. Oecologia 46:55-62.

GRAUL, W. D., J. TORRES, AND R. DENNEY. 1976. A species-ecosystem approach for nongame programs. Wildl. Soc. Bull. 4:79-80.

GRAY, L. 1979. The use of psychophysical unfolding theory to determine principal resource axes. Am. Nat. 114:695-706.

GRAY, L. 1981. Genetic and experiential differences affecting foraging behavior, p. 455-473. In A. C. Kamil and T. D. Sargent [eds.], Foraging behavior: ecological, ethological, and psychological approaches. Garland STPM Press, New York.

GRAY, L., AND J. A. KING. 1986. The use of multidimensional scaling to determine principal resource axes. Am. Nat. 127:577-592.

GRAY, R. D. 1987. Faith and foraging: a critique of the "paradigm argument from design," p. 69-140. In A. C. Kamil, J. R. Krebs, and H. R. Pulliam [eds.], Foraging behavior. Plenum Press, New York.

GREEN, H. W., AND F. M. JAKSIC. 1983. Food-niche relationships among sympatric predators: effects of level of prey identification. Oikos 40:151-154.

GREEN, L., E. B. FISHER, JR., S. PERLOW, AND L. SHERMAN. 1981. Preference reversal and self-control: choice as a function of reward amount and delay. Behav. Anal. Let. 1:244-256.

GREEN, P. 1985. Some results from the use of a long life radio transmitter package on corvids. Ringing and Migration 6:45-51.

GREEN, R. 1978. Factors affecting the diet of farmland Skylarks, Alauda arvensis. J. Anim. Ecol. 47:913-928.

GREEN, R. F. 1980. Bayesian birds: a simple example of Oaten's stochastic model of optimal foraging. Theor. Popul. Biol. 18:244-256.

GREEN, R. F. 1984. Stopping rules for optimal foragers. Am. Nat. 123:30-40.

GREEN, R. F. 1987. Stochastic models of optimal foraging, p. 273-302. In A. C. Kamil, J. R. Krebs, and H. R. Pulliam [eds.], Foraging behavior. Plenum Press, New York.

GREEN, R. H. 1971. A multivariate statistical approach to the Hutchinsonian niche: bivalve molluscs of central Canada. Ecology 52:543-556.

GREEN, R. H. 1974. Multivariate niche analysis with temporally varying environmental factors. Ecology 55:73-83.

GREEN, R. H. 1979. Sampling design and statistical methods for environmental biologists. John Wiley and Sons, New York.

GREENACRE, M. J. 1984. Theory and applications of correspondence analysis. Academic Press, New York.

GREENACRE, M., AND T. HASTIE. 1987. The geometric interpretation of correspondence analysis. J. Am. Stat. Assoc. 82:437-447.

GREENACRE, M. J., AND E. S. VRBA. 1984. Graphical display and interpretation of antelope census data in African wildlife areas, using correspondence analysis. Ecology 65:984-997.

GREENBERG, R. 1980. Demographic aspects of long-distance migration, p. 493-504. In A. Keast and E. S. Morton [eds.], Migrant birds in the Neotropics: ecology, behavior, distribution, and conservation. Smithsonian Instit. Press, Washington, D.C.

GREENBERG, R. 1981a. Dissimilar bill shapes in New World versus temperate forest foliage-gleaning birds. Oecologia 49:143-147.

GREENBERG, R. 1981b. Frugivory in some migrant tropical forest wood warblers. Biotropica 13:215-223.

GREENBERG, R. 1983. The role of neophobia in determining foraging specialization of some migratory warblers. Am. Nat. 122:444-453.

GREENBERG, R. 1984a. The winter exploitation systems of Bay-breasted and Chestnut-sided warblers in Panama. Univ. Calif. Publ. Zool. 116:1-107.

GREENBERG, R. 1984b. Differences in feeding neophobia in tropical migrant warblers, Dendroica castanea and D. pensylvanica. J. Comp. Psych. 98:131-136.

GREENBERG, R. 1984c. Neophobia in the foraging site selection of a neotropical migrant bird: an experimental study. Proc. Nat. Acad. Sci. 81:3778-3780.

GREENBERG, R. 1985. A comparison of foliage discrimination learning in a specialist and a generalist species of migrant warbler (Aves: Parulidae). Can. J. Zool. 63:773-776.

GREENBERG, R. 1986. Competition in migrant birds in the nonbreeding season, p. 281-307. In R. F. Johnston [ed.], Current ornithology. Vol. 3. Plenum Press, New York.

GREENBERG, R. 1987a. Development of dead leaf foraging in a tropical migrant warbler. Ecology 68:130-141.

GREENBERG, R. 1987b. Seasonal foraging specialization in the Worm-eating Warbler. Condor 89:158-168.

GREENBERG, R. 1987c. Social facilitation does not reduce neophobia in Chestnut-sided Warblers (Dendroica pensylvanica). J. Ethol. 5:7-10.

GREENBERG, R. 1989. Aversion to exposed sites, neophobia, and ecological plasticity in Song and Swamp sparrows. Can. J. Zool. 67:1194-1199.

GREENBERG, R., AND J. GRADWOHL. 1980. Leaf surface specializations of birds and arthropods in a Panamanian forest. Oecologia 46:115-124.

GREENE, H. W. 1982. Dietary and phenotypic diversity in lizards: why are some organisms specialized? p. 107-128. In D. Mossakowski and G. Roth [eds.], Environmental adaptation and evolution. Gustav Fischer, Stuttgart and New York.

GREENE, H. W., AND F. M. JAKSIC. 1983. Food-niche relationships among sympatric predators: effects of level of prey identification. Oikos 40:151-154.

GREENLAW, J. S. 1969. The importance of food in the breeding system of the Rufous-sided Towhee, Pipilo erythrophthalmus (L.). Ph.D. diss., Rutgers Univ., New Brunswick, NJ.

GREENLAW, J. S. 1976. Use of bilateral scratching behavior by emberizines and icterids. Condor 78:94-97.

GREENLAW, J. S., AND W. POST. 1985. Evolution of monogamy in Seaside Sparrows Ammodramus maritimus: tests of hypotheses. Anim. Behav. 33:373-383.

GREENSLADE, P. J. M. 1964. Pitfall trapping as a method for studying populations of Carabidae (Coleoptera). J. Anim. Ecol. 33:301-310.

GREENWOOD, P. J., AND P. H. HARVEY. 1978. Foraging and territory utilization of Blackbirds (Turdus merula) and Song Thrushes (Turdus philomelos). Anim. Behav. 26:1222-1236.

GREENWOOD, P. J., AND P. H. HARVEY. 1982. The natal and breeding dispersal of birds. Ann. Rev. Ecol. Syst. 13:1-21.

GREIG-SMITH, P. 1983. Quantitative plant ecology. Blackwell Scientific Publ., Oxford.

GREIG-SMITH, P. 1987. Aversion of starlings and sparrows to unexpected or unusual flavours and colouring in food. Ethology 74:155-163.

GREY, J. 1985. Behaviour of honeyeaters in relation to pollination and resource selection. M. Appl. Sc. thesis, Western Australian Institute of Technology.

GRINNELL, J., AND A. H. MILLER. 1944. The distribution of the birds of California. Pac. Coast Avif. No. 27.

GRINNELL, J., AND T. I. STORER. 1924. Animal life in the Yosemite. Contrib. Mus. Vert. Zool., Univ. California, Berkeley.

GROMADZKI, M. 1969. Composition of food of the starling, Sturnus vulgaris L., in agrocenoses. Ekol. Pol. Ser. A 17:287-311.

GROSS, A. O. 1949. Nesting of the Mexican Jay in the Santa Rita Mountains, Arizona. Condor 51:241-249.

GROSS, A. O. 1958. Life history of the Banaquit of Tobago Island. Wilson Bull. 70:257-279.

GRUBB, T. C., JR. 1975. Weather-dependent foraging behavior of some birds wintering in a deciduous woodland. Condor 77:175-182.

GRUBB, T. C., JR. 1977. Weather-dependent foraging behavior of some birds wintering in a deciduous woodland: horizontal adjustments. Condor 79:271-274.

GRUBB, T. C., JR. 1978. Weather-dependent foraging rates of wintering woodland birds. Auk 95:370-376.

GRUBB, T. C., JR. 1979. Factors controlling foraging strategies of insectivorous birds, p. 119-135. In J. G. Dickson, R. N. Connor, R. R. Fleet, J. C. Kroll, and J. A. Jackson [eds.], The role of insectivorous birds in forest ecosystems. Academic Press, New York.

GRUBB, T. C., JR. 1982a. On sex-specific foraging behavior in the White-breasted Nuthatch. J. Field Ornithol. 53:305-314.

GRUBB, T. C., JR. 1982b. Downy Woodpecker sexes select different cavity sites: an experiment using artificial snags. Wilson Bull. 94:577-579.

GRUBB, T. C., JR. 1987. Changes in the flocking behaviour of wintering English titmice with time, weather and supplementary food. Anim. Behav. 35:794-806.

GRUBB, T., AND L. GREENWALDT. 1982. Sparrows and a brushpile: foraging responses to different combinations of predation and energy costs. Anim. Behav. 30:367-370.

GUILLAUME, J., AND J. D. SUMMERS. 1970. Maintenance energy requirement of the rooster and influence of plane of nutrition on metabolizable energy. Can. J. Anim. Sci. 50:363-369.

GUNNARSSON, B. 1983. Winter mortality of spruce-living spiders: effect of spider interactions and bird predation. Oikos 40:226-233.

GURALNIK, D. B. [ed.]. 1970. Webster's new world dictionary of the American language. 2nd College ed. The World Publishing Co., New York.

GWINNER, E. 1986. Circannual rhythms in the control of avian migrations. Adv. Study Behav. 16:191-228.

GWINNER, E., AND D. CZESCHLIK. 1978. On the significance of spring migratory restlessness in caged birds. Oikos 30:364-372.

GWINNER, E., H. BIEBACH, AND I. V. KRIES. 1985. Food

availability affects migratory restlessness in caged Garden Warblers (Sylvia borin). Naturwissenschaften 72:51-52.

GWINNER, E., H. SCHWABL, AND I. SCHWABL-BENZINGER. 1988. Effects of food-deprivation on migratory restlessness and diurnal activity in the Garden Warbler Sylvia borin. Oecologia 77:321-326.

HAFTORN, S. 1956. Contribution to the food biology of tits, especially about storing of surplus food. Part IV. Det. Kgl. Norske Videnskabers Selskabs Skrifter 1-54.

HAGGSTROM, G. W. 1983. Logistic regression and discriminant analysis by ordinary least squares. J. Business Econ. Stat. 1:229-238.

HAILA, Y., J. TIAINEN, AND K. VEPSÄLÄINEN. 1986. Delayed autumn migration as an adaptive strategy of birds in northern Europe: evidence from Finland. Ornis Fenn. 63:1-9.

HAILMAN, J. P. 1960. A field study of the Mockingbird's wing-flashing behavior and its association with foraging. Wilson Bull. 72:346-357.

HAILMAN, J. P. 1973. Double-scratching and terrestrial locomotion in emberizines: some complications. Wilson Bull. 85:348-350.

HAILMAN, J. P. 1974. A stochastic model of leaf-scratching bouts in two emberizine species. Wilson Bull. 86:296-298.

HAILS, C. J. 1979. A comparison of flight energetics in hirundines and other birds. Comp. Biochem. Physiol. 63A:581-585.

HAINSWORTH, F. R. 1974. Food quality and foraging efficiency: the efficiency of sugar assimilation by hummingbirds. J. Comp. Physiol. 88:425-431.

HAINSWORTH, F. R., AND L. L. WOLF. 1976. Nectar characteristics and food selection by hummingbirds. Oecologia 25:101-113.

HAIRSTON, N. G., F. E. SMITH, AND L. D. SLOBODKIN. 1960. Community structure, population control, and competition. Am. Nat. 94:421-425.

HALSE, S. A. 1984. Food intake, digestive efficiency, and retention time in Spur-winged Geese Plectropterus gambensis. S. Afr. J. Wildl. Res. 14:106-110.

HAMM, S. L., AND S. J. SHETTLEWORTH. 1987. Risk-aversion in pigeons. J. Exp. Psychol., Anim. Behav. Processes 13:376-383.

HANOWSKI, J. M. 1982. Feeding niche and habitat relations of a black spruce bog bird community, NW Minnesota. M.S. thesis, Univ. of Minnesota.

HANSKI, I. 1978. Some comments on the measurement of niche metrics. Ecology 59:168-174.

HANSON, J. 1987. Tests of optimal foraging using an operant analogue, p. 335-362. In A. C. Kamil, J. R. Krebs, and H. R. Pulliam [eds.], Foraging behavior. Plenum Press, New York.

HANSSEN, I. 1979. A comparison of the microbiological conditions in the small intestine and caeca of wild and captive Willow Grouse (Lagopus lagopus lagopus). Acta Vet. Scand. 20:365-371.

HARDY, A. 1965. The living stream. Collins, London.

HARLOW, W. R. 1986a. Eigenvectors of a general matrix. Access 5:29-30.

HARLOW, W. R. 1986b. Determination of eigenvalues, eigenvectors of a general matrix. Access 5:35-37.

HARRIS, J. H. 1984. An experimental analysis of desert rodent foraging ecology. Ecology 65:1579-1584.

HARRIS, R. N. 1987. Density-dependent paedomorphosis in the salamander Notophthalmus viridescens dorsalis. Ecology 68:705-712.

HARRIS, W. F. 1986. The breeding ecology of the South Island fernbird in Otago wetlands. Ph.D. thesis, Univ. of Otago, Dunedin, New Zealand.

HARRISON, C. J. 1967. The double-scratch as a taxonomic character in the Holarctic emberizinae. Wilson Bull. 79:22-29.

HARRISON, C. J. O. 1967. Some aspects of adaptation and evolution in Australian Fan-tailed flycatchers. Emu 76:115-119.

HARTLEY, P. H. T. 1948. The assessment of the food of birds. Ibis 90:361-379.

HARTLEY, P. H. T. 1953. An ecological study of the feeding habits of the English titmice. J. Anim. Ecol. 22:261-288.

HARTSHORN, G. S. 1983. Plants, p. 118-157. In D. H. Janzen [ed.], Costa Rican natural history. Univ. of Chicago Press, Chicago.

HARVEY, P. H., AND R. J. PAXTON. 1981. The evolution of aposematic coloration. Oikos 37:391-396.

HASSELL, M. P. 1978. The dynamics of arthropod predator-prey systems. Princeton Univ. Press, Princeton, NJ.

HAUKIOJA, E., AND P. NIEMALA. 1977. Retarded growth of a geometrid larva after mechanical damage to leaves of its host tree. Ann. Zool. Fenn. 14:48-52.

HAYNES, J., AND M. MESLER. 1984. Pollen foraging by bumblebees: foraging patterns and efficiency on Lupinus polyphyllus. Oecologia 61:249-253.

HEATHCOTE, G. D. 1957a. The optimum size of sticky aphid traps. Plant Pathol. 6:104-107.

HEATHCOTE, G. D. 1957b. The comparison of yellow cylindrical, flat, and water traps and of Johnson traps for sampling aphids. Ann. Appl. Biol. 45:133-139.

HEATHCOTE, G. D., J. P. PALMER, AND R. L. TAYLOR. 1969. Sampling for aphids by traps and by crop inspection. Ann. Appl. Biol. 63:155-166.

HECHT, O. 1979. Light and colour reactions of Musca domestica under different conditions. Bull. Entomol. Soc. Am. 16:94-98.

HEINRICH, B. 1979a. Resource heterogeneity and patterns of movement in foraging bumblebees. Oecologia 40:235-245.

HEINRICH, B. 1979b. "Majoring" and "minoring" by foraging bumblebees, Bombus vagans: an experimental analysis. Ecology 60:245-255.

HEINRICH, B. 1979c. Foraging strategies of caterpillars: leaf damage and possible predator avoidance strategies. Oecologia 42:325-337.

HEINRICH, B., AND S. L. COLLINS. 1983. Caterpillar leaf damage and the game of hide-and-seek with birds. Ecology 64:592-602.

HEINSELMAN, M. L. 1963. Forest sites, bog processes, and peatland types in Glacial Lake Agassiz Region, Minnesota. Ecol. Monogr. 33:327-374.

HEISEY, D. M. 1985. Analyzing selection experiments with log-linear models. Ecology 66:1744-1748.

HEISTERBERG, J. F., C. E. KNITTLE, O. E. BRAY, D. F. MOTT, AND J. F. BESSER. 1984. Movements of radio-instrumented blackbirds and European Starlings among winter roosts. J. Wildl. Manage. 48:203-209.

HEITHAUS, E. R., AND T. H. FLEMING. 1978. Foraging movements of a frugivorous bat, Carollia perspicillata (Phyllostomatidae). Ecol. Monogr. 48:127-143.

HEITHAUS, E. R., T. H. FLEMING, AND P. A. OPLER. 1975. Foraging patterns and resource utilization in seven species of bats in a seasonal tropical forest. Ecology 56:841-854.

HELMS, C. W., AND B. SMYTH. 1969. Variation in major body components of the Tree Sparrow (Spizella arborea) sampled within the winter range. Wilson Bull. 81:280-292.

HENDRICKS, P. 1987. Habitat use by nesting Water Pipits (Anthus spinoletta): a test of the snowfield hypothesis. Arctic and Alpine Res. 19:313-320.

HENGEVELD, R. 1980. Polyphagy, oligophagy and food specialization in ground beetles (Coleoptera: Carabidae). Neth. J. Zool. 30:564-584.

HENSLEY, M. M., AND J. B. COPE. 1951. Further data on removal and repopulation of the breeding birds in a spruce-fir forest community. Auk 68:483-493.

HENTY, C. J. 1976. Foraging methods of the Song Thrush. Wilson Bull. 88:497-499.

HEPPNER, F. 1965. Sensory mechanisms and environmental cues used by the American Robin in locating earthworms. Condor 67:247-256.

HERD, R. M., AND T. J. DAWSON. 1984. Fiber digestion in

the Emu, Dromaius novaehollandiae, a large bird with a simple gut and high rates of passage. Physiol. Zool. 57:70-84.

HERRERA, C. M. 1976. A trophic diversity index for presence-absence food data. Oecologia 25:187-191.

HERRERA, C. M. 1978a. Ecological correlates of residence and nonresidence in a Mediterranean passerine bird community. J. Anim. Ecol. 47:871-890.

HERRERA, C. M. 1978b. Individual dietary differences associated with morphological variation in Robins, Erithacus rubecula. Ibis 120:542-545.

HERRERA, C. M. 1982. Defense of ripe fruits from pests: its significance in relation to plant-disperser interactions. Am. Nat. 120:218-241.

HERRERA, C. M. 1984a. A study of avian frugivores, bird-dispersed plants, and their interaction in mediterranean scrublands. Ecol. Monogr. 54:1-23.

HERRERA, C. M. 1984b. Adaptation to frugivory of Mediterranean avian seed dispersers. Ecology 65:609-617.

HERRERA, C. M., AND F. HIRALDO. 1976. Food niche and trophic relationships among European owls. Ornis Scand. 7:29-41.

HERRERA, C. M., AND M. RODRIGUEZ. 1979. Year-to-year site constancy among three passerine species wintering at a southern Spanish locality. Ringing and Migration 2:160.

HERTZ, P. E., J. V. REMSEN, JR., AND S. I. ZONES. 1976. Ecological complementarity of three sympatric parids in a California oak woodland. Condor 78:307-316.

HESPENHEIDE, H. A. 1966. The selection of seed size by finches. Wilson Bull. 78:191-197.

HESPENHEIDE, H. A. 1971. Food preference and the extent of overlap in some insectivorous birds, with special reference to the Tyrannidae. Ibis 113:59-72.

HESPENHEIDE, H. A. 1973a. A novel mimicry complex: beetles and flies. J. Entomol. 48:49-56.

HESPENHEIDE, H. A. 1973b. Ecological inferences from morphological data. Ann. Rev. Ecol. Syst. 4:213-229.

HESPENHEIDE, H. 1975a. Prey characteristics and predator niche width, p. 158-180. In M. L. Cody and J. M. Diamond [eds.], Ecology and evolution of communities. Harvard Univ. Press, Cambridge, MA.

HESPENHEIDE, H. 1975b. Selective predation by two swifts and a swallow in Central America. Ibis 117:82-99.

HEWLETT-PACKARD CO. 1984. HP41C Stat. Pac. Hewlett-Packard, Portable Computer Division, Corvallis, OR.

HEYER, W. R. 1974. Niche measurements of frog larvae from a seasonal tropical location in Thailand. Ecology 55:651-656.

HILDÉN, O. 1965. Habitat selection in birds: a review. Ann. Zool. Fenn. 2:53-75.

HILL, D. A., AND N. ELLIS. 1984. Survival and age related changes in the foraging behavior and time budget of Tufted Ducklings Aythya fuligula. Ibis 126:544-550.

HILL, D. C., E. V. EVANS, AND H. G. LUMSDEN. 1968. Metabolizable energy of aspen flower buds for captive Ruffed Grouse. J. Wildl. Manage. 32:854-858.

HILL, M. O. 1973. Reciprocal averaging: an eigenvector method of ordination. J. Ecol. 61:237-249.

HILL, M. O. 1974. Correspondence analysis: a neglected multivariate method. J. Roy. Stat. Soc., Ser. C 23:340-354.

HILL, M. O. 1979. DECORANA -- A Fortran program for detrended correspondence analysis and reciprocal averaging. Section of Ecology and Systematics, Cornell Univ., Ithaca, New York.

HILL, M. O., AND H. G. GAUCH. 1980. Detrended correspondence analysis: an improved ordination technique. Vegetatio 42:47-58.

HILTY, S. L. 1975. Notes on a nest and behavior of the Chestnut-crowned Anteater. Condor 77:513-514.

HILTY, S. L. 1980. Flowering and fruiting periodicity in a premontane rain forest in Pacific Colombia. Biotropica 12:292-306.

HILTY, S. L., AND W. L. BROWN. 1986. A guide to the birds of Colombia. Princeton Univ. Press, Princeton, NJ.

HILTY, S. L., T. A. PARKER III, AND J. SILLIMAN. 1979. Observations on Plush-capped Finches in the Andes with a description of the juvenal and immature plumages. Wilson Bull. 91:145-148.

HINDE, R. A. 1952. The behaviour of the Great Tit (Parus major) and some other related species. Behav. Suppl. 2:1-201.

HINDE, R. A., AND J. FISHER. 1951. Further observations on the opening of milk bottles by birds. Br. Birds 44:392-396.

HINDE, R. A., AND J. FISHER. 1972. Some comments on the republication of the two papers on the opening of milk bottles by birds. In P. H. Klopfer and J. P. Hailman [eds.], Function and evolution of behavior. Addison-Wesley, Reading, MA.

HIRONS, G., AND T. H. JOHNSON. 1987. A quantitative analysis of habitat preferences of Woodcock Scolopax rusticola in the breeding season. Ibis 129:371-381.

HIRONS, G. J. M., AND R. B. OWEN, JR. 1982. Radio tagging as an aid to the study of the Woodcock. Symp. Zool. Soc. Lond. 49:139-152.

HOAGLIN, D. C., F. MOSTELLER, AND J. W. TUKEY. 1983. Understanding robust and exploratory data analysis. John Wiley and Sons, New York.

HOBBS, J. N. 1954. Flame Robin "foot-pattering" feeding habit. Emu 54:278-279.

HOCKEY, P. A. R. 1984. Growth energetics of the African Black Oystercatcher, Haematopus moquini. Ardea 72:111-117.

HODGES, C. M. 1985. Bumble bee foraging: energetic consequences of using a threshold departure rule. Ecology 66:188-197.

HOEL, P. G., S. C. PORT, AND C. J. STONE. 1972. An introduction to stochastic processes. Houghton Mifflin, Boston, MA.

HOGG, R. V., AND A. T. CRAIG. 1978. Introduction to mathematical statistics. 4th ed. MacMillan, New York.

HOGSTAD, O. 1970. On the ecology of the Three-toed Woodpecker Picoides tridactylus (L.) outside the breeding season. Nytt. Mag. Zool. 18:221-227.

HOGSTAD, O. 1976. Sexual dimorphism and divergence in winter foraging behaviour of Three-toed Woodpeckers Picoides tridactylus. Ibis 118:41-50.

HOGSTAD, O. 1987a. Social rank in winter flocks of Willow Tits Parus montanus. Ibis 129:1-9.

HOGSTAD, O. 1987b. It is expensive to be dominant. Auk 104:333-336.

HOGSTAD, O. 1987c. Winter survival strategies of the Willow Tit Parus montanus. D. Phil. diss., Univ. of Trondheim, Norway.

HOGSTAD, O. 1988. Social rank and antipredator behavior of Willow Tits Parus montanus in winter flocks. Ibis 130:45-56.

HOLBERTON, R. L., K. P. ABLE, AND J. C. WINGFIELD. 1989. Status signalling in Dark-eyed Juncos, Junco hyemalis: plumage manipulations and hormonal correlates of dominance. Anim. Behav. 37:681-689.

HOLLANDER, M., AND D. A. WOLFE. 1973. Nonparametric statistical methods. John Wiley and Sons, New York.

HOLLING, C. S. 1959a. The functional response of invertebrate predators to prey density. Mem. Entomol. Soc. Can. 48:1-86.

HOLLING, C. S. 1959b. Some characteristics of simple types of predation and parasitism. Can. Entomol. 91:385-398.

HOLLING, C. S. 1965. The functional response of predators to prey density and its role in mimicry and population regulation. Mem. Entomol. Soc. Can. 45:1-60.

HOLLING, C. S. 1966. The functional response to invertebrate predators to prey density. Mem. Entomol. Soc. Can. 48:1-86.

HOLLIS, K. L. 1982. Pavlovian conditioning of signal-centered action patterns and autonomic behavior: a biological analysis of function. Adv. Study Behav. 12:1-64.

HOLM, E. 1985. The evolution of generalist and specialist species, p. 87-93. In E. S. Vrba [ed.], Species and speciation. Transvaal Museum Monogr. No. 4, Pretoria, South Africa.

HOLMES, R. T. 1966. Feeding ecology of the Red-backed Sandpiper (Calidris alpina) in arctic Alaska. Ecology 47:32-45.

HOLMES, R. T. 1980. Resource exploitation patterns and the structure of a forest bird community. Proc. XVII Int. Ornithol. Congr. 1056-1062.

HOLMES, R. T. 1986. Foraging patterns of forest birds: male-female differences. Wilson Bull. 98:196-213.

HOLMES, R. T. 1988. Community structure, population fluctuations, and resource dynamics of birds in temperate deciduous forests. Proc. XIX Int. Ornithol. Congr.:1318-1327.

HOLMES, R. T., AND F. A. PITELKA. 1968. Food overlap among coexisting sandpipers on northern Alaskan tundra. Syst. Zool. 17:305-318.

HOLMES, R. T., AND H. F. RECHER. 1986a. Search tactics of insectivorous birds foraging in an Australian eucalypt forest. Auk 103:515-530.

HOLMES, R. T., AND H. F. RECHER. 1986b. Determinants of guild structure in forest bird communities: an intercontinental comparison. Condor 88:427-439.

HOLMES, R. T., AND S. K. ROBINSON. 1981. Tree species preferences of foraging insectivorous birds in a northern hardwoods forest. Oecologia 48:31-35.

HOLMES, R. T., AND S. K. ROBINSON. 1988. Spatial patterns, foraging tactics, and diets of ground-foraging birds in a northern hardwoods forest. Wilson Bull. 100:377-394.

HOLMES, R. T., AND R. H. SAWYER. 1975. Oxygen consumption in relation to ambient temperatures in five species of forest-dwelling thrushes (Hylocichla and Catharus). Comp. Biochem. Physiol. 50A:527-531.

HOLMES, R. T., AND J. C. SCHULTZ. 1988. Food availability for forest birds: effects of prey distribution and abundance on bird foraging. Can. J. Zool. 66:720-728.

HOLMES, R. T., AND F. W. STURGES. 1975. Avian community dynamics and energetics in a northern hardwoods ecosystem. J. Anim. Ecol. 44:175-200.

HOLMES, R. T., C. P. BLACK, AND T. W. SHERRY. 1979a. Comparative population bioenergetics of three insectivorous passerines in a deciduous forest. Condor 81:9-20.

HOLMES, R. T., R. E. BONNEY, JR., AND S. W. PACALA. 1979b. Guild structure of the Hubbard Brook bird community: a multivariate approach. Ecology 60:512-520.

HOLMES, R. T., AND J. C. SCHULTZ, AND P. NOTHNAGLE. 1979c. Bird predation on forest insects: an exclosure experiment. Science 206:462-463.

HOLMES, R. T., T. W. SHERRY, AND S. E. BENNETT. 1978. Diurnal and individual variability in the foraging behavior of American Redstarts (Setophaga ruticilla). Oecologia 36:141-149.

HOLMES, R. T., T. W. SHERRY, AND F. W. STURGES. 1986. Bird community dynamics in a temperate deciduous forest: long-term trends at Hubbard Brook. Ecol. Monogr. 56:201-220.

HOLSON, P. R. 1987. Feeding neophobia: a possible explanation for the differential maze performance of rats reared in enriched or isolated environments. Psychol. and Behavior 38:191-201.

HOLT, D. W., L. J. LYON, AND R. HALE. 1987. Techniques for differentiating pellets of Short-eared Owls and Northern Harriers. Condor 89:929-931.

HOLTHUIJZEN, A. M. A., AND C. S. ADKISSON. 1984. Passage rate, energetics, and utilization efficiency of the Cedar Waxwing. Wilson Bull. 96:680-685.

HOOGE, P. N. 1989. Spatial use in the Acorn Woodpecker. M.A. thesis, Univ. of California, Berkeley.

HOOPER, R. G., AND H. S. CRAWFORD. 1969. Woodland habitat research for nongame birds. Trans. N. Am. Wildl. and Nat. Res. Conf. 34:201-207.

HOOPER, R. G., AND M. R. LENNARTZ. 1983. Roosting behavior of Red-cockaded Woodpecker clans with insufficient cavities. J. Field Ornithol. 54:72-76.

HORN, H. S. 1968. The adaptive significance of colonial nesting in the Brewer's Blackbird (Euphagus cyanocephalus). Ecology 49:682-694.

HOUSTON, A., C. CLARK, J. MCNAMARA, AND M. MANGEL. 1988. Dynamic models in behavioral and evolutionary ecology. Nature 332:29-34.

HOWARD, W. J. 1937. Bird behavior as a result of emergence of seventeen year locusts. Wilson Bull. 49:43-44.

HOWE, H. F. 1977. Bird activity and seed dispersal of a tropical wet forest tree. Ecology 58:539-550.

HOWE, H. F. 1979. Fear and frugivory. Am. Nat. 114:925-931.

HOWE, H. F. 1980. Monkey dispersal and waste of a neotropical fruit. Ecology 61:944-959.

HOWE, H. F. 1981. Dispersal of a neotropical nutmeg (Virola sebifera) by birds. Auk 98:88-98.

HOWE, H. F. 1983. Annual variation in a neotropical seed-dispersal system, p. 211-227. In S. L. Sutton, T. C. Whitmore, and A. C. Chadwick [eds.], Tropical rainforest: ecology and management. Special Publ. No. 2 of the British Ecol. Soc., Blackwell Scientific Publications, Oxford.

HOWE, H. F. 1984. Implications of seed dispersal by animals for tropical reserve management. Biol. Conserv. 30:261-281.

HOWE, H. F., AND G. F. ESTABROOK. 1977. On intraspecific competition for avian dispersers in tropical trees. Am. Nat. 111:817-832.

HOWE, H. F., AND J. SMALLWOOD. 1982. Ecology of seed dispersal. Ann. Rev. Ecol. Syst. 13:201-228.

HOWE, H. F., AND G. A. VANDE KERCKHOVE. 1979. Fecundity and seed dispersal of a tropical tree. Ecology 60:180-189.

HOWE, H. F., AND G. A. VANDE KERCKHOVE. 1980. Nutmeg dispersal by tropical birds. Science 210:925-927.

HOWE, H. F., AND G. A. VANDE KERCKHOVE. 1981. Removal of nutmeg (Virola suranimensis) crops by birds. Ecology 62:1093-1106.

HUBBELL, S. P., AND R. B. FOSTER. 1986. Commonness and rarity in a neotropical forest: implications for tropical tree conservation, p. 205-231. In M. E. Soule [ed.], Conservation biology. Sinauer Assoc., Sunderland, MA.

HUDDY, L. 1979. Social behaviour and feeding ecology of Scarlet Robins (Petroica multicolor). B. Sc. Hons. thesis, Univ. of New England, Armidale, NSW.

HUDSON, J. W., AND S. L. KIMZEY. 1966. Temperature regulation and metabolic rhythms in populations of the House Sparrow, Passer domesticus. Comp. Biochem. Physiol. 17:203-217.

HUEY, R. B., AND E. R. PIANKA. 1981. Ecological consequences of foraging mode. Ecology 62:991-999.

HUFFAKER, C. B., AND P. S MESSENGER, [eds.]. 1976. Theory and practice of biological control. Academic Press, New York.

HUNTER, M. D. 1987. Opposing effects of spring defoliation on late season oak caterpillars. Ecol. Entomol. 12:373-382.

HURLBERT, S. H. 1978. The measurement of niche overlap and some relatives. Ecology 59:67-77.

HURLBERT, S. H. 1984. Pseudoreplication and the design of ecological field experiments. Ecol. Monogr. 54:187-211.

HURST, G. A., AND W. E. POE. 1985. Amino acid levels and patterns in Wild Turkey poults and their food items in Mississippi, p. 133-143. In J. E. Kennamer and M. C. Kennamer [eds.], Proceedings 5th National Wild Turkey Symposium, Des Moines, IA.

HURTUBIA, J. 1973. Trophic diversity measurement in sympatric predatory species. Ecology 54:885-890.

HUSSELL, D. J. T., AND T. E. QUINNEY. 1987. Food

abundance and clutch size of Tree Swallows Tachycineta bicolor. Ibis 129:243-258.

HUTCHINSON, G.E . 1957. Concluding remarks. Cold Spring Harbor Symp. Quant. Biol. 22:415-427.

HUTTO, R. L. 1980. Winter habitat distribution of migratory land birds in western Mexico, with special reference to small, foliage-gleaning insectivores, p. 181-203. In A. Keast and E. S. Morton [eds.], Migrant birds in the Neotropics: ecology, behavior, distribution, and conservation. Smithsonian Inst. Press, Washington, D.C.

HUTTO, R. L. 1981a. Temporal patterns of foraging activity in some wood warblers in relation to the availability of insect prey. Behav. Ecol. Sociobiol. 9:195-198.

HUTTO, R. L. 1981b. Seasonal variation in the foraging behavior of some migratory western wood warblers. Auk 98:765-777.

HUTTO, R. L. 1985a. Habitat selection by nonbreeding, migratory land birds, p. 455-476. In M. L. Cody [ed.], Habitat selection in birds. Academic Press, Orlando, Florida.

HUTTO, R. L. 1985b. Seasonal changes in the habitat distribution of transient insectivorous birds in Southeastern Arizona: competition mediated? Auk 102:120-132.

HYSLOP, E. J. 1980. Stomach contents analysis -- a review of methods and their application. J. Fish. Biol. 17:411-429.

INOZEMTSEV, A. A., S. A. EZHOVA, S. L. PERESHKOL'NIK, AND G. I. FRENKIMA. 1980. Evaluation of the total effects of insectivorous birds on invertebrates in an oak-hornbeam forest. Sov. J. Ecol. 11:47-55.

ISAACSON, D. L., AND R. W. MADSEN. 1976. Markov chains: theory and applications. John Wiley and Sons, New York.

ISENHART, F. R., AND D. F. DESANTE. 1985. Observations of Scrub Jays cleaning ectoparasites from black-tailed deer. Condor 87:145-147.

ISLER, M. L., AND P. R. ISLER. 1987. The tanagers: natural history, distribution, and identification. Smithsonian Inst. Press, Washington, D.C.

IMSL. 1982. IMSL library reference manual, revision 9. IMSL Inc., Houston, Texas.

IVLEV, V. S. 1961. Experimental ecology of the feeding of fishes. Yale Univ. Press, New Haven, CT.

IWASA, Y., M. HIGASHI, AND N. YAMAMURA. 1981. Prey distribution as a factor determining the choice of optimal foraging strategy. Am. Nat. 117:710-723.

JACKSON, J. A. 1970. A quantitative study of the foraging ecology of Downy Woodpeckers. Ecology 51:318-323.

JACKSON, J. A. 1979. Tree surfaces as foraging substrates for insectivorous birds, p. 69-93. In J. G. Dickson, R. N. Conner, R. R. Fleet, J. C. Kroll, and J. A. Jackson [eds.], The role of insectivorous birds in forest ecosystems. Academic Press, New York.

JACKSON, J. A. 1985. A mutualistic feeding association between Boat-tailed Grackles and Pied-billed Grebes. Condor 87:147-148.

JACOBS, J. 1974. Quantitative measurement of food selection. Oecologia 14:413-417.

JAKSIC, F. M., AND H. E. BRAKER. 1983. Food-niche relationships and guild structure of diurnal birds of prey: competition versus opportunism. Can. J. Zool. 61:2230-2241.

JAMES, F. C. 1971. Ordinations of habitat relationships among breeding birds. Wilson Bull. 83:215-236.

JAMES, F. C. 1982. The ecological morphology of birds: a review. Ann. Zool. Fenn. 19:265-275.

JAMES, F. C. 1983. Environmental component of morphological differentiation in birds. Science 221:184-186.

JAMES, F. C., AND W. C. BOECKLEN. 1983. Interspecific morphological relationships and the densities of birds, p. 458-477. In D. R. Strong, Jr., D. Simberloff, L. G. Abele, and A. B. Thistle [eds.], Ecological communities: conceptual issues and the evidence. Princeton Univ. Press, Princeton, NJ.

JAMES, F. C., AND C. E. MCCULLOCH. 1985. Data analysis and the design of experiments in ornithology. Current Ornithol. 2:1-63.

JAMES, F. C., R. F. JOHNSTON, N. O. WAMER, G. J. NIEMI, AND W. J. BOECKLEN. 1984. The Grinnellian niche of the Wood Thrush. Am. Nat. 124:17-30.

JAMES, R. D. 1976. Foraging behavior and habitat selection of three species of vireos in southern Ontario. Wilson Bull. 88:62-75.

JAMES, R. D. 1979. The comparative foraging behaviour of Yellow-throated and Solitary vireos: the effect of habitat sympatry, p. 137-163. In J. G. Dickson, R. N. Conner, R. R. Fleet, J. C. Kroll, and J. A. Jackson [eds.], The role of insectivorous birds in forest ecosystems. Academic Press, New York.

JANSON, C. H. 1983. Adaptations of fruit morphology to dispersal agents in a neotropical forest. Science 219:187-189.

JANSON, C. H., E. W. STILES, AND D. W. WHITE. 1986. Selection on plant fruiting traits by brown capuchin monkeys: a multivariate approach, p. 83-92. In A. Estrada and T. H. Fleming [eds.], Frugivores and seed dispersal. Dr. W. Junk, Publishers, Dordrecht, The Netherlands.

JANZEN, D. H. 1966. Coevolution of mutualism between ants and acacias in Central America. Evolution 20:249-275.

JANZEN, D. H. 1980a. Heterogeneity of potential food abundance for tropical small land birds, p. 545-552. In A. Keast and E. S. Morton [eds.], Migrant birds in the Neotropics: ecology, behavior, distribution, and conservation. Smithsonian Inst. Press, Washington, D.C.

JANZEN, D. H. 1980b. When is it coevolution? Evolution 34:611-612.

JARMAN, P. J. 1971. Diets of large mammals in the woodlands around Lake Kariba, Rhodesia. Oecologia 8:157-178.

JEFFORDS, M. R., J. G. STERNBERG, AND G. P. WALDBAUER. 1979. Batesian mimicry: field demonstration of the survival value of pipevine swallowtail and monarch color patterns. Evolution 33:275-286.

JENKINS, J. M. 1979. Foraging behavior of male and female Nuttall Woodpeckers. Auk 96:418-420.

JENKINS, R. 1969. Ecology of three species of saltators in Costa Rica with special reference to their frugivorous diet. Ph.D. diss., Harvard Univ., Cambridge, MA.

JENKINS, S. H. 1975. Food selection by beavers: a multidimensional contingency table analysis. Oecologia 21:157-173.

JENNI, L. 1987. Mass concentrations of Bramblings Fringilla montifringilla in Europe 1900-1983: their dependence upon beechmast and the effect of snow-cover. Ornis Scand. 18:84-94.

JEUNIAUX, C. 1961. Chitinase: an addition to the list of hydrolases in the digestive tract of vertebrates. Nature 192:135-136.

JEUNIAUX, C., AND C. CORNELIUS. 1978. Distribution and activity of chitolytic enzymes in the digestive tract of birds and mammals, p. 542-549. In R. A. A. Muzzarelli and E. R. Pariser [eds.], Chitin and chitosans. Proc. 1st Int. Conf. on Chitin and Chitosans. MIT, Cambridge, MA.

JOHNSGARD, P. A. 1983. The hummingbirds of North America. Smithsonian Inst. Press, Washington, D.C.

JOHNSON, C. G. 1950. A suction trap for small airborne insects which automatically segregates the catch into successive hourly samples. Ann. Appl. Biol. 37:80-91.

JOHNSON, C. G., T. R. E. SOUTHWOOD, AND H. M. ENTWISTLE. 1957. A new method of extracting arthropods and molluscs from grassland and herbage with a suction apparatus. Bull. Entomol. Res. 48:211-218.

JOHNSON, D. H. 1980. The comparison of usage and availability measurements for evaluating resource preference. Ecology 61:65-71.

JOHNSON, D. H. 1981a. The use and misuse of statistics in wildlife habitat studies, p. 11-19. In D. E. Capen [ed.],

The use of multivariate statistics in studies of wildlife habitat. USDA For. Serv. Gen. Tech. Rep. RM-87, Fort Collins, CO.

JOHNSON, D. H. 1981b. How to measure habitat - a statistical perspective, p. 53-57. In D. E. Capen [ed.], The use of multivariate statistics in studies of wildlife habitat. USDA For. Serv. Gen. Tech. Rep. RM-87, Fort Collins, CO.

JOHNSON, E. A. 1977. A multivariate analysis of the niches of plant populations in raised bogs. II. Niche width and overlap. Can. J. Botany 55:1211-1220.

JOHNSON, E. J., AND L. B. BEST. 1982. Factors affecting feeding and brooding of Gray Catbird nestlings. Auk 99:148-156.

JOHNSON, E. J., L. B. BEST, AND P. A. HEAGY. 1980. Food sampling biases associated with the "ligature method." Condor 82:186-192.

JOHNSON, N. L., AND S. KOTZ. 1969. Discrete distributions. John Wiley and Sons, New York.

JOHNSON, E. V., G. L. MACK, AND D. Q. THOMPSON. 1976. The effects of orchard pesticide applications on breeding Robins. Wilson Bull. 88:16-35.

JOHNSON, R. A., M. F. WILLSON, J. N. THOMPSON, AND R. I. BERTIN. 1985. Nutritional values of wild fruits and consumption by migrant frugivorous birds. Ecology 66:819-827.

JOHNSON, S. F., AND J. A. GESSAMAN. 1973. An evaluation of heart rate as an indirect monitor of free-living energy metabolism, p. 44-54. In J. A. Gessaman [ed.], Ecological energetics of homeotherms. Utah Univ. Press, Logan.

JOHNSTON, T. D. 1982. The selective costs and benefits of learning: an evolutionary analysis. Adv. Study Behav. 12:65-106.

JONES, F. M. 1932. Insect coloration and the relative acceptability of insects to birds. Trans. Entomol. Soc. Lond. 80:345-386.

JORDANO, P., AND C. M. HERRERA. 1981. The frugivorous diet of Blackcaps Sylvia atricapilla in southern Spain. Ibis 123:502-507.

JUILLET, J. A. 1963. A comparison of 4 types of traps used for capturing flying insects. Can. J. Zool. 41:219-223.

KACELNIK, A., AND I. CUTHILL. 1987. Optimal foraging: just a matter of technique, p. 303-333. In A. C. Kamil, J. R. Krebs, and H. R. Pulliam [eds.], Foraging behavior. Plenum Press, New York.

KAGEL, J. H., L. GREEN, AND T. CARACO. 1986. When foragers discount the future: constraint or adaptation? Anim. Behav. 34:271-283.

KALE, M. R. 1964. Food ecology of the Wood Stork (Mycteria americanus) in Florida. Ecol. Monogr. 34:97-117.

KALELA, O. 1954. Populationsökologiche Gesichtspunkte zur Entstehung des Vogelzugs. Ann. Zool. Bot. Fennicae Vanamo 16:1-30.

KAM, M., A. A. DEGEN, AND K. A. NAGY. 1987. Seasonal energy, water, and food consumption of Negev Chukars and Sand Partridges. Ecology 68:1029-1037.

KAMIL, A. C. 1978. Systematic search by a nectar-feeding bird. J. Comp. Physiol. Psychol. 92:388-396.

KAMIL, A. C. 1988. Experimental design in ornithology, p. 313-346. In R. F. Johnston [ed.], Current ornithology. Plenum Press, New York.

KAMIL, A. C., AND R. P. BALDA. 1985. Cache recovery and spatial memory in Clark's Nutcracker (Nicifruga columbiana). J. Exp. Psychol. Anim. Behav. Processes 11:95-111.

KAMIL, A. C., AND S. I. YOERG. 1982. Learning and foraging behavior, p. 325-364. In P. P. G. Bateson and P. H. Klopfer [eds.], Perspectives in ethology. Plenum Press, New York.

KAMIL, A. C., AND S. I. YOERG. 1985. The effects of prey depletion on the patch choice of foraging Blue Jays. Anim. Behav. 33:1089-1095.

KAMIL, A. C., F. LINDSTROM, AND J. PETERS. 1985. The detection of cryptic prey by Blue Jays (Cyanocitta cristata). I. The effects of travel time. Anim. Behav. 33:1068-1079.

KAMIL, A. C., J. R. KREBS, AND H. R. PULLIAM [eds.]. 1987. Foraging behavior. Plenum Press, New York.

KAMIL, A. C., AND T. D. SARGENT [eds.]. 1981. Foraging behavior: ecological, ethological, and psychological approaches. Garland STPM Press, New York.

KAMIL, A. C., S. I. YOERG, AND K. C. CLEMENTS. In press. Rules to leave by: patch departure in foraging Blue Jays. Anim. Behav.

KANTAK, G. E. 1979. Observations on some fruit-eating birds in Mexico. Auk 96:183-186.

KARASOV, W. H. 1982. Energy assimilation, nitrogen requirement, and diet in free-living antelope ground squirrels, Ammospermophilus leucurus. Physiol. Zool. 55:378-392.

KARASOV, W. H. 1986. Energetics, physiology, and vertebrate ecology. Trends ecol. and evol. 1:101-105.

KARASOV, W. H. 1987. Nutrient requirements and the design and function of guts in fish, reptiles, and mammals, p. 181-191. In P. Dejours, L. Bolis, C. R. Taylor, and E. Weibel [eds.], Comparative physiology: life in water and on land. Fidia Research Series, IX-Liviana Press, Padova.

KARASOV, W. H., AND J. M. DIAMOND. 1983. Adaptive regulation of sugar and amino acid transport by vertebrate intestine. Am. J. Physiol. 245:G443-462.

KARASOV, W. H., AND J. M. DIAMOND. 1988. Interplay between physiology and ecology in digestion. Bioscience 38:602-611.

KARASOV, W. H., AND D. J. LEVEY. 1990. Digestive system tradeoffs and adaptations of passerine frugivorous birds. Physiol. Zool. (in press).

KARASOV, W. H., D. PHAN, J. M. DIAMOND, AND F. L. CARPENTER. 1986. Food passage and intestinal nutrient absorption in hummingbirds. Auk 103:453-464.

KARBAN, R. 1982. Increased reproductive success at high densities and predator satiation for periodical cicadas. Ecology 63:321-328.

KARR, J. R. 1971. Structure of avian communities in selected Panama and Illinois habitats. Ecol. Monogr. 41:207-233.

KARR, J. R. 1975. Production, energy pathways, and community diversity in forest birds, p. 161-176. In F. B. Golley and E. Medina [eds.], Tropical ecological systems: trends in terrestrial and aquatic research. Springer, New York.

KARR, J. R. 1976. Seasonality, resource availability, and community diversity in tropical bird communities. Am. Nat. 110:973-994.

KARR, J. R. 1979. On the use of mist nets in the study of bird communities. Inland Bird Banding 51:1-10.

KARR, J. R. 1980. Geographical variation in the avifaunas of tropical forest undergrowth. Auk 97:283-298.

KARR, J. R. 1981. Surveying birds with mist nets, p. 62-67. In C. J. Ralph and J. M. Scott [eds.], Estimating numbers of terrestrial birds. Stud. Avian Biol. 6. Cooper Ornithological Society.

KARR, J. R. 1989. Birds, p. 401-406. In H. L. Leith and M. J. A. Werger [eds.], Tropical rainforest ecosystems. Ecosystems of the world. Vol. 14B. Elseiver Scientific Publ. Co., Amsterdam.

KARR, J. R., AND K. E. FREEMARK. 1983. Habitat selection and environmental gradients: dynamics in the "stable" trophics. Ecology 64:1481-1494.

KARR, J. R., AND F. C. JAMES. 1975. Eco-morphological configurations and convergent evolution in species and communities, p. 258-291. In M. L. Cody and J. M. Diamond [eds.], Ecology and evolution of communities. Harvard Univ. Press, Cambridge, MA.

KARR, J. R., S. K. ROBINSON, J. G. BLAKE, AND R. O. BIERREGAARD, JR. In press. The avifauna of four neotropical forests. In A. Gentry [ed.], Four neotropical forests: a comparison of La Selva, Costa Rica; Barro Colorado Island, Panama; the minimum critical size of ecosystems area, Brazil; and Manu National Park, Peru. Yale Univ. Press, New Haven, CT.

KATTAN, G. 1988. Food habits and social organization of Acorn Woodpeckers in Colombia. Condor 90:100-106.

KATUSIC-MALMBORG, P., AND M. F. WILLSON. 1988. Foraging ecology of avian frugivores and some consequences for seed dispersal in an Illinois woodlot. Condor 90:173-186.

KEAST, A. 1968. Competitive interactions and the evolution of evological niches as illustrated by the Australian honeyeater genus Meliothreptus (Meliphagidae). Evolution 22:762-784.

KEAST, A. 1976. Ecolgoical opportunities and adaptive evolution on islands, with special reference to evolution in the isolated forest outliers of southern Australia. Proc. XVI Int. Ornith. Congr. 573-584.

KEAST, A. 1980. Migratory parulidae: what can species co-occurrence in the north reveal about ecological plasticity and wintering patterns?, p. 457-476. In A. Keast and E. S. Morton [eds.], Migrant birds in the Neotropics: ecology, behavior, distribution, and conservation. Smithsonian Inst. Press, Washington, D.C.

KEAST, A., AND E. S. MORTON [eds.]. 1980. Migrant birds in the Neotropics: ecology, behavior, distribution, and conservation. Smithsonian Inst. Press, Washington, D.C.

KEELER-WOLF, T. 1986. The Barred Antrshrike (Thamnophilus doliatus) on Trinidad and Tobago: habitat niche expansion of a generalist forager. Oecologia 70:309-317.

KEHOE, F. P., AND C. D. ANKNEY. 1985. Variation in digestive organ size among five species of diving ducks (Aythya spp.). Can. J. Zool. 63:2339-2342.

KELLER, M. K. 1978. Markov chains (^{u}107) and applications of matrix methods: fixed point and absorbing Markov chains (^{u}111). Modules and monographs in undergraduate mathematics and its applications project. EDC/UMAP. 55 Chapel St., Newton, Mass.

KENDALL, M. G., AND A. STUART. 1967. The advanced theory of statistics. Vol. 2. 2nd ed. Hafner, New York.

KENDEIGH, S. C. 1947. Bird population studies in the coniferous forest biome during a spruce budworm outbreak. Ontario Dept. of Lands and Forests, Biol. Bull. 1:1-100.

KENDEIGH, S. C. 1949. Effect of temperature and season on energy resources of the English Sparrow. Auk 66:113-127.

KENDEIGH, S. C. 1961. Energy of birds conserved by roosting in cavities. Wilson Bull. 73:140-147.

KENDEIGH, S. C. 1970. Energy requirements for existence in relation to size of birds. Condor 72:60-65.

KENDEIGH, S. C. 1979. Invertebrate populations of the deciduous forest: fluctuations and relations to weather. Illinois Biol. Monogr. No. 50.

KENDEIGH, S. C., V. R. DOL'NIK, AND V. M. GAVRILOV. 1977. Avian energetics, p. 127-204. In J. Pinowski and S. C. Kendeigh [eds.], Granivorous birds in ecosystems: their evolution, populations, energetics, adaptations, impact and control. IBP 12, Cambridge Univ. Press, England.

KENWARD, R. 1987. Wildlife radio tagging: equipment, field techniques and data analysis. Academic Press, London.

KENWARD, R. E., AND R. M. SIBLY. 1977. A Woodpigeon (Columba palumbus) feeding preference explained by a digestive bottleneck. J. Appl. Ecol. 14:815-826.

KEPHART, D. G., AND S. J. ARNOLD. 1982. Garter snake diets in a fluctuating environment: a seven-year study. Ecology 63:1232-1236.

KERFOOT, W. C., AND A. SIH. 1987. Predation -- direct and indirect impacts on aquatic communities. Univ. Press of New England, Hanover, NH.

KERLINGER, F. W. 1986. Foundations of behavioral research. Holt, Rinehart, and Winston, New York.

KESSEL, B. 1976. Winter activity patterns of Black-capped Chickadees in interior Alaska. Wilson Bull. 88:36-61.

KETTERSON, E. D. 1979. Aggressive behavior in wintering Dark-eyed Juncos: determinants of dominance and their possible relation to geographic variation in sex ratio. Wilson Bull. 91:371-383.

KETTERSON, E. D., AND V. NOLAN, JR. 1976. Geographic variation and its climatic correlates in the sex ratio of eastern-wintering Dark-eyed Juncos (Junco hyemalis hyemalis). Ecology 57:679-693.

KETTERSON, E. D., AND V. NOLAN, JR. 1982. The role of migration and winter mortality in the life history of a temperate-zone migrant, the Dark-eyed Junco, as determined from demographic analyses of winter populations. Auk 99:243-259.

KETTERSON, E. D., AND V. NOLAN, JR. 1983. The evolution of differential bird migration. Curr. Ornithol. 1:357-402.

KETTERSON, E. D., AND V. NOLAN, JR. 1985. Intraspecific variation in avian migration: evolutionary and regulatory aspects, p. 553-579. In M. A. Rankin [ed.], Migration: mechanisms and adaptive significance. Contrib. Marine Sci. suppl. vol. 27, Univ. of Texas Marine Sci. Inst., Port Aransas, Texas.

KETTLEWELL, H. B. D. 1955. Selection experiments on industrial melanin in the Lepidoptera. Heredity 9:323-342.

KETTLEWELL, H. B. D. 1956. Further selection experiments in industrial melanism in the Lepidoptera. Heredity 10:287-301.

KETTLEWELL, H. B. D. 1973. The evolution of melanism: the study of a recurring necessity. Clarendon Press, Oxford.

KILHAM, L. 1965. Differences in feeding behavior of male and female Hairy Woodpeckers. Wilson Bull. 77:134-145.

KILHAM, L. 1970. Feeding behavior of Downy Woodpeckers. I. Preference for paper birches and sexual differences. Auk 87:544-556.

KILHAM, L. 1972. Habits of the Crimson-crested Woodpecker in Panama. Wilson Bull. 84:28-47.

KILHAM, L. 1974. Debilitated condition of warblers encountered at Parc Daniel, Gaspe Peninsula. Can. Field-Nat. 8:223.

KILHAM, L. 1978. Sexual similarity of Red-headed Woodpeckers and possible explanations based on fall territorial behavior. Wilson Bull. 90:285-287.

KILHAM, L. 1979. Chestnut-colored Woodpeckers feeding as a pair on ants. Wilson Bull. 91:149-150.

KINCAID, W. B., AND G. N. CAMERON. 1982. Dietary variation in three sympatric rodents on the Texas coastal prairie. J. Mammal. 63:668-672.

KING, J. R. 1972. Adaptive periodic fat storage by birds. Proc. XV Int. Ornithol. Congr. 200-217.

KING, J. R. 1974. Seasonal allocation of time and energy resources in birds. In R. A. Paynter [ed.], Avian energetics. Publ. Nuttall Ornithol. Club 15:4-70.

KING, J. R., AND D. S. FARNER. 1966. The adaptive role of winter fattening in the White-crowned Sparrow with comments on its regulation. Am. Nat. 100:403-418.

KIRBY, B. A. 1963. The biochemistry of the insect fat body. Adv. Insect Physiol. 1-111-174.

KIRKWOOD, J. K. 1979. The partition of food energy for existence in the Kestrel (Falco tinnunculus) and the Barn Owl (Tyto alba). Comp. Biochem. Physiol. 63A:495-498.

KIRKWOOD, J. K. 1983. A limit to metabolisable energy intake in mammals and birds. Comp. Biochem. Physiol. 75A:1-3.

KISIEL, D. S. 1972. Foraging behavior of Dendrocopos villosus and D. pubescens in eastern New York State. Condor 74:393-398.

KLEINBAUM, D. G., L. L. KUPPER, AND H. MORGENSTERN. 1982. Epidemiologic research. Lifetime Learning Publications, Belmont, Calif.

KLIMKIEWICZ, M. K., AND C. S. ROBBINS. 1978. Standard abbreviations for common names of birds. N. Am. Bird Bander 3:16-25.

KLOMP, H. 1966. The dynamics of a field population of

the pine looper Bupalus piniarius L. (Lep. Geom.). Adv. Ecol. Res. 3:207-305.

KLOPFER, P. H. 1958. Influence of social interactions on learning rates in birds. Science (October):903-905.

KLOPFER, P. H. 1963. Behavioral aspects of habitat selection: the role of early experience. Wilson Bull. 75:15-22.

KLOPFER, P. H. 1965. Behavioral aspects of habitat selection: a preliminary report on stereotypy in foliage preferences of birds. Wilson Bull. 77:376-381.

KLOPFER, P. H. 1967. Behavioral stereotypy in birds. Wilson Bull. 79:290-300.

KLOPFER, P. H., AND R. H. MACARTHUR. 1960. Niche size and faunal diversity. Am. Nat. 94:293-300.

KLUYVER, N. H. 1957. Roosting habits, sexual dominance and survival in the Great Tit. Cold Spring Harbor Symp. Quant. Biol. 22:281-285.

KLUYVER, N. H. 1961. Food consumption in relation to habitat in breeding chickadees. Auk 78:532-550.

KNAPTON, R. W. 1980. Nestling foods and foraging patterns in the Clay-colored Sparrow. Wilson Bull. 92:458-465.

KNOKE, J. D. 1982. Discriminant analysis with discrete and continuous variables. Biometrics 38:191-200.

KOCH, R. F., A. E. COURCHESNE, AND C. T. COLLINS. 1970. Sexual differences in foraging behaviors of White-headed Woodpeckers. Bull. So. Calif. Acad. Sci. 69:60-64.

KODRIC-BROWN, A., AND J. H. BROWN. 1978. Influence of economics, interspecific competition, and sexual dimorphism on territoriality of migrant Rufous Hummingbirds. Ecology 59:285-296.

KODRIC-BROWN, A., J. H. BROWN, G. S. BYERS, AND D. F. GORI. 1984. Organization of a tropical island community of hummingbirds and flowers. Ecology 65:1358-1368.

KOENIG, W. D., AND R. L. MUMME. 1987. Population ecology of the cooperatively breeding Acorn Woodpecker. Princeton Univ. Press, Princeton, NJ.

KOENIG, W. D., AND F. A. PITELKA. 1981. Ecological factors and kin selection in the evolution of cooperative breeding in birds, p. 261-280. In R. D. Alexander and D. W. Tinkle [eds.], Selection and social behavior. Chiron Press, Massachusetts.

KOERSVELD, E. VAN. 1950. Difficulties in stomach analysis. Proc. X Int. Ornith. Congr. 592-594.

KOJIMA, K., AND S. MATSUOKA. 1985. Studies on the food habits of four sympatric species of woodpeckers. II. Black Woodpecker Dryocopus martius from winter to early spring. Tori 34:1-6.

KONTOGIANNIS, J. E. 1968. Effect of temperature and exercise on energy intake and body weight of the White-throated Sparrow Zonotrichia albicollis. Physiol. Zool. 41:54-64.

KOTB, A. R., AND T. D. LUCKEY. 1972. Markers in nutrition. Nutr. Abstr. Rev. 42:813-845.

KOZAK, A., D. D. MUNRO, AND J. H. G. SMITH. 1969. Taper functions and their application in forest inventory. For. Chron. 45:278-283.

KREBS, J. R. 1971. Territory and breeding density in the Great Tit, Parus major L. Ecology 52:2-22.

KREBS, J. R. 1973. Behavioral aspects of predation, p. 73-111. In P. P. G. Bateson and P. H. Klopfer [eds.], Perspectives in ethology. Vol. 1. Plenum Press, New York.

KREBS, J. R. 1974. Colonial nesting and social feeding as strategies for exploiting food resources in the Great Blue Heron (Ardea herodias). Behaviour 51:99-134.

KREBS, J. R. 1978. Optimal foraging: decision rules for predators, p. 23-63. In J. R. Krebs and N. B. Davies [eds.], Behavioural ecology: an evolutionary approach. Sinauer Associates, Sunderland, MA.

KREBS, J. R., AND R. J. COWIE. 1976. Foraging strategies in birds. Ardea 64:98-116.

KREBS, J. R., AND N. B. DAVIES. 1981. An introduction to behavioural ecology. Blackwell, Oxford.

KREBS, J. R., AND N. B. DAVIES. 1987. An introduction to behavioral ecology. 2nd ed. Sinauer Assoc., Sunderland, MA.

KREBS, J. R., J. T. ERICHSEN, M. I. WEBBER, AND E. L. CHARNOV. 1977. Optimal prey selection in the Great Tit (Parus major). Anim. Behav. 25:30-38.

KREBS, J. R., AND P. H. HARVEY. 1986. Busy doing nothing -- efficiently. Nature 320:18-19.

KREBS, J. R., AND R. H. MCCLEERY. 1984. Optimization in behavioural ecology, p. 91-121. In J. R. Krebs and N. B. Davies [eds.], Behavioural ecology: an evolutionary approach. Sinauer Assoc., Sunderland, MA.

KREBS, J. R., M. H. MACROBERTS, AND J. M. CULLEN. 1981. Flocking and feeding in the Great Tit Parus major - an experimental study. Ibis 114:507-530.

KREBS, J. R., J. C. RYAN, AND E. L. CHARNOV. 1974. Hunting by expectation or optimal foraging? A study of patch use by chickadees. Anim. Behav. 22:953-964.

KREBS, J. R., D. W. STEPHENS, AND W. J. SUTHERLAND. 1983. Perspectives in optimal foraging, p. 165-216. In A. H. Brush and G. A. Clark, Jr. [eds.], Perspectives in ornithology. Cambridge Univ. Press, New York.

KRZANOWSKI, W. J. 1980. Mixtures of continuous and categorical variables in discriminant analysis. Biometrics 36:493-499.

KUBAN, J. F., AND R. L. NEILL. 1980. Feeding ecology of hummingbirds in the highlands of the Chisos Mountains, Texas. Condor 82:180-185.

KUITUNEN, M., AND T. TORMALA. 1983. The food of Treecreeper (Certhia f. familiaris) nestlings in southern Finland. Ornis Fenn. 60:42-44.

KUNZ, T. H. 1987. Methods of assessing prey availability to insectivorous bats. In T. H. Kunz [ed.], Ecological and behavioral methods for the study of bats. Smithsonian Institution Press, Washington, D.C. [in press].

KUSHLAND, J. A. 1977. Growth energetics of the White Ibis. Condor 79:31-36.

KUSHLAN, J. A. 1983. Bill-shoving feeding behavior in Darwin's finches. J. Field Ornithol. 54:421-422.

KUTTNER, R. 1985. The poverty of economics. Atlantic Monthly (Feb.):74-84.

LAAKE, J. L., K. P. BURNHAM, AND D. R. ANDERSON. 1979. User's manual for program TRANSECT. Utah State Univ. Press, Logan.

LABISKY, R. F., AND W. L. ANDERSON. 1973. Nutritional responses of pheasants to corn, with special reference to high-lysine corn. Ill. Nat. Hist. Surv. Bull. 31:87-112.

LACK, D. 1947. Darwin's finches. Cambridge Univ. Press, Cambridge.

LACK, D. 1954. The natural regulation of animal numbers. Clarendon Press, Oxford.

LACK, D. 1966. Population studies of birds. Clarendon Press, Oxford.

LACK, D. 1968a. Bird migration and natural selection. Oikos 19:1-9.

LACK, D. L. 1968b. Ecological adaptations of reproduction in birds. Methuen, London.

LACK, D. 1971. Ecological isolation in birds. Harvard Univ. Press, Cambridge, Mass.

LACK, D. 1976a. Island biology. Blackwell Scientific Publ., Oxford.

LACK, D. 1976b. Island biology, illustrated by the land birds of Jamaica. Univ. of Calif. Press, Berkeley.

LACK, P. C. 1983. The movements of Palearctic landbird migrants in Tsavo East National Park, Kenya. J. Anim. Ecol. 52:513-524.

LACK, P. C. 1987. The structure and seasonal dynamics of the bird community in Tsavo National Park, Kenya. Ostrich 58:9-23.

LACK, D., AND E. LACK. 1958. The nesting of the Long-tailed Tit. Bird Study 5:1-19.

LACK, D., AND P. LACK. 1972. Wintering warblers in Jamaica. Living Bird 11:129-153.

LANCE, D. R., J. S. ELKINTON, AND C. P. SCHWALBE. 1987. Behaviour of late-instar gypsy moth larvae in high and low

density populations. Ecol. Entomol. 12:267-273.
LAND, H. C. 1963. A tropical feeding tree. Wilson Bull. 75:199-200.
LANDRES, P. B. 1980. Community organization of the arboreal birds in some oak woodlands of western North America. Ph.D. diss., Utah State Univ., Logan.
LANDRES, P. B. 1983. Use of the guild concept in environmental impact assessment. Environ. Manage. 7:393-398.
LANDRES, P. B., AND J. A. MACMAHON. 1980. Guilds and community organization: analysis of an oak woodland avifauna in Sonora, Mexico. Auk 97:351-365.
LANDRES, P. B., AND J. A. MACMAHON. 1983. Community organization of arboreal birds in some oak woodlands of western North America. Ecol. Monogr. 53:183-208.
LANGELIER, L. A., AND E. O. GARTON. 1986. Management guidelines for increasing populations of birds that feed on western spruce budworm. USDA Agric. Handb. 653, Washington, D.C.
LANYON, W. E. 1984. The systematic position of the Cocos Flycatcher. Condor 86:42-47.
LASIEWSKI, R. C., AND W. R. DAWSON. 1967. A reexamination of the relation between standard metabolic rate and body weight in birds. Condor 69:13-23.
LASKEY, A. R. 1957. Some Tufted Titmouse life history. Bird Banding 28:135-145.
LAURENT, J. L. 1986. Winter foraging behavior and resource availability for a guild of insectivorous gleaning birds in a southern alpine larch forest. Ornis Scand. 17:347-355.
LAURENZI, A. W., B. W. ANDERSON, AND R. D. OHMART. 1982. Wintering biology of Ruby-crowned Kinglets in the lower Colorado River Valley. Condor 84:385-398.
LAURSEN, K. 1978. Interspecific relationships between some insectivorous passerine species, illustrated by their diet during spring migration. Ornis Scand. 9:178-192.
LAVERTY, T. M., AND R. C. PLOWRIGHT. 1988. Flower handling by bumblebees: a comparison of specialists and generalists. Anim. Behav. 36:733-740.
LAWRENCE, E. S. 1985. Sit-and-wait predators and cryptic prey: a field study with wild birds. J. Anim. Ecol. 54:965-976.
LAWTON, M., AND C. F. GUINDON. 1981. Flock composition, breeding success, and learning in the Brown Jay. Condor 83:27-33.
LAWTON, M. F., AND R. O. LAWTON. 1986. Heterochrony, deferred breeding, and avian sociality. Curr. Ornithol. 4:187-222.
LEBART, L., A. MORINEAU, AND K. M. WARWICK. 1984. Multivariate descriptive statistical analysis: correspondence analysis and related techniques for large matrices. John Wiley and Sons, New York.
LEBRETON, J. D., AND N. YOCCOZ. 1987. Multivariate analysis of bird count data. Acta Oecol./Oecol. Gener. 8:125-144.
LECHOWICZ, M. J. 1982. The sampling characteristics of electivity indices. Oecologia 52:22-30.
LECK, C. F. 1971. Overlap in the diet of some Neotropical birds. Living Bird 10:89-106.
LECK, C. F. 1973. Observations of birds at Cecropia trees in Puerto Rico. Wilson Bull. 84:498-500.
LEDERER, R. J. 1972. The role of avian rictal bristles. Wilson Bull. 84:193-197.
LEDERER, R. J. 1977. Winter territoriality and foraging behavior of the Townsend's Solitaire. Am. Midl. Nat. 97:101-109.
LEDERER, R. J., AND R. CRANE. 1978. The effects of emetics on wild birds. N. Am. Bird Bander 3:3-5.
LEES, D. R., AND E. R. CREED. 1975. Industrial melanism in Biston betularia: the role of selective predation. J. Anim. Ecol. 44:67-83.
LEHIKOINEN, E. 1986. Winter ecology of passerines: significance of weight and size. Rep. Dept. Biol., Univ. of Turku (Finland), No. 14.

LEHIKOINEN, E. 1987. Seasonality of the daily weight cycle in wintering passerines and its consequences. Ornis Scand. 18:216-226.
LEIGH, E. G., JR. 1971. Adaptation and diversity. Freeman, Cooper and Co., San Francisco, CA.
LEIGH, E. G., JR., AND D. M. WINDSOR. 1982. Forest production and regulation of primary consumers on Barro Colorado Island, p. 111-112. In E. Leigh, Jr., A. S. Rand, and D. Windsor [eds.], The ecology of a tropical forest. Smithsonian Inst. Press, Washington, D.C.
LEIGH, T. F., D. GONZALEZ, AND R. VAN DEN BOSCH. 1970. A sampling device for estimating absolute insect populations in cotton. J. Econ. Entomol. 63:1704-1706.
LEIGHTON, M., AND D. R. LEIGHTON. 1983. Vertebrate responses to fruiting seasonality within a Bornean rain forest, p. 181-196. In S. L. Sutton, T. C. Whitmore, and A. C. Chadwick [eds.], Tropical rainforest: ecology and management. Special Publ. No. 2 of the British Ecol. Soc., Blackwell Scientific Publ., Oxford.
LEISLER, B. 1980. Morphological aspects of ecological specialization in bird genera. Okologia Vogel (Ecol. Birds) 2:119-220.
LEISLER, B., AND E. THALER. 1982. Differences in morphology and foraging behaviour in the Goldcrest Regulus regulus and the Firecrest R. ignicapillus. Ann. Zool. Fennici 19:277-284.
LEISLER, B., AND H. WINKLER. 1985. Ecomorphology. Curr. Ornithol. 2:155-186.
LEISLER, G. H., AND K. H. SIENBENROCK. 1983. Einnischung und interspezifische Territorialitat uberwinternder Steinschmatzer (Oenanthe isabellina, O. oenenthe, O. pleschanka) in Kenia. J. Ornith. 124:393-414.
LEONARD, D. E. 1964. Biology and ecology of Magicicada septendecim (L.) (Hemiptera: Cicadidae). J. New York Entomol. Soc. 72:19-23.
LEOPOLD, A. S. 1953. Intestinal morphology of gallinaceous birds in relation to food habits. J. Wildl. Manage. 17:197-203.
LEVEY, D. J. 1986. Methods of seed processing by birds and seed deposition patterns, p. 147-158. In A. Estrada and T. H. Fleming [eds], Frugivores and seed dispersal. Dr. W. Junk, Publ., Dordrecht, The Netherlands.
LEVEY, D. J. 1987a. Sugar-tasting ability and fruit selection in tropical fruit-eating birds. Auk 104:173-179.
LEVEY, D. J. 1987b. Seed size and fruit-handling techniques of avian frugivores. Am. Nat. 129:471-485.
LEVEY, D. J. 1988. Spatial and temporal variation in Costa Rican fruit and fruit-eating bird abundance. Ecol. Monogr. 58:251-269.
LEVEY, D. J., AND W. H. KARASOV. In press. Digestive responses of temperate birds switched to fruit or insect diets. Auk 106.
LEVEY, D. J., T. C. MOERMOND, AND J. S. DENSLOW. 1984. Fruit choice in neotropical birds: the effect of distance between fruits on preference patterns. Ecology 65:844-850.
LEVINGS, S. C., AND D. M. WINDSOR. 1984. Litter moisture content as a determinant of litter arthropod distribution and abundance during the dry season on Barro Colorado Island, Panama. Biotropica 16:125-131.
LEVINS, R. 1968. Evolution in changing environments. Monogr. Popul. Biol. No. 2. Princeton Univ. Press, Princeton, NJ.
LEWKE, R. E. 1982. A comparison of foraging behavior among permanent, summer, and winter resident bird groups. Condor 84:84-90.
LIEBERMAN, D., J. B. HALL, M. D. SWAINE, AND M. LIEBERMAN. 1979. Seed dispersal by baboons in the Shai Hills, Ghana. Ecology 60:65-75.
LIEBHOLD, A. M., AND J. S. ELKINTON. 1988. Techniques for estimating the density of late-instar gypsy moth (Lepidoptera: Lymantriidae), populations using frass drop and frass production measurements. Environ. Entomol. 17:381-384.

LIEM, K. F., AND L. S. KAUFMAN. 1984. Intraspecific macroevolution: functional biology of the polymorphic cichlid species Cichlasoma minckley, p. 203-215. In A. A. Echelle and I. Kornfield [eds.], Evolution of fish species flocks. Univ. of Maine Press, Orono, ME.

LIFSON, N., AND R. McCLINTOCK. 1966. Theory and use of the turnover rates of body water for measuring energy and material balance. J. Theor. Biol. 12:46-74.

LIGON, J. D. 1968a. Sexual differences in foraging behavior of two species of Dendrocopos woodpeckers. Auk 85:203-215.

LIGON, J. D. 1968b. Starvation of spring migrants in the Chiricahua Mountains, Arizona. Condor 70:387-388.

LIMA, S. L. 1983. Downy woodpecker foraging behavior: foraging by expectation and energy intake rate. Oecologia 58:232-237.

LIMA, S. L. 1984. Downy Woodpecker foraging behavior: efficient sampling in simple stochastic environments. Ecology 65:166-174.

LIMA, S. L. 1985. Sampling behavior of starlings foraging in simple patchy environments. Behav. Ecol. Sociobiol. 16:135-142.

LIMA, S. L. 1986. Predation risk and unpredictable feeding conditions: determinants of body mass in birds. Ecology 67: 377-385.

LIMA, S. L., AND T. J. VALONE. 1986. Influence of predation risk on diet selection: a simple example in the grey squirrel. Anim. Behav. 34:536-544.

LINDEN, H., AND M. WIKMAN. 1983. Goshawk predation of tetranoids: availability of prey and diet of the predator in the breeding season. J. Anim. Ecol. 52:953-968.

LINKER, H. M., F. A. JOHNSON, J. L. STIMAC, AND S. L. POE. 1984. Analysis of sampling procedures for corn earworm and fall armyworm (Lepidoptera: Noctuidae) in peanuts. Environ. Entomol. 13:75-78.

LLOYD, M., AND H. S. DYBAS. 1966a. The periodical cicada problem. I. Population ecology. Evolution 20:133-149.

LLOYD, M., AND H. S. DYBAS. 1966b. The periodical cicada problem. II. Evolution. Evolution 20:466-505.

LÖHRL, H. 1972. Zun Verhalten dis Wiessmaskenbaumhopfes (Phoeniculus bollei) (Bewegunsweise, Nahrungsaufnahme, und Sozialverhalten). J. Ornith. 113:49-52.

LOISELLE, B. A. 1987. Birds and plants in a neotropical rain forest: seasonality and interactions. Ph.D. diss., Univ. of Wisconsin, Madison.

LOISELLE, B. A. 1988. Bird abundance and seasonality in a Costa Rican lowland forest canopy. Condor 90:761-772.

LONGLAND, W. S. 1985. Comments on preparing owl pellets by boiling in NaOH. J. Field Ornithol. 56:277.

LOWE, V. P. W. 1980. Variation in digestion of prey by the Tawny Owl (Strix aluco). J. Zool. 192:283-293.

LOWTHER, P. E. 1975. Geographic and ecological variation in the family Icteridae. Wilson Bull. 87:481-495.

LOYN, R. H., R. G. RUNNALLS, G. Y. FORWARD, AND J. TYERS. 1983. Territorial Bell Miners and other birds affecting populations of insect prey. Science 221:1411-1413.

LUNDQUIST, R. W. 1988. Habitat use by cavity-nesting birds in the southern Washington Cascades. M.S. thesis, Univ. of Washington, Seattle.

MACARTHUR, R. H. 1958. Population ecology of some warblers of northeastern coniferous forests. Ecology 39:599-619.

MACARTHUR, R. H. 1959. On the breeding distribution pattern of North American migrant birds. Auk 76:318-325.

MACARTHUR, R. H. 1965. Patterns of species diversity. Biol. Rev. 40:510-533.

MACARTHUR, R. H. 1969. Species packing, and what interspecies competition minimizes. Proc. Nat. Acad. Sci. 64:1369-1371.

MACARTHUR, R. H. 1970. Species packing and competitive equilibrium for many species. Theor. Pop. Biol. 1:1-11.

MACARTHUR, R. H. 1972. Geographical ecology. Harper and Row, New York.

MACARTHUR, R. H., AND R. LEVINS. 1964. Competition, habitat selection, and character displacement in a patchy environment. Proc. Nat. Acad. Sci. 51:1207-1210.

MACARTHUR, R. H., AND R. LEVINS. 1967. The limiting similarity, convergence, and divergence of coexisting species. Am. Nat. 101:377-385.

MACARTHUR, R. H., AND J. W. MACARTHUR. 1961. On bird species diversity. Ecology 42:594-598.

MACARTHUR, R. H., J. W. MACARTHUR, AND J. PREER. 1962. On bird species diversity. II. Prediction of bird census from habitat measurements. Am. Nat. 96:167-174.

MACARTHUR, R. H., AND E. R. PIANKA. 1966. On optimal use of a patchy environment. Am. Nat. 100:603-609.

MACARTHUR, R. H., AND E. O. WILSON. 1967. The theory of island biogeography. Princeton Monogr. Pop. Biol. 1:1-203.

MACDONALD, D. W. 1978. Radio-tracking: some applications and limitations, p. 192-204. In B. Stonehouse [ed.], Animal marking: recognition marking of animals in research. University Park Press, Baltimore, MD.

MACDONALD, J. D. 1973. Birds of Australia. Reed, Sydney.

MACFADYEN, A. 1962. Soil arthropod sampling. Adv. Ecol. Res. 1:1-34.

MACLEAN, A. A. 1986. Age-specific foraging ability and the evolution of delayed breeding in three species of gulls. Auk 98:267-279.

MACLEAN, G. L. 1985. Roberts' Birds of Southern Africa. Voelcker Bird Book Fund, Cape Town.

MACLELLAN, C. R. 1958. Role of woodpeckers in control of the codling moth in Nova Scotia. Can. Entomol. 90:18-22.

MACLELLAN, C. R. 1959. Woodpeckers as predators of the codling moth in Nova Scotia. Can. Entomol. 91:673-680.

MACMAHON, J. A., D. J. SCHIMPF, D. C. ANDERSEN, K. G. SMITH, AND R. L. BAYN, JR. 1981. An organism-centered approach to some community and ecosystem concepts. J. Theor. Biol. 88:287-307.

MACROBERTS, M. H., AND B. R. MACROBERTS. 1976. Social organization and behavior of the Acorn Woodpecker in central coastal California. Ornith. Monogr. 21.

MCATEE, W. L. 1912. Methods for estimating the contents of bird stomachs. Auk 29:449-464.

MCATEE, W. L. 1932. Effectiveness in nature of the so-called protective adaptations in the animal kingdom, chiefly as illustrated by the food habits of nearctic brids. Smithsonian Misc. Coll. 85:1-201.

MCCAMBRIDGE, W. F., AND G. C. TROSTLE. 1972. The mountain pine beetle. USDA For. Serv. Pest Leaflet 2, Washington, DC.

MCCULLAGH, P., AND J. A. NELDER. 1983. Generalized linear models. Marcel Dekker, New York.

MCCUNE, B. 1987. Multivariate analysis in the PC-ORD system. Holcomb Res. Instit. Rep. No. 75, Butler Univ., Indiana.

MCDONALD, L. L. 1981. A discussion of robust procedures in multivariate analysis, p. 242-244. In D. E. Capen [ed.], The use of multivariate statistics in studies of wildlife habitat. USDA For. Serv. Gen. Tech. Rep. RM-87, Fort Collins, CO.

MCDONALD, L. L., AND B. F. J. MANLY. 1989. Calibration of biased sampling procedures, p. 467-483. In L. L. McDonald, B. F. J. Manly, J. A. Lockwood, and J. A. Logan [eds.], Estimation and analysis of insect populations. Springer Verlag, New York.

MCELLIN, S. M. 1979. Population demographies, spacing, and foraging behaviors of White-breasted and Pygmy nuthatches in ponderosa pine habitat, p. 301-330. In J. G. Dickson, R. N. Conner, R. R. Fleet, J. A. Jackson, and J. C. Kroll [eds.], The role of insectivorous birds in forest ecosystems. Academic Press, New York.

MCFARLAND, D. C. 1986a. The organization of a

honeyeater community in an unpredictable environment. Aust. J. Ecol. 11:107-120.

McFarland, D. C. 1986b. Determinants of feeding territory size in the New Holland Honeyeater Phylidonyris novaehollandiae. Emu 86:180-185.

McFarlane, R. W. 1976. Birds as agents of biological control. The Biologist 58:123-140.

McGill, A. R. 1970. Australian warblers. Bird Observers Club, Melbourne.

McIntyre, G. A. 1953. Estimation of plant density using line transects. J. Ecol. 41:319-330.

McKean, L. 1988. Foraging efficiency and cultural transmission of information between Gray-breasted Jay flock members. M.S. thesis, Univ. of Arizona, Tucson.

McKelvey, R. 1985. The metabolizable energy of chicken scratch, the rhizomes of Carex lyngbei and timothy grass to swans. Can. Wildl. Serv. Prog. Notes 152:1-3.

McKey, D. 1974. Adaptive patterns in alkaloid physiology. Am. Nat. 108:305-320.

McKey, D. 1975. The ecology of coevolved seed dispersal systems, p. 159-191. In L. E. Gilbert and P. H. Raven [eds.], Coevolution of animals and plants. Univ. of Texas Press, Austin.

McLean, J. A., A. Bakke, and H. Niemeyer. 1987. An evaluation of three traps and two lures for the ambrosia beetle Trypodendron lineatum (Oliv.) (Coleoptera: Scolytidae) in Canada, Norway, and West Germany. Can. Entomol. 119:273-280.

McLelland, J. 1979. Digestive system, p. 61-181. In A. S. King and J. McLelland [eds.], Form and function in birds. Vol. 1. Academic Press, London and New York.

McNair, J. N. 1982. Optimal giving-up times and the marginal value theorem. Am. Nat. 119:511-529.

McNamara, J. M. 1982. Optimal patch use in a stochastic environment. Theor. Popul. Biol. 21:269-288.

McNamara, J. M., and A. I. Houston. 1985. A simple model of information use in the exploitation of patchily distributed food. Anim. Behav. 33:553-560.

McNamara, J. M., and A. I. Houston. 1987a. A general framework for understanding the effects of variability and interruptions on foraging behaviour. Acta Biotheoretica 36:3-22.

McNamara, J. M., and A. I. Houston. 1987b. Partial preferences and foraging. Anim. Behav. 35:1084-1099.

McNaughton, S. J., and L. L. Wolf. 1970. Dominance and the niche in ecological systems. Science 167:131-139.

McPherson, J. M. 1987. A field study of winter fruit preferences of Cedar Waxwings. Condor 89:293-306.

Machlis, L. 1977. An analysis of the temporal patterning of pecking in chicks. Behaviour 63:1-70.

Madden, J. L. 1982. Avian predation of the woodwasp, Sirex noctilio F., and its parasitoid complex in Tasmania. Aust. J. Wildl. Res. 9:135-144.

Majer, J. D., and H. F. Recher. 1988. Invertebrate communities on Western Australian eucalypts: a comparison of branch clipping and chemical knockdown procedures. Aust. J. Ecol. 13.

Malaise, R. 1937. A new insect-trap. Entomol. Tidskr. 58:148-160.

Mallet, J., J. T. Longino, D. Murawski, A. Murawski, and A. Simpson de Gamboa. 1987. Handling effects in Heliconius: where do all the butterflies go? J. Anim. Ecol. 56:377-386.

Malone, C. R. 1965. Dispersal of plankton: rate of food passage in mallad ducks. J. Wildl. Manage. 29:529-533.

Mangel, M., and C. W. Clark. 1986. Towards a unified foraging theory. Ecology 67:1127-1138.

Manly, B. F. J. 1974. A model for certain types of selection experiments. Biometrics 30:281-294.

Manly, B. F. J. 1977. The estimation of the fitness function from several samples taken from a population. Biometrics 19:391-401.

Manly, B. F. J. 1985. The statistics of natural selection. Chapman and Hall, London.

Mannan, R. W., and E. C. Meslow. 1984. Bird populations and vegetation characteristics in managed and old-growth forests, northeastern Oregon. J. Wildl. Manage. 48:1219-1238.

Mannan, R. W., M. L. Morrison, and E. C. Meslow. 1984. Comment: the use of guilds in forest bird management. Wildl. Soc. Bull. 12:426-430.

Manuwal, D. A., and M. H. Huff. 1987. Spring and winter bird populations in a Douglas-fir forest sere. J. Wildl. Manage. 51:586-595.

Marcotullio, P. J., and F. B. Gill. 1985. Use of time and space by Chestnut-backed Antbirds. Condor 87:187-191.

Marcum, C. L., and D. O. Loftsgaarden. 1980. A nonmapping technique for studying habitat preferences. J. Wildl. Manage. 44:963-968.

Marden, J. H. 1984. Remote perception of floral nectar by bumblebees. Oecologia 64:232-240.

Mariani, J. M. 1987. Brown Creeper (Certhia americana) abundance patterns and habitat use in the southern Washington Cascades. M.S. thesis, Univ. Washington, Seattle.

Mariath, H. A. 1982. Experiments on the selection against different colors morphs of a twig caterpillar by insectivorous birds. Z. fur Tierpsychol. 60:135-145.

Markin, G. P. 1982. Abundance and life cycles of lepidoptera associated with an outbreak of the western spruce budworm (Lepidoptera: Tortricidae) in southeastern Idaho. J. Kans. Entomol. Soc. 55:365-372.

Marlatt, C. L. 1907. The periodical cicada. USDA Bur. Entomol. Bull. No. 71, Washington, D.C.

Marlatt, C. L. 1908. A successful seventeen-year breeding record for the periodical cicada. Proc. Entomol. Soc. Wash. 9:16-18.

Maron, J. L., and J. P. Myers. 1985. Seasonal changes in feeding success, activity patterns, and weights of non-breeding Sanderlings (Calidris alba). Auk 102:580-586.

Marquiss, M., and I. Newton. 1981. A radio-tracking study of the ranging behaviour and dispersion of European Sparrowhawks Accipiter nisus. J. Anim. Ecol. 51:111-133.

Marriott, R. W., and D. K. Forbes. 1970. The digestion of lucerne chaff by Cape Barren Geese, Cereopsis novacollandiae Latham. Aust. J. Zool. 18:257-263.

Marshall, J. T. 1960. Interrelations of Abert's and Brown Towhees. Condor 62:49-64.

Marston, N., D. G. Davis, and M. Gebhardt. 1982. Ratios for predicting field populations of soybean insects and spiders from sweep net samples. J. Econ. Entomol. 75:976-981.

Marston, N., C. E. Morgan, G. D. Thomas, and C. M. Ignotto. 1976. Evaluation of four techniques for sampling soybean insects. J. Kans. Entomol. Soc. 49:389-400.

Marti, C. D. 1974. Feeding ecology of four sympatric owls. Condor 76:45-61.

Martin, A. C., R. C. Gensch, and C. P. Brown. 1946. Alternative methods in upland gamebird food analysis. J. Wildl. Manage. 10:8-12.

Martin, A. C., H. S. Zim, and A. L. Nelson. 1951. American wildlife and plants. McGraw Hill, New York.

Martin, E. W. 1968. The effects of dietary protein on the energy and nitrogen balance of the Tree Sparrow (Spizella arborea arborea). Physiol. Zool. 41:313-331.

Martin, P., and P. Bateson. 1986. Measuring behaviour: an introductory guide. Cambridge Univ. Press, Cambridge.

Martin, T. E. 1980. Diversity and abundance of spring migratory birds using habitat islands on the Great Plains. Condor 82:430-439.

Martin, T. E. 1982. Frugivory and North American migrants in a neotropical second-growth woodland. Ph.D. diss., Univ. of Illinois, Champaign.

Martin, T. E. 1985a. Selection of second-growth woodlands by frugivorous migrating birds in Panama: an effect of fruit size and plant density? J. Trop. Ecol.

1:157-170.

Martin, T. E. 1985b. Resource selection by tropical frugivorous birds: integrating multiple interactions. Oecologia 66:563-573.

Martin, T. E. 1986. Competition in breeding birds: on the importance of considering processes at the level of the individual. Curr. Ornithol. 4:181-210.

Martin, T. E. 1987. Food as a limit on breeding birds: a life-history perspective. Ann. Rev. Ecol. Syst. 18:453-487.

Martin, T. E. 1988a. Resource partitioning and structure of animal communities. ISI Atlas of Science (Animal and Plant Sciences) 1:20-24.

Martin, T. E. 1988b. Processes organizing open-nesting bird assemblages: competition or nest predation? Evol. Ecol. 2:37-50.

Martin, T. E. 1988c. On the advantage of being different: nest predation and the coexistence of bird species. Proc. Nat. Acad. Sci. 85:2196-2199.

Martin, T. E., and J. R. Karr. 1986a. Temporal dynamics of neotropical birds with special reference to frugivores in second-growth woods. Wilson Bull. 98:38-60.

Martin, T. E., and J. R. Karr. 1986b. Patch utilization by migrating birds: resource oriented? Ornis Scand. 17:165-174.

Martinat, P. J. 1984. The effect of abiotic factors on saddled prominent, Heterocampa guttivitta (Walker), (Lepidoptera: Notodontidae) population biology. Ph.D. thesis, State Univ. of New York, Syracuse.

Martinat, P. J., C. C. Coffman, K. Dodge, R. J. Cooper, and R. C. Whitmore. 1988. Effect of Dimilin 25-W on the canopy arthropod community in a central Applachian forest. J. Econ. Entomol. 81:261-267.

Martinez del Rio, C., B. R. Stevens, D. E. Daneke, and P. T. Andreadis. 1988. Physiological correlates of preference and aversion for sugars in three species of birds. Physiol. Zool. 61:222-229.

Martinez del Rio, C., W. H. Karasov, and D. J. Levey. 1989. Physiological basis and ecological consequences of sugar preferences in Cedar Waxwings. Auk 106:64-71.

Masaki, S., and T. J. Walker. 1987. Cricket life cycles, p. 349-423. In M. K. Hecht, B. Wallace and G. Price [eds.], Evolutionary biology. Vol. 21. Plenum Press, New York.

Masman, D. 1986. The annual cycle of the Kestrel, Falco tinnunculus. Drukkerj van Denderen B. V., Groningen, The Netherlands.

Masman, D., and M. Klaasen. 1987. Energy expenditure during free flight in trained and free-living European Kestrels (Falco tinnunculus). Auk 104:603-616.

Mason, J. R., A. H. Arzt, and R. F. Reidinger. 1984. Comparative assessment of food preferences and aversions acquired by blackbirds via observational learning. Auk 101:796-803.

Mason, R. R. 1970. Development of sampling methods for the Douglas-fir tussock moth, Hemerocampa pseudotsugata (Lepidoptera: Lymantriidae). Can. Entomol. 102:836-845.

Mason, R. R. 1977. Sampling low density populations of the Douglas-fir tussock moth by frequency of occurrence in the lower tree crown. USDA For. Serv. Res. Pap. PNW-216, Portland, OR.

Mason, R. R. 1979. How to sample Douglas-fir tussock moth larvae. USDA Agric. Handb. 547. Washington, D.C.

Mason, R. R. 1987a. Frequency sampling to predict densities in sparse populations of the Douglas-fir tussock moth. For. Sci. 33:145-156.

Mason, R. R. 1987b. Nonoutbreak species of forest Lepidoptera, p. 31-57. In P. Barbosa and J. C. Schultz [eds.], Insect outbreaks. Academic Press, New York.

Mason, R. R., and W. S. Overton. 1983. Predicting size and change in nonoutbreak populations of the Douglas-fir tussock moth (Lepidoptera: Lymantriidae). Environ. Entomol. 12:799-803.

Mason, R. R., and T. R. Torgersen. 1983. Mortality of larvae in stocked cohorts of the Douglas-fir tussock moth, Orygia pseudotsugata (Lepidoptera: Lymantriidae). Can. Entomol. 115:1119-1127.

Mason, R. R., and T. R. Torgersen. 1987. Dynamics of a nonoutbreak population of the Douglas-fir tussock moth (Lepidoptera: Lymantriidae) in southern Oregon. Environ. Entomol. 16:1217-1227.

Mateos, G. G., and J. L. Sell. 1981. Metabolizable energy of supplemental fat as related to dietary-fat level and methods of estimation. Poul. Sci. 60:1509-1515.

Mathiak, H. A. 1938. A key to hairs of the mammals of southern Michigan. J. Wildl. Manage. 2:251-268.

Matsuoka, S., and K. Kojima. 1985. Studies on the food habits of four sympatric species of woodpeckers. I. Grey-headed Green Woodpecker Picus canus in winter. Tori 33:103-111.

Matthews, R. W., and J. R. Matthews. 1970. Malaise trap studies of flying insects in a New York mesic forest. I. Ordinal composition and seasonal abundance. J. New York Entomol. Soc. 78:52-59.

Mattocks, J. G. 1971. Goose feeding and cellulose digestion. Wildfowl 22:107-113.

Mattson, W. J., Jr. 1980. Herbivory in relation to plant nitrogen content. Ann. Rev. Ecol. Syst. 11:119-161.

Mattson, W. J., and N. D. Addy. 1975. Phytophagous insects as regulators of forest primary production. Science 190:515-522.

Mattson, W. J., F. B. Knight, D. C. Allen, and J. L. Feltz. 1968. Vertebrate predation on the jack-pine budworm in Michigan. J. Econ. Entomol. 61:229-234.

Maurer, B. A. 1984. Interference and exploitation in bird communities. Wilson Bull. 96:380-395.

Maurer, B. A. 1985a. On the ecological and evolutionary roles of interspecific competition. Oikos 45:300-302.

Maurer, B. A. 1985b. Avian community dynamics in desert grasslands: observational scale and hierarchical structure. Ecol. Monogr. 55:295-312.

Maurer, B. A. 1987. Scaling of biological community structure: a systems approach to community complexity. J. Theor. Biol. 127:97-110.

Maurer, B. A., and R. C. Whitmore. 1981. Foraging of five bird species in two forests with different vegetation structure. Wilson Bull. 93:478-490.

Maxson, S. J., and L. W. Oring. 1980. Breeding season time and energy budgets of the polyandrous Spotted Sandpiper. Behaviour 74:200-263.

May, R. M. 1973. Stability and complexity in model ecosystems. Monogr. Popul. Biol. 6. Princeton Univ. Press, Princeton, NJ.

May, R. M. 1975. Some notes on estimating the competition matrix, alpha. Ecology 56:737-741.

May, R. M. 1976. Models for two interacting populations, p. 49-70. In R. M. May [ed.], Theoretical ecolgoy, principles and applications. Blackwell Scientific Publ., Oxford.

May, R. M. 1979. Periodical cicadas. Nature 277:347-349.

Maynard, L. A., and J. K. Loosli. 1969. Animal nutrition. McGraw-Hill. New York.

Maynard-Smith, J. 1978. Optimization theory in evolution. Ann. Rev. Ecol. Syst. 9:31-56.

Mayr, E. 1974. Behavioral programs and evolutionary strategies. Am. Sci. 62:650-659.

Meanley, B. 1970. Methods of searching for food by the Swainson's Warbler. Wilson Bull. 82:228.

Mech, L. D. 1983. Handbook of animal radio-tracking. Univ. Minnesota Press, Minneapolis.

Medway, L. 1972. Phenology of a tropical rain forest in Malaya. Biol. J. Linn. Soc. 4:117-146.

Meents, J. K., J. Rice, B. W. Anderson, and R. D. Ohmart. 1983. Nonlinear relationships between birds and vegetation. Ecology 64:1022-1027.

Merriam Webster Co. 1986. Webster's ninth collegiate dictionary. Merriam Webster. Springfield, MA.

Meunier, M., and J. Bédard. 1984. Nestling foods of the

Savannah Sparrow. Can. J. Zool. 62:23-27.
MEYLAN, A. 1988. Spongivory in hawksbill turtles: a diet of glass. Science 239:393-395.
MICHENER, M. C., AND C. WALCOTT. 1966. Navigation of single homing pigeons: airplane observations by radiotracking. Science 154:410-413.
MIKKOLA, K. 1972. Behavioural and electrophysiological responses of night-flying insects, especially Lepidoptera, to near-ultraviolet and visible light. Ann. Zool. Fenn. 9:225-254.
MILES, D. B., AND R. E. RICKLEFS. 1984. The correlation between ecology and morphology in deciduous forest passerine birds. Ecology 65:1629-1640.
MILES, D. B., R. E. RICKLEFS, AND J. TRAVIS. 1987. Concordance of ecomorphological relationships in three assemblages of passerine birds. Am. Nat. 129:347-364.
MILLER, A. H. 1931. Notes on the song and territorial habits of Bullock's Oriole. Wilson Bull. 43:102-108.
MILLER, A. H. 1942. Habitat selection in higher vertebrates and its relation to intraspecific variation. Am. Nat. 76:25-35.
MILLER, A. H. 1956. Ecological factors that accelerate the formation of races and species of terrestrial vertebrates. Evolution 10:262-271.
MILLER, J. M., AND F. P. KEEN. 1960. Biology and control of the Western Pine Beetle. USDA For. Serv. Misc. Publ. No. 800, Washington, D.C.
MILLER, M. R. 1974. Digestive capabilities, gut morphology, and cecal fermentation in wild waterfowl (genus Anas) fed various diets. M.S. thesis, Univ. California, Davis.
MILLER, M. R. 1975. Gut morphology of Mallards in relation to diet quality. J. Wildl. Manage. 39:168-173.
MILLER, M. R. 1984. Comparative ability of Northern Pintails, Gadwalls, and Northern Shovelers to metabolize foods. J. Wildl. Manage. 48:362-370.
MILLER, M. R., AND K. J. REINECKE. 1984. Proper expression of metabolizable energy in avian energetics. Condor 86:396-400.
MILLER, R. G. 1981. Simultaneous statistical inference. Springer-Verlag, New York.
MILLER, R. S. 1967. Pattern and process in competition. Adv. Ecol. Res. 4:1-74.
MILTON, K., D. M. WINDSOR, D. W. MORRISON, AND M. A. ESTRIBI. 1982. Fruiting phenologies of two neotropical Ficus species. Ecology 63:752-762.
MINDELL, D. P., AND H. L. BLACK. 1984. Combined-effort hunting by a pair of Chestnut-mandibled Toucans. Wilson Bull. 96:319-321.
MINOT, E. O. 1981. Effects of interspecific competition for food in breeding blue and great tits. J. Anim. Ecol. 50:375-385.
MISSLEN, R., AND M. CIGRANG. 1986. Does neophobia necessarily imply fear or anxiety. Behav. Process 12:45-50.
MITCHELL, D. 1976. Experiments on neophobia in wild and laboratory rats: a reevaluation. J. Comp. Physiol. Psychol. 90:190-197.
MITCHELL, R. T. 1952. Consumption of spruce budworms by birds in a Maine spruce-fir forest. J. For. 50:387-389.
MIZELL, R. F. III, AND D. E. SCHIFFHAUER. 1987. Trunk traps and overwintering predators in pecan orchards: survey of species and emergence times. Florida Entomol. 70:238-244.
MOEED, A., AND M. J. MEADS. 1983. Vertebrate fauna of four tree species in Orongorongo Valley, New Zealand, as revealed by trunk traps. N. Z. J. Ecol. 6:39-53.
MOERMOND, T. C. 1979a. Habitat constraints on the behavior, morphology, and community structure of Anolis lizards. Ecology 60:152-164.
MOERMOND, T. C. 1979b. The influence of habitat structure on Anolis foraging behavior. Behaviour 70:147-167.
MOERMOND, T. C. 1986. A mechanistic approach to the structure of animal communities: Anolis lizards and birds. Zool. 26:23-27.
MOERMOND, T. C., AND J. S. DENSLOW. 1983. Fruit choice in neotropical birds: effects of fruit type and accessibility on selectivity. J. Anim. Ecol. 52:407-420.
MOERMOND, T. C., AND J. S. DENSLOW. 1985. Neotropical avian frugivores: patterns of behavior, morphology, and nutrition, with consequences for fruit selection, p. 865-897. In P. A. Buckley, M. S. Foster, E. S. Morton, R. S. Ridgely, and N. G. Smith [eds.], Neotropical ornithology. Ornithol. Monogr. No. 36.
MOERMOND, T. C., AND R. W. HOWE. In press. Ecomorphology of feeding: behavioral abilities and ecological consequences. Proc. XIX Intern. Ornithol. Congr.
MOERMOND, T. C., J. S. DENSLOW, D. J. LEVEY, AND E. SANTANA C. 1986. The influence of morphology on fruit choice in neotropical birds, p. 137-146. In A. Estrada and T. H. Fleming [eds.], Frugivores and seed dispersal. Dr. W. Junk Publ., Dordrecht, The Netherlands.
MOERMOND, T. C., J. S. DENSLOW, D. J. LEVEY, AND E. SANTANA C. 1987. The influence of context on choice behavior: fruit selection by tropical birds, p. 229-254. In M. L. Commons, A. Kacelnik, and S. J. Shettleworth [eds.], Quantitative analyses of behavior. Vol. VI: Foraging. Lawrence Earlbaum Assoc. Publ., Hillsdale, NJ.
MOLLER, A. P. 1983. Habitat selection and feeding activity in the Magpie, Pica pica. J. Ornithol. 124:147-161.
MONTGOMERIE, R. D., J. EADIE, AND L. D. HARDER. 1984. What do foraging hummingbirds maximize? Oecologia 63:357-363.
MOODY, D. T. 1970. A method for obtaining food samples from insectivorous birds. Auk 87:579.
MOOK, L. J., AND H. W. MARSHALL. 1965. Digestion of spruce budworm larvae and pupae in the Olive-backed Thrush, Hylocichla ustulata swainsoni (Tschudi). Can. Entomol. 97:1144-1149.
MOORE, D. M. 1972. Trees of Arkansas. 3rd ed. Arkansas Forestry Commission, Little Rock.
MOORE, F. R., AND P. A. SIMM. 1985. Migratory disposition and choice of diet by the Yellow-rumped Warbler (Dendroica coronata). Auk 102:820-826.
MOORE, F. R., AND P. A. SIMM. 1986. Risk-sensitive foraging by a migratory bird (Dendroica coronata). Experientia 42:1054-1056.
MOORE, J. 1983. Dietary variation among nestling Starlings. Condor 88:181-189.
MOREAU, R. E. 1972. The Palaearctic-African bird migration systems. Academic Press, New York.
MOREL, C. J., AND M. Y. MOREL. 1974. Recherches ecologiques sur une savane Sahelinee du Ferlo Septenrional, Senegal: influence de la secheresse de l'annee 1972-1973 sur l'avifaune. Terre Vie 28:95-123.
MORENO, J. 1981. Feeding niches of woodland birds in a montane coniferous forest in central Spain during winter. Ornis Scand. 12:148-159.
MORENO, J. 1984. Search strategies of Wheaters (Oenanthe oenanthe) and Stonechats (Saxicola torquata): adaptive variation in perch height, search time, sally distance and inter-perch move length. J. Anim. Ecol. 53:147-160.
MORENO, J., A. LUNDBERG, AND A. CARLSON. 1981. Hoarding of individual nuthatches (Sitta europaea) and Marsh Tits (Parus palustris). Holarctic Ecol. 4:263-269.
MORGAN, L. 1896. Habit and instinct. Arnold Publ. Ltd., London.
MORRILL, W. L. 1975. Plastic pitfall traps. Environ. Entomol. 4:596.
MORRIS, R. F. 1955. The development of sampling techniques for forest insect defoliators, with particular reference to the spruce budworm. Can. J. Zool. 33:225-294.
MORRIS, R. F. 1960. Sampling insect populations. Ann. Rev. Entomol. 5:243-264.
MORRIS, R. F. 1964. The value of historical data in population research, with particular reference to

Hyphantria cunea Drury. Can. Entomol. 96:356-368.
MORRIS, R. F., W. F. CHESHIRE, C. A. MILLER, AND D. G. MOTT. 1958. The numerical response of avian and mammalian predators during a gradation of the spruce budworm. Ecology 39:487-494.
MORRISON, D. F. 1976. Multivariate statistical methods. 2nd ed. McGraw-Hill, New York.
MORRISON, D. W., AND D. F. CACCAMISE. 1985. Ephemeral roosts and stable patches? A radiotelemetry study of communally roosting starlings. Auk 102:793-804.
MORRISON, M. L. 1980. Seasonal aspects of the predatory behavior of Loggerhead Shrikes. Condor 82:296-300.
MORRISON, M. L. 1981. The structure of western warbler assemblages: analysis of foraging behavior and habitat selection in Oregon. Auk 98:578-588.
MORRISON, M. L. 1982. The structure of western warbler assemblages: ecomorphological analysis of the Black-throated Gray and Hermit warblers. Auk 99:503-513.
MORRISON, M. L. 1984a. Influence of sample size and sampling design on analysis of avian foraging behavior. Condor 86:146-150.
MORRISON, M. L. 1984b. Influence of sample size on discriminant function analysis of habitat use by birds. J. Field Ornithol. 55:330-335.
MORRISON, M. L. 1988. On sample sizes and reliable information. Condor 90:275-278.
MORRISON, M. L., L. A. BRENNAN, AND W. M. BLOCK. 1989. Arthropod sampling methods in ornithology: goals and pitfalls, p. 484-492. In L. L. McDonald, B. F. J. Manly, J. A. Lockwood, and J. A. Logan [eds.], Estimation and analysis of insect populations. Springer-Verlag, New York.
MORRISON, M. L., R. D. SLACK, AND E. SHANLEY. 1978. Age and foraging ability relationships of Olivaceous Cormorants. Wilson Bull. 90:414-422.
MORRISON, M. L., I. C. TIMOSSI, K. A. WITH, AND P. N. MANLEY. 1985. Use of tree species by forest birds during winter and summer. J. Wildl. Manage. 49:1098-1102.
MORRISON, M. L., AND K. A. WITH. 1987. Intersexual and interseasonal resource partitioning in Hairy and White-headed Woodpeckers. Auk 104:225-233.
MORRISON, M. L., K. A. WITH, AND I. C. TIMOSSI. 1986. The structure of a forest bird community during winter and summer. Wilson Bull. 98:214-230.
MORRISON, M. L., K. A. WITH, I. C. TIMOSSI, AND K. A. MILNE. 1987a. Composition and temporal variation of flocks in the Sierra Nevada. Condor 89:739-745.
MORRISON, M. L., K. A. WITH, I. C. TIMOSSI, W. M. BLOCK, AND K. A. MILNE. 1987b. Foraging behavior of bark-foraging birds in the Sierra Nevada. Condor 89:201-204.
MORSE, D. H. 1967a. Foraging relationships of Brown-headed Nuthatches and Pine Warblers. Ecology 48:94-103.
MORSE, D. H. 1967b. Competitive relationships between Parula Warblers and other species during the breeding season. Auk 84:490-502.
MORSE, D. H. 1968. A quantitative study of foraging of male and female spruce-woods warblers. Ecology 49:779-784.
MORSE, D. H. 1970. Ecological aspects of some mixed-species foraging flocks of birds. Ecol. Monogr. 40:119-168.
MORSE, D. H. 1971a. The foraging of warblers isolated on small islands. Ecol. Monogr. 52:216-228.
MORSE, D. H. 1971b. The insectivorous bird as an adaptive strategy. Ann. Rev. Ecol. Syst. 2:177-200.
MORSE, D. H. 1973. The foraging of small populations of Yellow Warblers and American Redstarts. Ecology 54:346-355.
MORSE, D. H. 1974a. Foraging of Pine Warblers allopatric and sympatric to Yellow-throated Warblers. Wilson Bull. 86:474-477.
MORSE, D. H. 1974b. Niche breadth as a function of social dominance. Am. Nat. 108:818-830.
MORSE, D. H. 1976a. Variables affecting the density and territory size of breeding spruce-woods warblers. Ecology 57:290-301.
MORSE, D. H. 1976b. Hostile encounters among spruce-woods warblers (Dendroica, Parulidae). Anim. Behav. 24:764-771.
MORSE, D. H. 1977. The occupation of small islands by passerine birds. Condor 79:399-412.
MORSE, D. H. 1978a. Structure and foraging patterns of tit flocks in an English woodland. Ibis 120:298-312.
MORSE, D. H. 1978b. Populations of Bay-breasted and Cape May Warblers during an outbreak of the spruce budworm. Wilson Bull. 90:404-413.
MORSE, D. H. 1979. Habitat use by the Blackpoll Warbler. Wilson Bull. 91:234-243.
MORSE, D. H. 1980a. Behavioral mechanisms in ecology. Harvard Univ. Press, Cambridge, MA.
MORSE, D. H. 1980b. Foraging and coexistence of spruce-woods warblers. Living Bird 18:7-25.
MORSE, D. H. 1980c. Population limitation: breeding or wintering grounds?, p. 505-516. In A. Keast and E. S. Morton [eds.], Migrant birds in the Neotropics: ecology, behavior, distribution, and conservation. Smithsonian Inst. Press, Washington, D.C.
MORSE, D. H. 1981. Foraging speeds of warblers in large populations and in isolation. Wilson Bull. 93:334-339.
MORSE, D. H., AND R. S. FRITZ. 1987. The consequences of foraging for reproductive success, p. 443-455. In A. C. Kamil, J. R. Krebs, and H. R. Pulliam [eds.], Foraging behavior. Plenum Press, New York.
MORTON, E. S. 1973. On the evolutionary advantages and disadvantages of fruit eating in tropical birds. Am. Nat. 107:8-22.
MORTON, E. S. 1980a. Adaptations to seasonal changes by migrant land birds in the Panama Canal Zone, p. 437-453. In Keast, A., and E. S. Morton [eds.], Migrant birds in the neotropics: ecology, behavior, distribution, and conservation. Smithsonian Inst. Press, Washington, D.C.
MORTON, E. S. 1980b. The importance of migrant birds to the advancement of evolutionary theory, p. 555-557. In Keast, A., and E. S. Morton [eds.], Migrant birds in the neotropics: ecology, behavior, distribution, and conservation. Smithsonian Inst. Press, Washington, D.C.
MOSHER, J. A., AND P. F. MATRAY. 1974. Size dimorphism: a factor in energy savings for broad-winged hawks. Auk 91:325-341.
MOSS, R. 1973. The digestibility and intake of winter foods by wild ptarmigan in Alaska. Condor 75:293-300.
MOSS, R. 1974. Winter diets, gut lengths, and interspecific competition in Alaskan ptarmigan. Auk 91:737-746.
MOSS, R. 1977. The digestion of heather by Red Grouse during the spring. Condor 79:471-477.
MOSS, R. 1983. Gut size, body weight and digestion of winter food by grouse and ptarmigan. Condor 85:185-193.
MOSS, R., AND J. A. PARKINSON. 1972. The digestion of heather (Callusa vulgaris) by Red Grouse (Lagogus lagogus scoticus). Br. J. Nutr. 27:285-295.
MOSTELLER, F., AND J. W. TUKEY. 1977. Data analysis and regression. Addison-Wesley, Reading, MA.
MOUNTAINSPRING, S. R. 1987. Ecology, behavior, and conservation of the Maui Parrotbill. Condor 89:24-39.
MOUNTAINSPRING, S. R., AND J. M. SCOTT. 1985. Interspecific competition among Hawaiian forest birds. Ecol. Monogr. 55:219-239.
MUELLER-DOMBOIS, D., AND H. ELLENBERG. 1974. Aims and methods of vegetation ecology. John Wiley and Sons, New York.
MUGAAS, J. N., AND J. R. KING. 1981. Annual variation of daily energy expenditure by the Black-billed Magpie. Stud. Avian Biol. 5:1-78.
MUNN, C. A. 1985. Permanent canopy and understory flocks in Amazonia: species composition and population density, p. 683-712. In P. A. Buckley, M. S. Foster, E. S. Morton, R. S. Ridgely, and N. G. Smith [eds.], Neotropical ornithology. Ornithol. Monogr. No. 36.
MUNN, C. A. 1986. Birds that "cry wolf." Nature 319:143-

MUNN, C. A., AND J. W. TERBORGH. 1979. Multispecies territoriality in neotropical foraging flocks. Condor 81:338-347.
MUNZINGER, J. S. 1974. A comparative study of the energetics of the Black-capped and Carolina chickadees, Parus atricapillus and Parus carolinensis. Ph.D. diss., Ohio State Univ.
MURDOCH, W. W. 1969. Switching in general predators: experiments on predator specificity and stability of prey populations. Ecol. Monogr. 39:335-354.
MURDOCH, W. W., AND A. OATEN. 1975. Predation and population stability. Adv. Ecol. Res. 9:1-131.
MURPHY, M. T. 1986. Temporal components of reproductive variability in Eastern Kingbirds (Tyrannus tyrannus). Ecology 67:1483-1492.
MURPHY, M. T. 1987. The impact of weather on kingbird foraging behavior. Condor 89:721-730.
MURRAY, J. D. 1981. Introductory remarks (to a symposium on theories of biological pattern formation). Phil. Trans. Roy. Soc. Lond. B295:427-428.
MURRAY, K. G. 1986. Avian seed dispersal of neotropical gap-dependent plants. Ph.D. diss., Univ. of Florida, Gainesville.
MURRAY, K. G. 1987. Selection for optimal foraging fruit-crop size in bird-dispersed plants. Am. Nat. 129:18-31.
MURTON, R. 1971. Man and birds. Collins, London.
MUZTAR, A. J., S. J. SLINGER, AND J. H. BURTON. 1977. Metabolizable energy content of freshwater plants in chickens and ducks. Poult. Sci. 56:1893-1899.
MYERS, J. P. 1983. Commentary, p. 216-221. In A. H. Brush and G. A. Clark, Jr. [eds.], Perspectives in ornithology. Cambridge Univ. Press, Cambridge.
MYERS, J. P. 1984. Spacing behavior of nonbreeding shorebirds, p. 271-321. In J. Burger and B. L. Olla [eds.], Shorebirds: migration and foraging behavior. Plenum Press, New York.
MYRCHA, A., J. PINOWSKI, AND T. TOMEK. 1973. Energy balance of nestlings of Tree Sparrows Passer m. montanus and House Sparrows, Passer d. domesticus, p. 59-83. In S. C. Kendeigh and J. Pinowski [eds.], Productivity, population dynamics and systematics of granivorous birds. Warszawa: PWN-Polish Scientific Publishers.
NAGY, K. A. 1980. CO_2 production in animals: analysis of potential errors in the doubly labeled water method. Am. J. Physiol. R466-R473.
NAGY, K. A. 1983. The doubly labeled water ($^3H^{18}O$) method: a guide to its use. Univ. of California at Los Angeles Publ. 12-1417.
NAGY, K. A. 1987. Field metabolic rate and food requirement scaling in mammals and birds. Ecol. Monogr. 57:111-128.
NAGY, K. A. 1989. Field bioenergetics: accuracy of models and methods. Physiol. Zool. 62:237-252.
NAGY, K. A., R. S. SEYMOUR, A. K. LEE, AND R. BRAITHWAITE. 1978. Energy and water budgets in free-living Antechinus stuarti (Marsupialia: Dasyuridae). J. Mammal. 59:60-68.
NAGY, K. A., W. R. SIEGFRIED, AND R. P. WILSON. 1984. Energy utilization by free-ranging Jackass Penguins, Spheniscus demersus. Ecology 65:1648-1655.
NEFF, N. A., AND L. F. MARCUS. 1980. A survey of multivariate methods for systematics. New York: privately published.
NEILSON, M. J., AND J. A. MARLETT. 1983. A comparison between detergent and nondetergent analyses of dietary fiber in human foodstuffs, using high-performance liquid chromatography to measure neutral sugar composition. J. Agric. Food Chem. 31:1342-1347.
NESBITT, S. A., D. T. GILBERT, AND D. B. BARBOUR. 1978. Red-cockaded Woodpecker fall movements in a Florida flatwoods community. Auk 95:145-151.
NETER, J., W. WASSERMAN, AND M. H. KUTNER. 1983. Applied linear regression models. Richard D. Irwin, Homewood, IL.
NEU, C. W., C. R. BYERS, AND J. M. PEEK. 1974. A technique for analysis of utilization-availability data. J. Wildl. Manage. 38:541-545.
NEWLAND, C. E., AND R. D. WOOLLER. 1985. Seasonal changes in a honeyeater assemblage in Banksia woodland near Perth, Western Australia. N.Z. J. Zool. 12:631-636.
NEWTON, I. 1967. The adaptive radiation and feeding ecology of some British finches. Ibis 109:33-98.
NEWTON, I. 1975. Finches. Collins, Glasgow.
NICOLAI, V. 1986. The bark of trees: thermal properties, microclimate and fauna. Oecologia 69:148-160.
NIE, N. H., C. H. NULL, J. G. JENKINS, K. STEINBRENNER, AND D. H. BENT. 1975. SPSS: statistical package for the social sciences. 2nd ed. McGraw-Hill Book Co., New York.
NIEMI, G. J., AND J. M. HANOWSKI. 1984. Effects of a transmission line on bird populations in the Red Lake Peatland, Northern Minnesota. Auk 101:487-498.
NILES, D. M., S. A. ROWHER, AND R. D. ROBINS. 1969. An observation of midwinter nocturnal tower mortality of Tree Sparrows. Bird-Banding 40:322-323.
NIX, H. A. 1976. Environmental control of breeding, post-breeding dispersal and migration of birds in the Australian region. Proc. XVI Int. Ornith. Congr. 272-306.
NOLAN, V., JR., AND C. F. THOMPSON. 1975. The occurrence and significance of anomalous reproductive activities in two North American non-parasitic cuckoos Coccyzus spp. Ibis 117:496-503.
NOLAN, V., JR., AND D. P. WOOLDRIDGE. 1962. Food habits and feeding behavior of the White-eyed Vireo. Wilson Bull. 74:68-73.
NOON, B. R. 1981a. The distribution of an avian guild along a temperate elevational gradient: the importance and expression of competition. Ecol. Monogr. 51:105-124.
NOON, B. R. 1981b. Techniques for sampling avian habitats, p. 42-52. In D. E. Capen [ed.], The use of multivariate statistics in studies of wildlife habitat. USDA For. Serv. Gen. Tech. Rep. RM-87, Fort Collins, CO.
NOON, B. R. 1986. Summary: biometric approaches to modeling--the researcher's viewpoint, p. 197-201. In J. Verner, M. L. Morrison, and C. J. Ralph [eds.], Wildlife 2000: modeling habitat relationships of terrestrial vertebrates. Univ. Wisconsin Press, Madison.
NORBERG, R. A. 1978. Energy content of some spiders and insects on branches of spruce (Picea abies) in winter: prey of certain passerine birds. Oikos 31:222-229.
NORBERG, R. A. 1986. Treecreeper climbing: mechanics, energetics, and structural adaptations. Ornis Scand. 17:191-209.
NORBERG, U. M. 1979. Morphology of the wings, legs and tail of three coniferous forest tits, the Goldcrest, and the Treecreeper in relation to locomotor pattern and feeding station selection. Phil. Trans. Roy. Soc. Lond. B 287:131-165.
NORBERG, U. M. 1981. Flight, morphology and the ecological niche in some birds and bats. Symp. Zool. Soc. Lond. 48:173-197.
NORTON, D. W. 1970. Thermal regime of nests and bioenergetics of chick growth in the Dunlin (Chalidris alpina) at Barrow Alaska. M.S. thesis, Univ. of Alaska, College Park.
NORUSIS, M. J. 1986. SPSS/PC+ advanced statistics. SPSS Inc., Chicago, IL.
NOSKE, R. A. 1979. Co-existence of three species of treecreepers in northeastern New South Wales. Emu 79:120-128.
NOSKE, R. A. 1982. Comparative behaviour and ecology of some bark-foraging birds. Ph.D. thesis, Univ. of New England, Armidale, N.S.W., Australia.
NOSKE, R. A. 1985. Habitat use by three bark-foragers in eucalypt forests, p. 193-204. In J. A. Keast, H. F. Recher, H. A. Ford and D. Saunders [eds.], Birds of eucalypt forests and woodlands: ecology, conservation, and management. Surrey Beatty, Sydney.

NUE, C. W., C. R. BYERS, AND J. M. PEEK. 1974. A technique for analysis of utilization-availability data. J. Wildl. Manage. 38:541-545.

OATEN, A. 1977. Optimal foraging in patches: a case for stochasticity. Theor. Popul. Biol. 12:263-285.

OBST, B. S. 1986. Wax digestion in Wilson's Storm-Petrel. Wilson Bull. 98:189-195.

O'FARRELL, M. J., E. H. STUDIER, AND W. G. EWING. 1971. Energy utilization and water requirements of captive Myotis thysanodes and Myotis lucifugus (Chiroptera). Comp. Biochem. Physiol. 39A:549-552.

OHMART, C. P., AND D. L. DAHLSTEN. 1977. Biological studies of bud mining sawflies, Pleroneura spp. (Hymenoptera: Xyelidae), on white fir in the central Sierra Nevada of California. I. Life cycles, niche utilization, and interaction between larval feeding and tree growth. Can. Entomol. 109:1001-1007.

OHMART, C. P., AND D. L. DAHLSTEN. 1978. Biological studies of bud mining sawflies, Pleroneura spp. (Hymenoptera: Xyelidae), on white fir in the central Sierra Nevada of California. II. Larval distribution within tree crowns. Can. Entomol. 110:583-590.

OHMART, C. P., AND D. L. DAHLSTEN. 1979. Biological studies of bud mining sawflies, Pleroneura spp. (Hymenoptera: Xyelidae), on white fir in the central Sierra Nevada of California. III. Mortality factors of egg, larval, and adult stages and a partial life table. Can. Entomol. 111:883-888.

OKSANEN, J. 1983. Ordination of boreal heath-like vegetation with principal component analysis, correspondence analysis and multidimensional scaling. Vegetatio 52:181-189.

OPLER, P. A., G. W. FRANKIE, AND H. G. BAKER. 1980. Comparative phenological studies of treelet and shrub species in tropical wet and dry forests in the lowlands of Costa Rica. J. Ecol. 68:167-188.

ORIANS, G. H. 1966. Food of nestling Yellow-headed Blackbirds, Caribou Parklands, British Columbia. Condor 68:321-337.

ORIANS, G. H. 1969a. Age and hunting success in the Brown Pelican (Pelecanus occidentalis). Anim. Behav. 17:316-319.

ORIANS, G. H. 1969b. The number of bird species in some tropical forests. Ecology 50:783-801.

ORIANS, G. H. 1981. Foraging behavior and the evolution of discriminatory abilities, p. 389-405. In A. C. Kamil and T. D. Sargent [eds.], Foraging behavior: ecological, ethological, and psychological approaches. Garland STPM Press, New York.

ORIANS, G. H. 1983. Notes on the behavior of the Melodius Blackbird (Dives dives). Condor 85:453-461.

ORIANS, G. H. 1985a. Allocation of reproductive effort by breeding blackbirds, family Icteridae. Rev. Chil. Hist. Nat. 58:19-29.

ORIANS, G. H. 1985b. Blackbirds of the Americas. Univ. Washington Press, Seattle.

ORIANS, G. H., AND H. S. HORN. 1969. Overlap in foods and foraging in four species of blackbirds in the potholes of central Washington. Ecology 50:930-938.

ORMEROD, S. J. 1985. The diet of Dippers Cinclus cinclus and their nestlings in the catchment of the River Wye, Mid-Wales: a preliminary study of faecal analysis. Ibis 127:316-331.

OSTERHAUS, M. B. 1962. Adaptive modifications in the leg structure of some North American warblers. Am. Midl. Nat. 68:474-486.

OTVOS, I. S. 1965. Studies on avian predators of Dendroctonus brevicomis LeConte (Coleoptera: Scolytidae) with special reference to Picidae. Can. Entomol. 97:1184-1197.

OTVOS, I. S. 1970. Avian predation of the western pine beetle, p. 119-127. In R. W. Stark and D. L. Dahlsten [eds.], Studies on the population dynamics of the western pine beetle Dendroctonus brevicomis Leconte (Coleoptera: Scolytidae). Univ. California Div. Agric. Sci., Berkeley.

OTVOS, I. S. 1979. The effects of insectivorous bird activities in forest ecosystems: an evaluation, p. 341-374. In J. G. Dickson, R. N. Conner, R. R. Fleet, J. C. Kroll, and J. A. Jackson [eds.], The role of insectivorous birds in forest ecosystems. Academic Press, New York.

OTVOS, I. S., AND R. W. STARK. 1985. Arthropod food of some forest-inhabiting birds. Can. Entomol. 117:971-990.

OWEN, D. F. 1956. The food of nestling jays and magpies. Bird Study 3:257-265.

OWEN, M. 1975. An assessment of fecal analysis technique in waterfowl feeding studies. J. Wildl. Manage. 39:271-279.

OWEN, R. B., JR. 1970. The bioenergetics of captive Blue-winged Teal under controlled and outdoor conditions. Condor 72:153-163.

PACALA, S., AND J. ROUGHGARDEN. 1984. Control of arthropod abundance by Anolis lizards on St. Eustatius (Neth. Antilles). Oecologia 64:160-162.

PAGE, G. W., L. E. STENZEL, AND C. A. RIBIC. 1985. Nest site selection and clutch predation in the Snowy Plover. Auk 102:347-353.

PALADINES, O. L., J. T. REID, A. BEUSADOUN, AND B. D. H. VAN NIEKERK. 1964. Heat of combustion values of the protein and fat in the body and wool of the sheep. J. Nutr. 82:145-149.

PARKER, K. C. 1986a. Trunk vs. ground feeding in Cactus Wrens (Campylorhynchus brunneicapillus). Southwest. Nat. 31:111-114.

PARKER, K. C. 1986b. Partitioning of foraging space and nest sites in a desert shrubland bird community. Am. Midl. Nat. 115:255-267.

PARKER, T. A., III. 1982. Observations of some unusual rainforest and marsh birds in southeastern Peru. Wilson Bull. 94:477-493.

PARKER, T. A., III, AND J. P. O'NEILL. 1980. Notes on little known birds of the upper Urubamba Valley, southern Peru. Auk 97:167-176.

PARKER, T. A., III, AND S. A. PARKER. 1982. Behavioral and distributional notes on some unusual birds of a lower montane cloud forest in Peru. Bull. Br. Ornithol. Club 102:63-70.

PARKER, T. A., III, AND J. V. REMSEN, JR. 1987. Fifty-two Amazonian bird species new to Bolivia. Bull. Br. Ornithol. Club 107:94-107.

PARKER, T. A., III, J. V. REMSEN, JR., AND J. A. HEINDEL. 1980. Seven bird species new to Bolivia. Bull. Br. Ornithol. Club 100:160-162.

PARKER, T. A., III, T. S. SCHULENBERG, G. R. GRAVES, AND M. J. BRAUN. 1985. The avifauna of the Huancabamba region, northern Peru, p. 169-197. In P. A. Buckley, M. S. Foster, E. S. Morton, R. S. Ridgely, and F. G. Buckley [eds.], Neotropical ornithology. Ornithol. Monogr. No. 36.

PARRISH, J. R. 1988. Kleptoparasitism of insects by a Broad-tailed Hummingbird. J. Field Ornithol. 59:128-129.

PARTRIDGE, L. 1976a. Some aspects of the morphology of Blue Tits (Parus caeruleus) and Coal Tits (Parus ater) in relation to their behaviour. J. Zool. (Lond.) 179:121-134.

PARTRIDGE, L. 1976b. Field and laboratory observations on the foraging and feeding techniques of Blue Tits (Parus caeruleus) and Coal Tits (Parus ater) in relation to their habitats. Anim. Behav. 24:534-544.

PARTRIDGE, L. 1979. Differences in behaviour between Blue and Coal tits reared under identical conditions. Anim. Behav. 27:120-125.

PASZKOWSKI, C. A. 1982. Vegetation, ground, and frugivorous foraging of the American Robin. Auk 99:701-709.

PASZKOWSKI, C. A. 1984. Macrohabitat use, microhabitat use, and foraging behavior of the Hermit Thrush and Veery in a northern Wisconsin forest. Wilson Bull. 96:286-292.

PASZKOWSKI, C. A., AND T. C. MOERMOND. 1984. Prey

handling relationships in captive Ovenbirds. Condor 86:410-415.
PATON, D. C. 1982. The diet of the New Holland Honeyeater Phylidonyris novaehollandiae. Aust. J. Ecol. 7:279-298.
PATON, D. C. 1986. Food supply, population structure and behaviour of New Holland Honeyeaters Phylidonyris novaehollandiae in woodland near Horsham, Victoria, p. 219-230. In A. Keast, H. F. Recher, H. Ford, and D. Saunders [eds.], Birds of the eucalypt forest and woodlands. Surrey Beatty, Sydney.
PATON, D. C., AND F. L. CARPENTER. 1984. Peripheral foraging by territorial Rufous Hummingbirds: defense by exploitation. Ecology 65:1808-1819.
PATON, D. C., AND H. A. FORD. 1983. The influence of plant characteristics and honeyeater size on levels of pollination in Australian plants, p. 235-248. In C. E. Jones and R. J. Little [eds.], Handbook of experimental pollination ecology. Van Nostrand Reinhold, New York.
PAUL, H G. 1979. How to construct larval sampling equipment. USDA Agric. Handb. 545, Washington, D.C.
PAYNE, R. B. 1980. Seasonal incidence of breeding, moult, and local dispersal of Red-billed Firefinches Lagonosticta senegala in Zambia. Ibis 122:43-56.
PEARSON, D. L. 1977a. Ecological relationships of small antbirds in Amazonian bird communities. Auk 94:283-292.
PEARSON, D. L. 1977b. A pantropical comparison of bird community structure on six lowland forest sites. Condor 79:232-244.
PEET, R. K. 1974. The measurement of species diversity. Ann. Rev. Ecol. Syst. 5:285-307.
PENDERGAST, B. A., AND D. A. BOAG. 1971. Nutritional aspects of the diet of Spruce Grouse in central Alberta. Condor 73:437-443.
PENNEY, J. G., AND E. D. BAILEY. 1970. Comparison of the energy requirements of fledgling Black Ducks and American Coots. J. Wildl. Manage. 34:105-114.
PENRY, D. L., AND P. A. JUMARS. 1987. Modeling animal guts as chemical reactors. Am. Nat. 129:69-96.
PERDECK, A. C. 1964. An experiment on the ending of autumn migration in starlings. Ardea 52:133-139.
PERFECT, T. J., A. G. COOK, AND E. R. FERRER. 1983. Population sampling for planthoppers, leafhoppers (Hemiptera: Delphacidae and Cicadellidae) and their predators in flooded rice. Bull. Entomol. Res. 73:345-355.
PERRINS, C. 1979. British tits. Collins, Glasgow.
PERRY, M. C., G. H. HAAS, AND J. W. CARPENTER. 1981. Radio transmitters for Mourning Doves: a comparison of attachment techniques. J. Wildl. Manage. 45:524-527.
PERRY, J. N., AND J. BOWDEN. 1983. A comparative analysis of Chrysoperla carnea catches in light- and suction-traps. Ecol. Entomol. 8:383-394.
PETERMAN, R. M., W. C. CLARK, AND C. S. HOLLING. 1979. The dynamics of resilience: shifting stability domains in fish and insect systems. Symp. Br. Ecol. Soc. 20:321-341.
PETERS, S., W. SEARCY, AND P. MARLER. 1980. Species song discrimination in choice experiments with Swamp and Song sparrows. Anim. Behav. 28:393-404.
PETERS, W. D., AND T. C. GRUBB, JR. 1983. An experimental analysis of sex-specific foraging in the Downy Woodpecker Picoides pubescens. Ecology 64:1437-1443.
PETERSON, A. W., AND T. C. GRUBB, JR. 1983. Artificial trees as a cavity substrate for woodpeckers. J. Wildl. Manage. 47:790-798.
PETERSON, R. T. 1980. A field guide to the birds east of the Rockies. Houghton Mifflin Co., Boston, MA.
PETIT, D. R., K. E. PETIT, AND T. C. GRUBB, JR. 1985. On atmospheric moisture as a factor influencing the distribution of breeding birds in temperate deciduous forest. Wilson Bull. 97:88-96.
PETIT, L. J., AND D. R. PETIT. 1988. Continuous feeding from a spider's web by a Prothonotary Warbler. J. Field Ornith. 59:278-279.

PETTERSSON, B. 1983. Foraging behaviour of the Middle Spotted Woodpecker Dendrocopus medius in Sweden. Holarctic Ecol. 6:263-269.
PIANKA, E. R. 1973. The structure of lizard communities. Ann. Rev. Ecol. Syst. 4:53-74.
PIANKA, E. R. 1981. Competition and niche theory, p. 167-196. In R. M. May [ed.], Theoretical ecology: principles and applications. Blackwell Scientific Publ., London.
PIANKA, E. R. 1983. Evolutionary ecology. Harper and Row, New York.
PIELOU, E. C. 1975. Ecological diversity. John Wiley and Sons, New York.
PIELOU, E. C. 1977. Mathematical ecology. John Wiley and Sons, New York.
PIELOU, E. C. 1984. The interpretation of ecological data: a primer on classification and ordination. John Wiley and Sons, New York.
PIENKOWSKI, M. W. 1983. Changes in the foraging pattern of plovers in relation to environmental factors. Anim. Behav. 31:244-264.
PIENKOWSKI, M. W., AND P. R. EVANS. 1984. Migratory behavior of shorebirds in the western Palearctic, p. 73-123. In J. Burger and B. L. Olla [eds.], Shorebirds: migration and foraging behavior. Plenum Press, New York.
PIERCE, G. J., AND J. G. OLLASON. 1987. Eight reasons why optimal foraging theory is a complete waste of time. Oikos 49:111-118.
PIERCE, V., AND T. C. GRUBB, JR. 1981. Laboratory studies of foraging in four bird species of deciduous woodland. Auk 98:307-320.
PIEROTTI, R., AND C. ANNETT. 1987. Reproductive consequences of dietary specialization and switching in an ecological generalist, p. 417-442. In A. C. Kamil, J. R. Krebs, and H. R. Pulliam [eds.], Foraging behavior. Plenum Press, New York.
PIERPONT, N. 1986. Interspecific aggression and the ecology of woodcreepers (Aves: Dendrocolaptidae). Ph.D. diss., Princeton Univ., Princeton, NJ.
PIETREWICZ, A. T., AND A. C. KAMIL. 1977. Visual detection of cryptic prey by Blue Jays (Cyanocitta cristata). Science 195:580-582.
PIMM, S. L., M. L. ROSENZWEIG, AND W. MITCHELL. 1985. Competition and food selection: field tests of a theory. Ecology 66:798-807.
PINKOWSKI, B. C. 1977. Foraging behavior of the Eastern Bluebird. Wilson Bull. 89:404-414.
PINKOWSKI, B. C. 1978. Feeding of nestling and fledgling Eastern Bluebirds. Wilson Bull. 90:84-98.
PITELKA, F. A. 1961. Comments on types and taxonomy in the jay genus Aphelocoma. Condor 63:234-245.
PLACE, A. R., AND D. D. ROBY. 1986. Assimilation and deposition of dietary fatty alcohols in Leach's Storm-Petrel, Oceanodroma leucorhoa. J. Exp. Zool. 240:149-161.
PLEASANTS, B. Y. 1979. Adaptive significance of the variable dispersion pattern of breeding Northern Orioles. Condor 81:28-34.
POLLARD, E. 1979. Population ecology and change in range of the white admiral butterfly Ladoga camilla L. in England. Ecol. Entomol. 4:61-74.
POOLE, R. W. 1974. An introduction to quantitative ecology. McGraw-Hill, New York.
PORTER, M. L., M. W. COLLOPY, R. F. LABISKY, AND R. C. LITTELL. 1985. Foraging behavior of Red-cockaded Woodpeckers: an evaluation of research methodologies. J. Wildl. Manage. 49:505-507.
PORTER, W. P., AND P. A. MCCLURE. 1984. Climate effects on growth and reproduction potential in Sigmodon hispidus and Peromyscus maniculatus. Carnegie Mus. Nat. His. Spec. Publ. 16:173-181.
POTTER, E. F., AND R. DAVIS. 1974. Hermit Thrush practices foot-patting feeding behavior. Chat 38:95.
POTTINGER, R. P., AND F. J. LEROUX. 1971. The biology and dynamics of Lithocolletis blancardella (Lepidoptera: Gracillariidae) on apple in Quebec. Mem. Entomol. Soc.

Tabanidae as determined by Malaise trap collections. Mosq. News 31:509-512.

ROBERTSON, R. J. 1973. Optimal niche space of the Red-winged Blackbird. III. Growth rate and food of nestlings in marsh and upland habitat. Wilson Bull. 85:209-222.

ROBINSON, M. H. 1969. Defenses against visually hunting predators. Evol. Biol. 3:225-259.

ROBINSON, S. K. 1981. Ecological relations and social interactions of Philadelphia and Red-eyed vireos. Condor 83:16-26.

ROBINSON, S. K. 1985. The Yellow-rumped Cacique and its associates' nest pirates, p. 898-907. In P. A. Buckley, M. S. Foster, E. S. Morton, R. S. Ridgely, and F. G. Buckley [eds.], Neotropical ornithology. Ornithol. Monogr. No. 36.

ROBINSON, S. K. 1986. Three-speed foraging during the breeding cycle of Yellow-rumped Caciques (Icterinae: Cacicus cela). Ecology 67:394-405.

ROBINSON, S. K., AND R. T. HOLMES. 1982. Foraging behavior of forest birds: the relationships among search tactics, diet, and habitat structure. Ecology 63:1918-1931.

ROBINSON, S. K., AND R. T. HOLMES. 1984. Effects of plant species and foliage structure on the foraging behavior of forest birds. Auk 101:672-684.

ROBY, D. D., A. R. PLACE, AND R. E. RICKLEFS. 1986. Assimilation and deposition of wax esters in planktivorous seabirds. J. Exp. Zool. 238:29-41.

RODENHOUSE, N. L. 1986. Food limitation for forest passerine birds: effects of natural and experimental food reductions. Ph.D. thesis, Dartmouth College, Hanover, NH.

ROGERS, C. M. 1985. Foraging success and tree species use in the Least Flycatcher. Auk 102:613-620.

ROGERS, C. M., T. L. THEIMER, V. NOLAN, JR., AND E. D. KETTERSON. 1989. Does dominance determine how far Dark-eyed Juncos, Junco hyemalis, migrate into their winter range? Anim. Behav. 37:498-506.

ROGERS, L. E., W. J. HINDS, AND R. L. BUSCHBOM. 1976. A general weight vs. length relationship for insects. Ann. Entomol. Soc. Am. 69:387-389.

ROGERS, L. E., R. L. BUSCHBOM, AND C. R. WATSON. 1977. Length-weight relationships of shrubsteppe invertebrates. Ann. Entomol. Soc. Am. 70:51-53.

ROLAND, J., S. J. HANNON, AND M. A. SMITH. 1986. Foraging pattern of Pine Siskins and its influence on winter moth survival in an apple orchard. Oecologia 69:47-52.

ROOT, R. B. 1964. Ecological interactions of the Chestnut-backed Chickadee following a range extension. Condor 66:229-238.

ROOT, R. B. 1966. The avian response to a population outbreak of the tent caterpillar, Malacosoma constrictum (Stretch) (Lepidoptera: Lasiocampidae). Pan-Pacific Entomol. 42:48-53.

ROOT, R. B. 1967. The niche exploitation pattern of the Blue-gray Gnatcatcher. Ecol. Monogr. 37:317-350.

ROSCOE, J. T., AND J. A. BYARS. 1971. An investigation of the restraints with respect to sample size commonly imposed on the use of the chi-square statistic. J. Am. Stat. Assoc. 66:755-759.

ROSE, D. J. W. 1972. Times and size of dispersal flights by Cicadulina species (Homoptera: Cicadellidae), vectors of maize streak disease. J. Anim. Ecol. 41:495-506.

ROSENBERG, K. V., R. D. OHMART, AND B. W. ANDERSON. 1982. Community organization of riparian birds: response to an annual resource peak. Auk 99:260-274.

ROSENTHAL, G. A., AND D. H. JANZEN. 1979. Herbivores: their interactions with secondary plant metabolites. Academic Press, New York.

ROSENZWEIG, M. L. 1985. Some theoretical aspects of habitat selection, p. 517-540. In M. L. Cody [ed.], Habitat selection in birds. Academic Press, New York.

ROSKAFT, E., T. JARVI, M. BAKKEN, C. BECH, AND R. E. REINERTSEN. 1986. The relationship between social status and resting metabolic rate in Great Tits (Parus major) and Pied Flycatchers (Ficedula hypoleuca). Anim. Behav. 34:838-842.

ROTENBERRY, J. T. 1980a. Dietary relationships among shrubsteppe passerine birds: competition or oppportunism in a variable environment? Ecol. Monogr. 50:93-110.

ROTENBERRY, J. T. 1980b. Bioenergetics and diet in a simple community of shrubsteppe birds. Oecologia 46:7-12.

ROTENBERRY, J. T., AND J. A. WIENS. 1980a. Temporal variation in habitat structure and shrubsteppe bird dynamics. Oecologia 47:1-9.

ROTENBERRY, J. T., AND J. A. WIENS. 1980b. Habitat structure, patchiness, and avian communities in North American steppe vegetation: a multivariate analysis. Ecology 61:1228-1250.

ROTHSTEIN, S. I., J. VERNER, AND E. STEVENS. 1984. Radio-tracking confirms a unique diurnal pattern of spatial occurrence in the parasitic Brown-headed Cowbird. Ecology 65:77-88.

ROUGHGARDEN, J. 1974. Niche width: biogeographic patterns among Anolis lizards. Am. Nat. 108:429-442.

ROYAL STATISTICAL SOCIETY. 1986. The general linear interactive model (GLIM 3.77). Numerical Algorithms Group Ltd., Oxford.

ROYAMA, T. 1959. A device of an auto-cinematic food-recorder. Tori 15:172-176.

ROYAMA, T. 1970. Factors governing the hunting behaviour and selection of food by the Great Tit, Parus major. J. Anim. Ecol. 39:619-660.

RUBENSTEIN, D. I., R. J. BARNETT, R. S. RIDGELY, AND P. H. KLOPFER. 1977. Adaptive significance of mixed-species feeding flocks among seed-eating finches in Costa Rica. Ibis 119:10-21.

RUDD, W. G., AND R. L. JENSEN. 1977. Sweep net and ground cloth sampling for insects in soybeans. J. Econ. Entomol. 70:301-304.

RUDALL, K. M. 1963. The chitin/protein complexes of insect cuticles, p. 257-313. In J. W. L. Beaument, J. E. Treherne, and V. B. Wigglesworth [eds.], Advances in insect physiology, Vol. 1. Academic Press, New York.

RUDOLPH, S. G. 1982. Foraging strategies of American Kestrels during breeding. Ecology 63:1268-1276.

RUGGIERO, L. F., AND A. B. CAREY. 1984. A programmatic approach to the study of old-growth forest -- wildlife relationships, p. 340-345. In New forests for a changing world: Proc. 1983 Conv., Soc. Am. For., Portland, OR.

RUITER, L. DE. 1952. Some experiments on the camouflage of stick caterpillars. Behavior 4:222-232.

RUNDLE, W. D. 1982. A case for esophageal analysis in shorebird food studies. J. Field Ornithol. 53:249-257.

RUSTERHOLZ, K. A. 1981. Competition and the structure of an avian foraging guild. Am. Nat. 118:173-190.

RYPSTRA, A. L. 1984. A relative measure of predation on web-spiders in temperate and tropical forests. Oikos 43:129-132.

SABATIER, D. 1985. Saisonalite et determinisme du pic de fructification en forest guyanaise. Revue d'Ecologie 40:289-320.

SABINE, W. S. 1959. The winter society of the Oregon Junco: intolerance, dominance and the pecking order. Condor 61:110-135.

SABO, S. R. 1980. Niche and habitat relations in subalpine bird communities of the White Mountains of New Hampshire. Ecol. Monogr. 50:241-259.

SABO, S. R., AND R. T. HOLMES. 1983. Foraging niches and the structure of forest bird communities in contrasting montane habitats. Condor 85:121-138.

SABO, S. R., AND R. H. WHITTAKER. 1979. Bird niches in a subalpine forest: an indirect ordination. Proc. Nat. Acad. Sci. 76:1338-1342.

SAETHER, B. E. 1982. Foraging niches in a passerine bird community in a grey alder forest in Central Norway. Ornis. Scand. 13:149-163.

SAITOU, T. 1979. Ecological study of social organization in the Great Tit, Parus major L. II. Formation of the basic flocks. Misc. Rep. Yamashina Inst. Ornithol. 11:137-138.

SAKAI, H. F. 1987. Response of Hammond's and Western flycatchers to different aged Douglas-fir stands in northwestern California. M.S. thesis, Humboldt State Univ., Arcata, CA.

SAKAI, H. F. 1988. Breeding biology and behavior of Hammond's and Western flycatchers in northwestern California. Western Birds 19:49-60.

SALL, K. R. 1979. Caloric determination of (1) mealworm larvae, pupae, and adults, (2) stomach contents and feces of Myotis lucifugus. B.A. thesis, Boston Univ., Boston, MA.

SALOMONSEN, M. G., AND R. P. BALDA. 1977. Winter territoriality of Townsend's Solitaire (Myadestes townsendi) in a Pinon-Juniper-Ponderosa Pine ecotone. Condor 79:148-161.

SAMPLE, B. E. 1987. Effects of aerially applied and orally administered carbaryl on Wood Thrush nestlings. M.S. thesis, Univ. of Delaware, Newark.

SANDERS, C. J. 1970. Populations of breeding birds in the spruce-fir forests of northwestern Ontario. Can. Field-Nat. 84:131-135.

SARGENT, T. D. 1968. Cryptic moths: effects on background selections of painting the circumocular scales. Science 159:100-101.

SAS. 1982. SAS user's guide: statistics. SAS Instit., Inc., Cary, NC.

SAS. 1985. SAS user's guide: statistics. Version 5 edition. SAS Instit., Inc., Cary, NC.

SASVARI, L. 1985. Different observational learning capacity in juvenile and adult individuals of congeneric bird species. Z. Tierpsychol. 69:293-304.

SAUL, L. J., AND D. A. WASSMER. 1983. Red-bellied Woodpeckers roosting outside of cavities. Florida Field Nat. 11:50-51.

SAUNDERS, D. K., AND J. W. PARRISH. 1987. Assimilated energy of seeds consumed by Scaled Quail in Kansas. J. Wildl. Manage. 51:787-790.

SAVORY, C. J., AND M. J. GENTLE. 1976. Effects of dietary dilution with fibre on the food intake and gut dimensions of Japanese Quail. Br. Poult. Sci. 17:561-570.

SCHAEFFER, E. 1953. Contributions to the life history of the Swallow-Tanager. Auk 70:403-460.

SCHAEFFER, R. L., W. MENDENHALL, AND L. OTT. 1986. Elementary survey sampling. 3rd ed. Duxbury Press, Boston, MA.

SCHLEIDT, W. M. 1961. Reactionen von truthuhnern auf fliegende Raubrogogel und Versuche Zur Analyse ihrer AAM's. Z. Tierpsychol. 18:534-560.

SCHLUTER, D. 1981. Does the theory of optimal diets apply in complex environments? Am. Nat. 118:139-147.

SCHLUTER, D. 1982a. Seed and patch selection by Galapagos ground finches: relation to foraging efficiency and food supply. Ecology 63:1106-1120.

SCHLUTER, D. 1982b. Distributions of Galapagos ground finches along an altitudinal gradient: the importance of food supply. Ecology 63:1504-1517.

SCHLUTER, D. 1984. Feeding correlates of breeding and social organization in two Galapagos finches. Auk 101:59-68.

SCHMIDT, R. S. 1960. Predator behaviour and the perfection of incipient mimetic resemblance. Behaviour 16:149-158.

SCHNEIDER, K. J. 1984. Dominance, predation and optimal foraging in White-throated Sparrow flocks. Ecology 65:1820-1827.

SCHOENER, T. W. 1968. The Anolis lizards of Bimini: resource partitioning in a complex fauna. Ecology 49:704-726.

SCHOENER, T. W. 1970. Nonsynchronous spatial overlap of lizards in patchy habitats. Ecology 51:408-418.

SCHOENER, T. W. 1971a. Large-billed insectivores: a precipitous diversity gradient. Condor 73:154-161.

SCHOENER, T. W. 1971b. Theory of feeding strategies. Ann. Rev. Ecol. Syst. 2:369-404.

SCHOENER, T. W. 1974. Resource partitioning in ecological communities. Science 185:27-39.

SCHOENER, T. W. 1982. The controversy over interspecific competition. Am. Sci. 70:586-595.

SCHOENER, T. W. 1983. Field experiments on interspecific competition. Am. Nat. 122:240-285.

SCHOENER, T. W. 1986a. Overview: kinds of ecological communities--ecology becomes pluralistic, p. 467-479. In J. Diamond and T. J. Case [eds.], Community ecology. Harper and Row, New York.

SCHOENER, T. W. 1986b. Resource partitioning, p. 91-126. In J. Kikkawa and D. J. Anderson [eds.], Community ecology. Blackwell Scientific Publ., Melbourne.

SCHOENER, T. W. 1987. A brief history of optimal foraging ecology, p. 5-67. In A.C. Kamil, J. R. Krebs, and H. R. Pulliam [eds.], Foraging behavior. Plenum Press, New York.

SCHOWALTER, T. D., W. W. HARGROVE, AND D. A. CROSSLEY, JR. 1986. Herbivory in forested ecosystems. Ann. Rev. Entomol. 31:177-196.

SCHOWALTER, T. D., J. W. WEBB, AND D. A. CROSSLEY. 1981. Community structure and nutrient content of canopy arthropods in clearcut and uncut forest ecosystems. Ecology 62:1010-1019.

SCHUELER, F. W. 1972. A new method of preparing owl pellets: boiling in NaOH. Bird Banding 43:142.

SCHULENBERG, T. S. 1983. Foraging behavior, ecomorphology, and systematics of some antshrikes (Formicariidae: Thamnomanes). Wilson Bull. 95:505-521.

SCHULER, W. 1983. Responses to sugars and their behavioral mechanisms in the Starling (Sturnus vulgaris). Behav. Ecol. and Sociobiol. 13:243-251.

SCHULTER, D. 1981. Does the theory of optimal diets apply in complex environments? Am. Nat. 118:139-147.

SCHULTZ, J. C. 1983a. Habitat selection and foraging tactics of caterpillars in heterogeneous trees, p. 61-90. In R. F. Denno and M. S. McClure [eds.], Variable plants and herbivores in natural and managed systems. Academic Press, New York.

SCHULTZ, J. C. 1983b. Impact of variable plant defensive chemistry on susceptibility of insects to natural enemies, p. 37-54. In P. A. Hedin [ed.], Plant resistance to insects. Am. Chem. Soc. Symp. Ser. No. 208.

SCHULTZ, J. C., AND I. T. BALDWIN. 1982. Oak leaf quality declines in response to defoliation by gypsy moth larvae. Science 217:149-151.

SCHULTZ, J. C., P. C. NOTHNAGLE, AND I. T. BALDWIN. 1982. Seasonal and individual variation in leaf quality of two northern hardwoods tree species. Am. J. Bot. 69:753-759.

SCHWABL, H. 1983. Aus pragung und Bedentung des Teilzugverhaltens einer sudwestdeutschen population der Amsel Turdus merula. J. Orn. 124:101-116.

SCHWARTZ, P. 1964. The Northern Waterthrush in Venezuela. Living Bird 3:169-184.

SCHWARTZ, P. 1980. Some considerations of migratory birds, p. 31-34. In A. Keast and E. S. Morton [eds.], Migrant birds in the Neotropics: ecology, behavior, distribution, and conservation. Smithsonian Inst. Press, Washington, D.C.

SCHWARTZ, S. E., AND G. D. LEVINE. 1980. Effects of dietary fiber on intestinal glucose absorption and glucose tolerance in rats. Gastroenterology 79:833-836.

SEALY, S. G. 1979. Extralimital nesting of Bay-breasted Warblers: response to forest tent caterpillars? Auk 96:600-603.

SEALY, S. G. 1980. Reproductive responses of Northern Orioles to a changing food supply. Can. J. Zool. 58:221-227.

SEASTEDT, T. R., AND S. F. MACLEAN. 1979. Territory size and composition in relation to resource abundance in Lapland Longspurs breeding in arctic Alaska. Auk 96:131-

142.

SEDGWICK, J. A. 1987. Avian habitat relationships in pinyon-juniper woodland. Wilson Bull. 99:413-431.

SEIDEL, G. E., AND R. C. WHITMORE. 1982. Effect of forest structure on American Redstart foraging behavior. Wilson Bull. 94:289-294.

SELANDER, R. K. 1966. Sexual dimorphism and differential niche utilization in birds. Condor 68:113-151.

SERIE, J. R., AND R. E. JONES. 1976. Spring mortality of insectivorous birds in southern Manitoba. Prairie Nat. 8:33-39.

SERVELLO, F. A., AND R. L. KIRKPATRICK. 1987. Regional variation in the nutritional ecology of Ruffed Grouse. J. Wildl. Manage. 51:749-770.

SERVELLO, F. A., R. L. KIRKPATRICK, AND K. E. WEBB, JR. 1987. Predicting metabolizable energy in the diet of Ruffed Grouse. J. Wildl. Manage. 51:560-567.

SERVICE, M. W. 1977. A critical review of procedures for sampling populations of adult mosquitoes. Bull. Entomol. Res. 67:343-382.

SEVERINGHAUS, W. D. 1981. Guild theory development as a mechanism for assessing environmental impact. Environ. Manage. 5:187-190.

SHAPIRO, S. S., AND M. B. WILK. 1965. An analysis of variance test for normality (complete samples). Biometrika 52:591-611.

SHELDON, W. G. 1971. The food of the American Woodcock. Univ. of Mass. Press, Amherst, MA.

SHELTON, A. M., J. T. ANDALORO, AND C. W. HOY. 1983. Survey of ground-dwelling predaceous and parasitic arthropods in cabbage fields in upstate New York. Environ. Entomol. 12:1026-1030.

SHERMAN, P. W. 1988. The levels of analysis. Anim. Behav. 36:616-619.

SHERRY, D., M. AVERY, AND A. STEVENS. 1982. The spacing of stored food by Marsh Tits. Z. Tierpsychol. 58:153-162.

SHERRY, D. F. 1985. Food storage by birds and mammals. Adv. Study Behav. 15:153-187.

SHERRY, T. W. 1979. Competitive interactions and adaptive strategies of American Redstarts and Least Flycatchers in a northern hardwoods forest. Auk 96:265-283.

SHERRY, T. W. 1982. Ecological and evolutionary inferences from morphology, foraging behavior, and diet of sympatric insectivorous Neotropical flycatchers (Tyrannidae). Ph.D. diss., Univ. of California, Los Angeles.

SHERRY, T. W. 1983. Terenotriccus erythrurus (Mosqueitero Colirrufo, Tontillo, Ruddy-tailed Flycatcher), p. 605-607. In D. H. Janzen [ed.], Costa Rican natural history. Univ. of Chicago Press, Chicago, IL.

SHERRY, T. W. 1984. Comparative dietary ecology of sympatric insectivorous neotropical flycatchers (Tyrannidae). Ecol. Monogr. 54:313-338.

SHERRY, T. W. 1985. Adaptation to a novel environment: food, foraging, and morphology of the Cocos Island Flycatcher, p. 908-920. In P. A. Buckley, M. S. Foster, E. S. Morton, R. S. Ridgely, and F. G. Buckley [eds.], Neotropical ornithology. Orinithol. Monogr. No. 36.

SHERRY, T. W. 1986. Nest, eggs, and reproductive behavior of the Cocos Flycatcher (Nesotriccus ridgwayi). Condor 88:531-532.

SHERRY, T. W., AND R. T. HOLMES. 1985. Dispersion patterns and habitat responses of birds in northern hardwoods forests, p. 283-309. In M. L. Cody [ed.], Habitat selection in birds. Academic Press, New York.

SHERRY, T. W., AND R. T. HOLMES. 1988. Habitat selection by breeding American Redstarts in response to a dominant competitor, the Least Flycatcher. Auk 105:350-364.

SHERRY, T. W., AND L. A. MCDADE. 1982. Prey selection and handling in two Neotropical hover-gleaning birds. Ecology 63:1016-1028.

SHETTLEWORTH, S. J. 1972. The role of novelty in learned avoidance of unpalatable prey by chicks Gallus gallus. Anim. Behav. 20:29-35.

SHETTLEWORTH, S. J. 1983. Function and mechanism in learning. Adv. Anal. Behav. 3:1-39.

SHETTLEWORTH, S. J. 1984. Learning and behavioural ecology, p. 170-194. In J. R. Krebs and N. B. Davies [eds.], Behavioural ecology: an evolutionary approach. Sinauer Assoc., Sunderland, MA.

SHOEMAKER, D. K., AND D. T. ROGERS, JR. 1980. Food habits of some overwintering Fringillidae in Tuscalousa County, Alabama. Alabama Birdlife 28:4-29.

SHORT, H. L., AND E. A. EPPS, JR. 1976. Nutrient quality and digestibility of seeds and fruits from southern forests. J. Wildl. Manage. 40:283-299.

SHORT, H. L., AND L. C. DREW. 1962. Observations concerning behavior, feeding, and pellets of Short-eared Owls. Am. Midl. Nat. 67:424-433.

SHORT, L. L. 1973. Habits of some Asian Woodpeckers (Aves, Picidae). Bull. Am. Mus. Nat. Hist. 152:253-364.

SHORT, L. L. 1978. Sympatry in woodpeckers in lowland Malayan forest. Biotropica 10:122-133.

SHUBECK, P. P. 1976. An alternative to pitfall traps in carrion beetle studies. Entomol. News 87:176-178.

SIBBALD, I. R. 1976. A bioassay for true metabolizable energy in feeding stuffs. Poult. Sci. 55:303-308.

SIBLEY, C. G., AND J. E. AHLQUIST. 1985. The phylogeny and classification of the Australo-Papuan passerines. Emu 85:1-14.

SIBLY, R. M. 1981. Strategies of digestion and defecation, p. 109-139. In C. R. Townsend and P. Calow [eds.], Physiological ecology: an evolutionary approach to resource use. Sinauer Assoc., Sunderland, MA.

SIBLY, R. M., AND R. H. SMITH. 1985. Behavioral ecology: ecological consequences of adaptive behaviour. Blackwell Scientific Publ., Palo Alto, CA.

SIH, A. 1982. Optimal patch use: variation in selective pressure for efficient foraging. Am. Nat. 120:666-685.

SIH, A. 1987. Predators and prey lifestyles: an evolutionary and ecological overview, p. 201-224. In W. C. Kerfoot and A. Sih [eds.], Predation--direct and indirect impacts on aquatic communities. Univ. Press of New England, Hanover, NH.

SILKSTONE, B. E. 1987. The consequence of leaf damage for subsequent insect grazing on birch (Betula spp.): a field experiment. Oecologia 74:149-152.

SILVERIN, B., P-A. VIEBKE, AND J. WESTIN. 1984. Plasma levels of luteinizing hormone and steroid hormones in free-living winter groups of Willow Tits (Parus montanus). Hormones and Behavior 18:367-379.

SIMBERLOFF, D., AND W. BOECKLEN. 1981. Santa Rosalia reconsidered: size ratios and competition. Evolution 35:1206-1228.

SIMON, C. 1979. Evolution of periodical cicadas: phylogenetic inferences based on allozymic data. Syst. Zool. 28:22-39.

SINCLAIR, A. R. E. 1978. Factors affecting the food supply and breeding season of resident birds and movements of palaearctic migrants in a tropical African savannah. Ibis 120:480-497.

SKEAD, C. J. 1967. Sunbirds of South Africa. Balkema, Cape Town.

SKUTCH, A. F. 1945. Life history of the Allied Woodhewer. Condor 47:85-94.

SKUTCH, A. F. 1947. Life history of the Turquoise-browned Motmot. Auk 64:201-216.

SKUTCH, A. F. 1948. Life history notes on puff-birds. Wilson Bull. 60:81-97.

SKUTCH, A. F. 1954. Life histories of Central American birds. Pac. Coast Avif. No. 31.

SKUTCH, A. F. 1960. Life histories of Central American birds, II. Pac. Coast Avif. No. 34.

SKUTCH, A. F. 1969. Life histories of Central American birds, III. Pac. Coast Avif. No. 35.

SKUTCH, A. F. 1981. New studies of tropical American birds. Publ. Nuttall Ornithol. Club No. 19.

SLATER, P. J. B. 1974. The temporal pattern of feeding in

the Zebra Finch. Anim. Behav. 22:506-515.
SLOAN, N. F., AND A. C. COPPEL. 1968. Ecological implications of bird predators in the larch casebearer in Wisconsin. J. Econ. Entomol. 61:1067-1070.
SMITH, C. C., AND D. J. REICHMAN. 1984. The evolution of food caching by birds and mammals. Ann. Rev. Ecol. Syst. 15:329-351.
SMITH, C. R., AND M. E. RICHMOND. 1972. Factors influencing pellet egestion and gastric pH in the Barn Owl. Wilson Bull. 84:179-186.
SMITH, D. G., AND R. GILBERT. 1981. Backpack radio transmitter attachment success in Screech Owls (Otus asio). N. Am. Bird Bander 6:142-143.
SMITH, D. G., C. R. WILSON, AND H. H. FROST. 1972. The biology of the American Kestrel in central Utah. Southwest. Nat. 17:73-83.
SMITH, E. P. 1982. Niche breadth, resource availability, and inference. Ecology 63:1675-1681.
SMITH, H. R. 1985. Wildlife and the gypsy moth. Wildl. Soc. Bull. 13:166-174.
SMITH, H. S. 1939. Insect populations in relation to biological control. Ecol. Monogr. 9:311-320.
SMITH, J. N. M. 1974a. The food searching behaviour of two European thrushes. I. Description and analysis of search paths. Behaviour 48:276-302.
SMITH, J. N. M. 1974b. The food searching behaviour of two European thrushes. II. The adaptiveness of the search patterns. Behaviour 49:1-61.
SMITH, J. N. M., AND R. DAWKINS. 1971. The hunting behavior of individual Great Tits in relation to spatial variation in their food density. Anim. Behav. 19:695-706.
SMITH, J. N. M., P. R. GRANT, B. R. GRANT, I. J. ABBOTT, AND L. K. ABBOTT. 1978. Seasonal variation in feeding habits of Darwin's ground finches. Ecology 59:1137-1150.
SMITH, J. N. M., AND H. P. A. SWEATMAN. 1974. Food-searching behavior of titmice in patchy environments. Ecology 55:1216-1232.
SMITH, J. W., E. A. STADELBACHER, AND C. W. GANTT. 1976. A comparison of techniques for sampling beneficial arthropod populations associated with cotton. Environ. Entomol. 5:435-444.
SMITH, K. G. 1977. Distribution of summer birds along a forest moisture gradient in an Ozark watershed. Ecology 58:810-819.
SMITH, K. G. 1982. Drought-induced changes in avian community structure along a montane sere. Ecology 63:952-961.
SMITH, K. G., AND D. C. ANDERSON. 1982. Food, predation, and reproductive ecology of the Dark-eyed Junco in northern Utah. Auk 99:650-661.
SMITH, K. G., N. C. WILKINSON, K. S. WILLIAMS, AND V. B. STEWARD. 1987. Predation by spiders on periodical cicadas (Homoptera: Magicicada). J. Arachnol. 15:277-279.
SMITH, S. M. 1967. An ecological study of winter flocks of Black-capped and Chestnut-backed chickadees. Wilson Bull. 79:200-207.
SMITH, S. M. 1972. Roosting aggregations of Bushtits in response to cold temperatures. Condor 74:478-479.
SMITH, S. M. 1973. Factors directing prey-attack by the young of three passerine species. Living Bird 12:55-67.
SMITH, S. M. 1975. Innate recognition of coral snake pattern by a possible avian predator. Science 187:759-760.
SMITH, S. M. 1976. Ecological aspects of dominance hierarchies in Black-capped Chickadees. Auk 93:95-107.
SMITH, T. B. 1987. Bill size polymorphism and intraspecific niche utilization in an African finch. Nature 329:717-719.
SMITH, T. M., AND H. H. SHUGART. 1987. Territory size variation in the Ovenbird: the role of habitat structure. Ecology 68:695-704.
SMITH, W. J., AND F. VUILLEUMIER. 1971. Evolutionary relationships of some South American ground tyrants. Bull. Mus. Comp. Zool. 141:181-268.
SMYTHE, N., W. E. GLANZ, AND E. G. LEIGH, JR. 1982.
Population regulation in some terrestrial frugivores, p. 227-238. In E. Leigh Jr., A. S. Rand, and D. Windsor [eds.], The ecology of a tropical forest. Smithsonian Instit. Press, Washington, D.C.
SNEATH, P. H. A., AND R. R. SOKAL. 1973. Numerical taxonomy: the principles and practice of numerical classification. W. H. Freeman, San Francisco, CA.
SNEDECOR, G. W., AND W. G. COCHRAN. 1967. Statistical methods, 6th ed. Iowa State Univ. Press, Ames.
SNEDECOR, G. W., AND W. G. COCHRAN. 1980. Statistical methods, 7th ed. Iowa State Univ. Press, Ames.
SNOW, B. K., AND D. W. SNOW. 1971. The feeding ecology of tanagers and honeycreepers in Trinidad. Auk 88:291-322.
SNOW, D. W. 1962a. A field study of the Black and White Manakin, Manacus manacus, in Trinidad. Zoologica 47:65-104.
SNOW, D. W. 1962b. A field study of the Golden-headed Manakin, Pipra erythrocephala, in Trinidad. Zoologica 47:183-198.
SNOW, D. W. 1973. Distribution, ecology and evolution of the bellbirds (Procnias, Cotingidae). Br. Mus. Bull. Zool. 25:369-391.
SNOW, D. W. 1981. Tropical frugivorous birds and their food plants: a world survey. Biotropica 13:1-14.
SNOW, D. W., AND B. K. SNOW. 1980. Relationships between hummingbirds and flowers in the Andes of Colombia. Bull. Br. Mus. Nat. Hist. Zool. 38:105-139.
SNOW, D. W., AND B. K. SNOW. 1986. Some aspects of avian frugivory in a north temperate area relevant to tropical forest, p. 159-164. In A. Estrada and T. H. Fleming [eds.], Frugivores and seed dispersal. Dr. W. Junk Publ., Dordrecht, The Netherlands.
SNYDER, L. L. 1928. The summer birds of Lake Nipigon. Trans. Roy. Can. Inst. 16:251-277.
SOKAL, R. R., AND F. J. ROHLF. 1969. Biometry. W. H. Freeman and Co., San Francisco.
SOKAL, R. R., AND F. J. ROHLF. 1981. Biometry. 2nd ed. W. H. Freeman and Co., San Francisco.
SOLOMON, M. E., AND D. M. GLEN. 1979. Prey density and rates of predation by tits (Parus spp.) on larvae of codling moth (Cydia pomonella) under bark. J. App. Ecol. 16:49-59.
SOLOMON, M. E., D. M. GLEN, D. A. KENDALL, AND N. F. MILSOM. 1976. Predation of overwintering larvae of codling moth (Cydia pomonella (L.) by birds. J. Appl. Ecol. 13:341-351.
SORENSEN, A. E. 1981. Interactions between birds and fruits in a temperate woodland. Oecologia 50:242-249.
SORENSEN, A. E. 1984. Nutrition, energy and passage time: experiments with fruit preference in European Blackbirds (Turdus merula). J. Anim. Ecol. 53:545-557.
SÖRENSEN, U., AND G. H. SCHMIDT. 1987. Vergleichende Untersuchungen zum Beuteeintrag der Waldameisen (Genus: Formica, Hymenoptera) in der Bredstedter Geest (Schleswig-Holstein). J. Appl. Entomol. 103:153-176.
SOUTHWOOD, T. R. E. 1961. The number of species of insects associated with various trees. J. Anim. Ecol. 30:1-8.
SOUTHWOOD, T. R. E. 1977. Habitat, the templet for ecological strategies? J. Anim. Ecol. 46:337-366.
SOUTHWOOD, T. R. E. 1978. Ecological methods. Chapman and Hall, London.
SOUTHWOOD, T. R. E. 1980. Ecological methods, with particular reference to insect populations. 2nd ed. Chapman and Hall, London, New York.
SOUTHWOOD, T. R. E., AND H. N. COMINS. 1976. A synoptic population model. J. Anim. Ecol. 45:949-965.
SPAANS, A. L. 1977. Are starlings faithful to their individual wintering quarters? Ardea 65:83-87.
SPECTOR, W. S. 1958. Handbook of biological data. Saunders, Philadelphia.
SPERBER, I. 1985. Colonic separation mechanisms (CSM). A review. Acta Physiol. Scand. 124. Supplementum 542.

SPSS. 1986. SPSSX user's guide. 2nd ed. SPSS, Inc., Chicago, IL.
SRIVASTAVA, N., R. W. CAMPBELL, T. R. TORGERSEN, AND R. C. BECKWITH. 1984. Sampling the western spruce budworm: fourth instars, pupae, and egg masses. For. Sci. 30:883-892.
STACEY, P. B. 1981. Foraging behavior of the Acorn Woodpecker in Belize, Central America. Condor 83:336-339.
STADDON, J. E. R. 1983. Adaptive behavior and learning. Cambridge Univ. Press, New York.
STAIRS, G. R. 1985. Predation on overwintering codling moth populations by birds. Ornis Scand. 16:323-324.
STALLCUP, P. L. 1968. Spatio-temporal relationships of nuthatches and woodpeckers in ponderosa pine forests of Colorado. Ecology 49:831-843.
STALMASTER, M. V., AND J. A. GESSAMAN. 1982. Food consumption and energy requirements of captive Bald Eagles. J. Wildl. Manage. 46:646-654.
STALMASTER, M. V., AND J. A. GESSAMAN. 1984. Ecological energetics and foraging behavior of overwintering Bald Eagles. Ecol. Monogr. 54:407-428.
STARK, R. W. 1952. Analysis of a population sampling method for the lodgepole needle miner in Canadian Rocky Mountain Parks. Can. Entomol. 84:316-321.
STARK, R. W., AND D. L. DAHLSTEN [eds.]. 1970. Studies on the population dynamics of the western pine beetle Dendroctonus brevicomis LeConte (Coleoptera: Scolytidae). Univ. California Div. Agric. Sci., Berkeley.
STAUFFER, D. F., E. O. GARTON, AND R. K. STEINHORST. 1985. A comparison of principal components from real and random data. Ecology 66:1693-1698.
STEBBINGS, R. E. 1982. Radio tracking greater horseshoe bats with preliminary observations on flight patterns. Symp. Zool. Lond. 49:161-173.
STEEL, R. G. D., AND J. H. TORRIE. 1960. Principles and procedures of statistics. McGraw-Hill, New York.
STEEL, R. G. D., AND J. H. TORRIE. 1980. Principles and procedures of statistics. 2nd ed. McGraw-Hill, New York.
STEENHOF, K., AND M. N. KOCHERT. 1988. Dietary responses of three raptor species to changing prey densities in a natural environment. J. Anim. Ecol. 57:37-48.
STEPHEN, F. M., AND D. L. DAHLSTEN. 1976a. The temporal and spatial arrival pattern of Dendroctonus brevicomis in ponderosa pine. Can. Entomol. 108:271-282.
STEPHEN, F. M., AND D. L. DAHLSTEN. 1976b. The arrival sequence of the arthropod complex following attack by Dendroctonus brevicomis (Coleoptera: Scolytidae) in ponderosa pine. Can. Entomol. 108:283-304.
STEPHENS, D. W. 1985. How important are partial preferences? Anim. Behav. 33:667-669.
STEPHENS, D. W. 1987. On economically tracking a variable environment. Theor. Popul. Biol. 32:15-25.
STEPHENS, D. W., AND J. R. KREBS. 1986. Foraging theory. Princeton Univ. Press, Princeton, NJ.
STEPHENS, D. W., AND S. R. PATON. 1986. How constant is the constant of risk-aversion? Anim. Behav. 34:1659-1667.
STEPHENS, D. W., J. F. LYNCH, A. E. SORENON, AND C. GORDON. 1986. Preference and profitability: theory and experiment. Am. Nat. 127:533-553.
STEPHENSON, A. G. 1981. Flower and fruit abortion: proximate causes and ultimate function. Ann. Rev. Ecol. Syst. 12:253-281.
STEVENS, R. E., AND R. W. STARK. 1962. Sequential sampling for the lodgepole needle miner, Evagora milleria. J. Econ. Entomol. 55:491-494.
STEVENSON, J. 1933. Experiments on the digestion of food by birds. Wilson Bull. 45:155-167.
STEWARD, V. B. 1986. Bird predation on the 13-year periodical cicada (Homoptera: Cicadidae: Magicicada spp.) in an Ozark forest community, 1985. M.S. thesis, Univ. of Arkansas.
STEWARD, V. B., K. G. SMITH, AND F. M. STEPHEN. 1988a. Red-winged Blackbird predation on periodical cicadas (Cicadidae: Magicicada spp.): bird behavior and cicada responses. Oecologia 76:348-352.
STEWARD, V. B., K. G. SMITH, AND F. M. STEPHEN. 1988b. Predation by wasps on lepidopteran larvae in an Ozark forest canopy. Ecol. Entomol. 13:81-86.
STEWART, P. A. 1975. Cases of birds reducing or eliminating infestations of tobacco insects. Wilson Bull. 87:107-109.
STEWART, R. E., AND J. W. ALDRICH. 1951. Removal and repopulation of breeding birds in a spruce-fir forest community. Auk 68:471-482.
STEYSKAL, G. C. 1981. A bibliography of the Malaise trap. Proc. Entomol. Soc. Wash. 83:225-229.
STILES, E. W. 1978. Avian communities in temperate and tropical alder forests. Condor 80:276-284.
STILES, F. G. 1980. The annual cycle in a tropical wet forest hummingbird community. Ibis 122:322-343.
STILES, F. G. 1983a. Birds, p. 502-544. In D. H. Janzen [ed.], Costa Rican natural history. Univ. Chicago Press, Chicago, IL.
STILES, F. G. 1983b. The Rufous-rumped Antwren (Terenura callinota) in Costa Rica. Wilson Bull. 95:462-464.
STILES, F. G. 1985a. On the role of birds in the dynamics of neotropical forests, p. 49-59. In A. W. Diamond and T. Lovejoy (eds.), Conservation of tropical forest birds. International Council on Bird Preservation, Cambridge, England.
STILES, F. G. 1985b. Conservation of forest birds in Costa Rica: problems and perspectives, p. 141-168. In A. W. Diamond and T. Lovejoy [eds.], Conservation of tropical forest birds. International Council on Bird Preservation. Cambridge, England.
STILES, F. G. 1985c. Seasonal patterns and coevolution in the hummingbird-flower community of a Costa Rican subtropical forest, p. 757-787. In P. A. Buckley, M. S. Foster, E. S. Morton, R. S. Ridgely, and F. G. Buckley [eds.], Neotropical ornithology. Orinithol. Monogr. No. 36.
STILES, F. G., AND B. WHITNEY. 1983. Notes on the behavior of the Costa Rican Sharpbill (Oxyruncus cristatus frater). Auk 100:117-125.
STORK, N. E. 1987. Guild structure of arthropods from Bornean rain forest trees. Ecol. Entomol. 12:69-80.
STOREY, M. L., AND N. K. ALLEN. 1982. Apparent and true metabolizable energy of feedstuffs for mature, nonlaying Embden Geese. Poul. Sci. 59:1275-1279.
STREHL, C. E., AND J. WHITE. 1986. Effects of superabundant food on breeding success and behavior of the Red-winged Blackbird. Oecologia 70:178-186.
STRICKLAND, A. H. 1961. Sampling crop pests and their hosts. Ann. Rev. Entomol. 6:201-220.
STRONG, D. R., J. H. LAWTON, AND R. SOUTHWOOD. 1984. Insects on plants: community patterns and mechanisms. Harvard Univ. Press, Cambridge, MA.
STRONG, D. R., L. A. SZYSKA, AND D. SIMBERLOFF. 1979. Tests of community-wide character displacement against null hypotheses. Evolution 33:897-913.
STURGES, F. W., R. T. HOLMES, AND G. E. LIKENS. 1974. The role of birds in nutrient cycling in a northern hardwoods ecosystem. Ecology 55:149-155.
STURMAN, W. A. 1968. The foraging ecology of Parus atricapillus and P. rufescens in the breeding season, with comparisons with other species of Parus. Condor 70:309-322.
SUGDEN, L. G. 1973. Metabolizable energy of wild duck foods. Can. Wildl. Serv. Progress Notes No. 35.
SULLIVAN, K. A. 1984. The advantages of social foraging in Downy Woodpeckers. Anim. Behav. 32:16-22.
SUNSHINE, P., AND N. KRETCHMER. 1964. Intestinal disaccharidases: absence in two species of sea lions. Science 144:850-851.
SVARDSON, G. 1949. Competition and habitat selection in birds. Oikos 1:157-174.
SWANSON, G. A., AND J. C. BARTONEK. 1970. Bias

associated with food analysis in gizzards of Blue-winted Teal. J. Wildl. Manage. 34:739-746.

SWARTZMAN, G. L., AND S. P. KALUZNY. 1987. Ecological simulation primer. MacMillan, New York.

SWIHART, R. K., AND N. A. SLADE. 1986. The importance of statistical power when testing for independence in animal movements. Ecology 67:255-258.

SWINGLAND, I. R. 1984. Intraspecific differences in movement, p. 102-115. In I. R. Swingland and P. J. Greenwood [eds.], The ecology of animal movement. Clarendon Press, Oxford.

SZARO, R. C. 1976. Population densities, habitat selection, and foliage use by the birds of selected ponderosa pine forest areas in the Beaver Creek Watershed, Arizona. Ph.D. diss., Northern Arizona Univ., Flagstaff.

SZARO, R. C. 1986. Guild management: an evaluation of avian guilds as a predictive tool. Environ. Manage. 10:681-688.

SZARO, R. C., AND R. P. BALDA. 1979. Bird community dynamics in a ponderosa pine forest. Stud. Avian Biol. No. 3.

SZARO, R. C., AND R. P. BALDA. 1986. Relationships among weather, habitat structure, and ponderosa pine forest birds. J. Wildl. Manage. 50:253-260.

TACHA, T. C., P. A. VOHS, AND G. C. IVERSON. 1985. A comparison of interval and continuous sampling methods for behavioral observations. J. Field Ornithol. 56:258-264.

TAKEKAWA, J. Y. 1987. Energetics of Canvasbacks staging on an Upper Mississippi River pool during fall migration. Ph.D. diss., Iowa State Univ., Ames.

TAKEKAWA, J. Y., AND E. O. GARTON. 1984. How much is an evening grosbeak worth? J. For. 82:426-428.

TALLAMY, D. W., E. J. HANSENS, AND R. F. DENNO. 1976. A comparison of Malaise trapping and aerial netting for sampling a horsefly and deerfly community. Environ. Entomol. 5:788-792.

TANNER, J. T. 1942. The Ivory-billed Woodpecker. Res. Rep. No. 1, Nat. Aud. Soc., Washington, D.C.

TATNER, P. 1983. The diet of urban Magpies Pica pica. Ibis 125:90-107.

TATNER, P., AND D. M. BRYANT. 1986. Flight cost of a small passerine measured using doubly labeled water: implications for energetics studies. Auk 103:169-180.

TAYLOR, C. R., N. C. HEGLUND, AND G. M. O. MALOIY. 1982. Energetics and mechanics of terrestrial locomotion. J. Exp. Biol. 97:1-21.

TAYLOR, L. R. 1951. An improved suction trap for insects. Ann. Appl. Biol. 38:582-591.

TAYLOR, L. R. 1955. The standardization of air flow in insect suction traps. Ann. Appl. Biol. 43:390-408.

TAYLOR, L. R. 1962. The absolute efficiency of insect suction traps. Ann. Appl. Biol. 50:405-421.

TAYLOR, L. R. 1963. Analysis of the effect of temperature on insects in flight. J. Anim. Ecol. 32:99-117.

TAYLOR, L. R., AND E. S. BROWN. 1972. Effects of light-trap design and illumination on samples of moths in the Kenya highlands. Bull. Entomol. Res. 62:91-112.

TAYLOR, L. R., AND C. I. CARTER. 1961. The analysis of numbers and distribution in an aerial population of Macrolepidoptera. Trans. Roy. Entomol. Soc. Lond. 113:396-386.

TAYLOR, R. W. 1976. A submission to the enquiry into the impact on the Australian environment of the current woodchip industry programme. Hansard: 12 August 1976: 3724-3731.

TEATHER, K. L., AND R. J. ROBERTSON. 1985. Female spacing patterns in Brown-headed Cowbirds. Can. J. Zool. 63:218-222.

TELFORD, A. D., AND S. G. HERMAN. 1963. Chickadee helps check insect invasion. Audubon Mag. 65:78-81.

TEMPLE, S. A. 1977. Plant-animal mutualism: co-evolution with Dodo leads to near extinction of plant. Science 197:885-886.

TERBORGH, J. 1980a. Causes of tropical species diversity.

Proc. XVII Inter. Ornithol. Congr. 955-961.

TERBORGH, J. W. 1980b. The conservation status of Neotropical migrants: present and future, p. 21-30. In A. Keast and E. S. Morton [eds.], Migrant birds in the Neotropics: ecology, behavior, distribution, and conservation. Smithsonian Inst. Press, Washington, D.C.

TERBORGH, J. 1983. Five New World primates: a study in comparative ecology. Monographs in behavior and ecology, Princeton Univ. Press, Princeton, NJ.

TERBORGH, J. 1985. Habitat selection in Amazonian birds, p. 311-340. In M. L. Cody [ed.], Habitat selection in birds. Academic Press, Orlando, FL.

TERBORGH, J. 1986. Keystone plant resources in the tropical forest, p. 330-344. In M. E. Soule [ed.], Conservation biology. Sinauer Assoc., Sunderland, MA.

TERBORGH, J. W., AND J. R. FAABORG. 1980. Factors affecting the distribution and abundance of North American migrants in the eastern Caribbean region, p. 145-155. In A. Keast and E. S. Morton [eds.], Migrant birds in the Neotropics: ecology, behavior, distribution, and conservation. Smithsonian Inst. Press, Washington, D.C.

TERBORGH, J., AND S. K. ROBINSON. 1986. Guilds and their utility in ecology, p. 65-90. In J. Kikkawa and D. J. Anderson [eds.], Community ecology: pattern and process. Blackwell Scientific Publ., Melbourne, Australia.

TERBORGH, J., AND J. W. WESKE. 1969. Colonization of secondary habitats by Peruvian birds. Ecology 50:765-782.

TERBORGH, J., J. W. FITZPATRICK, AND L. EMMONS. 1984. Annotated checklist of bird and mammal species of Cocha Cashu Biological Station, Manu National Park, Peru. Fieldiana (Zoology) No. 21.

TERRILL, S. B. 1987. Social dominance and migratory restlessness in the dark-eyed Junco (Junco hyemalis). Behav. Ecol. Sociobiol. 21:1-11.

TERRILL, S. B. 1988. The relative importance of ecological factors in bird migration, p. 2180-2190. In H. Ouellet [ed.], Proc. XIX Inter. Ornithol. Congr. Vol. 2. University of Ottawa Press, Canada.

TERRILL, S. B. In press a. Ecophysiological aspects of movements by migrants in the wintering grounds. In E. Gwinner [ed.], Bird migration: physiology and ecophysiology. Springer-Verlag, Berlin.

TERRILL, S. B. In press b. The regulation of migratory behavior: interactions between exogenous and endogenous factors. J. Orn.

TERRILL, S. B. In press c. Evolutionary aspects of orientation and migration in birds. Experientia.

TERRILL, S. B., AND K. P. ABLE. 1988. Bird migration and terminology. Auk 105:205-206.

TERRILL, S. B., AND R. L. CRAWFORD. 1988. Additional evidence of nocturnal migration by Yellow-rumped Warblers in winter. Condor 90:261-263.

TERRILL, S. B., AND R. D. OHMART. 1984. Facultative extension of fall migration by Yellow-rumped Warblers (Dendroica coronata). Auk 101:427-438.

THALAU, H.-P., AND W. WILTSCHKO. 1987. Einflusse des Futterangebots auf die Tagesaktivitat von Trauerschnappern (Ficedula hypoleuca) auf dem Herbstzug. Cour. Forsch.-Inst. Senckenberg 97:67-73.

THOMAS, D. B., AND E. L. SLEEPER. 1977. The use of pitfall traps for estimating the abundance of arthropods, with special reference to the Tenebrionidae (Coleoptera). Ann. Entomol. Soc. Am. 70:242-248.

THOMAS, D. G. 1980. Foraging of honeyeaters in an area of Tasmanian sclerophyll forest. Emu 80:55-58.

THOMPSON, C. G., AND L. J. PETERSON. 1978. Rearing the Douglas-fir tussock moth. USDA Agric. Handb. 520, Washington, D.C.

THOMPSON, J. N., AND M. F. WILLSON. 1979. Evolution of temperate fruit/bird interactions: phenological strategies. Evolution 33:973-982.

THOMPSON, S. K. 1987. Sample sizes for estimating multinomial proportions. Am. Stat. 41:42-46.

THOMPSON, R. D., AND C. V. GRANT. 1968. Nutritive value

of two laboratory diets for starlings. Lab. Anim. Care 18:75-79.
THOMSON, J. D., W. P. MADDISON, AND R. C. PLOWRIGHT. 1982. Behavior of bumble bee pollinators of Aralia hispada Vent. (Araliaceae). Oecologia 54:326-336.
THORINGTON, R. W., JR., B. TANNENBAUM, A. TARAK, AND R. RUDRAN. 1982. Distribution of trees on Barro Colorado Island: a five hectare sample, p. 83-94. In E. Leigh, Jr., A. S. Rand, and D. Windsor [eds.], The ecology of a tropical forest. Smithsonian Inst. Press, Washington, D.C.
THORPE, W. H. 1963. Learning and instinct in animals. Methuen, London.
TICEHURST, M., AND R. REARDON. 1975. Malaise trap: a survey tool for collecting the adult stage of the gypsy moth, Lymantria dispar. Melsheimer Entomol. Ser. 18.
TIEBOUT, H. M. III. 1986. Downy Woodpecker feeds on insects in a spider's web. Wilson Bull. 98:319.
TINBERGEN, J. M. 1981. Foraging decisions in Starlings (Sturnus vulgaris L.) Ardea 69:1-67.
TINBERGEN, L. 1960. The natural control of insects in pinewoods. I. Factors influencing the intensity of predation by songbirds. Arch. Neer. Zool. 13:265-336.
TOFT, C. A. 1980. Feeding ecology of thirteen syntopic species of anurans in a seasonal tropical environment. Oecologia 45:131-141.
TOFT, C. A. 1985. Resource partitioning in amphibians and reptiles. Copeia 1985:1-21.
TOMBACK, D. F. 1975. An emetic technique to investigate food preferences. Auk 92:581-583.
TOMBACK, D. F. 1982. Dispersal of whitebark pine seeds by Clark's Nutcracker: a mutualism hypothesis. J. Anim. Ecol. 51:451-467.
TOPICK, C., N. M. HALVERSON, AND D. G. BROCKWAY. 1986. Plant associations and management guide for the Western Hemlock Zone, Gifford Pinchot National Forest. USDA For. Serv. PNW-R6-ECOL-230A, Portland, OR.
TORGERSEN, T. R., AND R. W. CAMPBELL. 1982. Some effects of avian predators on the western spruce budworm in north central Washington. Environ. Entomol. 11:429-431.
TORGERSEN, T. R., AND R. R. MASON. 1979. Predation and parasitization of Douglas-fir tussock moth egg masses. USDA Agric. Handb. 549, Washington, D.C.
TORGERSEN, T. R., AND R. R. MASON. 1987. Predation on egg masses of the Douglas-fir tussock moth (Lepidoptera: Lymantriidae). Environ. Entomol. 16:90-93.
TORGERSEN, T. R., R. R. MASON, AND H. G. PAUL. 1983. Predation on pupae of Douglas-fir tussock moth, Orygia pseudotsugata (McDunnough) (Lepidoptera: Lymantriidae). Environ. Entomol. 12:1678-1682.
TORGERSEN, T. R., R. W. CAMPBELL, N. SRIVASTAVA, AND R. C. BECKWITH. 1984a. Role of parasites in the population dynamics of the western spruce budworm (Lepidoptera: Tortricidae) in the northwest. Environ. Entomol. 13:568-573.
TORGERSEN, T. R., J. W. THOMAS, R. R. MASON, AND D. VAN HORN. 1984b. Avian predators of Douglas-fir tussock moth, Orygia pseudotsugata (McDunnough) (Lepidoptera: Lymantriidae) in southwestern Oregon. Environ. Entomol. 13:1018-1022.
TORTORA, R. D. 1978. A note on sample size estimation for multinomial populations. Am. Stat. 32:100-102.
TRAMER, E. J., AND T. R. KEMP. 1980. Foraging ecology of migrant and resident warblers and vireos in the highlands of Costa Rica, p. 285-296. In A. Keast and E. S. Morton [eds.], Migrant birds in the Neotropics: ecology, behavior, distribution, and conservation. Smithsonian Inst. Press, Washington, D.C.
TRAVIS, J. 1977. Seasonal foraging in a Downy Woodpecker population. Condor 79:371-375.
TRAYLOR, M. A., JR., AND J. W. FITZPATRICK. 1982. A survey of the tyrant flycatchers. Living Bird 19:7-45.
TRUMBLE, J. T., B. CARTWRIGHT, AND L. T. KOK. 1981. Efficiency of suction sampling for Rhinocyllus conicus and a comparison of suction and visual sampling techniques. Environ. Entomol. 10:787-792.
TRUMBLE, J. T., H. NAKAKIHARA, AND G. W. ZENDER. 1982. Comparisons of traps and visual searches of foliage for monitoring aphid (Heteroptera: Aphidae) population density of broccoli. J. Econ. Entomol. 75:853-856.
TRYON, P. R., AND S. F. MACLEAN. 1980. Use of space by Lapland Longspurs breeding in arctic Alaska. Auk 97:509-520.
TUKEY, J. W. 1980. We need both exploratory and confirmatory. Am. Stat. 34:23-25.
TURNBULL, A. L., AND C. F. NICHOLLS. 1966. A "quick trap" for area sampling of arthropods in grassland communities. J. Econ. Entomol. 59:1100-1104.
TURNER, A. K. 1982. Optimal foraging by the swallow (Hirundo rustica L.): prey size selection. Anim. Behav. 30:862-872.
TURNER, A. K. 1983a. Food selection and the timing of breeding of the Blue-and-white Swallow Notiochelidon cyanoleuca in Venezuela. Ibis 125:450-462.
TURNER, A. K. 1983b. Time and energy constraints on the brood size of swallows, Hirundo rustica, and Sand Martins, Riparia riparia. Oecologia 59:331-338.
TYE, A. 1981. Ground-feeding methods and niche separation in thrushes. Wilson Bull. 93:112-114.
UETZ, G. W., AND J. D. UNZICKER. 1976. Pitfall trapping in ecological studies of wandering spiders. J. Arachnol. 3:101-111.
ULFSTRAND, S. 1976. Feeding niches of some passerine birds in a south Swedish coniferous plantation in winter and summer. Ornis Scand. 7:21-27.
ULFSTRAND, S. 1977. Foraging niche dynamics and overlap in a guild of passerine birds in a south Swedish coniferous woodland. Oecologia 27:23-45.
U.S. FISH AND WILDLIFE SERVICE. 1980. Tensas River National Wildlife Refuge. USDI Fish and Wildl. Serv. Brochure. Tensas River Nat. Wildl. Ref.
VAN DER HEIJDEN, P. G. M., AND J. DE LEEUW. 1985. Correspondence analysis used complementary to loglinear analysis. Psychometrika 50:429-447.
VAN DER PIJL, L. 1969. Principles of dispersal in higher plants. Springer-Verlag, Berlin.
VAN DIJK, T. S. 1986. How to estimate the level of food availability in field populations of carabid beetles, p. 371-384. In P. J. DenBoer, M. L. Luff, D. Mossakowski, and F. Weber [eds.], Carabid beetles, their adaptations and dynamics. Gustav Fischer, Stuttgart, New York.
VAN HORNE, B., AND R. G. FORD. 1982. Niche breadth calculation based on discriminant analysis. Ecology 63:1172-1174.
VAN RIPER, C. III. 1984. The influence of nectar resources on nesting success and movement patterns of the Common Amakihi (Hemignathus virens). Auk 101:38-46.
VAN SCHAIK, C. P. 1986. Phenological changes in a Sumatran rain forest. J. Trop. Ecol. 2:327-347.
VAN SOEST, P. J., AND J. B. ROBINSON. 1976. The detergent system of analysis and its application to human foods. In P. James and O. Theander [eds.], Analysis of dietary fiber in food. M. Dekker, New York.
VAN SOEST, P. J., P. UDEN, AND K. F. WRICK. 1983. Critique and evaluation of markers for use in nutrition of humans and farm and laboratory animals. Nutr. Rep. Inter. 27:17-28.
VAN TYNE, J. 1948. Home range and duration of family ties in the Tufted Titmouse. Wilson Bull. 60:121.
VANDERMEER, J. H. 1972. Niche theory. Ann. Rev. Ecol. Syst. 3:107-132.
VANDERWALL, S. B., AND R. P. BALDA. 1981. Ecology and evolution of food-storage behavior in conifer-seed-caching corvids. Z. Tierpsychol. 56:217-242.
VARLEY, G. C., G. R. GRADWELL, AND M. P. HASSELL. 1973. Insect population ecology: an analytical approach. Blackwell Scientific Publ., Oxford.

VASSALLO, M. I., AND J. C. RICE. 1982. Ecological release and ecological flexibility in habitat use and foraging of an insular avifauna. Wilson Bull. 94:139-155.

VEHRENCAMP, S. L., AND L. HALPENNY. 1981. Capture and radio-transmitter attachment techniques for Roadrunners. N. Am. Bird Bander 6:128-132.

VERBEEK, N. A. M. 1975. Comparative feeding behavior of three coexisting Tyrannid flycatchers. Wilson Bull. 87:231-240.

VERNER, J. 1984. The guild concept applied to management of bird populations. Environ. Manage. 8:1-14.

VIA, J. M., JR. 1979. Foraging tactics of flycatchers in southwestern Virginia, p. 191-202. In J. G. Dickson, R. N. Conner, R. R. Fleet, J. A. Jackson, and J. C. Kroll [eds.], The role of insectivorous birds in forest ecosystems. Academic Press, NY.

VLACHONIKOLIS, I. G., AND F. H. C. MARRIOTT. 1982. Discrimination with mixed binary and continuous data. Appl. Stat. 31:23-31.

VONK, H. J., AND J. R. H. WESTERN. 1984. Comparative biochemistry and physiology of enzymatic digestion. Academic Press, New York.

VORIS, H. K., AND H. H. VORIS. 1983. Feeding strategies in marine snakes: an analysis of evolutionary, morphological, behavioral, and ecological relationships. Am. Zool. 23:411-425.

WAAGE, J. K. 1979. Foraging for patchily distributed hosts by the parasitoid, Nemeritis canescens. J. Anim. Ecol. 48:353-371.

WADDINGTON, K. D., AND B. HEINRICH. 1979. The foraging movements of bumblebees on vertical "inflorescences:" an experimental analysis. J. Comp. Physiol. 134:113-117.

WAGNER, J. L. 1981a. Visibility and bias in avian foraging data. Condor 83:263-264.

WAGNER, J. L. 1981b. Seasonal change in guild structure: oak woodland insectivorous birds. Ecology 62:973-981.

WAHLENBERG, W. G. 1946. Longleaf pine: its use, ecology, regeneration, protection, growth, and management. Charles Lathrop Pack Forestry Foundation, and USDA For. Serv., Washington, D.C.

WAIDE, R. B., AND J. P. HAILMAN. 1977. Birds of five families feeding from spider webs. Wilson Bull. 89:345-346.

WAITE, R. K. 1984a. Sympatric corvids: effects of social behavior, aggression and avoidance on feeding. Behav. Ecol. Sociobiol. 15:55-60.

WAITE, R. K. 1984b. Winter habitat selection and foraging behaviour in sympatric corvids. Ornis Scand. 15:55-62.

WAITE, T. A. 1987a. Vigilance in the White-breasted Nuthatch: effects of dominance and sociality. Auk 104:429-434.

WAITE, T. A. 1987b. Dominance-specific vigilance in the Tufted Titmouse: effects of social context. Condor 89:932-935.

WAITE, T. A., AND T. C. GRUBB, JR. In press. Diurnal caching rhythms in captive White-breasted Nuthatches. Ornis Scand.

WAKELEY, J. S. 1978. Factors affecting the use of hunting sits by Ferruginous Hawks. Condor 80:316-326.

WALDBAUER, G. P., AND W. E. LABERGE. 1985. Phenological relationships of wasps, bumblebees, their mimics and insectivorous birds in northern Michigan. Ecol. Entomol. 10:99-110.

WALDBAUER, G. P., AND J. K. SHELDON. 1971. Phenological relationships of some aculeate Hymenoptera, their dipteran mimics, and insectivorous birds. Evolution 25:371-382.

WALKER, T. J. 1978. Migration and re-migration of butterflies through north peninsular Florida: quantification with Malaise traps. J. Lepid. Soc. 32:178-190.

WALL, L. E. 1982. "Foot pattering" by Australian Ground Thrush. Tasmanian Nat. 70:8.

WALLNER, W. E. 1987. Factors affecting insect population dynamics: differences between outbreak and non-outbreak species. Ann. Rev. Entomol. 32:317-340.

WALSBERG, G. E. 1975. Digestive adaptation of Phainopepla nitens associated with the eating of mistletoe berries. Condor 77:169-174.

WALSBERG, G. E. 1977. Ecology and energetics of contrasting social systems in Phainopepla nitens (Aves: Ptiligonatidae). Univ. Calif. Publ. Zool. 108:1-63.

WALSH, H. 1978. Food of nestling Purple Martins. Wilson Bull. 90:248-260.

WALTZ, E. C. 1987. A test of the information-centre hypothesis in two colonies of Common Terns, Sterna hirundo. Anim. Behav. 35:48-59.

WARNER, A. C. I. 1981. Rate of passage of digesta through the gut of mammals and birds. Nutr. Abs. Rev. B51:780-820.

WARTENBERG, D., S. FERSON, AND F. J. ROHLF. 1987. Putting things in order: a critique of detrended correspondence analysis. Am. Nat. 129:434-448.

WATSON, A. [ed.]. 1970. Animal populations in relation to their food resources. Blackwell Scientific Publ., Oxford.

WAUGH, D. R. 1978. Predation strategies in aerial feeding birds. Ph.D. diss., Univ. of Stirling, Scotland.

WAUGH, D. R. 1979. The diet of Sand Martins in the breeding season. Bird Study 26:123-128.

WAUGH, D. R., AND C. J. HAILS. 1983. Foraging ecology of a tropical aerial feeding guild. Ibis 125:200-217.

WEATHERS, W. W., W. A. BUTTEMER, A. M. HAYWORTH, AND K. A. NAGY. 1984. An evaluation of the time-budget estimates of daily energy expenditure in birds. Auk 101:459-472.

WEGLARCZYK, G. 1981. Nitrogen balance and energy efficiency of protein disposition of the House Sparrow, Passer domesticus. Ekologia Pol. 29:519-533.

WEIGL, P. D., AND E. V. HANSON. 1980. Observational learning and the feeding behavior of the red squirrel Tamiasciurus hudsonicus: the ontogeny of optimization. Ecology 61:213-218.

WEISBERG, S. 1985. Applied linear regression. 2nd ed. John Wiley and Sons, New York.

WELTY, J. C. 1975. The life of birds. 2nd ed. W. B. Saunders Co., Philadelphia, PA.

WERNER, E. E. 1977. Species packing and niche complementarity in three sunfishes. Am. Nat. 111:553-578.

WERNER, E. E., AND J. F. GILLIAM. 1984. The ontogenetic niche and species interactions in size-structured populations. Ann. Rev. Ecol. Syst. 15:393-425.

WERNER, E. E., AND D. J. HALL. 1974. Optimal foraging and the size selection of prey by the bluegill sunfish (Lepomis macrochirus). Ecology 55:1042-1052.

WERNER, E. E., D. J. HALL, AND G. G. MITTELBACH. 1983a. An experimental test of the effects of predation risk on habitat use in fish. Ecology 64:1540-1548.

WERNER, E. E., G. G. MITTELBACH, D. J. HALL, AND J. F. GILLIAM. 1983b. Experimental test of optimal habitat use in fish: the role of relative habitat profitability. Ecology 64:1525-1534.

WERNER, T. K. 1988. Behavioral, individual feeding specializations by Pinaroloxias inornata, the Darwin's Finch of Cocos Island, Costa Rica. Ph.D. diss., Univ. of Massachusetts, Amherst.

WERNER, T. K., AND T. W. SHERRY. 1987. Behavioral feeding specialization in Pinaroloxias inornata, the "Darwin's Finch" of Cocos Island, Costa Rica. Proc. Natl. Acad. Sci. 84:5506-5510.

WESELOH, R. M. 1972. Field responses of gypsy moths and some parasitoids to colored surfaces. Ann. Entomol. Soc. Am. 65:742-746.

WESELOH, R. M. 1981. Relationships between colored sticky panel catches and reproductive behavior of forest tachinid parasitoids. Environ. Entomol. 10:131-135.

WESELOH, R. M. 1987. Accuracy of gypsy moth (Lepidoptera: Lymantriidae) population estimates based on counts of larvae in artificial resting sites. Ann. Entomol. Soc. Am. 80:361-366.

WEST, C. 1985. Factors underlying the late season

appearance of the lepidopterous leaf-mining guild on oak. Ecol. Entomol. 10:111-120.
WEST, G. C. 1960. Seasonal variation in the energy balance of the Tree Sparrow in relation to migration. Auk 77:306-329.
WEST, G. C. 1968. Bioenergetics of captive Willow Ptarmigan under natural conditions. Ecology 49:1035-1045.
WEST, G. C., AND J. S. HART. 1966. Metabolic responses of Evening Grosbeaks to constant and to fluctuating temperatures. Physiol. Zool. 39:171-184.
WESTERTERP, K. 1973. The energy budget of the Starling Sturnus vulgaris, a field study. Ardea 61:137-158.
WETHERBEE, J. 1968. The Southern Swamp Sparrow. In A. C. Bent and O. Austin [eds.], Life history of North American cardinals, grosbeaks, buntings, towhees, finches, sparrows, and allies. Dover Press, NJ.
WETHERWAX, P. B. 1986. Why do honeybees reject certain flowers? Oecologia 69:567-570.
WHEELER, D. E., AND S. C. LEVINGS. In press. The impact of the 1983 El Nino drought on the litter arthropods of Barro Colorado Island, Panama. In R. Arnett, Jr. [ed.], Advances in Myrmecology. E. J. Brill, Holland.
WHEELWRIGHT, N. T. 1983. Fruits and the ecology of Resplendent Quetzals. Auk 100:286-301.
WHEELWRIGHT, N. T. 1985. Fruit size, gape width, and the diets of fruit-eating birds. Ecology 66:808-818.
WHEELWRIGHT, N. T. 1986a. A seven-year study of individual variation in fruit production in tropical bird-dispersed tree families in the family Lauraceae, p. 19-35. In A. Estrada and T. H. Fleming [eds.], Frugivores and seed dispersal. Dr. W. Junk Publ., Dordrecht, The Netherlands.
WHEELWRIGHT, N. T. 1986b. The diet of American Robins: an analysis of U.S. Biological Survey records. Auk 103:710-725.
WHEELWRIGHT, N. T. 1988. Seasonal changes in food preferences of American Robins in captivity. Auk 105:374-378.
WHEELWRIGHT, N. T., AND G. H. ORIANS. 1982. Seed dispersal by animals: contrasts with pollen dispersal, problems with terminology, and constraints on coevolution. Am. Nat. 119:402-413.
WHEELWRIGHT, N. T., W. A. HABER, K. G. MURRAY, AND C. GUINDON. 1984. Tropical fruit-eating birds and their food plants: a survey of a Costa Rican lower montane forest. Biotropica 16:173-192.
WHELAN, C. J., R. T. HOLMES, AND H. R. SMITH. 1989. Bird predation on gypsy moth (Lepidoptera: Lymantridae) larvae: an aviary study. Environ. Entomol. 18:43-45.
WHITAKER, J. O., AND P. Q. TOMICH. 1983. Food habits of the hoary bat, Lasiurus cinereus, from Hawaii. J. Mammal. 64:151-152.
WHITE, S. C. 1974. Ecological aspects of growth and nutrition in tropical fruit-eating birds. Ph.D. diss., Univ. of Pennsylvania, Philadelphia.
WHITHAM, T. G., AND C. N. SLOBODCHIKOFF. 1981. Evolution by individuals, plant-herbivore interactions, and mosaics of genetic variability: the adaptive significance of somatic mutations. Oecologia 49:287-292.
WHITING, T. M., JR. 1979. Winter feeding niche partitioning by Carolina Chickadees and Tufted Titmice in East Texas, p. 331-340. In J. G. Dickson, R. N. Conner, R. R. Fleet, J. A. Jackson, and J. C. Kroll [eds.], The role of insectivorous birds in forest ecosystems. Academic Press, New York.
WHITMORE, R. C. 1981. Applied aspects of choosing variables in studies of bird habitats, p. 38-41. In D. E. Capen [ed.], The use of multivariate statistics in studies of wildlife habitat. USDA For. Serv. Gen. Tech. Rep. RM-87, Fort Collins, CO.
WHITTAM, T. G. 1977. Coevolution of foraging in Bombus and nectar dispensing in Chilopsis: a last-dreg theory. Science 197:593-596.
WICKLER, W. 1968. Mimicry in plants and animals. World University Library, London.
WIEDENMANN, R. N., AND K. N. RABENOLD. 1987. The effects of social dominance between two subspecies of Dark-eyed Juncos, Junco hyemalis. Anim Behav. 35:856-864.
WIENS, J. A. 1969. An approach to the study of ecological relationships among grassland birds. Ornithol. Monogr. 8:1-93.
WIENS, J. A. 1973. Patterns and process in grassland bird communities. Ecol. Monogr. 43:237-270.
WIENS, J. A. 1976. Population responses to patchy environments. Ann. Rev. Ecol. Syst. 7:81-120.
WIENS, J. A. 1977. On competition and variable environments. Am. Sci. 65:590-597.
WIENS, J. A. 1981. Scale problems in avian censusing, p. 513-521. In C. J. Ralph and J. M. Scott [eds.], Estimating numbers of terrestrial birds. Stud. Avian Biology 6.
WIENS, J. A. 1983. Avian community ecology: an iconoclastic view, p. 355-403. In A. H. Brush and G. A. Clark [eds.], Perspectives in ornithology. Cambridge Univ. Press, New York.
WIENS, J. A. 1984a. The place of long-term studies in ornithology. Auk 101:202-203.
WIENS, J. A. 1984b. Resource systems, populations, and communities, p. 397-436. In P. W. Price, C. N Slobodchikoff, and W. S. Gaud [eds.], A new ecology: novel approaches to interactive systems. John Wiley and Sons, New York.
WIENS, J. A., AND M. I. DYER. 1975. Rangeland avifaunas: their composition, energetics and role in the ecosystem, p. 146-182. In D. R. Smith [tech. coord.], Proc. Symposium on management of forest and range habitats for nongame birds. USDA For. Serv. Gen. Tech. Rep. WO-1, Washington, D.C.
WIENS, J. A., AND G. S. INNIS. 1974. Estimation of energy flow in bird communities: a population bioenergetics model. Ecology 55:730-746.
WIENS, J. A., AND J. T. ROTENBERRY. 1979. Diet niche relationships among North American grassland and shrubsteppe birds. Oecologia 42:253-292.
WIENS, J. A., AND J. T. ROTENBERRY. 1981. Habitat associations and community structure of birds in shrubsteppe environments. Ecol. Monogr. 51:21-41.
WIENS, J. A., AND J. T. ROTENBERRY. 1987. Shrub-steppe birds and the generality of community models: a response to Dunning. Am. Nat. 129:920-927.
WIENS, J. A., J. T. ROTENBERRY, AND B. VAN HORNE. 1986a. A lesson in the limitations of field experiments: shrubsteppe birds and habitat alteration. Ecology 67:365-376.
WIENS, J. A., J. F. ADDICOTT, T. J. CASE, AND J. DIAMOND. 1986b. Overview: the importance of spatial and temporal scale in ecological investigations, p. 145-153. In J. Diamond and T. J. Case [eds.], Community ecology. Harper and Row, New York.
WIENS, J. A., J. T. ROTENBERRY, AND B. VAN HORNE. 1987a. Habitat occupancy patterns of North American shrubsteppe birds: the effects of spatial scale. Oikos 48:132-147.
WIENS, J. A., B. VAN HORNE, AND J. T. ROTENBERRY. 1987b. Temporal and spatial variations in the behavior of shrubsteppe birds. Oecologia 73:60-70.
WIENS, J. A., S. G. MARTIN, W. R. HOLTHAUS, AND F. A. IWEN. 1970. Metronome timing in behavioral ecology studies. Ecology 51:350-352.
WIGGLESWORTH, V. B. 1965. The principles of insect physiology. Methuen, London.
WIJNANDTS, H. 1984. Ecological energetics of the Long-eared Owl. Ardea 72:1-92.
WIKLUND, C. 1975. Pupal colour polymorphism in Papilio machaon L. and the survival in the field of cryptic versus non-cryptic pupae. Trans. Roy. Entomol. Soc. Lond. 127:73-84.
WILEY, R. H. 1971. Cooperative roles in mixed flocks of

antwrens (Formicariidae). Auk 88:881-892.
WILKINSON, G. S. 1985. The social organization of the common vampire bat. I. Pattern and cause of association. Behav. Ecol. Sociobiol. 17:111-121.
WILKINSON, G. S., AND J. W. BRADBURY. 1988. Radio telemetry: techniques and analysis. In T. H. Kunz [ed.], Ecological and behavioral methods for the study of bats. Smithsonian Instit. Press, Washington, D.C.
WILLIAMS, B. K. 1981. Discriminant analysis in wildlife research: theory and applications, p. 59-71. In D. E. Capen [ed.], The use of multivariate statistics in studies of wildlife habitat. U.S.D.A. For. Serv. Gen. Tech. Rep. RM-87.
WILLIAMS, B. K., AND K. TITUS. In press. Assessment of sampling stability in ecological applications of discriminant analysis. Ecology.
WILLIAMS, C. S. 1938. Aids to the identification of mole and shrew hairs with general comments on hair structure and hair determination. J. Wildl. Manage. 2:239-250.
WILLIAMS, C. S., AND W. H. MARSHALL. 1938. Duck nesting studies, Bear River Migratory Bird Refuge, Utah, 1937. J. Wildl. Manage. 2:29-48.
WILLIAMS, D. A. 1976. Improved likelihood ratio tests for complete contingency tables. Biometrika 63:33-37.
WILLIAMS, J. B. 1980. Intersexual niche partitioning in Downy Woodpeckers. Wilson Bull. 92:439-451.
WILLIAMS, J. B., AND G. O. BATZLI. 1979a. Winter diet of a bark-foraging guild of birds. Wilson Bull. 91:126-131.
WILLIAMS, J. B., AND G. O. BATZLI. 1979b. Competition among bark-foraging birds in central Illinois: Experimental evidence. Condor 81:122-132.
WILLIAMS, P. L. 1988. Spacing behavior and related features of social organization in Northern Orioles of central coastal California. Ph.D. diss., Univ. California, Berkeley.
WILLIAMSON, M. 1981. Island populations. Oxford Univ. Press, London.
WILLIAMSON, P. 1971. Feeding ecology of the Red-eyed Vireo (Vireo olivaceus) and associated foliage-gleaning birds. Ecol. Monogr. 41:129-152.
WILLIS, E. O. 1960. A study of the foraging behavior of two species of ant-tanagers. Auk 77:150-170.
WILLIS, E. O. 1966. Competitive exclusion and birds at fruiting trees in western Colombia. Auk 83:479-480.
WILLIS, E. O. 1968. Studies of the behavior of Lunulated and Salvin's antbirds. Condor 70:128-148.
WILLIS, E. O. 1969. On the behavior of five species of Rhegmatorhina, ant-following antbirds of the Amazon Basin. Wilson Bull. 81:363-395.
WILLIS, E. O. 1972. The behavior of Plain-brown Woodcreepers, Dendrocincla fuliginosa. Wilson Bull. 84:377-420.
WILLIS, E. O. 1976. Seasonal changes in the invertebrate litter fauna on Barro Colorado Island, Panama. Rev. Brazil. Biol. 36:643-657.
WILLIS, E. O. 1979. Behavior and ecology of two forms of White-chinned Woodcreepers (Dendrocincla merula, Dendrocolaptidae) in Amazonia. Pap. Avulsos Zool. 33:27-66.
WILLIS, E. O. 1981a. Notes on the Slender Antbird. Wilson Bull. 93:103-107.
WILLIS, E. O. 1981b. Diversity in adversity: the behaviors of two subordinate antbirds. Arquiv. Zool. 30:159-234.
WILLIS, E. O. 1981c. Momotus momota and Barypthengus ruficapillus (Momotidae) as army ant followers. Ciência e Cultura 33:1636-1640.
WILLIS, E. O. 1982a. Ground-cuckoos (Aves, Cuculidae) as army ant followers. Rev. Brasil. Biol. 42:753-756.
WILLIS, E. O. 1982b. Notharchus puffbirds (Aves, Bucconidae) as army ant followers. Ciência e Cultura 34:777-782.
WILLIS, E. O. 1982c. Amazonian Bucco and Monasa (Bucconidae) as army ant followers. Ciência e Cultura 34:782-785.

WILLIS, E. O. 1982d. The behavior of Black-banded Woodcreepers (Dendrocolaptes picumnus). Condor 84:272-285.
WILLIS, E. O. 1982e. Malacoptila puffbirds (Aves, Bucconidae) as army ant followers. Ciência e Cultura 34:924-928.
WILLIS, E. O. 1982f. The behavior of Red-billed Woodcreepers (Hylexetastes perrotti). Rev. Brasil. Biol. 42:655-666.
WILLIS, E. O. 1983a. Jays, mimids, icterids, and bulbuls (Corvidae, Mimidae, Icteridae and Pycnonotidae) as ant followers. Gerfaut 73:379-392.
WILLIS, E. O. 1983b. Flycatchers, cotingas and drongos (Tyrannidae, Muscicapidae, Cotingidae and Dicruridae) as ant followers. Gerfaut 73:265-280.
WILLIS, E. O. 1983c. Trans-Andean Xiphorhynchus and relatives (Aves, Dendrocolaptidae) as army ant followers. Rev. Brasil Biol. 43:125-131.
WILLIS, E. O. 1983d. Cis-Andean Xiphorhynchus and relatives (Aves, Dendrocolaptidae) as army ant followers. Rev. Brasil Biol. 43:133-142.
WILLIS, E. O. 1984. Manakins (Aves, Pipridae) as army ant followers. Ciência e Cultura 36:817-823.
WILLIS, E. O. 1985a. East African Turdidae as safari ant followers. Gerfaut 75:140-153.
WILLIS, E. O. 1985b. Antthrushes, antpittas, and gnateaters (Aves, Formicariidae) as army ant followers. Rev. Brasil. Zool. 2:443-448.
WILLIS, E. O. 1986. West African thrushes as safari ant followers. Gerfaut 76:95-108.
WILLIS, E. O., AND Y. ONIKI. 1978. Birds and army ants. Ann. Rev. Ecol. Syst. 9:243-263.
WILLSON, M. F. 1966. Feeding ecology of the Yellow-headed Blackbird. Ecol. Monogr. 36:51-77.
WILLSON, M. F. 1970. Foraging behavior of some winter birds of deciduous woods. Condor 72:169-174.
WILLSON, M. F. 1971. Seed selection in some North American finches. Condor 73:415-429.
WILLSON, M. F. 1974. Avian community organization and habitat structure. Ecology 55:1017-1027.
WILLSON, M. F. 1986. Avian frugivory and seed dispersal in eastern North America. Curr. Ornithol. 3:223-279.
WILLSON, M. F., AND J. C. HARMESON. 1973. Seed preferences and digestive efficiency of Cardinals and Song Sparrows. Condor 75:225-234.
WILSON, L. T., AND A. P. GUTIERREZ. 1980. Within-plant distribution of predators on cotton: comments on sampling and predator efficiencies. Hilgardia 48:3-11.
WILSON, S. W. 1978. Food size, food type and foraging sites of Red-winged Blackbirds. Wilson Bull. 90:511-520.
WINKEL, W., AND D. WINKEL. 1980. Winter-Untersuchungen uber das nachtigen von Kohl-und Blaumeise (Parus major und P. caeruleus) in Kunstlichen Nisthohlen eines niedersachsischen Aufforstungsgebietes mit Japanischer Larche Larix leptolepis. Die Vogel. 101:47-61.
WINKLER, H., AND B. LEISLER. 1985. Morphological aspects of habitat selection in birds, p. 415-434. In M. L. Cody [ed.], Habitat selection in birds. Academic Press, New York.
WITHERS, P. C. 1983. Energy, water, and solute balance of the Ostrich Struthio camelus. Physiol. Zool. 56:568-579.
WITTENBERGER, J. F. 1980. Vegetation structure, food supply, and polygyny in Boblinks (Dolichnyx oryzivorus). Ecology 61:140-150.
WOINARSKI, J. C. Z. 1985. Foliage gleaners of the tree-tops: pardalotes, p. 165-175. In J. A. Keast, H. F. Recher, H. A. Ford and D. Saunders [eds.], Birds of the eucalypt forests and woodlands: ecology, conservation, management. Surrey-Beatty, Sydney.
WOINARSKI, J. C. Z., AND J. M. CULLEN. 1984. Distribution of invertebrates on foliage in forests of south-eastern Australia. Aust. J. Ecol. 9:207-232.
WOINARSKI, J. C. Z., AND R. R. ROUNSEVELL. 1983. Comparative ecology of pardalotes, including the Forty-

spotted Pardalote, Pardalotus quadragintus (Aves: Pardalotidae) in south-eastern Tasmania. Aust. Wildl. Res. 10:351-361.

WOLDA, H. 1979. Abundance and diversity of Homoptera in the canopy of a tropical forest. Ecol. Entomol. 4:181-190.

WOLDA, H., AND M. WONG. 1988. Tropical insect diversity and seasonality. Sweep-samples vs. light-traps. Proc. Kon. Nederl. Akad. Wetensch. Ser. C 91:203-216.

WOLF, L. L. 1975. Energy intake and expenditures in a nectar-feeding sunbird. Ecology 56:92-104.

WOLF, L. L., F. R. HAINSWORTH, AND F. G. STILES. 1972. Energetics of foraging: rate and efficiency of nectar extraction by hummingbirds. Science 176:1351-1352.

WOLF, L. L., F. G. STILES, AND F. R. HAINSWORTH. 1976. Ecological organization of a tropical, highland hummingbird community. J. Anim. Ecol. 45:349-379.

WONG, M. 1986. Trophic organization of understory birds in a Malaysian dipterocarp forest. Auk 103:100-116.

WOOD, A. G. 1986. Diurnal and nocturnal territoriality in the Grey Plover at Teesmouth, as revealed by radio telemetry. J. Field Ornithol. 57:213-221.

WOOD, B. 1979. Changes in numbers of over-wintering Yellow Wagtails, Motacilla flava, and their food supplies in a west African savanna. Ibis 121:228-231.

WOOD, D. L. 1972. Selection and colonization of ponderosa pine by bark beetles, p. 101-117. In Insect-plant relationships. Symp. Roy. Entomol. Soc. Lond., No. 6.

WOOLEY, J. B., JR., AND R. B. OWEN, JR. 1978. Energy costs of activity and daily energy expenditure in the Black Duck. J. Wildl. Manage. 42:739-745.

WOOLLER, R. D. 1982. Feeding interactions between sunbirds and sugarbirds. Ostrich 53:114-115.

WOOLLER, R. D., AND M. C. CALVER. 1981. Feeding segregation within an assemblage of small birds in the Karri Forest understory. Austr. Wildl. Res. 8:401-410.

WOOLLER, R. D., E. M. RUSSELL, M. B. RENFREE, AND P. A. TOWERS. 1983. A comparison of seasonal changes in the pollen loads of nectarivorous marsupials and birds. Aust. Wildl. Res. 10:311-317.

WORTHINGTON, A. 1982. Population sizes and breeding rhythms of two species of manakins in relation to food supply, p. 213-226. In E. Leigh, Jr., A. S. Rand, and D. Windsor [eds.], The ecology of a tropical forest: seasonal rhythms and long-term changes. Smithsonian Inst. Press, Washington, DC.

WORTHINGTON, A. H. 1983. Foraging ecology and digestive adaptations of two avian frugivores: Manacus vitellinus and Pipra mentalis (Pipridae). Ph.D. diss., Univ. of Washington, Seattle.

WOURMS, M. K., AND F. E. WASSERMAN. 1985. Prey choice by Blue Jays based on movement patterns of artificial prey. Can. J. Zool. 63:781-784.

WRIGHT, S. J. 1979. Competition between insectivorous lizards and birds in central Panama. Am. Zool. 19:1145-1156.

WRIGHT, S. J., AND C. C. BIEHL. 1982. Island biogeographic distributions: testing for random, regular, and aggregated patterns of species occurrence. Am. Nat. 119:345-357.

WUNDERLE, J. M., JR., AND T. G. O'BRIEN. 1986. Risk-aversion in hand-reared bananaquits. Behav. Ecol. Sociobiol. 17:371-380.

WUNDERLE, J. M., JR., M. SANTA CASTRO, AND N. FETCHER. 1987. Risk-averse foraging by bananaquits on negative energy budgets. Behav. Ecol. Sociobiol. 21:249-256.

WYKES, B. J. 1985. The jarrah forest avifauna and its re-establishment after bauxite mining. School of Biology, Bulletin No. 11, Western Australian Institute of Technology, Perth.

WYLES, S. J., J. G. KUNKEL, AND A. C. WILSON. 1983. Birds, behavior, and anatomical evolution. Proc. Natl. Acad. Sci. 80:4394-4397.

WYLLIE, I. 1981. The cuckoo. B. T. Batsford, London.

YAHNER, R. H. 1987. Use of even-aged stands by winter and spring bird communities. Wilson Bull. 99:218-232.

YATES, H. O., III. 1973. Attraction of cone and seed insects to six fluorescent light sources. USDA For. Serv. Res. Note SE-184.

YDENBERG, R. C. 1984. Great Tits and giving up times: decision rules for leaving patches. Behaviour 90:1-24.

YDENBERG, R. C., AND A. I. HOUSTON. 1986. Optimal tradeoffs between competing behavioural demands in the Great Tits. Anim. Behav. 34:1041-1050.

YEATON, R. I., AND M. CODY. 1977. Competitive release in island populations of Song Sparrows. Theor. Popul. Biol. 5:42-58.

YOUNG, A. M. 1971. Foraging for insects by a tropical hummingbird. Condor 73:36-45.

YOUNG, F. W., AND W. F. KUHFELD, 1986. The CORRESP procedure: experimental software for correspondence analysis, preliminary documentation. SAS Institute, Inc., Cary, NC.

YOUNGS, L. C., AND R. W. CAMPBELL. 1984. Ants preying on pupae of the western spruce budworm, Choristoneura occidentalis (Lepidoptera: Tortricidae), in eastern Oregon and western Montana. Can. Entomol. 116:1665-1669.

YOUNAN, E. G., AND F. P. HAIN. 1982. Evaluation of five trap designs for sampling insects associated with severed pines. Can. Entomol. 14:789-796.

ZACH, R., AND J. B. FALLS. 1976a. Bias and mortality in the use of tartar emetic to determine the diet of Ovenbirds (Aves: Parulidae). Can. J. Zool. 54:1599-1603.

ZACH, R., AND J. B. FALLS. 1976b. Do ovenbirds (Aves, Parulidae) hunt by expectation? Can. J. Zool. 54:1894-1903.

ZACH, R., AND J. B. FALLS. 1976c. Foraging behavior, learning, and exploration by captive Ovenbirds (Aves: Parulidae). Can. J. Zool. 54:1880-1893.

ZACH, R., AND J. B. FALLS. 1976d. Ovenbird (Aves: Parulidae) hunting behavior in a patchy environment: an experimental study. Can. J. Zool. 54:1863-1879.

ZACH, R., AND J. B. FALLS. 1979. Foraging and territoriality of male Ovenbirds (Aves: Parulidae) in a heterogeneous habitat. J. Anim. Ecol. 48:33-52.

ZAMMUTO, R. M., E. C. FRANKS, AND C. R. PRESTON. 1981. Factors associated with the interval between feeding visits in brood-rearing Chimney Swifts. J. Field Ornithol. 52:134-139.

ZANN, R., AND R. B. STRAW. 1984. A non-destructive method to determine the diet of seed-eating birds. Emu 84:40-42.

ZAR, J. L. 1974. Biostatistical analysis. Prentice-Hall, Englewood Cliffs, NJ.

ZAR, J. L. 1984. Biostatistical analysis. 2nd ed. Prentice-Hall, Englewood Cliffs, NJ.

ZARET, T. M., AND A. S. RAND. 1971. Competition in tropical stream fishes: support for the competitive exclusion principle. Ecology 52:336-342.

ZELAZNY, B., AND A. R. ALFILER. 1987. Ecological methods for adult populations of Oryctes rhinoceros (Coleoptera, Scarabaeidae). Ecol. Entomol. 12:227-238.

ZIMMERMAN, J. L. 1965. Bioenergetics of the dickcissel, Spiza americana. Physiol. Zool. 38:370-389.

ZIMMERMAN, M. 1981. Optimal foraging, plant density and the marginal value theorem. Oecologia 49:148-153.

ZIMMERMAN, M. 1983. Calculating nectar production rates: residual nectar and optimal foraging. Oecologia 58:258-259.

ZINK, R. 1982. Patterns of genic and morphological variation among sparrows of the genus Zonotrichia, Melospiza, Junco, and Passerella. Auk 99:632-650.

ZISWILER, V., AND D. S. FARNER. 1972. Digestion and the digestive system. In D. S. Farner and J. R. King [eds.], Avian biology. Vol. 2. Academic Press, New York.

ZONOV, G. B. 1967. (On the winter roosting of Paridae in Cisbaikal.) Ornitologiya 8:351-354. (In Russian; English translation by L. Kelso.)

ZUG, G. R., AND P. B. ZUG. 1979. The Marine Toad *Bufo marinus*: a natural history resume of native populations. Smithsonian Contrib. Zool. 284.

ZUSI, R. L. 1969. Ecology and adaptations of the Trembler on the island of Dominica. Living Bird 8:137-164.

ZUSI, R. L. 1978. Notes on song and feeding behaviour of *Orthonyx spaldingii*. Emu 78:156-157.